Part I

Aerosol Filtration Contributions

ADVANCES IN AEROSOL FILTRATION

edited by

Kvestoslav R. Spurny

Schmallenberg, Germany

LEWIS PUBLISHERS

Boca Raton Boston London New York Washington, D.C.

Acquiring Editor: Ken McCombs
Project Editor: Andrea Demby
Marketing Manager: Arline Massey
Cover design: Denise Craig
PrePress: Kevin Luong
Manufacturing: Carol Royal

Library of Congress Cataloging-in-Publication Data

Advances in aerosol filtration / edited by Kvetoslav R. Spurny.
p. cm.
Includes bibliographical references and index.
ISBN 0-87371-830-5 (alk. paper)
1. Filters and filtration. 2. Aerosols. 3. Gasses -- Cleaning.
I. Spurny, Kvetoslav.
TP156.F5A38 1997
660′.284245—dc21 97-27834
CIP

This book contains information obtained from authentic and highly regarded sources. Reprinted material is quoted with permission, and sources are indicated. A wide variety of references are listed. Reasonable efforts have been made to publish reliable data and information, but the author and the publisher cannot assume responsibility for the validity of all materials or for the consequences of their use.

Direct all inquiries to CRC Press LLC, 2000 Corporate Blvd., N.W., Boca Raton, Florida 33431.

Lewis Publishers is an imprint of CRC Press LLC

No claim to original U.S. Government works
International Standard Book Number 0-87371-830-5
Library of Congress Card Number 97-27834
Printed in the United States of America 1 2 3 4 5 6 7 8 9 0
Printed on acid-free paper

Foreword

Our monograph is generally dedicated to the great scientific progress in the field of aerosol science during the second part of the 20th century.

In 1995 we remembered the 100th anniversary of the birth of Nikolai Albertovich Fuchs as well as the 40th anniversary of the "birth" of his excellent book *The Mechanics of Aerosols*. In the foreword to this book, Fuchs explains "... the mechanics of aerosols deals with the movements of aerosol particles and clouds, the collision of particles and hydrodynamic disintegration of aggregates." These are the laws and rules of the aerosol mechanics, which are strongly involved in the theory of aerosol filtration. Another aerosol friend of mine, cofounder of basic aerosol and aerosol filtration research, first winner of the International Fuchs Memorial Award in Aerosol Science, Shal K. Friedlander used to divide aerosols into "good" and "bad" ones. The latter category, which is produced mainly in the form of industrial particulate emissions, has negative influence on human health, on ecological systems, the greenhouse effect, the ozone layer, etc. The "good" aerosols have enabled several new, highly effective technological productions of new, very clean, and highly dispersed materials. *Filtration* and *separation* processes play very important roles in several fields of this both in scientific and technological categories.

The filtration and separation of fine aerosols, i.e., the very effective gas-cleaning methods, belong in basic the requirements for atmospheric environmental activities, as well as into the processes of the production of new high-tech materials. The new results of aerosol filtration science will therefore have an important impact on further environmental and technological developments.

I'm happy that I was able — with great help from all contributors to this book and very effective support from the Lewis Publishers/CRC Press staff specifically, Susan Alfieri and Andrea Demby — to realize this publication.

Nikolai Albertovich Fuchs was my good friend, my great teacher and, last but not least, an excellent human character. He lived in a very difficult time. The communistic, antidemocratic, and antihuman system based on lies and oppressions was not able to disturb or break his character and change his human philosophy. He was, unfortunately, not able to live to the end of this inhuman system and to live as a free scientist and human being.

Kvetoslav R. Spurny

Dedication

The editor as well as the contributors would like to dedicate this monograph to the memory of a great aerosol scientist and humanist, cofounder of the basic aerosol sciences in the world.

Nikolai Albertovich Fuchs (1895–1982). Fuchs in Prague, Spring, 1966.

Aerosol Filtration Studies in the Fuchs' Laboratory*

A. A. Kirsch and I. B. Stechkina

The problem of air filtration by fibrous filters was one of the main branches of aerosol research by Prof. N. A. Fuchs in the newly organized laboratory in the Karpov Institute of Physical Chemistry in 1959. His interest in this problem had been suspended since the time he had studied the electrohydrodynamic method of fine polymer fiber production with his coworkers prior to his repression in 1937. Later this type of filer was named "filter FP" in the U.S.S.R. (Fuchs, 1979). According to Fuchs' opinion, the theory of filtration should thus be based on viscous flow through a system of cylinders at small Re, either determined by model experiments, or calculated theoretically (Fuchs, 1961, 1964).

We took an interest in his concept of modeling of filters and, during twenty years in his laboratory, we, collectively, step-by-step tried to approach the quantitative calculation of single fiber efficiency.

As the first such model we chose the "parallel fiber model" — a system of parallel, cylindrical, and regularly arranged fibers (Fuchs and Stechkina, 1963). Almost all properties of this model were treated theoretically and experimentally. We studied the viscous flow, the flow of rarefied gas, and particle deposition due to diffusion, interception, and inertia in parallel model filters with different packing densities (Kirsch and Fuchs, 1967). The parallel model with a known pattern of flow was a convenient model for study of the influence of a nonuniform arrangement of cylinders and their polydispersity, as well as the influence of gas slip on the pressure drop and the particle deposition. As a result of a set of works we could describe the diffusional and interception particle deposition in ordered systems of parallel cylinders. We also used this model for the study of the influence of an external electric field on the particle deposition, for modeling the process of building up a deposit on fibers, and the process of increasing of droplets on fibers at mist filtration, among others. However, the parallel model proved to be inadequate to represent the particle deposition process in commercially produced fibrous filters. Thus, in spite of a better understanding of the complex processes of aerosol deposition in fibrous filters, we were not able to reach our goal to calculate the collection efficiency of real filters.

Further investigation produced the "fan model filter," obtained by turning in the parallel filter model, all fiber rows, in their planes at an arbitrary angle. The theoretical treatment of the fan model was very difficult because it had a three-dimensional flow field. That caused us to pass into a semiempirical method. It was found that its properties were the same as those of real filters with a perfectly homogeneous structure. The "degree of inhomogeneity" of real filters was determined as the reverse ratio of its resistance and of the particle capture coefficient on its fiber to those in an equivalent fan model. This made it possible to calculate the filter collection efficiency from its resistance for a given size of particles depending on filter parameters, flow rate, temperature, pressure, etc. The method developed describes satisfactory experimental data obtained with different filters (Fuchs et al., 1973). In the later series of our collaborative works, we have shown the possibility of calculation of the high efficiency of HEPA filters with allowance for polydispersity of fibers and gas slip effect near ultrafine fibers (Kirsch, Stechkina, and Fuchs, 1973, 1974, 1975). Later we performed the comprehensive study of collection efficiency of different kinds of paper filters and published a review of the results about modeling and calculating of high efficiency filters (Kirsch and Stechkina, 1978). The work, started in the 1960s in Fuchs' laboratory, has been continued in the 1980s and 1990s in the Laboratory of Dispersed Systems in the Kurchatov Institute.

* See also *J. Aerosol Sci.*, 26, S61–S62, Suppl.1 (1995).

We have considered the questions of refinement of the calculation of capture coefficient, due to diffusion and interception at intermediate Knudsen numbers and at high gas temperature, and the questions of the effect of the filter structure on filter properties. We have begun the new line of investigation on the field of nonstationary filtration. The process of loading of fibrous filters was studied both theoretically and experimentally. The results made it possible to approach a solution of a new, very practical problem of optimization of different filter parameters (Stechkina and Kirsch, 1994; Kirsch and Lebedev, 1994).

In conclusion, it should be noted that the absence of fundamental hydrodynamic elements on a three-dimensional flow field and in a polydisperse fiber system restricts the further development of the aerosol filtration theory based on the single fiber element philosophy.

REFERENCES

Fuchs, N. A. (1961) Uspechi Mekhaniki Aerosolej (*Advances in the Mechanics of Aerosols,* in Russian) Izd. AN USSR (Publ. Acad. Sci.) Moscow.

Fuchs, N. A. and Stechkina, I. B. (1963) *Ann. Occup. Hyg.,* 6, 27–30.

Fuchs, N. A. (1964) *The Mechanics of Aerosols,* Pergamon Press, Oxford, U.K., p. 226.

Fuchs, N. A., Kirsch, A. A., and Stechkina, I. B. (1973) *Proc. Faraday Symp. Chem. Soc.,* 7, 143–156.

Fuchs, N. A. (1979) Gas and liquid filtration by fibrous filters. *Khimicheskaja Promishlennost* (Chemical Industry, Russian) 11, 48–51.

Kirsch, A. A. and Fuchs, N. A. (1967) *Ann. Occup. Hyg.,* 10, 23–30.

Kirsch, A. A. and Fuchs, N. A. (1968) *Ann. Occup. Hyg.,* 11, 299–304.

Kirsch, A. A., Stechkina, I. B., and Fuchs, N. A. (1973) *J. Aerosol Sci.,* 4, 287–293.

Kirsch, A. A., Stechkina, I. B., and Fuchs, N. A. (1974) *J. Aerosol Sci.,* 5, 39–45.

Kirsch, A. A., Stechkina, I. B., and Fuchs, N. A. (1975) *J. Aerosol Sci.,* 6, 119–124.

Kirsch, A. A. and Stechkina, I. B. (1978) The Theory of Aerosol Filtration with Fibrous Filters In: D. T. Shaw (Ed.), *Fundamentals in Aerosol Science*, John Wiley & Sons, New York, pp. 165–256.

Kirsch, A. A. and Lebedev, M. N. (1994) Proc. 12th ISCC Congress, Yokohama, Japan, pp. 53–57.

Stechkina, I. B. and Kirsch, A. A. (1994) Proc. 12th ISCC Congress, Yokohama, Japan, pp. 37–42.

The Editor

Prof. Dr. Kvetoslav Rudolf Spurny was Head of the Department of Aerosol Chemistry at the Fraunhofer Institute for Environmental Chemistry and Ecotoxicology in Germany from 1972 to 1988. He has retired, but is still working as an aerosol chemist. Prior to this, he was an environmental chemist at the Institute for Occupational Hygiene in Prague (1952 to 1956) and Head of the Department of Aerosol Sciences at the Czechoslovak Academy of Sciences in Prague (1957 to 1972). He was a visiting scientist at the National Center for Atmospheric Research, Boulder, CO (1966 to 1967) and visiting scientist at the Nuclear Research Center, Fontenay aux Roses, France, in 1969. Dr. Spurny obtained his Dipl. in Physics and Chemistry from Charles University, Prague, in 1948; a Ph.D. in chemistry at the same university in 1952; and a C.Sc. as a Candidate of Chemical Sciences at the Czechoslovak Academy of Sciences in Prague in 1964. Professor Spurny is a member of the American Chemical Society, American Association for the Advancement of Science, American Association of Aerosol Research, British Occupational Hygiene Society, the New York Academy of Sciences, and was president of the Association for Aerosol Research from 1983 to 1984. He has written three books on aerosols and over 150 original publications in aerosol physics and chemistry. He was recipient of the American David Sinclair Award in Aerosol Sciences in 1989.

Acknowledgments

The editor is extremely grateful to the authors for their excellent contributions. They have not only contributed to the broad information about the progress, state of the art, and new concepts and philosophy in the field of aerosol separation processes in filters, but they have also demonstrated their respect and admiration to the historical merits and contribution to the development of aerosol and aerosol filtration sciences by the excellent scientist and humanist Nikolai Albertovich Fuchs, whose 100th anniversary of his birth was celebrated in 1995.

The editor would like to thank very much the staff of Lewis Publishers/CRC Press for their substantial help with the realization of this book. For several of the authors English is not their mother tongue and without the great, patient, and highly professional support of the publishers, the realization of this book would not have been possible. All, authors and publishers, have done an excellent job for science and scientists.

Contributors

J. G. Alperovich
Laboratory of Physics of Anisotropic Materials
Academy of Sciences
St. Petersburg, Russia

Vladimir J. Alperovich
Laboratory of Physics of Anisotropic Materials
Academy of Sciences
St. Petersburg, Russia

Imre Báláshazy
Institute of Physics and Biophysics
University of Salzburg
Salzburg, Austria

V. V. Blagoveshschenskiy
Laboratory of Physics of Anisotropic Materials
Academy of Sciences
St. Petersburg, Russia

Denis Boulaud
CEA–SACLAY
Institut de Protection et de Sûreté Nucléaire
Gif Sur Yvette, France

Richard C. Brown
Health and Safety Laboratory
Sheffield, United Kngdom

E. D. Chunin
Laboratory of Physics of Anisotropic Materials
Academy of Sciences
St. Petersburg, Russia

Olga Filippova
Institut für Verbrennung und Gasdynamik
Universität Duisburg
Duisburg, Germany

H. Fissan
Process and Aerosol Measurement Technology Division
University of Duisburg
Duisburg, Germany

Leon Gradoń
Department of Chemical and Process Engineering
Warsaw University of Technology
Warsaw, Poland

Andreas Gutsch
Institut für Mechanische Verfahrenstechnik und Mechanik
Universität (TH)
Karlsruhe, Germany

Dieter Hänel
Institut für Verbrennung und Gasdynamik
Universität Duisburg
Duisburg, Germany

Thomas Heistracher
Institute of Physics and Biophysics
University of Salzburg
Salzburg, Austria

Wilhelm Höflinger
Institut für Verfahrens-, Brennstoff-, und Umwollttechnik TU
Vienna, Austria

Werner Hofmann
Institute of Physics and Biophysics
University of Salzburg
Salzburg, Austria

Chikao Kanaoka
Department of Civil Engineering
Kanzawa University
Kanazawa, Japan

V. V. Kashmet
Laboratory of Physics of Anisotropic Materials
Academy of Sciences
St. Petersburg, Russia

A. A. Kirsch
Laboratory of Dispersed Systems
Kurchatov Institute
Russian Research Center
Moscow, Russia

Jane F. Klassen
Duke Engineering & Services
Naperville, Illinois

Petros Koutrakis
Department of Environmental Health
School of Public Health
Harvard University
Boston, Massachusetts

Orest Lastow
Department of Working Environment
Lund Institute of Technology
Lund, Sweden

Ralf Maus
Institut für Mechanische Verfahrenstechnik und Mechanik
Universität (TH)
Karlsruhe, Germany

W. Mölter
Process and Aerosol Measurement Technology Division
University of Duisburg
Duisburg, Germany

Alexander M. Möllinger
Atmospheric Aerosol Observatory
La Pichotiere
Nocé, France

Toshihiko Myojo
National Institute of Industrial Health
Kawasaki, Japan

Vincent J. Novick
Aerosol Technology Group
Argonne National Laboratory
Argonne, Illinois

Jozef S. Pastuszka
Institute of Occupational Medicine and Environmental Health
Sosnowiec, Poland

Albert Podgórski
Department of Chemical and Process Engineering
Warsaw University of Technology
Warsaw, Poland

André Renoux
Laboratoire de Physique des Aérosols et de Transfert des Contaminations
Université Paris XII
Créteil, France

Rein André Roos
Atmospheric Aerosol Observatory
La Pichotière
Nocé, France

Michel Quintard
L.E.P.T.–ENSAM (UACNRS)
Esplanade des Arts et Métiers
Talence, France

Eberhard Schmidt
Institut für Mechanische Verfahrenstechnik und Mechanik
Universität (TH)
Karlsruhe, Germany

Michael Shapiro
Faculty of Mechanical Engineering
Technion — Israel Institute of Technology
Haifa, Israel

Constantinos Sioutas
Department of Civil Engineering
University of Southern California
Los Angeles, California

Kvetoslav R. Spurny
Schmallenberg, Germany

I. B. Stechkina
Laboratory of Dispersed Systems
Kurchatov Institute
Russian Research Center
Moscow, Russia

Mark Stoelinga
Atmospheric Aerosol Observatory
La Pichotière
Nocé, France

Gabriel I. Tardos
Department of Chemical Engineering
The City College of
The City University of New York
New York, NY

Chi Tien
Department of Chemical Engineering
National University of Singapore
Singapore

Stephen Whitaker
Department of Chemical Engineering and
Material Science
University of California at Davis
Davis, California

Contents

PART II: Aerosol Separation

ADVANCES IN AEROSOL FILTRATION

CHAPTER 1

The History of Dust and Aerosol Filtration

Kvetoslav R. Spurny

CONTENTS

The development of particulate filters has had several stimuli — the protection of workers from harmful dust, the call for a medical respirator, the protection of firefighters from smoke, protection in times of war and crimes, sampling of dust and aerosol particles, etc.

For this reason, it is no wonder that the history of dust and aerosol filtration is over 2000 years old. One of the most meritorious pioneers on the field of aerosol filtration science, Charles Norman

0-87371-830-5/98/$0.00+$.50

Davies (1910–1994), has described this historical development very extensively up to 1970 (Davies, 1949, 1973; Vincent and Kasper, 1994).

1.1 EARLY DEVELOPMENTS

1.1.1 Protection from Dust

The earliest records of air filtration date from Roman times. The problem of "industrial dust" is mentioned in the *Natural History* of Pliny the Elder (*ca* a.d. 50), and Julius Pollux (*ca* a.d. 150) describes the problem of the dust in mines in Egypt. Concerning the protection of the men, Plinius mentions loose bladders being used to prevent dust inhalation. Leonardo da Vinci (1452–1519) mentions wet cloths as a protection from fumes used in warfare. Agricola (Georg Bauer) describes in his book *De Re Metallica* (1556) the problem of dust in mines. He recommended ventilation as well as the use of primitive forms of respirators. The first observation about the health risk of dusts was done by Paracelsus (Theophrastus von Hohenheim). He mentioned new "dust diseases" in his book *Von der Bergsucht und anderen Bergkrankheiten* (*ca* a.d. 1560) (Davies, 1949).

One hundred years later Stockhausen published his *Libellus de Lythargyrill Fumio Noxto Morbifico* (Goslar, 1656) dealing with protection from rock dust. Bernadino Ramazzini (1633–1714) recognized in his *De Morbis Artificum* the need for better protection of the respiratory tract and cities dozens of occupations which were dangerous on account of dust. Nevertheless, no filtering respirator was recorded until the device of Brisé Faden (1814). It was a box filled with cotton which had a breathing tube to go in the mouth (Davies, 1949).

1.1.2 Medical Respirators

By 1860 Louis Pasteur (1822–1895) was demonstrating that putrefraction was due to airborne organisms. This experience was followed by needs for the filtration of air, i.e., for the development of medical respirators. The early equipment and their further improvement were very well described by Feldhaus (1929). At the beginning of the 20th century relatively effective respirators filled by cotton wool were available.

1.1.3 Smoke Filters

Another important field for the use of filters was the protection of firefighters from fire smoke. In the 19th century protective "wire masks" were already used in England, France, and Austria. An important filtering mask, the "smoke cap," was invented and produced in England by Sir Eyre Massey Shaw and Prof. John Tyndall, the well-known physicist. This filter was made of highly packed cotton wool. Tyndall was the first to be able to test qualitatively the performance of aerosol filters. His testing method was based on his "light-scattering" observations on smoke particles.

1.1.4 Chemical Warfare

Another field for the application of respirators and gas masks was, unfortunately, military technology. Gaseous and aerodisperse toxic chemicals were used during World War I. In this case a combination of a gas and aerosol filter was necessary. The harmful gases could be separated relatively well by adsorption on coarse granules of charcoal, whereas the aerodispersed phase was not removed. This problem was solved later (*ca* 1930) in Germany by using fibrous pads (Albrecht, 1931; Kaufman, 1936).

1.2 FIBROUS FILTERS

In July 1930, N. L. Hansen working in Copenhagen tried the powdering of a wood filter pad with colophony resin, and he found a very great increase in filtering efficiency. It was in fact the invention of the first electret filter, as the resin was an electrically insulating substance. By melting and solifying the resin in a strong electric field, electrically charged filter structures were produced.

The Hansen filter was used by the Danish Army for some years after 1931. At the time it was among the best smoke filters in the world. Later, resin filters were used by the Dutch, French, and Italian armies.

During World War II several types of fibrous materials in the form of soft pads, felt, or papers were tried. It had been found empirically that asbestos, mainly chrysotile and crocydolite, fibers were very effective filtering materials, because of the thinness of the single fibers. A mixture of about 20% asbestos with wool produced excellent pads in which almost all the filtering action was due to the asbestos. Asbestos-containing filters had been used for a long time as high-efficiency aerosol filters in respirators, gas masks, as technical air filters in ventilating systems, and also as very good liquid filters for cleaning, e.g., beverages, etc. Nevertheless their use as well as their production were stopped in the 1970s when it was clear that fine asbestos fibers, which are highly carcinogenic, are released from these filters and contaminate the filtering gases or liquids.

After this, asbestos was replaced by other fiber types, mainly by thin glass, carbon, ceramic as well as organic fibers. The filtration efficiency depends in other aspects also on the single-fiber diameter; efficiency increases with declining fiber diameter. Thin fibers (~1 μm) therefore have to be used in the production of high-efficiency filters.

At present, a relatively large assortment of very good, highly efficient glass as well as polymer fiber filters is commercially available. Almost ~50 aerosol filter producers in different countries in Europe, America, Asia, etc. are registered worldwide (Elsevier, 1995).

1.3 PORE FILTERS

In addition to fibrous structures, other porous media have also proved to be useful for separation of fine particles dispersed in gases and liquids. Gas or liquid flows slowly through a relatively thin layer of a microporous layer made of organic or inorganic material, mostly a dry gel. Particles are then separated on the solid single elements in a similar way as inside the fibrous filters. The pore filters are preferentially used as analytical filters for sampling dust, aerosol, or colloid probes for physical and chemical analysis.

The history of the development and application of analytical aerosol filters has been well documented in the review literature (Spurny, 1965/1966).

The collodial membrane filter (MF), the silver membrane filter (SMF), as well as the very precise polycarbonate filter (Nuclepore filter, NPF) are the most common and important filtration tools which allow the sampling and preconcentration of fine and very fine aerosol particles and enable a chemical characterization of the sampled particulates by very sensitive analytical methods. These methods can be used for bulk and for single-particle chemical analysis.

1.3.1 Celullose Membranes

Membranes of cellulose nitrate were first made for laboratory filtration purposes in the late 1800s and achieved considerable acceptance for bacteriological purposes in Russia and Germany (Spurny, 1965/1966; Gelman, 1965; Matteson, 1987). In 1916, Zsigmondy and Bachmann developed cellulose-derivate membranes and applied them to fundamental studies in colloid chemistry (Zsigmondy and Bachmann, 1916; 1918). Commercial production was started in Germany by the company Sartorius, in 1927 (Maier, 1958; 1959).

The German patents have been applied for commercial production of cellulose ester membrane filters in the U.S. and in Russia, since of about 1955. Collodial MF have been used in dust sampling and analysis since the beginning of the 1950s (Kruse, 1998; Reznik, 1951; Goetz, 1951; 1953; First and Silverman, 1953; Einbrodt and Maier, 1954; Spurny and Vondracek, 1957).

1.3.2 Metal Membranes

In 1964 the Selas Corporation, in the U.S. developed an MF of pure silver with relatively uniform porosity, the SMF (Matteson, 1987). One advantage of this SMF over MF lies in the ability to conduct electrically; furthermore, these filters are stable from –130 to +370°C.

1.3.3. Polycarbonate Filters

The conception of the polycarbonate filter is attributed to a research group at the General Electric Laboratories in California (Fleischer et al., 1965). The first applications of NPF for dust and aerosol sampling and analysis, as well as their first filtration theory, were published by Spurny et al. (1957; 1969). The NPF, because of its basically cylindrical pores, offers an advantage of high selectivity in filtration and particle sizing. Furthermore, this filter is very suitable for physical and chemical characterization of single particles by electron microscopical methods.

The characterizations and applications of aerosol analytical filters are described in detail elsewhere in this book.

1.4 EARLY AEROSOL FILTRATION THEORY

1.4.1 Browian Motion

The first recognition dealing with fine particle movement had already been done in the early 1800s when the botanist Robert Brown observed the irregular motion of minute particles suspended in a liquid (Brownian motion). Nevertheless, the role of the Brownian motion of fine particles dispersed in gases was mentioned much later when Bodaszewsky (1881) described the collision and adhesion of dust particles on vessel walls.

The surface of the fibers in the filter is a large area so that the number of collisions is very great, especially for smaller particles. When aerosol particle size is comparable with the mean free path of gas molecules (~0.06 μm in air), their Brownian motion is violent and collisions with neighboring surfaces are therefore frequent. Larger particles, e.g., greater than 0.3 μm, however, show negligible Brownian motion and filtration of these cannot be accounted for on this diffusion theory.

1.4.2 General Filtration Mechanics

A more general picture of the mechanics of aerosol filtration was developed in Germany in the 1920s (Freundlich, 1921; Engelhard, 1925). There was shown that a maximum degree of penetration occurred for aerosol particles between 0.1 and 0.2 μm radius. Larger and smaller particles were more efficiently filtered. The reason for the removal of large particles was not clearly understood at that time. A theory of the filtration of larger particles could be dated, e.g., from about 1850, when Stokes' law was first given.

However, the first clear idea of the process at work was included in the paper of Albrecht (1931). He considered the streamlines of air flowing past a cylindrical fiber. Large particles do not travel along the streamlines; owing to its mass, a larger particle tends to travel straight on when the airflow curves. These particles are then captured on the cylinder surface. Albrecht's theory was essentially improved by Sell (1931).

The first attempt to combine the theories of filtration of small particles, by Brownian motion, and of large particles, by impingement due to inertia, was made by Kaufmann (1936). Kaufmann had also already defined the mechanisms of direct interception, which he called "die Grundabscheidung." It was a purely geometric effect: if the center of the particle passes the surface of the fiber at a distance less than the particle radius, collision results.

It was Davies, who, in 1940, pointed out that theories could not be exact because they were founded on ideal, instead of viscous, fluid flow. He developed his own mathematical theory showing the important role of the single fiber and its diameter.

The next important improvement on the theory of aerosol filters was made by Langmuir (1942; 1962). He adopted Kaufmann's method of approach with a considerably refined technique. He considered two main separation mechanisms working together — particle diffusion and direct interception. Langmuir's chief advance lay in the way he avoided the difficulties of viscous flow hydrodynamics by adopting an ingenious method of approximation which made some allowance for the influence of the mutual proximity of fibers upon the flow field.

1.4.3 Isolated Fiber Theory

When the effects of neighboring fibers and the fiber packing density are disregarded, an isolated fiber efficiency can be calculated. Lamb's theoretical expression for the flow field around an isolated cylinder can be used for this purpose. Nevertheless, Lamb's theory was accurate at low values of the Reynolds number (Re). Davies (1950) and Davies and Peetz (1956) have improved and adopted Lamb's theory to be able to calculate the inertial particle deposition upon isolated fibers in viscous flow. Further important steps in the development of single-fiber theory were made by Friedlander (1958) and later by Japanese scientists Yoshioka et al. (1967; 1969a, b; 1972). These authors have included inertial and diffusional particle depositions at larger Re, and later also considered the effects of gravitational deposition and clogging.

1.5 RESISTANCE OF FILTERS

The term *resistance* is used for the pressure drop across a filter at some arbitrary rate of airflow. In early approximations the Darcy law for flow in porous systems was used. Later investigations showed that Darcy's law is not good enough for an exact description of flow resistance in a fiber filter. The first scientific description originated with Pich (1966) who related the pressure drop to the existing Knudsen number (Kn). All following theoretical solutions have strongly considered the bulk structure of the filter which they try to characterize by means of different geometric models (Davies, 1950; Finn, 1953; Happel, 1959; Stern et al., 1960).

Another very important improvement in the description of the airflow field through the fiber filter model was made by Kuwarbara in 1959. He solved the Navier–Stokes equations for flow transverse to a set of parallel cylinders. His filter model has been used by further theoretical computations as, e.g., by Pich (1966, 1971, 1987), Spielman and Goren (1968), Kirsch et al. (1971), etc.

1.6 NEW CONCEPTS OF FILTRATION

The first stage in the development of aerosol filtration theory ended with the realization of the importance of inertial deposition of particles upon filter fibers. After World War II a considerable amount of theoretical development took place, and it became apparent that the real commercially available fiber filters made of organic and glass wool fibers were not ideal for checking the fluid mechanical theories concerned with the mechanisms of filtration. The aerosol scientist began to use model filters. Kuwabara's brilliant theory started to be the most important base for future developments.

During the 1960s several very good experiments were performed which tried to confirm the penetration maximum, i.e., the maximal penetrating particle size for fiber filters (Whitby, 1965; Dyment, 1970, etc.). The existence of this penetration maximum was definitively confirmed at that time.

Further important improvements were made in the theoretical field by describing more exactly the partial separation mechanisms as well as their combinations. Very important contributions have come from the "Fuchs Russian School of Aerosol Filtration" since about the beginning of the 1960s (Fuchs and Stechkina, 1963; Kirsch and Fuchs, 1967, 1968; Stechkina et al., 1966, 1969, 1970). They have very much improved the theory of direct interception mechanisms, the combined effects of inertia and interception. They later proposed and used a new fiber filter model designated as the "fan model" (Kirsch and Fuchs, 1968). The fan model consists of a series of parallel layers of equidistant, parallel fibers, but successive layers are rotated in their one plane through random angles. Nevertheless, in the fan model the dispersion of fibers was perfectly homogeneous. This is not so in real filters in which inhomogeneities arise. For this purpose, Fuchs, Kirsch, and Stechkina had introduced an inhomogeneity factor in their model computations to bring the real filter into alignment with the equivalent fan model. Fuchs and co-workers have also improved the description of the action of the diffusional deposition of aerosols in fibrous filters (Stechkina and Fuchs, 1966; Kirsch and Fuchs, 1968).

At the end of the 1960s, Fuchs and his co-workers concluded that when the geometric parameters, fiber radius, packing density, and filter thickness are known, then it is possible to calculate the resistance and efficiency of a real fiber filter with sufficient accuracy for practical purposes. Unfortunately, this conclusion was to prove optimistic. Later developments have shown that there still were and still are nonnegligible differences between the structures of real fiber filters and mathematical models.

1.6.1 Electrical Forces in Filters

The aerosol particles, as well as the filter, or both can be electrically charged. This enhances the total collection efficiency of a filter. A charged particle is attracted toward an uncharged surface inside the filter by the image effects.

The first approximate theory of this action was published by Natanson (1957a, b) and by Havlicek (1961). Lundgren and Whitby (1965) demonstrated the electrical action of charged particles by experiments. If an electric field is applied across a fibrous filter, the fibers become polarized, and the particles will be polarized, too. The aerosol filtration under such conditions was described for the first time by Walkenhorst and Zebel (1964), and the theory was very much improved by Zebel (1965), 1969) and by Hochrainer (1969).

The most usual situation is the action of an electrically charged filter. As mentioned, Hansen had already built and used such a filter in 1930 (Davies, 1949). Further important applications of electrically charged and electret filters started in the 1950s (Van Orman and Endres, 1952; Gillespie, 1955; Walkenhorst, 1970).

The first studies of the charged and filtration electrokinetics of pore filters were published in 1961 by Spurny and Polydorova.

1.6.2 Clogging of Filters

While the particle deposit is building up and the filter is becoming clogged, through use, important changes may take place which considerably alter the filter's filtering characteristics.

While filtration statics describe the filtration parameters of a clean filter, the filtration kinetics is a time-dependent filter characteristic describing the filter clogging process.

The first experimental observations about clogging were published by Watson in 1946 and described by Leers (1957), Billings (1966), and Mohrmann (1970). Davies defined theoretically the increase of filter resistance and collection efficiency as results of clogging in 1970.

1.6.3 Adhesion of Particles in Filters

Solid particles are held in contact with filtering surfaces by van der Waals forces. When these forces are overcome by air drag, the deposited particles can be removed again into the gas stream. This effect had already been observed and described during the 1950s and 1960s (Jordan, 1954; Larsen, 1958; Löffler, 1968).

1.7 EARLY FILTER TESTING METHODS

One of the earliest pieces of apparatus described for testing dust respirators was that of Michaelis (1890). Thomas slag dust that had penetrated through the filter was sampled by an impinger and evaluated by microscopical particle counting. Better and more-sophisticated testing methods were developed in the U.S. during World War I (Fieldner et al., 1919). Tobacco smoke or silica dust were used as testing aerosols. The penetrated concentration was detected by a simple light-scattering method.

In 1937 Hill published a new apparatus — the "carbon smoke penetrometer." Carbon smoke came from a gas burner and its penetrated concentration was measured by a light transmission method. During the World War II the methylene blue test was developed and standardized in the U.K. by Walton (1940). This system would measure down to 0.01%. A second important test procedure at that time was the sodium flame test (Walton, 1941). The penetrated sodium chloride aerosol was detected by means of a flame photometer. The first application of a radioactive aerosol for filter testing was described in 1941 by Rodebush. He used radioactive triphenylphosphate smoke.

A very sensitive testing method was developed by Gucker et al. (1947). They used the first "optical particle counter" and were able to measure penetrations to 10^{-8}%. The filter testing techniques were very much improved after 1960, mainly in the 1980s, when computer-supported and automatic procedures were developed. Nevertheless, the testing principles and philosophies remained almost the same.

1.8 DEVELOPMENTS AND THEORIES AFTER THE 1970S

As mentioned, the "classical" period of aerosol filtration science was excellently documented and evaluated in the Davies monograph (1973) as well as in *The Mechanics of Aerosols* by Fuchs (1964). The further developments were very positively influenced by the facilities of modern computer techniques, which made it possible to solve complicated theoretical problems. There do exist two excellent and modern monographs describing development up until 1986 (Pich, 1987) and up until 1992 (Brown, 1993).

REFERENCES

Albrecht, F. (1931) Theoretische Untersuchungen übfer die Ablagerung von Staub aus der Luft und ihre Anwendung auf die Theorie der Staubfilter, *Phys. Zs.*, 32, 48.

Billings, C. E. (1966) Effects of Particle Accumulation in Aerosol Filtration. Report, W. M. Keck Laboratory, California Institute of Technology, Pasadena.

Bodaszewsky, L. U. (1881) Rauch und Dampf unter dem Mirkoskop, *Dinglers Polytech. J.*, 339, 325.

Brown, R. C. (1993) *Air Filtration,* Pergamon Press, Oxford, 1–22.

Davies, C. N. (1940) General Principles Underlying the Design of Particulate Filters, Report CDRD, Porton, U.K.

Davies, C. N. (1949) Fibrous filters for dust and smoke, in *Proc. 9th Int. Congress on Ind. Med.,* Simpkin Marshall, London, 162–196.

Davies, C. N. (1950) Viscous flow transverse to a circular cylinder, *Proc. Phys. Soc. B,* 63, 288–296.

Davies, C. N. (1952) The separation of airborne dust and particles, *Proc. Inst. Mech. Eng.,* 1B, 185–213.

Davies, C. N. (1973) *Air Filtration,* Academic Press, London, 1–171.

Davies, C. N. and Peetz, V. (1956) Impingement of particles on a transverse cylinder, *Proc. R. Soc. A,* 234, 269–295.

Dyment, J. (1970) Use of the Goetz Aerosol Spektrometer for measuring the penetration of aerosols through filters as a function of particle size, *J. Aerosol Sci.,* 1, 53–67.

Einbrodt, H. J. and Maier, K. H. (1954) Staub-Mikroskopie mit Membranfilter, *Staub,* 16, 246.

Elsevier (1995) *Profile of the Worldwide Filtration and Separation Industry,* Elsevier Advanced Technology, Elsevier, Oxford.

Engelhard, H. (1925) Atemschutz gegen Quecksilber, *Gasmaske,* 3, 156.

Feldhaus, G. M. (1929) Schutzmasken in vergangenen Jahrhunderten, *Gasmaske,* 1, 104.

Fieldner, A. C., Oberfall, G. C., Teague, M. C., and Lawrence, J. N. (1919) Methods of testing gasmasks and absorbents, *J. Ind. Eng. Chem.,* 2, 519.

Finn, R. K. (1953) Determination of the drag on a cylinder at low Reynolds numbers, *J. Appl. Phys.,* 24, 771–773.

First, M. W. and Siverman, L. (1953) Air sampling with membrane filters, *AMA Arch. Ind. Hyg. Occup. Med.,* 7, 1–13.

Fleischer, R. L., Price, P. B., and Walker, R. M. (1965) Tracks of charged particles in solids, *Science,* 149, 383–389.

Freundlich, H. (1921) *Kapilarchemie,* Berlin.

Friedlander, S. K. (1958) Theory of aerosol filtration, *Ind. Eng. Chem.,* 50, 1161–1164.

Fuchs, N. A. (1964) *The Mechanics of Aerosols,* Pergamon Press, Oxford.

Fuchs, N. A. and Stechkina, I. B. (1963) A note on the theory of fibrous aerosol filters, *Ann. Occup. Hyg.,* 6, 27–30.

Gelman, C. (1965) Microporous membrane technology, *Anal. Chem.,* 37, 29A.

Gillespie, T. (1955) The role of electric forces in filtration of aerosols by fiber filters, *J. Colloid Sci.,* 10, 299–314.

Goetz, A. (1951) Aerosol filtration with molecular filter membranes, *Chem. Eng. News,* 29, 193–198.

Goetz, A. (1953) Application of molecular filter membranes to the analysis of aerosols, *Am. J. Public Health,* 43, 150–157.

Gucker, F. T., O'Konski, C. T., Pickard, H. B., and Pitts, J. (1947) A photoelectric counter for colloidal particles, *J. Am. Chem. Soc.,* 69, 2422–2429.

Happel, J. (1959) Viscous flow relative to arrays of cylinders, *Am. Inst. Chem. Eng. J.,* 5, 174–177.

Havlicek, V. (1961) The improvement of the efficiency of fibrous dielectric filters by application of external electric field, *Ont. J. Air Water Pollut.,* 4, 225–236.

Hill, A. S. G. (1937) A photoelectric smoke penetrometer, *J. Sci. Inst.,* 14, 296.

Hochrainer, D. (1969) Zur Aerosolabscheidung an einer Einzelfaser unter dem Einfluß elektrischer Kräfte, *Staub Reinhalt. Luft,* 29, 67–70.

Jordan, D. W. (1954) The adhesion of dust particles, *Br. J. Appl. Phys.,* Suppl. 3, S194–S198.

Kaufmann, A. (1963) Die Faserstoffe für Atemschutzfilter, *Z. VDI,* 80, 593.

Kirsch, A. A. and Fuchs, N. A. (1967a) The fluid flow in a system of parallel cylinders perpendicular to the flow direction at small Reynolds numbers, *J. Phys. Soc. Jpn.,* 22, 1251–1255.

Kirsch, A. A. and Fuchs, N. A. (1967b) Studies on fibrous aerosol filters. Pressure drop in systems of parallel cylinders, *Ann. Occup. Hyg.,* 10, 23–30.

Kirsch, A. A. and Fuchs, N. A. (1968) Studies on fibrous aerosol filters. Diffusional deposition of aerosol in fibrous filters, *Ann. Occup. Hyg.,* 11, 299–304.

Kirsch, A. A., Stechkina, I. B., and Fuchs, N. A. (1971) Effect of gas slip on the pressure drop in a system of parallel cylinders of small Reynolds numbers, *J. Colloid Interface Sci.*, 37, 458–461.
Kruse, H. (1948) Ein neues Verfahren zur Bestimmung des Keimgehaltes der Luft, *Ges. Ing. H,* 7, 199.
Kuwabara, S. (1959) The forces experienced by randomly distributed parallel circular cylinders or spheres in viscous flow at small Reynolds numbers, *J. Phys. Soc. Jpn.*, 14, 527–532.
Langmuir, I. (1942) Report on Smokes and Filters. Section I. U.S. Office Sci. Res. Develop. No. 3460.
Langmuir, I. (1962) *The Collected Works of Irving Langmuir*, Pergamon Press, London.
Larsen, R. I. (1958) The adhesion and removal of particles attached to air filter surfaces, *Am. Ind. Hyg. Assoc. J.*, 19, 265–270.
Leers, R. (1957) Die Abscheidung von Schwebstoffen in Faserfiltern, *Staub,* 50, 402–417.
Löffler, F. (1968) Über die Haftung von Staubteilchen an Faser- und Teichenoberflächen, *Staub Reinhalt. Luft,* 28, 456–462.
Lundgren, D. A. and Whitby, K. T. (1965) Effect of particle electrostatic charge on filtration by fibrous filters, *And. Eng. Chem. Proc. Des. Dev.*, 4, 345–349.
Maier, K. H. (1958) Staubuntersuchung mit Hilfe von Membranfiltern. *VDI Ber.*, 26, 75.
Maier, K. H. (1959) Über die Verwendung von Membranfilters in der Luftuntersuchung, *Staub,* 19, 16–23.
Matteson, M. J. (1987) Analytical applications of filtration, in: Matteson, M. J. and Orr, C., Eds., *Filtration,* Marcel Dekker, New York, 629–673.
Michaelis, H. (1880) Prüfung der Wirksamkeit von Staubrespiratoren, *Z. Hyg.*, 9, 389.
Mohrmann, H. (1970) Beladung von Faserfiltern mit Aerosolen aus flüssigen Partikeln, *Staub Reinhalt. Luft,* 30, 317–321.
Natanson, G. (1957a) Diffusive deposition of aerosol particles flowing past a cylinder, *Dokl. Akad. Nauk USSR,* 112, 100–103.
Natanson, G. (1957b) Deposition of aerosol particles by electrostatic attraction upon a cylinder around which they are flowing, *Dokl. Akad. Nauk USSR,* 112, 696–699.
Pich, J. (1964) Impaction of aerosol particles in the neighbourhood of a circular hole, *Collect. Czech. Chem. Commun.*, 29, 2223–2227.
Pich, J. (1966a) Pressure drop of fibrous filters at small Knudsen numbers, *Ann. Occup. Hyg.*, 9, 23–27.
Pich, J. (1966b) Theory of aerosol filtration by fibrous and membrane filters, in C. N. Davies, Ed., *Aerosol Science,* Academic Press, London, 223–285.
Pich, J. (1969) Der Druckabfall der Faserfilter in molekularer Strömung, *Staub Reinhalt. Luft,* 29, 407–408.
Pich, J. (1971) Pressure characteristics of fibrous aerosol filters, *J. Colloid Interface Sci.*, 37, 912–917.
Pich, J. (1987) Gas filtration theory, in M. J. Matteson and C. Orr, Eds., *Filtration — Principles and Practices,* Marcel Dekker, New York, 1–132.
Ramazzini, B. (1940) *De Morbis Artificum,* translated by W. Cave, Wright, Chicago.
Reznik, J. B. (1951) Dust sampling by membrane filters, *Gig. Sanitar.,* (Russia), 10, 28–34.
Rodebush, W. H. (1941) Aerosol Filter Materials, U.S. Office of Sci. Res. and Develop. No. 58.
Sell, W. (1931) Staubabscheidung an einfachen Körpern und Luftfiltern, *VDI Ber.,* 247, 1–14.
Spielman, L. and Goren, S. L. (1968) Model for predicting pressure drop and filtration efficiency in fibrous media, *Environ. Sci. Technol.*, 2, 279–287.
Spurny, K. (1965/1966) Membranfilter in der Aerosologie 1–3; *Zbl Biol. Aerosol-Forsch.*, 12/13, 369–407; 3–56; 398–451.
Spurny, K. and Polydorova, M. (1961) Measurement of the electric surface charge of MF during filtration, *Collect. Czech. Chem. Commun.*, 26, 932–943.
Spurny, K. R. and Vondracek, V. (1957) Analytical methods for determination of aerosols by means of membrane ultrafilters, *Collect. Czech. Chem. Commun.*, 22, 22–31.
Spurny, K. R., Lodge, J. P., Frank, E. R., and Scheesley, D. C. (1969) Aerosol filtration by means of Nuclepore filters, *Environ. Sci. Technol.*, 3, 453–468.
Stechkina, I. B. and Fuchs, N. A. (1966) Studies of fibrous filters. Calculation of diffusional deposition of aerosols in fibrous filters, *Ann. Occup. Hyg.*, 9, 59–64.
Stechkina, I. B., Kirsch, A. A., and Fuchs, N. A. (1969) Studies on fibrous aerosol filters. Calculation of aerosol deposition in model filters in the range of maximum penetration, *Ann. Occup. Hyg.*, 12, 1–8.
Stechkina, I. B., Kirsch, A. A., and Fuchs, N. A. (1970) Effect of inertia on the capture coefficient of aerosol particles by cylinders at low Stokes' numbers, *Kolloidn. Zh.* (Russia), 32, 467–473.

Stern, S. C., Zeller, H. W., and Schekman, A. I. (1960) The aerosol efficiency and pressure drop of a fibrous filter at reduced pressures, *J. Colloid Sci.,* 15, 546–562.

Van Orman, W. T. and Endres, H. A. (1952) Self-charging electrostatic air filters, *ASHVE J. Sci.,* 157–163.

Vincent, J. A. and Kasper, F. (1995) C. N. Davies. In memoriam, *J. Aerosol Sci.,* 25, 1253–1268.

Walkenhorst, W. (1970) Reflections and research on the filtration of dust from gases with special consideration of electrical forces, *J. Aerosol Sci.,* 1, 225–242.

Walkenhorst, W. and Zebel, G. (1964) Über ein neues Schwebestoffilter hoher Abscheideleistung und geringerer Strömungswiderstandes, *Staub Reinhalt. Luft,* 24, 444–448.

Walton, W. H. (1940) The Methylene Blue Particulate Test for Respirator Containers, Porton Down Report Specification 1206.

Walton, W. H. (1941) The Sodium Flame Particulate Test for Respirator Containers, Porton Down Report 2465.

Watson, J. H. L. (1946) Filmless sample mounting for the electron microscope, *J. Appl. Phys.,* 17, 121–127.

Whitby, K. T. (1965) Calculation of the clean fractional efficiency of low media density filters, *Am. Soc. Heating, Refrig. Air Con. Eng. J.,* Sep. 56–65.

Yoshioka, N., Emi, H., and Fukashima, M. (1967) Filtration of aerosols by fibrous filters, *Kagaku Kogaku* [*Chem. Eng. Jpn.*], 31, 157–163.

Yoshioka, N., Emi, H., and Yasunami, M. (1969a) Filtration of aerosols through fibrous packed beds, *Kagaku Kogaku* [*Chem. Eng. Jpn.*], 33, 381–386.

Yoshioka, N., Emi, H., and Yasunami, M. (1969b) Filtration of aerosols through fibrous packed beds loaded with dust, *Kagaku Kogaku* [*Chem. Eng. Jpn.*], 33, 1013–1019.

Yoshioka, N., Emi, H., Kanaoka, C., and Ysasunami, M. (1972) Collection efficiency of aerosol by an isolated cylinder when gravity and inertia predominate, *Kagaku Kogaku* [*Chem. Eng. Jpn.*], 36, 313–319.

Zebel, G. (1965) Deposition of aerosol flowing past a cylinder fibre in a uniform electric field, *J. Colloid Sci.,* 20, 522–543.

Zebel, G. (1969) Zur Aerosolabscheidung and einer Einzelfaser unter dem Einfluß elektrischer Kräfte unter geringerer Strömungswiderstand, *Staub Reinhalt. Luft,* 29, 62–66.

Zsigmondy, R. and Bachmann, W. (1916) German Patents 329 060 l (May 9) and 329 117 (August 22).

Zsigmondy, R. and Bachmann W. (1918) Zellulosenitrat-Membranfilters. *Z. Anorg. Chem.,* 103, 119–128.

CHAPTER 2

Aerosol Filtration Science at the End of the 20th Century

Kvetoslav R. Spurny

CONTENTS

2.1 INTRODUCTION

The processes of separation and filtration of fine particles in gases are among the most important fields of aerosol mechanics. These processes deal with the removal of fine particles from a gas stream on macro- and micro-obstacles, whose geometry and structure are optimized in such a way that the particle collection efficiencies reach a desirable degree.

Filtration theories have been very well developed during the last 50 years, especially since the 1960s, when high-speed computer systems could be applied in this field. The same is also valid for the development of experimental methodology and for the development of new filters and separators. Very satisfactory filtration systems are available at the present time.

The history of air filters is well known and started almost 2000 years ago. Nevertheless, their forms and applications remained in the range of empirical knowledge for a long time. The first quantitative and theoretical considerations did not start until the 1920s. Rapid development occurred during and after World War II (see also Chapter 13).

0-87371-830-5/98/$0.00+$.50

For about the last 20 years, good fibrous and pore filters have been available and used. Filtration theories were also very successfully developed. The latest state of the art in this field is very well documented and evaluated in the monograph of Brown (1993). Nevertheless, there remain unsolved or partly solved problems in the field of aerosol filtration. This book tries to point out some new important contributions and new theoretical approaches.

2.2 CLASSICAL FILTRATION THEORIES

The classical filtration theories of aerosol particle separation in fibrous filters are principally based on the "single-fiber element" model. Several partial mechanisms are combined in integral particle collection on the surface of a cylindrical fiber (Figure 2.1). The total collection efficiency is defined by a superposition of these partial collection mechanisms. Therefore, the characteristic separation curves have penetration maxima or collection minima. The collection efficiency of the total filter is obtained by combining the single-fiber efficiencies with the macroscopic parameters of the whole filter.

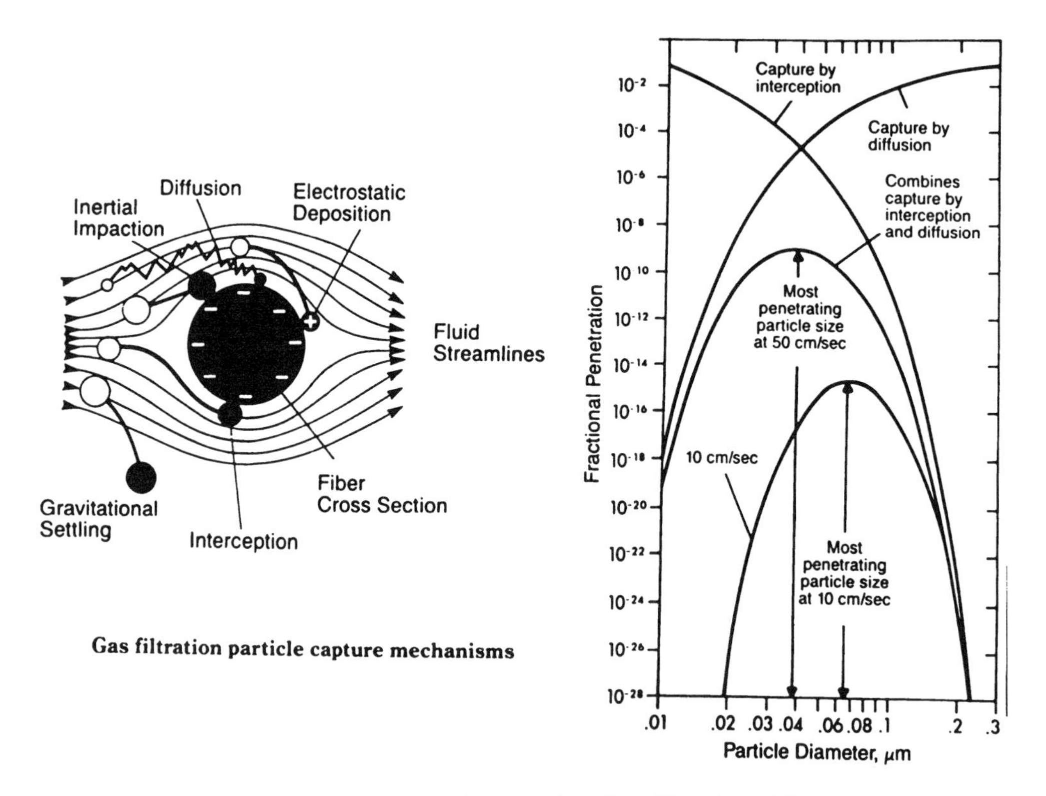

Figure 2.1 Separation mechanisms and filter characteristics, single-fiber element theory.

Particles separated on the surfaces of single-fiber elements are characterized by irregular distribution and often build dendrite structures (Figure 2.2). A new and very improved theory of particle separation and spatial distribution on cylindrical surfaces (single-fiber elements) was recently described by Rosner et al. (1995). By means of this theory, spatial particle deposition can be predicted by calculation procedures. Improved filtration theories, better definition and description of the filter structure, as well as the deposition mechanisms and their combinations, are of basic importance.

Figure 2.2 Particle deposited on the single fiber and the dendrites (Transmission electron micrographs.)

2.2.1 Irregularities in Filter Structures

Caused by manufacturing processes, the structure of fiber filter media is inhomogeneous. The real internal structure shows local fluctuations that influence filtration properties. Measurement procedures and computation methods have been recently developed and described by Mölter and Fißan (1995 and Chapter 13 of this book) which make it possible to consider these inhomogeneities in the estimation of the pressure drops and of the effective fiber diameter in the whole inhomogeneous fiber filter.

2.2.2 Model Improvements

The classical fiber filter models are based on the flow field of a viscous fluid and on the combination of the partial separation mechanisms. For further improvement, correction coefficients for the gas slip on the single fiber, considering, e.g., the diffusion (*Cd*), interception (*Cr*), etc., were introduced in the existing theories. These corrections lead to a much better fit of the theory with experimental data (Payet, 1992; Gougeon, 1994; Attoui, 1994 and 1995). Table 2.1 summarizes these correction factors.

Table 2.1 Original Parameters Characterizing the Behavior of Aerosol Particles in the Gas Flow through the Fiber Filter as Well as Their Values Corrected with Relation to the Gas Pressure, Gas, Slip, etc.

Parameters	Corrected Parameters
$\mathrm{Kn}_p = \dfrac{2\lambda_0}{D_p}$	$\mathrm{Kn}_{pc} = \dfrac{2\lambda_0}{D_p}\dfrac{P_0}{p} = \dfrac{P_0}{p}\mathrm{Kn}_p$
$\mathrm{Kn}_f = \dfrac{2\lambda_0}{D_f}$	$\mathrm{Kn}_{fc} = \dfrac{2\lambda_0}{D_f}\dfrac{P_0}{p} = \dfrac{P_0}{p}\mathrm{Kn}_f$
Cu	$Cu_c = 1 + \mathrm{Kn}_{pc}[1.246 + 0.42\exp(-0.87/\mathrm{Kn}_{pc})]$
Pe	$Pe_c = \dfrac{3D_f D_p U_0 \pi\mu}{k_B T Cu_c}$
Cd	$Cd_c = 1 + 0.388\,\mathrm{Kn}_{fc}\left[\dfrac{(1-\alpha)Pe_c}{\mathrm{Ku}}\right]^{1/3}$
Cr	$Cr_c = 1 + \dfrac{1.996}{R}\mathrm{Kn}_{fc}$
Cd′	$Cd'_c = \dfrac{1}{1 + 1.6\left[\dfrac{1-\alpha}{\mathrm{Ku}}\right]^{1/3} Pe_c^{-2/3}}$
Pe′	$\mathrm{Pe}'_c = \mathrm{Pe}_c Cd_c^{-2/3}$

See description in Attoui, 1995. (From Attoui, M. B., *Actes il Emes J. Etud. Aerosols,* Paris, 68–73, 1995.)

2.2.3 Filtration of Fibrous Aerosols

In the case when the shapes of aerosol particles may not be approximated by spheres, the problem of particle separation on a cylinder becomes more complicated. This requires the description of the trajectory of such a particle together with its rotation, as well as a new definition of the single-fiber efficiency. From practical point of view, the filtration of fibrous aerosols is of very great importance.

The complete movement of a fibrous particle in the airstream in the vicinity of a single cylinder can be described by a set of equations of particle translation and rotation (Gradoń et al., 1988; Grzybowski and Gradoń, 1995; see Chapter 10, this book) There is an important difference between the separation of spherical and fibrous particles (Figure 2.3).

Theoretical considerations as well as measurements have shown that the main separation mechanism for fibrous particles is direct interception. The total filter efficiency increases, therefore, with the increasing aspect ratio with respect to the length of the fibrous particle (Spurny, 1994; see Chapter 20, this book).

2.2.4 Adsorptive Aerosol Filters

Fibrous filters are also produced by using chemically activated carbon fibers. Such aerosol filters having high porosity as well as high specific surfaces can be used for a combined separation or collection of gaseous and particulate air pollutants (Figure 2.4) (Spurny, 1993; see Chapter 21, this book).

2.2.5 Loading Characteristics of Electrically Charged Filters

Electrically charged fibrous filters are commonly used as the filter of choice in industrial air-cleaning processes. The electric charge of the filter, also of the aerosol particles, can positively influence the filter characteristics. Nevertheless, the problem is relatively quite complex and depends on several time-dependent parameters — particle size, type, and charge, filter face velocity, and filter charge. The time-dependent behavior of such an electrically charged fibrous filter (electret filter) can be demonstrated well by its "loading characteristics." The initial filter penetration increases as the original filter charge is reduced by several neutralization processes. The penetration reaches a maximum. Then, because of heavy clogging, the penetration decreases and the pressure drop increases fast (Figure 2.5). This very complex process was studied more deeply recently by Walsh and Stenhouse (1997).

Their work has shown that the size, charge, and composition of aerosol particles have a significant effect on the loading characteristics of a mixed-fiber electrically active material. Smaller particles cause filter samples to become clogged more quickly, and as a result the filtration efficiency is also degraded more quickly. The charge of the particle also affects its clogging rate, with more-charged particles having greater clogging points. The effect of this charge on the clogging point causes filtration efficiency to be degraded more quickly by uncharged particles, implying that the mechanism of filter degradation is most certainly not charge neutralization. The composition of the loading particles also has a significant effect on loading behavior. If the loading of a filter sample is normalized, so that penetration is plotted as a function of available space occupied, the rate of filter degradation becomes independent of size and charge, with all data sets reaching a maximum penetration at approximately 65% of the dust-holding capacity.

2.3 MULTIFIBER FILTER MODELS

Most filtration models in the past considered only flow past a single fiber. At the end of the 1950s Happle and Kuwabara were the first to consider flow past various systems of cylinders in the flows. The influences of neighboring fibers were considered by imposing further artificial boundary conditions. The filtration theory was further advanced by solving the Navier–Stokes equations for flow around multifiber fibers by Brown in the later 1980s. Recently, Liu and Wang (1996) have substantially improved the multifiber filter model.

Flow fields around arrays of parallel and staggered fibers were computed by solving the incompressible steady-state Navier–Stokes equations numerically. They could show that viscous

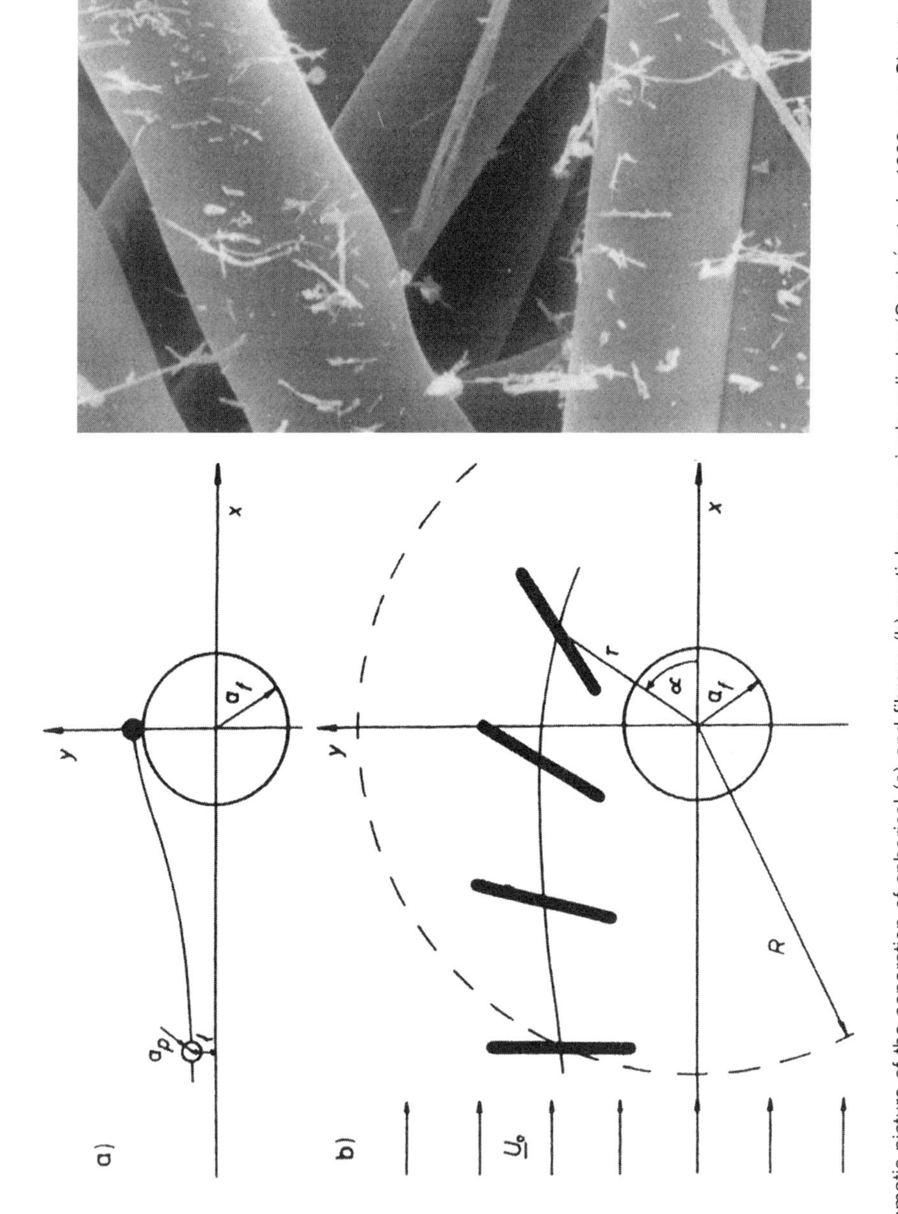

Figure 2.3 Schematic picture of the separation of spherical (a) and fibrous (b) particles on a single cylinder (Gradoń et al., 1988; see Chapter 10, this book). Scanning electron micrograph (right) of fibrous particles deposited in a fiber filter. (Gradoń, L. et al., *Chem. Eng. Sci.*, 43, 1253–1259, 1988. With permission.)

Figure 2.4 Scanning electron micrograph of the inner structure of a carbon fiber filter.

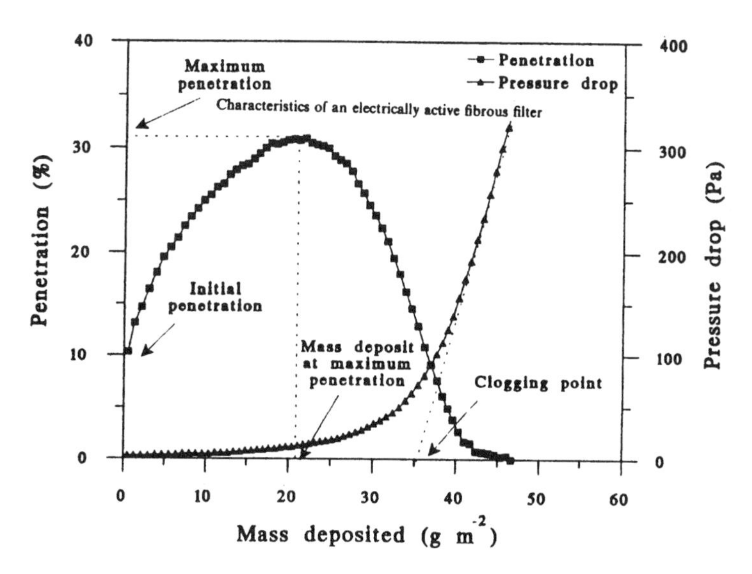

Figure 2.5 Behavior of time-dependent particle deposition in a fibrous electret filter. (From Walsh, D. C. and Stenhouse, J. I. T., *J. Aerosol Sci.,* 28, 307–321, 1997. With permission.)

flow fields around multifiber models becomes periodic immediately after the first fiber array downstream of the filer entrance until the last fiber array is reached (Figure 2.6).

In a next improvement, Filippova and Hänel (1996) used a numerical simulation of three-dimensional gas–particle flow through filters. This simulation of gas–particle flow through filters includes deposition of particles on the filter surfaces and their interactions with the flow field. The interaction results from the change of shape due to deposited particles, which influences again the

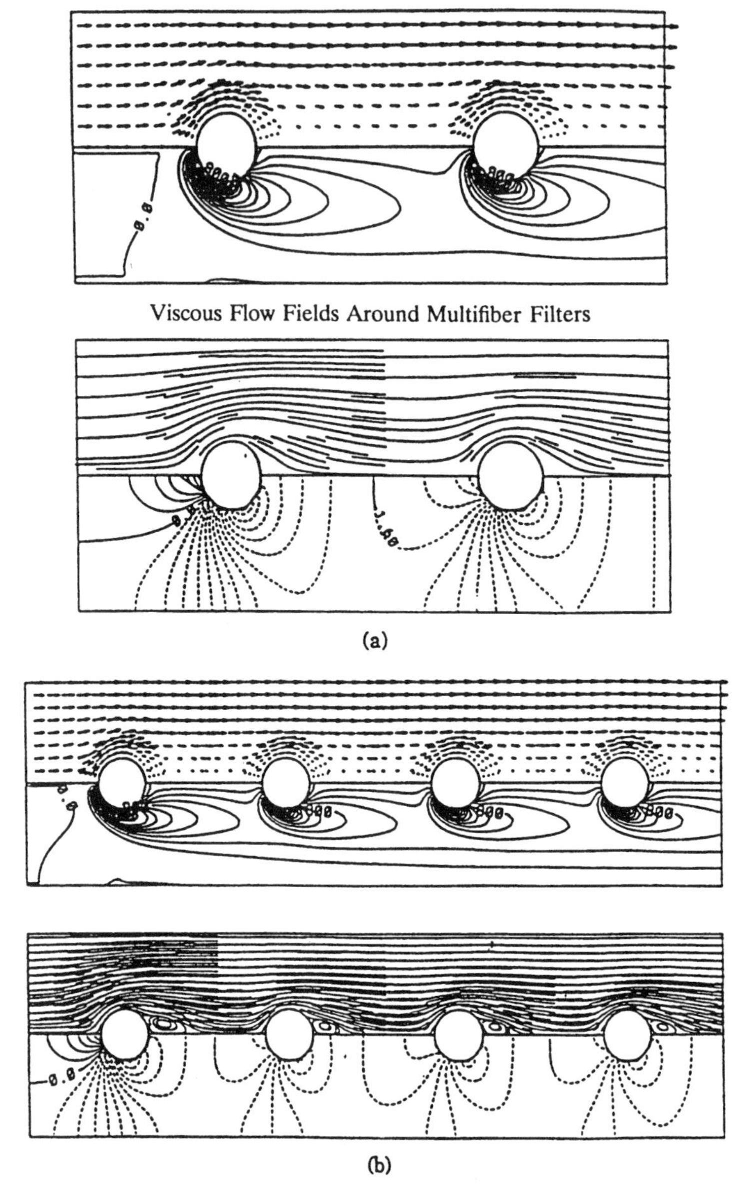

Figure 2.6 Viscous flow fields around arrays of circular fibers. (From Liu, Z. G. and Wang, P. K., *Aerosol Sci. Technol.,* 25, 375–391, 1996. With permission.) Velocity vectors and negative velocity contours (lower): Re = 10 (a) and Re = 20 (b).

flow field and the succedent particle deposition. The capability of this concept is demonstrated in Figure 2.7. Gas and particles flow through a model of a sieve filter. Particles are deposited on the fibers and the gas flow is visualized by the plotted streamlines (lowest picture). The filter exposure time increases in the perpendicular direction of the pictures (see also Chapter 9, this book).

2.4 INTEGRATED FILTRATION THEORIES

The classical "single-filter element" filtration theory is and will remain a very useful tool for prediction of particle separation in aerosol filters. Nevertheless, since about the 1990s important

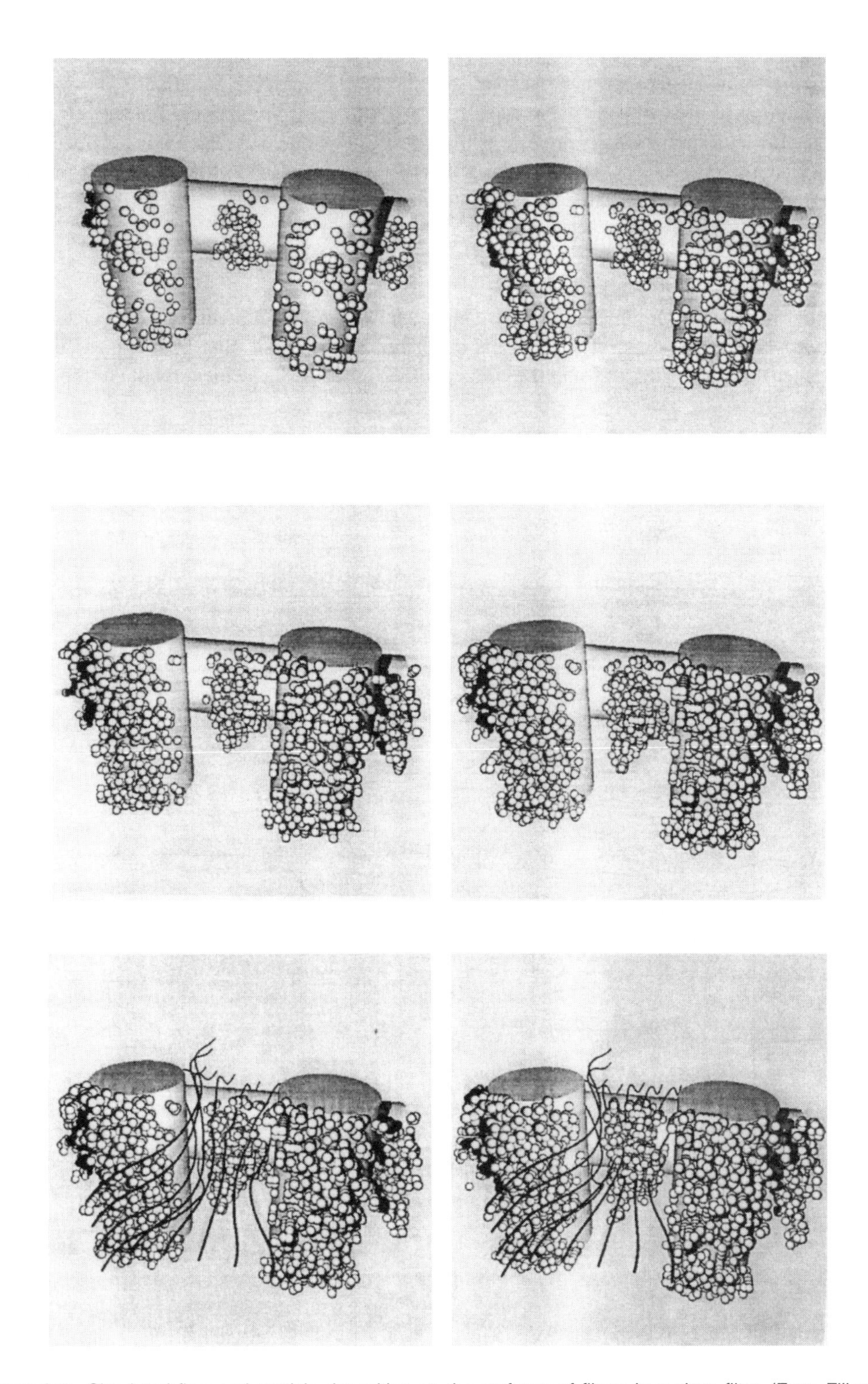

Figure 2.7 Simulated flow and particle deposition on the surfaces of fibers in a sieve filter. (From Filippova, O. and Hänel, D., *J. Aerosol Sci.,* 27 (Suppl. 1), 1996. With permission.)

progress has been achieved in the field of the general filtration theories based on an integrated porous filter model (Shapiro et al., 1991; Quintard and Whitaker, 1995; Shapiro, 1996, Chapters 5 and 6, this book). These methods consider the process of filtration in the whole structure of the porous media by using the particle transport and the particle capture equations. These enable a direct calculation of the cellular efficiencies.

Very useful and experimentally proved results have been obtained by using a dispersion–reaction model by Shapiro et al. (1991). This model provides a precise scheme for calculating an aerosol filtration length l_f, appearing in the collection efficiency equation

$$\eta = 1 - \exp(-L/l_f)$$

L is the filter processing length (Figure 2.8). This model is based on a precise physicomathematical formulation of the aerosol microtransport and deposition in the entire filter bed. The calculation of the parameter η is done without using the concept of the single-fiber element model of the classical filtration theory.

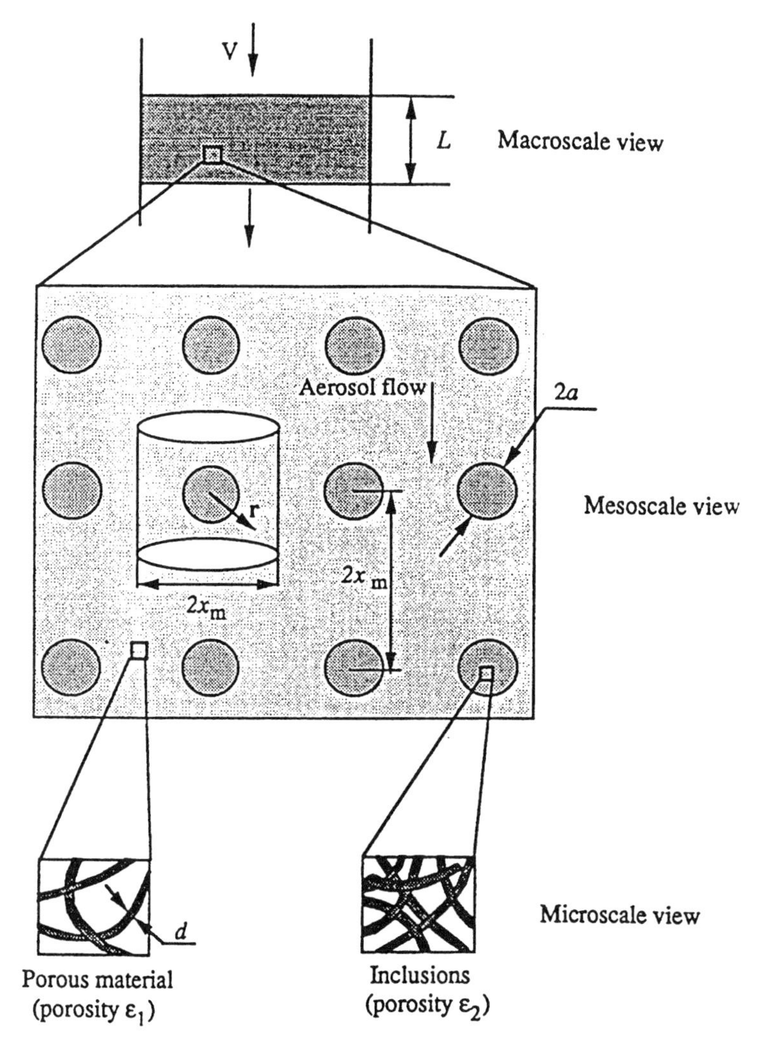

Figure 2.8 Spatially periodic microstructure of a porous filter. (From Shapiro et al., 1991; Shapiro, 1996; Chapter 5, this book; Dinariev and Shapiro, 1996.)

Another important advantage of these integrated theories can be the possibility of overcoming the problem of the nonhomogeneity of the porous filter structure. The effect of filter inhomogeneities is to cause a nonuniform pressure gradient. Shapiro (1996) has described an analytical model of such a filtration process. Calculations performed for spherical inclusions showed that they tend to decrease the aerosol filtration efficiency when their porosity is less than that of the surrounding filter medium.

2.5 FILTRATION PHYSICS AND CHEMISTRY

Separation of aerosol particles in a clean filter, designated as the filter statics and filter clogging, that is, the time-dependent particle deposition, designated as the filtration dynamics or the nonstatic filtration, are physical processes that characterize the use of aerosol filters in air cleaning and aerosol sampling. The above-mentioned partial separation mechanisms in filters belong also in filtration physics.

Particles deposited in the porous structure and on the surface of a real aerosol filter are in the majority of cases chemically active material. They may have very different chemical composition, can be partly volatile, and can undergo during the filter action or aerosol sampling chemical and physicochemical reactions and changes. Stable, as well as short-living and intermediate, new substances can be produced during the filter action. They may be harmless as well as harmful, inert as well as highly reactive. In such a way the *aerosol filtration chemistry* may play an important role in aerosol separation processes and should not be neglected. Nevertheless, there do exist only rare publications dealing with this problem and research needs in this field are broad. We have relatively more knowledge about these processes, concerning analytical aerosol filtration. They may produce important artifacts by changing the chemical composition of the sampled aerosol probe, may produce volatilization, etc.

The chemical composition of the particulate samples may therefore differ substantially from the chemical composition of the airborne particles (Appel et al., 1984; Matteson, 1986). Because of volatilization, aerosol mass concentration can be underestimated and, as a consequence of the sorption of gaseous pollutants, the mass concentration can be overestimated. These chemical artifacts can therefore erroneously decrease or increase the measured particulate concentrations in the air (Batterman et al., 1997).

Although the particle-removal mechanisms in filters are relatively well understood, the chemical reactions and processes in particulate deposits in filters have not been satisfactorily studied as yet. The reaction kinetics of small particles deposited in filters can be simple (e.g., absorption and adsorption of gases) as well as relatively complex. Particle oxidation is often a common chemical process in particulate filter deposits. Soot and other carbonaceous particles, including organic carbonaceous compounds, are important components of anthropogenic aerosols. After their separation and deposition in filters, they can undergo slow or fast oxidation. The oxidation rate can depend on the filter material as well as on the other components of the deposited aerosol (Lin and Friedlander, 1988) and is, of course, proportional to the partial pressure of the oxygen.

REFERENCES

Appel, B. R., Y. Tokiwa, M. Haik, and E. L. Kothny (1984) Artifact particulate sulfate and nitrate formation on filter media, *Atmos. Environ.,* 18, 409–416.

Attoui, M. (1994) Une Nouvelle Approche Experimentale de la Mesure de l'Efficacite des Filtres a Fibres a Basse Pression, these en science des aerosols, Universite de Paris XII, CEA Rapport.

Attoui, M. B. (1995) Efficacite des filtres a fibres a basse pression comparison theorie experience, *Actes 11 Emes J. Etud. Aerosols,* Paris, 68–73.

Batterman, S., I. Osask, and C. Gelman (1997) SO_2 sorption characteristics of air sampling filter media using a new laboratory test, *Atmos. Environ.*, 31, 1041–1047.

Brown, R. C. (1993) *Air Filtration,* Pergamon Press, Oxford, 1–272.

Dinariev, O. Y. and A. A. Shapiro (1996) Nonlocal accurate solutions of kinetic gas theory in porous medium, *J. Technol. Physi.* (Russia), 66, 24–34.

Filippova, O. and D. Hänel (1996) Numerical solution of particle deposition in filters, *J. Aerosol Sci.,* 27, (Suppl. 1), S627–S628.

Gougeon, R. (1994) Experimental Study of Stationary Filtration with Model Fiber Filters, these de l'Universite de Paris XII, Rapport.

Gradon, L., P. Grzybowski, and W. Pilaciski (1988) Analysis of motion and deposition of fibrous particles on single filter element, *Chem. Eng. Sci.,* 43, 1253–1259.

Grzybowski, P. and L. Gradon (1995) Filtering efficiency of stiff fibrous particles on single filter element, *J. Aerosol Sci.,* 26 (Suppl. 1), 725–726.

Lin, C. and S. K. Friedlander (1988) Soot oxidation in fibrous filters 1, 2, *Langmuir,* 4, 891–903.

Liu, Z. G. and P. K. Wang (1966) Numerical investigation of viscous flow fields around multifiber filter, *Aerosol Sci. Technol.,* 25, 375–391.

Matteson, M. J. (1986) Analytical applications of filtration, in *Filtration,* M. J. Matteson and C. Orr, Eds., Marcel Dekker, New York, 629–673.

Mölter, W. and H. Fißan (1995) Die Filtrationseigenschaften von HEPA-Glasfaser-Filtermedien, *Staub Reinhalt. Luft,* 55, 411–416 and 463–466.

Payet, S. (1992) Filtration Stationnaire et Dynamique des Aerosol Liquides Submicroniques, these de l'Universite de Paris XII, Rapport.

Quintard, M. and S. Whitaker (1995) Aerosol filtration: an analysis using the methods of volume averaging, *J. Aerosol Sci.,* 26, 1227–1255.

Rosner, D. E., P. Tandon, and A. G. Konstantopulous (1995) Local size distributions of particles deposited by internal impaction on cylindrical target in dust-loaden streams, *J. Aerosol Sci.,* 26, 1257–1279.

Shapiro, M. (1966) An analytical model for aerosol filtration by nonuniform filter media, *J. Aerosol Sci.,* 27, 263–280.

Shapiro, M., I. J. Kettner, and H. Brenner (1991) Transport mechanics and collection of submicrometer particles in fibrous filters, *J. Aerosol Sci.,* 22, 707–722.

Spurny, K. R. (1993) Aerosolfilter aus aktivierten Kohlenstoffasern, *Filtrieren Separieren* (Germany), 7, 34–36.

Spurny, K. R. (1994) Zur Filtration und Separation von Faserigen Mineralstäuben, *Filtrieren Separieren* (Germany), 8, 166–175.

Walsh, D. C. and J. I. T. Stenhouse (1997) The effect of particle size, charge, and composition on the loading characteristics of an electrically active fibrous filter material, *J. Aerosol Sci.,* 28, 307–321.

CHAPTER 3

Single-Fiber Collection Efficiency

Orest Lastow and Albert Podgorski

CONTENTS

0-87371-830-5/98/$0.00+$.50

3.1 INTRODUCTION

The collection efficiency of a single, clean fiber has been the subject of a number of studies. Knowledge about the deposition on single-fibers is of fundamental importance in the understanding of filtration mechanisms. The single-fiber is the smallest filtration element in a fibrous filter and the single-fiber results are often extrapolated to describe the whole filter. In this study a single-fiber is placed in an infinite space, i.e., no neighboring fibers are present and the collection efficiency is investigated.

Several results from calculations of the single-fiber collection efficiency using the potential flow model and the Lamb flow model have been published (Fuchs, 1964; Davies, 1973; Pulley and Walters, 1990). Another approach using a matrix of fibers has been developed by Kuwabara (1959). The Kuwabara approach is based on the Lamb flow model and takes into account the influence of the neighboring fibers. The potential flow model and the Lamb flow model do not represent a realistic flow around a cylindrical fiber in an infinite space. This investigation employs a new flow model, based on the Oseen approximation (Podgorski, 1993), see Appendix.

The traditional approach is to calculate separately one inertial collection efficiency for particles with zero diameter and one interception efficiency for particles with zero mass. The distinction and separation of the different efficiencies were made in order to simplify the calculations when computers were scarce, and also to save time. Diffusional deposition on fibers is normally calculated for an ensemble of particles moving in a concentration gradient. The diffusional collection efficiency is not valid for a single particle in a flow past a fiber. In the studies by Davies (1973), Hinds (1982), and Lee et al. (1993) the three efficiencies — interception efficiency, inertial collection efficiency, and diffusional efficiency — were added to obtain the total efficiency. The arithmetical sum of the efficiencies can, however, only be considered as a first approximation of the total efficiency (Hinds, 1982; Lee et al., 1993). The efficiencies are normally presented as functions of the Stokes number, Stk. One objective of this work is to calculate the combined inertial collection and interception efficiency as a function of particle diameter, fiber diameter, and air parameters. We also propose a general efficiency definition for any number of fibers and a combined collection efficiency definition for inertial collection, interception, and diffusion.

The collection efficiency of a single clean fiber depends on a number of particle and air parameters. When theoretical results are compared with experimental results it is important to know how variations in the ambient air conditions change the efficiency. Uncertainties in particle density will also give uncertainties in the efficiency. In this study, the influence of variations (10 to 100%) in the different parameters is calculated.

When a particle is captured on a fiber, it remains there as a result of a combination of different attraction forces, e.g., the van der Waals force. Other particles can then be captured on the earlier captured particle. The captured particles form chainlike agglomerates known as dendrites. Dendrites dramatically increase the collection efficiency of the fiber–dendrite system. Knowledge about filter loading mechanisms can help us to prolong the lifetimes of filters and to optimize their working conditions. This can result in economical savings for both the producer and the user of filters. A reliable model of the filtration process in a fibrous filter can be an important contribution to air cleaning in general. The growth of dendrites has been the subject of a number of investigations. Billings (1966) performed a number of systematic experimental observations of dendrites. A theoretical investigation of the phenomenon was made by Payatakes and Gradon (1980). They suggested a deterministic set of equations governing the growth of an idealized dendrite. In this study, a stochastic approach is used and serves as a complement to the deterministic approach. This approach is similar to that employed by Kanaoka et al. (1980), but in this case a single-fiber flow is used instead of the Kuwabara flow. In this chapter, the collection efficiency of a typical dendrite is calculated.

3.2 THEORY AND METHOD

3.2.1 Transport and Collection

When particles move through a fibrous filter they deviate from the air streamlines. This deviation is caused by a combination of the following mechanisms:

- Particle inertia when the streamlines bend;
- Diffusion due to Brownian motion of the particles;
- Electric forces due to charged particles, charged fibers, or external fields;
- Gravity;
- Other external forces.

In porous filters, the mechanism that removes particles from the air is interception with fibers. The interception of a homogeneous spherical particle by a cylindrical fiber is defined, in this chapter, in the following way:

- A particle is intercepted by a fiber when the distance from the center of mass of the particle to the fiber surface is equal or less than the radius of the particle.

The deviation mechanisms mentioned above are traditionally grouped together with interception and called deposition mechanisms or collection mechanisms (Fuchs, 1964; Davies, 1973; Hinds, 1982; Pulley and Walters, 1990; Lee et al., 1993). The only real deposition or collection mechanism is interception: the others are deviation mechanisms. The deviation mechanisms govern the particle trajectory in the airflow and can both increase and decrease the collection efficiency. The objective of this study is to calculate the combined inertial collection and interception efficiency as a function of particle diameter, fiber diameter, and air parameters.

3.2.2 FiFi

The calculations presented in this chapter were performed using an improved version of the personal computer code FiFi (Lastow and Bohgard, 1992). FiFi is a personal computer program which performs a real-time two-dimensional simulation of the dynamic deposition process of particles on one or two fibers (Figure 3.1). The details of the FiFi model will be discussed in the following sections.

FiFi can be used to simulate, not only the influence of a large number of independent parameters, but also the influence of dendrites on the collection efficiency. FiFi can also visualize the dependence of the particle dynamics on various parameters. Several types of electrical fields can be applied around the fiber.

FiFi is a stand-alone application which meets the Macintosh user-interface guidelines. Results can be exported as PICT or TEXT documents which can be imported by all major word processors, spreadsheet programs, and DTP-software. The user-friendly graphic human interface has made it possible even for nonexperts to use the software. FiFi is a very versatile and powerful program with many features:

- One or two fibers
- Two drag force models: Stokes's and Oseen's drag force;
- Six different flow models: potential flow, Lamb flow, Davies flow, Kuwabara flow, and single and double Podgorski flow;
- The flow can be illustrated by vectors or streamlines;

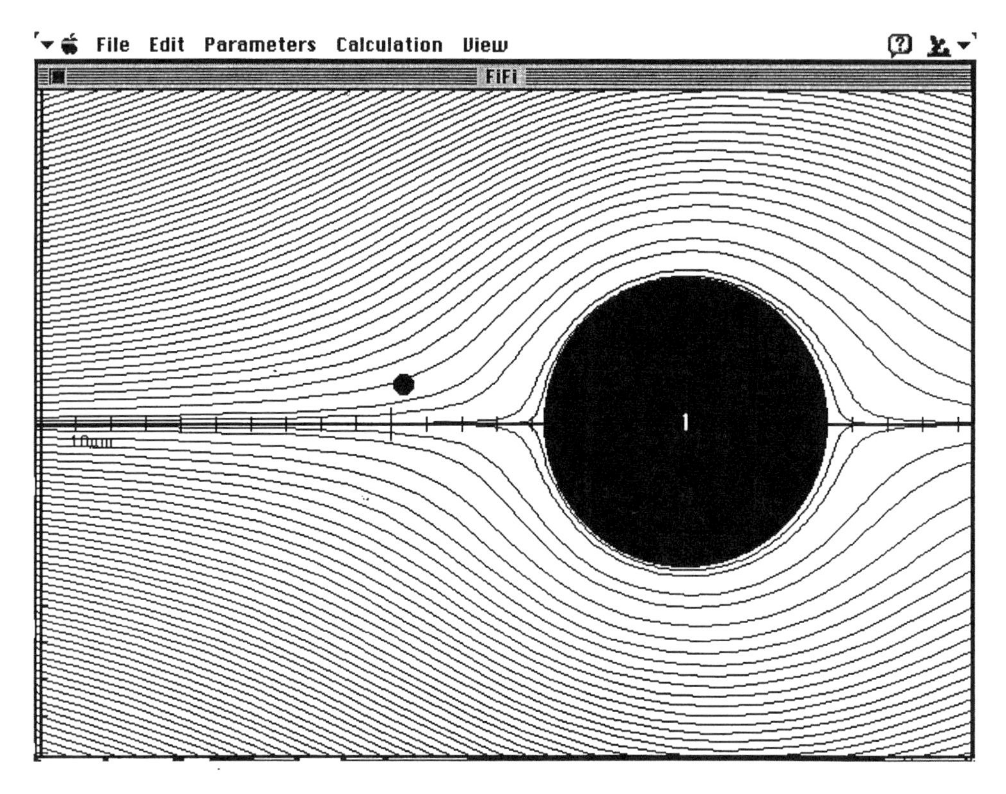

Figure 3.1 Simulation in progress performed by FiFi. The parameter values were V_0 = 0.1 m/s, d_p = 6 μm, and d_f = 80 μm. In the background the streamlines are shown. The flow is from left to right. The distance between two markings on the x-axis is 10 μm.

- Efficiency calculations;
- Three diffusion models;
- Gravity in any direction;
- Electrostatic force in any direction;
- Simulation of dendritic growth;
- Efficiency calculations for fibers with dendrites;
- Five variable particle parameters: density, diameter, charge, starting position, and starting velocity;
- The particle parameters can be randomly distributed according to uniform, Gaussian, or lognormal distributions;
- Four variable air parameters: density, viscosity, mean free path, velocity;
- Five variable fiber parameters: diameter, dielectric constant, position, charge, dipole moment;
- Integration accuracy can be set arbitrarily.

The number of drag force models, flow models, diffusion models, and particle parameter distributions can easily be increased.

The objective is to simulate the filtration process inside a fibrous filter. In modern fibrous filters the porosity is very high, close to 99%. In this kind of filter the single-fibers make up the basic filtration elements.

The Podgorski flow model is also derived for two fibers. FiFi implements this model which is the only analytical flow model for two fibers. The collection efficiency for a system of two fibers can thus be calculated. The traditional efficiency definition is, however, not applicable to a system of two fibers, which is why a new general definition must be found before calculations can be made. FiFi can also be used to simulate the growth of dendrites on a system of two fibers, see Figure 3.2.

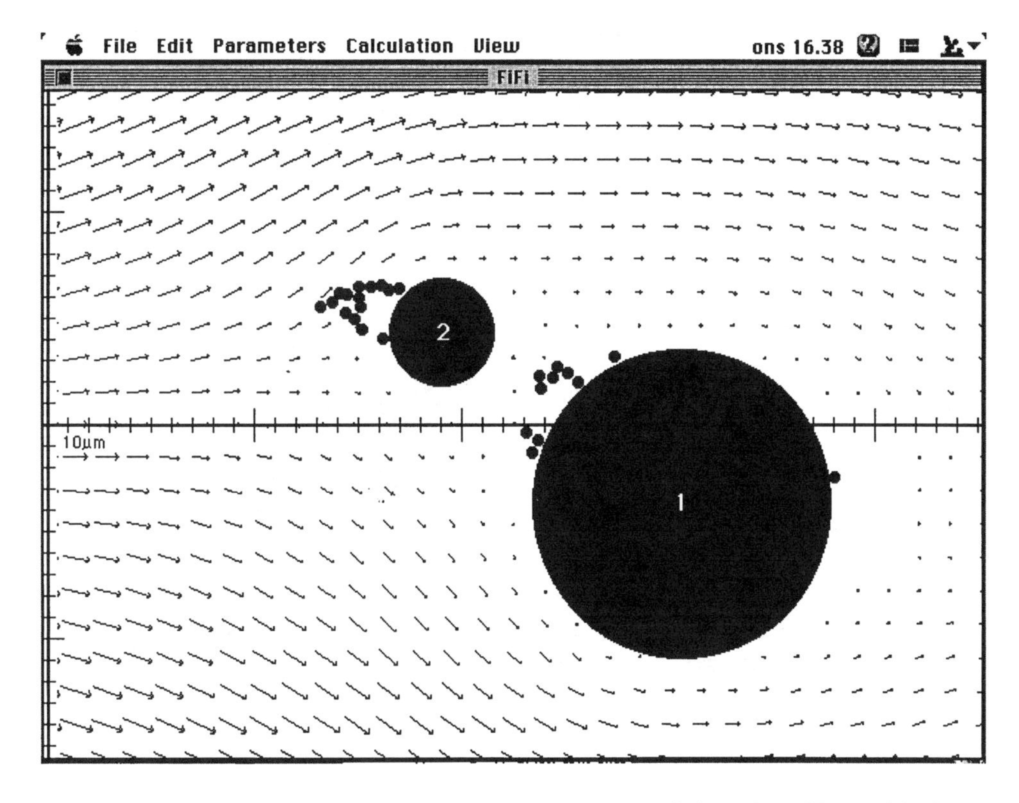

Figure 3.2 Typical dendrites. The fiber diameters are d_{f1} = 140 μm and d_{f2} = 50 μm. The particle diameter is d_p = 6 μm. The arrows are velocity vectors. The undisturbed air velocity was V = 0.1 m/s.

3.2.3 Flow Field Models

In the FiFi code the fiber is approximated by an infinitely long cylinder, perpendicular to the flow. The compressible, stationary, and linearly elastic flow past the fiber is governed by the Navier–Stokes equations.

$$\rho_a \dot{\boldsymbol{r}} \cdot \nabla \dot{\boldsymbol{r}} = -\nabla p + \eta \Delta \dot{\boldsymbol{r}} + \rho_a \boldsymbol{g} \tag{3.1}$$

Earlier calculations of the single-fiber efficiency have employed either the nonviscous flow approximation (potential flow) (see Figure 3.3A).

$$\eta \Delta \dot{\boldsymbol{r}} = 0 \rightarrow \rho_a \dot{\boldsymbol{r}} \cdot \nabla \dot{\boldsymbol{r}} = -\nabla p + \rho_a \boldsymbol{g} \tag{3.2}$$

or the noninertial creeping flow approximation derived by Stokes (Lamb flow), see Figure 3.3B.

$$\rho_a \dot{\boldsymbol{r}} = \rho_a \boldsymbol{g} = 0 \rightarrow 0 = -\nabla p + \eta \Delta \dot{\boldsymbol{r}} \tag{3.3}$$

The potential flow does not fulfill the boundary condition $V = 0$ at the fiber surface and does not exert any drag on the fiber. The potential flow is a good approximation of the flow far away from the fiber, but the error is considerable close to the fiber. The potential field can be used for very high Re_f. The Lamb flow is a better approximation close to the fiber and fulfills the surface boundary condition but, at large distances from the fiber, the velocity increases to infinity. This is called the

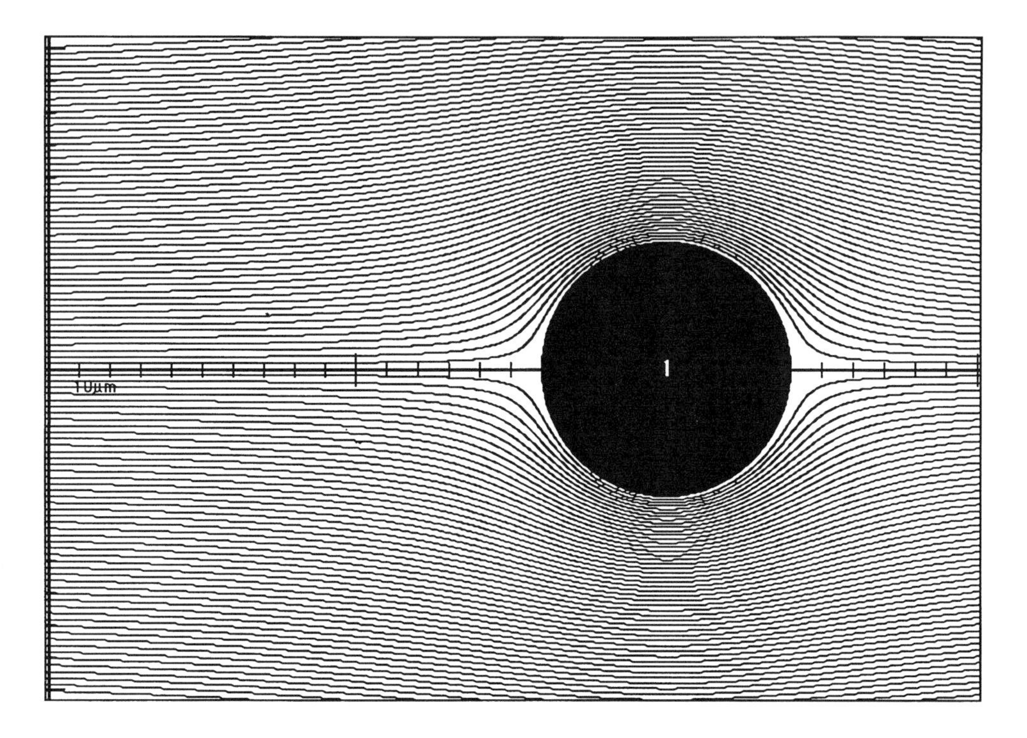

Figure 3.3A Potential flow.

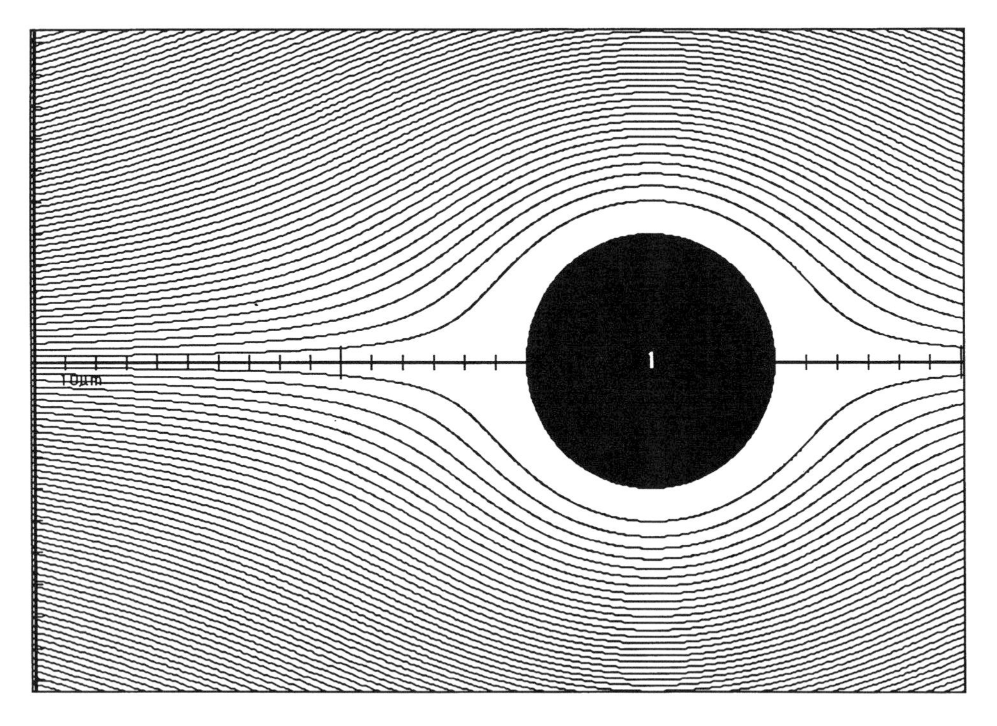

Figure 3.3B Lamb flow.

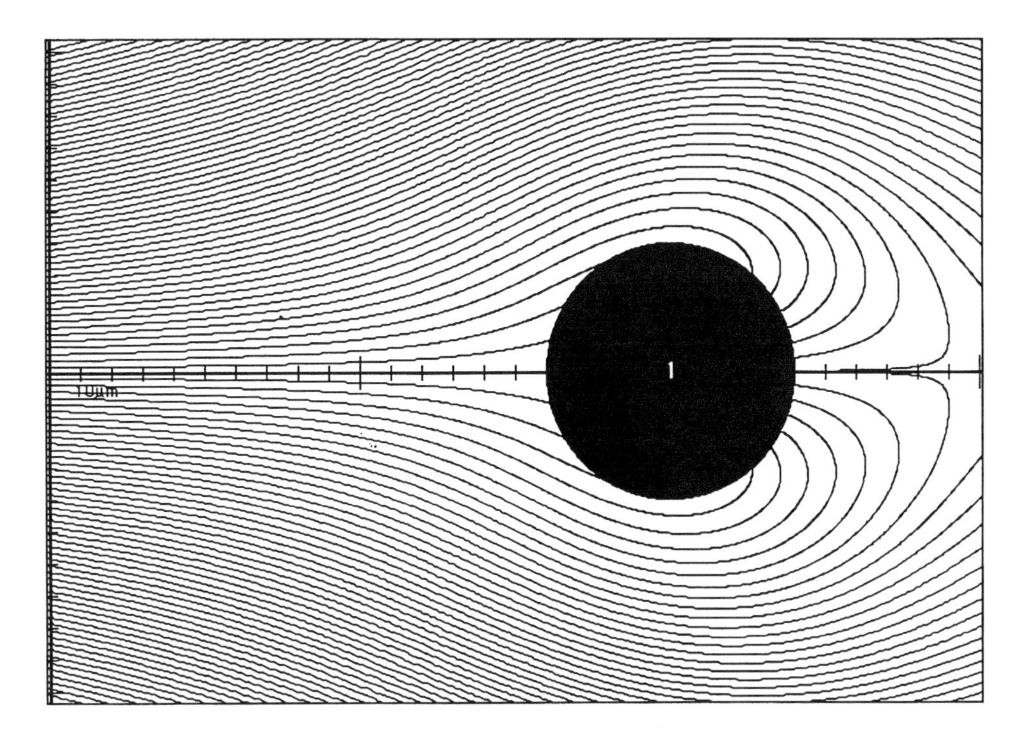

Figure 3.3C Davies flow.

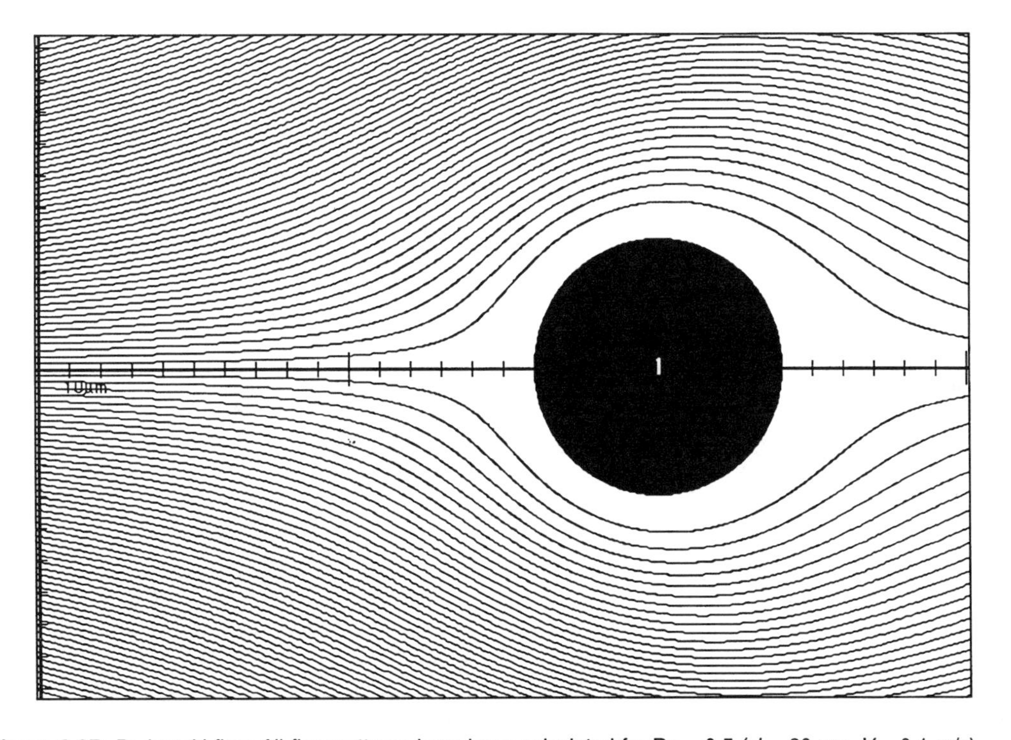

Figure 3.3D Podgorski flow. All flow patterns have been calculated for Re = 0.5 (d_f = 80 µm, V = 0.1 m/s).

Stokes paradox. Already at distances of about $20d_f$ the air velocity is 20% higher than the undisturbed velocity. To avoid these weaknesses, Davies (1949) suggested a flow model based on the linearized inertia approximation derived by Oseen (Happel and Brenner, 1991), see Figure 3.3C.

$$\rho_a \dot{r} \cdot \nabla \dot{r} = \rho_a V_0 \nabla \dot{r},\ \rho_a g = 0 \rightarrow \rho_a V_0 \nabla \dot{r} = -\nabla p + \eta \Delta \dot{r} \quad (3.4)$$

Davies used three terms in the expansion of the solution. This flow model starts to become unstable for $Re_f > 0.2$. For $Re_f < 0.2$, a reverse flow near the rear stagnation point can be seen and some streamlines cross the boundary of the fiber. The fourth flow model used has been developed by Podgorski (1993), see Appendix and Figure 3.3D. It is also based on the Oseen approximation, but the expansion may have an arbitrary number of terms. In this study ten terms were used. This flow model fulfills both the surface and infinity boundary conditions, and also takes into account the standing eddies behind the fiber. Wang (1991) solved the full stationary Navier–Stokes equations numerically and calculated E for $Re_f > 1$. This chapter deals with $Re_f < 1$.

3.2.4 Drag Force on Particles

The drag force model traditionally used is Stokes' drag force, based on the creeping flow approximation (Equation 3.3). For low Re_p ($Re_p < 1$) the Stokes' drag force shows good agreement with experimental data. For higher Re_p, a drag force based on Oseen's approximation shows better agreement with experimental results (Fuchs, 1964). Goldstein (1965) derived a drag force with six terms:

$$F_D = -\frac{3\pi\eta d_p}{C_c}\left(V - v_p\right)G$$

$$G = \left(1 + \frac{3}{16}Re_p - \frac{19}{1280}Re_p^2 + \frac{71}{20480}Re_p^3 - \frac{30179}{34406400}Re_p^4 + \frac{122519}{550502400}Re_p^5\right) \quad (3.5)$$

$$C_c = 1 + \frac{\lambda_a}{d_p}\left[2.514 + 0.800e^{-0.55 d_p/\lambda_a}\right]$$

When a particle is moving in an airflow the value of Re_p varies as a result of the velocity gradients, and it is difficult to say how many terms are necessary. After testing both drag force models for the different particle sizes and velocity gradients used in this work, no significant difference ($\Delta E < 0.1\%$) was found. Simulations also show that Re_p does not exceed 1 during transport towards the fiber. In this study, the Stokes drag force formula is used, i.e., (Equation 3.5) with $G = 1$.

$$F_D = -\frac{3\pi\eta d_p}{C_c}\left(V - v_p\right) \quad (3.6)$$

Both Stokes' drag force (Equation 3.6) and Goldstein's drag force (Equation 3.5) are fully valid only when the particle is at equilibrium, i.e., its acceleration is zero. Using (Equations 3.5 or 3.6) for a curvilinear motion imposes an additional error. This error is, however, less than 5% (Fuchs, 1964) and is neglected in this chapter.

The particle trajectories were calculated using the Fehlberg method (Fehlberg, 1970; Press, 1992). This is a fourth-order Runge–Kutta method with variable time step and automatic error control. To calculate the efficiency, an interpolation algorithm was used and all the efficiencies were calculated with four significant digits.

The cross sections of the particle and of the fiber are defined as mathematical circles. The particle is captured when the circles defining the fiber and the particle intersect. In the discrete algorithm this is checked at every time step. The distance the particle moves during one time step will cause an additional error. In this chapter, the particles used were assumed to be spherical and homogeneous and in the diameter range 1 to 12 μm.

3.2.5 Definition of Single-Fiber Efficiency

The single-fiber efficiency is defined as the ratio of the number of particles captured on a fiber to the number of particles passing the cross section of the fiber in an undisturbed field (Davies, 1973). It can be expressed as

$$E = \frac{h}{d_f}, \quad h = h_{\text{top}} - h_{\text{bottom}} \tag{3.7}$$

where h is the distance between the two limiting trajectories and d_f is the fiber diameter (Figure 3.4). The distance h is by definition given at a point of undisturbed field, i.e., mathematically at a distance $L = \infty$ from the fiber. Wang (1991) used $L_0 \approx 20d_f$ for $d_f = 15$ and 30 μm and $V_0 > 0.5$ m/s. For practical reasons, a distance $L = L_0$, where the disturbance of the field can be neglected, must be determined. To be able to compare different calculations of E there must be a common criterion for L_0.

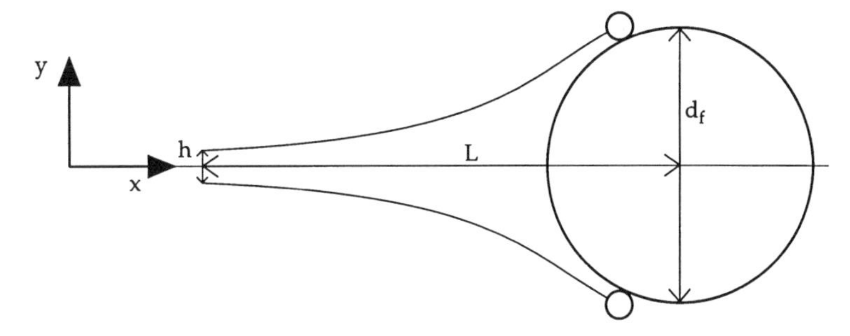

Figure 3.4 The limiting trajectories of particles collected on a fiber.

One way to determine the distance L_0 is to calculate the efficiency, E, as a function of L and use the criterion

$$\left|\frac{\partial E}{\partial L}\right| < \varepsilon \tag{3.8}$$

to determine L_0. The values of E with L as a parameter are plotted in Figure 3.5.

The curve in Figure 3.4 reaches a stable value, $|\partial E/\partial L| < 1 \times 10^{-8}$, at a distance equal to 40 to 50 fiber diameters from the fiber. This criterion is particle oriented and L_0 must be calculated for each particle diameter.

Another way is to calculate the disturbance of the field as a function of L. The disturbance is characterized as the ratio of the x and y velocity components.

$$\left|\frac{V_y(L, y_0)}{V_x(L, y_0)}\right| < \varepsilon \tag{3.9}$$

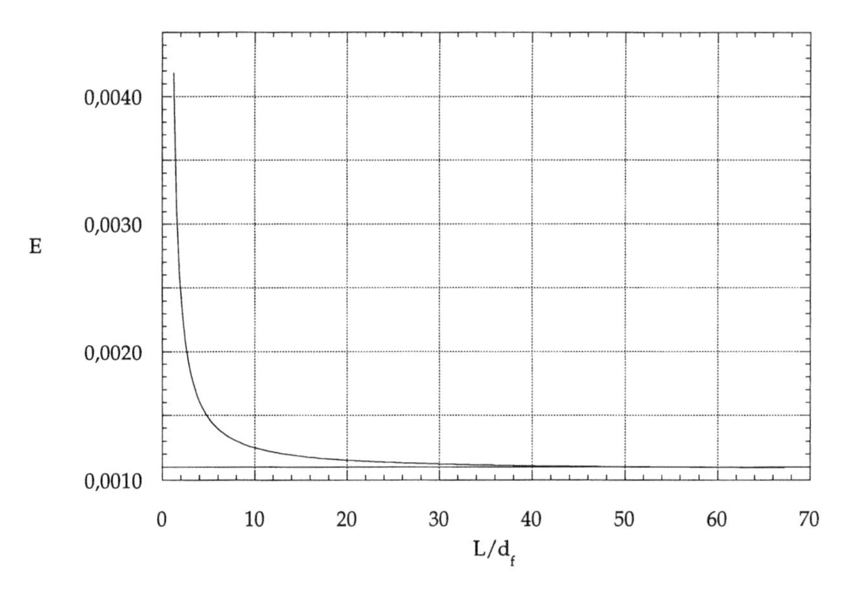

Figure 3.5 Efficiency as a function of upstream distance. Particle diameter d_p = 4 µm, fiber diameter d_f = 80 µm, and air velocity V = 0.1 m/s.

The second criterion (3.9) is flow dependent, i.e., is the same for all particles. The y_0 value is chosen to be of the same magnitude as the particle diameter, in this case y_0 = 1 µm. In the compromise between accuracy and speed of calculation a disturbance of $\varepsilon = 0.01\%$ was chosen. The advantage of a flow-dependent criterion is that it is only necessary to calculate L_0 once for each geometry and air velocity. Figure 3.6 shows L_0 for 0.01% disturbance. In this chapter the criterion (3.9) is used.

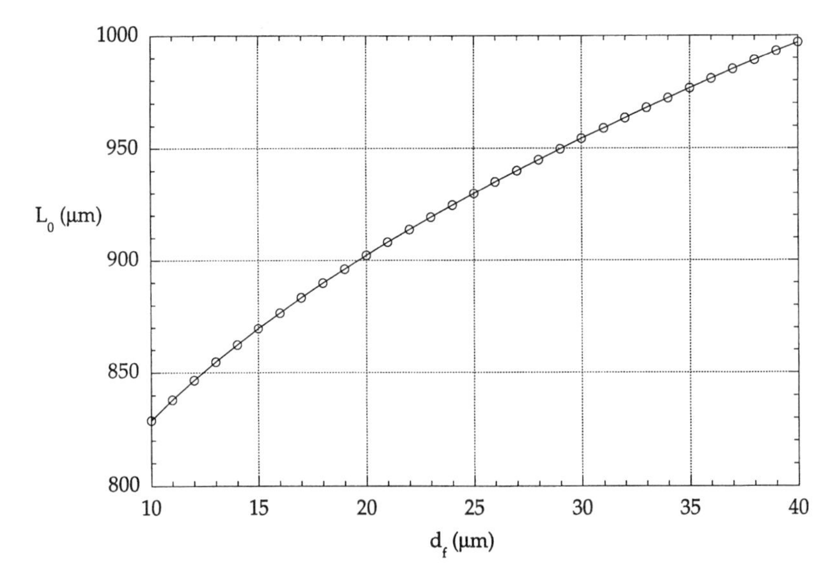

Figure 3.6 L_0 for 0.01% disturbance as a function of fiber diameter (V = 0.1 m/s).

3.2.6 Definition of Double-Fiber Efficiency

In the case of two fibers the efficiency definition must be modified. There are three different possible definitions.

$$E = \frac{h_1}{d_{f1}} + \frac{h_2}{d_{f2}} \tag{3.10}$$

$$E = \frac{h_1 + h_2}{d_{f1} + d_{f2}} \tag{3.11}$$

In Equations 3.10 and 3.11 the two single efficiencies are added in two different ways. Neither of the definitions includes any information on the distance between the fibers and the orientation of the fiber system relative to the flow field. The third definition uses an equivalent single-fiber diameter in the denominator.

$$E = \frac{h_1 + h_2}{d_{\text{eqv}}} \tag{3.12}$$

The equivalent diameter, d_{eqv}, is the diameter of a single-fiber which experiences the same drag force as the total drag force acting on the system of two fibers. When the two fibers are very far apart, Equation 3.12 reduces to the sum of the two single-fiber efficiencies. This definition includes information on the relative position of the fibers and relates the efficiency to the pressure drop. This definition can be used for any number of fibers and is equivalent to Equation 3.7 in the case of a single-fiber. The collection efficiency of more than one fiber will not be further investigated in this paper.

3.2.7 Diffusional Efficiency

For submicron particles, the diffusion due to Brownian motion has a major influence on the particle trajectories. For large particles, the diffusion can be neglected and the trajectory will be deterministic. For submicron particles a stochastic motion must be added to the deterministic trajectory. FiFi can simulate the combined deterministic and stochastic motion of particles. Figure 3.7 shows the result from one simulation. All particles were released from the same point. The particle with a deterministic trajectory was captured, while only 20% of the particles with combined deterministic and stochastic trajectory were captured.

Traditionally, the diffusional collection efficiency is calculated for an ensemble of particles moving in a concentration gradient (Fuchs 1964; Davies, 1973; Hinds, 1982; Lee et al., 1993). This diffusional collection efficiency definition is not valid for a single particle in a flow past a fiber. In the case of stochastic motion of a single particle, a new probability definition of the collection efficiency must be used. In the deterministic case all particles released within the region h will be captured. In the case of combined deterministic and stochastic motion, only a fraction of the particles will be captured. We define a probability, Ω, such that a particle released within the region h will be captured. In the purely deterministic case the probability is $\Omega = 1$. In the combined deterministic and stochastic case the probability cannot be $\Omega = 1$ for any h. We must therefore limit the probability to $\Omega < 1$. The new stochastic distance h_s can then be defined by the desired probability, Ω, and can be determined by solving Equation 3.13.

$$\Omega = \int_{h_{\text{bottom}}}^{h_{\text{top}}} f_y(y) \Pr(\text{captured} \mid y) dy, \quad h_s = h_{\text{top}} - h_{\text{bottom}} \tag{3.13}$$

The factor $f_y(y)$ is the probability density of the release position of the particles. In this case we used the uniform distribution, $f_y(y) = 1/h$. Pr(captured | y) is the conditional probability that a

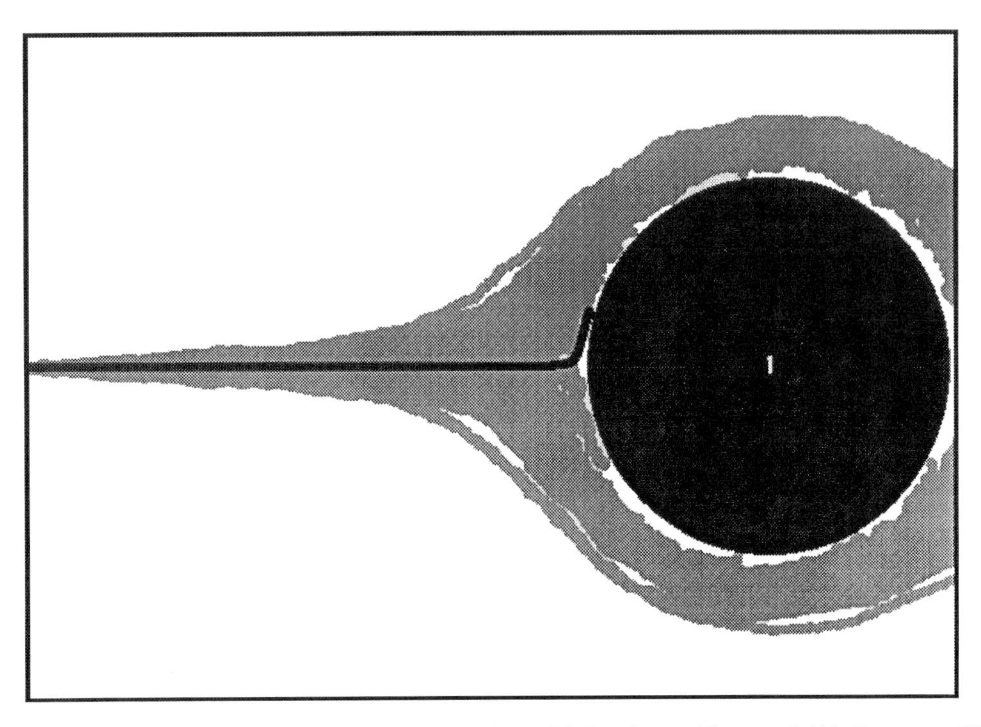

Figure 3.7 The black line is the trajectory of a purely deterministic trajectory. The gray field is the group of the combined deterministic and stochastic particle trajectories. The fiber diameter is d_f = 20 μm and the particle diameter is d_p = 0.5 μm. The undisturbed air velocity is V = 0.1 m/s.

particle will be captured when released from the position y. The conditional probability can be determined by releasing a large number of particles from a position y and calculating the ratio between captured particles and the total number of released particles. In the case shown in Figure 3.7 the conditional probability is Pr = 0.2. The conditional probability depends strongly on L. In the purely deterministic case the conditional probability will be Pr = 1; $h_{\text{bottom}} \leq y \leq h_{\text{top}}$ and Pr = 0; $h_{\text{bottom}} > y > h_{\text{top}}$. In the case of combined deterministic and stochastic motion there is only one value of h_s that satisfies Equation 3.13. In the deterministic case all values of h_s smaller than the value of h according to the traditional definition 3.7 will satisfy 3.13. The maximum value of h_s that satisfies 3.13 coincides with the definition 3.7. The definition in Equation 3.13 can thus be used as a general definition of the single-fiber collection efficiency for both diffusional and non-diffusional particle transport. The definition can also be used in the case of turbulent flow.

3.3 RESULTS AND DISCUSSIONS

The flow past a fiber and the drag on a particle depend explicitly on air velocity, air viscosity, air density, the mean free path in air, relative humidity, air pressure, and air temperature. The particle motion depends on the particle density and particle diameter. The calculations presented in this chapter were performed using the standard values in Table 3.1. The standard values for λ_a, ρ_{a0}, η_0, and P_0 are those suggested by Jennings (1988).

Table 3.1 Standard Values

T_0 = 293.15 K (20°C)	ρ_{a0} = 1.2045 kg/m^3	η_0 = 18.1920 × 10^{-6} kg/(m · s)
λ_a = 6.5430 × 10^{-8} m	ρ_p = 1000 kg/m^3	P_0 = 1.0325 × 10^5 Pa
V_0 = 0.1 m/s	d_p = 1–12 μm	d_f = 10–80 μm
g = 0	RH = 0%	

3.3.1 Flow Fields

Since most earlier calculations of single-fiber efficiency have been made using either the potential field or the Lamb field, calculations were made using these field models in order to compare them with the Podgorski model. The results in Table 3.2 also include results from calculations made using the Davies field model.

Table 3.2 E_0 **Calculated Using the Podgorski Field and** ΔE **Calculated for Various Other Flow Field Models**

	Podgorski Field, E_0, d_f				Lamb Field, ΔE, d_f			
	10 μm	20 μm	40 μm	80 μm	10 μm	20 μm	40 μm	80 μm
d_p = 1 μm	2.190×10^{-3}	6.735×10^{-4}	2.135×10^{-4}	7.126×10^{-5}	–21%	–26%	–33%	–41%
2 μm	8.524×10^{-3}	2.669×10^{-3}	8.542×10^{-4}	2.862×10^{-4}	–22%	–27%	–34%	–42%
4 μm	3.650×10^{-2}	1.131×10^{-2}	3.585×10^{-3}	1.191×10^{-3}	–26%	–31%	–37%	–45%
8 μm	2.202×10^{-1}	7.141×10^{-2}	2.049×10^{-2}	6.113×10^{-3}	–30%	–38%	–45%	–52%

	Davies Field, ΔE, d_f				Potential Field, ΔE, d_f			
	10 μm	20 μm	40 μm	80 μm	10 μm	20 μm	40 μm	80 μm
d_p = 1 μm	1237%	9799%	67535%	403428%	7197%	10354%	14687%	20431%
2 μm	238%	2377%	16735%	101102%	4592%	6494%	8251%	10159%
4 μm	61%	472%	3855%	24423%	2363%	4324%	63551%	7178%
8 μm	11%	73%	591%	4675%	618%	1349%	3062%	5629%

A relative difference in efficiency, ΔE, is calculated. E_0 is the efficiency calculated using the Podgorski model.

$$\Delta E = \frac{E - E_0}{E_0} \tag{3.14}$$

The potential field results show very large deviations compared with the Podgorski field. The Lamb model gives smaller and more constant deviations. The Davies flow also shows large deviations compared with the Podgorski flow. This may appear strange since the Davies flow is based on the same approximation as the Podgorski field. According to Davies (Davies, 1949), the flow model is not accurate for $Re_f > 0.2$, see Figure 3.3. Calculations made by FiFi show that even for $Re_f = 0.06$ (d_f = 10 μm) a reverse flow near the rear stagnation point is evident and some streamlines cross the boundary of the fiber. This gives a higher efficiency for small particles. It can be observed that ΔE is less for large particles. This can be explained by the fact that large particles are less sensitive to the reverse flow vortexes on the back of the fiber.

3.3.2 Gravity

ΔE was calculated using gravity ($g = 9.81$ m/s^2) directed in three different directions: perpendicular to the airflow and fiber ($\alpha = 90°$), perpendicular to the fiber and against the main flow direction ($\alpha = 180°$), and perpendicular to the fiber and along the main flow direction ($\alpha = 0°$). The results are compared with zero-gravity efficiencies. For $\alpha = 90°$, gravity has most influence for large d_f and small d_p. ΔE is positive. This can be explained by the fact that when large particles reach the stagnation area in front of the fiber they settle down past the fiber. Small particles, on the other hand, pass through the stagnation area and follow the air up along the fiber and settle on the top of the fiber. For $\alpha = 180°$, the efficiency decreases dramatically. For d_f = 40 μm and d_f = 80 μm, the efficiency drops to zero. As a particle enters the stagnation area in front of the fiber, the drag force decreases until it equals the gravity force. When the particle reaches equilibrium in the main flow direction ($V_{px} = 0$), it is swept away by the perpendicular component of the airflow,

V_{ay}. For $\alpha = 0°$, the efficiency is increased by over 500%. In the remainder of this chapter gravity is not taken into account since the strong influence of gravity would obscure the influence of the other parameters.

3.3.3 Single-Fiber Efficiency

The single-fiber efficiency calculated using the standard parameter values in Table 3.1, as a function of d_p for different d_f, is shown in Figure 3.8. In most studies of filtration, the efficiency, E, is calculated or measured as a unique function of the Stokes number (Fuchs, 1964; Davies, 1973; Pulley and Walters, 1990; Wang, 1991). The Stk value is a purely inertial number containing no information on the interception of particles. Normally, the inertial collection is considered to be a function of Stk, and the interception a function of the interception number, d_p/d_f (Fuchs, 1964). If the combined inertial collection and interception is being investigated, it is not sufficient to plot the efficiency against the Stk number since one Stk number can correspond to an unlimited number of interception numbers. Results expressed as a function of Stk cannot be compared unless two of the three parameters — air velocity, particle diameter, and fiber diameter — are the same. In Figure 3.9, E is plotted as a function of d_p/d_f.

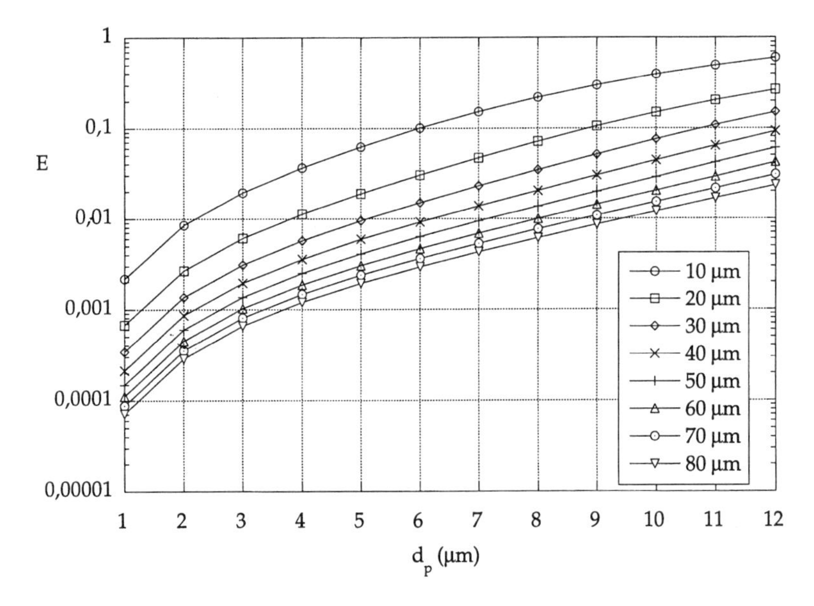

Figure 3.8 E calculated for d_p = 1 to 12 μm and d_f = 10 to 80 μm.

E plotted in a log–log diagram for different values of d_p/d_f shows a set of lines, one line for each value of d_p. The lines can be fitted to straight lines. The slope, k, and vertical offset, m, of the lines are plotted in Figure 3.10 and 3.11.

For particles of d_p = 1 to 12 μm and d_f ranging from 10 to 80 μm, E can be expressed as

$$^{10}\log(E) = k\ ^{10}\log\left(\frac{d_p}{d_f}\right) + m \tag{3.15}$$

or simplified, as

$$E = 10^m\left(\frac{d_p}{d_f}\right)^k \tag{3.16}$$

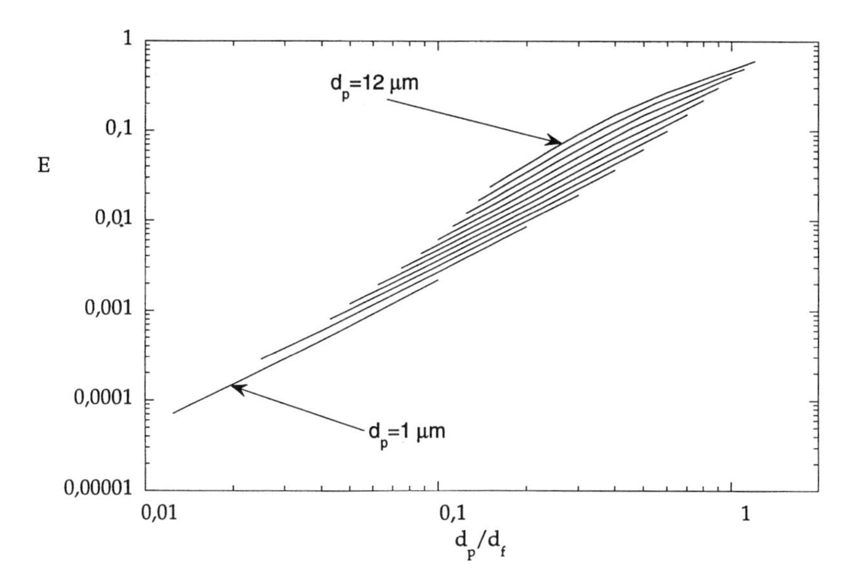

Figure 3.9 E calculated for d_p = 1 to 12 µm and d_f = 10 to 80 µm and plotted as a function of d_p/d_f. The first line from the left (bottom line) is d_p = 1 µm, the second is d_p = 2 µm, and so on.

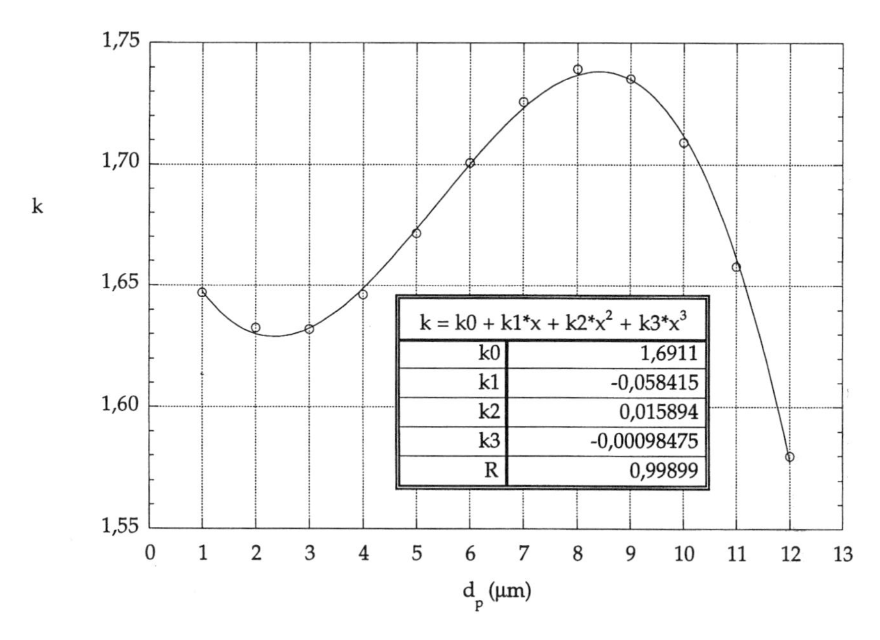

Figure 3.10 The slope, k, of the lines in Figure 3.9, as a function of d_p. The points are fitted to a third-order polynomial curve. R is the correlation coefficient.

The slope, k, and the vertical offset, m, are functions of particle diameter, particle density, air velocity, air viscosity, air density, mean free path in air, air pressure, air temperature, and relative humidity

$$\begin{aligned} k &= k\left(d_p, \rho_p, V, \eta, \rho_a, \lambda_a, P, T, \mathrm{RH}\right) \\ m &= m\left(d_p, \rho_p, V, \eta, \rho_a, \lambda_a, P, T, \mathrm{RH}\right) \end{aligned} \tag{3.17}$$

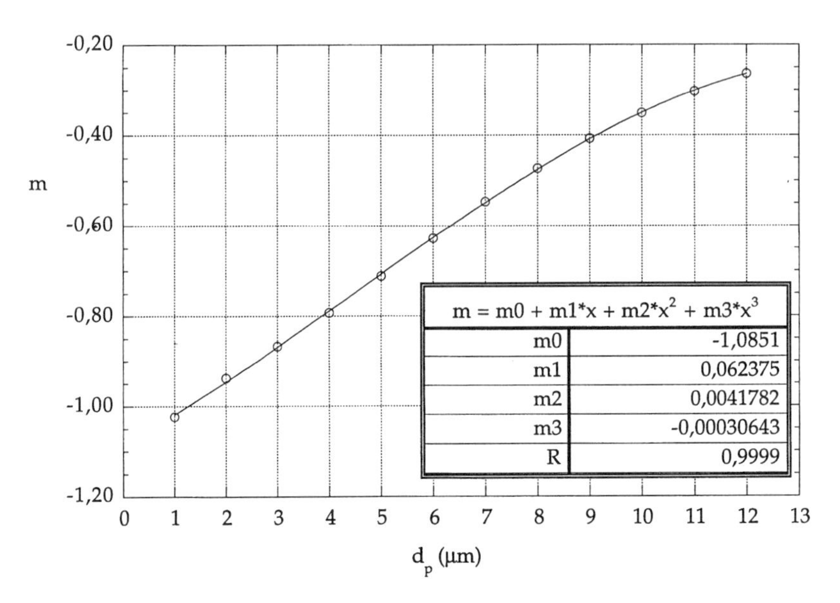

Figure 3.11 The vertical offset, *m*, of the lines in Figure 3.9 as a function of d_p. The points are fitted to a third-order polynomial curve. *R* is the correlation coefficient.

For a set of constant air parameters, k and m as a function of d_f can be fitted to a third-order polynomial function.

$$
\begin{aligned}
k &= k_0 + k_1 d_p + k_2 d_p^2 + k_3 d_p^3 \\
m &= m_0 + m_1 d_p + m_2 d_p^2 + m_3 d_p^3
\end{aligned}
\tag{3.18}
$$

The coefficients k_n and m_n are given in the tables in Figures 3.10 and 3.11.

Upon comparing efficiencies calculated using Equation 3.16 with results from the calculations using FiFi, a difference is observed. The relative difference is quantified using the definition in Equation 3.14. For particle diameters smaller than 8 μm, ΔE is less than 5%, and for $d_p < 10$ μm ΔE is less than 10%. For particle diameters of $d_p = 20$ μm ΔE increases to 20%. This must be borne in mind when using Equation 3.16.

3.3.4 *k-m* Analysis

A simplified and less accurate expression can be used instead of Equation 3.16. In the particle diameter range 1 to 8 μm, k can be approximated to a constant value of $k = 1.67$. The relative deviation, s, from this value is 2.5%.

$$
s = \frac{k_{\max} - k_{\min}}{\bar{k}} \tag{3.19}
$$

$\bar{k}$ is the mean value of k, and $k_{\max}$ and $k_{\min}$ denote the maximum and minimum values of k, respectively.

The function of m can be approximated to a straight line,

$$
m = m_0 + m_1 d_p \tag{3.20}
$$

with $m_0 = -1.1$ and $m_1 = 0.079$, giving the correlation coefficient $R = 0.9996$. This gives us

$$E = 10^{(m_0 + m_1 d_p)} \left(\frac{d_p}{d_f} \right)^k \tag{3.21}$$

In this simplified formula, the exponent k is a constant. Using Equation 3.21, one must accept values of ΔE in the range of 10 to 15%.

We define an upper particle diameter limit, d_{pl}, for which the correlation coefficient, R, for the linear curve fitting of m is 0.99 and the relative maximum deviation, s, from the mean value of k is 10% or less. For the standard values the upper particle diameter limit is d_{pl} = 12 μm. The calculation of the functions of $k(d_p)$ and $m(d_p)$ and the determination of d_{pl} is referred to in this study as k-m analysis.

3.3.5 Particle and Air Parameters

The dimensionless functions $k(d_p)$ and $m(d_p)$ for 10%, 50%, and 100% variations in particle density, air velocity, air viscosity, air density, and mean free path in air have been investigated. The calculations were performed for d_p = 1, 2, 4, and 8 μm and d_f = 10, 20, 40, and 80 μm. The functions $k(d_p)$ and $m(d_p)$ were also investigated for air temperatures of 10, 50, and 100°C and relative humidities of 50 and 100%. The upper particle diameter limit d_{pl} was calculated to determine the range of applicability of Equation 3.21.

Air viscosity, air density, and mean free path in air depend on the temperature and the relative humidity. The new values for η, ρ_a, and λ_a when the relative humidity is 50% and 100% are taken from Jennings (1988).

A standard temperature of 20°C is used. The values of λ_a, ρ_a, and η for T = 10, 50, and 100°C were calculated according to

$$\frac{\lambda_a}{\lambda_{a_0}} = \frac{T}{T_0} \qquad \frac{\rho_a}{\rho_{a_0}} = \frac{T_0}{T} \qquad \frac{\eta}{\eta_0} = \sqrt{\frac{T}{T_0}} \tag{3.22}$$

T_0, λ_{a0}, ρ_{a0}, and η_0 are the standard values and T, λ_a, ρ_a, and η are the corresponding values of 10, 50, and 100°C.

The various parameters can be divided into three groups:

- Parameters which, when increased, do not change d_{pl} by more than 1 μm (relative humidity, mean free path, and air temperature);
- Parameters which, when increased, decrease d_{pl} (air velocity, particle density, and air density);
- Parameters which, when increased, increase d_{pl} (air viscosity).

A k-m analysis will be performed for each group of parameters separately.

(1) Relative Humidity, Mean Free Path, and Air Temperature

ΔE was calculated for RH = 50% and 100%. In both cases, ΔE was less than 1%. For variations (10 to 50%) in the mean free path, ΔE did not exceed 1%. Equation 3.21 is applicable in the whole particle diameter range, i.e., d_{pl} = 12 μm.

ΔE is positive (ΔE = 1 to 3%) for $T < 20$°C and negative for $T > 20$°C. When the temperature is increased to T = 50°C, the efficiency falls for large values of d_p to $\Delta E = -10$%. For $T > 20$°C, ΔE is almost independent of d_f, but increases with increased d_p. In the case of increased air temperature (T = 50, 100°C), d_{pl} is the same, d_{pl} = 12 μm, (s < 7% and R = 1.00) as for T = 20°C. For T = 10°C d_{pl} falls to d_{pl} = 11 μm (s < 6.6% and R = 0.99).

(2) Air Velocity, Particle Density, and Air Density

When one of the parameter's air velocity, particle density, or air density is increased, the efficiency is increased; i.e., ΔE is positive. The shape of the lines, when E is plotted against d_p/d_f (Figures 3.12 to 3.14), is typical for this group of parameters. In this group the applicability of Equation 3.21 decreases when the parameter is increased. The lines are less straight (R < 0.99) and s > 10%, for increased particle diameter.

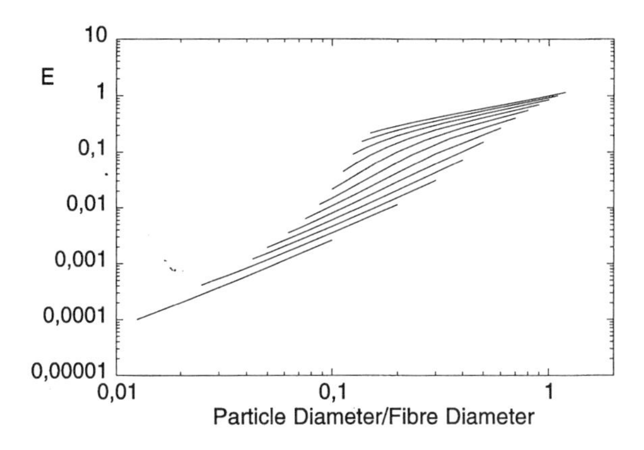

Figure 3.12 *E* as a function of d_p/d_f for V = 0.2 m/s.

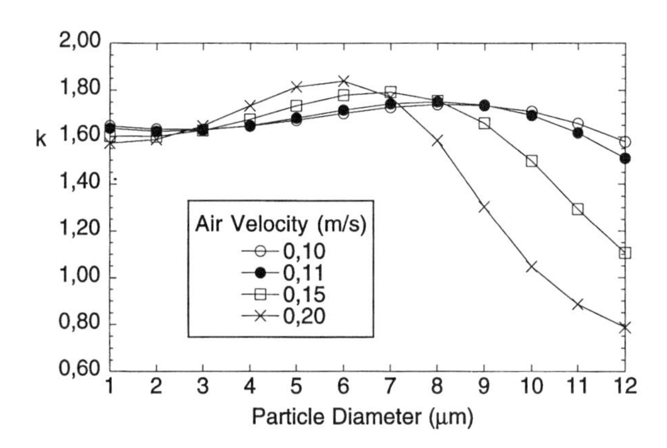

Figure 3.13 *k* as a function of d_p for V = 0.10, 0.11, 0.15, and 0.20 m/s.

As the air velocity is increased ΔE is positive and increases with increased d_p. The highest value of ΔE is reached for $d_f \approx$ 20 μm. When V is increased to V = 0.11 m/s, ΔE is 17% for large d_p but only 3% for small d_p. For V = 0.11 m/s, the upper particle diameter limit is d_{pl} = 11 μm (s = 7.9% and R = 0.99). It then falls very rapidly to d_{pl} = 5 μm (s = 8.0% and R = 1.00) for V = 0.15 m/s. For V = 0.20 m/s, the value of d_{pl} has fallen to d_{pl} = 4 μm (s = 9.8% and R = 0.99). For d_p > 7 μm, m flattens out and reaches an almost constant value of 0.

As the particle density is increased, ΔE is positive and increases with increasing d_p. As in the case of air velocity, the highest values of ΔE are reached for $d_f \approx$ 20 μm. When ρ_p is increased to ρ_p = 1100 kg/m³, ΔE is 9% for large d_p but less than 1% for small d_p. For ρ_p = 1100 kg/m³, d_{pl} is

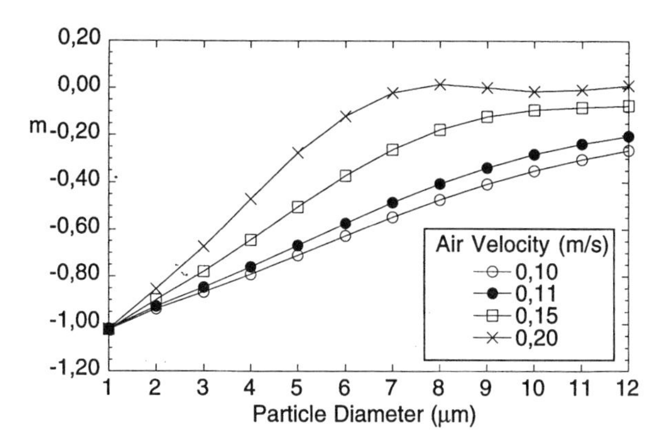

Figure 3.14 *m* as a function of d_p for V = 0.10, 0.11, 0.15, and 0.20 m/s.

still high, d_{pl} = 11 μm (s = 7.4% and R = 0.99). For ρ_p = 1500 kg/m^3, d_{pl} falls to d_{pl} = 7 μm (s = 9.4% and R = 1.00). For ρ_p = 2000 kg/m^3, d_{pl} is very low, d_{pl} = 4 μm (s = 6.5% and R = 1.00). For large particles, d_p > 9 μm, m reaches an almost constant value of –0.07.

As the air density is increased, ΔE is positive and increases with increased d_p and increased d_f. When ρ_a is increased to ρ_a = 1.325 kg/m^3, ΔE is 6% for large d_p and large d_f but only 2% for small d_p and small d_f. The applicability of Equation 3.21 decreases with increasing air density although not as rapidly as for increasing air velocity and particle density. For ρ_a = 1.325 kg/m^3, d_{pl} is 11 μm (s = 6.5% and R = 1.00). For ρ_a = 1.807 kg/m^3, d_{pl} remains high, d_{pl} = 11 μm (s = 9.2% and R = 0.99) and does not fall much for ρ_a = 2.809 kg/m^3, d_{pl} = 10 μm (s = 7.4% and R = 0.99).

(3) Air Viscosity

As the air viscosity is increasing, ΔE is negative. For small d_p, ΔE decreases most significantly for large d_f. For large d_p, ΔE decreases most significantly for small d_f. When η is increased 10% to $\eta = 20.011 \times 10^{-6}$ kg/(m · s), ΔE is –11% for large d_p, and small d_f but only –2% for small d_p and small d_f.

In the case of increasing air viscosity the lines are straighter and the applicability of Equation 3.21 increases (Figures 3.15 to 3.17). For all the increased values of air viscosity, d_{pl} is at least 12 μm (s < 4% and R = 1.00).

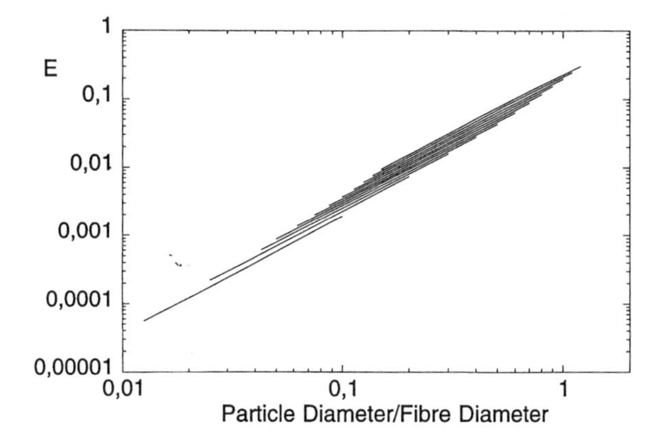

Figure 3.15 E as a function of d_p/d_f for $\eta = 36.384 \times 10^{-6}$ kg/(m · s).

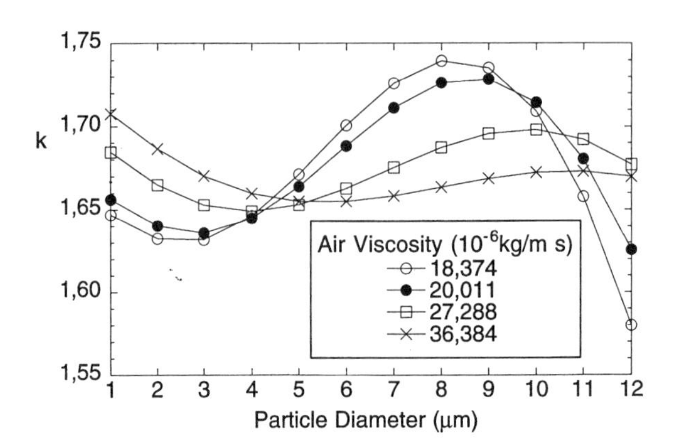

Figure 3.16 k as a function of d_p for η = 18.374, 20.011, 27.288, and 36.384 × 10^{-6} kg/(m · s).

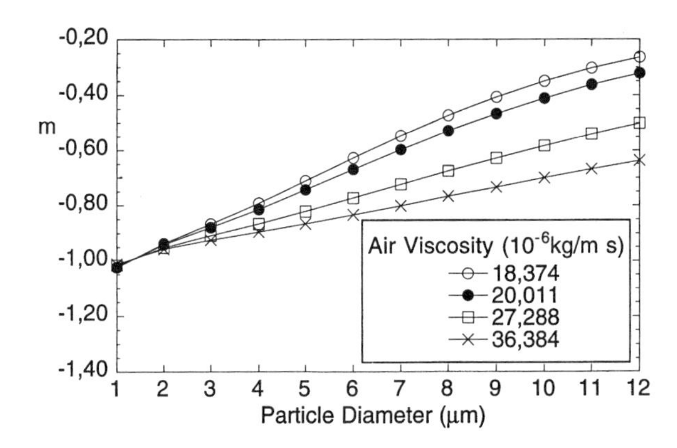

Figure 3.17 m as a function of d_p for η = 18.374, 20.011, 27.288, and 36.384 × 10^{-6} kg/(m · s).

3.3.6 Aerodynamic Equivalent Particle Diameter

The possibility of using the equivalent aerodynamic diameter to account for variations in particle density was examined. New particle diameters for increased densities ρ_p = 1010 to 1500 kg/m^3 were calculated according to

$$d_{ae} = d_p \sqrt{\frac{\rho_p}{\rho_0}}$$
$$C_c(d_p) \approx C_c(d_{ae}) \qquad (3.23)$$

The collection efficiency of particles with the equivalent aerodynamic diameter and unity density were compared with collection efficiencies of particles with diameters of d_p = 1 to 8 μm and densities of ρ_p = 1010 to 1500 kg/m^3.

The value of ΔE is positive and approximately of the same magnitude as the relative change in particle density. ΔE is smaller for larger d_p. The conclusion is that in this type of study it is not preferable to use the equivalent aerodynamic diameter to compensate for non-unity densities.

3.4 DENDRITES

A large number of simulations of the dendritic growth was performed. The simulations were performed under the standard conditions and only the fiber and particle diameters were varied. Particles with diameters d_p = 1 to 8 μm were released randomly upstream of the fiber and after each capture the collection efficiency, E, was recalculated according to the definition in Equation 3.7. The fiber diameters were d_f = 10 to 80 μm. The simulation was terminated when 30 particles had been captured. For each combination of fiber and particle diameter, 20 simulations were performed. The influence of the dendrites on the flow was neglected.

In most cases, there was only one dendrite on the cross section of the fiber. The number of dendrites increased with the fiber–particle size ratio, d_f/d_p, but even for the case of d_f = 80 μm and d_p = 1 μm, less than 50% of the simulations performed showed two dendrites or more. Less than 10% showed three dendrites and none showed more than three. This can be explained by the locally very high collection efficiency of the dendrite. The morphology of some typical dendrites is shown in Figure 3.18. The dendrites are more slender and fewer in number per cross section than those presented by Kanaoka et al. (1980).

The collection efficiency as a function of number of collected particles was investigated for different combinations of particle and fiber diameters. The collection efficiency increases rapidly with each captured particle, see Figure 3.19.

After 30 captured particles the value of E increased by about two orders of magnitude. The shape of the bold line in Figure 3.19 is typical for the average efficiency increase as a function of number of captured particles. In Figure 3.20 the results from calculations made for eight different combinations of particle and fiber diameters are shown.

3.5 DISCUSSION

When comparing calculated efficiencies with experimental results, it is vital to know the conditions under which the experiment was performed. Gravity can increase E considerably, which is why the orientation of the flow and the fiber should be considered when designing experiments. The variation in efficiency due to variations in air parameters and particle density depends strongly on particle size. A 10% variation in air viscosity may cause a 2 to 12% variation in efficiency. An increased temperature of 50°C decreases the efficiency by 3 to 11%. Thus, in technical applications, e.g., air filtration of hot gases, the filtration efficiency may increase if the gas is cooled prior to filtration. A 10% increase in air velocity may give a 3 to 17% increase in efficiency. For a 1-μm particle, a 10% increase in particle density may cause less than a 1% change in E, but a 9% increase for an 8-μm particle. A 10% increase in air density will only cause a 2 to 6% increase in the efficiency. Using the aerodynamic diameter to compensate for variations in density will give an error of the same magnitude as the variation in density. When setting up a filtration experiment, even small variations in fiber orientation, particle density, air velocity, and air viscosity may cause considerable variations in the results. Variations in the mean free path and humidity will only cause very small changes in the results.

Equation 3.16 describes the combined inertial collection and interception efficiency. For particles $d_{pl} < 10$ μm the error is less than 10%, compared with the results from calculations using FiFi. The simplified version, Equation 3.21, is easier to use but can, in some cases, give an error of up to 15%. When the various parameters are varied, the applicability of Equation 3.21 is changed. The applicability is, in some cases, increased and, in some cases, decreased. The results of the *k-m* analysis are summarized in Table 3.3.

The *k-m* analysis shows that the simplified version, Equation 3.21, can be used for particle diameters up to 10 μm, for an increase of up to 50% in air temperature, air density, and air viscosity. When the particle density or air velocity is increased by more than 50%, d_{pl} falls rapidly to below

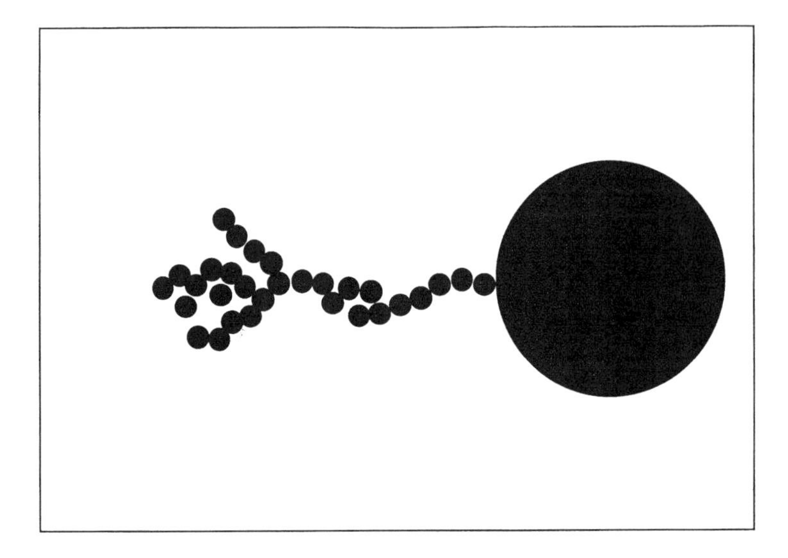

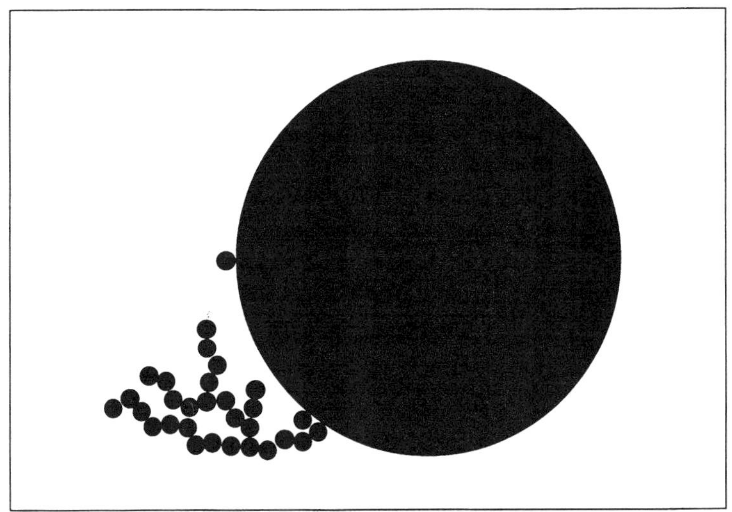

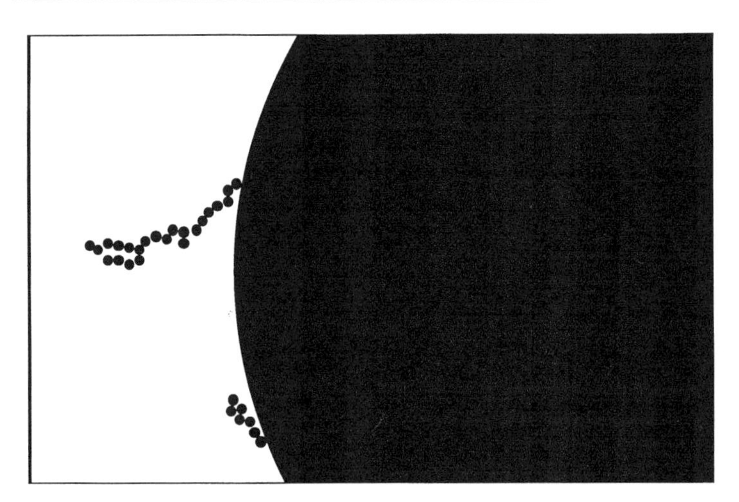

Figure 3.18 Typical dendrites. In total, 30 particles are captured on each fiber. Top: d_f = 10 μm and d_p = 1 μm. Middle: d_f = 20 μm and d_p = 1. Bottom: d_f = 80 μm and d_p = 1 μm.

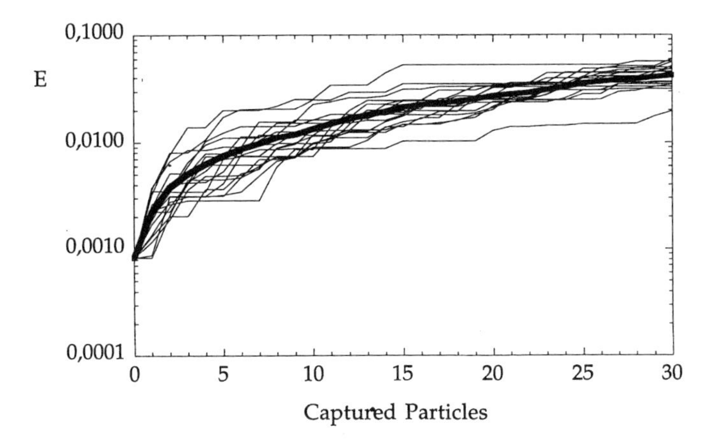

Figure 3.19 The collection efficiency, E, as a function of the number of captured particles; 20 simulations are shown. The bold line is the average value of E. The particle diameter was 1 μm and the fiber diameter was 20 μm.

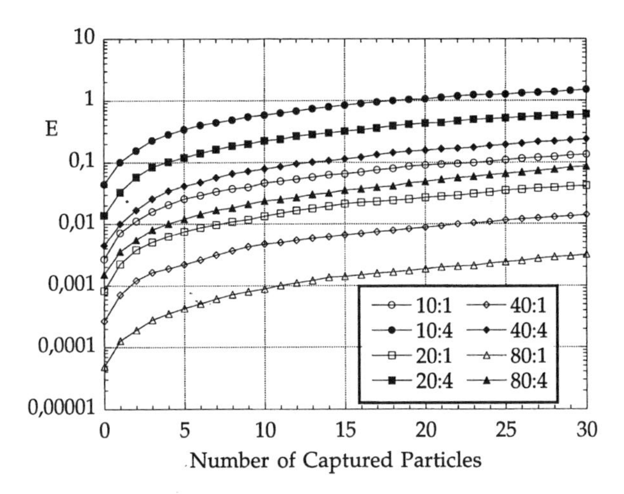

Figure 3.20 The average collection efficiency, E, as a function of number of captured particles. The average is taken from a total of 20 simulations. The particle diameters, d_p, are 1 and 4 μm and the fiber diameters, d_f, are 10, 20, 40, and 80 μm. In the figure the different d_p and d_f combinations are denoted (d_f:d_p).

5 μm. It can be seen that for large particles and high air velocity or high particle density, the lines become more parallel and straighter but with a different slope than for small particles. For air temperature, air density, and air viscosity no such tendencies can be seen.

Table 3.3 Limiting Particle Diameter, d_{pl} (μm), for Variations in the Various Parameters

ρ (kg/m³)	d_{pl}	V(m/s)	d_{pl}	ρ_a (kg/m³)	d_{pl}	η (kg/(m · s))	d_{pl}	T (°C)	d_{pl}
1000	12	0.10	12	1.204	12	18.374×10^{-6}	12	10	11
1100	11	0.11	11	1.325	11	20.011×10^{-6}	12	20	12
1500	7	0.15	5	1.807	11	27.288×10^{-6}	12	50	12
2000	4	0.20	4	2.209	10	36.384×10^{-6}	12	100	12

The results from FiFi and Equation 3.21 are the result of a numerical implementation of a theoretical model. Whether or not they present a useful description of reality is yet to be experimentally verified.

NOMENCLATURE

Boldface denotes vectors.

C_c	Cunningham correction factor
d	diameter (m)
d_{eqv}	equivalent drag diameter (m)
E	fiber collection efficiency
E_{err}	iteration error in E
ΔE	relative efficiency difference, $\Delta E = \frac{E - E_0}{E_0}$
$\mathbf{F}_D$	drag force (N)
$\mathbf{g}$, g	gravity (m/s^2)
G	drag force factor
h	distance between the two limiting trajectories (m)
h_{top}, h_{bottom}	vertical position of top and bottom limiting trajectories (m)
Δh	absolute error in h (m)
m	y offset (m)
L	upstream distance from the center of the fiber (m)
k	slope
n	number of iterations
N	number of terms in flow velocity function, Equation 3.A10
P	static air pressure (Pa)
Pr	conditional probability
p	dynamic air pressure (Pa)
$\mathbf{r}$	position vector (m)
R	correlation coefficient
R_f	fiber radius (m)
Re_f	Reynolds number for fiber, $\mathrm{Re}_f = \frac{\rho_a V d_f}{\eta}$
Re_p	Reynolds number for particle, $\mathrm{Re}_p = \frac{\rho_a \lvert V - v_p \rvert d_p}{\eta}$
RH	relative humidity (%)
Stk	Stokes number for fiber, $\mathrm{Stk} = \frac{V \rho_p d_p^2 C_c}{18 \eta d_f}$
s	relative deviation
T	air temperature (K) or (C°)
$\mathbf{V}$, V	air velocity (m/s)
V_t	tangential velocity component (m/s)
v_p	particle velocity (m/s)
x, y	coordinates (m)
α	angle between gravity and direction of flow
ε	constant, Equations 3.8 and 3.9
η	air viscosity (Ns/m^2)
κ	constant, Equation 3.A1
λ_a	air mean free path (m)
ρ_0	unit density (1000 kg/m^3)

ρ	density (kg/m^3)
Ω	probability
ω	vorticity, Equation 3.A1

Subscripts

0	standard/reference value
k	kth iteration step
x, y	x and y components of a vector
p	particle
f	fiber
a	air
n, m	summation indices
1	limit
eqv	equivalent
s	stochastic

ACKNOWLEDGMENTS

The authors wish to express their gratitude to Dr. Leon Gradoń for valuable comments and discussions. This work was supported by the Swedish Work Environment Fund and the Lund Institute of Technology.

REFERENCES

Billings, C. E. (1966) Ph.D. dissertation, California Institute of Technology, Pasadena.
Davies, C. N. (1949) *Proc. Phys. Soc.*, LXIII, 288–296.
Davies, C. N. (1973) *Air Filtration,* Academic Press, London.
Fehlberg, E. (1970) *Computing,* 6, 61–71.
Fuchs, N. A. (1964) *The Mechanics of Aerosols,* Dover, New York.
Goldstein, S. (1965) *Modern Developments in Fluid Dynamics,* Dover, New York.
Happel, J. and Brenner, H. (1991) *Low Reynolds Number Hydrodynamics,* Kluwer Academic Publishers, Dordrecht.
Hinds, W. C. (1982) *Aerosol Technology,* John Wiley & Sons, New York.
Imai, I. (1954) *Proc. R. Soc. A,* 244, 141–160.
Jennings, S. G. (1988) *J. Aerosol Sci.,* 19(2), 159–166.
Kanaoka, C., Emi, H., Myojo, T. (1980) *J. Aerosol Sci.,* 11, 377–389.
Kuwabara, S. (1959) *J. Phys. Soc. Jpn.,* 14, 527–532.
Lamb, H. (1932) *Hydrodynamics,* Dover, New York.
Lastow, O. and Bohgard, M. (1992) *J. Aerosol Sci.,* 23, Suppl. 1, S105–S108.
Lee, K. W. and Ramamuthi, M., (1993) in *Aerosol Measurement,* Willeke, K. and Baron, P. A., Eds., Van Nostrand Reinhold, New York.
Payatakes, A. C. and Gradon, L. (1980) *Chem. Eng. Sci.,* 35, 1083–1096.
Podgorski, A. (1993) *J. Aerosol Sci.,* 24, Suppl. 1, S277–S278.
Press, W. H. (1992) *Numerical Recipes in C,* Cambridge University Press, Cambridge.
Pulley, R. A. and Walters, J. K. (1990) *J. Aerosol Sci.,* 21, 6, 733–743.
Wang, P. K. (1991) *Aerosol Sci. Technol.,* 15, 149–155.

APPENDIX

We consider a two-dimensional steady-state, incompressible flow past a circular cylinder with a diameter d_f in an infinite space. The direction of the uniform flow velocity V_0, coincides with the Ox axis of the Cartesian coordinate system, Oxy, fixed in the center of the fiber. Using the equation of continuity and eliminating the pressure from Oseen's equation of motion (3.4) we express (3.4) in terms of vorticity, ω.

$$\Delta\omega - 2\kappa\frac{\partial\omega}{\partial x} = 0$$
$$\kappa = \frac{V_0\rho_a}{2\eta}, \quad \omega = \frac{\partial V_y}{\partial x} - \frac{\partial V_x}{\partial_y} \tag{3.A1}$$

Following Imai (1954), we consider the velocity field $V_0 = [V_x, V_y]$ as a sum of a uniform flow, $[V_0, 0]$, and a perturbation velocity $W = [W_x, W_y]$ due to the fiber.

$$V_x = V_0 + W_x \tag{3.A2}$$

$$V_y = V_y \tag{3.A3}$$

We introduce a complex variable $z = x + iy$ and its conjugate $\bar{z} = x - iy$. Similarly, we define a complex perturbation velocity.

$$\overline{W} = W_x - iW_y \tag{3.A4}$$

This allows us to express ω in terms of $\bar{z}$

$$\omega = 2i\frac{\partial\overline{W}}{\partial\bar{z}} \tag{3.A5}$$

and hence, after inserting Equation 3.A5 into 3.A1, the equation of motion can be expressed by means of $\overline{W}$. The boundary conditions for the flow are

$$r \to \infty: \quad V_x \to V_0, \quad V_y \to 0, \quad W_x = W_y = 0 \tag{3.A6}$$

$$r \to \frac{d_f}{2}: \; V_x = V_y = 0, \; W_x = -V_0, \; W_y = 0 \tag{3.A7}$$

We introduce polar cylindrical coordinates, r, θ, so that

$$r = \sqrt{x^2 + y^2}, \quad \tan\theta = \frac{y}{x} \tag{3.A8}$$

The equation of motion (3.A1 + 3.A5) can be divided into a harmonic and an inharmonic part and then integrated analytically. The general solution, obeying the boundary conditions (3.A6), has the form

$$\overline{W} = \sum_{n=1}^{\infty} \left\{ \frac{a_n}{r^n} e^{-in\theta} + b_n e^{\kappa x} \left[K_{n-1}(\kappa r) e^{i(n-1)\theta} + K_n(\kappa r) e^{-in\theta} \right] \right\} \tag{3.A9}$$

K_n are modified Bessel functions of the second kind and nth order. Because of the flow symmetry with respect to the x-axis, it can be shown that the constants a_n and b_n are real. Expanding the exponents and separating real and imaginary parts, we obtain

$$\begin{aligned} V_x &= V_0 + \sum_{n=1}^{\infty} \left\{ \frac{a_n \cos(n\theta)}{r^n} + b_n e^{\kappa x} \left[K_{n-1}(\kappa r) \cos((n-1)\theta) + K_n(\kappa r) \cos(n\theta) \right] \right\} \\ V_y &= \sum_{n=1}^{\infty} \left\{ \frac{a_n \sin(n\theta)}{r^n} - b_n e^{\kappa x} \left[K_{n-1}(\kappa r) \sin((n-1)\theta) - K_n(\kappa r) \sin(n\theta) \right] \right\} \end{aligned} \tag{3.A10}$$

To determine the constants a_n and b_n, we use the second boundary condition for $r = d_f/2$ (3.A7) which should be fulfilled over the entire surface of the fiber, i.e., for any angle θ. This angle is present explicitly in the exponential term $e^{-m\theta}$ and implicitly in the inharmonic term $e^{\kappa m}$, since $x = r\cos\theta$. To obtain the harmonic form we use the following lemma:

$$e^{\kappa x} = e^{\kappa r \cos\theta} \sum_{m=-\infty}^{\infty} I_m(\kappa r) e^{-im\theta} \tag{3.A11}$$

Where I_m are modified Bessel functions of the first kind and mth order. We now limit ourselves to the finite series with N terms. In the general solution (3.A10) the sum limited to N terms contains $2N$ multiples of the angle θ: $-N\theta$, $-(N-1)\theta\ldots$, $\theta\ldots$, $-\theta,0,\theta,2\theta\ldots$, $(N-1)\theta$. Inserting 3.A11 into 3.A9 for the fiber surface ($r = d_f/2 = R_f$) and retaining N terms, we get, after simplifications,

$$\overline{W} = \sum_{n=1}^{N} \left\{ \frac{a_n}{r^n} e^{-n\theta} + b_n \sum_{m=-N}^{N-1} e^{im\theta} \left[K_{n-1}(\kappa R_f) I_{n-m-1}(\kappa R_j) + K_n(\kappa R_f) I_{n+m}(\kappa R_f) \right] \right\} \tag{3.A12}$$

Expressing the exponents by trigonometric functions and using the boundary conditions (3.A7), we obtain the following set of $2N$ linear algebraic equations

$$\begin{aligned} &m = -N,\ldots,(N-1);\; m \neq 0;\; \sum_{n=1}^{N} \frac{a_n}{r^n} + b_n \left[K_{n-1}(\kappa R_f) I_{n-m-1}(\kappa R_f) + K_n(\kappa R_f) I_{n+m}(\kappa R_f) \right] = 0 \\ &m = 0;\; \sum_{n=1}^{N} \left[K_{n-1}(\kappa R_f) I_{n-m-1}(\kappa R_f) + K_n(\kappa R_f) I_{n+m}(\kappa R_f) \right] = -V_0 \end{aligned} \tag{3.A13}$$

The solution of Equation 3.A13 gives us the constants a_n and b_n. Having determined the constants, the velocity field can be calculated from 3.A10. These series are rapidly convergent, and for conditions normally present in fibrous filters four to six terms should give satisfactory accuracy. In this chapter $N = 10$ was used. Using ten terms, 3.A10 can be used up to $Re_f = 10$. In this case, 3.A10 also describes the standing eddies behind the fiber (Podgorski, 1993).

CHAPTER 4

Stationary and Nonstationary Filtration of Liquid Aerosols by Fibrous Filters

Denis Boulaud and André Renoux

CONTENTS

0-87371-830-5/98/$0.00+$.50

4.1 INTRODUCTION

The experts know that the first theory on aerosol filtration was published by Albrecht (1931), but one has to wait until after World War II to observe systematic studies, especially those by Langmuir (1942). During the 1960s, theoretical works took into account the main mechanisms: Brownian diffusion, interception and inertial impaction, and the flow field around a single fiber with adjacent fibers obtained by Kuwabara (1959). Since this date, experimental and theoretical works have been extensively developed and many papers have been published on filtration.

Broadly speaking, the behavior of a filter over time can be divided into two phases:

- In the first phase, one can assume that changes occurring in filter structure due to deposition of particles are sufficiently negligible so that filter efficiency is unaffected. Particles making contact with a fiber are captured without modification of filtration mechanisms. During this "stationary phase" the penetration P and the pressure drop across the filter Δp do not change with time.
- Gradual buildup of particles at the fiber surface causes secondary effects, such as formation of aggregates with solid aerosols or bridges with liquid aerosols. This results in changes in P and Δp during this "nonstationary" phase of filtration.

Various studies devoted to the stationary phase have given rise to several models which allow prediction of the "instantaneous" penetration of a filter, as one can see in the first part of this chapter. In contrast, nonstationary filtration remains poorly understood, notably in the case of liquid where some authors (Billard et al., 1963; Mohrmann and Marchlewitz, 1974; Accomazzo et al., 1984) have reported anomalies, such as increases in filter penetration P as the pressure drop Δp across the filter rises, unlike what is seen with solid aerosols. Yet, the theoretical explanations of this phenomenon remain entirely qualitative; so, it appeared very important to us to perform in our laboratories an experimental and theoretical study of the nonstationary filtration of submicronic liquid aerosols. This will be developed in the second part of this chapter.

4.2 STATIONARY FILTRATION

4.2.1 Introduction

Filter performance is generally described in terms of overall efficiency E and pressure drop Δp. The overall collection efficiency is the ratio of the number of particles collected by the filter to the number of particles going into the medium:

$$E = \frac{n_2 - n_1}{n_2}$$

where n_2 is the concentration upstream and n_1 is the concentration downstream of the filter. The penetration P, defined by $P = 1 - E$, is also used to characterize the performance of a filter. It is generally expressed in the form of an exponential:

$$P = \exp\left(\frac{-4\alpha h \eta}{\pi(1-\alpha)D_f}\right)$$

where α represents the filter solidity, h the thickness of the medium, D_f the diameter of the fiber, and η the single-fiber efficiency. Modeling filtration by fibrous filters involves studying the mechanisms responsible for collection of particles by the fibers. In the region of maximum penetrating particle size and when there are no external forces acting, there are three mechanisms: Brownian diffusion, direct interception, and inertial impaction.

One of the major problems in modeling fibrous filters is the description of the complex structure of these media. A filter is generally composed of a random mixture of polydispersed fibers, where the finest have the task of catching the particles, whereas the largest give to the filter its strength. Thus, for the theoretical description of the filtration process, this structure is idealized in order to simplify the calculations connected with the description of the flow in the filter. Models have been worked out in this way: the simplest considers a single fiber placed perpendicular to the flow; unfortunately, it does not permit taking account either of the effect of proximity of adjacent fibers or of the intersections of fibers which disrupt the flow in the medium. Kuwabara's cell model (1959) is more sophisticated since it takes account of the effect of adjacent fibers by describing the filter as a succession of parallel fibers placed perpendicular to the flow. However, the comparison between the theoretical results from numerical calculations based on these models and the results of experiments relating to industrial filters always reveals a discrepancy connected with an insufficient description of the complexity of the fibrous structure. In order to reduce these differences between theory and experiment, Gougeon (1994) used another method which consists, on the experimental level, of using special fibrous filters with clearly defined structural characteristics, such as monodispersed fibers, the absence of a binding agent, etc. Such filters are manufactured by the Bernard Dumas Company (France) and are called the "formettes."

In this research into the stationary phase of filtration, Gougeon (1994) has measured the penetration of these special filters in the regimes of diffusion, interception, and inertia. A comparison of her experimental results with theoretical models from the literature enables her to define model forming, describing the efficiency of fiber filters over a wide range of filtration velocities and particle diameters.

4.2.2 Theroretical Background

(1) Diffusion and Interception Regimes

The single-fiber theory (Happel's 1959, and Kuwabara's, 1959, cell models), in which a single fiber is studied without neglecting the effect of the adjacent fibers (which is not the case with the isolated-fiber theory), is one of the most widely used theories because it has the enormous advantage of being simple. It permits arriving at an analytical expression for the stream function, from which the streamlines and the particle capture efficiency can be easily calculated.

It is demonstrated by Kirsch and Fuchs (1967) and by Yeh (1972) that between the two cell models, Kuwabara's is more representative of the flow around the fiber in the case of low Reynolds numbers.

Stechkina and Fuchs (1966) are the first to use this approach to deal with the problem of Brownian diffusion. They model the single-fiber efficiency for this mechanism (η_d) through the following equation:

$$\eta_d = 2.9 \text{ Ku}^{-1/3} \text{ Pe}^{-2/3}, \tag{4.1}$$

in which

$$\text{Ku} = \frac{-\ln \alpha}{2} - \frac{3}{4} + \alpha - \frac{\alpha^2}{4}, \tag{4.2}$$

is the Kuwabara's hydrodynamic coefficient and Pe, the Peclet number.

Friedlander (1958) uses a boundary layer approach, similar to that commonly used in heat and mass transfer analysis to determine the single-fiber efficiency in the case of a single fiber placed transverse to the flow. To consider the flow in the immediate neighborhood of the surface, as it is the boundary layer which is of importance, the Lamb's flow field is used. In the case of pure diffusion (very small particles) the single-fiber efficiency η is proportional to $B\mathrm{Pe}^{-2/3}$:

$$\eta \cong B^{1/3}\,\mathrm{Pe}^{-2/3}, \tag{4.3}$$

where B is a function of Re. In the case of pure interception the efficiency is approximately given by

$$\eta = 2BR^2, \tag{4.4}$$

with R the interception parameter.

The author correlates the experimental results of Chen (1955) and Wong et al. (1956) by multiplying both sides of Equations 4.3 and 4.4 by $R\mathrm{Pe}$ and obtains an equation given by

$$Y = 6X + 3X^3 \tag{4.5}$$

where

$$Y = \eta R\mathrm{Pe} \tag{4.6a}$$

and

$$X = (R^3\mathrm{Pe}B)^{1/3} \tag{4.6b}$$

From this former analysis, Friedlander and Pasceri (1962) recommend a semiempirical correlation (see also Friedlander, 1967):

$$Y = 1.3X + 0.7X^3 \tag{4.7a}$$

$$\eta = 1.3\mathrm{Pe}^{-2/3} + 0.7R^2 \tag{4.7b}$$

Later, using a boundary layer theory associated to the Kuwabara flow field, Lee and Liu (1982a) combine the effects of interception and diffusion by adding the single-fiber efficiencies of these two mechanisms, η_r and η_d, respectively. They obtain the following equations:

$$\eta = \eta_d + \eta_r$$
$$\eta_d = 2.6\left(\frac{1-\alpha}{\mathrm{Ku}}\right)^{1/3}\mathrm{Pe}^{-2/3} \tag{4.8}$$

and

$$\eta_r = \left(\frac{1-\alpha}{\mathrm{Ku}}\right)^{1/3}\frac{R^2}{(1+R)} \tag{4.9}$$

In another article, these same authors, Lee and Liu (1982b), amend this equation in line with their experimental data:

$$\eta_d = 1.6\left(\frac{1-\alpha}{\mathrm{Ku}}\right)^{1/3} \mathrm{Pe}^{-2/3} \tag{4.10}$$

and

$$\eta_r = 0.6\left(\frac{1-\alpha}{\mathrm{Ku}}\right)\frac{R^2}{(1+R)} \tag{4.11}$$

The interpretation of their experimental results (Lee and Liu, (1982b) in these two regimes is done using the method of representation proposed by Friedlander (1958) and detailed above. They show that the results of their experiments can be correlated to two dimensionless parameters $Y = \eta\ (2\ \mathrm{Pe}R/\sqrt{1+R})$ and $X = ((1-\alpha)/\mathrm{Ku})^{1/3}\ \mathrm{Pe}^{1/3}\ (R/\sqrt{1+R})$ found by multiplying the two equations (4.10 and 4.11) by $\mathrm{Pe}R/\sqrt{1+R}$.

They then obtain an universal curve in the following form:

$$Y = 1.6X + 0.6X^3 \tag{4.12}$$

This correlation was developed in the case of the fluid being considered as continuous, i.e., in the case of fiber diameter clearly greater than the mean free path, λ, of the carrier gas molecules. However, very high efficiency filters have fibers with a diameter generally less than a micron. Consequently, the description of the flow around the fiber must take account of the slip flow effects which are greater the finer the fiber. The degree of slip is given by the Knudsen number relative to the fiber:

$$\mathrm{Kn}_f = 2\lambda/D_f \tag{4.13}$$

This approach is followed by Liu and Rubow (1990) who, from Pich's work (1966), bring to Equations 4.10 and 4.11 two correction terms which take account of the slip flow effect. Then, η_d and η_r take the following form:

$$\eta_d = 1.6\left(\frac{1-\alpha}{\mathrm{Ku}}\right)^{1/3} \mathrm{Pe}^{-2/3} C_d \tag{4.14}$$

and

$$\eta_r = 0.6\left(\frac{1-\alpha}{\mathrm{Ku}}\right)\frac{R^2}{(1+R)} C_r \tag{4.15}$$

with

$$C_d = 1 + 0.388\ \mathrm{Kn}_f\left(\frac{(1-\alpha)\mathrm{Pe}}{\mathrm{Ku}}\right)^{1/3} \tag{4.16}$$

and

$$C_r = 1 + \frac{1.996\ \mathrm{Kn}_f}{R} \tag{4.17}$$

Revertifng to this last correlation and adjusting it to their own experimental data, Payet (1991) and Payet et al. (1992) introduce a new correction relating solely to the mechanism of diffusion and propose the following correlation:

$$\eta_d' = 1.6\left(\frac{1-\alpha}{\text{Ku}}\right)^{1/3} \text{Pe}^{-2/3} C_d C_d' \tag{4.18}$$

and

$$\eta_r = 0.6\left(\frac{1-\alpha}{\text{Ku}}\right)\frac{R^2}{(1+R)} C_r \tag{4.19}$$

with

$$C_d' = \frac{1}{1+\eta_d} \tag{4.20}$$

η_d being given by Equation 4.14.

Rao and Faghri (1988) use numerical techniques in order to develop a model which is not limited to a viscous flow (low Reynolds number, Re), but can also be used on a laminar flow. The authors consider the filter as an inline array of parallel cylinders, placed transverse to the flow and resolve the full Navier–Stokes equations with the assumption of periodic, fully developed flow. Their model, obtained without making the boundary layer approximation which is only theoretically valid for high Peclet numbers (Pe > 100), allows the effect of interference from adjacent fibers to be taken into consideration. For four values for the filter solidity they obtain two expressions for the single-fiber diffusion efficiency in accordance with Peclet's range:

$$\text{for the range Pe} < 50 \quad \eta_d = 4.89\left(\frac{1-\alpha}{\text{Ku}}\right)^{0.54} \text{Pe}^{-0.92} \tag{4.21}$$

$$\text{for } 100 < \text{Pe} < 300 \quad \eta_d = 1.8\left(\frac{1-\alpha}{\text{Ku}}\right)^{1/3} \text{Pe}^{-2/3} \tag{4.22}$$

(2) Inertial Impaction Regime

According to the authors, the characteristic geometric parameter of obstacle X, which appears in the definition of the Stokes number, can be represented either by radius R_f or by the diameter of the fiber D_f. The Stokes number then takes on the following two forms:

$$\text{If } X = R_f, \text{ then Stk} = \left(\rho_p D_p^2 \text{Cu} U\right)/\left(9\mu D_f\right) \tag{4.23a}$$

$$\text{If } X = D_f, \text{ then Stk} = \left(\rho_p D_p^2 \text{Cu} U\right)/\left(18\mu D_f\right) \tag{4.23b}$$

in which ρ_p represents the particle density, Cu the Cunningham coefficient (function of the Knudsen number of the particle, Kn_p), U the fluid velocity, and μ the dynamic viscosity of the fluid.

From a mathematical point of view, the calculation for the single-fiber efficiency of a fiber goes through the resolution of the Navier–Stokes equations (with various approximations) for an incompressible flow. Depending on the value of the Reynolds number relative to the fiber Re_f ($Re_f = (D_f U \rho_l)\mu$), which indicates the relative extent of the inertia forces and of the viscosity forces, a distinction is generally made between three flow regimes.

- At a low Reynolds number ($Re_f < 0.2$), with the extent of the inertia forces being relatively low, the flow around the fiber is purely viscous. Disturbance of the streamlines due to the fiber being present starts well upstream of the obstacle.
- For high Reynolds numbers ($Re_f > 1000$), the inertia forces are predominant and the flow around the cylinder is considered as ideal, nonviscous, and nonrotational. In this potential flow rating, the streamlines come very close to the fiber before deviating strongly in order to get around it.
- In the intermediate range of $Re_f (1 < Re_f < 1000)$, the flow is considered as transitory.

The nature of the flow around the fiber exerts considerable influence on the path which the aerosol is going to follow. In fact, a particle, in a potential flow, is more likely to be caught in the fiber due to the more sudden change of direction of the flow lines which, indeed, come closer to the fiber than in the case of a viscous flow.

In the following paragraphs only flows at low and intermediate Reynolds numbers are taken into consideration. Potential flow is too far from our experimental conditions to be developed here.

Model of Inertial Impaction in Viscous Flow

Case of an Isolated Fiber — Calculation of the collection efficiencies η_i and η_{ir} due, respectively, to pure inertial impaction and the couple of impaction and interception, in the case of a viscous flow is more difficult on account of the great complexity of the corresponding velocity domain.

Davies (1952) established the following equation for Re = 0.2:

$$\eta_{ir} = 0.16(R + (0.5 + 0.8R)\text{Stk} - 0.105R\text{Stk}^2) \tag{4.24}$$

in which R is the interception parameter and Stk is defined by Equation 4.23a.

A change to this equation is proposed by the same author in order to take account of the effects of interference from adjacent fibers. Coefficient 0.16 is then replaced by the expression ($0.16 + 10.9\alpha - 17\alpha^2$) in which α represents the filter solidity.

Case of a Fiber System — Yeh and Liu (1974) calculate the single-fiber efficiency η_i by basing their velocity domain calculations on Kuwabara's cell model (1959) which seems the most appropriate for $Re_f < 1$.

The authors do not give any analytical expression for η_i but they compare their experimental and theoretical results with the model developed by Stechkina et al. (1969).

Stenhouse (1975) proposes an equation for η_i, a function of R, η_r, α, and Stk (Equation 4.23a) which shows discontinuity. It is given by

$$\eta_i = (1 + R - \eta_r)(1 - J - 1) \text{ for } J \geq 1.0 \tag{4.25}$$

$$\eta_i = 0 \text{ for } J < 1.0 \tag{4.26}$$

$$J = 0.45 + 1.4\alpha + (1.3 + 0.5 \log_{10} \alpha)\text{Stk} \tag{4.27}$$

$$\eta_r = \frac{1}{2\,\text{Ku}}\left[(1+R)^{-1} - (1+R) + 2\left(1 + R\ln(1+R) + \alpha\left(-2R^2 - 0.5R^4 + 0.5R^5\right)\right)\right] \tag{4.28}$$

In one of their articles, Conder and Liew (1989) compare Yeh and Liu's model (1974) with that of Stenhouse (1975), concluding that the two models describe similarly the efficiencies obtained from experiments.

Empirical Correlations — In articles regarding experimental research carried out on the inertial deposition of particles in filters, the results are generally expressed in the form of graphs. Due to this, few analytical expressions describing the single fiber efficiency relative to impaction are available. Only Friedlander and Pasceri (1967), correlating data from experiments arising from the work of Wong and Johnstone (1953), express η_i with the equation:

$$\eta_i = 0.075\ \text{Stk}^{6/5} \tag{4.29}$$

for $0.8 < \text{Stk} < 2$, $\text{Re} < 1$ and $R < 0.2$. Here the Stokes number is calculated from Equation 4.23b.

The wide spread of the points from Wong and Johnstone's experiments (1953) for Stokes numbers below 0.5 indicates that the impaction mechanism is not appreciable in this domain. The domain of validity in this relationship being relatively limited, the authors consider their correlation as an attempt at model forming.

Models of Inertial Impaction in Transition Flow

Contrary to flows of the viscous and potential type for which Navier–Stokes equations can be simplified to find analytical solutions for the velocity domain, the intermediate regime requires complete resolution of these equations.

Landahl and Hermann (1949) base their calculations for velocity domain on Thom's model (1953) for $\text{Re}_f = 10$. They then adjust the efficiencies found with an empirical equation, solely the function of the Stokes number expressed by Equation 4.23b, given by

$$\eta_i = \frac{\text{Stk}^3}{\left(\text{Stk}^3 + 0.77\ \text{Stk}^3 + 0.22\right)} \tag{4.30}$$

Suneja and Lee (1974) calculate the collision efficiencies of spherical particles impacting on a cylindrical fiber for an intermediate domain of Re_f (1 to 60). From their results, expressed in graphs, they deduce an analytical expression for unit collection efficiency, a function of Stk (Equation 4.23b), Re, and R, given by

$$\eta_i = \frac{1}{\left(1 + \dfrac{1.53 - 0.23\ln\text{Re} + 0.0167(\ln\text{Re})^2}{\text{Stk}}\right)} + \frac{2R}{3\text{Stk}} \quad \text{for Re} < 500 \tag{4.31}$$

Iilas and Douglas (1989), after resolving the Navier–Stokes equations numerically, propose an equation based on a statistical analysis for all the single-fiber efficiencies by inertia.

This equation, a function of the Stokes number, given by Equation 4.23b, and of the Reynolds number Re_f, is defined by

$$\eta_i = \frac{\text{Stk}^3 + \dfrac{1.622\times10^{-4}}{\text{Stk}}}{1.031\ \text{Stk}^3 + \left(1.14 + 0.04044\ln\text{Re}_f\right)\text{Stk}^2 + 0.01479\ln\text{Re}_f + 0.2013} \tag{4.32}$$

for $30 \le \text{Re}_f \le 40$ and $0.07 \le \text{Stk} \le 5.0$.

4.2.3 Experimental

(1) Description of the Experimental Setup

To study the spectral efficiency of filters, Payet (1991) and Gougeon (1994) use the experimental setup, sketched in Figure 4.1, which consists essentially of two parts. The first concerns the production of the monodispersed liquid aerosol by a method originally proposed by Liu and Pui (1974). The second permits measuring of the concentrations of particles downstream and upstream of the filter tested.

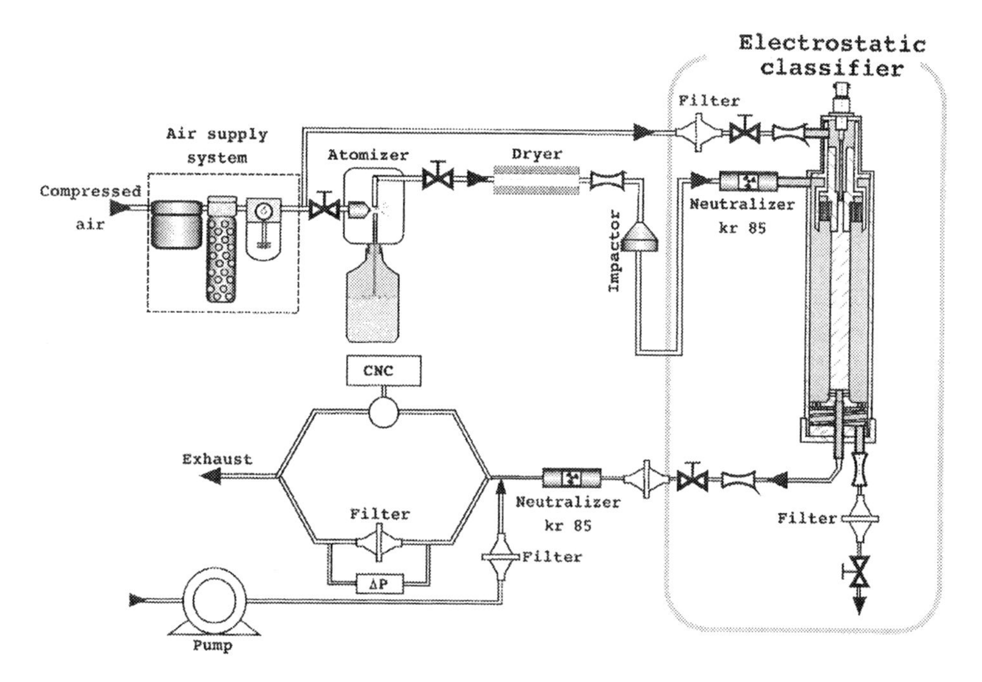

Figure 4.1 Experimental setup.

Initially, the dried and purified compressed air feeds a Collison-type atomizer containing a solution of DES (diethyl sebaçate) diluted in ethanol. The droplets produced, diluted with dry air, are then evaporated in a dryer. A micro-orifice impactor removes multiple charge-carrying particles with diameter above 1.1 μm. The polydisperse dry aerosol thus obtained then goes through a neutralizer (Kr85 source) where the particles are brought to the Boltzmann equilibrium. This makes their electric mobility a unique function of size. This property is used in the differential mobility analyzer (DMA) to select an aerosol of the precise diameter. The monodispersed aerosol thus produced then passes through the filter to be tested. The concentration is measured downstream and upstream of the filter with a continuous flow condensation nucleus counter (CNC) (Model 3020 from Thermo System Incorporated, St. Paul, MN). The pressure drop of the fibrous medium is monitored regularly by means of a linear water manometer. For further details, refer to Gougeon's work (1994).

(2) Filter Characteristics

The spectral penetration of fiber filters are measured, in the size range from 0.02 to 1 μm. The filters used for this research are "formettes" No. 106475 and 110 × 475, manufactured by the Bernard Dumas Company, which have clearly defined structural characteristics such as the diameter of the fiber D_f, the weight by unit surface g, and the solidity α. For comparison, an industrial HEPA filter from the PALL Company was also tested. The characteristics of these different types of media are listed in Table 4.1; for further details, refer to the work of Gougeon (1994).

Table 4.1 Structural Characteristics of the Different Filters

Fibers Filter	Formette 106475	Formette 110 × 475	HEPA Filter
Manufacturer	Bernard Dumas Company	Bernard Dumas Company	PALL Company
Referred to as	Model filter "F"	Model filter "C"	HEPA
D_f (μm)	0.65 ± 0.045	2.7 ± 0.1	1 to 1.5
h (mm)	0.78 ± 0.05	0.3 ± 0.05	0.2 ± 0.05
g ($g \cdot m^{-2}$)	76.9 ± 1	78 ± 1	40 ± 1
ρ_f ($g \cdot cm^{-3}$)	2.657 ± 0.005	2.657 ± 0.005	2.5 ± 0.05

The fiber diameter of the HEPA filter was determined from Davies' equation (1952):

$$\frac{\Delta p}{\mu h U_o} = \frac{16\alpha^{3/2}}{R_f^2} \quad \text{and} \quad D_f = 8\sqrt{\mu h \alpha^{3/2}\frac{U_o}{\Delta p}} \tag{4.33}$$

For low values of the frontal velocity, U_o, Δp is a linear function of U_o, the frontal velocity (Darcy, 1856). The term $U_o/\Delta p$ was determined from measurements of pressure drop as a function of velocity. D_f was found to be 1.3 μm. However, as the parameters α and h are marred by uncertainties, it is estimated in fact that D_f is between 1 and 1.3 μm.

In fact, in determining the pressure drop, Equation 4.33 favors coarse fibers which have a larger contact area but are not involved in the process of particles collection, to the detriment of the finer fibers which have a greater collection capability and are therefore of greater interest to us. Due to this, when making comparisons between theory and experiment, Gougeon (1994) assumes the fiber diameter of the HEPA filter to be between 1 and 1.3 μm.

The test aerosol was obtained by spraying a solution of DES in ultrapure ethanol at concentrations of 0.1% and 1% by volume. The experiments were carried out at filtration velocities from 2 to 8 cm s^{-1} for the diffusional regime and from 20 to 140 cm s^{-1} for the inertial domain.

(3) Results of Experiments

Figure 4.2 shows the penetration of the model filter "F" obtained by Gougeon (1994) as a function of the particle diameter at filtration velocities from 2 to 8 cm s^{-1}. In this domain where Brownian diffusion and direct interception are predominant, all the curves show a maximum which moves toward small diameters as U_o increases.

From 0.04 to 0.2 μm, the rise in frontal velocity creates an increase in the penetration of the filter, while this effect weakens appreciably above 0.2 μm. This is easily explained by the fact that the interception mechanism is purely geometric and therefore independent of the velocity, which is not the case with the diffusion mechanism where η_d decreases with the velocity.

Figure 4.3 shows the spectral penetration of the model filter "C" obtained by Gougeon (1994) for filtration velocities from 10 to 140 cm s^{-1}. Over the range of particle sizes studied, 0.1 to 1 μm, the increase in the frontal velocity (contrary to the previous regime) causes a drop in penetration which is the more marked the larger the particle diameter. These results agree with the theory since the larger the particles and the faster the velocity, the greater is the inertial mechanism.

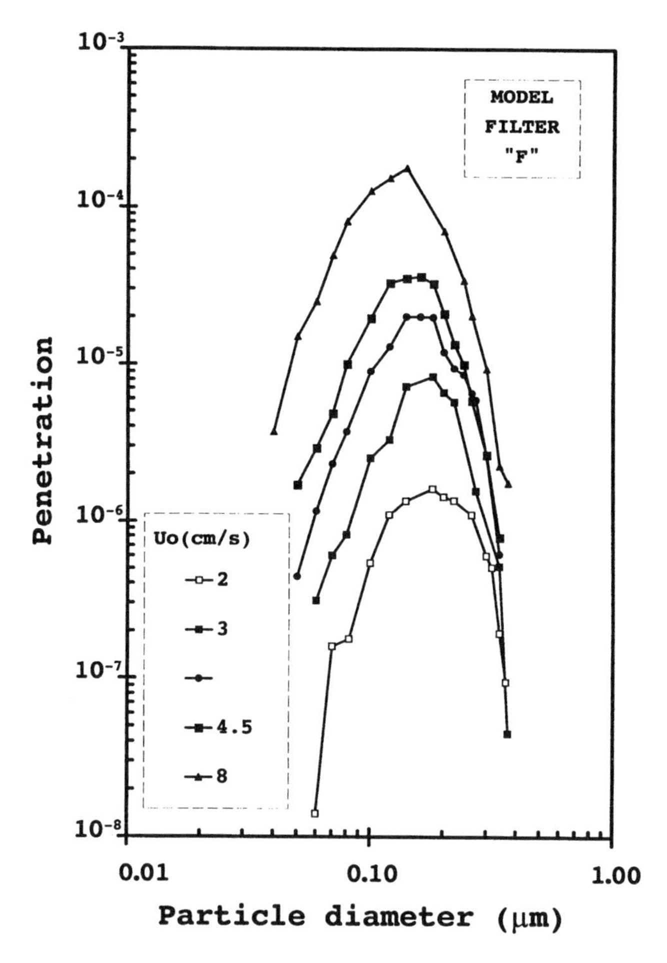

Figure 4.2 Penetration of the model filter "F" as a function of particle diameter for different frontal velocities.

4.2.4 Discussion

From the Figures 4.2 and 4.3, Gougeon (1994) deduces the experimental unit collection efficiency η_{exp} through the equation:

$$\eta_{exp} = \frac{\pi(1-\alpha)D_f \ln P_{exp}}{4\alpha h} \tag{4.34}$$

(1) Model Filter

Diffusional Regime — Figure 4.4 shows the unit diffusion collection efficiency η_d as a function of Peclet number Pe for the model filter "F" with the smallest fiber diameter. For Peclet numbers below 50, the points in her experiments are close to the correlation of Payet et al. (1992). For very low Peclet numbers such as 1 to 3, the experimental values of η_d tend to become constant. This is due to the limits of her experimental arrangement. In fact, in this domain, the data points for η_d correspond to very low experimental penetrations, in the order of 10^{-9}, due to very low filtration

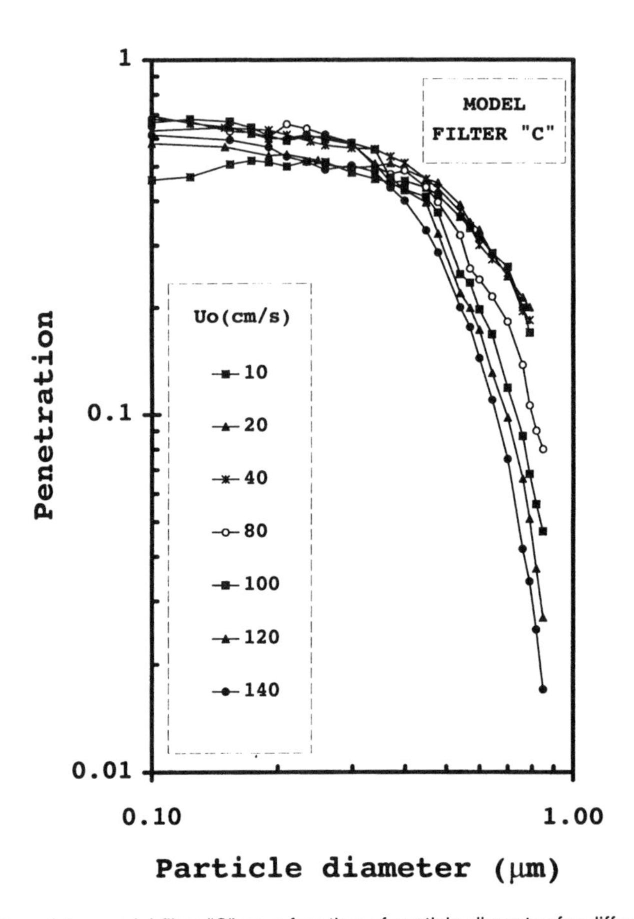

Figure 4.3 Penetration of the model filter "C" as a function of particle diameter for different frontal velocities.

velocities 0.6 cm s^{-1} which require very long counting times for particle concentrations downstream of the filter. In spite of this, the statistical error remains high (see error bars). Above Pe = 50, the difference between the correlation of Payet et al. (1992) and the results of the experiments originates from the role that interception plays in this domain.

Diffusion and Interception Regimes — In the new system of coordinates X and Y (Equation 4.12) developed by Friedlander and repeated by Lee and Liu (1982b), Gougeon rewrites the correlations of Liu and Rubow (1990) and Payet et al. (1992) in accordance with the calculations done in Table 4.2.

Figure 4.5 shows the values from her experiments for η_{exp} found with model filter "F," with the three models by replacing η with η_{exp} in the expression for Y. The curves clearly illustrate the various changes made to Lee and Liu's basic model (1982b) and show how well the theory agrees with the experiments by using the Payet et al. correlation (1992) for the diffusional regime ($X < 1$) as well as Liu and Rubow's model (1990) for the interception ($X > 1$).

Figures 4.6 and 4.7 compare the theoretical penetrations calculated from the three models and the experimental penetration with the model filters "F" and "C" as a function of particle diameter at filtration velocities equal to 3 and 4 cm s^{-1}, respectively. These curves show the influence of the slip effect which is considerable for the model filter "F" since it has the smallest fiber diameter. In both instances, Payet's correlation best describes the penetration in the diffusional regime and Liu and Rubow's model describes well the results of Gougeon's experiments in the domain where interception is predominant. Note that the penetrations vary in a large domain from 3×10^{-7} to 6×10^{-1}, agreeing satisfactorily with the models.

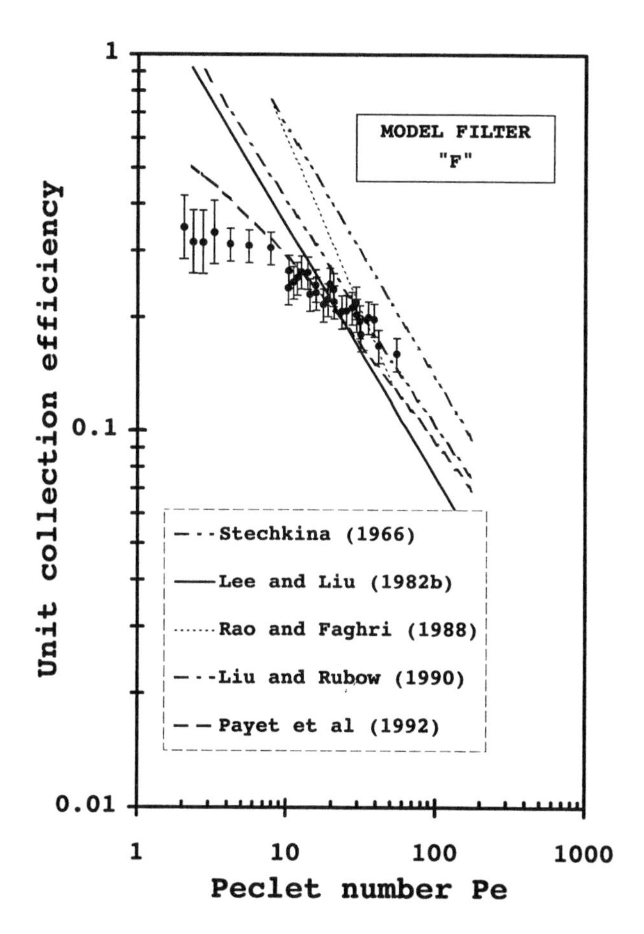

Figure 4.4 Comparison the collection due to diffusion between Gougeon's experimental results (1994) and works of different authors.

Table 4.2 Expressions of Coordinates *X* and *Y* with Lee and Liu's model (1982b), Corrected by Liu and Rubow (1990) and by Payet et al. (1992)

	X	Y
Lee and Liu (1982b)	$\left(\frac{1-\alpha}{Ku}\right)^{1/3} Pe^{1/3} \frac{R}{\sqrt{1+R}}$	$\eta Pe \frac{R}{\sqrt{1+R}}$
Liu and Rubow (1990)	$\left(\frac{1-\alpha}{Ku}\right)^{1/3} Pe^{1/3} C_d^{-1/2} \sqrt{\frac{R^2 C_r}{1+R}}$	$\eta Pe C_d^{-3/2} \sqrt{\frac{R^2 C_r}{1+R}}$
Payet et al. (1992)	$\left(\frac{1-\alpha}{Ku}\right)^{1/3} Pe^{1/3} \left(C_d C_d'\right)^{-1/2} \sqrt{\frac{R^2 C_r}{1+R}}$	$\eta Pe \left(C_d C_d'\right)^{-3/2} \sqrt{\frac{R^2 C_r}{1+R}}$

Inertial Regime — Gougeon (1994) has shown previously that η_d is best represented by Payet's correlation (4.18) and that η_r is described well by Liu and Rubow's Equation (4.15). Consequently, the single-fiber efficiency for inertial impaction is determined by

$$\eta_i = \eta_{exp} - (\eta_d \text{ (Payet)} + \eta_r \text{ (Liu and Rubow)}) \tag{4.35}$$

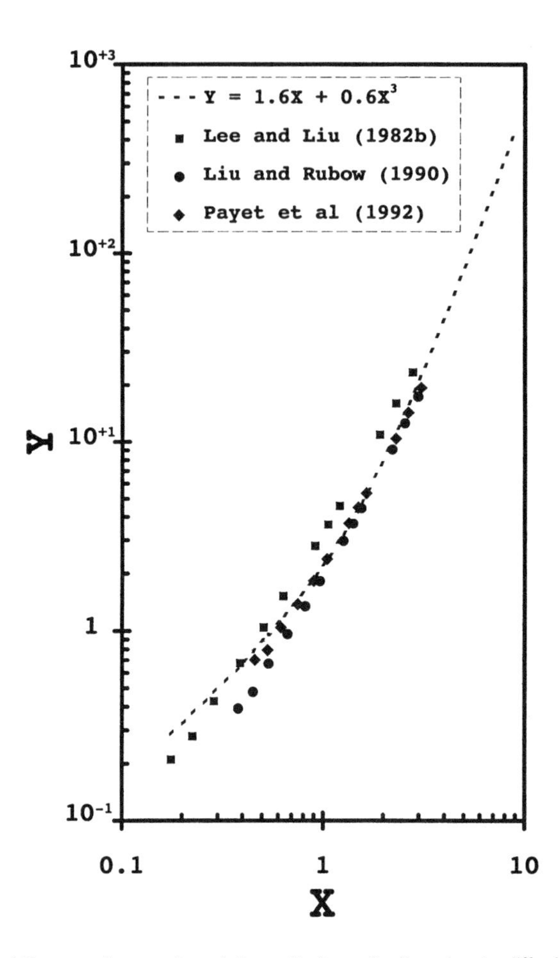

Figure 4.5 Correlation of Gougeon's experimental results for collection due to diffusion and interception in the form of Equation 4.12 (Lee and Liu, 1982b) corrected by Liu and Rubow (1990) and Payet et al. (1992) (see also Table 2.4).

Figure 4.8 compares the experimental values for η_i calculated from Equation 4.35 with two models for the impaction in intermediate and viscous flows, as a function of the Stokes number defined in Equation 4.23a.

We have not taken into account the models established on potential flow which are too far away from Gougeon's experimental conditions. Nevertheless, it can be considered that the lower limit of the models on transition flow can be linked to the viscous situation.

Gougeon (1994) noted a lack of agreement between her experiments and the theories described previously. With regard to Suneja and Lee's model (1974), higher values are to be expected since this relationship is determined for a transitory flow, therefore for higher Reynolds numbers ($1 \leq Re_f \leq 60$) than those ($Re_f \leq 0.25$) in our study. Concerning the work of Friedlander, although the relationship, which has been adapted with our Stokes number (Equation 4.23a), has the same domain of validity as the values from her experiments, she does not find any agreement.

In light of the unsatisfactory nature of the above comparison between theory and experiment, Gougeon (1994) developed an empirical correlation based on the single-fiber efficiencies from experiments using the model filter "C."

$$\eta_i = 0.0334 Stk^{3/2} \tag{4.36}$$

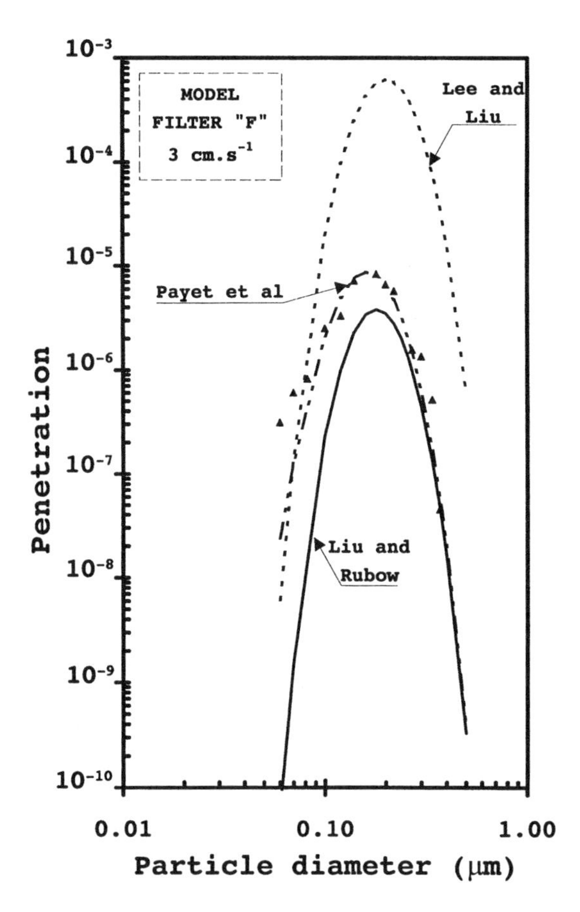

Figure 4.6 Penetration as a function of particle diameters. Comparison between Gougeon's experimental results (1994) and the correlation of Lee and Liu (1982b) corrected by Liu and Rubow (1990) and Payet et al. (1992). Experimental results are obtained with the model filter "F" and a frontal velocity of 3 cm s^{-1} (high slip effect).

for $0.5 \leq \mathrm{Stk} \leq 4.1$ and $0.0263 \leq \mathrm{Re}_f \leq 0.25$. This correlation can be extrapolated to higher Stokes numbers:

$$\eta_i = \frac{0.0334\ \mathrm{Stk}^{3/2}}{1 + 0.0334\ \mathrm{Stk}^{3/2}} \tag{4.37}$$

The correlation coefficient is 0.98 due to the well-defined structure of the model filter "C." It is important, therefore, to study the behavior of a real filter, the HEPA filter, under the same hydraulic conditions.

(2) HEPA Filter

Figure 4.9 shows the penetration of the HEPA filter at a frontal velocity of 5 cm s^{-1} in the particle size range of 0.02 to 0.7 µm. The major difficulty in interpreting the results relating to the HEPA filter lies in the difficulty in estimating the real diameter of the fibers from which it is made. Gougeon (1994) assumes here that the fiber diameter of the HEPA filter is between 1 and 1.3 µm. Therefore, she calculates the theoretical penetration of this filter for these two values of D_f. In the

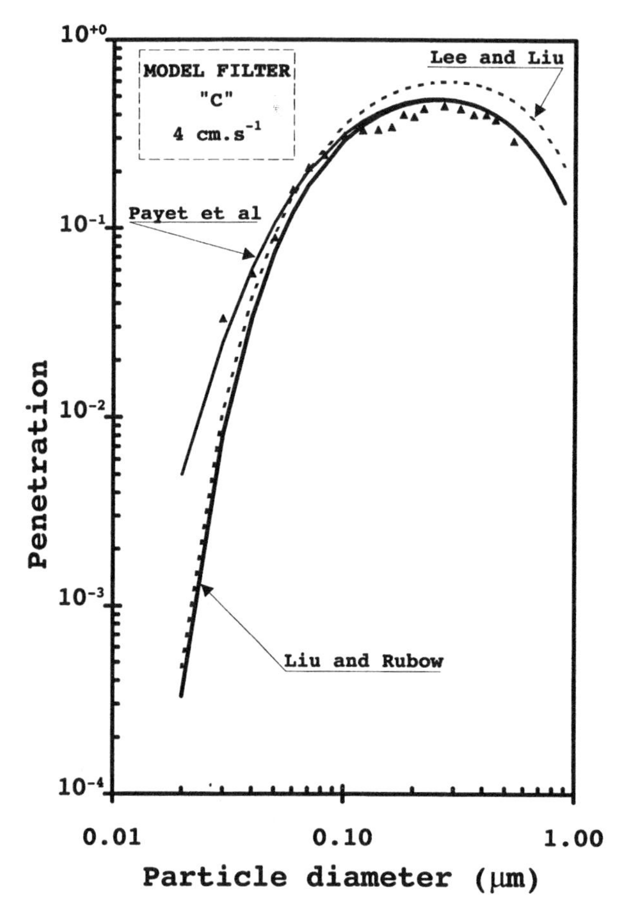

Figure 4.7 Penetration as a function of particle diameters. Comparison between Gougeon's experimental results (1994) and the correlation of Lee and Liu (1982b) corrected by Liu and Rubow (1990) and Payet et al. (1992). Experimental results are obtained with the model filter "C" and a frontal velocity of 4 cm s^{-1} (low slip effect).

diffusional and interception regimes the theory (Payet's correlation and Liu and Rubow's model) describes well the results when this interval is reduced between 1 and 1.1 μm.

Figure 4.10 shows the penetration of the HEPA filter at a frontal velocity of 60 cm s^{-1} in the size range of 0.1 to 1 μm. In this case, too, the experimental results agree well with the theoretical curves if the interval of D_f is reduced between 1 and 1.1 μm. The discrepancy observed for the particle diameters between 0.1 and 0.2 μm is systematic and is enhanced as the filtration velocity increases. The fact that this difference only appears for small particle sizes leads us to think that model-forming the Brownian diffusion, using Payet's correlation, underestimates the real collection of aerosols by this mechanism at high velocity. A possible explanation of this underestimating is the synergy of the diffusion and inertial impaction mechanisms. In fact, the works of Fernandez de la Mora and Rosner (1981; 1982) and Ramarao et al. (1994), based on a numerical simulation for the collection of aerosols by fibers, show that the inertia of the particles causes an increase in their concentration in the vicinity of the fiber. This local increase would bring about an increase in the collection of aerosols by Brownian diffusion. Interaction would then occur between these two mechanisms in the vicinity of the minimum efficiency of the filters and that is something which is not taken into consideration in our model.

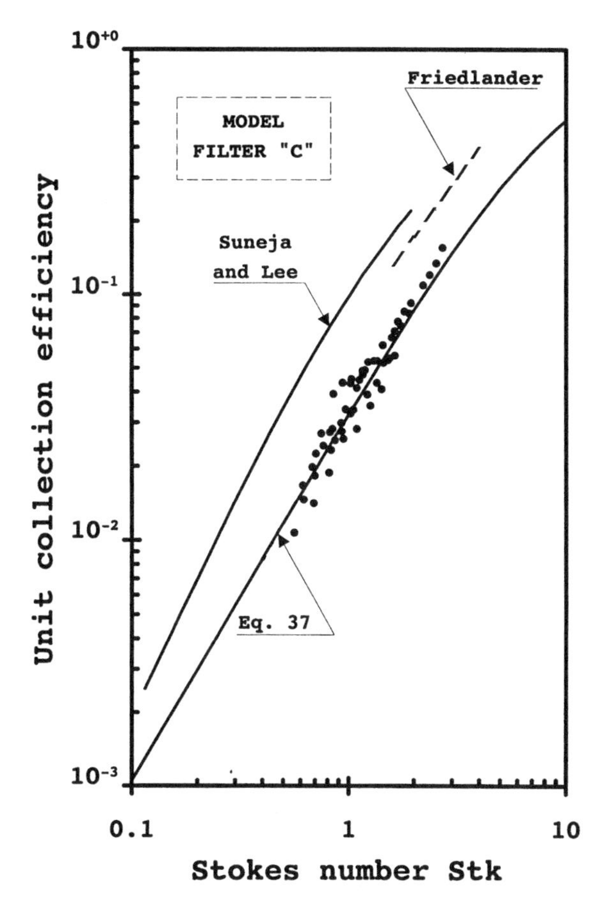

Figure 4.8 Comparison of the collection due to inertia with Gougeon's experimental results (1994) and correlations of different authors.

4.2.5 Conclusion

In this first part we have drawn up, on one hand, a review on research work carried out within the framework of the stationary filtration of aerosols by fiber filters. On the other hand, from the Payet (1992) and Gougeon (1994) works using special filters called "formettes" with clearly defined structural characteristics (fiber diameter, thickness, density), we have been able to arrive at precise results from their experiments with regard to penetration measurements. These results have allowed us to select two equations from among the models which appear in literature on the subject. One comes from the early work of Friedlander improved by different authors (Lee and Liu, 1982a; Liu and Rubow, 1990; Payet et al., 1992, and the other comes from Liu and Rubow, 1990), describing, respectively, the diffusion and interception mechanisms in strong slip effect regimes (small D_f) or weak slip effect regimes (large D_f). In the domain of inertial impaction, Gougeon (1994) proposes a new empirical correlation whereby it is possible to describe the capture efficiency of a fiber by inertial impaction as a function of the Stokes number in a viscous flow regime (low Reynolds number). These three relationships have been checked and then validated on an industrial HEPA filter.

Stationary filtration having now been clearly described, it was possible to study nonstationary filtration of liquid aerosols by a filter in the process of clogging, and that will be the subject of the second part of this chapter.

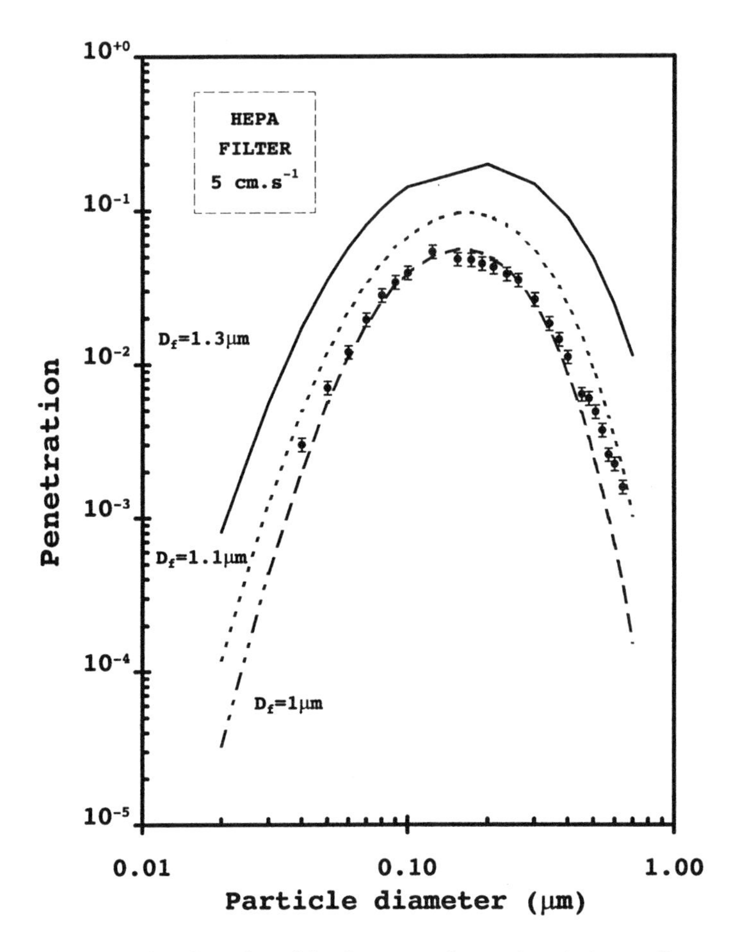

Figure 4.9 Penetration as a function of particle diameters. Comparison between Gougeon's experimental results (1994) and theoretical prediction using the correlations of Liu and Rubow (1990) and Payet et al. (1992) for η. Experimental results are obtained with the HEPA filter and a frontal velocity of 5 cm s^{-1}.

4.3 NONSTATIONARY FILTRATION

4.3.1 Introduction

In the filtration theory, two phases are generally distinguished. In the first one, as we just have seen, the stationary phase, it is assumed that one particle captured by a fiber remains fixed (the collision efficiency is unity) and that the collected particles do not disturb the filtration process. Thus, only the fibers are responsible for the liquid or solid aerosol capture. The basic filtration parameters, the filter efficiency E, and the pressure drop Δp, are therefore time independent.

In fact, these two assumptions are verified only in the initial stage of the filtration process, i.e., as the collected particle concentration is low. When this concentration increases, a time variation of E and Δp is observed because of structural changes in filter and the filtration process becomes nonstationary. In this second phase, the particle accumulation, in or upon the filter surface, induces an increase in the pressure drop Δp which is of special importance for the industry because of the filter "lifetime."

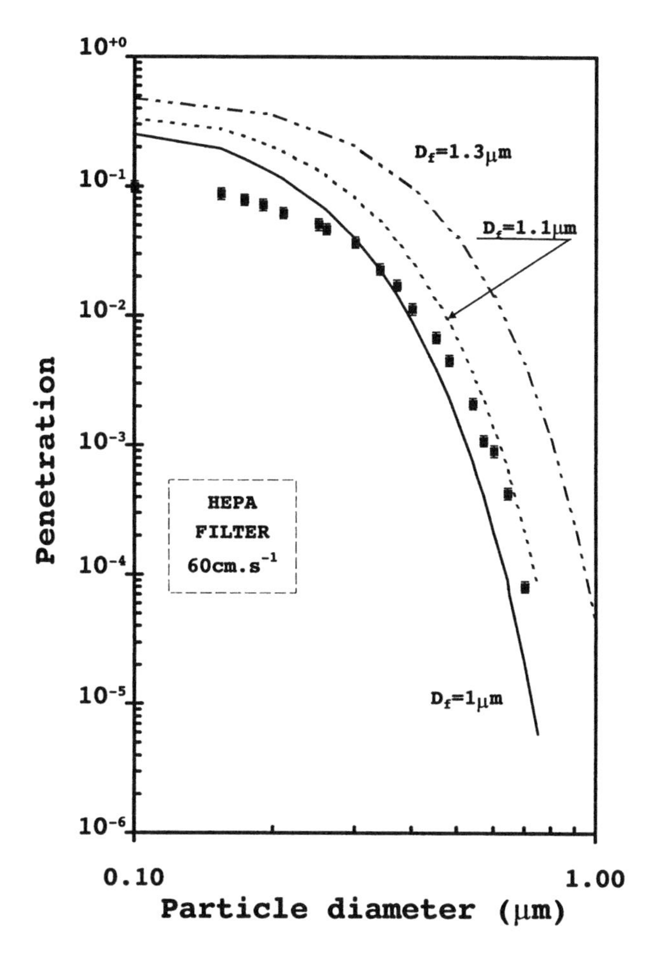

Figure 4.10 Penetration as a function of particle diameters. Comparison between Gougeon's experimental results (1994) and theoretical prediction using Equation 4.37 for η. Experimental results are obtained for the HEPA filter and a frontal velocity of 60 cm s^{-1}.

The filter efficiency variation, during filter loading, depends on the aerosol nature. During filter loading with solid aerosol, the incident particles depose preferentially on arrested particles rather than on uncovered fiber (Leers, 1957; Billings, 1966) forming aggregates at the fiber surface called "dendrites" because of their treelike appearance. Those dendrites are efficient in particle capture (as well as fibers) since this is the way they grow. The larger they grow, the greater is their ability to arrest more particles and so the filter efficiency increases. The dendrite formation by interception, inertial impaction, or diffusion is well known (Payatakes and Tien, 1976; Kanaoka et al., 1986) and their growth is well simulated by numerical calculations (Cai, 1990).

The increase in E and in Δp when clogging with solid particles is well described by different analytical equations.

In the case of filter clogging with liquid particles, few studies have been dedicated to this problem, the majority being of an experimental character. Recently, experimental and theoretical works carried out in our laboratories have tried to understand the different mechanisms involved in nonstationary filtration of liquid aerosols.

These works performed first by Payet (1991) in the diffusional and interception regimes and second by Gougeon (1994) in the interception and inertial regimes are developed hereafter. We emphasize the results and discussion, the materials and experimental methods having been described in the first part of this chapter.

4.3.2 Diffusional and Interception Regimes

(1) Experimental Results

Reproducibility — Payet et al. (1992) verified experimental reproducibility by measuring, under the same experimental conditions, the penetration of ten HEPA filters by monodisperse articles of DOP. The 95% confidence interval for particle diameters of 0.1 to 0.3 μm is given in Table 4.3.

Table 4.3 95% Confidence Interval for the Penetration of Ten HEPA Filters for DOP Particles (Frontal Velocity 3.65 cm s^{-1}; Δp_i = 100 Pa)

	Particle Diameter (μm)	
	0.1	0.3
95% confidence interval	8×10^{-3} to 1.85×10^{-2}	1.1×10^{-3} to 3.3×10^{-3}

Experimental uncertainty, defined as

$$\frac{\text{maximum value} - \text{minimum value}}{\text{mean value}} \tag{4.38}$$

was about 35%.

Effect of the Nature of the Aerosol — In the experiments of Payet et al. (1992) the HEPA filter was clogged with DES particles at a velocity of 2 cm s^{-1}. Measurements of penetration were taken alternately with monodisperse 0.15-μm particles of liquid DES and of solid sodium chloride up to 2.5 times the initial pressure drop across the filter (Δp_i = 60 Pa).

For both natures there is an initial period of improved collection followed by its decline. This is a typical well-known phenomenon corresponding to the preclogging of the filter by liquid aerosol which increases the collection efficiency due to the clogging of "microleaks."

The curves in Figure 4.11 show that the penetration was essentially the same whatever the test aerosol used. Consequently, for a filter clogged with liquid aerosol, the penetration of solid or liquid aerosols increases with a rise in pressure drop above 1.25 Δp_i.

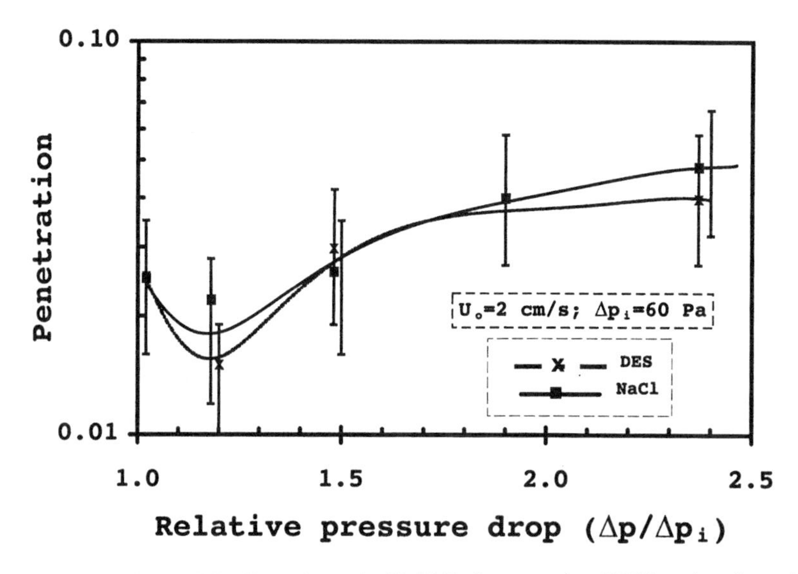

Figure 4.11 Penetration of the HEPA filter, clogged with DES, for aerosols of DES and sodium chloride.

Effect of Viscosity — The curves in Figure 4.12 show the experimental change in spectral penetration of three HEPA filters as a function of the increase in pressure drop when the filters are clogged with liquid particles of DOP, DES, or TBP (Tributylphosphate).

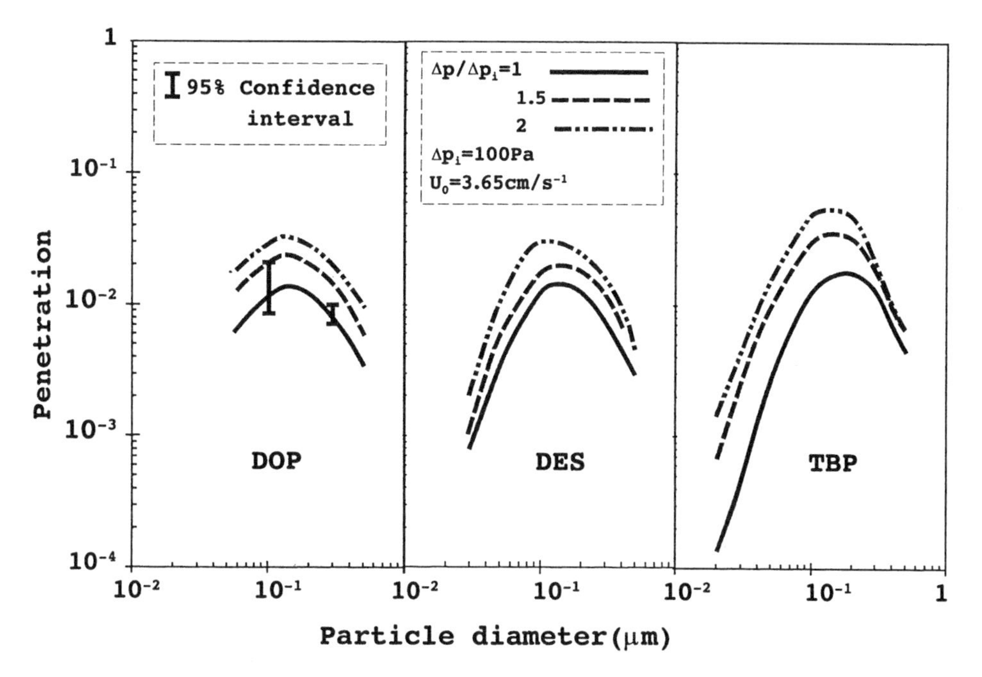

Figure 4.12 Variations of the penetration of the HEPA filter with particles diameters, as a function of the pressure drop for three liquid aerosols (DOP, DES and TBP).

These curves reveal increased penetration P of the filter medium during clogging, i.e., as the pressure drop increases, in contrast to what occurs with solid aerosols. The increase in P was particularly great around the maximum penetration, which seemed to shift toward the finer particles during clogging: 0.15 μm for a clean filter and 0.12 μm for the same filter clogged at $2\Delta p_i$. This suggests that the velocity increases within the filter as the pressure drop rises. Note that the variation in P as a function of $\Delta p/\Delta p_i$ was essentially the same for all the aerosols used to clog the filter.

The results concerning relative penetration (ratio of the penetration of the clean filter to that of the clogged filter) for three particle diameters are given in Table 4.4.

Table 4.4 Variation of the Relative Penetration of the HEPA Filters, Clogged with DOP, as a Function of the Relative Pressure Drop ($\Delta p/\Delta p_i$) for three particle diameters (frontal velocity = 3.65 cm s^{-1}; Δp_i = 100 Pa)

	$\Delta p/\Delta p_i$				
D_p (μm)	1	1.5	2.0	2.5	3
0.06	1	1.9	2.8	4.0	4.5
0.12	1	1.9	2.3	3.0	3.7
0.30	1	1.75	2.3	3.4	—

The same experiments were performed by Payet et al. (1992) with mineral oils. Figure 4.13 presents the relative variations in maximum penetration of the HEPA filter as a function of the

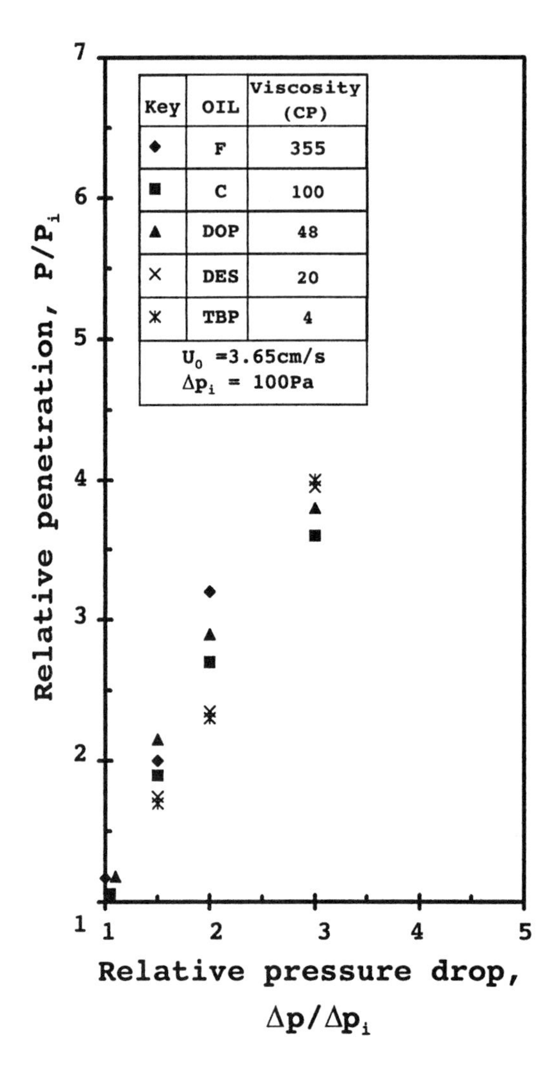

Figure 4.13 Relative maximum penetration of the HEPA filter as a function of the relative pressure drop for different oils.

increase in initial pressure drop, when the filter is clogged with oils C or F. The results obtained with DOP, DES, and TBP are also indicated in this figure.

It can be seen that these variations in maximum penetration are very similar, the precision of the measurement being about 35%. Approximately the same data scatter was seen for measurements at 1.5 and $3\Delta p_i$.

As these various liquids have different viscosities (from 4 to 355 cP), it can be deduced that viscosity does not play a significant role in the increase in penetration of a filter clogged by a liquid aerosol.

Test with Substantial Clogging — Changes in filter penetration were monitored up to an increase in pressure drop of $11\Delta p_i$, using DES particles at a face velocity of 2 cm s^{-1}. The measurements of relative penetration for various particle diameters are given in Table 4.5.

Two remarks can be made on the basis of these data:

- The increase in penetration of the filter was essentially the same for all selected particle sizes (0.12 to 0.34 μm) up to four times the initial pressure drop, Δp_i; hence, at $2\Delta p_i$, the penetration was

Table 4.5 Relative Penetration of the HEPA Filters, Clogged with DES, as a Function of the Relative Pressure Drop ($\Delta p/\Delta p_i$), for Different Particle Diameters (Frontal Velocity = 2 cm s^{-1}; Δp_i = 60 Pa)

	$\Delta p/\Delta p_i$								
D_p (μm)	1	2	3	4	5	8	9	10	11
0.12	1	1.5	3.9	4.8	10	17	18	17.7	20.8
0.15	1	2.2	4	4.7	8.6	13	14.6	15.5	18
0.20	1	2.0	4.4	5.1	4.0	8.8	10	9.7	14.3
0.26	1	2.1	2.9	4.3	3.8	7.3	10.8	8.7	—
0.34	1	2.0	3.2	2.0	3.3	10	10.9	7.1	—

increased by a factor of 2 or 2.5, whereas the increase was fourfold at $3\Delta p_i$ (except for particles larger than 0.25 μm, for which the relative penetration was lower);

- The relative penetration of the filter decreased with increasing particle size above $5\Delta p_i$; a potential explanation of this effect is a rise in interstitial velocity which affects the process of aerosol capture by diffusion, η_d, and impaction, η_i; indeed, the decrease in η_d for the finer particles (0.12, 0.15, 0.20 μm) may explain the increase in relative penetration; the increasing contribution of η_i for the largest particles (0.26, 0.34 μm) may slow the rise in relative penetration of the filter between 5 and $9\Delta p_i$, and decrease it above $10\Delta p_i$.

(2) Discussion

Background — Electron microscope observations by Liew and Conder (1985) suggest that the distribution of liquid in a filter medium essentially depends on filter solidity, α.

Hence, when α is less than 0.04, the interstices between the fibers are too large to be completely filled by liquid, which therefore forms large drops (of about 100 μm) at the intersection of two or more fibers. When α is greater than 0.04, liquid wetting of the fibers assumes several forms:

1. A film uniformly coating the fibers.
2. Droplets forming liquid bridges between several fibers.

The consequences of this could be

1. An increase in the apparent fiber diameter.
2. A decrease in the "number" or the "length" of "efficient" fibers in the filter.

As observed by Liew and Conder (1985), this last point is due to the fact that, since the distance between fibers and between points of intersection is small, the liquid forms irregular pools or patches spanning several fibers. In this condition, a significant proportion of the smaller interstices between the fibers is filled with collected liquid and not available for gas flow. The fibers bounding these filled interstices are not available for aerosol collection.

Furthermore, addition of liquid to the filter would decrease porosity, thus resulting in an increase in interstitial velocity.

Modeling — Payet (1991) has shown that the potential increase in fiber diameter has no real effect on the single-fiber efficiency. The increase in P is not due to a change in D_f, but rather to

1. The decrease in the "length" of "efficient" fibers in the filter, and
2. The increase in interstitial velocity.

Another possible explanation for the decrease in collection efficiency may be the increased reentrainment effect. With the increase in Δp, there should be a corresponding increase in the interstitial velocity which, in turn, may cause an increase in reentrainment. This explanation was very simple to reject because when the flow through the clogged filter was free of particles, the concentration downstream the filter was zero, too, and there is no evidence of the reentrainment effect for these experimental conditions.

Furthermore, when the ratio $\Delta p/\Delta p_i$ reached a value of 10, the interstitial velocity increased only a factor of two and remains too weak (4 cm s^{-1}) to induce a reentrainment effect.

So, only the two first explanations (1) and (2) were retained by Payet (1991) to develop her model. To estimate the fraction of fibers still available in the clogged filter, it is assumed that all wetted fibers cease to participate in the capture of incident aerosols, due to the formation of bridges between fibers, resulting in a drop in the length of efficient fibers and hence a reduction in the effective filtration surface area.

This reduction can be calculated as follows. If V_ℓ denotes the volume of liquid retained in the filter S_f the initial filtration surface area, ℓ the average liquid film thickness in the filter, the wetted fiber surface, S_F' is given by

$$S_F' = V_\ell/\ell \tag{4.39}$$

The fraction of fiber surface rendered ineffective is

$$S_F'/S_F = V_\ell/(S_F\ell) \tag{4.40}$$

From observations using an optical microscope we can consider that the liquid is spread through all the filter by capillarity and the liquid film thickness can be taken as the same as the filter thickness, h.

$$S_F'/S_F = V_\ell/(S_F h) \tag{4.41}$$

So the new solid fraction efficient for filtration, α', can then be expressed as follows:

$$\alpha' = \alpha(1 - V_\ell/(S_F h) \tag{4.42}$$

In this relation $(1 - V_\ell/(S_F h)$ represents the new fraction of surface efficient for filtration. It should be recalled that α represents the ratio of the fiber volume V_f to the total volume of the filter V_F. The volume of liquid V_ℓ is determined from the mass of liquid aerosol deposited, m_o, since

$$V_\ell = m_o/\rho_\ell, \tag{4.43}$$

where ρ_ℓ denotes the mass per unit volume of the liquid constituting the aerosol.

To calculate the new interstitial velocity U it is necessary to estimate the new porosity of the filter ε', from the following equation:

$$\varepsilon' = 1 - \alpha'', \tag{4.44}$$

where the term α'' represents the ratio of the volume of fibers plus the volume of the deposited liquid to the total filter volume, i.e.,

$$\alpha'' = \frac{V_\ell + V_f}{V_F} \tag{4.45}$$

Now, the overall penetration P of a fibrous filter in stationary filtration is expressed as

$$P = \exp\left[\frac{-4\alpha' h}{\pi(1-\alpha')D_f}\right], \tag{4.46}$$

and the Peclet number is calculated with the new interstitial velocity U:

$$U = \frac{U_o}{\varepsilon'} = \frac{U_o}{(1-\alpha'')} \tag{4.47}$$

Figures 4.14 and 4.15 illustrate the comparison of the experimental results with those obtained from the mass of liquid particles collected by the HEPA filter, for a given change in pressure drop across the filter.

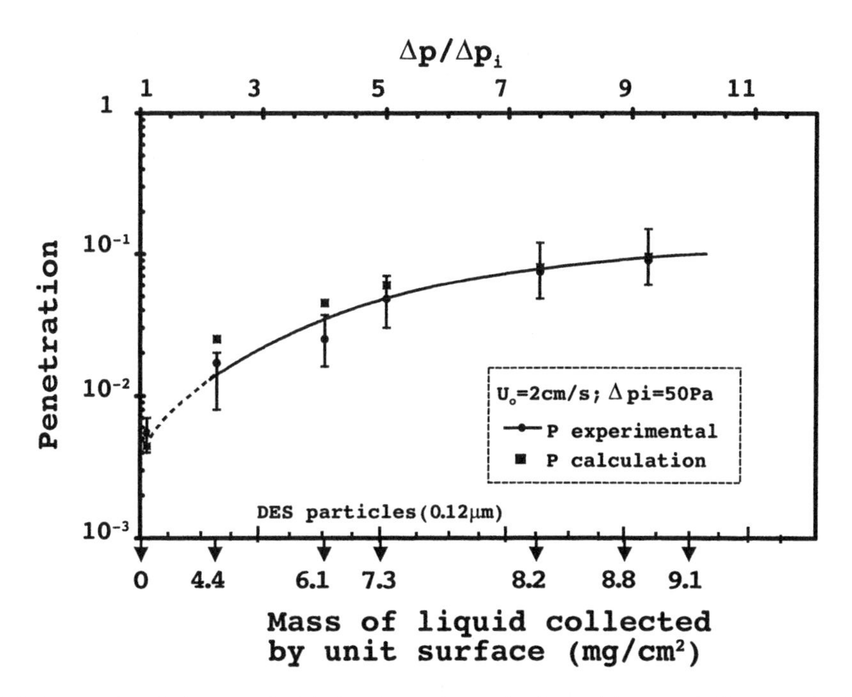

Figure 4.14 Comparison between calculated and experimental values of the penetration as a function of the relative pressure drop or as a function of the mass of liquid collected by unit initial surface of filtration.

From Table 4.6, which gives the relation between the variation of pressure drop ($\Delta p/\Delta p_i$) and the mass collected in the filter for a frontal velocity of 2 cm s^{-1}, it is possible to represent Figures 4.14 and 4.15 as the variations of penetration as a function of collected mass by unit initial surface of filtration.

Figures 4.14 and 4.15 show the changes in penetration calculated as a function of $\Delta p/\Delta p_i$, as well as the penetration derived from experiments on filter clogging with DES particles. Experimental results presented on Figures 4.14 and 4.15 have been obtained using one filter. On these two figures, it is not possible to observe the improvement of the collection efficiency for a $\Delta p/\Delta p_i$ smaller than 1.25 (see Figure 4.11). Payet et al. (1992) have chosen to study the increase of the penetration for values of $\Delta p/\Delta p_i$ corresponding to those of the Table 4.6, i.e., for $\Delta p/\Delta p_i$ greater than 2. So, in the range of $\Delta p/\Delta p_i$ comprised between 1 and 2 the experimental points are bound by a dashed line.

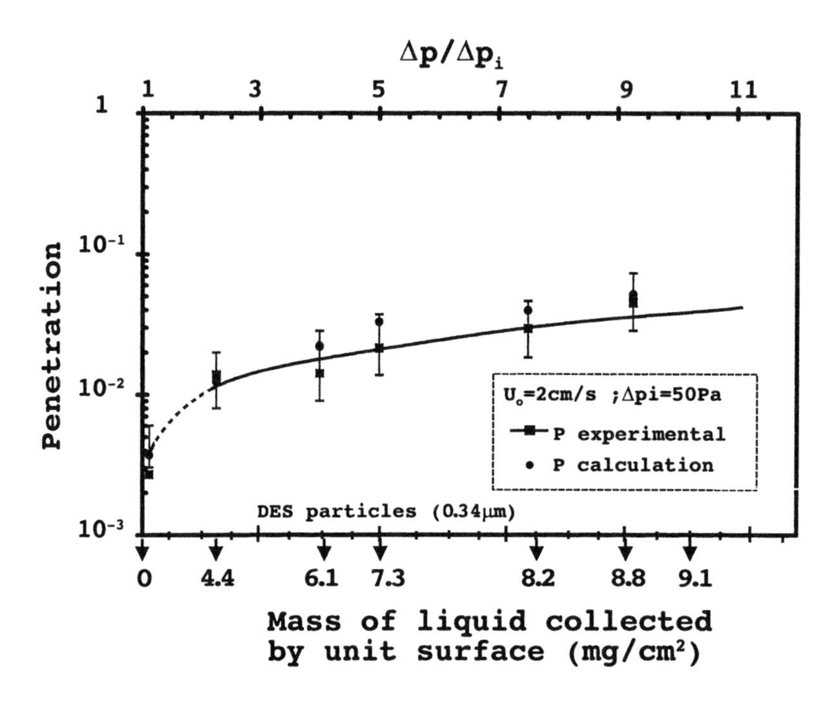

Figure 4.15 Comparison between calculated and experimental values of the penetration as a function of the relative pressure drop or as a function of the mass of liquid collected by unit initial surface of filtration.

Table 4.6 Relation Between the Relative Pressure Drop ($\Delta p/\Delta p_i$) and the Mass and the Volume of Liquid Aerosol Collected in the Filter (Frontal Velocity is 2 cm s^{-1})

$\Delta p/\Delta p_i$	m_o (g)	V_1 (cm^3)
2.3	0.064	0.070
4	0.088	0.107
5	0.106	0.116
7.5	0.119	0.130
9	0.128	0.140
10	0.132	0.145

Note: m_o: mass of liquid aerosols in the filter; V_1: volume of liquid aerosols in the filter; S_i: initial surface of filtration is 14.5 cm^2.

It can be seen that the predictions of the model of Payet (1991) are satisfactory. Indeed, the calculated values of P for different particle sizes (from 0.12 to 0.34 μm) are very close to the experimental values, even for large increases in pressure drop across the filter. Very similar agreements not reported here have been observed by Payet (1991) for particles of 0.15, 0.25, and 0.3 μm diameter.

In conclusion, the model, developed by Payet (1991) and Payet et al. (1992) provides a good understanding of the behavior of HEPA filters clogged with liquid aerosols. From simple formulae involving only the volume of liquid contained in the filter, which decreases the filtration surface area, and the filter porosity, it is possible to estimate correctly the changes in the penetration of these filters during clogging in the particle size range of 0.08 to 0.4 μm. Application of Payet's model shows that the increase in penetration can be explained in part by an increase in interstitial velocity and in part by a decrease in the "length" of "efficient fibers."

4.3.3 Interception and Inertial Regimes

(1) Experimental Results

Figures 4.16 and 4.17 obtained by Gougeon (1994) represent the HEPA filter penetration as a function of relative pressure drop $\Delta p/\Delta p_i$ at filtration velocities from 5 to 80 cm s^{-1} for particle sizes equal to 0.1 and 0.6 μm, respectively. For small particles (0.1 μm), the penetration increases in the first stages of clogging and tend to stabilize, except for 5 cm s^{-1} where it always rises. For larger particles (0.6 μm), all the curves show a maximum, which is the more marked, the faster the velocity. This maximum appears at smaller clogging rates as the filtration velocity increases. At 10 cm s^{-1}, the maximum of the curve corresponds to $\Delta p/\Delta p_i = 3.2$, and at 60 cm s^{-1} it corresponds to $\Delta p/\Delta p_i = 2$. For 0.6 μm particles, when the filter is clean ($\Delta p/\Delta p_i = 1$), the filter penetration is the lower, the faster the velocity. This phenomenon is real for all the clogging rates. This shows the dependence of inertial impaction on factors such as filtration velocity and particle size.

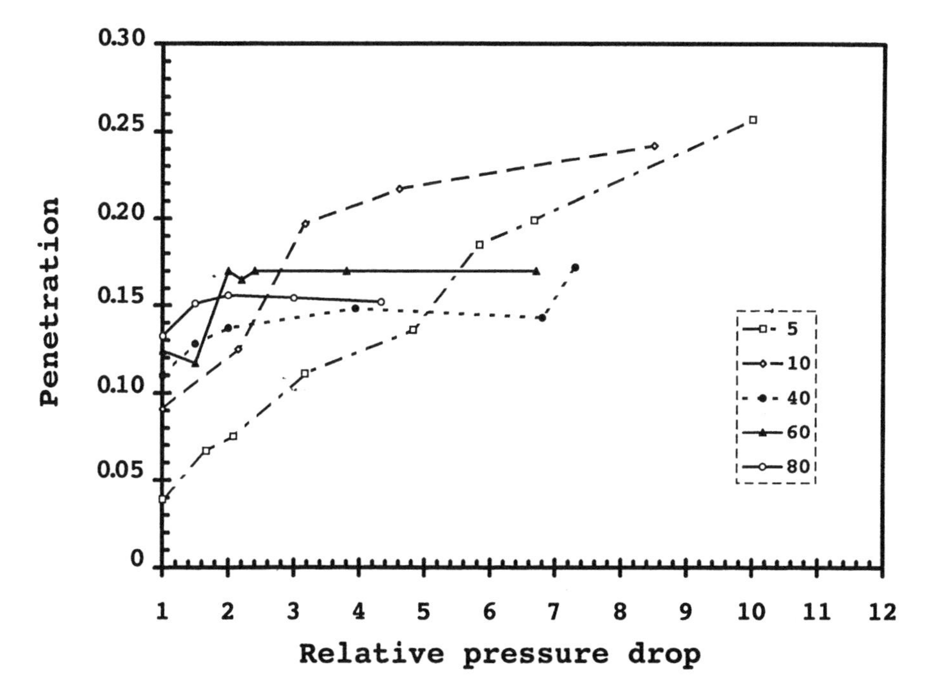

Figure 4.16 Penetration of the HEPA filter as a function of relative pressure drop $\Delta p/\Delta p_i$ for particle size of 0.1 μm and for different frontal velocities (cm/s).

(2) Discussion

Gougeon (1994) and Gougeon et al. (1994) observed the surface and the inner of the fiber filter clogged with liquid particles with a confocal light-scanning microscope. This microscope is able to change its focal plane, so this enables them to extend the depth of the microscope focus and to see the particles of all recorded layers. The clogging aerosol was glycerine tagged with fluorescent uranine. The liquid particles show a fluorescence like uranine, which has its maximum excitation at about 490 nm, lying in the green region of visible light.

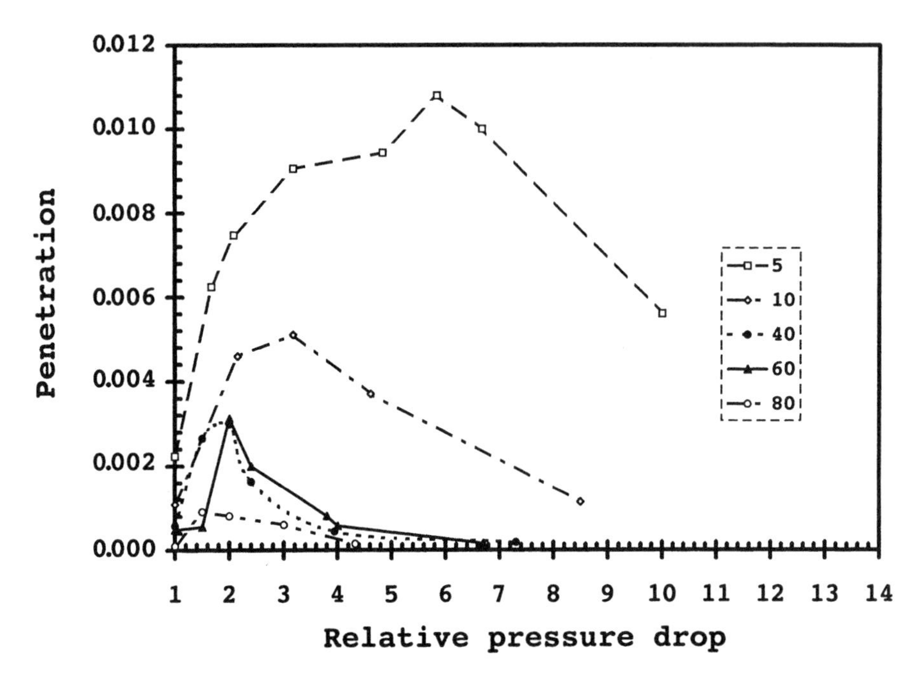

Figure 4.17 Penetration of the HEPA filter as a function of relative pressure drop $\Delta p/\Delta p_i$ for particle size of 0.6 µm and for different frontal velocities (cm/s).

For clogging with weak mass of aerosols, intersections and interstices between fibers are filled with liquid to form small droplets. As the clogging rate increases, those droplets grow up to form large surfaces of liquid (in order of 100 μm^2) located in the fiber intersections. Gougeon never observed a film uniformly coating the fibers.

As previously observed by Payet et al. (1992) in the diffusional and interception regimes, the increase in the filter penetration, Figure 4.16, is well predicted by the model developed by Payet (1991) for the wet structure of fibrous filters. The two main assumptions are

- Increase in interstitial velocity,
- Decrease in the fiber "length" available to collect particles.

Nevertheless, the increase in the interstitial velocity alone does not explain the decrease in the penetration observed at high velocities in the interception and inertial regimes.

A further mechanism has to be taken into account. It concerns the inertial impaction on the liquid surface formed at high loading rates and located in the fiber intersections (Gougeon, 1994).

4.3.4 Conclusion

The influence of the liquid collected particles on filter performances has been described in the case of fibrous filters, loaded in the diffusional, interception, and inertial regimes. Based mainly on the results obtained by Payet (1991) and Gougeon (1994), the following conclusions can be drawn:

- In the diffusional and interception regimes, the penetration increases with the loading rates of the filter;
- In the inertial regime, the penetration first increases and then decreases over a certain quantity of particles deposited in the filter.

Microscopical observations on the filters loaded with glycerine show at high loading rates a liquid distribution in the form of large liquid surfaces located in the fiber intersections of the filter.

The penetration and pressure drop results can be explained by taking into account the wet fibrous filter structure.

The increase in the filter penetration in the diffusional and interception regimes is well predicted in using the model developed in stationary filtration and adapted by Payet (1991) to the wet surface of the fibrous filters as follows:

- Increase in the interstitial velocity.
- Decrease in the fiber "length" available to collect particles.

The increase in the interstitial velocity alone does not explain the decrease in the penetration observed at high velocities in the interception and inertial regimes. A further mechanism must be taken into account. It concerns the inertial impaction on the liquid surfaces formed at high loading rates and located in the fiber intersections.

NOMENCLATURE

C_d	correction term for the diffusion slip flow effect
C_d'	correction term for the diffusion slip flow effect (Payet et al. correction, 1992)
Cr	correction term for the interception slip flow effect
Cu	Cunningham coefficient
D	diffusion coefficient
D_f	fiber diameter
D_p	particle diameter
E	overall efficiency of the filter
g	weight by unit surface of filter
h	thickness of the filter
Kn_f	Knudsen number of the fiber ($\mathrm{Kn}_f = 2\lambda/\mathrm{D}_f$)
Kn_p	Knudsen number of the particle ($\mathrm{Kn}_p = 2\lambda/\mathrm{D}_p$)
Ku	Kuwabara hydrodynamic factor (Equation 4.2)
ℓ	liquid film thickness
m_o	mass of liquid aerosol in the filter
n_1	number of particles downstream of the filter
n_2	number of particles upstream of the filter
P	penetration of the filter
P_{exp}	experimental penetration of the filter
Δp	pressure drop across the filter
Δp_i	initial pressure drop across the filter
Pe	Peclet number ($\mathrm{Pe} = UD_p/D$)
R	interception parameter ($R = D_p/D_f$)
Re_f	Reynolds number of the fiber
R_f	radius of the fiber
S_F	initial surface of filtration
S_F'	wetted fiber surface
Stk	Stokes number ($\mathrm{Stk} = 2\,\tau U/X$)
U_o	frontal velocity of filtration
U	interstitial velocity of filtration
V_ℓ	volume of liquid in the filter
X	geometric characteristics parameter of the fiber

Greek Symbols

α	filter solidity
α'	solid fiber fraction efficient for filtration
ε	filter porosity
ε'	filter porosity for a clogged filter
η	total single-fiber efficiency
η_d	single-fiber efficiency for diffusion
η_{exp}	experimental single-fiber efficiency
η_d'	single-fiber efficiency for diffusion (corrected by Payet et al., 1992)
η_i	single-fiber efficiency for impaction
η_r	single-fiber efficiency for interception
λ	mean free path of the carrier gas molecules
μ	dynamic viscosity of the fluid
ν	kinematic viscosity of the fluid
ρ_1	volumetric weight of the fluid
ρ_p	volumetric weight of the particle
τ	relaxation time of the particle

REFERENCES

Accomazzo, M. A., Rubow, K. L., and Liu, B. Y. H. (1984). *Solid State Technol.,* 141.

Albrecht, F. (1931). *Phys. Z.,* 32, 48.

Billard, F., Madelaine, G., and Pradel, J. (1963). Colloque sur la Pollution Radioactive des Milieux Gazeux, Saclay, 12–15 Novembre 1963. Presses Universitaires de France, Paris, 415.

Billings, C. E. (1966). Ph.D. Thesis, California Institute of Technology, Pasadena.

Cai, J. (1990). *Proceedings of the 3rd International Aerosol Conference,* Kyoto, Sept. 24–27, Pergamon, Oxford, 715.

Chen, C. Y. (1955). *Chem. Rev.,* 55, 595.

Conder, J. R. and Liew, T. P. (1989). *J. Aerosol Sci.,* 20, 45.

Darcy, H. P. G. (1856). *Les Fontaines Publiques de la Ville de Dijon,* Victor Dalmont Publ., Paris.

Davies, C. N. (1952). *Proc. Inst. Mech. Eng. London,* 1B, 185.

Davies, C. N. and Peetz, C. V. (1956). *Proc. R. Soc. London, Set. A,* 234, 269.

Fernandez de la Mora, J. and Rosner, D. E. (1981). *Physicochem. Hydrodyn.,* 2, 1.

Fernandez de la Mora, J. and Rosner, D. E. (1982). *J. Fluid Mech.,* 125, 379.

Friedlander, S. K. (1958). *Ind. Eng. Chem.,* 50, 1161.

Friedlander, S. K. (1967). *J. Colloid Int. Sci.,* 23, 157.

Friedlander, S. K. with help of Pasceri, R. E. (1967), in *Biochemical and Biological Engineering,* N. Blakenbrough, Ed., Academic Press, London.

Gougeon R. (1994). Thèse de l'Université Paris XII, 26 Sept. 1994. Rapport CEA-R-5684, 224 pp.

Gougeon, R., Boulaud, D., Fissan, H., Lange, R., and Renoux, A. (1994). *J. Aerosol Sci.,* 25(1), S209.

Happel, J. (1959). *Am. Inst. Chem. Eng. J.,* 5, 174.

Iilas, S. and Douglas, P. L. (1989). *Chem. Eng. Sci.,* 44, 1, 81.

Kanaoka, C., Emi, H., Hiragi, S., and Myojo, T. (1986). In *Aerosols, Formation and Reactivity,* 2nd Int. Aerosol Conf., Berlin, Pergamon, 674.

Kirsch, A. A. and Fuchs, N. A. (1967). *J. Phys. Soc. Jpn.,* 22(5), 1251.

Kuwabara, S. (1959). *J. Phys. Soc. Jpn.,* 14(4), 527.

Landahl, H. and Hermann, K. (1949). *J. Colloid Sci.,* 4, 103.

Langmuir, I. (1942). In Section I. U.S. Office of Scientific Research and Development, 865, part IV.

Lee, K. W. and Liu, B. Y. H. (1982a). *Aersol Sci. Technol.,* 1, 35.

Lee, K. W. and Liu, B. Y. H. (1982b). *Aersol Sci. Technol.,* 1, 147.

Leers, R. (1957). *Staub,* 17, 402.

Liew, T. P. and Conder, J. R. (1985). *J. Aerosol Sci.,* 16, 497.
Liu, B. Y. H. and Pui, D. Y. H. (1974). *J. Colloid Interface Sci.,* 47, 155.
Liu, B. Y. H. and Rubow, K. L. (1990). *Proc. of the 5th World Filtration Congress,* Nice, 5–8 Juin 1990. Société Française de Filtration, Paris, 3, 112.
Mohrmann, H. and Marchlewitz, W. (1974). *Staub Reinhalt. Luft,* 3, 91.
Payatakes, A. C. and Tien, C. (1976). *J. Aerosol Sci.,* 7, 85.
Payet, S. (1991). Thèse de l'Université Paris XII, 4 Oct. 1991. Rapport CEA-R-5589, 150 p.
Payet, S., Boulaud, D., Madelaine, G., and Renoux, A. (1992). *J. Aerosol Sci.,* 23, 723.
Pich, J. (1966). In *Aerosol Science,* C. N. Davies, Ed., Academic Press, New York.
Ramarao, B. V., Tien, C., and Mohan, S. (1994). *J. Aerosol Sci.,* 25, 295.
Rao, N. and Faghri, M. (1988). *Aerosol Sci. Technol.,* 8, 133.
Stechkina, I. B. and Fuchs, N. A. (1966). *Ann. Occup. Hyg.,* 9, 59.
Stechkina, I. B., Kirsch, A. A., and Fuchs, N. A. (1969). *J. Colloid Interface Sci.,* 31, 97.
Stenhouse, J. I. T. (1975). *Filt. Sep.,* 12 (May/June), 268.
Suneja, S. K. and Lee, C. H. (1974). *Atmos. Environ.,* 8, 1081.
Thom. A. (1933). *Proc. R. Soc. London,* A141, 651.
Wong, J. B. and Johnstone, H. F. (1953). Tech. Rep. No. 11, Eng. Expt. Station, University of Illinois, Urbana.
Wong, J. B., Ranz, W. E., and Johnstone, H. F. (1956). *J. Appl Phys.,* 27, 161.
Yeh, H. C. (1972). Ph.D. Thesis, University of Minnesota, Minneapolis.
Yeh, H. C. and Liu, B. Y. H. (1974. *J. Aerosol. Sci.,* 5, 191.

CHAPTER 5

Modeling of Particle Transport and Collection in Aerosol Flow Devices

Michael Shapiro

CONTENTS

5.1 INTRODUCTION

Principles of aerosol transport mechanics are widely used in modeling performance of various filtration, sampling, and aerosol processing devices and to predict their main parameters of engineering interest. These parameters include aspiration and sampling efficiency, losses in sampling tubes, efficiency of aerosol processing reactors, collection efficiency of porous filters, electrostatic precipitators, etc. The basic means of analysis is the aerosol transport equation (Friedlander, 1977).

$$\frac{\partial c_w}{\partial t}+\frac{\partial}{\partial \mathbf{R}}\left\{\left[\mathbf{U}(\mathbf{R},t)+\mathbf{m}_w\mathbf{F}(\mathbf{R},t)\right]c_w-\mathbf{D}_w\frac{\partial c_w}{\partial \mathbf{R}}\right\}+\dot{N}_w=0 \tag{5.1}$$

0-87371-830-5/98/$0.00+$.50

governing spatial (**R**) and temporal (t) evolution of concentration, $c_w = c_w(\mathbf{R},t)$ of a certain population w of aerosol particles. The latter possess diffusivity $\mathbf{D}_w$, and mobility $\mathbf{m}_w$, move in the flow velocity field $\mathbf{U}(\mathbf{R},t)$, and experience external forces $\mathbf{F}(\mathbf{R},t)$. In addition, $\dot{N}_w$ describes the rate of gain (loss) of the considered population due to aerosol transformations, that may occur within the device (e.g., evaporation, condensation, coagulation, etc.). This equation is supplemented by the appropriate initial and boundary conditions, usually specified on solid surfaces existing within each given device and, possibly, in the inlet flow region.

Even in circumstances where the aerosol transport problems posed for Equation 5.1 are linear, their solutions may be obtained in very few cases, pertaining to overly simplistic approximations of the flow fields within the devices and their geometry. The latter factor is most important for porous filters, which are normally characterized by a complicated (oftentimes, fractal) interstitial structure. Obtaining exact solutions for the convective–diffusive transport problems, posed for polydisperse aerosols in such systems, is a formidable task even for modern supercomputers.

On the other hand, efficient modeling of aerosol transport and deposition processes in such devices necessitates development of simplifying approaches, capable of sufficiently and accurately predicting their performance, while adequately accounting for the geometric structure of the complicated system and the physical processes occurring therein. Such an approach, which is applicable to a variety of physical situations and geometries, is described below.

5.2 CONVECTIVE DISPERSION OF AEROSOLS

Fortunately, in spite of the apparent complexity of the processes described above, in many circumstances, a short time after entering a device aerosol particles achieve an asymptotic spatial distribution within a characteristic local (bounded) region, dependent on the geometry of each specific device. This region is the cross-sectional area of a sampling tube, aerosol reactor, electrostatic precipitator, etc. For porous filters it is a macroscopically small volumetric element, however, well representing the interstitial structure of the filter (see Section 5.8). This asymptotic aerosol state is characterized by a balance (prevailing within the characteristic region) between the flux components of the particles due to diffusion, convection, and external forces, acting upon the particles, as well as due to aerosol transformations. The process of interaction between this local aerosol state and the global convective aerosol transport in the direction, say, x along the principal airflow through the device, is called *convective dispersion.* A general theory of convective dispersion phenomena (Shapiro and Brenner, 1986; Brenner and Edwards, 1993) is aimed at establishing the principles of transport and evolution of an *averaged* aerosol concentration, $\bar{c}_w$. The averaging is performed over the characteristic region, described above, so that $\bar{c}_w$ remains a function of x. Therefore, $\bar{c}_w$ may be used to calculate the global parameters of engineering interest, characterizing the overall performance of the aerosol device.

5.3 AXIAL AEROSOL TRANSPORT PROPERTIES

Consider pulses of monodisperse aerosol particles, instantaneously introduced from a particle generator into the clean air, flowing through a certain aerosol device (e.g., porous filter), with the particle concentration being monitored at two stations within the device (see Figure 5.1), or beyond it. While moving through the device, some of the particles will be captured inside due to several deposition mechanisms. Those of the particles which "survive," will disperse in longitudinal and lateral directions due to flow velocity gradients, Brownian diffusion, and, possibly, external forces, acting on the particles within the device. The time, t, which it takes the center of mass of the pulse to cover the distance L between the two stations, is related to the mean aerosol velocity of the particle

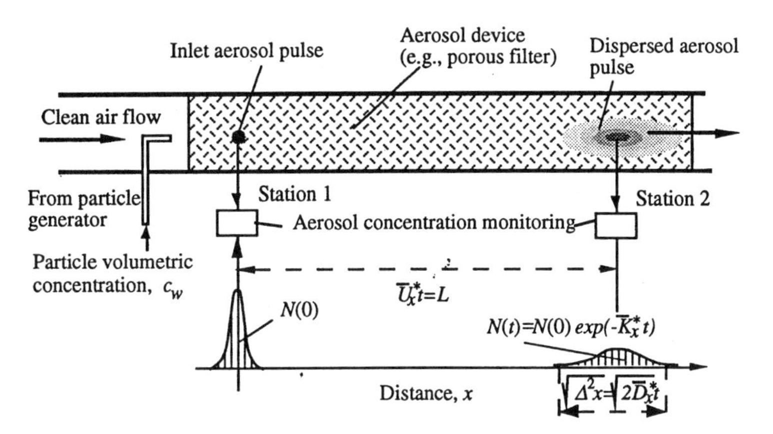

Figure 5.1 Definitions and experimental procedure for measuring effective axial aerosol transport properties.

$$\overline{U}_x^* t \sim L \,. \tag{5.2}$$

Measurements of the total number of particles passing the inlet and outlet stations, respectively, $N(0)$ and $N(t)$, yield the value of the aerosol volumetric deposition coefficient (Shapiro and Brenner, 1986)

$$\overline{K}_x^* \sim \frac{1}{t} \ln \frac{N(t)}{N(0)} \,. \tag{5.3}$$

Finally, the axial spread of the aerosol pulse (longitudinal mean square displacement, $\Delta^2 x$), obtained via measurements of the time-dependent particle concentrations at the two stations, yields the aerosol axial dispersivity $\overline{D}_x^*$ (Fuchs, 1964; Brenner and Edwards, 1993)

$$\Delta^2 x \sim 2\overline{D}_x^* t \,. \tag{5.4}$$

Practical implementations of the above measurement method involve simultaneous monitoring the aerosol concentrations at the two stations, aimed at determination of the time t, appearing in Equation 5.2, the mean square displacement $\Delta^2 x$, and the total amount of particles $N(0)$ and $N(t)$ within the inlet and outlet pulses. For such measurements noninvasive optical methods may be employed, similar to those used in bolus dispersion analysis (Gebhart et al., 1988).

5.4 CALCULATION OF THE EFFECTIVE TRANSPORT COEFFICIENTS

The coarse-scale aerosol transport coefficients, described in the previous section, will be given here rigorous mathematical definitions, and a method for their calculation will be described. Suppose, that the geometry of an aerosol device enables the following decomposition of aerosol particle position vector $\mathbf{R}$

$$\mathbf{R} = \mathbf{q} + \mathbf{i}_x x,$$

where x is the coordinate parallel to the principal aerosol flow direction, and $\mathbf{q}$ is the variable lying within the device cross section. In the simplest case of flow in a tube of any construct cross-sectional geometry, x is directed along the tube axis and $\mathbf{q} = (r, \theta)$, with r and θ being polar coordinates within its cross section (see Figure 5.2).

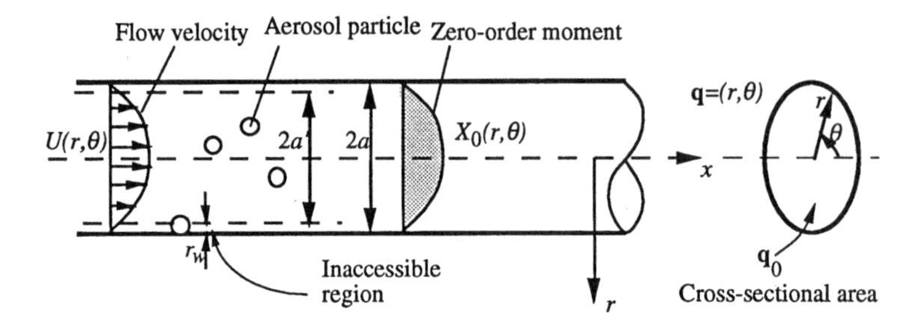

Figure 5.2 Schematic of motion of aerosol particles in a tubular device.

Consider a fully developed flow of a clean air flowing in a device with the velocity $\mathbf{U} = \mathbf{i}_x U(\mathbf{q})$. Suppose that an aerosol cloud of monodisperse aerosol particles is introduced into the flow where it is transported by convection and diffusion. Define the following *axial* moments, via the aerosol concentration distribution $c_w(t, \mathbf{q}, x)$:

$$M_{xn}(t) = \int_{\mathbf{q}_0} dq \int_{-\infty}^{\infty} x^n c_w(t, \mathbf{q}, x) dx, \quad n = 0, 1, 2, \ldots, \tag{5.5}$$

where $\mathbf{q}_0$ denotes the cross-sectional area. Explicitly, the zero-order moment $M_{x0}(t)$ gives the total amount of aerosol particles, present at time t in the device. These moments are used to define *mean axial* displacement

$$\bar{x}(t) = M_{x1}(t)/M_{x0}(t), \tag{5.6}$$

and *mean square axial* displacement of the aerosol cloud:

$$\Delta^2 x \equiv \overline{(\Delta x)^2} = \frac{1}{M_{x0}} \int_{\mathbf{q}_0} d\mathbf{q} \int_{-\infty}^{\infty} (x - \bar{x})^2 c_w(t, \mathbf{q}, x) dx = \frac{M_{x2}}{M_{x0}} - \frac{M_{x1}^2}{M_{x0}^2}, \tag{5.7}$$

in accordance with the intuitive definitions of these quantities given in the previous section. They may be used to define (Shapiro and Brenner, 1986) the phenomenological coefficients, characterizing *mean axial* aerosol transport:

$$\bar{U}_x(t) = \frac{d\bar{x}}{dt} = \frac{d}{dt}\left(\frac{M_{x1}}{M_{x0}}\right), \tag{5.8a}$$

$$\bar{D}_x(t) = \frac{1}{2}\frac{d}{dt}\overline{(\Delta x)^2} = \frac{1}{2}\frac{d}{dt}\left(\frac{M_{x2}}{M_{x0}} - \frac{M_{x1}^2}{M_{x0}^2}\right), \tag{5.8b}$$

$$\bar{K}_x(t) = \frac{d}{dt}\ln M_{x0}\,. \tag{5.8c}$$

These coefficients are, respectively, the mean axial aerosol velocity, dispersivity (effective axial diffusivity), and the mean axial aerosol deposition coefficient. Generally, they depend both on time and on the initial concentration distribution within the cloud $c_{w0}(\mathbf{q}, x)$. However, in fairly general circumstances and for sufficiently long times these coefficients may be shown to achieve *constant* values, marked by asterisks following:

$$\left(\overline{U}_x^*, \overline{D}_x^*, \overline{K}_x^*\right) = \lim_{t\to\infty}\left(\overline{U}_x, \overline{D}_x, \overline{K}_x\right), \tag{5.8d}$$

which are independent of $c_{w0}(\mathbf{q}, x)$. These coefficients may be also shown to govern long-time, area-averaged aerosol concentration

$$\overline{c}_w(t,x) = \int_{\mathbf{q}_0} c_w(t,\mathbf{q},x)d\mathbf{q}, \tag{5.9}$$

via the effective axial transport equation

$$\frac{\partial \overline{c}_w}{\partial t} + \frac{\partial}{\partial x}\left(\overline{U}_x^* \overline{c}_w - \overline{D}_x^* \frac{\partial \overline{c}_w}{\partial x}\right) + \overline{K}_x^* \overline{c}_w = 0. \tag{5.10}$$

As indicated in Equation 5.5, calculation of the effective axial transport coefficients requires knowledge of the aerosol concentration field $c_w(t, \mathbf{q}, x)$. Below, a method will be described for determination of the coefficients $\overline{U}_x^*, \overline{D}_x^*, \overline{K}_x^*$ without calculating this overly extensive, usually unavailable information.

5.5 TRANSPORT AND DEPOSITION OF AEROSOLS IN A CIRCULAR TUBE

Consider a cloud of monodisperse submicrometer aerosol transported by diffusion and convection in a tube (see Figure 5.2) of a circular cross section of radius a. The aerosol concentration is governed by the following equation:

$$\frac{\partial c_w}{\partial t} + U(r)\frac{\partial c_w}{\partial x} = D_w\left[L(c_w) + \frac{\partial^2 c_w}{\partial x^2}\right] \quad 0<r<a,\ \ 0<\theta<2\pi,\ \ -\infty<x<\infty, \tag{5.11}$$

where L *is the differential operator*

$$L = \frac{1}{r}\frac{\partial}{\partial r}\left(r\frac{\partial}{\partial r}\right) + \frac{1}{r^2}\frac{\partial^2}{\partial \theta^2}. \tag{5.12}$$

Moreover, $U(r)$ is given by the fully developed Poiseuille air velocity profile

$$U(r) = 2\overline{U}\left(1 - \frac{r^2}{a^2}\right), \tag{5.13}$$

with $\overline{U}$ being the mean air velocity. We will assume the aerosol Brownian diffusivity D_w related to the aerosol radius r_w by the Stokes–Einstein formula:

$$D_w = \frac{kTC}{6\pi\mu r_w}, \tag{5.14}$$

where μ is the air dynamic viscosity, kT is the Boltzmann factor and C is the Cunningham's slip correction coefficient.

The concentration is subjected to the following boundary and initial conditions

$$c_w = 0 \text{ at } r = a' \equiv a - r_w,\ c_w \text{ — continuous in } \mathbf{q}_0, \tag{5.15a,b}$$

$$x^n c_w \to 0, \text{ as } |x| \to \infty,\ n = 0, 1, 2, \ldots, \tag{5.16}$$

$$c_w = c_{w0}(r, \theta, x) \text{ at } t = 0. \tag{5.17}$$

Condition 5.15a implies that any aerosol particle, which approaches the tube wall to a distance comparable to its radius, sticks to the wall and never reentrains the flow.

We will use the above posed mathematical problem (5.11 through 5.17) to calculate the long-time behavior of the moments M_{xn}, $n = 0,1,2$ of the aerosol distribution. Toward this goal, define the comparable local moments by

$$P_{xn}(\mathbf{q},t) = \int_{\mathbf{q}_0} x^n c_w(t,\mathbf{q},x)dx, \quad n = 0,1,2,\ldots \tag{5.18}$$

To obtain the equation governing P_{xn}, multiply both sides of Equation 5.11 by x^n, $n = 0,1,2\ldots$ and integrate over the infinite x-domain using boundary conditions (5.16) to get the following equations

$$\frac{\partial P_{x0}}{\partial t} - D_w L(P_{x0}) = 0\ , \tag{5.19}$$

$$\frac{\partial P_{x1}}{\partial t} - D_w L(P_{x1}) = U(r)P_{x0}\ , \tag{5.20}$$

$$\frac{\partial P_{x2}}{\partial t} - D_w L(P_{x2}) = 2\left[U(r)P_{x1} + D_w P_{x0}\right]. \tag{5.21}$$

In addition, each of P_{xn} satisfies the following boundary conditions

$$P_{xn} = 0 \text{ at } r = a',\ r_w,\ P_{xn} \text{ — continuous in } \mathbf{q}_0 \tag{5.22}$$

easily obtainable from 5.18 and 5.15a,b.

The solution for the problem posed for the *zero-order* moment may be sought in the form of the eigenfunction expansion

$$P_{x0}(t,r,\theta) = \sum_{k=0}^{\infty} A_k X_k(r,\theta)\exp(-\lambda_k t) \tag{5.23}$$

where $X_k(r, \theta)$, $k = 0,1,2,\ldots$ are eigenfunctions, satisfying the following eigenvalue problem:

$$D_w L(X) = \lambda X, \tag{5.24}$$

$$X = 0 \text{ at } r = a',\ X \text{ — continuous in } \mathbf{q}_0. \tag{5.25}$$

The coefficients A_k ($k = 0,1,2,\ldots$) appearing in Equation 5.23 are determined from the appropriate initial condition, imposed on P_{x0}, which may be obtained by introducing Equation 5.17 into 5.18 with $n = 0$.

We will be interested in the asymptotic long-time solution for the local moment P_{x0}, i.e., valid for the time $t >> a^2/D_w$. This time characterizes the aerosol transport within the tube cross section; specifically, by this time the aerosol cloud effectively fills the tube cross-sectional area. However, the latter time should be significantly less than the characteristic time $L/\overline{U}$ of the aerosol axial transport, with L being the tube length. The asymptotic solution for P_{x0} is given by the leading-order term in expansion 5.23, namely

$$P_{x0}(t,r,\theta) \cong A_0 X_0(r)\exp(-\lambda_0 t)(1 + \exp),\ t >> a^2/D_w, \tag{5.26a,b}$$

where exp means terms decaying exponentially fast with time, $X_0(r)$ is a θ-independent leading eigenfunction, given by

$$X_0(r) = J_0\left(\gamma_0 \frac{r}{a'}\right), \tag{5.27}$$

with J_0 being the zero-order Bessel function of the first kind, and $\gamma_0 = 2.4048$ is the smallest positive root of the transcendental equation $J_0(\gamma) = 0$. The corresponding smallest eigenvalue is given by

$$\lambda_0 = D_w \frac{\gamma_0^2}{a'^2} \tag{5.28}$$

First-Order Moment — The equation governing asymptotic first-order moment is obtained by substituting 5.26a into 5.20:

$$\frac{\partial P_{x1}}{\partial t} - D_w L P_{x1} \cong U(r) A_0 X_0(r)\exp(-\lambda_0 t)(1+\exp)\ . \tag{5.29}$$

We will construct the long-time solution of 5.29 subject to the boundary conditions (5.22) in the form:

$$P_{x1} \cong A_0\left[X_0(r)\overline{U}^* t + B(r)\right]\exp(-\lambda_0 t)(1+\exp)\ , \tag{5.30}$$

where $B(r)$ is a certain function and $\overline{U}^*$ is a constant, both to be determined. Introduce 5.30 into 5.29, 5.22 to obtain the following θ-independent problem governing B-field:

$$-D_w\left[\frac{1}{r}\frac{\partial}{\partial t}\left(r\frac{\partial B}{\partial r}\right)\right] + B\lambda_0 = \left[U(r) - \overline{U}^*\right]X_0(r),\quad 0 < r < a'\ , \tag{5.31}$$

$$B = 0 \text{ at } r = a',\ B \text{ — finite at } r = 0. \tag{5.32}$$

To determine the requisite value of $\overline{U}^*$, multiply both sides of Equation 5.30 by $rX_0(r)$, integrate over the interval $0 < r < a'$, and use Equation 5.24 to obtain that the resulting integral over the left-hand side is equal to zero. Hence, the corresponding integral over the right-hand side is also zero, i.e.,

$$\int_0^{a'} \left[U(r) - \overline{U}^*\right]X_0^2(r)rdr = 0\ . \tag{5.33}$$

This provides an expression for the yet unknown value of $\overline{U}^*$:

$$\overline{U}^* = \frac{1}{N_0^2}\int_0^{a'} U(r)X_0^2(r)rdr\,, \tag{5.34}$$

where N_0^2 is given by

$$N_0^2 = \int_0^{d} X_0^2(r)rdr\,. \tag{5.35}$$

It may be shown that Problem 5.31, 5.32 with the value of $\overline{U}^*$ obtained in 5.34, possesses a solution, which is determined to within an additive term $B_0X_0(r)$, where B_0 is an arbitrary constant. This ambiguity will, however, be shown to have no effect on the values of the aerosol effective transport coefficients.

Second-Order Moment — The equation governing the asymptotic long-time second-order moment is obtained by substitution of the solutions 5.26a, 5.30 into Equation 5.21:

$$\frac{\partial P_{x2}}{\partial t} - D_w L(P_{x2}) \cong A_0\left\{U(r)\left[X_0(r)\overline{U}^* t + B(r)\right] + D_w X_0(r)\right\}\exp(-\lambda_0 t)(1+\exp)\,, \tag{5.36}$$

which is to be solved subject to the boundary conditions 5.22. We will construct the long-time solution for the latter problem in the following trial form:

$$P_{x2} \cong A_0\left[X_0(r)\left(\overline{U}^*\right)^2 t^2 + 2\overline{U}^* tB(r) + 2\overline{D}^* tX_0(r) + H(r)\right]\exp(-\lambda_0 t)(1+\exp)\,, \tag{5.37}$$

where $\overline{D}^*$ is a constant to be determined and H is an r-dependent function. Introduce the above solution into Equations 5.21 and 5.22 to obtain the following problem for the H-field

$$-D_w\left[\frac{1}{r}\frac{\partial}{\partial r}\left(r\frac{\partial H}{\partial r}\right)\right] + H\lambda_0 = 2\left\{\left[\left(D_w - \overline{D}^*\right)x_o(r) + \left[U(r) - \overline{U}^*\right]B(r)\right]\right\}0 < r < a'\,, \tag{5.38a}$$

$$H = 0 \text{ at } r = a',\ H \text{ — finite at } r = 0 \tag{5.38b}$$

Multiply both sides of Equations 5.38a by $rX_0(r)$, integrate over the interval $0 < r < a'$, and use Equation 5.24 to obtain that the resulting integral over the left-hand side vanishes. This yields the following expression for $\overline{D}^*$:

$$\overline{D}^* = D_w + \frac{1}{N_0^2}\int_0^{a'}\left[U(r) - \overline{U}^*\right]X_0(r)B(r)rdr\,. \tag{5.39}$$

Axial Transport Phenomenological Coefficients — Integrate Expressions 5.26a, 5.30, and 5.37 for the local moments over the tube radius to obtain the asymptotic long-time total moments (compare Equation 5.5)

$$M_{x0}(t) \cong A_0\overline{X}\exp(-\lambda_0 t)(1+\exp) \tag{5.40a}$$

$$M_{x1}(t) \cong A_0\left[\overline{X}\,\overline{U}^* t + \overline{B}\right]\exp(-\lambda_0 t)(1+\exp)\,, \tag{5.40b}$$

$$M_{x2}(t) \cong A_0\left[\overline{X}\left(\overline{U}^*\right)^2 t^2 + 2\overline{U}^* t\overline{B} + 2\overline{D}^* t\overline{X} + \overline{H}\right]\exp(-\lambda_0 t)(1+\exp)\,, \tag{5.40c}$$

wherein

$$\overline{X} = \int_0^{a'} X_0(r)rdr,\ \ \overline{B} = \int_0^{a'} B(r)rdr,\ \ \overline{H} = \int_0^{a'} H(r)rdr\,. \tag{5.41}$$

Finally, substitute expressions 5.40a,b,c for the total moments into Equations 5.8a,b,c,d to obtain

$$\overline{K}_x^* = \lambda_0,\ \ \overline{U}_x^* = \overline{U}^*,\ \ \overline{D}_x^* = \overline{D}^*, \tag{5.42}$$

where $\lambda_0, \overline{U}^*, \overline{D}^*$ are, respectively, given by Equations 5.28, 5.34, 5.39.

Discussion — The explicit expressions for the mean axial aerosol velocity and dispersivity are obtainable from 5.31, 5.34, 5.39 in terms of the Bessel functions. These quantities are easily shown to possess the following scaling representations:

$$\overline{U}_x^* = \overline{U} f_1\left(\frac{r_w}{a}\right),\ \ \overline{D}_x^* = D_w + \frac{\overline{U}^2 a^2}{D_w} f_2\left(\frac{r_w}{a}\right), \tag{5.43a,b}$$

where f_1 increases and f_2 decreases with r_w/a. In particular for point particles ($r_w/a = 0$), one has (Shapiro and Brenner, 1986)

$$f_1(0) = 1.2812,\ f_2(0) = 0.00125. \tag{5.44}$$

One can see that $f_1(0) > 1$, which is true for all values of r_w/a. This means that on average the aerosol moves along the x-axis faster than does the air. This may be explained by two reasons: first, the low-velocity region, $a' < r < a$ is unaccessible to the aerosol particles due to their final size (see Figure 5.2). Hence, they are transported in the faster-moving streamlines. Second, due to the aerosol depletion from the region adjacent to the wall, the long-time radial distribution $X_0(r)$ has a maximum at the tube axis (Figure 5.2). Hence, a major portion of the aerosol particles is transported near the axis, where the velocity is larger. Mathematically, this also follows from 5.34, where $U(r)$ is seen to be weighted in favor of small r.

Practically for all important cases, involved transport of submicrometer aerosols, $r_w/a \ll 1$. Therefore, f_1 and f_2 do not differ much from their limiting values (5.44a,b).

Expression 5.43b shows that the effective axial dispersivity consists of the Brownian diffusivity of the particles, D_w, and the so-called convective part, which in inversely proportional to D_w. This part stems from the gradient of the velocity field $dU(r)/dr$, which acts to increase the axial spread of the aerosol cloud. $f_2(0)$ is less than $f_2 = 1/48$, which value obtains in the case where the particles do not deposit on (repelled from) the wall (Taylor, 1953). The fact that the dispersivity of the depositing particles is smaller than their inert counterparts also follows from the observation made above (see $X_0(\mathbf{q})$ in Figure 5.2) that the major part of a depositing aerosol flows near the center of the tube, where the radial velocity gradients (promoting aerosol dispersion) are smaller.

The value of the dispersivity given by Expression 5.39 is unaffected by the nonuniqueness of the solution of the $B(r)$ field. This may be shown by considering any two solutions B, B' of 5.31

and 5.32, related by $B(r) = B'(r) + B_0X_0(r)$, with B_0 being any constant. Substituting this expression into 5.39 and using 5.33, one obtains that the term containing B_0 vanishes. It is seen that all three effective aerosol axial transport coefficients are independent of the initial aerosol concentration distribution, the information on which is implicitly embodied in A_0. They, however, depend upon the aerosol diameter $2r_w$, external forces, and the flow regime prevailing within the device.

5.6 AXIAL AEROSOL TRANSPORT EQUATION

Having obtained the scaling relationships (5.28 and 5.43a,b) of the axial aerosol transport coefficients, one can use them to estimate the relative magnitudes of the different terms in the axial aerosol transport Equation 5.10. We will consider the frequently encountered situation where the axial dispersivity is dominated by the convective term, i.e.,

$$D_w << \frac{\overline{U}^2 a^2}{D_w} f_2\left(\frac{r_w}{a}\right), \tag{5.45}$$

which may be expressed in a nondimensional form, as

$$\mathrm{Pe}_a = \frac{\overline{U}a}{D_w} >> f_2^{-1/2} \text{ or } \mathrm{Pe}_a >> 30 . \tag{5.46}$$

Consider now a tube of a finite length L. Equation 5.10 is valid when each portion of aerosol particles entering the tube reaches the asymptotic state (when its zero-order moment is described by 5.26a) well before the tube outlet. This requires that L should be greater than the asymptotic dispersion length L_∞, given by (compare 5.26b)

$$L >> L_\infty = \overline{U}\frac{a^2}{D_w} \text{ or } \mathrm{Pe}_a \frac{a}{L} \cong \frac{\overline{D}_x^*}{\overline{U}_x^* L} << 1 . \tag{5.47a,b}$$

Rewrite Equation 5.10 in terms of the dimensionless variables $\bar{t} = t\,\overline{U}/L$, $\bar{x} = x/L$, and assume $a' \approx a$ to obtain

$$\frac{\partial \bar{c}_w}{\partial \bar{t}} + f_1 \frac{\partial \bar{c}_w}{\partial \bar{x}} - \mathrm{Pe}_a \frac{a}{L} f_2 \frac{\partial^2 \bar{c}_w}{\partial \bar{x}^2} + \gamma_0^2 \frac{L}{a\mathrm{Pe}_a}\bar{c}_w = 0 . \tag{5.48}$$

One can see that in circumstances, where 5.47b is valid, the diffusive term in Equation 5.48 is small. Neglecting this term, one obtains that the steady-state (i.e., $\partial \bar{c}_w/\partial t = 0$) average aerosol concentration exponentially decreases along the tube axis:

$$\bar{c}_w = \bar{c}_{w0} \exp\left(-\overline{K}_x^* \frac{x}{\overline{U}_x^*}\right) = \bar{c}_{w0} \exp\left(-\gamma_0^2 \frac{x}{a\mathrm{Pe}_a}\right), \tag{5.49}$$

where $\bar{c}_{w0}$ is the inlet aerosol concentration.

We will also note that the axial aerosol transport equation is also valid when the average aerosol concentration is defined via a more general formula (Shapiro and Brenner, 1986)

$$\bar{c}_w(t,x) = \frac{1}{\| \mathbf{q}_0 \|} \int_{\mathbf{q}_0} g(r,\theta) c_w(t,x,r,\theta) d\mathbf{q} , \tag{5.9a}$$

where g is any nonnegative, appropriately normalized weighting function and $\|\mathbf{q}_0\|$ is the cross-sectional area. This definition will prove useful for calculation of the aerosol deposition velocity in electrostatic precipitators.

Table 5.1 Effective Axial Diffusivities $\bar{D}_x^*$ and Asymptotic Dispersion Length, L_∞, of Aerosol Particles Flowing in a Circular Tube of Radius a = 1 mm

	$\bar{U}$ = 10 mm/s			$\bar{U}$ = 1 mm/s		
r_w, μm	$\bar{D}_x^*/D_w$	$\bar{D}_x^*$, cm²/s	L_∞, cm	$\bar{D}_x^*/D_w$	$\bar{D}_x^*$, cm²/s	L_∞, cm
0.001	1.074	0.014	0.769	1.01	0.013	0.077
0.005	44.6	0.0236	18.9	1.445	$7.66 \cdot 10^{-4}$	1.89
0.01	636	0.0890	71.4	7.30	$1.02 \cdot 10^{-3}$	7.14
0.02	9,660	0.348	278	96.6	$3.48 \cdot 10^{-3}$	27.8
0.05	$2.7 \cdot 10^5$	1.836	1,480	2,706	0.018	148
0.1	$2.58 \cdot 10^6$	5.68	4,550	$2.58 \cdot 10^4$	0.0568	455
0.2	$1.77 \cdot 10^7$	14.9	11,990	$1.77 \cdot 10^5$	0.149	1,190
0.5	$1.64 \cdot 10^8$	45.2	36,200	$1.64 \cdot 10^6$	0.452	3,620

Table 5.1 presents the values of the effective aerosol diffusivity and asymptotic dispersion length of diffusional aerosol particles flowing with velocity $\bar{U}$ = 10 mm/s and 1 mm/s in a tube of 1 mm radius. One can see that $\bar{D}_x^*$ and L_∞ rapidly increase with particle radius. For small r_w, when the characteristic Peclet number $\mathrm{Pe}_a = \bar{U}a/D_w < 1$, the effective axial diffusivity is dominated by the aerosol molecular diffusivity (see 5.43b). Therefore, for small particle diameters the effective axial diffusivity decreases with r_w, which is for larger r_w followed by an increase of $\bar{D}_x^*$. This effect is best observed for small air velocities (e.g., 1 mm/s in Table 5.1).

It should be mentioned that the effective aerosol axial diffusivity is significantly affected by the tube bends. This is especially true for measurements performed for long tubes, in which case the dispersion depends on the bending arrangement (e.g., radius of curvature). In particular, the airflow within coiled tubes is affected by the Coriolis force, which leads to the air recirculation streams within the tube cross-section. These secondary flows insignificantly affect the airflow parameters (e.g., mean air velocity), but can reduce the effective axial diffusivity (Shlafstein et al., 1994).

5.7 AEROSOL COLLECTION FROM TURBULENT FLOWS IN ELECTROSTATIC PRECIPITATORS

5.7.1 Parallel Plate Geometry

In this section we outline the implementation of the general scheme for calculation of the effective axial transport properties of charged micrometer particles moving in a turbulent flow through a model one-stage electrostatic precipitator (Shapiro et al., 1988) (Figure 5.3a). The cross-sectional domain $\|\mathbf{q}_0\|$ is chosen here as the distance between the plates $0 < y < b$, ($b >> r_w$). Accordingly, the local coordinate $\mathbf{q}$ is taken as y, whereas x serves as a global coordinate along the airflow direction. The particle transport equation in this case is

$$\frac{\partial c_w}{\partial t} + U(y)\frac{\partial c_w}{\partial x} + u_e \frac{\partial c_w}{\partial y} = \frac{\partial}{\partial y}\left[D_w(y) \frac{\partial c_w}{\partial y} \right], \tag{5.50}$$

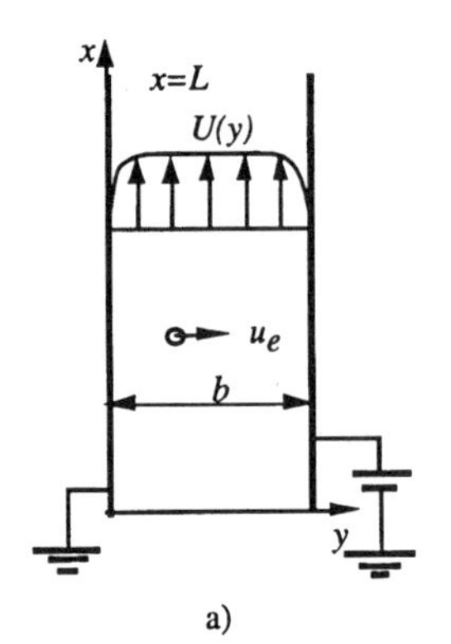

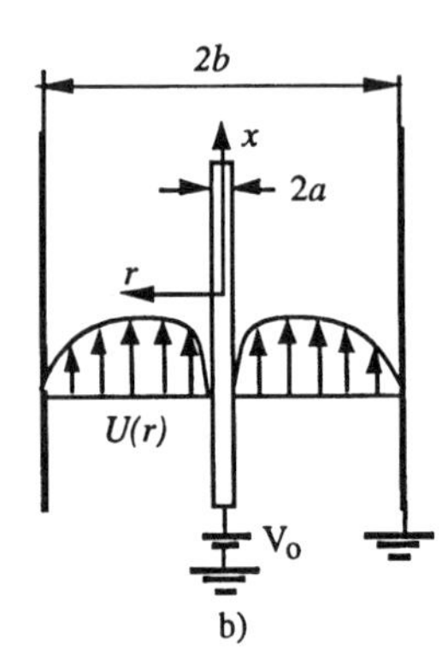

Figure 5.3 Schematic particle motion in an electrostatic precipitators. (a) Parallel plate geometry; (b) cylindrical geometry.

where $u_e = qEm_w$ is the electrostatic drift velocity of a particle possessing a constant charge q in an electrostatic field of intensity E, expressed via the mobility $m_w = C/(6\pi\mu r_w)$. Furthermore, $\overline{U}\,(y)$ is the time-averaged turbulent air velocity, and $D_w(y)$ is the turbulent diffusivity of the particles. In 5.50 the aerosol turbulent diffusive transport in the axial direction is neglected in comparison with the axial convective transport. The space-charge effect in Equation 5.50 is also neglected, albeit may be accounted for, using available data on typical aerosol concentration distributions within electrostatic precipitators (White, 1963).

The boundary conditions for the aerosol concentration c_w are chosen in the form:

$$j_y = u_e \frac{\partial c_w}{\partial y} - D_w(0)\frac{\partial c_w}{\partial y} = 0 \text{ at } y = 0\,, \tag{5.51a}$$

$$c_w = 0 \text{ at } y = b, \tag{5.51b}$$

which assume a zero lateral aerosol flux component on the grounded plate $y = 0$, and an infinitely fast precipitation of particles on the charged plate $y = b$.

The turbulent velocity profile is assumed to be fully developed and may be approximated by the logarithmic formula (Schlichting, 1968)

$$\frac{U(y/b)}{u_*} = \left\{ \begin{array}{ll} \bar{y}\,\mathrm{Re}_t, & 0 \le \bar{y} \le 5/\mathrm{Re}_t \\ 5.04\ln(\bar{y}\;\mathrm{Re}_t) - 3.96, & 5/\mathrm{Re}_t < \bar{y} \le 30/\mathrm{Re}_t \\ 2.5\ln(\bar{y}\;\mathrm{Re}_t) + 5.5, & 30/\mathrm{Re}_t < \bar{y} \le 120/\mathrm{Re}_t \\ 2.5\;\ln\;120 + 5.5, & 120/\mathrm{Re}_t < \bar{y} \le 0.5 \end{array} \right\} \tag{5.52}$$

where $\mathrm{Re}_t = u_* b/\nu$ is the turbulent Reynolds number and u_* is the wall friction velocity. The expression for the particle turbulent diffusivity is due to Sehmel (1970):

$$\overline{D}_w(\bar{y}) = \frac{D(y/b)}{D_{wo}} = \left\{ \begin{array}{ll} S + 0.275\bar{y}^{1.1}\tau_+^{1.1}\,\mathrm{Re}_t^{0.1}, & 0 \le \bar{y} \le \alpha_1/\mathrm{Re}_t \\ S + 0.25\bar{y}^3\,\mathrm{Re}_t^2, & \alpha_1/\mathrm{Re}_t \le \bar{y} \le 20/\mathrm{Re}_t \\ S + 10\bar{y}, & 20/\mathrm{Re}_t \le \bar{y} \le 10 \\ S + 1, & 10 \le \bar{y} \le 0.5 \end{array} \right\} \tag{5.53}$$

In the above $D_t = 0.4u_*h$, where h is the thickness of the y-dependent region of the turbulent diffusivity, taken here as $0.1b$; $S = D_{w0}$ is the particle constant Brownian diffusivity and $S = D_{w0}/D_t$, and τ_+ is the particles dimensionless relaxation time, given by

$$\tau_+ = \frac{2}{9}\frac{\rho_w}{\rho}\left(\frac{r_w u_*}{\nu}\right)^2, \tag{5.54}$$

with ρ_w, ρ being the particle and the air densities. Moreover, according to Sehmel (1970) the constant $\alpha_1 = (11\tau_+^{1.1})^{1.9}$.

The theory described in Section 5.5 can be extended to incorporate the particle drift within the local space, i.e., across the gap (Shapiro et al., 1988). The effective axial transport coefficients $\overline{U}_x^*, \overline{D}_x^*, \overline{K}_x^*$ were calculated for the above expressions for the turbulent velocity and diffusivity of the particles. The computational results showed that the average axial aerosol velocity only slightly (by less than 10%) exceeds the mean flow velocity. All the data obtained for the mean axial aerosol dispersivity showed that it is sufficiently small, so that Condition 5.47b is satisfied and the axial diffusion term in Equation 5.48 may be neglected. Under this condition the steady-state axial distribution of the average aerosol concentration along the precipitator is given by Equation 5.49. This may be used to calculate the precipitation efficiency:

$$\eta \equiv 1 - \frac{\bar{c}_w(0)}{\bar{c}_w(L)} = 1 - \exp\left(-\frac{\overline{K}_x^* L}{\overline{U}_x^*}\right). \tag{5.55}$$

One can also characterize aerosol precipitation by the deposition velocity (Friedlander, 1977)

$$k = \frac{1}{\bar{c}_b}\left(u_e c_w - D_w \frac{\partial c_w}{\partial y}\right)_{y=0}, \tag{5.56}$$

where $\bar{c}_b(x)$ is the bulk aerosol concentration (compare Equation 5.9a)

$$\bar{c}_b(x,t) = \frac{1}{U_0 b}\int_0^b U(y)c(x,y,t)dy\,. \tag{5.57}$$

Integrate the steady-state version of Equation 5.50 wherein the axial diffusion term is dropped, across the width b to rewrite Equation 5.56 for the deposition velocity in the form

$$k = \overline{K}_x^*\, b/\overline{U}_x^*\,. \tag{5.58}$$

The dimensionless effective aerosol axial deposition coefficient is plotted in Figure 5.4 vs. the electrostatic Peclet number $\mathrm{Pe}_e = u_e b/D_t$.

For strong electric fields ($\mathrm{Pe}_e > 10$) the axial deposition rate weakly depends on the turbulent diffusivity model and approaches the curve $\overline{K}_x^* = u_e^2/4D_t$, derived for $D_w = D_t =$ constant. On the other hand, for small Pe_e choice of $D_w(y)$ governs the deposition rate. The curve obtained for $D_w = D_t$ approaches the constant value $\overline{K}_x^* = 2.5D_t/b^2$, which greatly overestimates the deposition rate. This turbulent diffusivity model does not account for the damping influence of the wall on the aerosol transport.

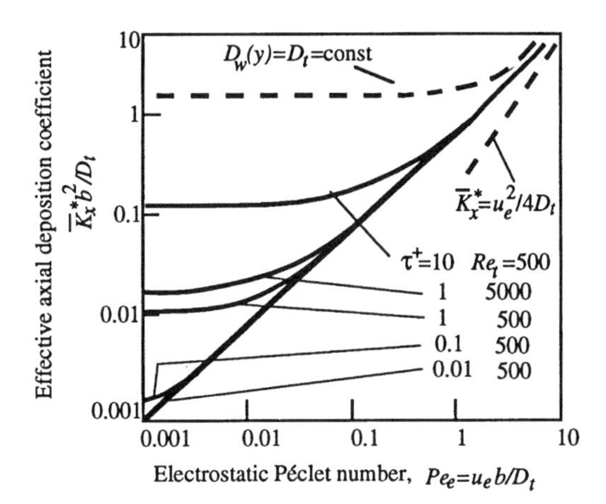

Figure 5.4 Dimensionless axial aerosol deposition coefficient in the parallel plate electrostatic precipitator. Bold broken curves: D_w = const. Solid curves: $D_w(y)$ is given by Sehmel (1970). The lowermost solid curve corresponds to the Deutsch (1922) formula.

The set of curves computed using the Sehmel diffusivity model yields much lower $\overline{K}_x^*$. The asymptotic values for $\overline{K}_x^*$ which were computed for $Pe_e \to 0$, are determined by the ratio S of the particle molecular and turbulent diffusivities, as well as the relaxation time.

It may be seen from Figure 5.4 that each of the curves calculated with the Sehmel turbulent diffusivity model may be characterized by the following three regions:

1. Very small Pe_e numbers, where the influence of the electric force is negligible, as compared with that of the particle diffusion. Here $\overline{K}_x^*$ may be approximately expressed by the formula:

$$\overline{K}_x^* = \gamma(\tau_+, S, \mathrm{Re}_t) D_t / b^2, \tag{5.59}$$

where $\gamma(\tau_+, S, \mathrm{Re}_t)$ is a coefficient primarily dependent on particle inertia. For example, for $\tau_+ = 1$ and $\mathrm{Re}_t = 500$ this region is approximately described by $Pe_e < 0.01$ and $\gamma = 0.015$.

2. Intermediate Pe_e, where the electric transport prevails over the diffusive transport in the vicinity of the collecting wall, but is still negligible far from the wall, where particle transport is governed by the turbulent diffusion coefficient D_t. This turbulent diffusion mechanism brings particles sufficiently close to the collecting wall, where the electric force constitutes its sole collection mechanism. One can see from Figure 5.4 that in this region $\overline{K}_x^*$ is given by the straight line

$$\overline{K}_x^* b^2/D_t = Pe_e = u_e b/D_t \text{ or } \overline{K}_x^* = u_e/b, \tag{5.60}$$

and independent of diffusion. This result, together with $\overline{U}_x^* \cong U_0$ and Formula 5.49, yields

$$\eta = 1 - \exp\left(-\frac{u_e L}{U_0 a}\right) \tag{5.61}$$

which is the Deutsch (1922) formula. The data in Figure 5.4 establishes the validity range of 5.61 in terms of the inertial properties of the particles and the turbulent Reynolds number. One can see that for $\tau_+ = 0.1$ and $\mathrm{Re}_t = 500$ this intermediate region is approximately described by $0.0025 < Pe_e < 1$.

3. High Peclet numbers, where particle transport due to electric force is comparable or even exceeds turbulent diffusion transport everywhere in the flow. In contrast with the previous two regions,

diffusion here plays a negative role in particle deposition by reducing particle transport toward the collecting wall. In particular, for very large Pe_e one can use the expression $\overline{K}_x^* = u_e^2 / 4D_t$ to obtain the following formula for the precipitation efficiency (Leonard et al., 1980)

$$\eta = 1 - \exp\left(-\frac{u_e^2 L}{4D_t \overline{U}_x^*}\right), \tag{5.62}$$

which is independent of the distance b separating the plates.

The effect of the turbulent Reynolds number is to increase the intensity of the turbulent pulsations in the vicinity of the wall, which increases the particle diffusivity. This leads to a considerable augmentation of the collection rate for low Pe_e, in which case particle diffusion promotes the deposition. For high Pe_e the effect of Re_t is negligible since in this case η is governed by the bulk turbulent diffusivity.

In all cases $\overline{D}_x^*$ was found to be small, so that Condition 5.47b was satisfied, which provides validation of Formula 5.55 for the precipitation efficiency.

Equation 5.62 may be rewritten via the Deutsch parameter $Z = u_e L / U_0 b$ in the form

$$\eta = 1 - \exp(-\Omega Z), \tag{5.63}$$

where Ω is a correction factor given by

$$\Omega = \frac{\overline{K}_x^* U_0}{\overline{U}_x^* u_e} . \tag{5.64}$$

Clearly, Ω represents the extent to which the collection rate, as predicted by Equation 5.55 differs from that established by the Deutsch formula. This factor is plotted in Figure 5.5 as a function of the electrostatic Peclet number for two models of the particle turbulent diffusivity. Clearly, the Deutsch equation is described by the line $\Omega = 1$. It provides the lower bound for the precipitation efficiency, calculated for all models.

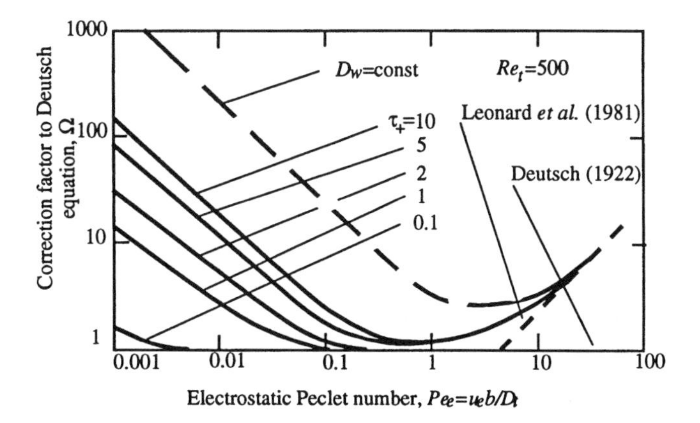

Figure 5.5 Correction factor to the Deutsch equation vs. electrostatic Peclet number.

Similarly to the above conclusions drawn on $\overline{K}_x^*$, one can identify for each curve three regions:

1. Weak electric force (small Pe_e), where particle deposition is promoted by diffusion. Here Ω increases with decreasing Pe_l for both turbulent diffusivity models;

2. Intermediate Pe_e where the effects of diffusion and the electric force are comparable. In this range η is given by the Deutsch formula. Figure 5.5 establishes the ranges of its applicability. Although the turbulent diffusion coefficient does not explicitly appear in this formula, its influence is manifested in the dependence of its applicability range on the particle inertial properties;
3. Large Peclet number, i.e., $Pe_e > 10$, in which region both diffusivity models yield the same expression $\Omega = Pe_e/4$.

5.7.2 Cylindrical Geometry

Consider aerosol particles moving in the flow between two coaxial cylinders $b < r < a$, $(a >> b)$. This geometry will be regarded as modeling the flow in a cylindrical electrostatic precipitator. An electric voltage is applied to the inner cylinder (electrode), while the outer cylinder is grounded (see Figure 5.3b). Aerosol axisymmetric convective turbulent transport equation in this case is

$$\frac{\partial c_w}{\partial t} + \frac{\partial j_x}{\partial x} + \frac{1}{r}\frac{\partial (r j_r)}{\partial r} = 0 , \tag{5.65}$$

where the particle flux components are, respectively, given by

$$j_x = U(r)c_w,\ j_r = w(r)c_w - D_w(r)\frac{\partial c_w}{\partial r} . \tag{5.66}$$

In the above $w(r)$ is the electrostatic drift velocity, which consists of two components:

$$w(r) = m_w\left(f_c + f_d\right) = \frac{w_c}{r/b} - \frac{w_d}{(r/b)^3} , \tag{5.67}$$

respectively, representing coulombic and dielectrophoretic forces (Shapiro et al., 1989). The characteristic velocities appearing in 5.67 are

$$w_c = m_w q \frac{V_0}{b\ln(b/a)} , \quad w_d = 4\pi\varepsilon_0 m_w \frac{r_w^3}{b^3}\gamma_w\left(\frac{V_0}{\ln(b/a)}\right)^2 , \tag{5.68}$$

where m_w is the particle mobility and $\gamma_w = (\varepsilon_w - \varepsilon_0)/(\varepsilon_w + 2\varepsilon_0)$.

Boundary conditions for Equation 5.65 are formulated as follows. On the surface of the outer cylinder, particles are collected and never reentrained into the flow. Accordingly,

$$c_w = 0 \text{ at } r = b. \tag{5.69}$$

On the surface of the inner cylinder, the particle flux is zero when the total electric force repels them from the wall:

$$j_r(a) = w(a)c_w - D_w(a)\frac{\partial c_w}{\partial r} = 0 \ \text{ at } r = a, \ \text{ when } w_c > w_d \frac{b^2}{a^2} . \tag{5.70}$$

On the other hand, when the total electric force attracts the particles toward the inner cylinder, they are collected on it:

$$c_w = 0 \ \text{ at } r = a, \ \text{ when } w_d < w_\alpha \frac{b^2}{a^2} . \tag{5.71}$$

In contrast with the parallel plate geometry the problem is now characterized by two characteristic Peclet numbers: coulombic, Pe_c, i.e., associated with w_c and dielectrophoretic, Pe_d, associated with w_d:

$$\mathrm{Pe}_c = w_c b/D_t,\ \mathrm{Pe}_d = w_d b/D_t \tag{5.72}$$

We will use expressions 5.52 and 5.53 (with y measured from the outer cylinder wall) for the average velocity and turbulent diffusivity, respectively. The inner cylinder will be considered thin (the ratio $a = \alpha/b \ll 1$), so that its influence both on the velocity and the turbulence is negligible.

As in the case of the parallel plate precipitator, here the effective axial velocity only slightly exceeds the average flow velocity. The nondimensional effective axial deposition coefficient in the absence of the dielectrophoretic force, i.e., where $\mathrm{Pe}_a = 0$ is plotted in Figure 5.6. The range of Pe_c shown in Figure 5.6 may be divided in three regions corresponding to weak, moderate, and strong electric force. In the latter region ($\mathrm{Pe}_c \gg 10$) $\overline{K}_x^*$ was found to weakly depend on the outer cylinder radius b. This is in contrast with the parallel plate precipitator, where the collection rate is independent of the distance between the plates. For regions of small and moderate Pe_c the conclusions drawn for the parallel plate case equally apply here. In particular, in the range of intermediate Pe_c the collection rate is described by the formula

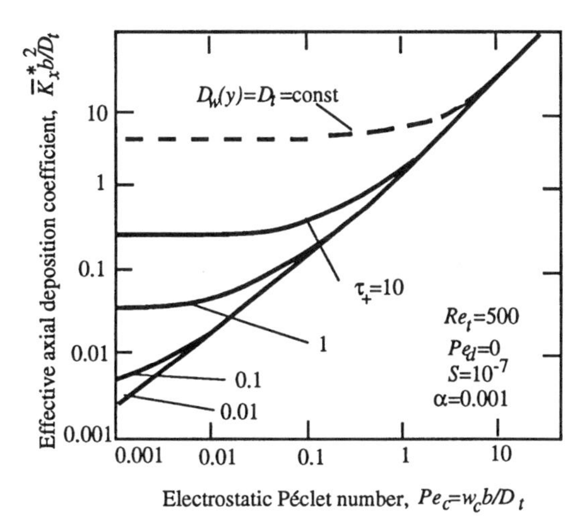

Figure 5.6 Dimensionless axial aerosol deposition coefficient in the cylindrical electrostatic precipitator. Bold broken curves: D_w = const. Solid curves: $D_w(y)$ is given by Sehmel (1970). The lowermost solid curve corresponds to the Deutsch (1922) formula.

$$\overline{K}_x^* = 2w_c/b\ , \tag{5.73}$$

and independent of the turbulent diffusivity. Introducing the effective particle drift velocity $w_e = 2w_c$ and employing the result $\overline{U}_x^* \approx U_0$, one can use 5.49 to obtain the Deutsch formula

$$\eta = 1 - \exp\left(-\frac{w_e L}{U_0 b}\right). \tag{5.74}$$

This is valid for cylindrical precipitators with expression $w_e = 2w_c$ for the particle effective drift velocity, which may be rewritten in the form

$$w_e = 2m_w q \frac{V_0}{b \ln(b/a)} \tag{5.75}$$

The validity range of the Deutsch formula in the present case, formulated in terms of the correction factor Ω may be found in Shapiro et al. (1989).

Deposition in the Presence of the Dielectrophoretic Force — When both coulombic and dielectrophoretic forces act upon the particles, they may be collected both on the inner and the outer cylinder, provided that

$$\mathrm{Pe}_a > \mathrm{Pe}_c\,\alpha^2. \tag{5.76}$$

This case is quantitatively described in Figure 5.7, which shows the effective axial deposition coefficient vs. particle coulombic Peclet number for several values of Pe_d. For large Pe_c and relatively low values of Pe_d, i.e., those for which Condition 5.76 is not fulfilled, no particles are collected on the inner cylinder. Corresponding curves plotted for $\mathrm{Pe}_d = 10^{-7}$ and $\mathrm{Pe}_c > 0.1$ practically coincide with those calculated for $\mathrm{Pe}_d = 0$. This indicates that in cases where inequality (5.76) is not fulfilled, the dielectrophoretic force has a negligible influence on particle collection. In a precipitator characterized by $\alpha = 0.001$, $\mathrm{Pe}_d = 10^{-7}$, Condition (5.76) is valid when $\mathrm{Pe}_c < 0.1$. For the latter value of Pe_c one observes in Figure 5.7 an instantaneous increase of $\overline{K}_x^*$. For lower values of Pe_d analogous

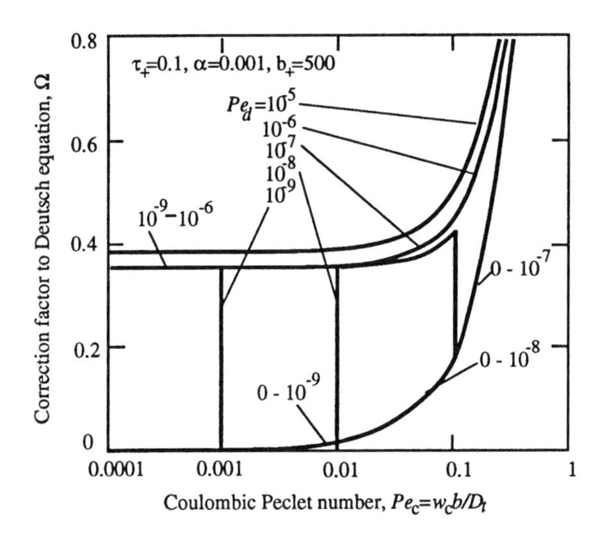

Figure 5.7 Effective deposition rate coefficient vs. particle coulombic Peclet number for several values of the dielectrophoretic Peclet number.

jumps occur at lower values of Pe_c. These sharp changes of the particle collection rate are clearly attributed to the particle deposition on the inner cylinder. This phenomenon may be interpreted in terms of the particle electric charge, by rewriting Pe_c in the form

$$\mathrm{Pe}_c = \frac{q}{q_0} = q\frac{m_w V_0}{D_t \ln \alpha^{-1}}\ . \tag{5.77}$$

This expression allows evaluation of the electric charges on particles of given sizes (given m_w), for which the dielectrophoretic deposition on the inner cylinder is important. For $2r_w = 10$ μm, $\alpha = 0.001$, and the electric field in the vicinity of the inner cylinder of the order of air break-down field (~30 kV/cm), the characteristic charge $q_0 = 1.6 \cdot 10^4$ elementary charges (e). For such circumstances and the precipitator parameters listed above, the dielectrophoretic capture mode prevails for $q <$

900*e*. Particles with such charges form a buildup on the inner electrode. However, this electric charge is significantly lower than the saturation charge acquired by a 10–μm particle by the ion bombardment charging (Marietta and Swan, 1976). Therefore, normally, fully charged particles are collected on the outer cylinder surface.

Figure 5.8 shows the collection efficiency plotted vs. the distance parameter Z accounting for the dielectrophoretic capture mechanism. The Deutsch formula is seen to yield the lowest collection efficiency. For low Pe_c the dielectrophoretic force has the most profound influence on the particle collection, i.e., it increases the efficiency by 150 to 250% in the range $0.05 < Z < 1$. For moderate Pe_c the effect of the dielectrophoretic capture is to increase η by 10 to 30%.

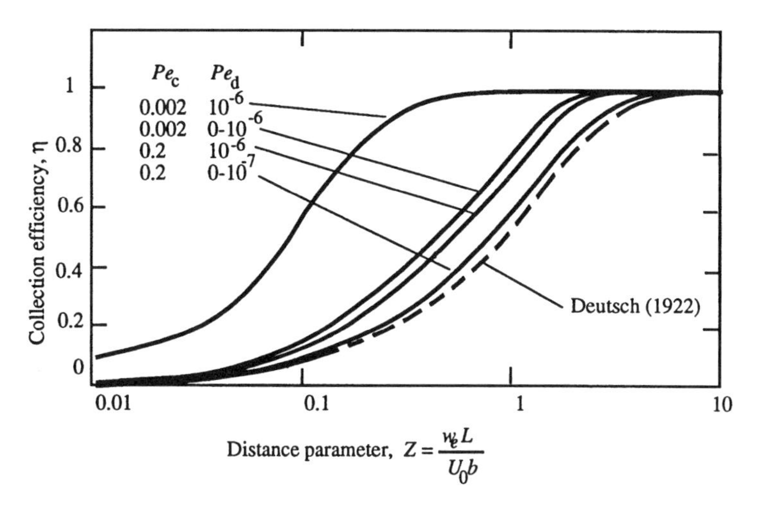

Figure 5.8 Particle collection efficiency vs. distance parameter.

5.8 AEROSOL TRANSPORT AND COLLECTION IN POROUS FILTERS

5.8.1 Dispersion–Reaction Theory

Aerosol filtration by porous filters can be treated as a convective dispersion phenomenon (Shapiro and Brenner, 1990a). The filtration process is viewed on a macroscale or coarse scale (much larger than the mean collector size), where the filter material is replaced by an effective homogeneous medium. Motion and collection of aerosol particles on the macroscale is characterized in terms of aerosol concentration, averaged over a certain representative volume, including several collector elements. The objective of this approach is calculation of the effective coarse-scale aerosol transport properties governing particle collection rate.

The above goal may be achieved by several methods, known in the literature as homogenization techniques. Application of these methods requires knowledge of the microscale geometric structure of the filter material. This may be specified either in the deterministic or statistical way. In the latter case, the microscale geometry is defined in terms of appropriate distribution functions — e.g., two- or three-point distribution function (Streider and Aris, 1970), collector size distribution, etc. In contrast with the above, deterministic geometric models describe the porous filter material as having a certain *a priori specified* microstructure. One way to specify the latter is to assume that the geometry of the material is spatially periodic. That is, a representative volume chosen to best represent the material structure constitutes a unit cell, which is indefinitely reproduced in space. In this case several homogenization techniques (Bensoussan et al., 1978; Brenner, 1980; Adler, 1992) may be applied to calculate the effective coarse-scale transport coefficient governing the average aerosol concentration.

It must be noted that the representative volume, say τ_0, no matter how large it may be, always has a finite size. Any consistent homogenization method should unambiguously address the questions (1) "what is the material structure beyond τ_0" and (2) "what are the laws of aerosol transport across its boundaries." These questions are important for formulation of the aerosol transport problem in the *whole, effectively infinite filter material.* Such a formulation constitutes the basis of any rigorous implementation of the homogenization methods.

Classical filtration theory (Pich, 1966; Davis, 1973) treats aerosol transport only within a specified representative or unit bed element (UBE), i.e., Happel–Kuwabara-type spherical or cylindrical cells. Accordingly, this theory does not address question (1) and introduces several assumptions to answer question (2) (i.e., particle redistribution at the entrance to the cell, etc.). The method of volume averaging (Crapiste et al., 1986) which is allegedly applicable to an arbitrary porous microstructure, in fact, utilizes the assumption of spatial periodicity in any specific application. However, similarly to the classical filtration theory, this method formulates and solves the problem of aerosol microscale transport only within the representative volume τ_0.

We will here describe the dispersion reaction model of aerosol filtration by porous filters (Shapiro and Brenner, 1990a). A necessary prerequisite for implementation of the model is the requirement of a spatially periodic filter structure (Figure 5.9). The spatial periodicity implies that

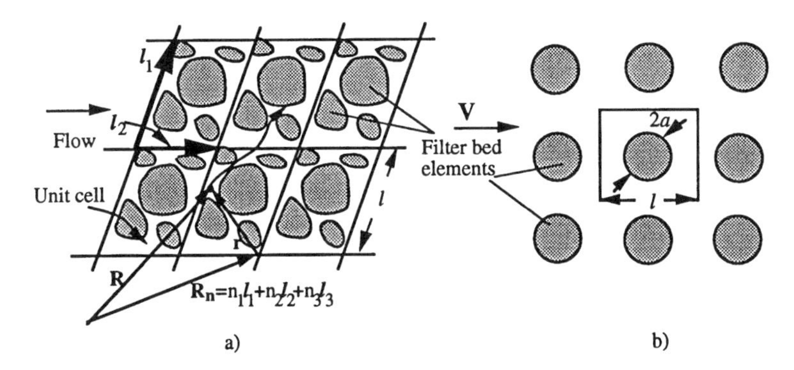

Figure 5.9 (a) Schematic of convective–diffusive motion of an aerosol particle in a generic spatially periodic model of a porous filter. $\mathbf{l}_1$, $\mathbf{l}_2$, $\mathbf{l}_3$— basic lattice vectors, n_1, n_2, n_3 — integer numbers. $\mathbf{R} = \mathbf{R}_n + \mathbf{r}$, where $\mathbf{R}_n$ is a global-space discrete coordinate, describing coarse-scale particle position, and $\mathbf{r}$ is a local-space intracell particle location. (b) Square lattice array of fibers.

a general position vector may be decomposed into the sum:

$$\mathbf{R} = \mathbf{R}_n + \mathbf{r},$$

where $\mathbf{R}_n = n_1\mathbf{l}_1 + n_2\mathbf{l}_2 + n_3\mathbf{l}_3$ is the discontinuous global unbounded vector, describing location of the **n**th unit cell and **r** is the continuous intracell position vector. This decomposition is very essential, since it enables reduction of the problem of aerosol transport and deposition, formulated in the effectively infinite domain to several boundary problems to be solved within a single unit cell. In particular, due to the spatially periodic filter geometry it follows that the airflow velocity field and the fields of forces (electric, magnetic, etc.) that may act on aerosol particles are also spatially periodic, i.e., dependent on the intracell variable **r**:

$$\mathbf{v} = \mathbf{v}(\mathbf{r}), \; \mathbf{f} = \mathbf{f}(\mathbf{r}). \tag{5.78}$$

Furthermore, if an account is taken for short-range particle-collector dispersion forces or hydrodynamic interactions, these will be also spatially periodic, and, hence, can be incorporated into the model.

Although aerosol filtration normally is a (quasi-) steady process, the basic properties of the effective (coarse-scale) aerosol transport and deposition are best studied for the following model *unsteady* problem. We will assume that a cloud of aerosol particles is instantaneously introduced into the infinite interstitial domain V_f of the filter material, where a clean air is flowing. Aerosol transport is described by the following continuity equation:

$$\frac{\partial c}{\partial t} + \nabla \cdot \mathbf{J} = 0 \text{ in } V_f , \tag{5.79}$$

where $\nabla = \partial/\partial\mathbf{r}$ is the intracell gradient operator, and $\mathbf{J}$ is the convective–diffusive flux vector, possessing the following constitutive form:

$$\mathbf{J} = [\mathbf{v}(\mathbf{r}) + m_w\mathbf{f}(\mathbf{r})]c_w - D_w\nabla c_w \tag{5.80}$$

Implicit in Equation 5.80 is the assumption that $R_p \equiv r_w/a << 1$, so that no wall effects are included in Equation 5.79. The latter can be included by replacing the undisturbed (by the presence of the aerosol phase) fluid velocity $\mathbf{v}(\mathbf{r})$ by the aerosol particle velocity $\mathbf{u}(\mathbf{r})$, which by virtue of the chosen geometric model will also be a spatially periodic function. In the latter case m_w and D_w are not position-independent scalars, but rather position-dependent, spatially periodic second-rank tensors $\mathbf{m}_w(\mathbf{r})$, $\mathbf{D}_w(\mathbf{r})$ (Brenner and Gaydos, 1977).

At the collector surfaces the particle concentrtion obeys the boundary condition

$$c_w = 0 \text{ on } S_p', \tag{5.81}$$

formulated on the hypothetical surface S_p' lying in the interstitial fluid domain, and which stands out from the collector surface S_p by a distance equal to the particle radius r_w. Another necessary condition is the requirement of continuity of the aerosol concentration across the boundaries $\partial\mathbf{r}_0$ separating any two adjacent unit cells:

$$c_w, \nabla c_w \text{ — continuous across } \partial\mathbf{r}_0. \tag{5.82}$$

In addition to the boundary conditions (5.81 and 5.82) the aerosol concentration is required to satisfy the attenuation condition

$$c_w \to 0 \text{ as } |\mathbf{R}| \to \infty, \tag{5.83}$$

at large distances from the location at which the aerosol cloud has been originally introduced.

It may be shown that there exists a unique solution of the problem (5.79 to 5.83) subject to the initial condition

$$c_w(\mathbf{R_n},\mathbf{r}, 0) = c_{w0}(\mathbf{R_n},\mathbf{r}), \tag{5.84}$$

where c_{w0} is any sufficiently smooth function. However, either computational or analytical solution of this problem in an infinite filter domain is a formidable task. The objective of the present model is to describe spatial distribution and temporal evolution of the coarse-scale aerosol concentration, defined by averaging c_w over the unit cell interstitial volume τ_f:

$$\bar{c}_w(\mathbf{R_n},t) = \int_{\tau_f} c_w(\mathbf{R_n},\mathbf{r},t)d^3\mathbf{r}. \tag{5.85}$$

This average aerosol concentration clearly depends upon the initial condition c_{w0}, i.e., the way the aerosol has been introduced in the filter material. Therefore, we will focus on calculation of the effective transport coefficients governing $\bar{c}_w$. These include $\bar{\mathbf{U}}^*, \bar{\mathbf{D}}^*, \bar{K}^*$, which, respectively, represent the aerosol effective (coarse-scale) velocity vector, diffusivity dyadic, and the volumetric deposition-rate coefficient. These properties appear in the coarse-scale (effective medium) aerosol transport equation (compare Equation 5.10)

$$\frac{\partial \bar{c}_w}{\partial t} + \bar{\mathrm{U}}^* \cdot \bar{\nabla} \bar{c}_w - \bar{\mathrm{D}}^* : \bar{\nabla}\bar{\nabla} \bar{c}_w + \bar{K}^* \bar{c}_w = 0 , \tag{5.86}$$

written for $\bar{c}_w$ in the effectively infinite coarse-scale filter domain. In the above $\bar{\nabla} = \partial/\partial\bar{\mathbf{R}}$ is the coarse scale gradient, taken with respect to the macroscopic variable $\bar{\mathbf{R}} = \mathbf{R_n}$ (Bensoussan et al., 1978). The coefficients $\bar{\mathbf{U}}^*, \bar{\mathbf{D}}^*, \bar{K}^*$ are defined by the following asymptotic formulas

$$\bar{K}^* = -\lim_{t\to\infty} \frac{d}{dt} \ln M_0(t), \quad \bar{\mathbf{U}}^* = \lim_{t\to\infty} \frac{d\bar{\mathbf{R}}}{\mathrm{d}t}, \quad \bar{\mathbf{D}}^* = \frac{1}{2} \lim_{t\to\infty} \overline{(\Delta \mathbf{R})^2} , \tag{5.87a,b,c}$$

where $M_0, \bar{\mathbf{R}}, \overline{(\Delta\mathbf{R})^2}$, are, respectively, the total aerosol number (mass), mean displacement, and mean square displacement. These are given by appropriate generalizations of the respective formulae 5.6 and 5.7

$$\bar{\mathbf{R}} = \mathbf{M}_1/M_0, \quad \overline{(\Delta\mathbf{R})^2} = \mathbf{M}_2/M_0 - \mathbf{M}_1\mathbf{M}_1/M_0^2 , \tag{5.88a,b}$$

with the moments M_0, $\mathbf{M}_1$, $\mathbf{M}_2$ being given by the generic relations

$$\mathbf{M}_m(t) = \int_{\tau_f} \sum_{\mathbf{n}} \mathbf{R}_{\mathbf{n}}^m c_w(\mathbf{R}_{\mathbf{n}}, \mathbf{r}, t) d^3\mathbf{r}, \quad m = 0,1,2,\dots . \tag{5.88c}$$

The long-time limits appearing in Definitions 5.87 mean $t >> t_\infty$ where $t_\infty = l^2/D_w$ is the time period in which aerosol particles achieve a certain asymptotic unit cell distribution. This time and the corresponding entrance length $\bar{V}l^2/D_w$ may be evaluated, for given size l of the aerosol particle and the filter porosity.

Calculation of the coarse-scale aerosol transport coefficients, as defined above via the concentration c_w, does not require explicit knowledge of this quantity. Rather, a variant of the method of moments (Brenner, 1980) can be employed to express $\bar{\mathbf{U}}^*, \bar{\mathbf{D}}^*, \bar{K}^*$ via the solutions of several boundary value problems, defined within one unit cell. These include (Shapiro and Brenner, 1990a)

1. Characteristic eigenvalue problem (EVP):

$$(\nabla \cdot \mathbf{j} + \lambda I)\phi = 0 \; (\mathbf{r} \in \tau_f), \tag{5.89}$$

$$\phi = 0 \; \mathbf{r} \in S_p', \; \phi \text{ — spatially periodic,} \tag{5.90a,b}$$

in which $\mathbf{j}$ denotes the partial differential operator $\mathbf{j} = \mathbf{v} - D_w\nabla$, and I is the identity operator.

2. Adjoint EVP:

$$(\mathbf{v} \cdot \nabla + D_w\nabla^2 - \lambda I)\psi = 0 \; (\mathbf{r} \in \tau_f), \tag{5.91}$$

$$\psi = 0 \ (\mathbf{r} \in S_p'), \ \psi \text{ — spatially periodic.} \tag{5.92a,b}$$

Using the *leading order solutions* ψ_0, ϕ_0, λ_0 of these EVPs, one can express the effective aerosol velocity, dispersivity, and the effective volumetric deposition coefficient in the form:

$$\overline{K}^* = -\lambda_0 \, , \tag{5.93}$$

$$\overline{\mathbf{U}}^* = \int_{\tau_f} D_w\left[\left(\phi_0 \nabla \psi_0 - \psi_0 \nabla \phi_0\right) + \phi_0 \psi_0 \mathbf{v}\right] d^3\mathbf{r} \, , \tag{5.94}$$

and

$$\overline{\mathbf{D}}^* = D_w \int_{\tau_f} \phi_0 \psi_0 \nabla \mathbf{B}^\dagger \cdot \nabla \mathbf{B} d^3\mathbf{r} \, . \tag{5.95}$$

The cellular vector field **B** appearing in the latter equation is determined by the solution of the following time-independent unit cell boundary value problem:

$$\left(\nabla \cdot \mathbf{j} + \lambda_0 I\right)\left(\phi_0 \mathbf{B}\right) = -\phi_0 \overline{\mathbf{U}}^* \ \text{ in } \ \tau_f \, , \tag{5.96}$$

subject to the boundary conditions

$$\mathbf{B} \text{ — finite on } S_p' \tag{5.97}$$

$$[[\mathbf{B}]] = [[\mathbf{r}]], \ [[\nabla \mathbf{B}]] = 0. \tag{5.98}$$

In the above [[..]] denotes the change in the value of any function across equivalent points lying on opposite faces of the unit cell. In particular, spatially periodic functions ϕ, ψ which solve EVPs 5.89, 5.90 and 5.91, 5.92 have $[[\phi]] = [[\phi]] = [[\nabla\psi]] = [[\nabla\psi]] = 0$.

Elementary dimensional analysis reveals that the three coarse-scale aerosol transport coefficients depend functionally upon the following dimensionless microscale parameters:

1. Characteristic Peclet number, $\text{Pe} = 2a\,\overline{V}\,/D_w$;
2. Collector Reynolds number $\text{Re} = 2a\rho\,\overline{V}\,/\mu$, with ρ and μ the respective air density and viscosity, where the scalar $\overline{V} \equiv |\,\overline{\mathbf{V}}\,|$;
3. Aerosol particle interception parameter $R_p = r_w/a$;
4. Bed porosity, $\varepsilon = \tau_f/\tau_0$ (where τ_f, τ_0 are the interstitial and superficial volumes of the unit cell).

5.8.2 Coarse-Scale Aerosol Transport Coefficients

Here we present the results obtained only for the square array case (Figure 5.9b) where the mean air velocity is parallel to the principal lattice axes. In such circumstances the mean aerosol particles velocity $\overline{\mathbf{U}}^*$ is parallel to the mean velocity vector $\overline{\mathbf{V}}$ of the air. The scalars $\overline{U}^*$ and $\overline{V}$ represent the respective aerosol and air *speeds*. The flow velocity field in this array was computed by Edwards et al. (1990). Figure 5.10a displays the functional dependence of the dimensionless aerosol velocity $\overline{U}^* / \overline{V}$ of effectively point-size aerosol particles ($R_p = 0$) upon the filter bed porosity. For (low) Peclet numbers of order unity (in which case the convective and diffusive particle transport mechanisms are comparable everywhere within the unit cell), particle deposition on the

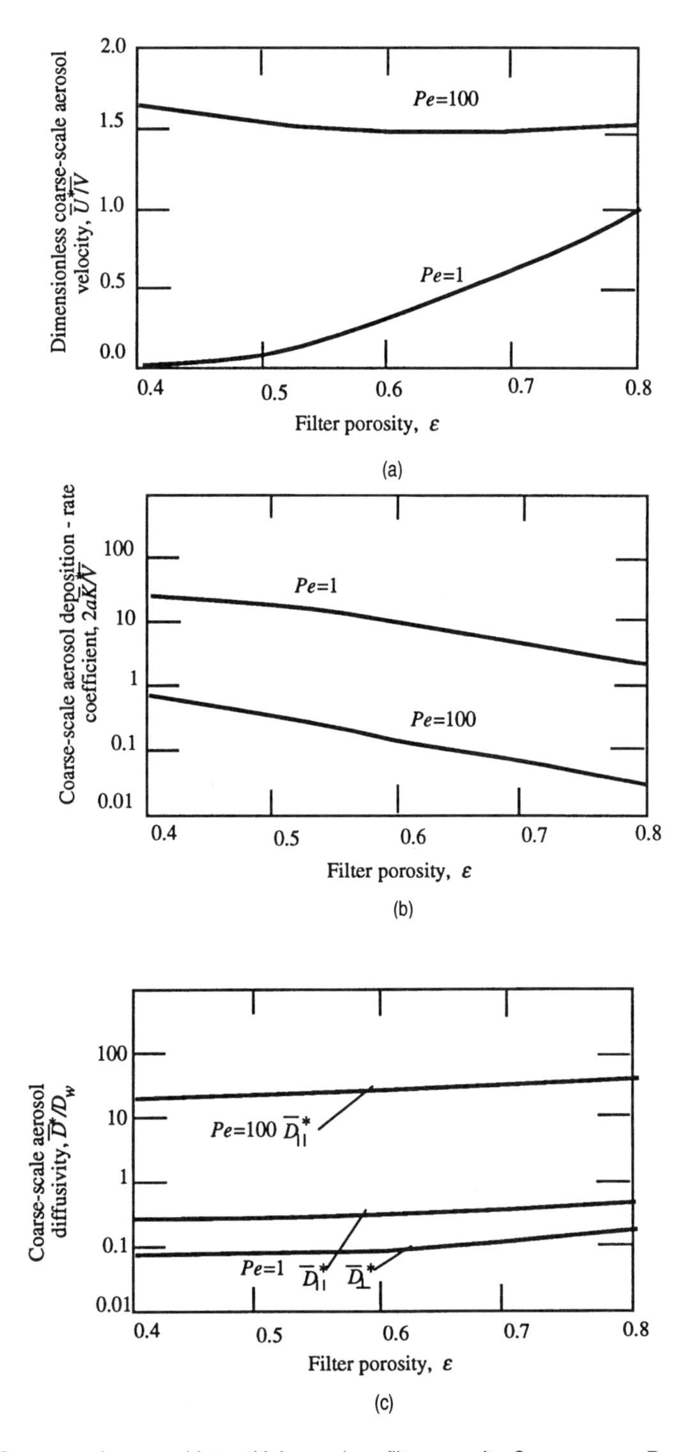

Figure 5.10 (a) Coarse-scale aerosol interstitial speed vs. filter porosity. Square array, $R_p = 0$. (b) Coarse-scale aerosol deposition rat coefficient vs. filter porosity. Square array, $R_p = 0$. (c) Coarse-scale longitudinal (||) and lateral (⊥) aerosol dispersivity. Square array, $R_p = 0$.

collector surfaces reduces the aerosol concentration in the bulk, far from the collectors, where the local interstitial velocity is higher (owing to the zero velocity, no-slip boundary condition on the collector surface). As a result, the mean aerosol speed $\overline{U}^*$ is less than the comparable mean interstitial airspeed $\overline{V}$ within the filter bed.

A different situation prevails at large Peclet numbers, in which circumstances a thin diffusional boundary layer exists proximate to the collector surfaces. Aerosol concentration within this thin layer, where the local interstitial air velocity is low due to the no-slip boundary condition, is depleted as a result of removal of particles. However, far from the collector the aerosol concentration remains virtually unchanged. Therefore, the major portion of the aerosol is transported by the faster-moving streamlines. Consequently, for large Peclet numbers the average aerosol particle velocity $\overline{U}^*$ exceeds the mean air velocity $\overline{V}$. Since the large Peclet number regime usually prevails during the filtration of submicrometer aerosols with $2r_w$ = 0.1 to 1 μm, it may generally be concluded that the phenomenon of enhanced mean aerosol velocity obtains for most filtration operating conditions. In all cases, however, $\overline{U}^* / \overline{V} \rightarrow 1$ when $\varepsilon \rightarrow 1$.

As may be seen in Figure 5.10a, the effect of increasing bed porosity is to diminish the difference between the large and small Peclet number coarse-scale aerosol velocities. For high-porosity fibrous filters, where the porosity exceeds 0.9, the effect of Pe upon the mean aerosol velocity is negligible. However, the effect of Peclet number is certainly significant for packed granular filters, where the porosity may be as low as 0.35.

Figure 5.10b shows the dependence of the nondimensional coarse-scale aerosol deposition-rate coefficient upon the bed porosity. As may be seen $\overline{K}_x^*$ is largely determined by the Peclet number, whereas the effect of decreasing porosity is to diminish $\overline{K}_x^*$ for all Peclet numbers, as the deposition surface to volume ratio increases. This implies that coarser particles and more loosely packed filters possess smaller values of $\overline{K}_x^*$ than do their opposite members. This is, of course, true in the absence of the interception collection mechanism (see discussion in Section 5.8.3).

Figure 5.10c shows the dependence of the longitudinal $\overline{D}_{\parallel}^*$ and lateral $\overline{D}_{\perp}^*$ components (with respect to the mean airflow direction) of the mean aerosol dispersivity dyadic. For low Peclet number regime (Pe = 1) longitudinal dispersivity is comparable to the molecular diffusivity and exceeds the lateral component by the factor of about three. Both $\overline{D}_{\parallel}^*$ and $\overline{D}_{\perp}^*$ slightly increase with increasing filter porosity. For the high Peclet number aerosol transport regime (corresponding to relatively large particle sizes) the longitudinal diffusivity is significantly larger than its low Pe number counterpart, which effect is attributed to the convective dispersion mechanism (Edwards et al., 1991). On the other hand, the lateral component of the mean aerosol diffusivity was found to be almost independent of the Peclet number. Accordingly, for Pe = 100, $\overline{D}_{\perp}^*$ (which is not shown in Figure 5.10c) does not differ significantly from the one calculated for Pe = 1.

5.8.3 Aerosol Filtration Length

Equation 5.86 provides the coarse-scale description of the interstitial aerosol transport and deposition processes. As specified in the definitions of the effective aerosol transport coefficients, they exist (as time-dependent quantities) a long time after the introduction of the aerosol cloud within the filter medium. The timescale t_∞ on which Equation 5.86 is valid may be explicitly evaluated for each interstitial filter geometry. It may be generally shown that

$$t_\infty = O(l^2/D_w), \tag{5.99}$$

where l is the characteristic size of the unit cell. The latter is related with the collector size a, specifically, $l \sim a(1-\varepsilon)^{-\nu}$, where $\nu = {}^1/_2$ for fibrous and ${}^1/_3$ for granular beds. Using this, one can rewrite the condition of applicability (compare 5.26b) of the dispersion/reaction model in the form:

$$t >> a^2(1-\varepsilon)^{-2\nu}/D_w, \tag{5.100}$$

Normally, aerosol filtration is a steady process, occurring in a filter with a finite length *L*. In this case the unsteady Equation 5.86 reduces to the following steady form:

$$\overline{\mathbf{U}}^* \cdot \overline{\nabla}\overline{c}_w - \overline{\mathbf{D}}^* : \overline{\nabla}\overline{\nabla}\overline{c}_w + \overline{K}^*\overline{c}_w = 0\,, \tag{5.101}$$

to be solved in the whole filter domain. We will use Condition 5.100 to evaluate the range of applicability of the above steady equation. Each portion of the aerosol particles entering the filter will propagate with the velocity of order $\overline{V}$ and by the time specified in 5.100 will cover the distance of order $\overline{V}a^2(1-\varepsilon)^{-2\nu}/D_w$. This distance should be much less than *L*, which yields

$$\overline{V}a^2(1-\varepsilon)^{-2\nu}/D_w \ll L\,. \tag{5.102}$$

This condition describes the situation, where the characteristic intracell particle equilibration time is much less than the aerosol residence time within the filter bed. Condition 5.102 is satisfied for submicrometer aerosols, provided that the amount of collector layers in the filter is sufficiently large.

For a given macroscale filter geometry, and the coarse-scale aerosol transport coefficients evaluated form 5.89 to 5.95, Equation 5.101 may be solved by standard analytical or numerical methods to determine the amount of aerosol collected within the filter. The boundary conditions for this equation include continuity of the aerosol flux at the inlet and the Dankwertz condition (Levenspiel, 1962; see below Equations 5.105, 5.106) at the filter outlet.

A significant simplification of Equation 5.101 may be achieved in the case where the filter bed is isotropic and the macroscale external forces are applied parallel to the average airflow direction, say $\overline{x}$. In this case Equation 5.101 reduces to the following one-dimensional equation:

$$\overline{U}_f \frac{\partial \overline{c}_f}{\partial \overline{x}} - \overline{D}_f \frac{\partial^2 \overline{c}_f}{\partial \overline{x}^2} + \overline{K}_f \overline{c}_f = 0\,, \tag{5.103}$$

wherein $\overline{x}$ is the coarse-scale coordinate along the airflow direction and $\overline{c}_f$ is the cross-sectionally averaged aerosol concentration, governed by the *scalar* mean axial transport coefficient:

$$\overline{U}_f = \overline{U}_x^*,\ \ \overline{D}_f = \overline{D}_{xx}^*,\ \ \overline{K}_f = \overline{K}^*, \tag{5.104}$$

Equation 5.103 may be solved subject to the boundary conditions:

$$U_0 c_{\text{in}} = \overline{U}_f \overline{c}_f - \overline{D}_f \frac{d\overline{c}_f}{d\overline{x}} \quad \text{at } \overline{x} = 0\,, \tag{5.105}$$

$$\frac{d\overline{c}_f}{d\overline{x}} = 0 \quad \text{at } \overline{x} = L\,. \tag{5.106}$$

Two limiting cases may be outlined:

1. Small filter face velocity, i.e.,

$$\varepsilon_f \equiv \overline{K}_f \overline{D}_f / \overline{U}_f^2 \gg 1\,, \tag{5.107}$$

which yields the following expression for the filter efficiency:

$$\eta \equiv 1 - \frac{c_{out}}{c_{in}} = 1 - 2\frac{\overline{U}_f}{\sqrt{\overline{K}_f \overline{D}_f}} \exp\left[-\left(\frac{\overline{K}_f}{\overline{D}_f}\right)^{1/2} L\right]. \tag{5.108}$$

2. Large filter face velocity:

$$\varepsilon_f = \overline{K}_f \, \overline{D}_f / \overline{U}_f^2 << 1 , \tag{5.109}$$

which gives

$$\eta = 1 - \exp\left[-\frac{\overline{K}_f}{\overline{U}_f} L\right]. \tag{5.110}$$

In this case, describing the majority of practically important situations, filtration may be characterized by the filtration length, defined by (Leers, 1957)

$$l_f = \frac{-L}{\ln(1-\eta)} . \tag{5.111}$$

This in combination with 5.110 provides the following expression for l_f

$$l_f = \frac{\overline{U}_f}{\overline{K}_f} . \tag{5.112}$$

The computational results shown in Figure 5.10a–c suggest the following scaling relationships:

$$\overline{K}_f = O\left(\overline{V}/a\mathrm{Pe}\right),\ \overline{U}_f = O\left(\overline{V}\right),\ \overline{D}_f = O\left(D_w \mathrm{Pe}\right), \tag{5.113}$$

valid for Pe >> 1. Introduction of 5.113 into Condition 5.109 yields

$$\varepsilon_f = O(\mathrm{Pe})^{-1} << 1, \tag{5.114}$$

which is obviously satisfied for aerosols with $2r_w = 0.1$ to 1 μm and a wide range of the filtration operating parameters. This condition describes the situation where the aerosol filtration can be characterized in terms of the aerosol filtration length. For smaller particles, however, Condition 5.109 may not be fulfilled. Indeed for Pe = 1 one can evaluate (for $\varepsilon = 0.8$) $\varepsilon_f = O(1)$, in which circumstances Equation 5.108 could be more appropriate for describing the aerosol filtration.

Figure 5.11 shows the characteristic filtration length l_f calculated for several particle interception parameters. It reveals that l_f increases with increasing bed porosity and the Peclet number. The influence of the interception parameter (increasing R_p), is to decrease l_f which corresponds to enhancement of the aerosol collection rate for beds with thinner fibers. Similar dependences of the characteristic filtration length upon the filter solidity are observed for smaller Peclet number values, namely, Pe = 100 and Pe = 10, respectively (Shapiro et al., 1991). The relative influence of the interception collection mechanism was found to be more pronounced for larger Peclet numbers, corresponding to larger aerosol particles. This accords generally with experimental observations,

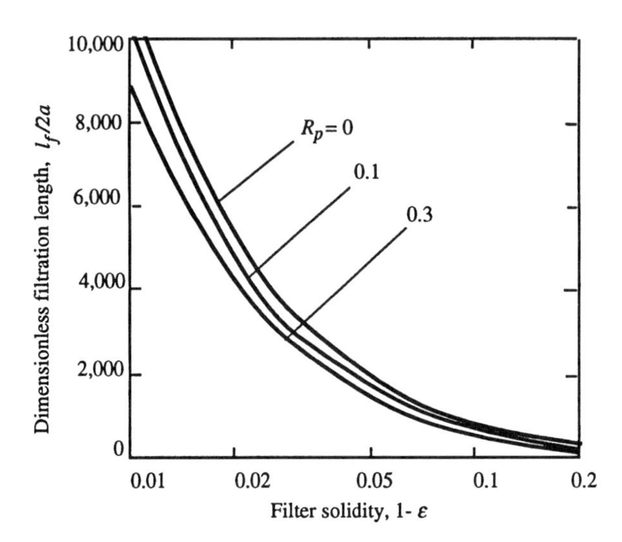

Figure 5.11 Filtration length vs. filter solidity. Square array. Pe = 1000.

which reveal that the diffusion particle capture mechanism becomes less efficient with increasing aerosol particle size.

5.8.4 Comparison with Analytical and Experimental Results

Figures 5.12 to 5.14 depict the characteristic filtration length vs. particle diameter (the latter being roughly proportional to the particle Peclet number) for three different airspeeds $\overline{U}$. These figures compare the l_f values derived from the dispersion/reaction model with those calculated from the semiempirical model of Stechkina et al. (1969), Stechkina and Fuchs (1966), Kirsch and Fuchs (1966), based upon the UBE approach, as well as with the experimental results of Lee (1977), Lee and Liu (1982), obtained for Dacron filters of 11 μm fiber diameter. For all calculations performed, the characteristic filtration lengths computed for the square array were observed to be larger than for the staggered array (all other parameters being equal). As may be seen from Figure 5.12, which shows data for a superficial air velocity of $\overline{V} = 1$ cm/s, the experimental results of Lee (1977) are satisfactorily described by the present model. On the other hand, the theory of Stechkina et al. (1969) predicts much lower values of l_f, thereby significantly overestimating the aerosol collection rate for the given filter.

Note that the data presented in Figure 5.12 for $\overline{V} = 1$ cm/s are characterized by relatively small values of the Peclet number, namely Pe ≤ 3000, corresponding to the case of a relatively strong influence of the diffusional over the interceptional deposition mechanism. Figure 5.13 shows a similar comparison for $\overline{V} = 3$ cm/s, characterized by larger values of the Peclet number, in which situation the relative effect of the interceptional collection mechanism is stronger. The figure indicates that our model displays a good agreement with the experimental data of Lee (1977), collected for the aerosol particles with diameters below 0.3 μm. Characteristic filtration lengths measured by Lee (1977) for 0.1 to 0.2 μm particles lie between the theoretical predictions of Stechkina et al. (1969) and those of the present model for the staggered array case.

Figure 5.14 similarly compares the numerical, alternate theoretical, and experimental data at larger values of the superficial air velocity, namely $\overline{V} = 10$ cm/s, corresponding to the case where the interceptional collection mechanism dominates over diffusion. One can see that the experimental values of l_f for particles with diameters smaller than about 0.1 μm are well described by our results for the staggered array. However, values of l_f experimentally measured for larger particles are better correlated by the model of Stechkina et al. (1969).

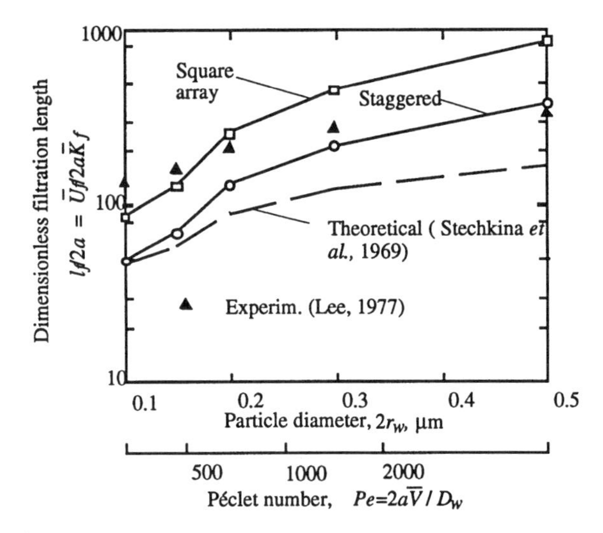

Figure 5.12 Filtration length calculated for $2a = 11$ µm, $\varepsilon = 0.849$. Air velocity 1 cm/s. Hollow symbols and solid lines = present calculations, filled symbols = experimental data of Lee (1977) for Dacron fiber filters.

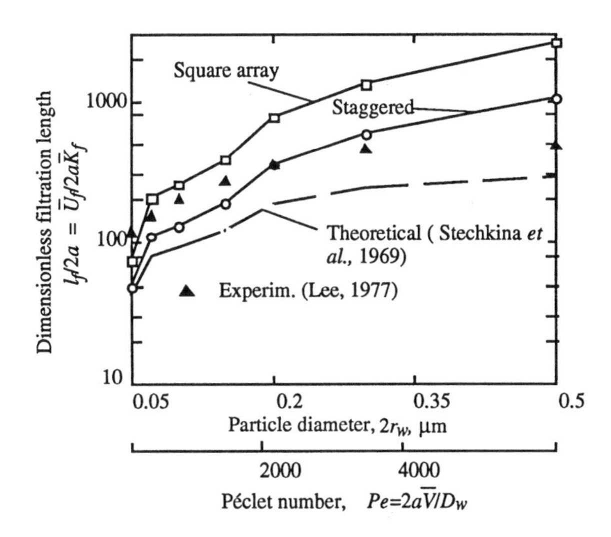

Figure 5.13 As in Figure 5.12. Air velocity 3 cm/s.

Figures 5.12 through 5.14 suggest that the values of l_f calculated by the present model for the *staggered array* better accord with the experimental data than do the comparable square array results. This agreement is better for smaller (diffusional) particles, than for larger particles, the deposition of which is governed mainly by interception. Schematics of these two geometric models are shown in Figure 5.15, where the square array corresponds to the zero angle θ between the mean airflow velocity and the lattice axis, whereas for the staggered array $\theta = 45°$. For very fine particles, aerosol collection is governed solely by diffusion and is, hence, independent of the microstructure geometry. In this situation the two geometric models yield virtually identical results. Collection of larger aerosol particles, which more closely adhere to the interstitial velocity streamlines (weak Brownian diffusion) is influenced by the collector shielding (see Figure 5.15). For the square array each successive collector is shielded by its closed predecessor, and is thus exposed to a depleted aerosol particle stream. For large Pe particle diffusion from the open streamlines into the interior of the depleted zone is slow, being limited by the available length of the unit cell. This length is

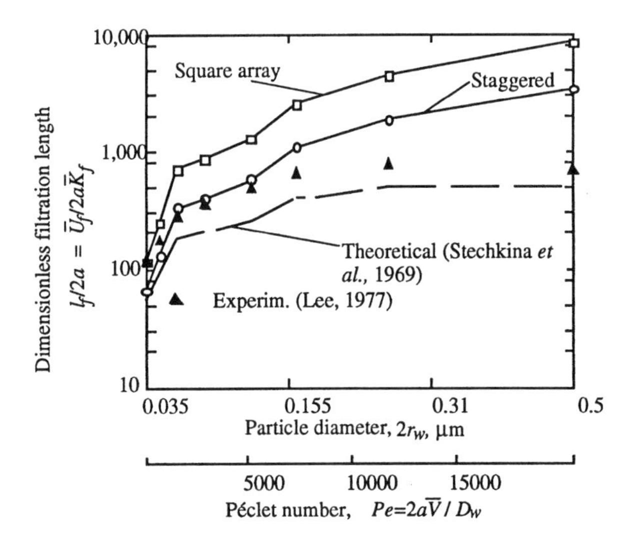

Figure 5.14 As in Figure 5.12. Air velocity 10 cm/s.

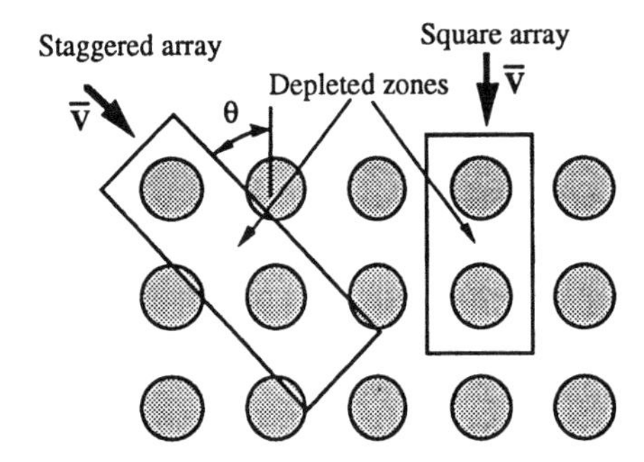

Figure 5.15 Collector shielding for different flow orientations.

larger for the staggered array. Therefore, more particles will enter the depleted zone in this array by diffusion and, hence, will deposit upon the subsequent collector element before passing to the next cell. This observation explains why l_f for the staggered array is less than that for the square array.

Another mechanism affecting the collection rate resides in the "tortuosity" of the interstitial flow pattern (characterizing the relative intensity of the lateral air velocity component with respect to the longitudinal component). The staggered array exhibits a more tortuous flow path than does the square array, leading to a better intracell mixing (Edwards et al., 1991). This further explains the more-efficient aerosol collection by the former array, discussed in the preceding paragraph.

Several remarks are in order concerning the classical filtration model of Stechkina et al. (1969). One of the elements characterizing the latter model is the implicit assumption of *perfect mixing* of the aerosol stream before entering the (simplified, cylindrical envelope) unit cell, as if no collector shielding exists between the successive cells. The present calculations show that the aerosol concentration distribution downstream of each collector element is not uniform, especially for large Pe and relatively low bed porosities, which result essentially contradicts the basic mixing assumption of the UBE model. On the other hand, for highly porous fibrous filters (with $\varepsilon \geq 0.99$) the distance between the adjacent fibers and hence particle mixing length is very large. In the latter case the use of the *ad hoc* mixing assumption is justified and the UBE model may better agree with experimental data (Lee and Liu, 1982). The present dispersion/reaction model may be further used to establish the validity range and the limitations of the simplified UBE approach.

It appears that the *ad hoc* assumption of perfect intracell mixing is the main factor underlying the relatively good agreement of the UBE model with the experimental data of Lee (1977) for *large* (interceptional) particles, and concomitant poor agreement for small (diffusional) particles. Such an implicit mixing assumption may be viewed as attempting to reflect the disordered pore structure of real porous filters. The dispersion/reaction model does not, however, utilize this assumption, nor does it assume any degree of geometric randomness in a stochastic (nondeterministic) sense. Instead, the aerosol concentration and flux are assumed continuous across the boundaries separating adjacent unit cells. However, such cells may be chosen such as to reflect some degree of local (intracell) geometric disorder, expressed, perhaps, by including several collectors within one cell, or by having several particle shapes and/or sizes present in the cell unit. This may be accounted for within the framework of our model (for which the underlying theory requires only spatial periodicity, but not necessarily a simple unit cell configuration composed of but a *single* collector). However, obtaining solutions for these more realistic (albeit spatially periodic) configurations requires larger computational efforts. The limited calculations performed thus far suggest that these more elaborate geometric lattice models of the filter bed would be expected to agree better with the experimental measurements over the wide range of aerosol particle diameters.

5.9 OTHER APPLICATIONS AND SUMMARY

The proposed approach calls for a novel procedure for testing of aerosol filtration devices. In the usually employed testing method, the filtration efficiency of given aerosol particles in (various) *steady-state* conditions is measured. A more fundamental testing procedure constitutes monitoring the time-dependent characteristics of pulses of monodisperse aerosol particles passing through the filtration devices, as described in Section 5.3. This information is shown to directly yield the effective axial aerosol transport coefficients, as well as the collection efficiency. These coefficients, especially the aerosol dispersivity, are sensitive to changes (nonuniformities) in the porous structure of the filter and the interstitial flow field. Therefore, the data obtained by the proposed testing method may be used to investigate the microstructure of the filter media, in particular, the inhomogeneities, either occurring during the process of their production, or arising from the filter contamination by the deposited particles.

The solution scheme for calculation of the effective aerosol transport properties, outlined in Section 5.5, may be generalized to include a variety of physical processes, and geometries, which may be decomposed into a *local, bounded space* (i.e., tube cross section) and a *global, unbounded space,* corresponding to the main flow direction. The various situations are encompassed by the general theory of convective–diffusive transport of reactive species (Shapiro and Brenner, 1986; Brenner and Edwards, 1993). In particular, the following processes and phenomena may be modeled by the considered scheme:

1. External forces acting on aerosols in the local and global spaces. These may include electric (as in Section 5.7), magnetic, gravitational forces, as well as thermophoretic, photophoretic, and diffusiophoretic forces acting on aerosol particles.
2. Devices of different cross-sectional and longitudinal geometries.
3. Transport in time-periodic flows (e.g., occurring in lungs during various breathing patterns), or in time-periodic fields of forces (Shapiro and Brenner, 1990b).
4. Kinetics of particles reentrainment from the wall surface. Particles captured on the walls (of the tube, or of the filtering elements) may be characterized by an appropriate surface concentration, governing the reentrainment flux. The general solution scheme may be extended to account for this effect. As a result, the dynamics of particle accumulation on surfaces may be described. This problem is related to contamination of porous filters, where the above approach can provide the time-dependent and spatially nonhomogeneous distribution of aerosol particles in the filtering media.

5. Transport in other globally discontinuous structures, where the present analyses are applicable, in particular, in spatially periodic models of the coherent air motions (Grass, 1971) existing in turbulent flows in ducts and in the atmosphere.
6. Transport of polydisperse aerosols, undergoing size transformation process (coagulation, evaporation, condensation, etc.). These processes may be modeled by generalization of the solution method, described above, to multicomponent systems possessing discrete (Shapiro and Adler, 1994) and continuous (Iosilevsky and Brenner, 1993) aerosol size distributions.
7. Transport and sedimentation of nonspherical aerosol particles (Brenner, 1981) (see Figure 5.4b), (i.e., fibrous, or porous agglomerates). Such particles perform rotational motion, which is generally coupled with their translational motion. The convective dispersion theory applied to this problem is based upon the Fokker–Planck equation, describing particle translational–rotational motion. The objectives of this approach are calculation of the effective translational properties of the aerosol particles. Accordingly, the orientational space serves in this case as a local space, over which the averaging is performed. Comparable global space is the position **R** of the particle center.

REFERENCES

P. M. Adler, *Porous Media. Geometry and Transports,* Butterworth-Heinemann, Boston, 1992.

J. Bensoussan, L. Lions, and G. Papanicolau, *Asymptotic Analysis for Periodic Structures,* North Holland, Amsterdam, 1978.

H. Brenner, Dispersion resulting from flow through spatially periodic porous media, *Phil. Trans. R. Soc. Lond.,* A297, 81, 1980.

H. Brenner, Taylor dispersion in systems of sedimenting nonspherical Brownian particles II. Homogeneous ellipsoidal particles, *J. Colloid Interface Sci.,* 80, 548, 1981.

H. Brenner and D. A. Edwards, *Macrotransport Processes,* Butterworth-Heinemann, Boston, 1993.

H. Brenner and L. J. Gaydos, The constrained Brownian movement of spherical particles in cylindrical pores of comparable radius: models of the diffusive and convective transport of solute molecules in membranes and porous media, *J. Colloid Interface Sci.,* 58, 312, 1977.

G. H. Crapiste, E. Rotstein, and S. Whitaker, A general closure scheme for the method of volume averaging, *Chem. Eng. Sci.,* 41, 227, 1986.

C. N. Davies, *Air Filtration,* Academic Press, London, 1973.

W. Deutsch, Bewegung and Ladung der Elektrozotaetstraeger in Zylinderkondensator, *Ann. Phys.,* 68, 335, 1922.

D. A. Edwards, M. Shapiro, P. Bar-Yoseph, and M. Shapira, The effect of Reynolds number upon the apparent permeability of spatially periodic arrays of cylinders, *Phys. Fluids A,* 2, 45, 1990.

D. A. Edwards, M. Shapiro, H. Brenner, and M. Shapira, Dispersion of inert solutes in spatially periodic two-dimensional model porous media. *Transp. Porous Media,* 6, 337, 1991.

S. K. Friedlander, *Smoke, Dust and Haze: Fundamentals of Aerosol Behavior,* Wiley, New York, 1977.

N. A. Fuchs, *The Mechanics of Aerosols,* Pergamon Press, New York, 1964.

J. Gebhart, G. Heigwer, J. Heyder, C. Roth, and W. Stahlhoften, The use of light scattering photometry in aerosol medicine, *J. Aerosol Med.,* 1, 89, 1988.

A. G. Grass, Structural features of turbulent flow over rough and smooth boundaries, *J. Fluid Mech.,* 50, 233, 1971.

G. Iosilevsky and H. Brenner, Taylor dispersion in systems containing a continuous distribution of reactive species, *Int. J. Non-linear Mech.,* 28(1), 69, 1993.

A. A. Kirsch and N. A. Fuchs, Studies on fibrous aerosol filters — III. Diffusional deposition of aerosols in fibrous filters, *Ann. Occup. Hyg.,* 11, 299, 1966.

K. W. Lee, Filtration of submicron aerosols by fibrous filters, Ph.D. thesis; University of Minnesota, Minneapolis, 1977.

K. W. Lee and B. Y. H. Liu, Experimental study of aerosol filtration by fibrous filters, *Aerosol Sci. Technol.,* 1, 35, 1982.

R. Leers, Die Abscheidung von Schwebstoffen in Faserfiltern, *Staub,* 17, 402, 1957.

G. M. Leonard, M. Mitchner, and S. A. Self, Particle transport in electrostatic precipitators, *Atmos. Environ.,* 14, 1289, 1980.

O. Levenspiel, *Chemical Reaction Engineering,* Wiley, New York, 1962.

M. G. Marietta and G. W. Swan, Particle diffusion in electrostatic precipitators, *Chem. Eng. Sci.,* 31, 795, 1976.

J. Pich, Theory of aerosol filtration by fibrous and membrane filters, in *Aerosol Science,* C. N. Davies, Ed., Academic Press, London, 1966, 223.

H. Schlichting, *Boundary Layer Theory,* McGraw-Hill, New York, 1968.

G. A. Sehmel, Experimental study on the effect of turbulent diffusion on the precipitation efficiency, *J. Geophys. Res. A,* 75, 1766, 1970.

M. Shapiro and P. M. Adler, Dispersion et transports couplés de solutés multiples en milieus poreux, *Entropie,* 184/185, 57, 1994.

M. Shapiro and H. Brenner, Taylor dispersion of chemically reactive species: irreversible first-order reactions in bulk and on boundaries, *Chem. Eng. Sci.,* 41, 1417, 1986.

M. Shapiro and H. Brenner, Dispersion of a chemically reactive solute in a spatially periodic model of a porous media, *Chem. Eng. Sci.,* 43, 551, 1988.

M. Shapiro and H. Brenner, Dispersion/reaction model of aerosol collection by porous filters, *J. Aerosol Sci.,* 21, 97, 1990a.

M. Shapiro and H. Brenner, Taylor dispersion in the presence of time-periodic convection phenomena. Part II. Transport of transversely oscillating Brownian particles in a plane Poiseuille flow, *Phys. Fluids A,* 2, 1744, 1990b.

M. Shapiro, A. Oron, and C. Gutfinger, A dispersion model for electrostatic precipitation from turbulent flows, *Physicochem. Hydrodyn.,* 10, 471, 1988.

M. Shapiro, A. Oron, and C. Gutfinger, Electrostatic precipitation of charged particles from a turbulent flow between coaxial cylinders, *J. Colloid Interface Sci.,* 127, 401, 1989.

M. Shapiro, H. Brenner, and I. J. Kettner, Transport mechanics and collection of aerosol particles in fibrous filters, *J. Aerosol Sci.,* 22, 707, 1991.

M. Shlafstein, M. Fichman, and M. Shapiro, Effective axial transport coefficients of submicrometer particles in tube flows, in *Proc. of 4th Internat. Aerosol Conf.,* R. C. Flagan, Ed., Los-Angeles, Aug. 29 to Sept. 2, 1994, 701–702.

I. B. Stechkina and N. A. Fuchs, Studies on fibrous aerosol filters — I. Calculation of diffusional deposition of aerosol in fibrous filters, *Ann. Occup. Hyg.,* 9, 59, 1966.

I. B. Stechkina, A. A. Kirsch, and N. A. Fuchs, Studies on fibrous filters — IV. Calculation of aerosol deposition in model filters in the range of maximum penetration, *Ann. Occup. Hyg.,* 12, 1, 1969.

W. Streider and R. Aris, *Variational Methods Applied to Problems of Diffusion and Reaction,* Springer, Berlin, 1970.

G. I. Taylor, Dispersion of soluble matter in solvent flowing slowly through a tube, *Proc. R. Soc. London Ser. A,* 219, 186, 1953.

H. J. White, *Industrial Electrostatic Precipitators,* Addison-Wesley, New York, 1963.

CHAPTER 6

The Analysis of Aerosol Filtration Using the Method of Volume Averaging

Michel Quintard and Stephen Whitaker

CONTENTS

6.1 INTRODUCTION

The process of filtration of noncharged, submicron particles represents and example of transport in porous media that can be analyzed using the method of volume averaging. In this chapter we illustrate how one can derive the local volume-averaged particle transport equation and how one can determine the effective coefficients that appear in that equation. Of particular interest is the capture coefficient that allows one to predict the filter efficiency for a homogeneous filter.

The particle continuity equation is represented in terms of the *first correction* to the Smoluchowski equation. This takes particle inertia into account for small Stokes numbers and leads to a filter efficiency that contains a minimum in the efficiency as a function of the particle size. This allows one to identify the *most-penetrating particle size* and this is often a key element in the design of a filter. Comparison of the theory with experimental data indicates that the first correction to the Smoluchowski equation gives reasonable results for the most-penetrating particle size and

0-87371-830-5/98/$0.00+$.50

for smaller particles; however, results for larger particles clearly indicate the need to extend the Smoluchowkski equation to include higher-order corrections for the inertia of the particles.

The process of filtration takes place in hierarchical porous media (Cushman, 1990) and we have illustrated this in Figure 6.1. In order to design a filter, one needs a particle transport equation in which the porosity heterogeneities have been *spatially smoothed.* This suggests the use of the first averaging volume shown in Figure 6.1 along with the method of large-scale averaging (Quintard and Whitaker, 1987, 1988, 1990ab; Plumb and Whitaker, 1988ab, 1990ab). Large-scale averaging requires the use of local volume-averaged equations that are associated with the second averaging volume shown in Figure 6.1. These equations are sometimes referred to as the *Darcy-scale* transport equations, and they represent the point in the hierarchial process at which the governing differential equations and boundary conditions are *joined.* These boundary conditions are imposed at the γ-σ interface which is illustrated in the third volume contained in Figure 1.6 where we have identified the fibers as the σ-phase and the fluid as the γ-phase. The governing equation for the fluid velocity in the γ-phase will be taken to be Stokes' equations, while the governing equation for the particle concentration is represented by a Fokker–Planck equation for the probability density function. This idea is suggested in the last volume illustrated in Figure 6.1 where we have identified the particles as the κ-phase and the pure fluid as the β-phase.

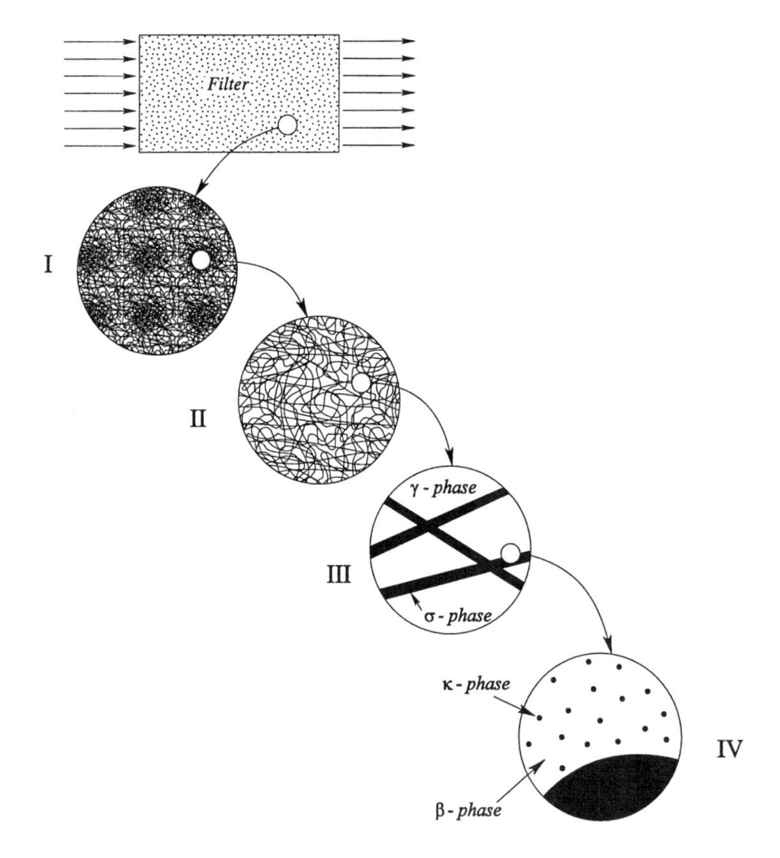

Figure 6.1 Hierarchical view of the filtration process.

Our starting point for a theoretical derivation of the filter collection efficiency is a pore-scale description of the particle transport which must be subjected to volume averaging in order to obtain a filter transport equation. If one assumes pure Brownian diffusion, particle transport can be viewed as equivalent to the process of convective–diffusion in a porous medium with heterogeneous chemical reaction. However, it has been recognized for some time that the existence of a *most-penetrating particle size* is the result of a complex interaction between Brownian diffusion and inertial effects. Because of this, it is important to make use of a more complete description of the particle transport process that takes into account the effect of particle inertia.

6.2 PARTICLE MOTION

Our description of the motion of the particles begins with the single particle Langevin equation

$$\text{B.C.1} \quad \tilde{n}_p = \underbrace{-\langle n_p \rangle^\gamma}_{\text{source}}, \quad \text{at the } \gamma - \sigma \text{ in} \tag{6.1}$$

in which m_p is the mass of the particle and $\mathbf{v}_p$ is the velocity of the particle. The first term on the right-hand side of Equation 6.1 describes the Stokes drag on a single isolated particle with c_s representing the Cunningham correction factor (Tien, 1989). The use of this form for the force indicates that we are ignoring particle-particle interactions and the fluid mechanical complications that arise when a particle approaches a solid surface (Peters and Ying, 1991). These effects should certainly be taken into account; however, in our first study of this process (Quintard and Whitaker, 1995) every effort was made to keep the analysis as simple as possible.

It is convenient to express Equation 6.1 as

$$\frac{d\mathbf{v}_p}{dt} = -\gamma\left(\mathbf{v}_p - \mathbf{v}\right) + \Gamma(t) \tag{6.2}$$

in which γ is inversely proportional to the Stokes number

$$\gamma = \frac{3\pi\mu d_p}{m_p c_s} \sim \text{St}^{-1} \tag{6.3}$$

Here St represents the Stokes number to be defined later. There are many particle transport processes for which the Stokes number is small compared with 1 and this so-called *high friction limit* represents an important case for filtration. The Fokker–Planck equation associated with the stochastic process described by Equation 6.2 takes the form (Risken, Sec. 10.1, 1989):

$$\frac{\partial W_p}{\partial t} + \frac{\partial}{\partial \mathbf{r}}\left(\mathbf{v}_p W_p\right) - \frac{\partial}{\partial \mathbf{v}_p}\left[\gamma\left(\mathbf{v}_p - \mathbf{v}\right)W_p\right] = \left(\frac{\gamma k T}{m_p}\right)\frac{\partial}{\partial \mathbf{v}_p} \cdot \left[\frac{\partial}{\partial \mathbf{v}_p}\left(W_p\right)\right] \tag{6.4}$$

The dependent variable represents a probability density function

$$W_p = W_p\,(t, \mathbf{r}, \mathbf{v}_p) \tag{6.5}$$

and the particle number density is given by

$$n_p = \int_{-\infty}^{+\infty} W_p\left(t, \mathbf{r}, \mathbf{v}_p\right) d\mathbf{v}_p \tag{6.6}$$

The solution of Equation 6.4 for the high friction limit is based on matrix-continued fraction methods (Risken, Sec. 10.4, 1989) which yield

$$\frac{\partial n_p}{\partial t} = \nabla \cdot \left\{ \left[-\mathbf{v} + D_p \nabla \right] n_p \right\} + \gamma^{-1} \nabla \cdot \left\{ \left[\mathbf{v} \cdot \nabla \mathbf{v} - (\nabla \cdot \mathbf{v}) D_p \nabla \right] n_p \right\}$$
$$+ \gamma^{-3} \nabla \cdot \left\{ [\cdots] n_p \right\} + \mathbf{O}\left(\gamma^{-5}\right) + \cdots \tag{6.7}$$

in which D_p is the Brownian diffusivity. It is important to keep in mind that this result *does not take into account* the complex fluid mechanics that occur when a particle approaches a solid surface. If only the first term on the right-hand side of Equation 6.7 is retained we have the Smoluchowski equation (Gardiner, Sec. 6.4, 1990) and under these circumstances the mean motion of the particles follows the fluid streamlines. As we shall see in subsequent paragraphs, this leads to a situation that cannot predict a crucial characteristic of many filtration processes; thus we will retain the second term on the right-hand side of Equation 6.7 so that our particle transport equation takes the form:

$$\frac{\partial n_p}{\partial t} + \nabla \cdot \left\{ \left[\mathbf{v} - \gamma^{-1} \mathbf{v} \cdot \nabla \mathbf{v} \right] n_p \right\} = \nabla \cdot \left(D_p \nabla n_p \right), \quad \text{in the } \gamma \text{ - phase} \tag{6.8}$$

Equation 6.8 represents the first smoothing process in the hierarchy of averaging processes illustrated in Figure 6.1. The next step in this process requires that we form the local volume average of Equation 6.8.

6.3 VOLUME AVERAGING

At this point we are ready to express the complete transport problem under consideration, and we list the governing differential equations and boundary conditions as

$$\frac{\partial n_p}{\partial t} + \nabla \cdot \left\{ \left[\mathbf{v} - \gamma^{-1} \mathbf{v} \cdot \nabla \mathbf{v} \right] n_p \right\} = \nabla \cdot \left(D_p \nabla n_p \right) \tag{6.9}$$

$$\text{B.C. 1} \quad n_p = 0, \text{ at the } \gamma\text{–}\sigma \text{ interface} \tag{6.10}$$

$$0 = \nabla p + \rho \mathbf{g} + \mu \nabla^2 \mathbf{v} \tag{6.11}$$

$$\text{B.C. 2} \quad \mathbf{v} = 0, \text{ at the } \gamma\text{–}\sigma \text{ interface} \tag{6.12}$$

$$\nabla \cdot \mathbf{v} = 0 \tag{6.13}$$

It should be clear that we have already used Equation 6.13 with Equation 6.7 in order to simplify that result to Equation 6.8, and we will need to use Equation 6.13 again in our analysis of the convective transport term in Equation 6.9. The boundary condition represented by Equation 6.10 must be thought of as a limiting case which will create an upper bound for the filter efficiency.

The method of volume averaging begins by associating with every point in space (in both the γ-phase and the σ-phase) an averaging volume that we denote by $\mathscr{V}$. Such a volume is illustrated

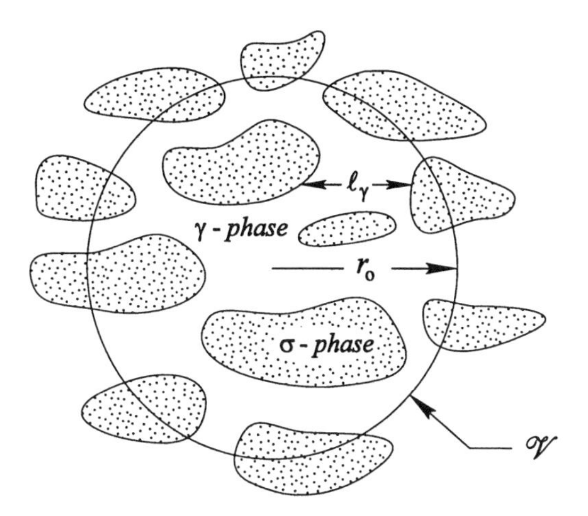

Figure 6.2 Averaging volume.

in Figure 6.2 where we have identified the radius of the averaging volume by r_0, and the characteristic length of the γ-phase by ℓ_γ. We will make use of two averages in our analysis of Equation 6.9 and the first of these is the superficial volume average which can be expressed as

$$\langle \psi_\gamma \rangle = \frac{1}{\mathcal{V}} \int_{V_\gamma} \psi_\gamma dV \tag{6.14}$$

Here ψ_γ is any function associated with the γ-phase and V_γ is the volume of the γ-phase contained within the averaging volume, $\mathcal{V}$. In addition to the superficial average, we will also make use of the intrinsic volume average that is defined by

$$\langle \psi_\gamma \rangle^\gamma = \frac{1}{V_\gamma} \int_{V_\gamma} \psi_\gamma dV \tag{6.15}$$

These two averages are related by

$$\langle \psi_\gamma \rangle = \varepsilon_\gamma \langle \psi_\gamma \rangle^\gamma \tag{6.16}$$

The average velocity is often represented in terms of the *superficial* velocity, while the average particle concentration is typically represented in terms of an *intrinsic* average. To avoid confusion between these two averages we will always make use of the nomenclature indicated in Equations 6.14 through 6.16. When we form the volume average of Equation 6.9 we will encounter averages of gradients and we will need to convert these to gradients of averages by means of the spatial averaging theorem (Howes and Whitaker, 1985), which we represent as

$$\langle \nabla \psi_\gamma \rangle = \nabla \langle \psi_\gamma \rangle + \frac{1}{\mathcal{V}} \int_{A_{\gamma\sigma}} \mathbf{n}_{\gamma\sigma} \psi_\gamma dA \tag{6.17}$$

Here $A_{\gamma\sigma}$ represents the area of the fluid–fiber interface contained within the averaging volume and $\mathbf{n}_{\gamma\sigma}$ represents the unit normal vector directed *from* the γ-phase *toward* the σ-phase.

6.4 PARTICLE TRANSPORT EQUATION

We begin the analysis of the particle transport process by expressing the superficial average of Equation 6.9 as

$$\left\langle \frac{\partial n_p}{\partial t} \right\rangle + \left\langle \nabla \cdot \left\{ \left[\mathbf{v} - \gamma^{-1} \mathbf{v} \cdot \nabla \mathbf{v} \right] n_p \right\} \right\rangle = \langle \nabla \cdot D_p \nabla n_p \rangle \tag{6.18}$$

and note that the first term can be written as

$$\left\langle \frac{\partial n_p}{\partial t} \right\rangle = \frac{\partial \langle n_p \rangle}{\partial t} \tag{6.19}$$

since V_γ is independent of time. The convective transport term in Equation 6.18 requires the use of the averaging theorem which leads to

$$\langle \nabla \cdot [\mathbf{v} - \gamma^{-1} \mathbf{v} \cdot \nabla \mathbf{v}] n_p \rangle = \nabla \cdot \langle [\mathbf{v} - \gamma^{-1} \mathbf{v} \cdot \nabla \mathbf{v}] n_p \rangle \tag{6.20}$$

on the basis of the no-slip condition given by Equation 12.6. The diffusive term on the right-hand side of Equation 6.18 provides us with

$$\langle \nabla \cdot D_p \nabla n_p \rangle = \nabla \cdot \langle D_p \nabla n_p \rangle + \frac{1}{\mathcal{V}} \int_{A_{\gamma\sigma}} \mathbf{n}_{\gamma\sigma} \cdot D_p \nabla n_p dA \tag{6.21}$$

Use of Equations 6.19 through 6.21 along with Equation 6.16 in the form

$$\langle n_p \rangle = \varepsilon_\gamma \langle n_p \rangle^\gamma \tag{6.22}$$

allows us to express the *superficial* average particle transport equation as

$$\begin{aligned} &\underbrace{\varepsilon_\gamma \frac{\partial \langle n_p \rangle^\gamma}{\partial t}}_{\text{accumulation}} + \underbrace{\nabla \cdot \langle \mathbf{v} n_p \rangle}_{\text{convection}} - \underbrace{\gamma^{-1} \nabla \cdot \langle [\mathbf{v} \cdot \nabla \mathbf{v}] n_p \rangle}_{\text{particle inertia correction}} \\ &= \underbrace{\nabla \cdot \left\{ D_p \left[\nabla \langle n_p \rangle + \frac{1}{\mathcal{V}} \int_{A_{\gamma\sigma}} \mathbf{n}_{\gamma\sigma} n_p dA \right] \right\}}_{\text{diffusion}} + \underbrace{\frac{1}{\mathcal{V}} \int_{A_{\gamma\sigma}} \mathbf{n}_{\gamma\sigma} \cdot D_p \nabla n_p dA}_{\text{particle capture}} \end{aligned} \tag{6.23}$$

Here we have identified the correction to the Smoluchowski equation as the *particle inertia correction,* and it is this term that plays a key role in the determination of the most-penetrating particle size in the filtration process.

In order to eliminate the point value of the particle concentration in the diffusion term, we use the following decomposition for the particle density

$$n_p = \langle n_p \rangle^\gamma + \tilde{n}_p \tag{6.24}$$

and later we will use the velocity decomposition given by

$$\mathbf{v} = \langle \mathbf{v} \rangle^\gamma + \tilde{\mathbf{v}} \tag{6.25}$$

One can use Equation 6.24 in the diffusion term and follow the type of analysis given by Whitaker (Sec. 2, 1986a) or Quintard and Whitaker (Sec. II, 1993) to obtain

$$\nabla\cdot\left\{D_p\left[\nabla\langle n_p\rangle+\frac{1}{\mathcal{V}}\int_{A_{\gamma\sigma}}\mathbf{n}_{\gamma\sigma}n_p\,dA\right]\right\}=\nabla\cdot\left\{D_p\left[\nabla\langle n_p\rangle^{\gamma}+\frac{1}{\mathcal{V}}\int_{A_{\gamma\sigma}}\mathbf{n}_{\gamma\sigma}\tilde{n}_p\,dA\right]\right\} \tag{6.26}$$

One can also use the decomposition given by Equation 6.24 in order to express the particle capture term as

$$\frac{1}{\mathcal{V}}\int_{A_{\gamma\sigma}}\mathbf{n}_{\gamma\sigma}\cdot D_p\nabla n_p\,dA=-\left(\nabla\varepsilon_{\gamma}\right)\cdot D_p\nabla\langle n_p\rangle^{\gamma}+\frac{1}{\mathcal{V}}\int_{A_{\gamma\sigma}}\mathbf{n}_{\gamma\sigma}\cdot D_p\nabla\tilde{n}_p\,dA \tag{6.27}$$

Substitution of Equations 6.26 and 6.27 in Equation 6.23 leads to

$$\begin{aligned}&\varepsilon_{\gamma}\frac{\partial\langle n_p\rangle^{\gamma}}{\partial t}+\nabla\cdot\langle\mathbf{v}n_p\rangle-\gamma^{-1}\nabla\cdot\left\langle[\mathbf{v}\cdot\nabla\mathbf{v}]n_p\right\rangle\\&=\nabla\cdot\left\{\varepsilon_{\gamma}D_p\left[\nabla\langle n_p\rangle^{\gamma}+\frac{1}{V_{\gamma}}\int_{A_{\gamma\sigma}}\mathbf{n}_{\gamma\sigma}\tilde{n}_p\,dA\right]\right\}\\&-\left(\nabla\varepsilon_{\gamma}\right)\cdot D_p\nabla\langle n_p\rangle^{\gamma}+\frac{1}{\mathcal{V}}\int_{A_{\gamma\sigma}}\mathbf{n}_{\gamma\sigma}\cdot D_p\nabla\tilde{n}_p\,dA\end{aligned} \tag{6.28}$$

and in order to complete the averaging procedure we would like to express the convective transport in terms of $\langle n_p\rangle^{\gamma}$ and $\tilde{n}_p$. One can follow Carbonell and Whitaker (1983) in order to represent the traditional convective transport as

$$\langle\mathbf{v}n_p\rangle=\underbrace{\varepsilon_{\gamma}\langle\mathbf{v}\rangle^{\gamma}\langle n_p\rangle^{\gamma}}_{\substack{\text{traditional}\\\text{convective transport}}}+\underbrace{\langle\tilde{\mathbf{v}}\tilde{n}_p\rangle}_{\substack{\text{traditional}\\\text{dispersive transport}}} \tag{6.29}$$

The so-called inertial contribution to the convective transport is algebraically more complex, and it is convenient to use only the decomposition given by Equation 6.24 to obtain

$$\left\langle(\mathbf{v}\cdot\nabla\mathbf{v})n_p\right\rangle=\underbrace{\langle\mathbf{v}\cdot\nabla\mathbf{v}\rangle\langle n_p\rangle^{\gamma}}_{\substack{\text{particle}\\\text{inertia convection}}}+\underbrace{\left\langle(\mathbf{v}\cdot\nabla\mathbf{v})\tilde{n}_p\right\rangle}_{\substack{\text{particle}\\\text{inertia dispersion}}} \tag{6.30}$$

On the basis of Equations 6.29 and 6.30 we can express the volume averaged particle transport equation as

$$\begin{aligned}&\underbrace{\varepsilon_{\gamma}\frac{\partial\langle n_p\rangle^{\gamma}}{\partial t}}_{\text{accumulation}}+\nabla\cdot\underbrace{\left(\varepsilon_{\gamma}\langle\mathbf{v}\rangle^{\gamma}\langle n_p\rangle^{\gamma}-\gamma^{-1}\langle\mathbf{v}\cdot\nabla\mathbf{v}\rangle\langle n_p\rangle^{\gamma}\right)}_{\text{convection}}+\nabla\cdot\underbrace{\left(\langle\tilde{\mathbf{v}}\tilde{n}_p\rangle-\gamma^{-1}\langle(\mathbf{v}\cdot\nabla\mathbf{v})\tilde{n}_p\rangle\right)}_{\text{dispersion}}\\&=\underbrace{\nabla\cdot\left\{\varepsilon_{\gamma}D_p\left[\nabla\langle n_p\rangle^{\gamma}+\frac{1}{V_{\gamma}}\int_{A_{\gamma\sigma}}\mathbf{n}_{\gamma\sigma}\tilde{n}_p\,dA\right]\right\}}_{\text{diffusion}}+\underbrace{\frac{1}{\mathcal{V}}\int_{A_{\gamma\sigma}}\mathbf{n}_{\gamma\sigma}\cdot D_p\nabla\tilde{n}_p\,dA}_{\text{particle capture}}\end{aligned} \tag{6.31}$$

Here we have imposed the simplification

$$\left(\nabla\varepsilon_\gamma\right)\cdot D_p\nabla\langle n_p\rangle^\gamma << \frac{1}{\mathcal{V}}\int_{A_{\gamma\sigma}} \mathbf{n}_{\gamma\sigma}\cdot D_p\nabla\tilde{n}_p dA \tag{6.32}$$

and the justification is given by Quintard and Whitaker (1995).

Before moving on to the closure problem, we note that the volume-averaged form of Equation 6.11 is given by (Whitaker, 1986b)

$$\langle\mathbf{v}\rangle = -\frac{\mathbf{K}}{\mu}\cdot\left(\nabla\langle p\rangle^\gamma - \rho\mathbf{g}\right), \quad \text{Darcy's law} \tag{6.33}$$

in which $\langle\mathbf{v}\rangle$ represents the *superficial* volume-averaged velocity and $\langle p\rangle^\gamma$ represents the *intrinsic* volume-averaged pressure. The volume-averaged continuity equation can be expressed either in terms of the superficial average velocity

$$\nabla\cdot\langle\mathbf{v}\rangle = 0 \tag{6.34}$$

or in terms of the intrinsic average velocity

$$\nabla\cdot(\varepsilon_\gamma\langle\mathbf{v}\rangle^\gamma) = 0 \tag{6.35}$$

Here we have made use of the nomenclature given by Equations 6.14 and 6.15 and the relation between the two averages indicated by Equation 6.16.

In order to obtain a closed form of Equation 6.31, we need to develop the boundary value problem for $\tilde{n}_p$. In a complete analysis of the filtration process we need to take porosity variations into account via the method of large-scale averaging (Quintard and Whitaker, 1987); however, in the development of the closure problem it is permissible to consider a local homogeneous region in which variations of ε_γ can be ignored. Under these circumstances we can divide Equation 6.31 by ε_γ to obtain

$$\begin{aligned}&\frac{\partial\langle n_p\rangle^\gamma}{\partial t} + \nabla\cdot\left(\langle\mathbf{v}\rangle^\gamma\langle n_p\rangle^\gamma - \gamma^{-1}\langle\mathbf{v}\cdot\nabla\mathbf{v}\rangle^\gamma\langle n_p\rangle^\gamma\right) + \nabla\cdot\left(\langle\tilde{\mathbf{v}}\tilde{n}_p\rangle^\gamma - \gamma^{-1}\langle(\mathbf{v}\cdot\nabla\mathbf{v})\tilde{n}_p\rangle^\gamma\right)\\ &= \nabla\cdot\left\{D_p\left[\nabla\langle n_p\rangle^\gamma + \frac{1}{V_\gamma}\int_{A_{\gamma\sigma}}\mathbf{n}_{\gamma\sigma}\tilde{n}_p dA\right]\right\} + \frac{\varepsilon_\gamma^{-1}}{\mathcal{V}}\int_{A_{\gamma\sigma}}\mathbf{n}_{\gamma\sigma}\cdot D_p\nabla\tilde{n}_p dA\end{aligned} \tag{6.36}$$

With this *intrinsic* form of the particle transport equation we are ready to begin the derivation of the closure problem.

6.5 CLOSURE PROBLEM

In order to develop the governing differential equation for $\tilde{n}_p$ we recall Equation 6.8

$$\frac{\partial n_p}{\partial t} + \nabla\cdot\left\{\left[\mathbf{v} - \gamma^{-1}\mathbf{v}\cdot\nabla\mathbf{v}\right]n_p\right\} = \nabla\cdot\left(D_p\nabla n_p\right) \tag{6.37}$$

and remember Equation 6.24 so that Equation 6.36 can be subtracted from Equation 6.37 to obtain

$$\frac{\partial \tilde{n}_p}{\partial t} + \nabla \cdot \left\{ \left[\mathbf{v} - \gamma^{-1}\mathbf{v}\cdot\nabla\mathbf{v} \right] n_p - \left[\langle \mathbf{v} \rangle^\gamma - \gamma^{-1}\langle \mathbf{v}\cdot\nabla\mathbf{v} \rangle^\gamma \right] \langle n_p \rangle^\gamma \right\}$$
$$- \nabla \cdot \langle \tilde{\mathbf{v}}\tilde{n}_p \rangle^\gamma + \gamma^{-1}\nabla \cdot \langle (\mathbf{v}\cdot\nabla\mathbf{v})\tilde{n}_p \rangle^\gamma = \nabla \cdot \left(D_p \nabla \tilde{n}_p \right) \tag{6.38}$$
$$- \nabla \cdot \left[\frac{D_p}{V_\gamma} \int_{A_{\gamma\sigma}} \mathbf{n}_{\gamma\sigma} \tilde{n}_p \, dA \right] - \frac{\varepsilon_\gamma^{-1}}{\mathcal{V}} \int_{A_{\gamma\sigma}} \mathbf{n}_{\gamma\sigma} \cdot D_p \nabla \tilde{n}_p \, dA$$

The second and third terms in this result can be arranged as

$$\left(\mathbf{v} - \gamma^{-1}\mathbf{v}\cdot\nabla\mathbf{v} \right) n_p - \left(\langle \mathbf{v} \rangle^\gamma - \gamma^{-1}\langle \mathbf{v}\cdot\nabla\mathbf{v} \rangle^\gamma \right) \langle n_p \rangle^\gamma$$
$$= \left(\mathbf{v} - \gamma^{-1}\mathbf{v}\cdot\nabla\mathbf{v} \right) \tilde{n}_p - \left[\left(\langle \mathbf{v} \rangle^\gamma - \mathbf{v} \right) - \left(\langle \mathbf{v}\cdot\nabla\mathbf{v} \rangle^\gamma - \mathbf{v}\cdot\nabla\mathbf{v} \right) \right] \langle n_p \rangle^\gamma \tag{6.39}$$

If we neglect variations of the porosity in the closure problem, we can use the various forms of the continuity equation to obtain

$$\nabla \cdot \left(\langle \mathbf{v} \rangle^\gamma - \mathbf{v} \right) = 0 \tag{6.40}$$

and this allows us to substitute Equation 6.39 into Equation 6.38 and obtain the following transport equation for $\tilde{n}_p$

$$\frac{\partial \tilde{n}_p}{\partial t} + \nabla \cdot \left[\left(\mathbf{v} - \gamma^{-1}\mathbf{v}\cdot\nabla\mathbf{v} \right) \tilde{n}_p \right] + \underbrace{\left[\tilde{\mathbf{v}} - \gamma^{-1}\left(\mathbf{v}\cdot\nabla\mathbf{v} - \langle \mathbf{v}\cdot\nabla\mathbf{v} \rangle^\gamma \right) \right] \cdot \nabla \langle n_p \rangle^\gamma}_{\text{source}}$$
$$\underbrace{- \left\{ \nabla \cdot \left[\gamma^{-1}\left(\mathbf{v}\cdot\nabla\mathbf{v} - \langle \mathbf{v}\cdot\nabla\mathbf{v} \rangle^\gamma \right) \right] \right\} \langle n_p \rangle^\gamma}_{\text{source}} \underbrace{- \nabla \cdot \langle \tilde{\mathbf{v}}\tilde{n}_p \rangle^\gamma + \gamma^{-1}\nabla \cdot \langle (\mathbf{v}\cdot\nabla\mathbf{v})\tilde{n}_p \rangle^\gamma}_{\text{nonlocal convective transport}} \tag{6.41}$$
$$= \nabla \cdot \left(D_p \nabla \tilde{n}_p \right) \underbrace{- \nabla \cdot \left[\frac{D_p}{V_\gamma} \int_{A_{\gamma\sigma}} \mathbf{n}_{\gamma\sigma} \tilde{n}_p \, dA \right]}_{\text{nonlocal diffusive transport}} \underbrace{- \frac{\varepsilon_\gamma^{-1}}{\mathcal{V}} \int_{A_{\gamma\sigma}} \mathbf{n}_{\gamma\sigma} \cdot D_p \nabla \tilde{n}_p \, dA}_{\text{particle capture}}$$

Here we see terms representing

1. The classic effects of accumulation, local convection, and local diffusion;
2. Nonlocal convection and nonlocal diffusion;
3. Sources proportional to $\nabla \langle n_p \rangle^\gamma$ and $\langle n_p \rangle^\gamma$;
4. Particle capture.

We use the word *nonlocal* to describe those terms which involve integrals of $\tilde{n}_p$, and one can draw upon previous studies (Carbonell and Whitaker, 1984; Whitaker, 1986a; Quintard and Whitaker, 1993; Quintard and Whitaker, 1994a) to argue that these terms are negligible. The analysis consists of comparing the nonlocal terms with the associated local terms and demonstrating that the former are smaller than the latter by a factor of ℓ_γ / L where L represents the characteristic length associated with $\langle n_p \rangle^\gamma$. This occurs because the nonlocal terms involved the derivatives of average quantities while the local terms always contain the derivatives of point quantities. Because ℓ_γ/L is always small compared to one, the nonlocal terms can be neglected and Equation 6.41 simplifies to

$$\frac{\partial \tilde{n}_p}{\partial t} + \nabla \cdot \left[\left(\mathbf{v} - \gamma^{-1}\mathbf{v} \cdot \nabla \mathbf{v}\right)\tilde{n}_p\right] + \underbrace{\left[\tilde{\mathbf{v}} - \gamma^{-1}\left(\mathbf{v} \cdot \nabla \mathbf{v} - \langle \mathbf{v} \cdot \nabla \mathbf{v}\rangle\right)\right] \cdot \nabla \langle n_p \rangle^\gamma}_{\text{source}}$$
$$\underbrace{-\left\{\nabla \cdot \left[\gamma^{-1}\left(\mathbf{v} \cdot \nabla \mathbf{v} - \langle \mathbf{v} \cdot \nabla \mathbf{v}\rangle^\gamma\right)\right]\right\}\langle n_p \rangle^\gamma}_{\text{source}} = \nabla \cdot \left(D_p \nabla \tilde{n}_p\right) - \frac{\varepsilon_\gamma^{-1}}{\mathcal{V}} \int_{A_{\gamma\sigma}} \mathbf{n}_{\gamma\sigma} \cdot D_p \nabla \tilde{n}_p dA \tag{6.42}$$

The last term in this result is also a nonlocal term; however, it is not negligible since it represents the rate at which particles are captured per unit volume.

Use of the boundary condition given by Equation 6.10 and the decomposition represented by Equation 6.24 leads to the following boundary condition:

$$\text{B.C.1} \quad \tilde{n}_p = \underbrace{-\langle n_p \rangle^\gamma}_{\text{source}}, \quad \text{at the } \gamma - \sigma \text{ interface} \tag{6.43}$$

and in order to determine the $\tilde{n}_p$-field in some local, representative region, we are forced to accept the spatially periodic model of a porous medium and impose the following periodicity condition.

$$\text{Periodicity: } \tilde{n}_p(\mathbf{r} + \ell_i) = \tilde{n}_p(\mathbf{r}), \; i = 1,2,3 \tag{6.44}$$

In addition, when the length-scale constraints indicated by Equations 6.28 and 6.29 are valid, Carbonell and Whitaker (1984) have shown that the average of the spatial deviation can be set equal to zero, and we express this idea as

$$\langle \tilde{n}_p \rangle^\gamma = 0 \tag{6.45}$$

Strictly speaking, we need an initial condition for $\tilde{n}_p$ to complete our problem statement; however, both the volume-averaged equation given by Equation 6.36 and the closure equation represented by Equation 6.42 can be treated as quasi steady; thus the initial condition for both $\langle n_p \rangle^\gamma$ and $\tilde{n}_p$ can be ignored.

6.6 QUASI-STEADY CLOSURE PROBLEM

The quasi-steady closure problem is given by

$$\nabla \cdot \left[\left(\mathbf{v} - \gamma^{-1}\mathbf{v} \cdot \nabla \mathbf{v}\right)\tilde{n}_p\right] + \underbrace{\left[\tilde{\mathbf{v}} - \gamma^{-1}\left(\mathbf{v} \cdot \nabla \mathbf{v} - \langle \mathbf{v} \cdot \nabla \mathbf{v}\rangle\right)\right] \cdot \nabla \langle n_p \rangle^\gamma}_{\text{source}}$$
$$\underbrace{-\left\{\nabla \cdot \left[\gamma^{-1}\left(\mathbf{v} \cdot \nabla \mathbf{v} - \langle \mathbf{v} \cdot \nabla \mathbf{v}\rangle^\gamma\right)\right]\right\}\langle n_p \rangle^\gamma}_{\text{source}} = \nabla \cdot \left(D_p \nabla \tilde{n}_p\right) - \frac{\varepsilon_\gamma^{-1}}{\mathcal{V}} \int_{A_{\gamma\sigma}} \mathbf{n}_{\gamma\sigma} \cdot D_p \nabla \tilde{n}_p dA \tag{6.46a}$$

$$\text{B.C.1} \quad \tilde{n}_p = \underbrace{-\langle n_p \rangle^\gamma}_{\text{source}}, \quad \text{at the } \gamma - \sigma \text{ interface} \tag{6.46b}$$

$$\text{Periodicity: } \tilde{n}_p(\mathbf{r} + \ell_i) = \tilde{n}_p(\mathbf{r}),\ i = 1,2,3 \tag{6.46c}$$

$$\text{Average: } \langle \tilde{n}_p \rangle^\gamma = 0 \tag{6.46d}$$

The form of this boundary value problem suggests a representation for $\tilde{n}_p$ given by

$$\tilde{\boldsymbol{n}}_p = \mathbf{b} \cdot \nabla \langle n_p \rangle^\gamma - s\langle n_p \rangle^\gamma \tag{6.47}$$

in which **b** and s are referred to as the closure variables or the *mapping variables* since they map the sources onto the spatial deviation concentration. One can draw upon a series of studies associated with the closure problem (see, for example, Quintard and Whitaker, 1993) to conclude that the vector **b** and the scalar s are determined by two boundary value problems that are analogous to the problem given by Equations 6.46. These problems are given by Quintard and Whitaker (1995), and here we will make use of their results in order to express the closed form of Equation 6.31 as

$$\begin{aligned} &\varepsilon_\gamma \frac{\partial \langle n_p \rangle^\gamma}{\partial t} + \nabla \cdot \left[\left(\langle \mathbf{v} \rangle - \gamma^{-1} \langle \mathbf{v} \cdot \nabla \mathbf{v} \rangle \right) \langle n_p \rangle^\gamma \right] - (\mathbf{d} + \mathbf{u}) \cdot \nabla \langle n_p \rangle^\gamma \\ &= \nabla \cdot \left(\mathbf{D}^* \cdot \nabla \langle n_p \rangle^\gamma \right) - k_{eff} \langle n_p \rangle^\gamma \end{aligned} \tag{6.48}$$

It is important to recognize that this is a *superficial average* transport equation. This means that each term represents a certain quantity *per unit volume of the porous medium* and not per unit volume of the fluid phase. This is obvious for the accumulation term; however, it can be confusing for the particle capture term. The various coefficients that appear in Equation 6.48 are defined as

$$\mathbf{u} = \frac{1}{\mathcal{V}} \int_{A_{\gamma\sigma}} \mathbf{n}_{\gamma\sigma} \cdot D_p \nabla \mathbf{b} dA \tag{6.49}$$

$$\mathbf{d} = -D_p \left[\frac{1}{\mathcal{V}} \int_{A_{\gamma\sigma}} \mathbf{n}_{\gamma\sigma} s dA \right] + \left[\langle \tilde{\mathbf{v}} s \rangle - \gamma^{-1} \langle (\mathbf{v} \cdot \nabla \mathbf{v}) s \rangle \right] \tag{6.50}$$

$$\mathbf{D}^* = \varepsilon_\gamma D_p \left(\mathbf{I} + \frac{1}{\mathcal{V}} \int_{A_{\gamma\sigma}} \mathbf{n}_{\gamma\sigma} \mathbf{b} dA \right) - \langle \tilde{\mathbf{v}} \mathbf{b} \rangle - \gamma^{-1} \langle (\mathbf{v} \cdot \nabla \mathbf{v}) \mathbf{b} \rangle \tag{6.51}$$

$$k_{eff} = \frac{1}{\mathcal{V}} \int_{A_{\gamma\sigma}} \mathbf{n}_{\gamma\sigma} \cdot D_p \nabla s dA \tag{6.52}$$

In order to develop a complete understanding of the filtration process, one needs to make use of these effective coefficients in Equation 6.48 to determine $\langle n_p \rangle^\gamma$ as a function of the system parameters. The interested reader can find a discussion of **d**, **u**, and D* in the work of Quintard and Whitaker (1995); however, the capture coefficient k_{eff}, dominates the filtration process and in the next section we use theoretical values of k_{eff} to determine filter efficiencies that are compared with experiment.

6.7 DETERMINATION OF THE CAPTURE COEFFICIENT

The simplest model of a fibrous porous medium is a regular array of cylinders such as that shown in Figure 6.3. For most practical cases, the Reynolds number for flow in fibrous filters is

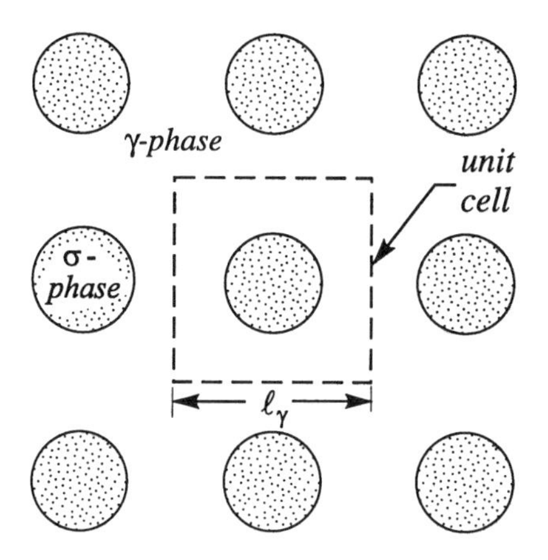

Figure 6.3 Spatially periodic array of cylinders.

less than one; thus we can use Stokes' equations so determine the fluid velocity field. For the case of a *macroscopically uniform flow* one can argue that

$$\langle \mathrm{v} \cdot \nabla \mathbf{v} \rangle = 0 \tag{6.53}$$

and Equation 6.48 simplifies to

$$\varepsilon_\gamma \frac{\partial \langle n_p \rangle^\gamma}{\partial t} + \nabla \cdot \left(\langle \mathbf{v}_p \rangle \langle n_p \rangle^\gamma \right) - (\mathbf{d} + \mathbf{u}) \cdot \nabla \langle n_p \rangle^\gamma = \nabla \cdot \left(\mathbf{D}^* \cdot \nabla \langle n_p \rangle^\gamma \right) - k_{eff} \langle n_p \rangle^\gamma \tag{6.54}$$

Calculated values of **d** and **u** are given by Quintard and Whitaker (1995), and they show that these terms may contribute as much as 10% to the convective transport in Equation 6.54. While this is not negligible, it is not a key issue in terms of the comparison of the theory with experiments. Thus, we will discard these terms, along with the dispersive transport so that our superficial volume-averaged transport equation takes the form:

$$\varepsilon_\gamma \frac{\partial \langle n_p \rangle^\gamma}{\partial t} + \nabla \cdot \left(\langle \mathbf{v} \rangle \langle n_p \rangle^\gamma \right) = -k_{eff} \langle n_p \rangle^\gamma \tag{6.55}$$

This result will be quasi steady when the following constraint is satisfied

$$k_{eff}\, t^* >> 1 \tag{6.56}$$

Here t^* is a characteristic process time that is generally taken to be infinite for filtration processes. Under these circumstances one can use the continuity equation given by Equation 6.34 in order the express Equation 6.55 as.

$$\langle \mathbf{v} \rangle \cdot \nabla \cdot \langle n_p \rangle^\gamma = -k_{eff} \langle n_p \rangle^\gamma \tag{6.57}$$

One should think of this result as being a reasonable approximation for homogeneous filters; however, real filters are heterogeneous and the terms that have been discarded in going from Equation 6.54 to Equation 6.57 will be retained in future studies of heterogeneous porous media. In order to determine k_{eff} for the system shown in Figure 6.3, one must first solve the Stokes equations for a spatially periodic system and then solve the closure problem (Quintard and Whitaker, 1995) to determine the capture coefficient as defined by Equation 52.6.

6.8 CELLULAR EFFICIENCY

In order to present our results for k_{eff} in a traditional form, we will write Equation 6.57 as

$$\langle v_x \rangle \frac{d\langle n_p \rangle^\gamma}{dx} = -k_{eff} \langle n_p \rangle^\gamma \tag{6.58}$$

and note that for our unit cell calculations, or any homogeneous porous filter, we have

$$\mathbf{i} \cdot \langle \mathbf{v}_p \rangle = \mathbf{i} \cdot \langle \mathbf{v} \rangle = \langle v_x \rangle \tag{6.59}$$

We can solve Equation 6.58 in order to represent the change in particle concentration that takes place across a unit cell as

$$\frac{\langle n_p \rangle^\gamma \big|_{x=0} - \langle n_p \rangle^\gamma \big|_{x=\ell_\gamma}}{\langle n_p \rangle^\gamma \big|_{x=0}} = -1 - e^{\left(k_{eff}\ell_\gamma / \langle v_x \rangle\right)} \tag{6.60}$$

It is convenient to define the left-hand side of this result as the *cellular efficiency,* η_c, in order to distinguish it from the *single-fiber efficiency,* and this leads to

$$\eta_c = 1 - e^{-(k_{eff}\ell_\gamma / \langle v_x \rangle}, \text{ cellular efficiency} \tag{6.61}$$

To illustrate the general nature of the solutions for the cellular efficiency, η_c, calculations were carried out for the parameters listed in Table 1.6. The illustrative values for η_c are shown in Figure 6.4 as a function of the particle diameter. The Stokes number is defined by

Table 6.1 Physical Properties

Particle diameter	d_p = 0.5 and 0.1 μm
Fiber diameter	$2a$ = 0.5 μm
Particle density	ρ_p = 4.0 g/cm^3
Temperature	283.0 K
Viscosity	1.8 × 10^{-5} Pa s
Porosity	ε_γ = 0.95

$$\mathrm{St} = \frac{\langle v_x \rangle^\gamma \, \rho_p d_p^2 c_s}{18\mu \ell_\gamma} \tag{6.62}$$

and the curve for St = 0 represents the purely diffusive case that does not exhibit a minimum. This means that the Smoluchowski equation for the particle concentration cannot be used to determine

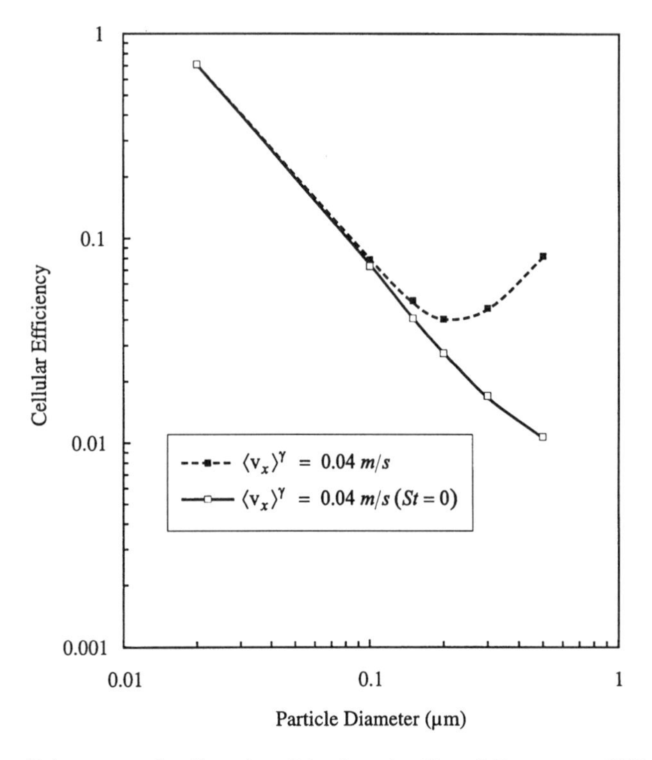

Figure 6.4 Cellular efficiency as a function of particle diameter ($2a$ = 0.5 μm, ρ_p = 4000 kg/m³, T = 283 K, μ = 1.8 · 10^{-5} Pa s, ε_γ = 0.05).

the most-penetrating particle size. The curve for finite Stokes numbers, which range from 10^{-4} to almost 10^{-1},indicates a minimum cellular efficiency for particle diameters on the order of 1 to 2 μm. In thinking about the results shown in Figure 6.4, one must be careful to remember that the additional convective transport represented by the coefficients **u** and **d** has been neglected along with the dispersive transport. Inclusion of these effects would change the values presented in Figure 6.4, but not the conclusion that the *corrected* Smoluchowski equation exhibits a minimum in the cellular efficiency as a function of particle diameter.

6.9 COMPARISON WITH EXPERIMENT

If a filter can be thought of as a series of unit cells, the cellular efficiency determined on a theoretical basis can be determined with the filter efficiency determined on an experimental basis. This approach neglects the heterogeneities illustrated in Figure 6.1 and the inclusion of the effects of those heterogeneities will be the subject of future studies. In Figure 6.5 we have shown a comparison with the work of Lee and Liu (1982) for an average velocity given by $\langle v \rangle$ = 0.1 m/s, and there one can see attractive agreement between theory and experiment for particle diameters that are equal to or smaller than the diameter of the most-penetrating particle. For larger particles, the agreement diminishes and this would appear to confirm our suspicions concerning the importance of the higher-order terms in Equation 6.7. In Figure 6.6 the comparison between theory and experiment in shown for $\langle v \rangle$ = 0.03 m/s and we again see reasonable agreement for the smaller particles. The comparison for an even smaller velocity given by $\langle v \rangle$ = 0.01 m/s is shown in Figure 6.7 and here it becomes apparent that the theory is less reliable at lower velocities. We have no explanation for this observation; however, one must remember that the model illustrated in Figure 6.3 cannot possibly capture all the characteristics of a real filter and further studies using more

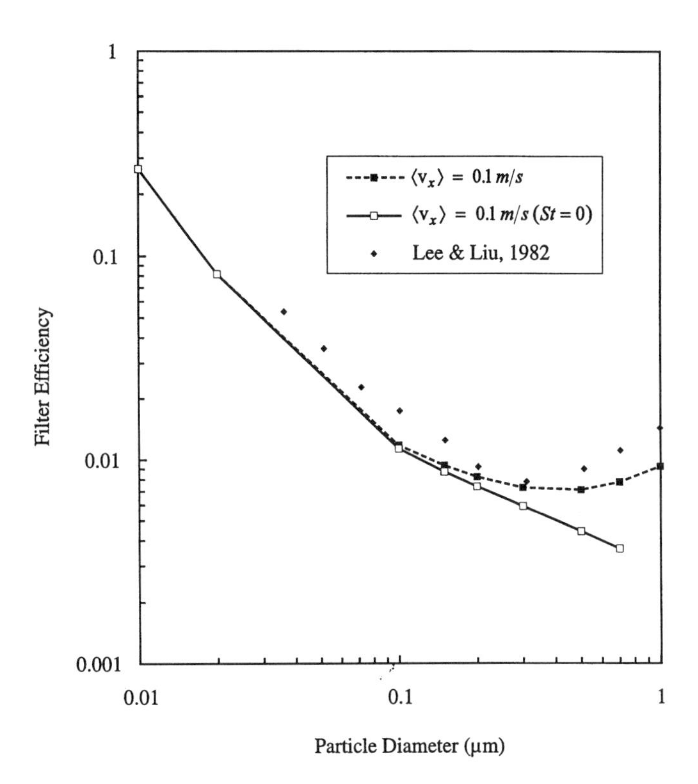

Figure 6.5 Comparison with laboratory experiments ($2a = 11$ μm, $\rho_p = 1000$ kg/m³, $T = 278$ K, $\mu = 1.83 \cdot 10^{-5}$ Pa s, $\varepsilon_\gamma = 0.05$).

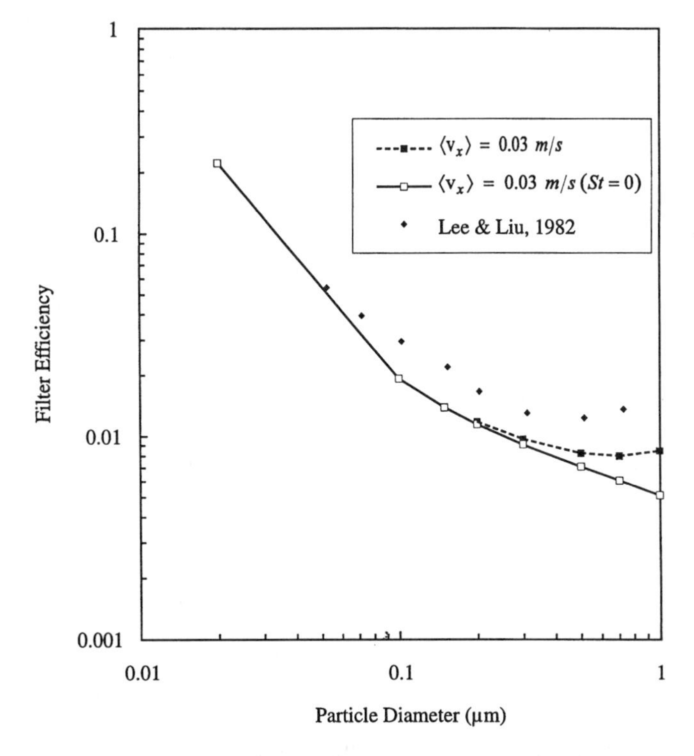

Figure 6.6 Comparison with laboratory experiments ($2a = 11$ μm, $\rho_p = 1000$ kg/m³, $T = 278$ K, $\mu = 1.83 \cdot 10^{-5}$ Pa s, $\varepsilon_\gamma = 0.849$).

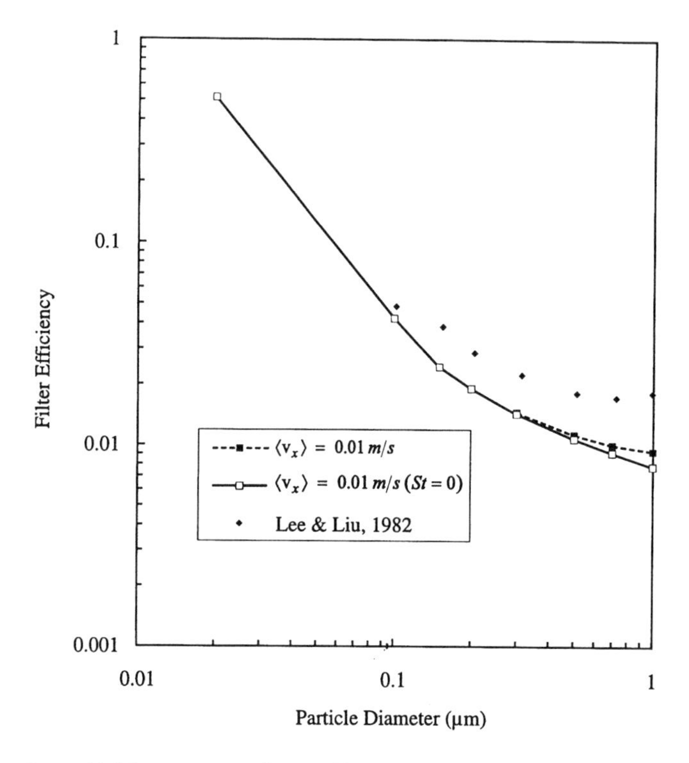

Figure 6.7 Comparison with laboratory experiments ($2a$ = 11 μm, ρ_p = 1000 kg/m³, T = 278 K, $\mu = 1.83 \cdot 10^{-5}$ Pa s, ε_γ = 0.849).

complex unit cells are certainly in order. In addition, the influence of local heterogeneities must be determined and that will be the objective of a subsequent study.

6.10 CONCLUSIONS

In this work we have used the first correction to the Smoluchowski equation to describe the effects of particle inertia, and the resulting particle transport equation has been used to develop a local volume average transport equation that includes the effects of nontraditional convective transport, dispersion, and particle capture. A spatially periodic model of a fibrous filter has been used, along with two closure problems, to calculated the effective coefficients that appear in the volume-averaged transport equation. This leads to a direct calculation of the cellular efficiency, and results were determined for a unit cell containing a single fiber that is orthogonal to the mean flow field. The results are in reasonably good agreement with experiments for the most-penetrating particle size; however, higher-order corrections need to be included in the Smoluchowski equation in order to predict the behavior of large particles.

NOMENCLATURE

Roman Letters

a	fiber radius, m
$A_{\gamma\sigma}$	area of the γ–σ interface contained within the averaging volume, m²
$\mathbf{b}$	a closure variable that maps $\nabla\langle n_p\rangle^\gamma$ onto $\tilde{n}_p$, m

c_s	slip coefficient
d_p	particle diameter, m
$\mathbf{d}$	a velocity-like coefficient, m/s
D_p	Brownian diffusivity, m²/s
D*	dispersion tensor, m²/s
$\mathbf{F}_r(t)$	Brownian or random force, N
$\mathbf{g}$	gravity vector, m/s²
K	Darcy's law permeability tensor, m²
k_{eff}	effective rate coefficient for particle capture, s^{-1}
ℓ_γ	characteristic length for a unit cell, m
ℓ_i	$i = 1,2$, and 3, lattice vectors, m
L	characteristic length for volume-averaged quantities, m
m_p	mass of a particle, kg
n_p	particle density, number/m³
$\tilde{n}_p$	spatial deviation of the particle of the particle, number/m³
$\langle n_p \rangle$	superficial average particle concentration, number/m³
$\langle n_p \rangle^\gamma$	intrinsic average particle concentration, number/m³
$\mathbf{n}_{\gamma\sigma}$	unit normal vector pointing from the γ-phase toward the σ-phase
p	fluid pressure, N/m²
$\langle p \rangle^\gamma$	intrinsic average pressure in the γ-phase, m³
$\mathbf{r}$	position vector, m
s	a closure variable that maps $\langle n_p \rangle^\gamma$ onto $\tilde{n}_p$
St	$\langle v_x \rangle^\gamma \rho_p d_p^2 c_s / 18 \mu \ell_\gamma$, the Stokes number
t	time, s
$\mathbf{u}$	a velocity-like coefficient, m/s
$\mathcal{V}$	averaging volume, m³
V_γ	volume of the γ-phase contained in the averaging volume, m³
$\mathbf{v}$	fluid velocity vector, m/s
$\langle \mathbf{v} \rangle^\gamma$	intrinsic average fluid velocity, m/s
$\langle \mathbf{v} \rangle$	$\varepsilon_\gamma \langle \mathbf{v} \rangle^\gamma$, superficial average fluid velocity, m/s
$\tilde{\mathbf{v}}$	$\mathbf{v} - \langle \mathbf{v} \rangle^\gamma$, spatial deviation fluid velocity, m/s

Greek Letters

μ	fluid viscosity, Ns/m²
π	3.1416.
ε_γ	porosity
γ	$3\pi\mu d_p / m_p c_s$, s^{-1}
ρ	fluid density, kg/m³
ρ_p	particle density, kg/m³

REFERENCES

Barrère, J., Gipouloux, O., and Whitaker, S. (1992). On the closure problem for Darcy's law, *Transp. Porous Media,* 76, 209–222.

Carbonell, R. G. and Whitaker, S. (1983). Dispersion in pulsed systems II: theoretical developments for passive dispersion in porous media, *Chem. Eng. Sci.,* 38, 1795–1802.

Carbonell, R. G. and Whitaker, S. (1984). Heat and mass transfer in porous media, in *Fundamentals of Transport Phenomena in Porous Media,* J. Bear and M. Y. Corapcioglu, Eds., Martinus Nijhoff Publishers, Dordrecht, The Netherlands, 123–198.

Cushman, J. H. (1990). *Dynamics of Fluids in Hierarchical Porous Media,* Academic Press, London.

Gardiner, C. W. (1990). *Handbook of Stochastic Methods,* Springer-Verlag, New York.

Howes, F. A. and Whitaker, S. (1985). The spatial averaging theorem revisited. *Chem. Eng. Sci.,* 40, 1387–1392.

Lee, K. W. and Liu, B. Y. H. (1982). Theoretical study of aerosol filtration by fibrous filters, *Aerosol Sci. Technol.,* 1, 147–161.

Peters, M. H. and Ying, R. (1991). Phase-space diffusion equations for single Brownian particle motion near surfaces, *Chem. Eng. Commun.,* 108, 165–185.

Plumb, O. A. and Whitaker, S. (1988a). Dispersion in heterogeneous porous media I: local volume averaging and large-scale averaging, *Water Resour. Res.,* 24, 913–926.

Plumb, O. A. and Whitaker, S. (1988b). Dispersion in heterogeneous porous media II: predictions for stratified and two-dimensional spatially periodic systems, *Water Resour. Res.,* 24, 927–938.

Plumb, O. A. and Whitaker, S. (1990a). Diffusion, adsorption, and dispersion in porous media: small-scale averaging and local volume averaging, in *Dynamics of Fluids in Hierarchical Porous Media,* J. H. Cushman, Ed., Academic Press, New York, chap. 5.

Plumb, O. A. and Whitaker, S. (1990b). Diffusion, adsorption, and dispersion in porous media: the method of large-scale averaging in *Dynamics of Fluids in Hierarchical Porous Media,* J. H. Cusman, Ed., Academic Press, New York, chap. 6.

Quintard, M. and Whitaker, S. (1987). Écoulement monophasique en milieu poreux: effet des hétérogénéités locales, *J. Méc. Théor. Appl.,* 6, 691–726.

Quintard, M. and Whitaker, S. (1988). Two-phase flow in heterogeneous porous media: the method of large-scale averaging, *Transp. Porous Media,* 3, 357–413.

Quintard, M. and Whitaker, S. (1990a). Two-phase flow in heterogeneous porous media I: the influence of large spatial and temporal gradients, *Transp. Porous Media,* 5, 341–379.

Quintard, M. and Whitaker, S. (1990b). Two-phase flow in heterogeneous porous media II: numerical experiments for flow perpendicular to a stratified system, *Transp. Porous Media,* 5, 429–472.

Quintard, M. and Whitaker, S. (1993). One and two-equation models for transient diffusion processes in two-phase systems, in *Advances in Heat Transfer,* Vol. 23, Academic Press, New York, 369–465.

Quintard, M. and Whitaker, S. (1994a). Convection dispersion, and interfacial transport of contaminants: homogeneous porous media, *Adv. Water Resour.,* 17, 221–239.

Quintard, M. and Whitaker, S. (1994b). Transport in ordered and disordered porous media I: the cellular average and the use of weighting functions, *Transp. Porous Media,* 14, 163–177.

Quintard, M. and Whitaker, S. (1994c). Transport in ordered and disordered porous media II: generalized volume averaging, *Transp. Porous Media,* 14, 179–206.

Quintard, M. and Whitaker, S. (1994d). Transport in ordered and disordered porous media III: closure and comparison between theory and experiment, *Transp. Porous Media,* 15, 31–49.

Quintard, M. and Whitaker, S. (1994e). Transport in ordered and disordered porous media IV: computer generated porous media, *Transp. Porous Media,* 15, 51–70.

Quintard, M. and Whitaker, S. (1994f). Transport in ordered and disordered porous media V: geometrical results for two-dimensional systems, *Transp. Porous Media,* 15, 183–196.

Quintard, M. and Whitaker, S. (1995). Aerosol filtration: an analysis using the method of volume averaging, *J. Aerosol Sci.,* 26, 1227–1255.

Risken, H. (1989), *The Fokker-Planck Equation: Methods of Solution and Applications,* Springer-Verlag, New York.

Rubinstein, J. and Mauri, R. (1986). Dispersion and convection in periodic porous media, *SIAM J. Appl. Math.,* 1018–1023.

Sanchez-Palencia, E. (1980). Non-homogeneous media and vibration theory, *Lecture Notes Phys.,* 127, Springer-Verlag, New York.

Tien, C. (1989). *Granular Filtration of Aerosols and Hydrosols,* Butterworth, Boston.

Whitaker, S. (1986a). Transport processes with heterogeneous reaction, in *Concepts and Design of Chemical Reactors*, S. Whitaker and A. E. Cassano, Eds., Gordon and Breach Publishers, New York, 1–94.

Whitaker, S. (1986b). Flow in porous media I: a theoretical derivation of Darcy's law, *Transp. Porous Media,* 1, 3–25.

CHAPTER 7

The Efficiency of Clarification and Resource of Fiber Filter Predicted by Modeling Its Structure with a Packet of Plane Nets

Vladimir J. Alperovich, J. G. Alperovich, E. D. Chunin, V. V. Blagoveshchenskiy, and V. V. Kashmet

CONTENTS

Abstract — A model of the structure of low-density fiber filters prepared from thin monodisperse fibers is suggested. The model correlates very well with experimental data on filter porosity; the efficiency of clarification of fiber filter is calculated depending on its structure. The data obtained were compared with those of calculations made using other models and with experimental data. The resource of fiber filter is estimated and the optimal, in terms of resource and efficiency of clarification, configuration of a column of fiber filter with different apparent density is determined.

0-87371-830-5/98/$0.00+$.50

7.1 INTRODUCTION

Wide application of fiber filters (ff) in modern technologies depends on two factors: (1) the efficiency of clarification value $j = 1 - C/C_0$ (C and C_0 are concentration of particles after and before application of ff) being the same, the efficiency and resource of ff significantly exceed those of filters with a different structure, e.g., nuclear filter and polymer membranes (screen principle of action) and deep-bed ceramic filters; (2) thin and rather cheap mineral fibers with diameter $d < 1$ mkm making it possible to clarify aerosol and hydrosol (designated below as *as* and *hs*, respectively) of submicron particles. As a result, ff came to be unavoidable in many clarification processes.

There is a method worked out by Langmuir for theoretical calculation of j for a given configuration of and parameters of filtration.[6] Having been further modified by other authors,[7-11] it has now as main points the following: (1) calculation of the capture efficiency of a single fiber to remove particles from the flow; (2) for a real filter taking into account the influence on j of fiber located close by. As far as submicron particles are concerned, this approach puts emphasis on two main mechanisms of particle capture comparable in importance. These are contact and adhesion to the fiber caused, first, by particle deviation due to diffusion and, second, by a decrease of distance between the flow trajectory and the fiber to less than $2r$ (r being the particle radius). No parameters of ff structure are taken into consideration despite the fact that ff geometry is of prime importance in the process of filtration. This explains why several competing methods are now in use based on the principles of j calculation[7-11] giving the results which differ up to tenfold. Thus, the calculation values typical of clarification of aerosol with $2r = 1$ mkm, using ff made of fibers with $d = 10$ mkm, give values of j ranging from 0.1 to 0.9.[12] This is why j is determined, as a rule, using empirical equations, e.g., Reference 13. However, the advent of ff prepared from fibers with $d < 1$ mkm imposes significant restrictions on the applicability of empirical equations. All this speaks in favor of a model in which structural parameters of ff would be chosen related to the main parameters of filtration. Given below is a model showing a relationship between structural characteristics of ff and the efficiency of clarification and resource. Because of their complexity some aspects of the problem remain beyond the scope of this chapter: (1) modeling of aerodynamics (hydrodynamics) of ff: (2) modeling of ff prepared from polydisperse fibers. (3) filtration of polydisperse *as* (*hs*), and (4) rather dense ff with the ratio $\rho_f/\rho_a > 5$, where ρ_a is an apparent density of ff and ρ_f is a density of the fiber (this will not significantly restrict the use of the above model since in the majority of cases ratio $\rho_f/\rho_a < 5$ holds).

7.2 MODEL OF PLANE SQUARE NETS AS AVERAGED DESCRIPTION OF FF REAL STRUCTURE

It is typical of a real structure of ff that the angles formed by intersection of a precipitated fiber and the X and Y *axes*, as well as coordinates of its gravity center (X*c*, Y*c*), Figure 7.1a, are random values: $0° < a < 90°$, (X*c*, Y*c*) being located with equal probability at any point within the filter contours (chaotic arrangement). A further analysis will require several assumptions: (1) fibers have the length L (m), diameter (mkm), and density ρ_f, all expressed as mean values; (2) filter fibers lie parallel to the plane *XOY*, Figure 1a (under the condition of plane); (3) The fibers in layers are parallel and at the same distance away from each other.

While assumption (2), as shown by Corte and Callmes,[14] rests on realistic grounds, assumption (3) leads to formalization of a real picture and its validity needs confirmation. The equal distance between fibers, provided they are parallel, is dependent on the condition that the density of ff is constant throughout the whole area, i.e., ρ_a = const. (X, Y, Z). The uniformity of density is inversely proportional to the diameter of the fibers. Experimental data supports this (see Figure 2).[15] The explanation for this follows. In a real ff the constancy of density is solely due to equal spacing in the plane *XOY* of fiber gravity centers. In the case of irregular arrangement of fibers, a local increase

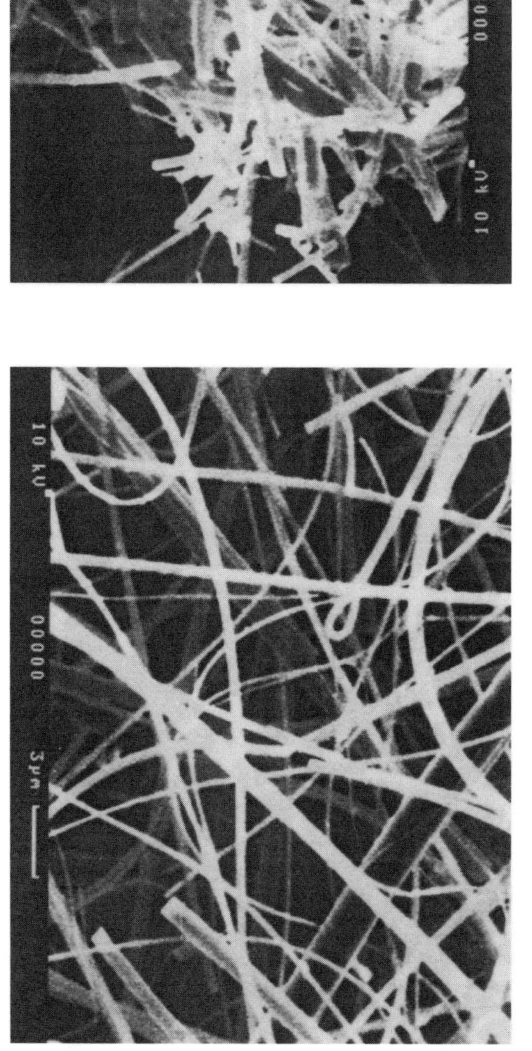

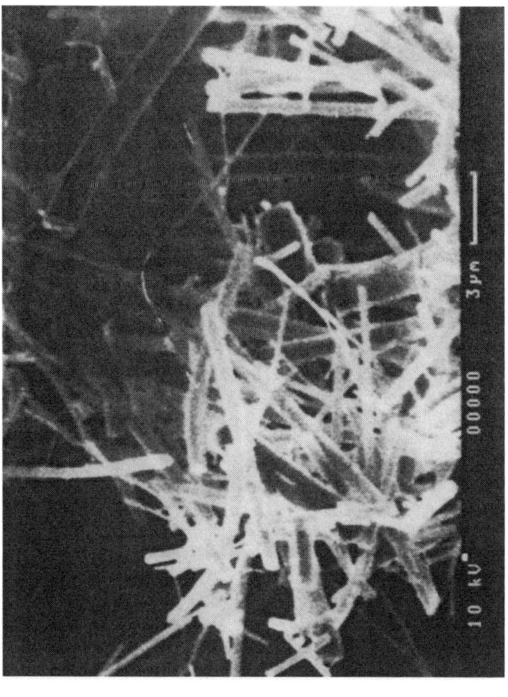

Figure 7.1 Microphoto of ff, d = 0.35 mkm, ρ_a = 0.08 g/cm^3. (a) View of ff in the plane perpendicular to the direction of ff growth. Chaotic arrangement of filter fibers is due to a random angle formed by a fiber and to random location of the center of any fiber mass within ff contour. (b) View of ff in the plane parallel to the direction of ff growth. Part of the fiber length is parallel to the axis of ff growth.

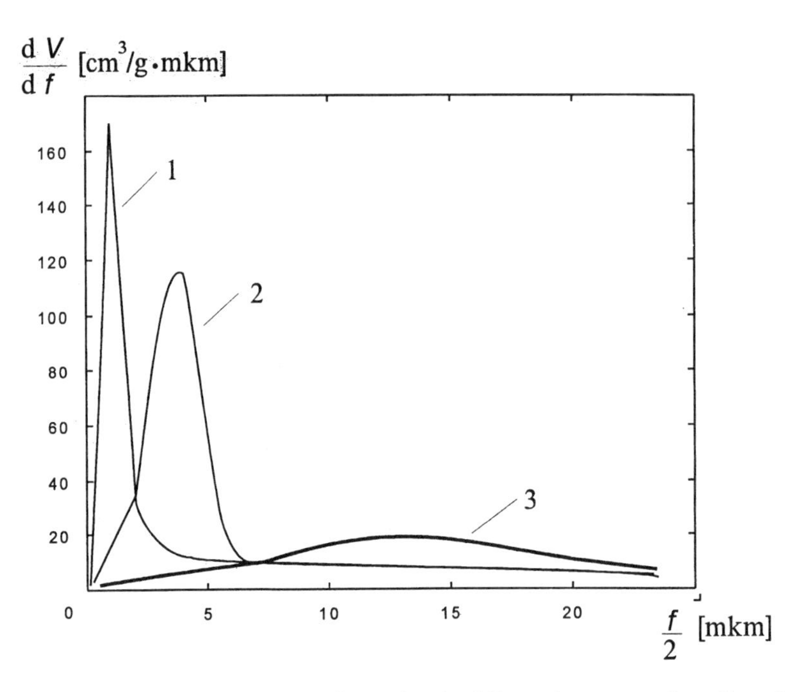

Figure 7.2 Distribution of ff pore volume according to its size.[5] Curve 1 corresponds to fiber d = 0.35 mkm, curve 2 to fiber d = 0.65 mkm, curve 3 to fiber d = 20 mkm.

of ρ_a within the filter area occurs when part of the fiber length reaches the upper (lower) layer. This is equivalent to a local increase of the z-component, see Figure 7.1b (z-component is a ratio of a total vertical projection of all fibers to their total length). A decrease of ρ_a within the filter area is equivalent to its increase in the part of the area located close by. At the initial stage of preparation of ff the orientation of fibers with respect to each other is arbitrary and the z-component mainly depends on the forces of reciprocal adhesion of fibers. Adhesion prevents gravitational force F_{gr} from arranging fibers horizontally.

$$F_{\text{gr}} \propto M \cos(XOZ), \tag{7.1}$$

where M is weight of fiber, (XOZ) is the angle made by the fiber and plane XOY. Frictional force F_{fr} is capable of fixing the fiber at an angle to the plane XOY, in which case it is proportional to d^2:

$$F_{\text{fr}} \propto M \cos(XOZ) \mathrm{A_c}, \tag{7.2}$$

where $\mathrm{A_c} \propto d^2$ is the area of fiber contact. A comparison of 7.1 and 7.2 leads to a suggestion that at the initial stage of ff preparation a mechanism responsible for a decrease of the z-component proportionally to d^2 comes into action.

7.2.1 Determination of the Free Size of ff Cell Designated as *f*

See Figures 7.3a and 7.4. Let us consider a filter $2d$ thick coinciding in form with an elementary cell (Figure 7.4). V_v stands for the volume of the ff free space (void) and V_f and M_f for its overall volume and weight. The volume of glue agent is negligible, as compared with the overall volume of V_f; hence,

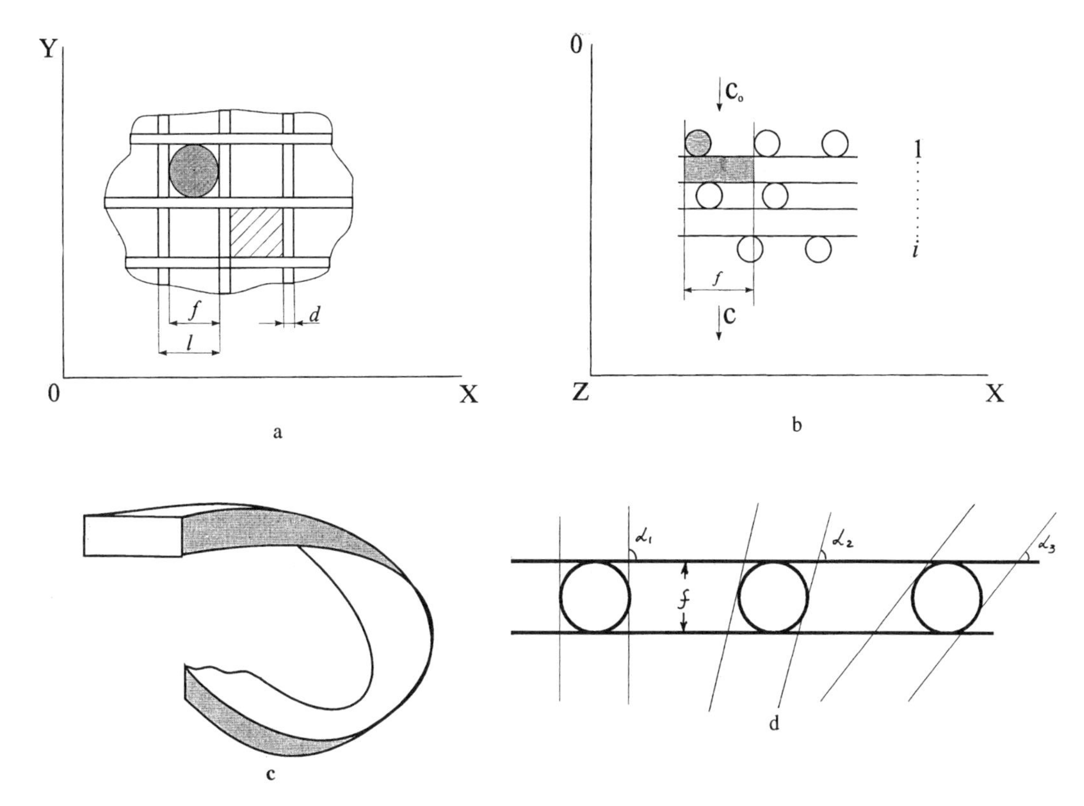

Figure 7.3 (a) Model of ff structure built using a square net. The hashed is an elementary cell; f = cell-free size corresponding to a maximum diameter of a circle within cell limits; d = fiber thickness; l = overall size of cell. (b) Cell column with j equal to the value of a whole ff. (x) Spiral-like form of cellulose fiber. (d) Independence of f value on the angle of intersection of two adjacent layers.

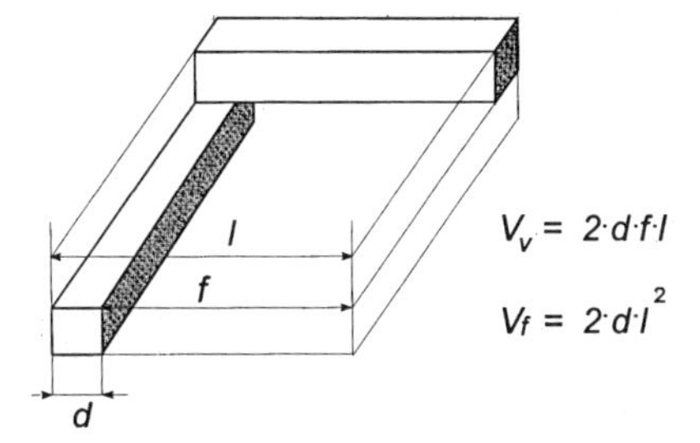

Figure 7.4 View of elementary cell to be used in calculation of parameter f.

$$M_f = \rho_f (V_f - V_v) = V_f \rho_a, \tag{7.3}$$

and

$$\frac{V_v}{V_f} = \left(1 - \frac{\rho_a}{\rho_f}\right). \tag{7.4}$$

It follows from Figure 7.4 that $V_v = 2dfl$, and $V_f = 2dl^2$, which leads to

$$\frac{V_v}{V_f} = \frac{f}{(f+d)} \, . \tag{7.5}$$

A comparison with Equation 7.4 gives

$$f = d\left(\frac{\rho_f}{\rho_a} - 1\right) \tag{7.6}$$

The latter differs from Equation 7.7[16]:

$$f = a_1 \exp(-a_2\rho_a) \, , \tag{7.7}$$

where a_1 and a_2 are empirical coefficients. The data[1,16,17] in Figure 7.5 were treated according to Equation 7.6 for comparison, and the root-mean-square deviation ψ for regression performed following Equation 7.6 and 7.7 is given. The slope of the curves is expressed by the values below those of d found for cellulose fibers.[14] Such a discrepancy can be ascribed to the fact that cellulose

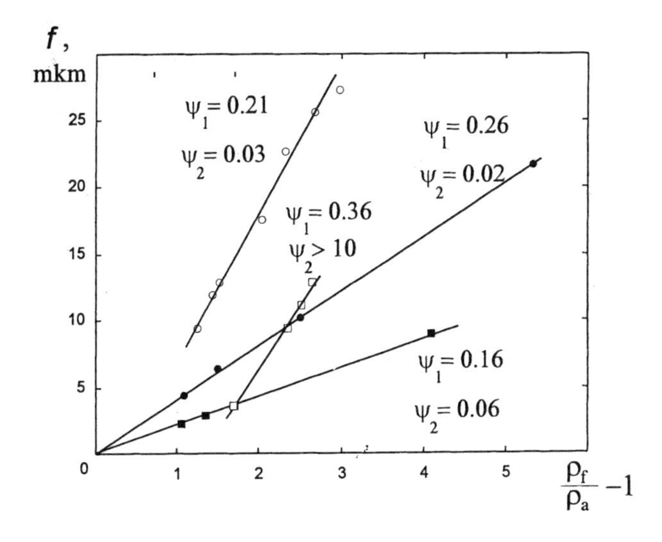

Figure 7.5 Treatment of the dependence on f = fun(ρ_a), obtained by means of a bubble test using Equation 7.6 of the model.[1,16,17] ψ = standard deviation of regression, ψ_1 = treatment by Equation 7.6. ψ_2, the same by Equation 7.7.

fibers have a spiral form, which gives the ff z-component, Figure 7.3c. And this was not taken into account when deriving Equation 7.6 (rectangular form of fibers does not disturb the quantitative correlation among ρ_a, d, and f, provided a maximum size of fiber is taken for d). This explains why in practice it is preferable to use polyempirical Equation 7.6′:

$$f = b_1\left(\frac{\rho_f}{\rho_a/b_1} - 1\right), \tag{7.6′}$$

where $b_1 \sim 1$ for synthetic fibers and ~0.25 for cellulose; b_2 is a coefficient in which ff microdefects increasing the parameter f are taken into account. $1.0 < b_2 < 1.5$. In addition, the suggested plane square nets model correlates very well with experimental data about impregnation of fibrous material.[18]

7.2.2 Estimation of the number of ff Layers Responsible for the Given Error in Determination of *f*

Let the error be designated as u:

$$u = \frac{\Delta F}{f}, \tag{7.8}$$

where F is an unknown true mean value of the cell-free size for a given ff; f is the value to be used. The number of local values of $f(X, Y, Z)$ necessary to attain the given u is determined according to Equation 7.9[19]

$$\frac{t}{\sqrt{n}} = \frac{\Delta f}{s}, \tag{7.9}$$

where s is the root-mean-square deviation and t is the Student criterion. However, it would be erroneous to take as n the number of elementary cells N in some ff volume, because in reality only a number of cells one above the other, as shown in Figure 7.3b, gives an adequate characteristic of ff. The curves showing distribution of pores by their radii[15] allow us to estimate s as $0.8 < s < 1.2$. Then, for example, for ff having $f = 10$ mkm with a degree of certainty 0.95 at $n > 20u < 0.1$, and at $n > 400u < 0.01$, i.e., ff with $d = 1$ mkm will be 0.02 to 0.4 thick, which seems quite reasonable. *Corollary:* if assumption (3) holds, the value f does not depend on the angle of intersection of two adjacent layers (Figure 7.3d). This corollary allows us to consider the square net $2d$ thick as being equivalent to a $2d$-filter.[14]

7.3 PREDICTION OF *J* OF FF BY MODELING ITS STRUCTURE WITH A PACKET OF PLANE NETS

Given below are the assumptions to be made for filtration (initial stage) through low-density ff.

1. The case under consideration is capture of particles from *as* (*hs*) diluted so that the volume coagulation of particles can be neglected; coagulation takes place only on the fiber surface; particles do not interfere with the direction of the flow within ff and evenly concentrate in the flow section.
2. The particle size is $2r \ll f$; particles are spherical in form and readily adhere to each other and to the filter irrespective of the material it is made of. Capture of particles is made possible because of their sorption on the ff surface.
3. The only mechanism of transportation of particles from the centerline of flow to the fiber is their direct ingress toward it, avoiding a diffusion mechanism.
4. A dynamics of passage of medium through ff is such that every cell in a cell column (Figure 7.3b) can be used as an independent filter penetrated by randomly located particles.
5. The impact of coulomb force is considered to be negligible.

7.3.1 Substantiation of Assumptions

The fact that the screen mechanism of particle capture at the initial (determining the course of the process) stage of filtration past ff can be ignored — assumption 2 — can be taken as proved in Reference 12, p. 19. Another important point is that calculation of the Brownian diffusion of particles from the flow to the fiber cell is neglected — assumption 3. As was shown by Fuchs,[20] displacement of a particle whose size is $2r = 0.1$ mkm and density 1 g/cm^3 in a gas stream under pressure of 1 atm/s is equal to 21 mkm, $2r = 1$ to 5 mkm. At a usual rate of filtration of gases of about 1 m/s and with thickness of $2d$-layer (ff cell) about 50 mkm the duration of particle presence in the cell is about 5×10^{-5} s. Since f is tens of micrometers, it can be concluded that the impact of diffusion in the transportation of particles from the centerline of the flow to the fiber is negligible. For *hs* the value f will be even smaller due to a sharp decrease of the diffusion coefficient. Thus, in terms of physics, assumption 3 is valid. A third peculiarity of ff is random location of particles in the cell at the initial stage of filtration — assumption 4. Under condition $2r << f$ (assumption 2) the particle trajectory does not depend on the ratio $2r/f$; hence, j could probably be considered as not depending on this ratio either. This, however, is not consistent with what happens in practice (see Table 7.1) and shows that geometric construction of ff is not enough to predict the value j. In

Table 7.1 Main Parameters of Filtering Materials with Different Types of Structure

Type of Clarification Max s.p. $D(p)$, mkm	No. No.	S.ch. mkm	H, mm	f, mkm	Max s.p.	Q, t/h·m^2	T, 1
	1	$d = 75$	—	25/35	20	10.3	1
$D(p) \geq 3.0$	2	$d = 20$	0.39	27	3.0	~20	~1
	3	~100/200	3.0	20	10	21.2	—
	4	$d = 0.4/0.6$	0.6	21	2.5	~30	~6
$3.0 > D(p) > 0.3$	5	$d = 1.4$	3.0	43	1/3	120	11
	6	$d = 0.35$	3.0	12	0.7	12.3	8
	7	—	0.01	0.5	0.5	42	0
$D(p) < 0.3$	8	—	0.01	0.01	0.01	0.03	—

Note: Q = output (water); pressure differential $\Delta p = 1$ atm; T calculated value of resource for filter area 1 cm^2 at $C_0 = 1 \times 10^{-6}$ (cm^3/cm^3) and particle size $r < d$. S. ch = size characteristics of structure (fiber or granule diameter); H = filter thickness; f = pore size; Max s.p. = maximal size of particles penetrating throughout filter. 1 = bronze ceramics[1]; 2 = filter paper[2]; 3 = filter of baked fluoroplastic granules[3]; 4,[2] 5, and 6 = ff consisting of glass fibers; 7[4] and 8[5] = nuclear filters. The values Q and T for No. 2 and 4 are arbitrary; only their ratio is true.

other words, characteristic features of the passage of medium through ff should also be taken into account. The influence of this factor on j is most significant in the case of dense medium (*hs*), while for *as* with the pore size of several micrometers a leading role in the transportation of particles from the flow centerline to the fiber is rightly ascribed to electrostatic charges on particles and ff. It is known that $\log(k) \sim R$, where $k = 1 - j$ is the breakthrough coefficient, R is the resistance of ff;[1] however, to build an adequate model R must be calculated independently of j. Since the hydrodynamics of ff is far from simple, only the qualitative aspect of the problem will be considered here. It is one of the structural features of ff that the pore size f significantly exceeds the arbitrary depth $h = d$. This is why it would be wrong to simulate ff using a bundle of capillaries and expect the output of $Q \sim f^2$. Nevertheless, the Poiseuille equation prompts that $Q \sim \Delta p$, but this condition is not satisfied either, because a passage of *hs* through a porous body with $f < 10$ mkm does not correspond to the Poiseuille equation.[4,21-23] By using the model suggested by the authors, this can be explained as follows. When ff f has the size of several micrometers, in liquid whose viscosity is not zero, a stationary distribution of velocities around streamlined fiber with the size d is hardly possible. This leads us to propose that (1) flow past any fiber is characterized by vibration; and (2) even in close proximity to a fiber there is a weak deviation of stream creating conditions causing particles to collide with fiber. Hence, the smaller the *f*, the larger the percentage of particles colliding with ff fibers, i.e., the larger the value *j*. This causes non-Poiseuille type of liquid flow as well as increases *j* with a decrease of *f* which has the form of a complex function. The fact that flow is a process characterized by vibrations can be confirmed by acoustic vibrations which arise as *hs* penetrates through ff. Indeed, as it follows from Table 7.1, vibrations with amplitude approximating 10 to 30 nVt and frequency about 1 kHz are recorded for ff No. 5. Thus, assumption 4 can exist as an hypothesis. Assumption 5 stems from our understanding of the mechanism of adsorption responsible for capture of a particle from *hs* and its retention on fiber: the percentage of desorbed particles depends on the disperse but not electrostatic forces. Dielectric permeability of *hs* being very high, the impact of coulomb forces in particle capture can be neglected. This, however, is not correct for *as*, in which case the impact of coulomb forces is taken into account by means of the empirical coefficient E_{ff}, see Equation 7.11, without disturbing *j*.

7.3.2 Calculation of Parameter *j* for ff

Due to arbitrary rotation of fibers at the initial stage of ff preparation, elementary cells in a cell column appear to be shadowed only a little. The area that happened to be in the aerodynamic (hydrodynamic) shadow does not exceed $4d^2$, which can be neglected if the ff density is less than $(4\ (d^2/l^2) \sim 1/1000)$. Consider the capture of particles by an elementary cell of ff. Let the number of particles likely to collide with cell fiber be determined as a ratio of the area of fiber forming a cell to its total area, see Figure 7.3a:

$$j = 1 - k = 1 - \frac{C_{i+1}}{C_i} = \frac{(dl + df)}{l^2}, \tag{7.10}$$

with $\rho_f/\rho_a < 5 \sim 2dl/l^2$, where k is the coefficient of breakthrough of particles before and after penetration through cell, respectively; $l = d + f$, Figure 7.3a. Then, taking into account Equation 7.6, deduce

$$\frac{C_{i+1}}{C_i} = 1 - 2\frac{\rho_a}{\rho_f} \tag{7.11}$$

Now, if we suppose, in addition to the above assumption, that the action of flow particles do not deviate, that making contact with fiber they easily adhere to it, and that cells upstream do not cast shade on those downstream, we will have a hypothetical case where a breakthrough coefficient k_{min} will be the minimum for all types of filtration through the given ff:

$$k_{min} = \left(1 - 2\frac{\rho_a}{\rho_f}\right)^m, \tag{7.12}$$

where m is the number of layers,

$$m = \frac{H}{d} - 1, \tag{7.13}$$

where H is ff thickness. Subtraction of one unit is dictated by the fact that *any* two layers located near each other form a 2d-filter. To consider real ff we introduce the parameter E_{ff} showing the capturing efficiency and retention of particles by fibers in a real process.

$$E_{\text{ff}} = \frac{d_{\text{ef}}}{d}, \tag{7.14}$$

where d_{ef} is effective diameter of the fiber, mkm, determined by Equations 7.12 and 7.13:

$$d_{\text{ef}} = \frac{H}{\dfrac{\log(k)}{\log\left(1 - 2\dfrac{\rho_a}{\rho_f}\right) + 1}}, \tag{7.15}$$

where k is a real breakthrough coefficient for given ff.

The preferred ff, allowing us to obtain a given value j, with a minimum E_{ff}, as in this case it provides j for a minimum thickness and improves the aerodynamic (hydrodynamic) characteristic of ff. The results of experiments show E_{ff} for *hs*[24] and *as*[25] calculated by Equation 7.14 to be in the range 10 (cellulose) to 100 (glass fiber) and 2 (cellulose) to 30 (glass fiber), respectively. Changing d to d_{ef} obtains the major equation:

$$\log(k) = H\frac{\log\left(1 - 2\dfrac{\rho_a}{\rho_f}\right)}{d_{\text{ef}}} - \log\left(1 - 2\frac{\rho_a}{\rho_f}\right) \tag{7.16}$$

or

$$\log(k) = A\left(\frac{H}{d_{\text{ef}}} - 1\right). \tag{7.17}$$

Now compare it with calculation of k carried out by Reference 11:

$$\log(k) = 0.87\frac{\rho_a Hq}{\rho_f \pi e} \tag{7.18}$$

q is a total coefficient of particle capture by a single cylinder, $q = q(R) + q(D) + q(D + R) + q(\text{St})$, where $q(R)$ is capture of particle due to contact. $q(D)$ due to diffusion, $q(D + R)$ due to contact plus diffusion, and $q(\text{St})$ due to inertia; $q(R) = (2x)^{-1}[(1 + Z)^{-1} - 1 - Z + 2(1 + Z)\ln(1 + Z)]$; $q(D) = 2.9x^{-1/3}\text{Pe}^{-2/3} + 0.62\ \text{Pe}^{-1}$; $q(D + R) = (1.24x\text{Pe})^{-1/2}Z^{2/3}$; $q(\text{St}) = ((2x)^{-2}(29.6 - 28(\rho_a/\rho_f)^{0.62}\ Z^2 - 27.5Z^{2,8})\text{St}$, St is Stokes number, Pe is Peclet number, $Z = r/d$, $x = -1.15\log(\rho_a/\rho_f) - 0.52$, $e = F_{f\text{-}1}/F_r$, where $F_{f\text{-}1} = 4\pi/x$ is force acting on a unit of the fiber length in a fanlike filter, $F_r = \Delta p\pi d^2/((\rho_a/\rho_f)\ yvH)$ the same for real ff; Δp, y, and v are pressure differential, viscosity of medium, and velocity of *as* (*hs*) being filtered. Table 7.2 shows the ratio of k logarithms calculated by Equations 7.14 and 7.15 to those obtained experimentally in Reference 1.

As it follows from Table 7.2, there is a good correlation between calculations carried out by Equation 7.17 and made in the experiment. That Equation 7.17 holds true is confirmed by the treatment of dependence of $\log(k) \sim H$: slope of curve $\log(k) \sim H$ was 1.41[1] and 1.31 (Equation 7.17), i.e., the discrepancy does not exceed 7%. Determining the dependence of j on ρ_a/ρ_f by

Table 7.2 Values of Breakthrough Coefficient for Cellulose ff with Different Density $v = 0.083$ m/s, $H = 1$ mm, $r/d \sim 0.2$

NN	ρ_a, g/cm³	Δp, Pa	~log(k)	log(k)/B_1	log(k)/B_2
1	0.20	11	1.19	1.00	0.40
2	0.29	33	2.05	1.07	0.12
3	0.40	66	3.44	1.13	0.04

Note: k and Δp are the breakthrough coefficient and pressure difference obtained by Reference 1; B_1 and B_2 are values of log(k) calculated by Equations 7.17 and 7.18, respectively. Calculation of B_1 was made taking a value of ρ_f equal to 1.50 g/cm³,[26] d_{ef} = 98 mkm.

Equation 7.18 the evidence is obtained that a relationship between B_2 and ρ_d/ρ_f is much stronger, compared with what was found in the experiment. Besides, in accordance with Equation 7.18, j must grow with an increase of v for particles with $2r > 1$ mkm, while in reality the case is just the opposite. All this suggests calculation of j should be made taking into consideration the ff structure.

7.4 CALCULATION OF THE RESOURCE *T* OF FF — OPTIMAL RESOURCE OF MULTILAYER EFF

7.4.1 Model of ff plugging — Calculation of *T* or ff

Consider stoppage of pores in two types of filters — (1) nuclear and (2) ff — whose action is based on the screen and deep-bed principles, respectively. Let the clarified medium be *hs* with $C_o = 1 \times 10^{-6}$ (cm³/cm³). The resource T of ff is the volume of *hs* (*as*) which passes through ff whose area is 1 cm² until the latter is completely plugged. The calculation of T given below can be used for comparison of different types of ff, because a number of factors remain beyond the scope of consideration. For (1) take $2r = 0.5$ mkm $- f$, $d = 0.5$ mkm, the percent of open area being 10%.[4] Suppose a complete stoppage of filter by particles occurs with plugging of all of its pores; the fiber is equal in size to the size of particles. Let N_p be the quantity of particles in 1 cm³ and N_f the number of pores in nuclear filter; then

$$T = \frac{N_f}{N_p} = \frac{4}{\left(0.1\pi d^2\right)} : \frac{3C_o}{4\pi r^3} \sim 3 \text{ ml (No. 7, Table 7.1)}. \tag{7.19}$$

To build a model of ff plugging consider plugging of the first two layers of ff from the inlet of *as* (*hs*) into the filter. This seems reasonable since only these layers come in contact with medium whose particles have concentration C_o, the highest for this case, C_o, and the volume of filtrate passing through these layers is T. A peculiarity of this process is that part of each cake adds to the previous one, making it grow vertically, and part decreases the size of the pore opening resulting in an increase of T. When the filtration process starts $f >> 2r$ and the rate of plugging is determined by j calculated from Equation 7.17. But as the fibers grow thicker due to adhesion of particles, assumptions 2 through 4 of Section 7.3 fail to be satisfied, and j decreases, which also leads to an increase of T of ff. Since $f >> 2d$, the cake from the underlying layers will reach, due to a vertical rate component, the cell in question and this is likely to find reflection in a steplike shape of the curve j–S, where S is the degree of plugging of ff area, %; growth of cake from the downstream to the upstream cell decreases T. To make the calculation less complex, let a vertical rate component of cake growth be zero and variations of j be neglected. See Figure 7.6b. First, calculate T for a single-pore T_f, and then for ff with the area 1 cm².

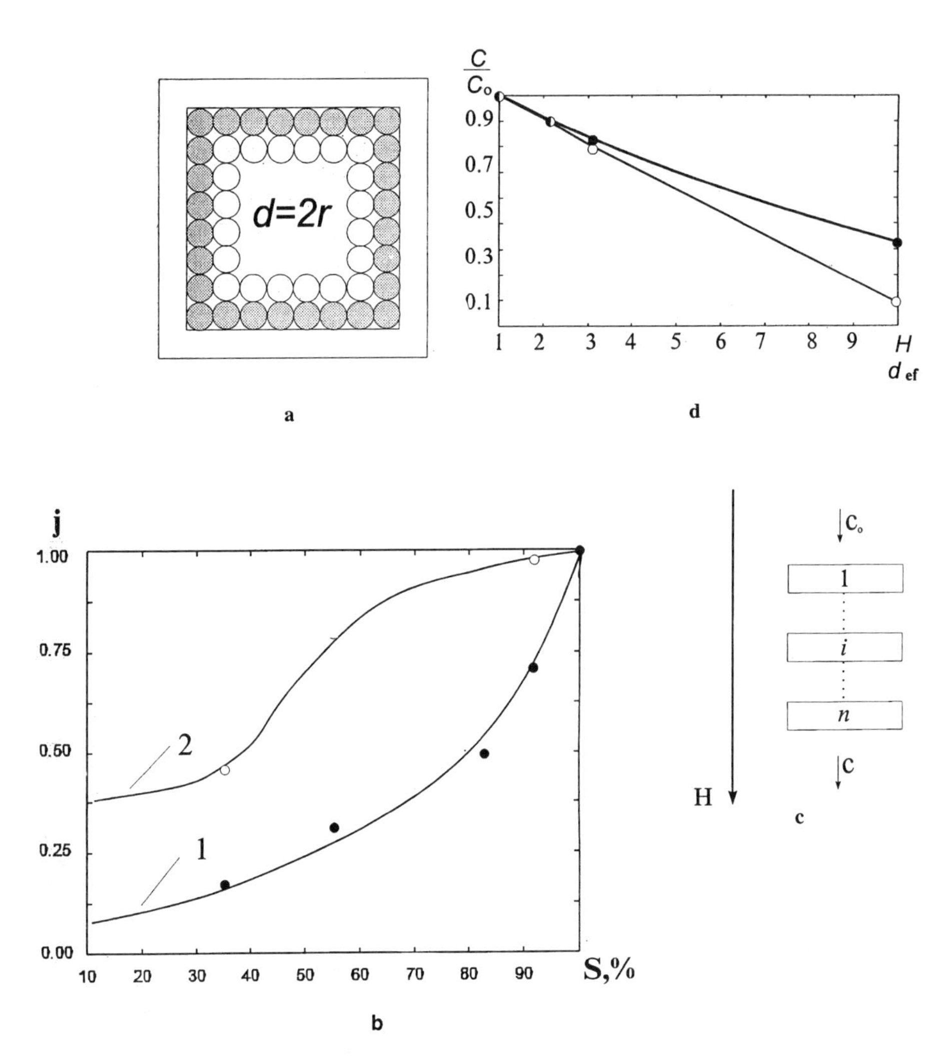

Figure 7.6 (a) Elementary cell plugged with particles $2r = d$; i is a number of the monolayer; vertical rate of cake growth is zero. (b) The dependence of the efficiency of ff clarification (j) on the degree of plugging of its area (S). Curve 1 corresponds to Reference 12, p. 75, curve 2 to Equation 7.25. (c) Left side: Schematic representation of multilayer filter for determination of its optimal $\rho_a(i + 1) > \rho_a(i)$. Right side: The dependence of C_H/C_o on the ff thickness H; curve 1 corresponds to ff with f = const, curve 2 to optimal multilayer ff (Equation 7.25).

$$T_f = \frac{w}{C_o} \sum_{i=1}^{n} \frac{N_i}{j_i}, \tag{7.20}$$

where $w = (4/3)\pi r^3$ is particle volume, C_o is volume concentration on the ff inlet, N_i the number of particles in the ith monolayer (Figure 7.6a), j_i a value of j when the ith monolayer is being filled,

$$j_i = 1 - \left[1 - \frac{\rho_f}{\rho_a / b_2} (1 + 2i) \right]^2 \tag{7.21}$$

when $2r = d$; b_2 is coefficient used in Equation 7.6′, Figure 7.6b. Summing up of i is continued until the number of layers reaches n and the pore is plugged completely. To obtain T, T_f should be multiplied by the number of cells $U = S_{ff}/S_c$, where S_{ff} = 1 cm^2; $S_c = l^2$, Figure 7.3a. If E_{ff} = 100, then with r = 0.25 mkm, T ~ 8 l; r = 0.5 mkm, T ~ 10 l, which corresponds to No. 6 and 5, Table 7.1. The value E_{ff} being the same, T for a ceramic surface decreases about ten times (the above assumptions taken into account), because the number of surface pores is approximately 0.1 of those of ff (see Table 7.1, No. 1).

7.4.2 Optimization of *T* for Multilayer ff

Comparing the quantity of filtrate having passed through a filter with a total volume of void, a conclusion is made — note the conditionality of the calculations — that not more than 10% of maximum value of T is utilized. The problem is that a rather significant time gap exists between stoppage of the upper and underlying layers. Such an approach contributes to a better understanding of the preference for polydisperse ff: each layer has a number of cells with different values of f, the size of some being $f_{max} >> f$, which is why stoppage of the first layer is delayed. Although polydisperse ff have been used in practice rather a long time, it is still not easy to attain a desirable value of j. To overcome this difficulty the use of ff with a decreasing f is recommended. Then it will be possible to calculate a step decreasing f so that ff resource could be maximum. For example, consider a column of ff 0.1 mm thick, as in Figure 7.6c, prepared of fibers with diameter d and having different density. The first ff has a minimum $\rho_a = \rho_a(1)$ and is plugged by being penetrated by T 1 of filtrate at a mean concentration C_1. The number of particles trapped in the upper layer of the first filter is N_1. To make an approximate calculation, let the capacitance of any of the underlying ff particle capture be the same, i.e., $N_1 = N_1 = N$, since the number of "particle seats" in cell N_c decreases almost proportionally with the cell area S_c, and a total number of cells increases inversely with S_c. It follows that to attain optimal values of T and j, the upper layers of each of the column ff must capture N particles. For such a column of filters the following equation holds:

$$C_n = C_o\, k_1\, k_2 \,\dots\, k_i\,, \tag{7.22}$$

where k_i is the mean breakthrough coefficient of the ith ff during filtration: i changes from 1 to n, n being the number of layers of ff.

$$k_i = \frac{C_i}{C_{i-1}} \tag{7.23}$$

C_i is a mean concentration of particles in a filtrate after it passed through the ith ff. A total number of particles captured by ff is proportional to the number of particles N captured by the upper layer; this is why the optimal sequence of ff will be the following:

$$j_{i+1} C_i = j_i C_{i-1},\; j_i = 1 - k_i\,. \tag{7.24}$$

A comparison of Equations 7.24 and 7.22 will finally give:

$$i_{i+1} = \frac{j_i}{1 - j_i} \tag{7.25}$$

Given the value j_1, we can calculate C_i/C_o for any ith ff. Figure 7.6c shows the dependence of C_H/C_o on the ff thickness for monolayer ff and ff consisting of n layers, each having j_i calculated by Equation 7.25.

7.5 CONCLUSION

The model of ff structure suggested is in good correlation with experimental data about ff porosity (Figure 7.5) and impregnation of fibrous material.[18] The analysis of dynamics of *as* (*hs*) filtration made using this model showed that Brownian motion is not to be regarded as a mechanism of transportation of particle to fiber from the stream centerline. Calculation of the efficiency of clarification j is made by Equation 7.17 where the member $(1 - 2\rho_a/\rho_f)$ describes structural peculiarities of ff. This is a principal difference of the model under consideration from that of Langmuir[6] and other authors.[7-13,15] Equation 7.17 allows prediction of the dependence of log(k) on H ff, and the dependence of j on ff density ρ_a, Table 7.2. The net model made it possible to estimate the resource of ff as equal to that of the two upper layers. The curve of dependence of ff efficiency j on the amount of plugged cake calculated by Equation 7.21 is qualitatively same as in the experiment (Figure 7.66). With all conditionality of calculation, the present approach allowed us to explain the relation between resource of filters having different structural types, Table 7.1. In terms of resource and of efficiency of clarification, the optimal density of each ff in the column of filters with different density ρ_a, Equation 7.25 is determined.

ACKNOWLEDGMENTS

The authors are grateful to I. B. Menina for translation and correction of the text and D. A. Kurdyukov for computer design of the chapter.

REFERENCES

1. Puzyryov, S. A., *Paper and Cardboard as Filtering Materials,* Moscow, 1970 (in Russian).
2. Kanarskiy, A. V., Platistina, N. V., Aleshin, V. A. et al., *Cellulose, Paper, Cardboard,* 10, 12–13, 1990 (in Russian).
3. Fluoroplast porous plates. Specs. 6-051741-85.
4. Barashenkov, V. S., *The New Professions of Heavy Ions,* Moscow 1970, 81 (in Russian).
5. Bildukevich, A. V. and Kaputckiy, F. N., *Colloid J.,* 48(4), 780–783 (in Russian).
6. Langmuir, I., *The Collected Works,* L., Vol. 10, 1961.
7. Davies, C. N., *Proc. Inst. Mech. Eng.,* 1B, 182, 1952.
8. Friedlander, S. K., *J. Colloid Interface Sci.,* 23, 157, 1967.
9. Whitby, K. T., *ASHRAE J.,* 7(9), 56, 1965.
10. Torgeson, V. L., General Mills, Inc., Mineapolis 13, Minnesota, 1961.
11. Kirsh, A. A. and Fuchs, N. A., *Ann. Occup. Hyg.,* 10, 23, 1967.
12. Uzhov, V. N. and Mjagkov, B. I., Industrial gases clarification by filters, Moscow, 1970 (in Russian).
13. Nevolin, V. F., *Cellulose, Paper, Cardboard,* 15, 9–11, 197.
14. Corte, H. and Callmes, O., *Formation and Structure of Paper,* L., Vol, 1, 1962, 351–369.
15. Irtegova, L. F., Ways of preparation of glass-asbestos filtering material for aggressive fluids. Synopsis of thesis for degree of Cand., *Tech. Sci. Leningrad,* 1978 (in Russian).
16. Corte, H., *Filt. Sep.,* 9, 10, 1966.
17. Tishenko, A. F., and Lohmachev, V. F., *Paper Industry,* 2, 13–15, (in Russian).
18. Starov, V. N., Churaev, N. V., and Panasuk, A. L., *Colloid J.,* 44(2), 271–276 (in Russian).
19. Hudson, D. J., *Statistics,* Geneva, 1964.
20. Fuchs, N. A., *Mechanics of Aerosol,* Moscow, 1955 (in Russian).
21. Pfeiter, P., Anvir, D., and Farin, D., *J. Chem. Phys.,* 73, 3558–3566, 1983.
22. de Gennes, P. G., Jr., *Fluid Mech.,* 136, 136, 189, 1983.

23. Halperin, B. I., Feng, S., and Sen, P., *N. Phys. Rev. Lett.,* 53, 2391, 1985.
24. Kanarskii, A. V., Platistina, N. V., and Fljate, D. M., *Lesnoi, J.,* 3, 84–87, (in Russian).
25. Pirumov, A. I., Provolovitch, O. V., Kanarskiy, V. A. et al., *Vodosnabzh. Sanit. Tekh.,* 12, 9–10, 1989, (in Russian).
26. Kanarskii, A. V., *Kinds of Filtering Paper and Cardboard,* Moscow, 1991 (in Russian).

CHAPTER 8

Airflow through Filters — Beyond Single-Fiber Theory

Richard C. Brown

CONTENTS

8.1 PRINCIPLES OF AIRFLOW THROUGH FILTERS

Early attempts to study airflow through filters considered the flow around a single fiber, an important approach because its simplicity allows easy numerical computation of filter properties.

0-87371-830-5/98/$0.00+$.50

As a description of flow pattern, however, it is being superseded by more sophisticated approaches, and in the text below studies that take account of filter structure will be described. A summary of the fundamental nature of the flow through filters is a necessary introduction.

In porous media turbulent flow is not favored because the structural elements prevent the development of vortices larger than the interstitial spaces. Small vortices are readily damped by viscous drag, and so the flow in porous media is usually Stokes flow, flow dominated by viscosity. In this situation the governing equations are relatively simple, and analytical solutions are frequently possible. The criterion for Stokes flow is expressed in terms of the Reynolds number, Re, where

$$\mathrm{Re} = \frac{d\rho U}{\eta} \tag{8.01}$$

η is the coefficient of viscosity, ρ is the density of the air, and d is a characteristic length in the filter, usually identified with the fiber diameter. A low value of Re, which is the case in most filters, suggests viscosity-dominated flow. The value at which the Stokes flow approximation is a good one depends on the size of the interstices, and this in turn depends on the packing fraction of the filter, the quotient of the volume of the fibers and the volume of the filter as a whole. The packing fraction, c, which is the most fundamental parameter giving information of filter structure, is normally a few percent. In this situation Stokes flow occurs provided that $\mathrm{Re} \ll 1$. The approximation ceases to apply if the air velocity is high, or if either the fiber diameter or the interfiber spacing is large.

A general characteristic of Stokes flow theory is that the pressure drop associated with it is proportional to the flow velocity. This relationship was observed in practice by Darcy (Davies, 1973) and is equivalent to Ohm's law in electrical theory. It is predicted by all Stokes flow theories, single fiber or many fiber, and it does not depend on the filter structure other than by the criterion above. Earlier theories of airflow such as potential flow models (Albrecht, 1931) and isolated fiber models (Davies, 1950), do not fulfill this requirement.

The pressure drop depends on η, d, U, and h, the filter thickness.

$$\Delta p = \frac{4U\eta h f(c)}{d^2} \tag{8.2}$$

$f(c)$ is a function of the packing fraction, the form of which depends on the particular model considered. The simple relationship in Equation 8.2 requires that the air in contact with the fiber is stationary. If the fiber diameter is small compared with the mean free path of air molecules (0.063 μm for oxygen molecules in air at 20°C, 760 mmHg), this ceases to hold and aerodynamic slip (see later) is the consequence.

8.2 CELL THEORY

In cell theory, the simplest credible quantitative approach, the problem is reduced to that of finding the airflow around a single fiber, considered to be typical of the filter as a whole. This fiber, and therefore any fiber, is assumed to lie perpendicular to the airflow and to have associated with it a region of space in the form of a concentric cylinder, the ratio of the volume of the fiber to the volume of this space being equal to the packing fraction of the filter.

In conditions of Stokes flow, the stream function, ψ, obeys the biharmonic equation

$$\nabla^4\psi = 0 \tag{8.3}$$

This equation is easy to solve in the cell geometry, and requires four boundary conditions, two being that the normal and tangential components of the velocity vanish at the fiber surface. The third is that the mean fluid velocity across the widest part of the cell is the same as that through the entire filter being modeled. The fourth is rather more arbitrary. Kuwabara (1959) assumed that the vorticity of the fluid vanished at the surface of the cell, whereas Happel (1959) assumed that the tangential stress vanished. Other possibilities exist for the boundary conditions, but the resultant stream function always takes the form:

$$\psi = \left(Ar + Br^{-1} + Cr\ln\left(\frac{r}{R}\right) + Dr^{3} \right)\sin(\theta) \tag{8.4}$$

where the constants A, B, C, and D vary according to the theory. Of the two theories Kuwabara's is found to give better agreement with experiment and is probably even now the most popular model of airflow through a filter. A typical flow pattern obtained with this model is illustrated in Figure 8.1.

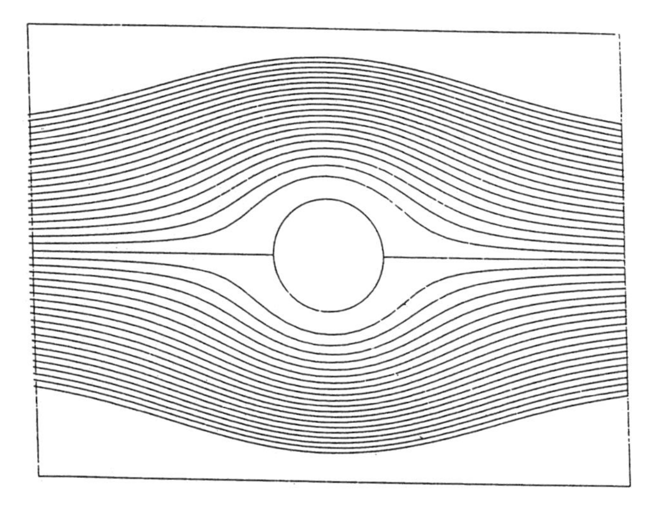

Figure 8.1 Illustration of streamlines based on Kuwabara model. (British Crown Copyright; by permission of Her Majesty's Stationery Office.)

8.2.1 Strengths and Weaknesses of Cell Theory

The great strength of cell theory is its simplicity, which has enabled it to form the basis of many simple numerical calculations of filter properties. However, a rigorous theory of pressure drop or particle capture would require the flow problem to be solved for each individual fiber, and the results of this to be averaged; whereas in the cell theory the conditions in which fibers in the filter find themselves are averaged; and then the flow equation is solved, a much easier though less satisfactory approach. This approximation is fair so far as pressure drop is concerned, and evidence for this is given by Jackson and James (1986) who showed that a reasonable correlation existed between experimentally observed pressure drop and packing fraction for a large range of filters.

The pressure drop across a filter can be calculated from the viscous drag at the fiber surface. Particle capture by interception depends critically on the flow pattern here and although this will vary from fiber to fiber, the contribution of airflow to each of the two phenomena varies in much the same way. This means that, although cell theory alone does not give a complete solution, the particle capture efficiency by interception can be related to the quotient of pressure drop and macroscopic fluid velocity (Brown and Wake, 1991). There is a weaker correlation between scaled pressure drop and capture efficiency by diffusional deposition (Kirsch and Fuchs, 1968), although neither process can be understood entirely in terms of single-fiber theory. In inertial impaction the model is at its weakest, because there is no satisfactory way of specifying the initial condition for a particle undergoing motion in which its own inertia is critical. Cell theory gives no account of airflow in the interstitial regions of the filter, and a particle may have its trajectory significantly perturbed by a fiber that does not capture it (Brown, 1985).

The extent of the approximations involved in cell theory are illustrated by Figure 8.2, an electron micrograph of a normal fibrous filter, which has an irregular structure that cell theory cannot account for. The fibers of the filter are not all perpendicular to the macroscopic flow direction whatever that direction might be, and the packing fraction of the filter is a function of position. Other filters may be complicated by a range of fiber diameters or by fibers that are not circular in cross section. No theory exists which gives a good account of all of these problems, although a number of refinements exist.

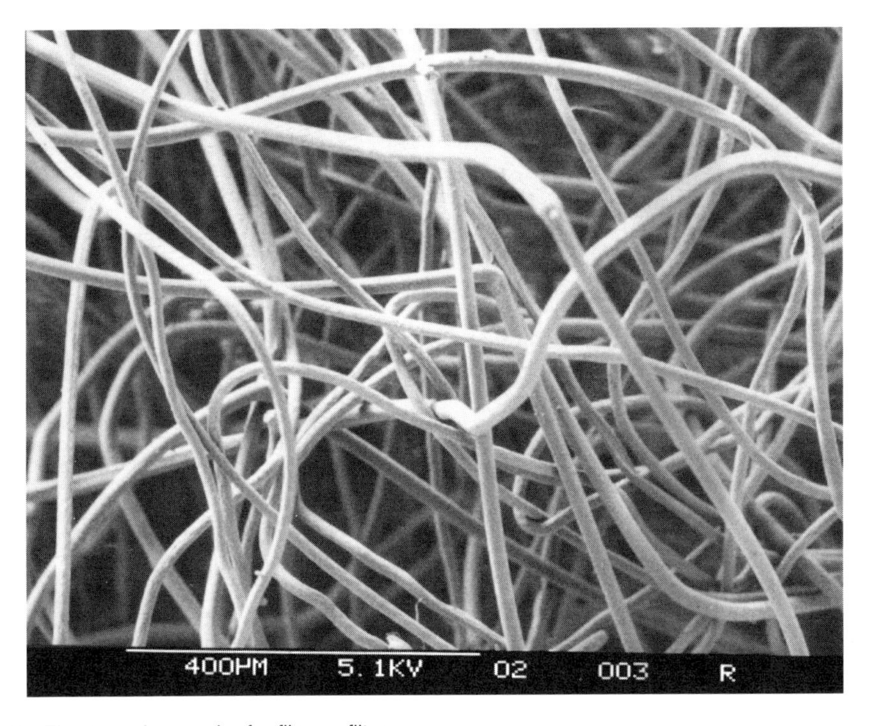

Figure 8.2 Electron micrograph of a fibrous filter.

8.3 FLOW THROUGH ROWS OF PARALLEL FIBERS

The simplest theory that accounts for the influence of the nearest neighbor fibers on the flow pattern and which describes a structure corresponding to a physically realizable model is that of flow perpendicular to a single infinite row of parallel, uniformly spaced fibers, a form in which the flow pattern, illustrated in Figure 8.3, is calculated by means of a complex variable approach (Miyagi, 1958). A general solution of the Stokes flow equation is written down relating the complex velocity, W, to the complex coordinate, w

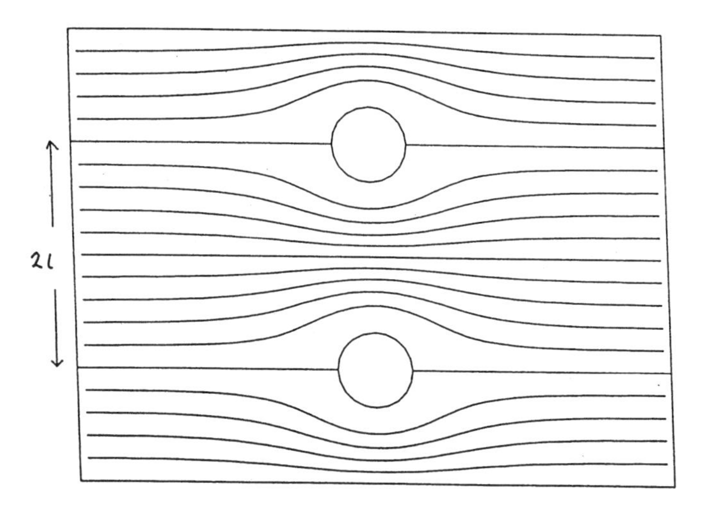

Figure 8.3 Streamlines calculated on the basis of the complex variable approach. (British Crown Copyright; by permission of Her Majesty's Stationery Office.)

$$\frac{W}{U}=\frac{U_x+U_y}{U}=1+a_0\left[\ln\left(2\sinh(\overline{w})\right)+\ln\left(2\sinh(w)\right)-(w+\overline{w})\coth(w)\right]+\sum_{n=0}^{\infty}a_{2n}$$
$$\times\left[\frac{d^{2n-1}}{d\overline{w}^{2n-1}}\coth(\overline{w})-(w+\overline{w})\frac{d^{2n}}{dw^{2n}}\coth(w)\right]+\sum_{n=1}^{\infty}b_{2n}\frac{d^{2n-1}}{dw^{2n-1}}\coth(w) \tag{8.5}$$

where

$$w=\frac{\pi}{2l}(x+iy) \tag{8.6}$$

The illustration of flow pattern shows that the flow deviates from uniformity over a distance comparable with the fiber spacing, a fact that can be deduced from the form of fundamental solutions of the biharmonic equation for periodic structures. This means that the theory will give a good description of airflow through filters provided that the interlayer spacing is not significantly smaller than the interfiber spacing within a layer. In fact, pressure drop calculated by this theory is comparable with that given by the cell theory if the packing fraction is assumed to be that of a stack of such layers individually separated by a distance equal to the fiber spacing within a layer.

8.4 FLOW THROUGH ARRAYS OF PARALLEL FIBERS

A regular array of parallel fibers is the simplest theoretical model corresponding to anything that could serve as a depth filter. A number of models with structures having various degrees of symmetry are shown in Figure 8.4, all of which would be seen as equivalent by the theories discussed above. The two simplest of these structures require three parameters for a full description, giving, respectively, the scale, the packing fraction, and the ratio of the interfiber spacing to the interlayer spacing. The two structures of lower symmetry require a fourth parameter, giving the extent of stagger between adjacent fiber layers. In contrast, the cell model requires only two.

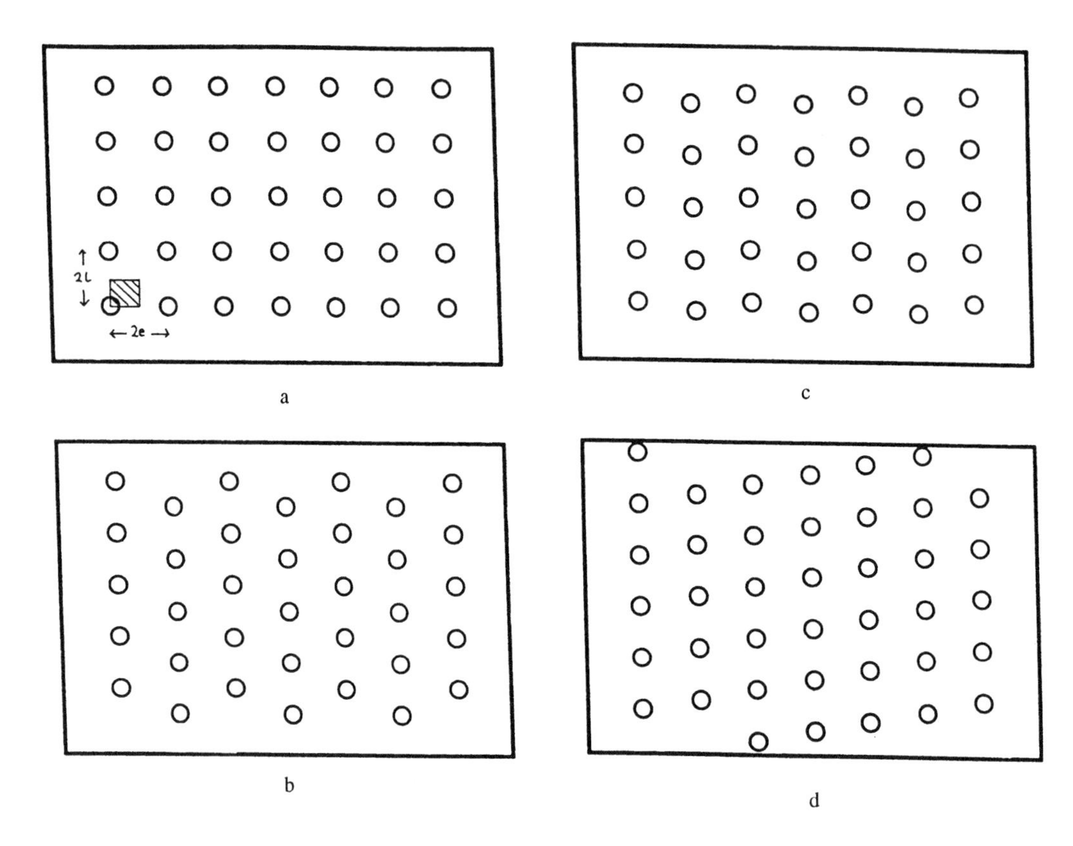

Figure 8.4 Arrays used in the variational and other approaches: (a) channel model (the irreducible element is shaded); (b) staggered model; (c) limited stagger model 1; (d) limited stagger model 2. (British Crown Copyright; by permission of Her Majesty's Stationery Office.)

Stokes flow has upstream–downstream symmetry, and so it will be sufficient to obtain a description of the flow in the irreducible symmetry element shown in the figure. Four approaches to this will be summarized: extensions of the cell model, numerical methods, a variational method, and a method based on the boundary element approach.

8.4.1 Extension of Cell Model to Account for Adjacent Fibers

The biharmonic equation applied to the cell model described above was easy to solve because the boundary conditions were set at points where the radial coordinate had one of two values: the fiber radius or the cell radius. In this situation only the simplest solution of Equation 8.3 needs to be considered, since this is sufficient to satisfy all of the boundary conditions. However, at the boundary of the irreducible symmetry element, shown in Figure 8.4, the radial coordinates vary continuously. It is, therefore, necessary to use a series involving those higher-order solutions of Equation 8.3 that have the required symmetry. In polar coordinates the nth term of a series of which Equation 8.4 gives the zeroth-order term is

$$\psi_n = (A_n r^{2n-1} + B_n r^{1-2n} + C_n r^{3-2n} + D_n r^{2n+1})\sin((2n-1)\theta) \quad (8.7)$$

One method of finding a higher-order solution is that of Sangani and Acrivos (1982) who calculated the coefficients of the series by requiring the boundary conditions to be met at a fixed number of points on the surface of the lowest symmetry element of the array. The solution in effect deals with a prismatic cell rather than a cylindrical one.

An alternative solution is that of Drummond and Tahir (1984) who wrote down the series for the stream function centered around a number of fibers considered to interact significantly with the fiber at the origin, rather than for just this fiber. They then found the values of the appropriate coefficients by ensuring consistency of the pressure drop at the surface of the fiber at the origin.

8.4.2 Numerical Methods

The flow pattern through arrays like those shown in Figures 8.4 can be solved by finite element or finite difference methods, in which a mesh is set up and the solution of the flow equations is carried out at each point in the mesh (Fardi and Liu, 1992). A number of methods can be used, and the one chosen by the authors above is a finite difference solution of the components of the simplified Navier–Stokes equation.

$$\rho(\underline{U} \cdot \nabla)\underline{U} = \eta\nabla^2\underline{U} - \nabla p \tag{8.8}$$

along with that of the equation of continuity in the fluid velocity. The pressure drop and velocity field are solved discretely at the center point of each mesh element, in the form of simultaneous equations relating the parameters to those at each contiguous element. The problem needs to be solved only in the irreducible symmetry zone, provided that appropriate boundary conditions are applied at the surface. In this approach there is no fundamental limit to the complexity that can be dealt with, but the equation must be solved numerically at every mesh point throughout the irreducible zone, and the size of this increases as the symmetry of the array decreases. The advantages of this approach and all of the approaches dealing with arrays are that the flow pattern in the interstitial regions is realistic and that there is no ambiguity about boundary conditions.

8.4.3 Variational Method

A third approach to the problem is the direct use of Helmholtz's principle, by finding the flow pattern that gives rise to the lowest rate of dissipation of energy by viscous drag, Θ. This parameter can be expressed in terms of the stream function:

$$\Theta = \eta\iint\left(\frac{\partial^2\psi}{\partial x^2} - \frac{\partial^2\psi}{\partial y^2}\right)^2 + 4\left(\frac{\partial^2\psi}{\partial x\partial y}\right)^2 dxdy \tag{8.9}$$

The biharmonic equation is the Euler equation of the functional in Equation 8.9.

Solution by the variational method requires a trial stream function containing adjustable parameters, and the periodic symmetry of the fiber array is exploited by the use of a Fourier series with the coefficients as variational parameters. For the channel array this function takes its simplest form

$$\psi = Uy + Ul\sum_{n=1}^{\infty}\sum_{k=0}^{\infty} a_{nk}\sin\left(\frac{n\pi y}{l}\right)\cos\left(\frac{k\pi x}{e}\right) \tag{8.10}$$

where $2l$ is the interfiber spacing within a layer and $2e$ is the interlayer spacing. The advantage of this model over purely numerical methods is that a continuous stream function throughout the filter is obtained in a simple analytical form. The pressure drop through a channel filter is given by

$$\Delta p = \frac{U\eta h l^2\pi^2}{2}\sum_{n=1}^{\infty}\sum_{k=0}^{\infty} a_{nk}^2\left(\frac{n^2}{l^2} + \frac{k^2}{e^2}\right)^2 q_k \tag{8.11}$$

where q_k takes the value 4 or 2 depending on whether k is zero or not.

Calculations using the variational model Brown (1984, 1993a) applied to the four structures illustrated in Figure 8.4 predict that the pressure drop is not particularly sensitive to fiber arrangement when the two distances, 1 and e, are equal. However, when e is reduced, so that the filter is compressed in the direction of airflow, the several structures behave very differently. The drag caused by each row of fibers decreases in the case of the channel structure and increases in the case of the staggered structure, as shown in Figure 8.5. The behavior of the arrays with limited stagger is intermediate as expected. The theoretical calculations based on the variational model agree well with experimental observations using model structures, made by Kirsch and Fuchs (1967). Observation of the behavior of the pressure drop of real filters when they are compressed, interpreted with respect to this, may give some information on their structure, although the structural changes in practice will be more complicated than those described above.

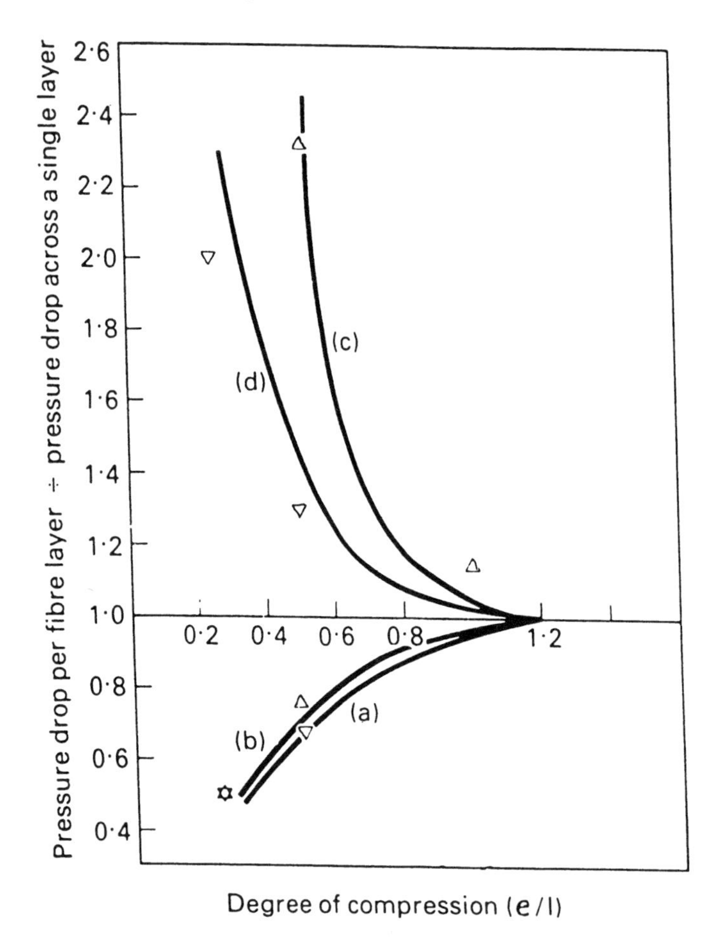

Figure 8.5 Variation of pressure drop for fiber arrays: (British Crown Copyright; by permission of Her Majesty's Stationery Office.) △▽ channel model theory; a,b channel model experiment; △▽ staggered model theory; c,d staggered model experiment.

8.4.4 Boundary Element Method

A further method is the boundary element method (Hildyard et al., 1985) which, like the variational method, does not require a mesh throughout the irreducible symmetry zone. In this method the biharmonic equation is split into two equations, by way of the vorticity, ω.

$$\nabla^2\psi = -\omega \tag{8.12}$$

The result of this procedure is expressed in the form of two integral equations:

$$\nu(\underline{r})\psi(\underline{r}) = \int_c \psi(\underline{s})G_1'(\underline{r},\underline{s}) - \psi'(\underline{s})G_1(\underline{r},\underline{s}) + \omega(\underline{s})G_2'(\underline{r},\underline{s}) - \omega'(\underline{s})G_2(\underline{r},\underline{s})d\underline{s} \tag{8.13}$$

$$\nu(\underline{r})\omega(\underline{r}) = \int_c \omega(\underline{s})G_1'(\underline{r},\underline{s}) - \omega'(\underline{s})G_1(\underline{r},\underline{s})d\underline{s} \tag{8.14}$$

where $\underline{r}$ and $\underline{s}$ are position vectors, and G_1 and G_2 are the fundamental free-space solutions of Laplace's equation and the biharmonic equation, respectively. Both integrals are carried out over a contour C, which is the complete boundary of the irreducible symmetry element, the fiber surface, and the lines of symmetry. The function ν takes the value 2π within the region of interest, zero outside, and π on the boundary except at sharp corners. The boundary is split up into a number of discrete units, reducing the problem to one of the solution of simultaneous equations. Flow patterns and pressure drop obtained by this method resemble those obtained with the variational method.

8.5 AIRFLOW NONPERPENDICULAR TO FIBERS

The situation of airflow parallel to fibers has received less interest because in such a situation a fiber array would not serve as a particularly good filter, although particle capture by diffusional deposition and by electric forces would still occur. Nevertheless, Figure 8.2 illustrates that in real filters regions of parallel flow will exist, and there will also be flow at an orientation that is neither parallel nor perpendicular to the fibers, a situation that can be described by solving for the two orthogonal directions separately and simply adding the solutions.

In cell theory the solution of flow parallel to a fiber is quite straightforward. Only a single component of flow exists, and the boundary conditions are that the air velocity is zero at the fiber surface and that the velocity gradient is zero at the cell surface. The result is that the velocity in the z-direction, parallel to the fiber, is

$$U_z(r) = \frac{-Uc\left(r^2 - R^2 - 2\frac{R^2}{c}\ln\left(\frac{r}{R}\right)\right)}{2R^2\left[-\frac{1}{2}\ln(c) - 0.75 + c\frac{c^2}{4}\right]} \tag{8.15}$$

There are equivalents in the more sophisticated theories; for example, the theory of Sangani and Acrivos has its counterpart in the theory of Sparrow and Loeffler (1959) for flow parallel to the fibers, whereas Drummond and Tahir (1984) apply their own model to flow parallel to the fibers.

When the variational model is applied to flow parallel to the fibers, the velocity flow field is expressed in terms of a Fourier series in x and y, and the rate of dissipation of energy, which depends only on U_z, can be obtained from a rather simpler calculation than that needed for flow parallel to the array, with the boundary condition of stagnant air at the fiber surface. Velocity contours for such a flow field are shown in Figure 8.6.

To some extent the effect of fiber inclination can be considered in the single-fiber model that adds a continuum term to take account of the other fibers in the filter. In this approach, due to Spielman and Goren (1969), the Stokes flow problem is solved for a single fiber, but account is taken of the surrounding fibers by adding to the governing equation a term representing a homogeneous resistive medium, simulating the filter as a whole, giving

$$\nabla p = \eta\nabla^2\underline{U} - \eta\underline{\underline{K}}\cdot\underline{U} \tag{8.16}$$

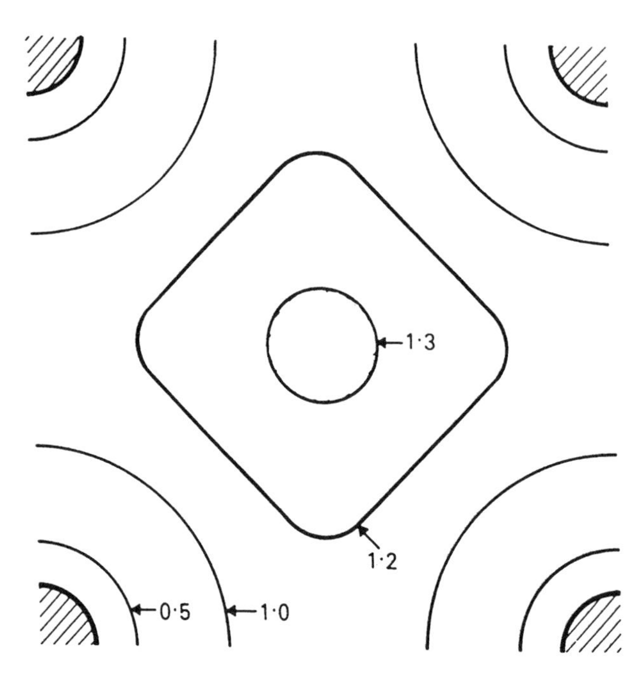

Figure 8.6 Constant velocity contours calculated for flow parallel to fibers on the basis of the variational model. (British Crown Copyright; by permission of Her Majesty's Stationery Office.)

The form of this additional term depends on the arrangement that the other fibers are assumed to take. The authors consider a number of arrangements, from that of parallel fibers to that of completely random fibers, the latter giving rise to a scalar resistance term; but in all of the calculations of flow, the fiber considered in detail is assumed to lie perpendicular to the macroscopic flow direction.

8.5.1 Pressure Drop at Arbitrary Inclination

All of the above theories predict a pressure drop for flow parallel to the fiber axes close to one half of that in the direction perpendicular to the axes. In the case of the cell theory of Kuwabara, the result is exact. Drummond and Tahir demonstrated that the result is correct to $O(c^2)$ for a square and to $O(c^4)$ for a staggered array in which each fiber is equidistant from its nearest neighbors in its own row and those in adjacent rows. Calculations based on the variational model and the boundary element model give results consistent with the above, being almost exact when the packing fraction of the filter is low, but poorer at high packing fractions.

The pressure drop corresponding to flow in an arbitrary direction through any filter that is not completely isotropic will involve a tensor term like that on the right-hand side of Equation 8.16. The array in Figure 8.4a is isotropic in two directions. Textile fibers in filters produced by carding have a preferred direction of fiber alignment, which is normally perpendicular to the direction of airflow through the filter. The fibers in filters produced by laying from air or liquids usually have a direction of least alignment, and this lies parallel to the direction of airflow. The former structure is closer to the parallel arrays described so far.

Since filter resistance is described by a tensor, the orientation of the pressure drop vector will not, in general, be parallel to the macroscopic flow direction. This has special importance with respect to semimicroscopic descriptions of airflow (Schweers and Loeffler, 1994). If, however, there are two equivalent directions, as there are in square fiber arrays, the pressure drop in macroscopic flow in those two directions will be the same. It follows that the pressure drop caused by macroscopic flow in any direction in the plane formed by the two equivalent directions will be the same and will be parallel to the flow. This is a two-dimensional example of the well-known physical principle that properties that depend on linear response are completely isotropic in systems of cubic symmetry. Examples of the latter are refractive index and relative permittivity, which are isotropic in crystals of cubic symmetry.

8.5.2 Elimination of Spurious Order in Model Filters

Although the arrays of fibers considered above give credible values for pressure drop in Stokes flow, the simple geometric relationship between the position of fibers in successive layers would preclude their use as filters. For example, a particle interacting by interception only, and not captured by the initial layer of fibers in the array in Figure 8.4a, would not be captured at all, however many layers it encountered. The sample particle challenging filters of Structures 8.4b and c would stand a chance of being captured by the second layer, but not by any subsequent layers.

A real filter has a structure approaching random, and even the concept of fiber layer is not very well defined. None of the structures described so far has randomness, but two at least can be structured so that no two layers are identical with respect to the coordinates normal to the direction of macroscopic flow. The first of these is obtained by particular combinations of flow in two directions through the square array described above, and is illustrated in Figure 8.7. The second is Structure 8.4d for which the flow pattern is illustrated in Figure 8.8. The latter requires a transformation of coordinates, both in the trial function and in the dissipation integral, for its calculation (Brown, 1993a and b); and the introduction of the degree of stagger as an additional parameter describing the structure results in a doubling of the number of coefficients required to describe the flow, relative to the number required for the channel or simple staggered models.

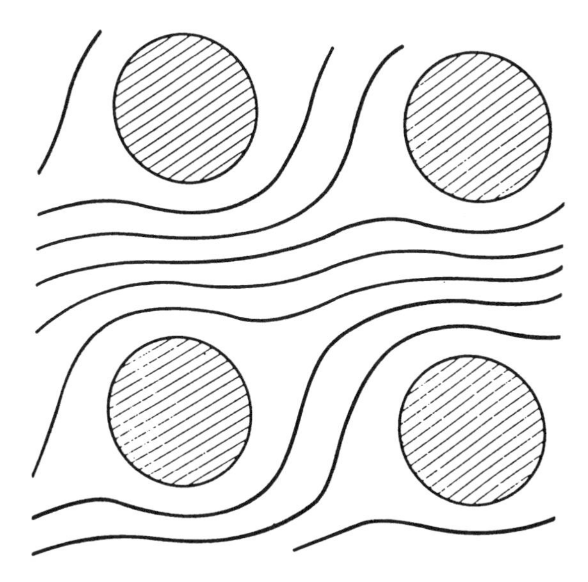

Figure 8.7 Flow at arbitrary inclination through channel array of square symmetry. (British Crown Copyright; by permission of Her Majesty's Stationery Office.)

8.6 EFFECT OF FIBER PROPERTIES

8.6.1 Effect of Fiber Shape

The simple cell model and the modifications described so far are all limited to fibers of circular cross section, whereas in practice fibers of other cross section are common. Rectangular fibers can be produced from the fibrillation of a sheet. Fibers produced by extrusion from a dilute polymer

Figure 8.8 Flow through limited stagger array 2. (British Crown Copyright; by permission of Her Majesty's Stationery Office.)

solution tend to dry from the surface inward, so that their cross section has a higher perimeter/area ratio than that of a circle, which often results in the fibers assuming a shape not unlike that of a railway line. More conventional textile fibers can be produced from spinnerets to give a noncircular section; in particular nylon is often produced in a trilobal form (Moncrieff, 1975).

A finite difference solution of the flow around a fiber of rectangular cross section within a regular array of similar fibers has been carried out by Fardi and Liu (1992). The stream function varies more rapidly in the region of the vertex of the rectangle than at other points and so the mesh has to be much finer there.

The problem can be solved for the variational model or the boundary element model in a very simple way, because the only difference between this and the flow around a fiber or circular cross section is in the position of the points where boundary conditions are applied. These have to be more closely spaced in the region of the vertex, but, as in the numerical solution, there is no easy formula for optimizing their position.

In Figure 8.9 the flow pattern around a fiber of square cross section is shown. There is no simple expression for the pressure drop of such a filter as a function of its packing fraction, because of the nonanalytical form of the solution of the flow equations, but for these fibers, as for cylindrical fibers, the pressure drop increases rapidly as the packing fraction becomes high. A fundamental theory of Stokes flow (Hill and Power, 1956) states that the resistance of a body is less than that of any other body that it can enclose and, therefore, less than that of any body enclosing it; and this means that the resistance of square fibers must lie between that of fibers whose cross section is the inscribed circle and that of fibers whose cross section is the circumscribed circle. Calculations based on the variational model indicate that the pressure drop of square fibers is similar to that of a square array of cylindrical fibers with the same surface area, which is in basic agreement with the philosophy of the Kozeny–Carman equation (Carman, 1937), though the latter applies strictly only to densely packed filters. This simple relationship does not apply to fibers with reentrant cross sections like that of the trilobal fiber.

In practice it is found that filters with fibers of rectangular cross section, such as that shown in Figure 11.1c (Chapter 11 on electrostatic filters) tend to suffer a larger increase in pressure drop with compression than do filters with cylindrical fibers, although it is possible that one effect of compression might be a change in orientation of the fibers with respect to the flow direction.

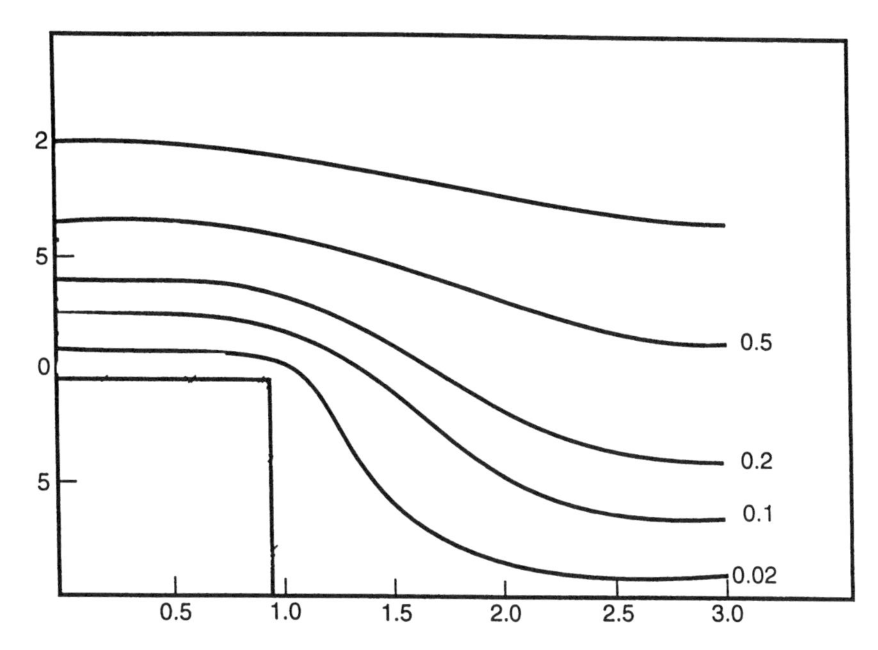

Figure 8.9 Flow around a fiber with a square cross section.

8.6.2 Effect of Fiber Size Range

Real filters are not normally made from completely monodisperse fibers but from fibers with a range of diameters. The range of sizes in carded textile filters may be relatively small, with the coefficient of variation well below unity; but certain filters are made specifically from fibers with a bimodal size range. For example, fine fibers, needed for a high level of filtration efficiency, may be too weak to be formed into a filter by themselves, and may need to be supported by much coarser fibers, which give mechanical strength but which contribute little to filtration performance.

The problem of a mixture of two fiber sizes in a single layer was solved by Kirsch and Stechkina (1973), using the complex variable approach previously applied to a filter of identical fibers. They found that the pressure drop across such a filter was well approximated by the pressure drop of monodisperse fibers with a diameter equal to that of the arithmetical mean of the two fiber types, provided that there was a factor of less than about three between their diameters. The flow pattern observed experimentally for a model filter of the same geometry was reasonably well described by the calculation. The implications of this are most appropriate to a filter containing a relatively small spread in fiber diameters.

The problem can also be tackled by other approaches, and Figure 8.10 shows the calculated flow pattern through the irreducible symmetry zone of a square array of fibers of which the fundamental unit is a four-fiber complex, each of which can be chosen to have a diameter independent of the other three. In the figure three are chosen to be equal. A larger zone could be described by this method, and the calculation time would increase with the number of fibers chosen but the conceptual difficulty would not. The observation of Kirsch and Stechkina is confirmed by this model, but the zone in Figure 8.10 has the ability to describe mixtures of two fiber types with various proportions; 1:3, 1:1, 3:1. The 1:1 mixture can be made in three different geometries according to which pairs of fibers are chosen to be similar, but the calculated pressure drop is insensitive to this aspect of structure when the packing fraction is relatively small, which is consistent with previous observations made on arrays with various degrees of stagger. Of particular interest is the pressure drop of a fiber mixture as a function of the mass of each fiber component, as this is the variable normally chosen in practice to describe the composition of a mixture. (A filter made of equal masses of fine and coarse fibers will,

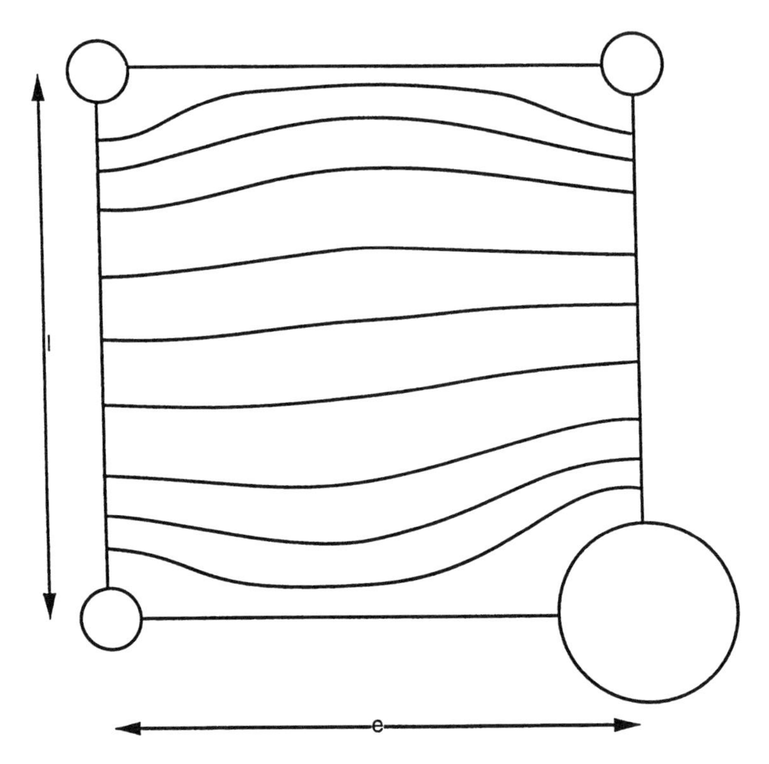

Figure 8.10 Flow through the irreducible symmetry element of a filter comprising a regular 3:1 mixture of fibers of two sizes.

of course, have a greater length per unit volume of the fine.) The calculated pressure drop of such mixtures is well approximated by a simple linear relationship.

8.6.3 Effect of Microscopic Fiber Clumping

The effect of heterogeneous structure on a small scale can be investigated by a model in which parallel fibers lying in a row consist of pairs closer together than each is to its next nearest neighbor, like the fibers shown in a single row in Figure 8.11. The smallest unit of this is a pair of fibers rather than a single fiber. Kirsch and Fuchs (1967) showed, using the complex variable approach, that the pressure drop across a row of paired fibers was invariably less than that across a uniform row with fiber spacing equal to the mean of the above.

If a row of paired fibers is seen as part of an array, then the geometric relationship between pairs of fibers in a row and those in the next row becomes important. Calculations based on the variational model, applied to the simplest example, in which each pair of fibers lies in front of and behind identical pairs, corresponding to the channel array of single fibers, show that as the fibers of a pair are brought closer together the pressure drop falls, at a faster rate than that for a single row. In the case of the staggered array, shown in Figure 8.11, pairing of fibers initially causes an increase in pressure drop, but increased clumping of fiber pairs causes a reduction in pressure drop, particularly at small packing fraction, as parts of the filter become more open.

8.7 AIRFLOW AT FINITE REYNOLDS NUMBER

In practice the criterion of low Reynolds number is usually fulfilled in fibrous filters, although in coarse-fibered filters operated at high velocities this may cease to be the case. Isolated fiber theory takes account of fluid inertia by necessity because there is no Stokes flow solution for flow

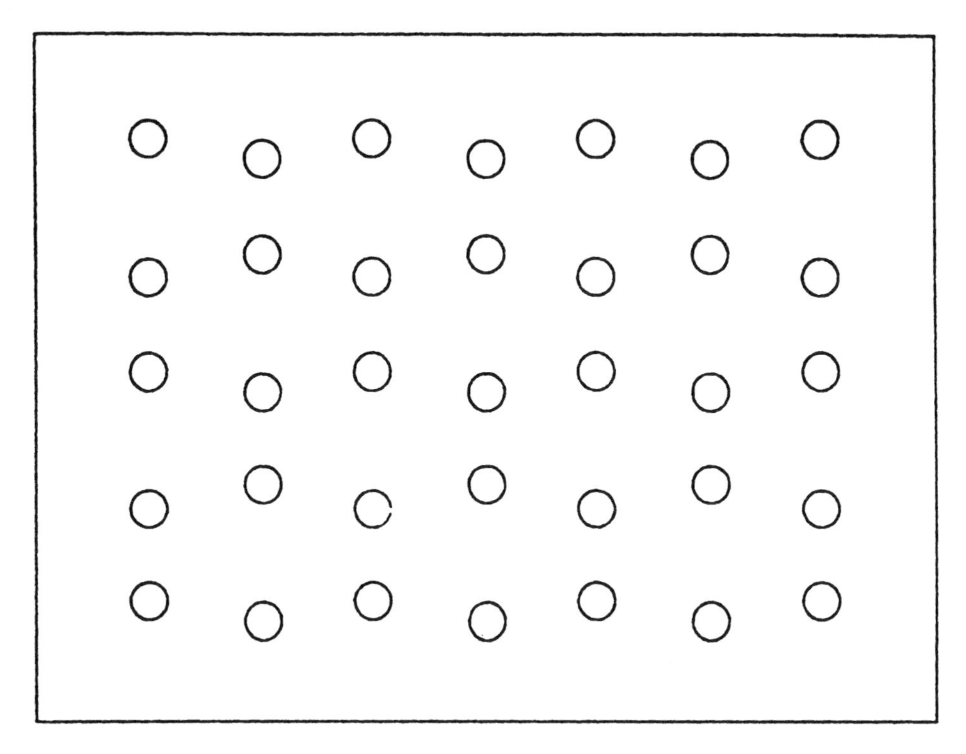

Figure 8.11 Model structure of paired filters.

around an isolated fiber. However, the functional dependence of fiber drag on Reynolds number is quite different from that observed in practice, and so this theory must still be considered inapplicable. The flow pattern through an array is highly dependent on structure even at finite Reynolds number.

It is relatively straightforward to generalize numerical methods to the finite Re situation (Fardi and Liu, 1992). The complex variable method was extended by Miyagi (1958), with the result that the pressure drop can be written as a series in fluid velocity of which the first term is the Darcy's law term and the second a cubic term. A similar calculation has been carried out for the variational method (Brown, 1986), with the same result. Dimensional arguments can be used to show that the cubic term is independent of fiber diameter, and results fitted to the experimental data of Jefferies (1974) on porous foams support this (Brown 1988). A point not revealed by the complex variable approach, which deals with a single layer of fibers only, is that this term is highly dependent on structure. Unlike the Darcy's law term it is very different for the channel array and the staggered array, a fact borne out by experiment (Brown, 1986). This means that the coefficient of this term observed experimentally may give information on the structure of a filter.

A cubic term is small at small Reynolds number, and so it is likely that fluid inertia will affect the flow pattern even at values of Re where the effect on pressure drop will be negligible. The upstream–downstream symmetry of Stokes flow may be destroyed to the extent of the formation of weak vortex on the downstream surface of the fiber. However, symmetrical vortices can occur in Stokes flow where fibers are in close proximity (Davies et al., 1976) and it is likely that this is more important in determining the flow pattern than inertial effects.

8.8 AERODYNAMIC SLIP

A detailed account of the nature of aerodynamic slip with reference to flow through fibrous filters has been given by Pich (1971), who specified a number of regimes according to the value of the Knudsen number, Kn, where

$$\mathrm{Kn} = \frac{\lambda}{R} \tag{8.17}$$

Of these regimes only one, the slip regime, where Kn < 9.25, is amenable to relatively straightforward analysis. In this situation molecular effects can be adequately described by an adjustment of the boundary condition of zero tangential flow to one in which the air velocity at the surface is proportional to the tangential stress.

$$U_\theta \cong 0.998\ \mathrm{Kn}\left(\frac{\partial U_\theta}{\partial r} + \frac{1}{r}\frac{\partial U_r}{\partial \theta} - \frac{U_\theta}{r}\right)_{r=R} \tag{8.18}$$

The effect of this is to allow more air to pass close to the fibers, which generally reduces the pressure drop and increases the filtration efficiency.

This approach can be applied to the cell theories, to the complex variable theory, and to any of the more complicated approaches described above. The effect on pressure drop for small Kn obeys the following relationship for cell theory and complex variable theory, in which an analytical approach is possible

$$\frac{1}{\Delta p} = \frac{1}{\Delta p_0} + \frac{R^2}{4\eta c U h} g(c)\mathrm{Kn} \tag{8.19}$$

If applied to the Kuwabara model the multiplying term is (Pich, 1966)

$$g(c) = 1.0 - 2c + c^2. \tag{8.20}$$

For the complex variable model, in which an interlayer spacing similar to that of the interfiber spacing is assumed, the value is

$$g(c) = 1.5 - \frac{\pi c}{2}, \tag{8.21}$$

although the values of pressure drop predicted by the three models are very similar indeed. This implies that the response of filters to aerodynamic slip conditions is rather more structure dependent than the pressure drop in non-slip conditions.

8.9 MACROSCOPIC FLOW PATTERN THROUGH FILTERS

The problem of macroscopic flow has been mentioned with reference to the continuum theory of flow, but it merits study in its own right. If flow is considered at a scale large compared with individual fibers or interfiber spaces but small compared with the size of the filter itself, the filter can be considered to be a uniformly resistive medium. Two-dimensional flow at this level obeys Laplace's equation, and potential flow is a good description whatever the geometry.

$$\nabla^2\psi = 0 \tag{8.22}$$

This means that flow problems in this regime can be solved by electrical analogue methods such as conducting paper in two dimensions or conducting electrolyte in three.

Two simple examples of interest are pleated filters (Brown, 1983) and filters that are restricted by having supporting grids close to them. In the case of a filter with a regular two-dimensional grid, the stream function for the flow within the filter itself has a simple two-dimensional solution.

$$\psi = Uy + Ul\sum_{n=1}^{\infty} d_n \sin\left(\frac{n\pi y}{l}\right)\exp\left(-\frac{n\pi x}{l}\right) \tag{8.23}$$

The coefficients $\{d_n\}$ can be obtained by Fourier analysis on the presumed flow profile at the position of the grid, and the flow pattern for a thick filter is shown in Figure 8.12. The occlusion of the filter will increase its effective resistance, although no part of the filter receives zero fluid flow.

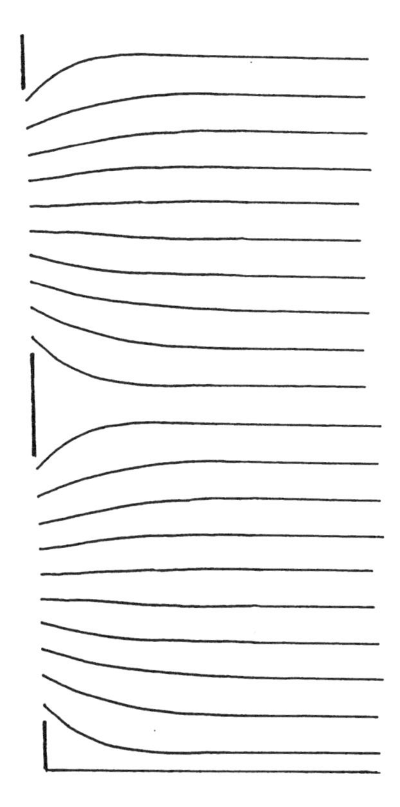

Figure 8.12 Calculated flow pattern for macroscopic flow through a filter supported by a grid.

8.10 SEMIMICROSCOPIC MODELS OF AIRFLOW

The models of airflow at a microscopic level have shown that the complexity of calculation increases considerably as the symmetry of the model is reduced. There is no fundamental limit to the degree of complexity that can be accommodated, but the computation time soon becomes prohibitive. Moreover, if a solution of very high complexity is obtained, it is not certain that we are very much the wiser for it. Approximations are needed that contain more detail than single-fiber theory but which are simple enough to illuminate general aspects of filter behavior.

If a local resistance in a filter can be defined, then the macroscopic flow problem can be solved, either by taking small elements of filter and matching the flow at the boundaries (Lajos, 1985) or by defining a spatially varying resistance and finding the flow pattern by variational method (Brown,

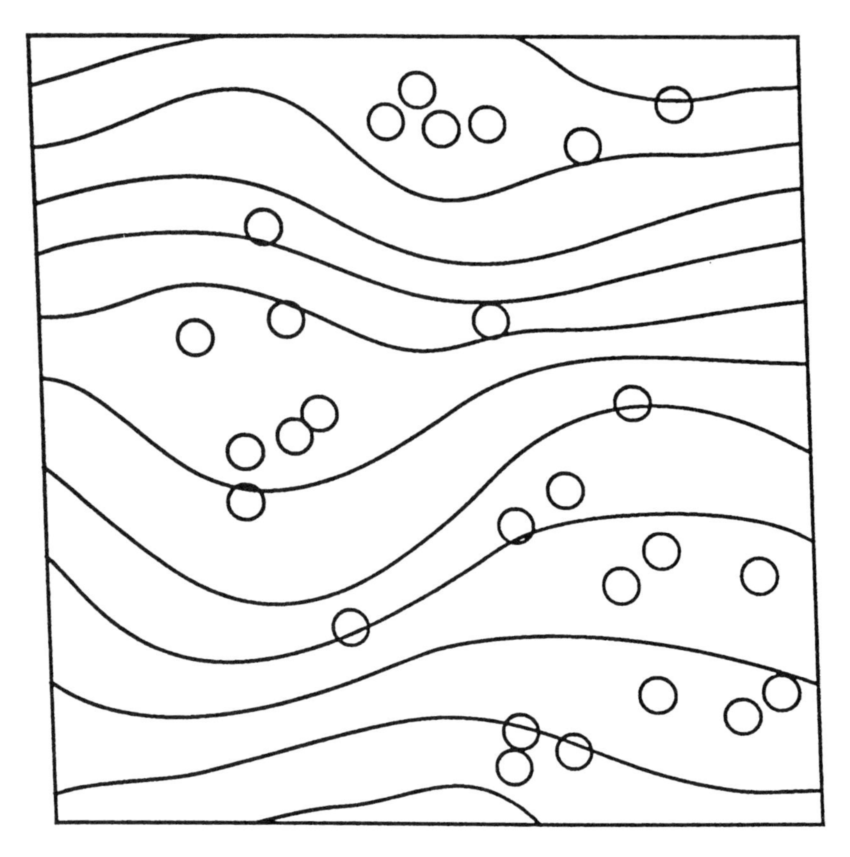

Figure 8.13 Section through two-dimensional filter illustrating fibers and semimicroscopic flow pattern. (British Crown Copyright; by permission of Her Majesty's Stationery Office.)

1994) using Fourier coefficients as variational parameters. An example of variable density and the flow through it are shown for a two-dimensional structure in Figure 8.13.

A serious problem with both approaches is that the picture obtained depends on the scale at which the filter is subdivided. If the filter is taken as a whole, the flow will be uniform and all structure will be lost. If the flow is studied at microscopic level all details will be retained, but the problem becomes intractable. It is necessary to seek a level that will contain the essential details of the flow but will be within the scope of ordinary computation. The local flow velocity obtained with this, combined with a single fiber or regular array approach, should then be capable of enabling the fiber and the filter properties to be described.

The problem of scale in a description of filter structure is illustrated by Schweers and Löffler (1994) who plotted cumulative distributions of the packing fraction observed in cubic elements of filter as a function of size. Packing fraction depends on the number of fibers counted and so is strictly a discrete variable. Complete randomness would result in a Poisson distribution, which for the values of parameters studied is well approximated by a normal distribution in which mean and variance in the number if fibers counted are equal. This functional form was observed by these authors, the variance in packing fraction being roughly proportional to the field size raised to the power $-^1/_2$ as expected. Vaughan and Brown (1996), using a basically similar method of two-dimensional sections, studied a filter in which the variance exceeded the mean when the field size was large, signifying a measure of clumping of the fibers.

Any study of airflow through a filter of heterogeneous structure will indicate that flow is greater through regions where fibers are fewer, as illustrated in Figure 8.13, and this will inevitably reduce the likelihood of particle–fiber contact. In detailed calculations of filtration efficiency, Schweers and Loeffler (1994) used a semimicroscopic approach to the calculation of flow velocity and direction through cubic regions in a filter, and then a single-fiber approach within these regions. Filtration efficiency tended to be reduced below that of a homogeneous filter, but in addition its dependence on the appropriate dimensionless parameters became weaker. This was particularly marked for capture by inertial impaction, in which simple theories predict an abrupt increase in single-fiber efficiency when the Stokes number reaches a value of the order of unity.

8.11 CONCLUSIONS

A study of airflow beyond single-fiber approaches needs to take detailed account of filter structure. Flow through regular arrays is relatively easy to describe, and this allows experimental results on model filters to be interpreted. However, the computational difficulty increases as the structure becomes less regular. A complete semimicroscopic method is needed, but the dependence of the solution on the choice of scale needs to be resolved, and the connection between flow at a larger scale and microscopic flow around fibers, which is necessary for the description of particle capture, needs to be placed on a secure footing.

Filter structure merits study in its own right, and any means of obtaining information on this is useful. Studies of airflow have revealed that pressure drop in the Darcy's law regime is not particularly sensitive to structure other than by the packing fraction, unless the filter is compressed. However, the pressure drop at air velocities where Darcy's law ceases to apply, and the pressure drop in the situation of aerodynamic slip are more structure sensitive, and quantitative observations of these effects may give information on certain aspects of structure.

Although the fiber arrays studied so far, in particular those that have reduced symmetry, represent an advance on single-fiber theory, none could be considered accurately to represent even the structures of simplified model filters, because it is impossible to construct a filter without points of fiber–fiber contact. Description of these requires a three-dimensional approach. Most of the descriptions of microscopic three-dimensional flow are aimed at granular filters, and are not readily applicable to real fibrous filters. It is necessary, for further progress, that three-dimensional models of fibrous filters be developed.

REFERENCES

Albrecht, F. (1931) Theoretische Untersuchungen uber die ablegerung von Staub und Luft ihre Anwendungauf die Theorie der Staubfilter, *Phys. Z.,* 32, 48.

Brown, R. C. (1983) The use of the variational principle in the solution of Stokes flow problems in fibrous filters, *J. Phys. D.,* 16, 743–754.

Brown, R. C. (1984) A many-fibre model of airflow through a fibrous filter, *J. Aerosol Sci.,* 15(5), 583–593.

Brown, R. C. (1985) A model of filter behaviour including the effects of fibre shape and filter structure, paper presented at International Symposium on Air Pollution Abatement by Air Filtration and Related Methods (Respiratory Protection), Dantest, Copenhagen.

Brown, R. C. (1986) A many fibre theory of airflow through a fibrous filter — II: fluid inertia and fibre proximity, *J. Aerosol Sci.,* 17(4), 685–697.

Brown, R. C. (1988) An account of depth filtration in air cleaning with particular reference to electrically charged filter material, in *Ventilation '88,* J. H. Vincent, Ed., Pergamon, Oxford.

Brown, R. C. (1993a) Theory of airflow through filters modelled as arrays of parallel fibres, *Chem. Eng. Sci.,* 48(20), 3535–3543.

Brown, R. C. (1993b) *Air Filtration: An Integrated Approach to the Theory and Applications of Fibrous Filters,* Pergamon, Oxford.

Brown, R. C. (1994) Theory of Stokes flow at a semi-microscopic level through fibrous filters of non-uniform structure, *Staub Reinhalt. Luft,* 54, 59–65 (in English).

Brown, R. C. and Wake, D. (1991) Air Filtration by interception — theory and experiment, *J. Aerosol Sci.*, 22(2), 181–186.

Carman, P. C. (1937) Fluid flow through granular beds, *Trans. Inst. Chem. Eng.*, 15, 150–166.

Davies, A. M. J., O'Neill, M. E., Dorrepaal, J., and Ranger, K. B. (1976) Separation from the surface of two equal spheres in Stokes flow, *J. Fluid Mech.*, 77, 625–644.

Davies, C. N. (1950) Viscous flow transverse to a circular cylinder, *Proc. Phys. Soc.*, B63, 288–296.

Davies, C. N. (1973) *Air Filtration,* Academic Press, London.

Drummond, J. E. and Tahir, M. I. (1984) Laminar, viscous flow through regular arrays of parallel circular cylinders, *Int. J. Multiphase Flow,* 10, 515–540.

Fardi, B. and Liu, B. Y. H. (1992) Flow field and pressure drop of filters with rectangular fibres, *Aerosol Sci. Technol.*, 17, 36–44.

Happel J. (1959) Viscous flow relative to arrays of cylinders, *A.I.Che.E.J.*, 5(2), 174–177.

Hildyard, M. L., Ingham, D. B., Heggs, P. J., and Kelmanson, M. A. (1985) Integral equation solution of viscous flow through a fibrous filter, in *Boundary Elements,* VII, C. A. Brebbia and G. Maier, Eds., Springer-Verlag, Berlin.

Hill, J. R. and Power, G. (1956) Extremum principles for slow viscous flow and the approximate calculation of drag, *Q. J. Mech. Appl. Math.*, 9, 313–319.

Jackson, G. W. and James, D. F. (1986) The permeability of porous media, *Can. J. Chem. Eng.*, 64, 364–374.

Jefferies, R. A. (1974) Fibrous material for the filtration of gases, Publication No. S14, Shirley Institute, Manchester, U.K.

Kirsch, A. A. and Fuchs, N. A. (1967) Studies on fibrous aerosol filters II. Pressure drops in systems of parallel cylinders, *Ann. Occup. Hyg.*, 10, 23–30.

Kirsch, A. A. and Fuchs, N. A. (1968) Studies on fibrous aerosol filters — III. Diffusional deposition of aerosol in fibrous filters, *Ann. Occup. Hyg.*, 11, 299–304.

Kirsch, A. A. and Stechkina, I. B. (1973) Pressure drop and diffusional deposition of aerosol in polydisperse model filter, *J. Coll. Int. Sci.*, 43, 10–16.

Kuwabara, S. (1959) The forces experienced by randomly distributed parallel circular cylinders or spheres in a viscous flow at small Reynolds numbers, *J. Phys. Soc. Jpn.*, 14, 527–532.

Lajos, T. (1985) The effect of inhomogeneity on flow in fibrous filters, *Staub Reinhalt. Luft,* 45(1), 19–22 (in English).

Miyagi, T. (1958) Viscous flow at low Reynolds numbers past an infinite row of equal circular cylinders, *J. Phys. Soc. Jpn.*, 13(5), 493–496.

Moncrieff, R. W. (1975) *Man-Made Fibres*, Newnes-Butterworth, London.

Pich, J. (1966) in *Aerosol Science,* Davies, C. N., Ed., Academic Press, London.

Pich, J. (1971) Pressure characteristics of fibrous aerosol filters, *J. Coll. Int. Sci.*, 37(4), 912–917.

Sangani, A. S. and Acrivos, A. (1982) Slow flow past periodic arrays of cylinders with application to heat transfer, *Int. J. Multiphase Flow,* 8(3), 193–306.

Schweers, E. and Loeffler, F. (1994) Realistic modelling of the behaviour of fibrous filters through a consideration of fibre structure, *Powder Technol.*, 80, 191–206.

Sparrow, E. M. and Loeffler, A. L. (1959) Longitudinal laminar flow between circular cylinders arranged in regular array, *A.I.Ch.E.J.*, 5(3), 325–330.

Spielman, I. and Goren, S. L. (1969) Model for predicting pressure drop and filtration efficiency in fibrous media, *Environ. Sci. Technol.*, 2, 279–287.

Vaughan, N. P. and Brown, R. C. (1996) Observations of the microscopic structure of fibrous filter, *Filtration and Separation,* (refereed papers), 33(5), 741–748.

CHAPTER 9

Numerical Simulation of Gas-Particle Flow in Filters by Lattice Bhatnagar–Gross–Krook Model

Olga Filippova and Dieter Hänel

CONTENTS

Abstract — This chapter deals with the numerical simulation of two- and three-dimensional gas–particle flow through filters. Typical for filter flows is the deposition of particles on the surfaces of filter material, resulting in change of the shapes of surfaces, sometimes in a complex irregular manner, which in interaction with the flow field influences the next particle deposition. The investigations have shown that the lattice Bhatnagar–Gross–Krook approach performs very well for such interactive flow around complex surfaces. The new kind of boundary conditions fits with the second-order accuracy the hydrodynamic no-slip conditions on the surface of arbitrary form lying between the nodes of regular lattice. This simulates in an effective way the viscous incompressible flow with complex boundaries at small Reynolds numbers. The motion of the solid phase is calculated depending on the particle size by means of a Lagrangian approach or by convective–diffusion equation. This solution concept is demonstrated for gas–particle flow through a model of a fibrous filter for the cases of diffusional and inertial deposition. The obstruction of the filter vs. time is considered.

0-87371-830-5/98/$0.00+$.50

9.1 INTRODUCTION

Filters of a different kind are used for extraction of a solid phase from a gas phase. The geometric and physical properties of two-phase flow in realistic filter configurations are usually so complex that in practice filter performance is estimated by empirical or semiempirical relations only. However, for prediction and improvement of filter performance, detailed knowledge of internal two-phase flow and, in particular, of its interaction with the filter material is essential. Modern computational fluid dynamics offers a promising way to achieve more-detailed information about filter flow, at least for model problems close to realistic configurations.

The present chapter deals with the numerical simulation of gas–particle flow through filters. A numerical solution concept is presented which appears well suited to study the fluid flow, the particle transport, and deposition in three-dimensional (3-D) models of fibrous filters. Particle–gas flow over rigid surfaces causes deposition of particles on the surfaces. The change of shape due to deposited particles changes the flow field which influences again the next particle deposition. Such surface effects determine the collection efficiency of filters, which is one of the important parameters of the flow in filters.

The computation of this interactive flow problem results in an iterative procedure between fluid flow and particle transport and deposition. The simulation of the filter flow requires, therefore, modeling and solution of three essential parts, of the particle motion, of the deposition, and of the fluid flow.

Two different typical sizes of particles are considered, very small nanoparticles with diameter on the order of 10 nm and large particles with diameter on the order of 1 μm. The motion of large particles is appropriately described by the Lagrangian approach, i.e., by the integration of the equation of motion for each single particle. This method enables consideration of particle motion under inertia and any other external force. The motion of nanoparticles is described by the convective–diffusion equation for particle concentration.

The deposition of particles is of complex nature; it depends in principle on intermolecular forces. Its detailed description is not the task of this study; therefore, a rather simple model is used, which assumes any large particle as deposited if it has touched the solid wall or another deposited particle. For nanoparticles the condition of zero particle concentration at the surfaces of obstacles describes well the diffusional particle mass flux to them. This flux is proportional to the speed of growth of the deposited layer whose boundary is considered now as a smooth surface.

The motion of the fluid, which is the carrier phase, determines in essence the particle transport and deposition and requires the high-accuracy calculation of a viscous flow around 3-D surfaces. The fluid flow is characterized by small speeds and small dimensions, resulting in small values of Reynolds numbers and very small Mach numbers. Particular difficulties arise from the 3-D irregular surfaces caused by deposited large particles. Conventional Navier–Stokes solvers which are able to solve the flow around complex geometries (as solvers on adaptive unstructured grids) become rather expensive because of the high resolution necessary near the deposit, while due to the small Reynolds number there are no regions of high gradients of the hydrodynamical values. An efficient alternative for the description of this kind of flow was found to be the lattice Bhatnagar–Gross–Krook (LBGK) model. The conditions on complex, rigid boundaries can be satisfied on a fixed lattice with second-order accuracy with a currently developed boundary-fitting procedure for the LBGK model.

Lattice gas methods proposed in the pioneering work of Frisch, Hasslacher, and Pommeau[1] in 1986 can be defined as low Mach number approximations of kinetic gas dynamics; thus, they approximate the Navier–Stokes equations for nearly incompressible flow in the continuum limit. The lattice gas concepts can be realized either by statistical methods (Boolean gas, lattice gas automata, LGA[1,2] analogous to Monte Carlo methods or by solving an equation for the rate of a velocity distribution function, e.g., a Boltzmann equation[3,4] or an approximation of it, like the BGK model,[5,6] (below LBGK) used here. Essential advantages, compared with solutions of the macroscopic conservation laws, are seen in their simple algorithms, allowing parallel implementation, intrinsic stability, and the ability to deal with complex geometric boundaries.

Several attempts were made to apply LGA to the problems of particle deposition. In Reference 7 the deposition of particles on a wall in a 2-D semi-infinite flow was considered. The two-phase flow was described with the LGA model developed earlier for fluid[2] in which some modifications were introduced. The possibility of LGA application to the problems of dendrite formation was noted by Fissan.[8] But LGA have some essential drawbacks, such as statistical noise in the results and very complex extension to 3-D. All these drawbacks were removed in the LBGK model proposed by Qian et al.[5] First results with this method were obtained by the present authors in References 9 and 10. Results were achieved for particle deposition on 2-D and 3-D geometries using the assumption of small particle concentration in the flow.

The full extension of the LBGK method in combination with a Lagrangian approach for the particle phase for the problems of inertial particle deposition in filters is presented in this chapter. In combination with the convective–diffusion equations the LBGK model is able to describe the problems of pure diffusional deposition. In the following sections the details of the numerical approach are briefly outlined. The capability of the solution concept is demonstrated by a 2-D and 3-D computation of gas–particle flow through a model of a filter.

9.2 METHODS OF SOLUTION

9.2.1 LBGK Model for Fluid Flow

The computation of the fluid phase is based on the LBGK model proposed by Qian et al.[5] The main idea of lattice gas methods is the creation of a fictitious microscopic world of molecular dynamics in which all is discrete: the space, the time, and the possible molecular velocities. Fictitious "molecules" of the same mass can move only along the links of some regular lattice and collide in the nodes with the conservation of mass and momentum. By analogy with the gas kinetic theory one can introduce the discrete set of the distribution function $f(t,\vec{r},\vec{c})$ which can be interpreted as the probability of the finding some "molecule" moving with the speed $\vec{c}$ at the time t in the node of the lattice with coordinates $\vec{r}$.

The rate of change of the distribution function $f(t,\vec{r},\vec{c})$ due to transport and collisions is described by the BGK model, well known in gas kinetic theory:

$$\frac{\partial f}{\partial t}+\vec{c}\frac{\partial f}{\partial \vec{r}}=\omega\left(f^{\mathrm{eq}}-f\right) \tag{9.1}$$

where $f^{eq}(t,\vec{r},\vec{c})$ is the equilibrium distribution function and ω is a collision frequency, depending on the viscosity.

The BGK equation (Equation 9.1) is discretized in a Cartesian lattice with additional diagonal links. The "molecular" speeds have the directions along the links i to the neighboring nodes (8 links + 1 for zero speed in 2-D; 14 links + 1 for zero speed in 3-D, Figure 9.1). The Cartesian components of the speeds have unit values, which are chosen so that the CFL condition CFL = $c_k \cdot \Delta t/\Delta r_k = 1 (k = 1, 2, 3)$ is satisfied. The distributions function is updated for each node and each discrete speed i by the discretized BGK model, Equation 9.1:

$$f_{pi}\left(t+\Delta t,\vec{r}+\vec{c}_{pi}\Delta t\right)=\left(1-\omega\Delta t\right)f_{pi}\left(t,\vec{r}\right)+\omega\Delta t f_{pi}^{eq}\left(t,\vec{r}\right) \tag{9.2}$$

The index p is the square modulus of the "molecular" speeds $\vec{c}_{pi}$, $p =$ 0, 1, 2 in 2-D and $p =$ 0, 1, 3 in 3-D.

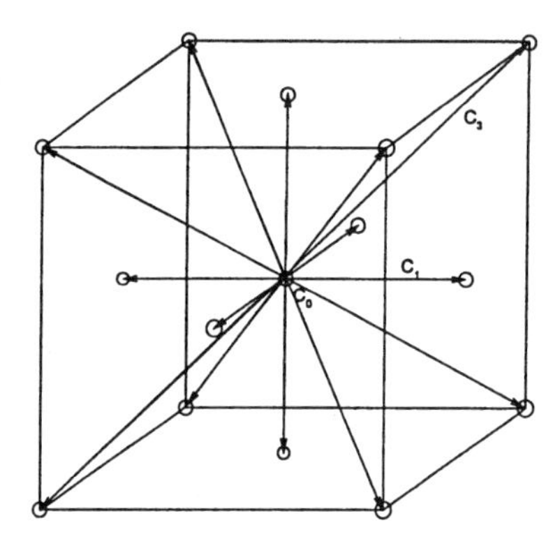

Figure 9.1 Cubic lattice with 15 possible velocities.

The equilibrium distribution function f_{pi}^{eq} is derived from the Maxwellian equilibrium function for the discrete model and for small Mach numbers $M = u/c_s$, where c_s is the isothermal speed of sound. By using Taylor series expansion, the discrete equilibrium function results in:[5]

$$f_{pi}^{eq}(t,r) = t_p\rho\left(1 + \frac{c_{pia}u_a}{c_s^2} + \frac{u_a u_b}{2c_s^2}\cdot\left(\frac{c_{pia}c_{pib}}{c_s^2} - \delta_{ab}\right)\right) \tag{9.3}$$

Cartesian coordinates are represented by a and b, assuming implied summation for repeated indices. The weighting coefficients t_p and the dimensionless speed of sound c_s are obtained from the moments of the equilibrium distribution function. For the cubic lattice of dimension d (denoted $DdQb$) the values of t_p are the following:[5]

$$D2Q9: t_0 = \frac{4}{9},\ t_1 = \frac{1}{9},\ t_2 = \frac{1}{36};\ D3Q15: t_0 = \frac{2}{9},\ t_1 = \frac{1}{9},\ t_3 = \frac{1}{72} \tag{9.4}$$

The dimensionless speed of sound c_s is equal to $1/\sqrt{3}$. The macroscopic quantities are achieved by taking the moments of f for each node.

$$\rho = \sum_i f_i,\quad \rho\vec{v} = \sum_i \vec{c}_i f_i,\quad p = c_s^2 \cdot \rho \tag{9.5}$$

With a Chapman–Enskog expansion

$$f_{pi} = f_{pi}^{eq} + \varepsilon f_{pi}^{(1)} + \varepsilon^2 f_{pi}^{(2)} + \cdots$$

using multiscale technique,[11] one can prove that this discrete lattice gas model approximates the following fluid equations

$$\partial_t\rho + \partial_\alpha(\rho u_\alpha) + O(\varepsilon^3) = 0 \tag{9.6}$$

$$\rho\partial_t(u_\alpha) + \rho u_\beta\partial_\beta(u_\alpha) + \partial_\alpha(c_s^2\rho) - \nu\partial_\beta\left(\partial_\beta(\rho u_\alpha) + \partial_\alpha(\rho u_\beta)\right) + O(\varepsilon^3) = 0 \tag{9.7}$$

The similarity parameter ε corresponds to a Knudsen number $\varepsilon = l/L$ where l is the length of the unit link and L is the characteristic length of the fluid flow.

The equation of state is given by

$$p = c_s^2 \rho \tag{9.8}$$

The dimensionless shear viscosity ν is defined as function of the collision frequency ω by

$$\nu = \frac{1}{6}\left(\frac{2}{\omega} - 1\right)$$

As far as the filter flows can be considered as incompressible ones, we will prove that in the stationary case Equations 9.6 and 9.7 approach the incompressible Navier–Stokes equation. In the case of small particle concentration, when the particle phase influences the fluid phase only through the changed form of the boundary, one can consider the motion of the carrier phase also as stationary. Consider the flow with small deviations in density $\rho = \rho_0 + M^a r$. Here M is the Mach number $M = u/c_s$. With the assumption of the steady-state flow one can rewrite Equations 9.6 through 9.8 in the following form

$$\rho_0 \partial_\alpha u_\alpha + \partial_\alpha \left(M^a r u_\alpha\right) + O\left(\varepsilon^3\right) = 0 \tag{9.9}$$

$$\rho_0 u_\beta \partial_\beta(u_\alpha) + \partial_\alpha(p) - \nu\rho_0 \Delta u_\alpha - \nu \partial_\beta u_\alpha \cdot \partial_\beta \rho - \nu u_\alpha \Delta\rho + M^a r u_\beta \partial_\beta u_\alpha + O(\varepsilon^3) = 0 \tag{9.10}$$

Then, the order of the terms in Equations 9.9 and 9.10 is the following:

$$M\varepsilon,\ M^{a+1}\varepsilon,\ \varepsilon^3 \tag{9.11}$$

$$M^2\varepsilon,\ M^a\varepsilon,\ \frac{M^2\varepsilon}{\mathrm{Re}},\ \frac{M^{a+2}\varepsilon}{\mathrm{Re}},\ \frac{M^{a+2}\varepsilon}{\mathrm{Re}},\ M^{a+2}\varepsilon,\ \varepsilon^3 \tag{9.12}$$

The first three terms in 9.12 correspond to the Navier–Stokes equation for incompressible fluid. The other are the "waste" of this artificial model.

It can be shown easily that for the stationary case the last term in Equation 9.9 is in the order of $M\varepsilon^3$ and in 9.10 in the order of $O(M^2,M^a)\varepsilon^3$ instead of ε^3.

Then the order of terms in the Equations 9.11 and 9.12 is the following:

$$M\varepsilon,\ M^{a+1}\varepsilon,\ M\varepsilon^3 \tag{9.13}$$

$$M^2\varepsilon,\ M^a\varepsilon,\ \frac{M^2\varepsilon}{\mathrm{Re}},\ \frac{M^{2+a}\varepsilon}{\mathrm{Re}},\ M^{a+2}\varepsilon,\ M^a\varepsilon^3,\ M^2\varepsilon^3 \tag{9.14}$$

Here $\mathrm{Re} \sim M/\varepsilon$. For Mach numbers $M \sim \varepsilon$, $a = 2$ and $M \sim \varepsilon^2$, $a = 1.5$ the ratio between truncated and leading terms in the incompressible Navier–Stokes equations (Equations 9.13 and 9.14) is $O(\varepsilon^2)$. Accordingly, the solution obtained from the discrete BGK model approximates the solution of incompressible Navier–Stokes equations (fluid density ρ_0) with second-order accuracy in space, if the calculated deviation in density $\rho - \rho_0 \sim M^a$ which defines the field of pressure according to the Equation 9.8, is of the prescribed order a. This order must be checked during the calculations.

9.2.2 Boundary-Fitting Concept

Solid bodies in the computational domain are represented with the set of nodes of the grid ("rigid" nodes) that are lying inside of the bodies. The "stream and collide" procedure, i.e., the redefinition of the distribution functions and shift of the new values to the neighboring nodes according to Equation 9.2 produces hydrodynamic values in every "fluid" node of equidistant Cartesian grid.

Two different types of boundary conditions are used usually. The first one is bouncing-back of the distribution function in the rigid nodes. This corresponds to a no-slip boundary lying halfway between the wall nodes and the nearest fluid nodes.[12] The other way is to set the distribution function in the rigid boundary nodes equal to the equilibrium distribution function with zero velocity and density extrapolated from the flow and include boundary nodes into the relaxation procedure of Equation 9.2. In this case the boundary passes exactly through the rigid nodes. Both boundary conditions have a common drawback: they are only of first-order accuracy; thus, in Cartesian grids any curved boundary is approximated as a "stepbody." To increase the resolution near the boundaries one has to increase the number of nodes in the whole computational domain (or use local refinement near the bodies).

A new improved boundary formulation has been developed using the fact that fluid flow in filters is characterized by relatively small Re numbers (Re $\leq O(10)$). The thickness of the boundary layer is on the order of the characteristic length of the body. For these kinds of flows it is possible to conserve the accuracy of solution on the equidistant Cartesian grid with the new kind of boundary conditions described below.

Consider a curved boundary lying between the nodes of the equidistant lattice with the size of the link ε as sketched in Figure 9.2. According to the above estimations $\partial_x\rho$, $\partial_y\rho$ are in the order of $O(M^a)$.

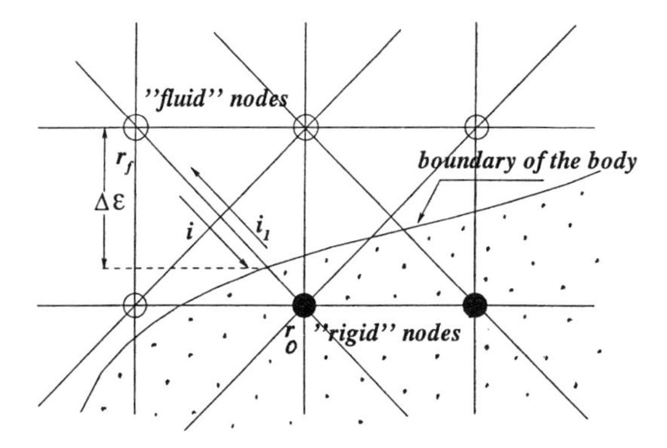

Figure 9.2 Computational mesh and geometric relations for fitting of solid boundaries.

Introduce the notations "incoming" and "outcoming" distribution functions in the fluid node

$$f_{pi}^{\text{in}}(\vec{r},t) = f_{pi}(\vec{r},t)$$

$$f_{pi}^{\text{out}}(\vec{r},t) = f_{pi}(\vec{r},t) + \omega\left(f_{pi}^{eq}(\vec{r},t) - f_{pi}(\vec{r},t)\right)$$

The relationship between incoming and outcoming distribution functions in the fluid node close to the boundary along the link that connects fluid node $\vec{r}$ and rigid node $\vec{r}_0$ is prescribed to be the following:

$$f_{pi_1}^{\text{in}}(\vec{r}) = f_{pi}^{\text{out}}(\vec{r})(1-\omega_i) + a_1\omega_i f_{pi}^{\text{eq}}(\vec{r}_0) + a_2\omega_i f_{pi}^{\text{eq}}(\vec{r}) \tag{9.15}$$

where

$$\vec{c}_{pi_1} = -\vec{c}_{pi}$$

with the assumption that

$$a_1 \cdot a_2 = 0, \quad a_1^2 + a_2^2 = 1$$

The "equilibrium distribution function" in the rigid nodes $f_{pi}^{\text{eq}}(\vec{r}_0)$ (which, of course, has no physical sense) is defined as

$$f_{pi}^{\text{eq}}(\vec{r}_0) = t_p\rho(\vec{r})\left(1 + \frac{u_\alpha(\vec{r}_0)c_{pi\alpha}}{c_s^2} + \frac{u_\alpha(\vec{r})u_\beta(\vec{r})}{2c_s^2}\left(\frac{c_{pi\alpha}c_{pi\beta}}{c_s^2} - \delta_{\alpha\beta}\right)\right) \tag{9.16}$$

where $u(\vec{r}_0)$ is the value of u extrapolated from the node $\vec{r}$ and the value of density and velocity in the square-order term are taken from the node $\vec{r}$.

The adjusting parameter ω_i is connected with the distance $\Delta\sqrt{c_{pix}c_{pix} + c_{piy}c_{piy}}\,\varepsilon$ to the boundary along the link i. When $\omega_i = \omega$ and $a_2 = 0$, Equation 9.15 is the Krook equation on the boundary crossing the node $\vec{r}_0$ ($\Delta_i = 1$).

Taking into account that

$$f_{pi}^{\text{eq}}(\vec{r}) = t_p\rho(\vec{r})\left(1 + \frac{u_\alpha(\vec{r})c_{pi\alpha}}{c_s^2} + \frac{u_\alpha(\vec{r})u_\beta(\vec{r})}{2c_s^2}\left(\frac{c_{pi\alpha}c_{pi\beta}}{c_s^2} - \delta_{\alpha\beta}\right)\right) \tag{9.17}$$

and that in the stationary case the first term in Chapman–Enskog expansion is

$$f_{pi}^{(1)}(\vec{r}) = -\frac{1}{\omega}\frac{\partial f_{pi}^{\text{eq}}(\vec{r})}{\partial x_\alpha}c_{pi\alpha} = -\frac{t_p\rho}{\omega c_s^2}\frac{\partial u_\beta}{\partial x_\alpha}c_{pi\alpha}c_{pi\beta} + O(M^a) \tag{9.18}$$

one can obtain from Equation 9.15 with accuracy $O(M\varepsilon^2)$

$$\begin{aligned} f_{pi_1}^{\text{in}}(\vec{r}) &= f_{pi_1}^{\text{eq}}(\vec{r}) + \varepsilon f_{pi_1}^{(1)} = f_{pi}^{\text{eq}}(\vec{r}) - 2t_p\rho(\vec{r})\frac{u_\alpha c_{pi\alpha}}{c_s^2} + \varepsilon f_{pi}^{(1)} \\ &= (1-\omega_i)\left(f_{pi}^{\text{eq}}(\vec{r}) + \varepsilon f_{pi}^{(1)}(1-\omega)\right) \\ &\quad + a_1\omega_i t_p\rho(\vec{r})\left(1 + \frac{u_\alpha(\vec{r}_0)c_{pi\alpha}}{c_s^2} + \frac{u_\alpha(\vec{r})u_\beta(\vec{r})}{2c_s^2}\left(\frac{c_{pi\alpha}c_{pi\beta}}{c_s^2} - \delta_{\alpha\beta}\right)\right) \\ &\quad + a_2\omega_i t_p\rho(\vec{r})\left(1 + \frac{u_\alpha(\vec{r})c_{pi\alpha}}{c_s^2} + \frac{u_\alpha(\vec{r})u_\beta(\vec{r})}{2c_s^2}\left(\frac{c_{pi\alpha}c_{pi\beta}}{c_s^2} - \delta_{\alpha\beta}\right)\right) \end{aligned} \tag{9.19}$$

The value of velocity in the rigid node $\vec{r}_0$ is prescribed by linear extrapolation from the node $\vec{r}$ through the zero value on the boundary:

$$\vec{u}(\vec{r}_0) = \frac{\Delta\varepsilon - \varepsilon}{\Delta\varepsilon}\vec{u}(\vec{r})$$

Taylor expansion of the velocity gives

$$\vec{u}(\vec{r}_0) = \frac{\Delta - 1}{\Delta}\partial_\alpha \vec{u}(\vec{r})\Delta\varepsilon c_{pi\alpha} + O(M\varepsilon^2)$$

Then in the case $a_1 = 1$, $a_2 = 0$ one can obtain from the Equation 9.19 the following relationship between ω_i and Δ

$$-2 + \frac{\omega_i}{\Delta} + \frac{\omega_i + \omega - \omega\omega_i}{\Delta\omega} + O(M^{a-1}\varepsilon) = 0$$

Then

$$\begin{cases} \omega_i = \omega(2\Delta - 1) \\ a_1 = 1,\ a_2 = 0 \end{cases} \tag{9.20}$$

In the case $a_1 = 0$, $a_2 = 1$ one can obtain from the Equation 9.19 the following relationship between ω_i and Δ

$$((\omega_i + \omega - \omega\omega_i) - 2\omega\Delta) + O(M^{a-1}\varepsilon) = 0$$

Then

$$\begin{cases} \omega_i = \omega\dfrac{2\Delta - 1}{1 - \omega} \\ a_1 = 0,\ a_2 = 1 \end{cases} \tag{9.21}$$

where ω is a relaxation parameter in the fluid nodes.

For reasons of numerical stability the boundary conditions are taken in the form Equation 9.20 when $\Delta \geq 0.5$ and in the form Equation 9.21 when $\Delta < 0.5$.

These boundary formulations approximate the no-slip conditions with second-order accuracy at the exact position of any boundary cutting the link between a fluid and a rigid node. The boundary conditions are derived for the stationary case. This restriction is sufficient for gas–particle flow in filters since the typical timescale of the flow is much smaller then the timescale of the growth of the deposit.[14] By analogy with the method of artificial compressibility we are coming to these solutions through nonphysical time–space using the usual stream-and-collide procedure with the time-averaged distribution function in Equation 9.2 and boundary conditions set as for the limiting case $\partial_t = 0$.

9.2.3 Particle Motion and Deposition

Two different typical sizes of particles are considered, nanoparticles with diameter $O(10\text{ nm})$ and large particles with diameter $O(1\ \mu\text{m})$. In the absence of the external forces, such as electrical forces, etc., the main mechanism of deposition for nanoparticles is their diffusion and for the large ones the inertia motion. Therefore, the motion of particles is simulated in different ways. For large particles the Lagrangian approach is preferable, while for nanoparticles a convective–diffusion equation for particle concentration is solved.

(1) Continuum Approach for Fine Particles

The convective–diffusion equation for particle concentration in the known flow field has the following form:

$$\frac{\partial c}{\partial t} + \operatorname{div}\left(c\vec{u}_{\text{gas}}\right) = D\Delta c \tag{9.22}$$

where D is the coefficient of particle diffusion.

In the case of the deposition of the fine particles the surface of the deposit can be considered as smooth ("dust cake") and the process of filter obstruction can be considered in 2-D if the initial filter flow was two dimensional.

Consider the flow in the 2-D computational cell with the periodic boundary conditions in y-direction. Then the boundary conditions for c are periodic conditions on the upper and lower boundaries of the computational cell, constant concentration at the entrance and zero concentration on the surfaces of obstacles. The field of concentration in the known flow field is calculated up to steady state as far as, with the assumption of the small particle concentration in the flow, one can consider the process of filter obstruction as quasi stationary. Also in the case of high packing density of the particles in the deposit one can neglect the porosity of the growing deposited layer and consider it as the rigid body.

Diffusion particle mass flux to the surface

$$j_n = -D\frac{\partial c}{\partial n}$$

is proportional to the rate of the growth of the deposited layer

$$U_n = \frac{j_n}{c_{\text{dep}}}$$

Consider the equation of the surface of fiber with deposited layer in polar coordinates (Figure 9.3)

$$r(t,\varphi) = \sum_{k=1}^{n} a_k(t)\cos\left(\varphi(k-1)\right)$$
$$i = 1,\ldots,n, \;\; a_1(t=0) = r_0(\varphi), \;\; a_j(t=0) = 0, \;\; j = 2,\ldots,n \tag{9.23}$$

The speed of the growth of deposited layer in n points φ_i in the directions r_i, $i = 1, \ldots, n$ is the following

$$\frac{\partial r}{\partial t}\left(t_0, \varphi_i\right) = -\frac{D}{c_{\text{dep}}}\frac{\dfrac{\partial c}{\partial n}}{\cos\left(\vec{r}_i\vec{n}\right)}$$

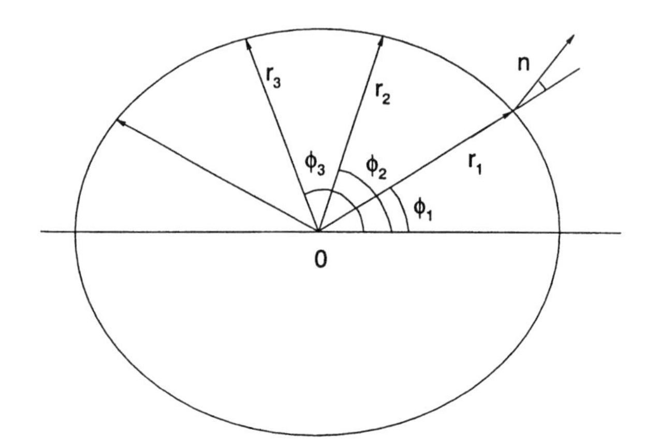

Figure 9.3 The surface of a filter in polar coordinates.

Then the derivatives of $a'_{k'}(t_0)$ can be found from the system of equations

$$-\frac{D}{c_{\text{dep}}}\frac{\frac{\partial c}{\partial n}}{\cos(\vec{r}_i\vec{n})}=\sum_{k=1}^{n} a'_k(t_0)\cos(\varphi_i(k-1)) \quad i=1,\ldots,n \tag{9.24}$$

With the new values of $a'_k(t_0 + \Delta t) = a_k(t_0) + a_{k'}(t_0)\,\Delta t$ the equation of the surface (Equation 9.23) is known at the new time $t_0 + \Delta t$. The nodes of the lattice which are lying inside of the new boundary alter from the fluid to the rigid ones. The flow field is recalculated according to the new boundary, and the procedure repeats again.

(2) Lagrangian Approach for Large Particles

Particles with the diameter in the order of 1 μm are considered as solid spheres in dilute concentration. They are treated as individual particles, and their motion is described by the Lagrangian approach of the equation of motion.

$$\frac{d\vec{v}_p}{dt}=-g\left(1-\frac{\rho_g}{\rho_p}\right)\vec{i}+\frac{\vec{F}_d}{m_p}+\frac{\vec{F}_{\text{ext}}}{m_p} \tag{9.25}$$

The drag force $\vec{F}_d$ is defined by

$$\vec{F}_d=\frac{1}{2}\rho_g c_d \pi r_p^2(\vec{v}_g-\vec{v}_p)|\vec{v}_g-\vec{v}_p|,\quad c_d=f(\mathrm{Re}_p),\quad \mathrm{Re}_p=\frac{2r_p|\vec{v}_g-\vec{v}_p|}{\nu} \tag{9.26}$$

where the indexes refer to the gas (g) and the particle (p). r_p is the particle radius, ν is the gas viscosity, and g is the gravity constant.

The drag of spherical particles is linear in velocity and is described by the Stokes formula $c_d = 24/\mathrm{Re}_p$ for Re_p less than 0.1. The drag coefficient c_d for the higher Re_p can be calculated using the equation of Morsi and Alexander.[16]

The numerical integration of the initial value problem, Equation 9.25, is performed in the same Cartesian grid as used for the LBGK model, starting from an initial position and velocity of the single particle equal to the gas velocity at the entrance. A four-step Runge–Kutta scheme is used with the second-order interpolation of the values of gas velocity from the nodes of equidistant Cartesian grid where they were calculated with the LBGK approach.

In the absence of external forces $\vec{F}_{ext}$ the main mechanism of deposition of particles on surfaces is their inertia motion. The inertia effect depends on the density ration ρ_p/ρ_g between solid phase and gas. For large density ratio, particles decelerate much slower in the boundary layers than the gas and touch or hit on the rigid surface.

Deposition of particles is assumed when a particle contacts a rigid wall or a previously deposited particle; then the particle is stopped and forms a part of a new surface. In the frame of the LBGK approach this means that each node of the grid lying inside of a deposited particle alters from fluid node to a rigid node.

In consequence, particles deposited on the rigid surfaces change their contours partially in an irregular and complex manner; also dendrite-like formations of particles can appear. The deposited layer changes the form of obstacles and, accordingly, the fluid field, which influences to the next particle deposition.

9.3 COMPUTATIONAL EXAMPLES FOR FILTER FLOW

The geometry of real filters is 3-D and complex, usually unknown in details. The model configurations with simplified geometry enable the analyses of the essential features of the internal two-phase flow and interactions, useful for the prediction and improvement of filter performance.

9.3.1 Diffusional Deposition of Fine Particles

The classical problem of the growth of deposited layer on the periodical grid of fibers[15] was considered. The gas–particle flow was computed for a periodic cell including three cylinders placed one behind the other at Re = 1, related to the diameter of one cylinder and at Sc = 15. Although the Re number is small, there are recirculation zones between the cylinders which grow together with the growth of the deposited layer. As long as diffusional particle mass fluxes from the recirculation zones are very small, the whole collection efficiency of the filter does not change remarkably during the considered time of its obstruction. On Figure 9.4a the particle concentration field and streamlines around filter with deposited layer at $T = 11$ where T is a scaled time $T = 3c_{entr}t/10c_{dep}$ in microseconds. On Figure 9.4b the growth of deposited layer at time $T = 11$ is shown. The collection efficiency as the ratio of deposited particle mass flux to incoming particle mass flux vs. a time T is plotted in Figure 9.4c. The change in the field of pressure is shown at times $T = 0$ and $T = 11$. The well-known effect that the deposited layer is growing more on the sides of the fibers[15] is shown here.

9.3.2 Inertial Deposition of Large Particles

To check the validity of the present model for filter simulation comparison was done with values of collection efficiency obtained experimentally for a periodic grid of fibers vs. Stokes number.[17] Here collection efficiency of the single fiber C_{eff} was determined as the quotient of the number of particles actually removed and the number that would be removed by a 100% efficient fiber ($C_{eff} = y/r_f$) (Figure 9.5). The range of experimental data is well predicted with the present model.

In the present study, gas–particle flow with deposition is computed for a 3-D sieve filter model. The integration domain, Figure 9.6, is a periodical mesh node of the sieve. Crossing fibers are represented by two circular cylinders, one perpendicular behind the other. The diameter of the cylinders is set to $D = 20$ μm; the distance between two parallel cylinders equals the diameter.

At the boundaries in y- and z-direction, periodic boundary conditions are assumed, so that globally an infinitely extended grid of crossing fibers is simulated. The inflow and outflow plane

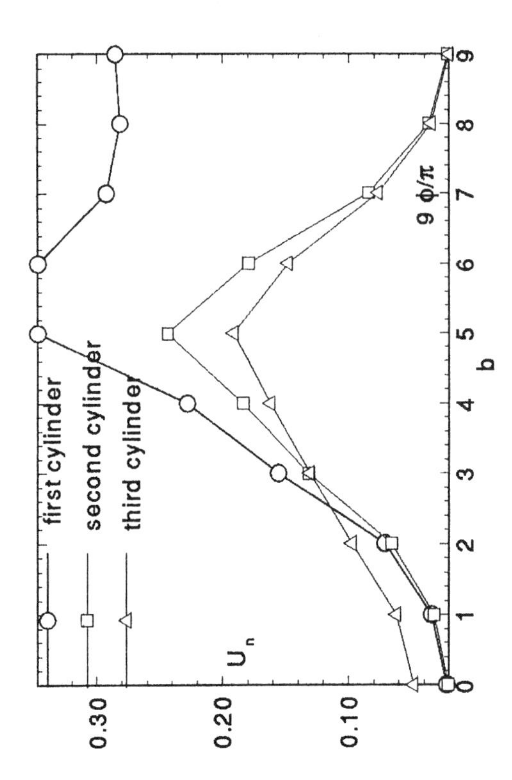
first cylinder
second cylinder
third cylinder
U n
0.30
0.20
0.10
0
1
2
3
4
5
6
7
8
9
9 φ/π
b

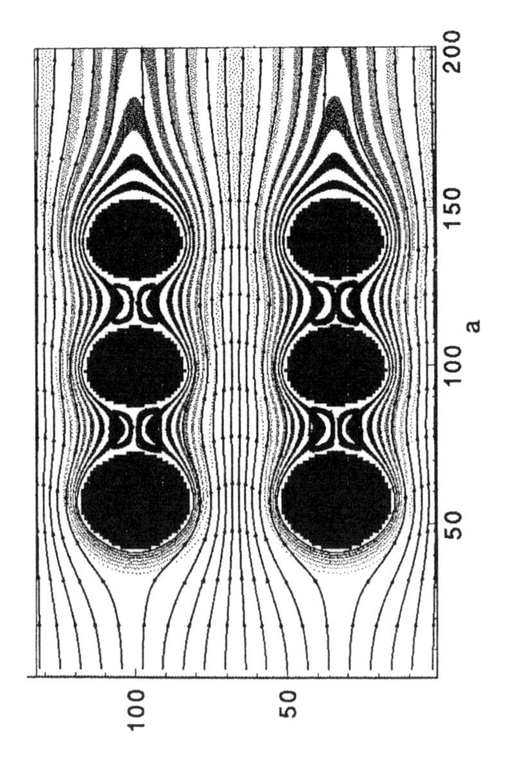
100
50
50
100
150
200
a

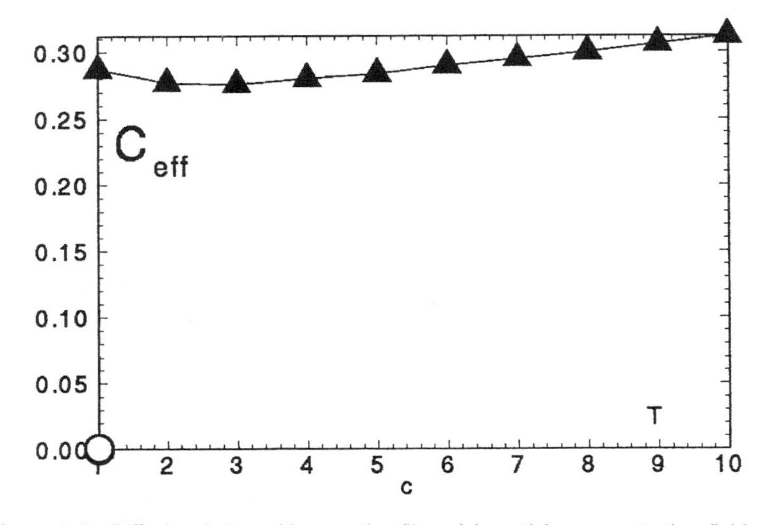

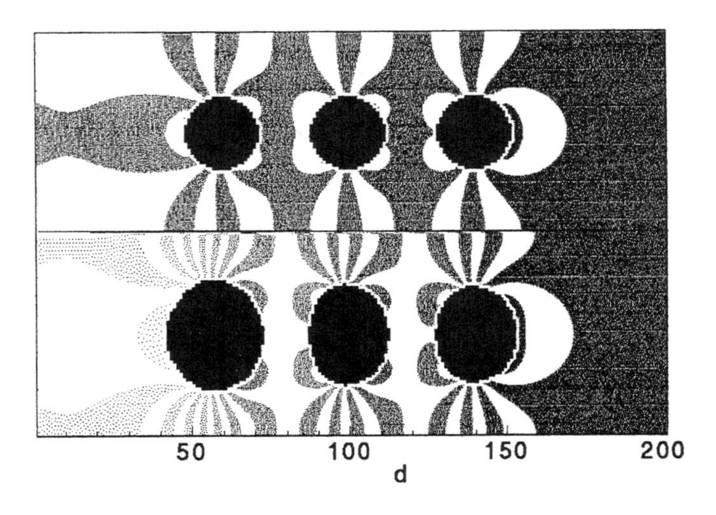

Figure 9.4 Diffusional deposition on the filter: (a) particle concentration field and streamlines around filter with deposited layer; (b) the growth of deposited layer; (c) collection efficiency; (d) change in the field of pressure. (Re = 1, Sc = 15).

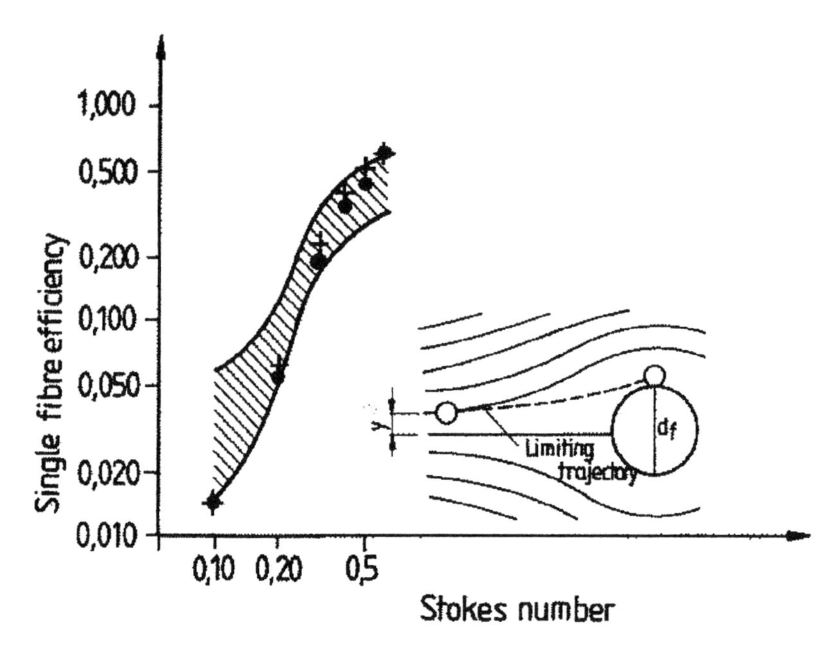

Figure 9.5 Single-fiber efficiency vs. Stokes number; hatched area = experimental data[17] (Re = 0.5–1); solid circles = present calculations, Re = 1; crosses = present calculations, Re = 0.5.

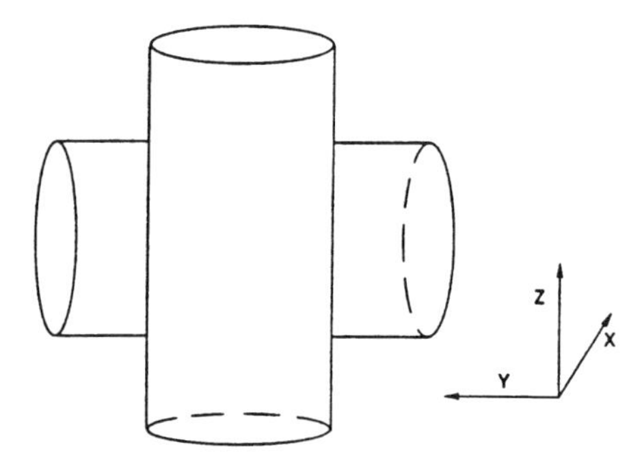

Figure 9.6 Arrangement of cylinders in the integration domain (periodical cell of a sieve).

are at positions x = const. The fluid flow is calculated by the LBGK model; the Reynolds number of the fluid flow is assumed to be $\mathrm{Re} = \frac{2u_{\mathrm{entr}} D}{3\nu_g}$ because of the second cylinder. The whole computational domain is a cubic box, discretized by a Cartesian mesh of step size of 1 μm with 81 nodes in inflow (x-) direction and 41 × 41 nodes across ($\varepsilon = 0.025$).

The cylinders inside the domain are represented as a set of rigid nodes. The distribution functions coming out from the rigid nodes are defined by the boundary-fitting conditions as described before. In the nodes of the entrance the values for the distribution functions are given by the equilibrium distribution functions with hydrodynamic values $u_{i=1} = u_{\mathrm{entrance}}$, $v_{i=1} = 0$, $w_{i=1} = 0$, $\rho_{i=1} = \rho_{i=2}$. In the nodes of the exit the distribution functions are set as equilibrium distribution functions with extrapolated hydrodynamic quantities, i.e., with $u_{i=i\max} = u_{i=i\max-1}$, $v_{i=i\max} = v_{i=i\max-1}$, $w_{i=i\max} = w_{i=i\max-1}$, $\rho_{i=i\max} = \rho_{\mathrm{exit}}$. A new solution for the distribution function at the next time step is achieved from

Equation 9.2. The new hydrodynamic values p and ν are obtained by the moments of the distribution functions according to Equation 9.5.

In the first step, the LBGK model is used to calculate the steady-state solution of the fluid flow around the crossing cylinders at $M_{entr} = 0.0087$, $a = 2$. This solution serves as initial flow field for the particle transport.

The computation of gas–particle flow is initiated by admitting solid particles to the steady-state fluid flow. Particles are considered as spheres of diameter of $d = 1.24$ μm with a density of ρ_p. The density ratio between solid and fluid phases are assumed to be $\rho_p/\rho_g = 656$ and $\rho_p/\rho_g = 2624$, respectively. The Stokes number is defined as $\mathrm{St} = (d/D)^2 (\rho_p/\rho_g) \cdot \frac{u_{entr} \cdot D}{\nu_g \cdot 18}$ and has the values of St = 0.14 and St = 0.56, respectively.

To study the mechanism of deposition and influence of the deposited particles on the flow a big parcel (2000 particles) is injected into the flow with constant concentration over the inlet boundary. Each single particle is randomly placed at the entrance with a velocity equal to the gas velocity. The motion of particle is computed by the Lagrangian approach (Equation 9.25 and 9.26). If the center of a particle approaches to a distance less than a particle radius to the surface of cylinders or less than particle diameter to the centers of another deposited particles, it is assumed to be deposited. Then the fluid nodes that are lying inside of the particle volume will be the rigid ones. The particles are represented as spheres and form a new boundary, which is described in the frame of boundary fitting. After the deposition of ten particles, the flow field around the new rigid surfaces is recalculated up to the steady state.

The following figures show results of the calculations of flow with particle deposition for a Stokes number of St = 0.14. The changed form of the filter surface due to particle deposition after passage of the parcel and streamlines around the filter with deposited layer is shown in Figure 9.7 for two periodical cells in the y-direction. Here spheres represent rigid nodes of the grid that are inside of the deposited particles. The influence of the deposited particles on the fluid flow is taken into account in the calculations presented in Figure 9.7. To estimate this influence, the same particle flux was calculated, but neglecting the change in hydrodynamics. The surface with streamlines around the clean filter (starting from the same positions as streamlines in Figure 9.7) is shown in Figure 9.8. To study quantitatively the difference between the two cases, the collection efficiency C_{eff} defined as the ratio between the number of deposited particles and the total number of injected particles is plotted in Figure 9.9 vs. parcels number m — parcel m consists of particles with numbers from $(m - 1) \cdot 200 + 1$ up to $m \cdot 200$ (Figure 9.9). From Figure 9.9 one can see the difference (more than 30%) between the values of collection efficiency predicted with the different models. In the case of the changed hydrodynamics pressure drop vs. time $t/\Delta t_p$ where Δt_p is the time between particle injection is shown in Figure 9.10.

When the Stokes number is higher (St = 0.56) the number of deposited particles is greater. The calculations were repeated with these data and the results are presented in similar form in Figures 9.11 to 9.14. Figure 9.11 represents the corresponding streamlines around the filter with deposited layer after the passage of the parcel. Figure 9.12 shows the form of the deposited layer after the passage of the parcel in the case of unchanged hydrodynamics and streamlines around the clean filter. The corresponding collection efficiency is plotted in Figure 9.13. As in the case St = 0.14 the value of collection efficiency predicted with the model of changed hydrodynamics is lower than with the model of unchanged hydrodynamics, but the difference between them is smaller. For the lower values of Stokes number the effect of inertia forces is weaker, and the change in hydrodynamic field is more important for particle motion. The change in pressure drop vs. time during the particle deposition is given in Figure 9.14. Due to the greater number of deposited particles, the pressure drop at the end of deposition is higher than the pressure drop in the case of St = 0.14.

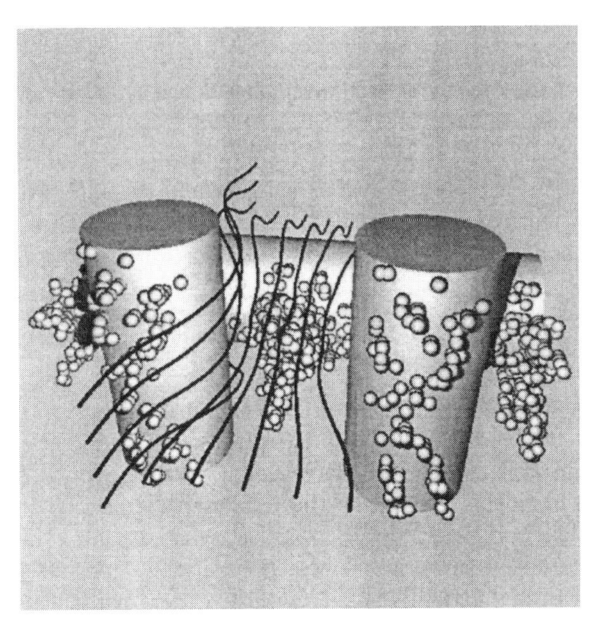

Figure 9.7 Deposited layer on the surfaces after the passage of 2000 particles (changed hydrodynamics) and streamlines around the new boundary (Re = 1, St = 0.14).

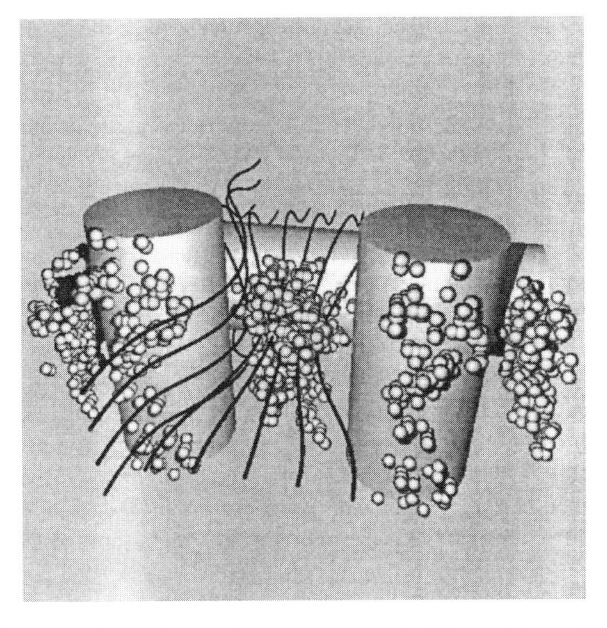

Figure 9.8 Deposited layer on the surfaces after the passage of 2000 particles (unchanged hydrodynamics) and streamlines around clean filter (Re = 1, St = 0.14).

Computations of gas–particle flow in the model filter described above have shown that the influence of changed surfaces on the hydrodynamics and, accordingly, on the next particle deposition is very important and has to be taken into consideration for the prediction of filter performance.

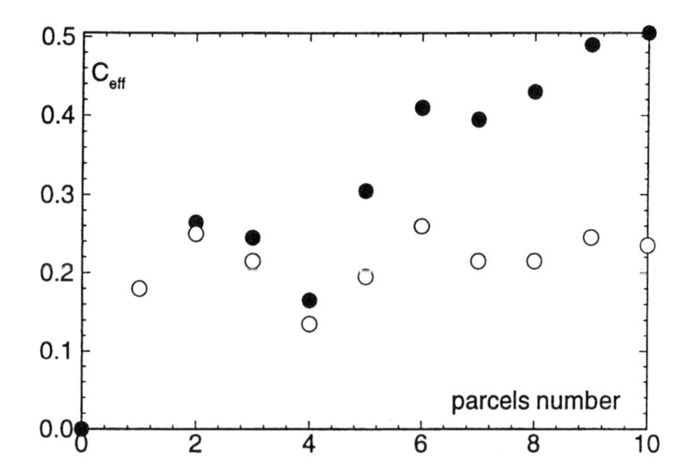

Figure 9.9 Collection efficiency vs. parcels number. Solid circles = unchanged hydrodynamics; open circles = changed hydrodynamics. (Re = 1, St = 0.14).

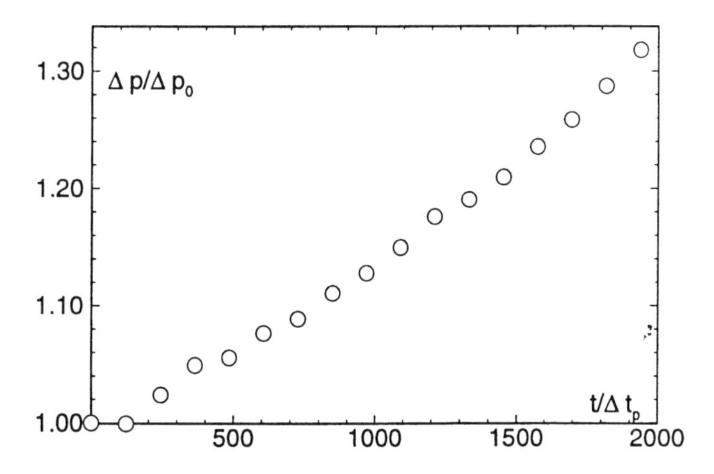

Figure 9.10 Change in pressure drop vs. time. (Re = 1, St = 0.14).

9.4 CONCLUSIONS

The new lattice gas concepts on the molecular basis offer a promising way to deal with incompressible fluid flows around complex geometries. The LBGK model used here leads to rather simple and efficient algorithms; a special strength is the flexible way of implementing boundary conditions with high accuracy. The capability of this concept in combination with a continuum and Lagrangian approach for particle motion is demonstrated for gas–particle flow through filterlike arrangements of bodies in two and three dimensions. The forms of filter surfaces, as well as their change due to deposited particles, can be very complex, and probably the lattice gas models are the simplest and most effective methods to calculate fluid flow around such geometries.

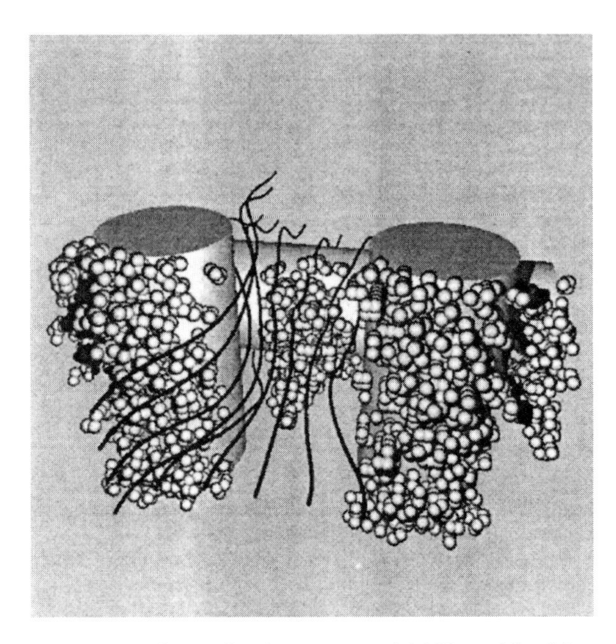

Figure 9.11 Deposited layer on the surfaces after the passage of 2,000 particles (changed hydrodynamics) and streamlines around the new boundary. (Re = 1, St = 0.56).

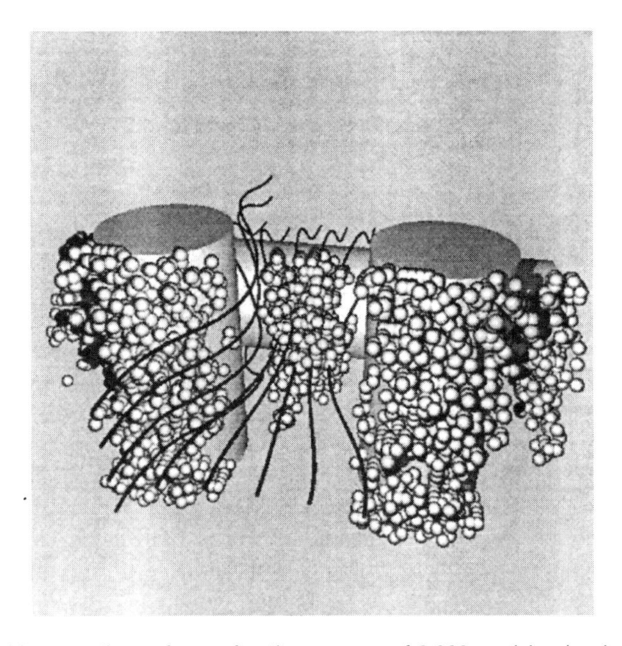

Figure 9.12 Deposited layer on the surfaces after the passage of 2,000 particles (unchanged hydrodynamics) and streamlines around clean filter. (Re = 1, St = 0.56).

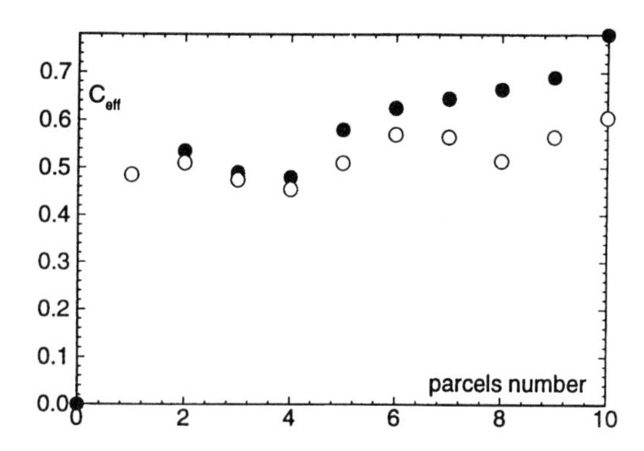

Figure 9.13 Collection efficiency vs. parcel number. Solid circles = unchanged hydrodynamics; open circles = changed hydrodynamics. (Re = 1, St = 0.56).

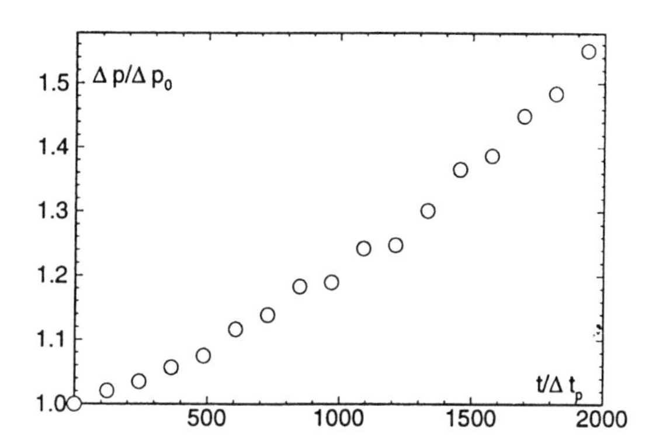

Figure 9.14 Change in pressure drop vs. time. (Re = 1, St = 0.56).

REFERENCES

1. U. Frisch, B. Hasslacher, and Y. Pommeau, Lattice gas automata for the Navier-Stokes equation, *Phys. Rev. Lett.,* 56, 1505, 1986.
2. D. d'Humieres and P. Lallemand, Numerical simulations of hydrodynamics with lattice gas automata in two dimensions, *Complex Syst.,* 1, 599, 1987.
3. G. McNamara and G. Zanetti, Use of the Boltzmann equation to simulate lattice-gas automata, *Phys. Rev. Lett.,* 61, 2332, 1988.
4. F. Higuera, S. Succi, and R. Benzi, Lattice gas dynamics with enhanced collisions, *Europhys. Lett.,* 9, 345, 1989.
5. Y. H. Qian, D. d'Humieres, and P. Lallemand, Lattice BGK models for Navier-Stokes equation, *Europhys. Lett.,* 17(6), 479, 1992
6. S. Chen, Z. Wang, X. Shan, and G. D. Doolen, Lattice Boltzmann computational fluid dynamics in three dimensions, *J. Stat. Phys.,* 68, 379, 1992.
7. J.-C. Toussant, J.-M. Debierre, and L. Turban, Deposition of particles in a two-dimensional lattice gas flow, *Phys. Rev. Lett.,* 68(13), 2027, 1992.
8. H. Fissan, private communication, 1993.
9. O. Filippova and D. Hänel, Simulation of gas-particle flow by lattice gas methods, in *Numerical Methods for Fluid Dynamics,* K. W. Morton, M. J. Baines, Eds., Clarendon Press, Oxford, 1995, 407–414.

10. O. Filippova and D. Hänel, Computation of viscous flow by lattice-gas methods, in *Proc. of 6th International Symposium on CFD,* Lake Tahoe, NV, 1995, 306.
11. U. Frisch, D. d'Humieres, B. Hasslacher, P. Lallemand, Y. Pomeau, and J.-P. Rivet, Lattice gas hydrodynamics in two and three dimensions, *Complex Syst.,* 1, 649, 1987.
12. I. Ginzbourg and P. M. Adler, Boundary flow condition analysis for the three-dimensional lattice Boltzmann model, *J. Phys. II* (Paris), 4, 191, 1994.
13. D. P. Ziegler, Boundary conditions for lattice Boltzmann simulations, *J. Stat. Phys.,* 71, 1171, 1993.
14. R. C. Brown, *Air Filtration. An Integrated Approach to the Theory and Application of Fibrous Filters,* Pergamon Press, Oxford, 1993.
15. Fuchs, N. A., *Mechanics of Aerosols,* Pergamon Press, Oxford, 1964.
16. S. A. Morsi and A. J. Alexander, An investigation of particles trajectories in two-phase flow system, *J. Fluid Mech.,* 55, 193, 1972.
17. K. C. Fan, B. Wamsley, and J. W. Gentry, The effect of Stokes and Reynolds numbers on the collection efficiency of grid filters, *J. Coll. Int. Sci.,* 65(1), 162, 1978.

Chapter 10

Mechanics of a Deformable Fibrous Aerosol Particle: General Theory and Application to the Modeling of Air Filtration

Albert Podgórski and Leon Gradoń

CONTENTS

10.1 INTRODUCTION

Dispersion aerosols are formed by the grinding or atomization of solids or liquids and by the transfer of powders into a state of suspension through the action of airflow or vibration. Condensation aerosols are formed when supersaturated vapor condenses and as a result of reactions between gases leading to the formation of nonvolatile products, such as metal oxides or soot. The difference between aerosols with liquid- and solid-dispersed phases is apparent in that the liquid particles are spherical and after the coalescence of several particles the result is also a spherical particle. Solid particles have the most varied shapes and coagulate to form more or less aggregates. The irregular shape of particles results also from their crystalline nature or cakes structure of the primary material.

0-87371-830-5/98/$0.00+$.50

According to Fuchs (1964) it is convenient to divide aerosol particles into three classes, taking into account their shape determined by the relative dimensions along the three axes perpendicular to one another: (1) isometric particles, for which all three dimensions are roughly the same (spherical, compact); (2) particles having much greater lengths in two dimensions than in the third one (flakes, disks); (3) particles with relatively great length in only one dimension (needles, fibers). The particles of the third class will be an object of our interest in this chapter.

The shape of an aerosol particle is one of the important factors influencing its behavior in carrier gas and therefore affecting the form of mathematical description of motion (Figure 10.1). For the first case mentioned, that is, for a spherical particle, it is sufficient to consider translation of the mass center which is described by the single (in the vector form) ordinary differential equation (ODE). For the third class of elongated (fibrous) particles we have to take into account their rotation. In this case, the resultant form of equations of motion is also dependent on particle mechanical properties. For a stiff fiber its motion may be still described by the set of ODEs for translation of the mass center and additionally for rotation around main axes. Otherwise, when a fibrous particle is deformable, instead of analyzing motion of the entire object we have to consider motion of its subsequent elements, which leads to the set of partial differential equations (PDEs).

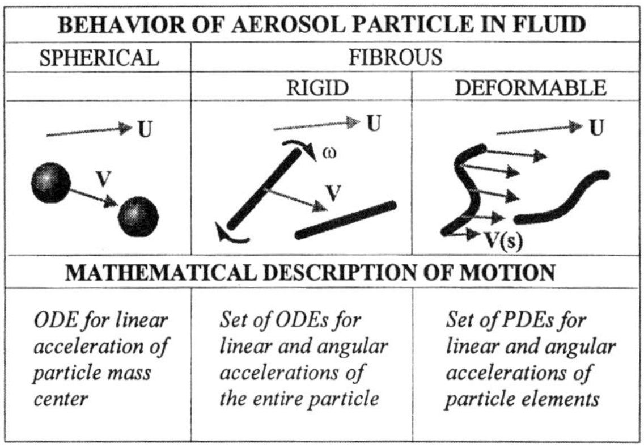

BEHAVIOR OF AEROSOL PARTICLE IN FLUID		
SPHERICAL	FIBROUS	
	RIGID	DEFORMABLE
U, V	U, ω, V	U, V(s)
MATHEMATICAL DESCRIPTION OF MOTION		
ODE for linear acceleration of particle mass center	*Set of ODEs for linear and angular accelerations of the entire particle*	*Set of PDEs for linear and angular accelerations of particle elements*

Figure 10.1 Schematic comparison of behavior and mathematical description of motion in fluid for spherical, and stiff and deformable fibrous aerosol particles.

Recently, an increasing interest in the analysis of fibrous aerosol particle behavior has been observed. Since motion of an elongated, deformable aerosol particle is a complex phenomenon in which translation, rotation, deformation, and oscillation may occur simultaneously and which may be affected by several factors, our understanding of the problem is, as yet, far from complete. On the contrary, the importance of the analysis of fibrous particle behavior is clear in light of the fact that such aerosols are known to be very dangerous to human health, causing some diseases of the lung such as fibrosis and cancer.

Many aspects of elongated aerosol particles have been investigated in the last three decades. The influence of particle size on its settling velocity was studied by Timbrell (1965), and Odgen and Walton (1975) considered the effect of fiber orientation. Gailly and Krushkal (1984) and Griffith (1988) studied the effects of fiber orientation on the drag force. Diffusion of fibrous particles was investigated by Gentry et al. (1988) and Asgharian et al. (1988). Other studies concern electrostatic problems for elongated aerosol particles, such as charge distribution on the fiber (Wang et al., 1988), and deposition by coulombic attraction, (Schamberger et al., 1990; Grzybowski and Gradoń, 1992). The behavior of fibrous particles in the turbulent flow was considered by Shapiro and Goldenberg (1993) and Bernstein and Shapiro (1994). All those investigations were concerned with the special case of the nondeformable particle. The model of a rigid body can be a reasonable approximation for many kinds of aerosols, e.g., glass fibers, asbestos fibers. The problem is even more complex when a particle is deformable. The second limiting case, opposite to the model of stiff body, is a model of perfectly flexible fiber. Theoretical work in this area started with the paper by Hinch (1976), and then the theory was extended by Podgórski and Gradoń (1990).

The aim of this chapter is to present a general description of the mechanics of a fibrous particle in a three-dimensional space which would be valid for arbitrary flow structure of the carrier gas and for any elastic properties of the particle material. These generalized equations will find their exemplification in a particular process of fibrous particle filtration. Filtration in a structure of fibrous filters is a process used extensively to remove suspended particulate matter from the gas stream. The results of fundamental studies delivered information being used in the design of more-efficient filtration equipment and respirators. The previous research concerned mainly the filtration of aerosols with spherical particles or with a compact structure. Detailed information about the influence of particulate mechanisms on filtration efficiency was obtained from the analysis of particle motion around a single filter element. Preliminary studies of filtration of stiff and perfectly flexible fibrous aerosol particles were done by Gradoń et al. (1988; 1989), Gradoń and Podgórski (1990), Podgórski (1994a and b), and Podgórski et al. (1995). In this chapter such analysis based on generalized equations of particle motion is presented. Deposition efficiencies computed for flexible and stiff fibers and for spherical particles are compared.

10.2 GENERAL DESCRIPTION OF DEFORMABLE FIBROUS AEROSOL PARTICLE MOTION

10.2.1 Geometric and Kinematic Preliminaries

We will consider motion of an elongated aerosol particle having length L and cross section diameter d, such that the slenderness ratio $\beta \equiv L/d >> 1$. The particle is assumed to be inextensible (L is constant), but it may undergo some deformation (bending, torsion). Let us introduce stationary Cartesian coordinate system $Ox_1x_2x_3$ (Figure 10.2). The vector $\mathbf{x}$ having components x_i, $i = 1,2,3$,

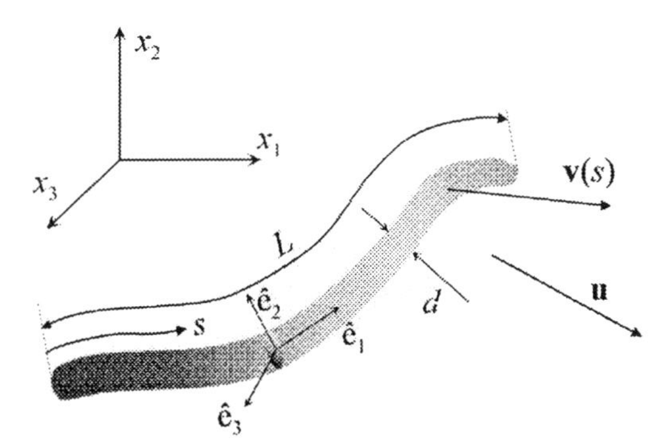

Figure 10.2 Definition of the coordinate systems and main symbols for analysis of fibrous particle motion.

denotes position of a material point along the particle major axis in the Cartesian coordinates. The length of arc measured along this axis from one chosen end of particle is denoted by s. For an infinitesimal linear element of arc we have

$$ds^2 = dx_i \, dx_i, \qquad (10.1)$$

where summation convention for repeated indices was introduced. Hence, the condition for local inextensibility may be written as

$$\left(\frac{\partial x_i}{\partial s}\right)\left(\frac{\partial x_i}{\partial s}\right) = 1 . \tag{10.2}$$

Taking successively the two following derivatives of the above formula with respect to s we get the next useful constrains:

$$\left(\frac{\partial x_i}{\partial s}\right)\left(\frac{\partial^2 x_i}{\partial s^2}\right) = 0 , \tag{10.3}$$

$$\left(\frac{\partial x_i}{\partial s}\right)\left(\frac{\partial^3 x_i}{\partial s^3}\right) = -\left(\frac{\partial^2 x_i}{\partial s^2}\right)\left(\frac{\partial^2 x_i}{\partial s^2}\right) . \tag{10.4}$$

Let us denote by $\mathbf{v}$ the vector of local linear velocity of a material point lying on the particle major axis. In Cartesian coordinates it has the components:

$$v_{x_i} \equiv \frac{\partial x_i}{\partial t} , \tag{10.5}$$

where t denotes time. Taking partial derivatives of Equation 10.2 with respect to time, we obtain kinematic constrains for local linear velocity and acceleration:

$$\left(\frac{\partial x_i}{\partial s}\right)\left(\frac{\partial v_{x_i}}{\partial s}\right) = 0 , \tag{10.6}$$

$$\left(\frac{\partial^2 v_{x_i}}{\partial s \partial t}\right)\left(\frac{\partial x_i}{\partial s}\right) = -\left(\frac{\partial v_{x_i}}{\partial s}\right)\left(\frac{\partial v_{x_i}}{\partial s}\right) . \tag{10.7}$$

In addition to the fixed Cartesian coordinates, we will use the local orthonormal coordinate system moving with particle major axis. The local vector basis of this system constitute: $\hat{\mathbf{e}}_1$ = unit vector locally tangent to the particle major axis, $\hat{\mathbf{e}}_2$ = unit normal, and $\hat{\mathbf{e}}_3$ = unit vector binormal to this axis (Figure 10.2). The unit tangent vector $\hat{\mathbf{e}}_1$ can be represented in the Cartesian basis as

$$\hat{\mathbf{e}}_1 = \frac{\partial x_i}{\partial s} \hat{\mathbf{i}}_{x_i} , \tag{10.8}$$

where $\hat{\mathbf{i}}_{xi}$ are unit vectors of the Cartesian system. The two remaining versors $\hat{\mathbf{e}}_2$ and $\hat{\mathbf{e}}_3$ may be defined in various ways, but their components should fulfill conditions of orthonormality of the $\hat{\mathbf{e}}_i$ basis:

$$\hat{\mathbf{e}}_i \cdot \hat{\mathbf{e}}_j = \delta_{ij} , \tag{10.9}$$

$$\hat{\mathbf{e}}_3 = \pm \hat{\mathbf{e}}_1 \times \hat{\mathbf{e}}_2 , \tag{10.10}$$

where δ_{ij} is Kronecker's symbol, dot $(\cdot)$ denotes scalar product and cross $(\times)$ = vector product. For example, we may choose:

$$\hat{\mathbf{e}}_2 = \frac{1}{K}\frac{\partial^2 x_i}{\partial s^2}\hat{\mathbf{i}}_{x_i}, \tag{10.11}$$

$$\hat{\mathbf{e}}_3 = \frac{1}{K}\varepsilon_{kij}\frac{\partial x_i}{\partial s}\frac{\partial^2 x_j}{\partial s^2}\hat{\mathbf{i}}_{x_k}, \tag{10.12}$$

where $K = [(\partial^2 x_i/\partial s^2)(\partial^2 x_i/\partial s^2)]^{1/2}$ and ε_{ijk} is permutation (Levi–Civita) symbol. Having these two coordinate systems defined, we can express any vector, for example, local particle velocity $\mathbf{v}$, in the basis $\hat{\mathbf{i}}_{xi}$, or $\hat{\mathbf{e}}_i$:

$$\mathbf{v} = v_{x_i}\hat{\mathbf{i}}_{x_i} = v_i\hat{\mathbf{e}}_i, \tag{10.13}$$

where v_{xi} are Cartesian, and v_i = local components of $\mathbf{v}$. We define the vector of local curvature of particle main axis, κ, having the components κ_i in the basis $\hat{\mathbf{e}}_i$, as

$$\frac{\partial\hat{\mathbf{e}}_i}{\partial s} = \boldsymbol{\kappa}\times\hat{\mathbf{e}}_i = \varepsilon_{kji}\kappa_j\hat{\mathbf{e}}_k, \tag{10.14}$$

Thus, the main curvatures (κ_2 and κ_3) and torsion (κ_1) of the center line are

$$\kappa_1 = \hat{\mathbf{e}}_3\cdot\frac{\partial\hat{\mathbf{e}}_2}{\partial s} = -\hat{\mathbf{e}}_2\cdot\frac{\partial\hat{\mathbf{e}}_3}{\partial s}, \tag{10.15}$$

$$\kappa_2 = \hat{\mathbf{e}}_1\cdot\frac{\partial\hat{\mathbf{e}}_3}{\partial s} = -\hat{\mathbf{e}}_3\cdot\frac{\partial\hat{\mathbf{e}}_1}{\partial s}, \tag{10.16}$$

$$\kappa_3 = \hat{\mathbf{e}}_2\cdot\frac{\partial\hat{\mathbf{e}}_1}{\partial s} = -\hat{\mathbf{e}}_1\cdot\frac{\partial\hat{\mathbf{e}}_2}{\partial s}. \tag{10.17}$$

Similarly, the vector of local angular velocity ω is defined by

$$\frac{\partial\hat{\mathbf{e}}_i}{\partial t} = \boldsymbol{\omega}\times\hat{\mathbf{e}}_i = \varepsilon_{kji}\omega_j\hat{\mathbf{e}}_k, \tag{10.18}$$

and, hence, its local components are as follows:

$$\omega_1 = \hat{\mathbf{e}}_3\cdot\frac{\partial\hat{\mathbf{e}}_2}{\partial t} = -\hat{\mathbf{e}}_2\cdot\frac{\partial\hat{\mathbf{e}}_3}{\partial t}, \tag{10.19}$$

$$\omega_2 = \hat{\mathbf{e}}_1\cdot\frac{\partial\hat{\mathbf{e}}_3}{\partial t} = -\hat{\mathbf{e}}_3\cdot\frac{\partial\hat{\mathbf{e}}_1}{\partial t}, \tag{10.20}$$

$$\omega_3 = \hat{\mathbf{e}}_2\cdot\frac{\partial\hat{\mathbf{e}}_1}{\partial t} = -\hat{\mathbf{e}}_1\cdot\frac{\partial\hat{\mathbf{e}}_2}{\partial t}. \tag{10.21}$$

Comparing second order mixed derivatives $\partial^2\hat{\mathbf{e}}_i/\partial s\partial t$ computed by differentiation of Equation 10.14 with respect to t, and Equation 10.18 with respect to s, we find kinematic equations relating ω and κ in local coordinates:

$$\frac{\partial \omega_k}{\partial s} - \frac{\partial \kappa_k}{\partial t} - \varepsilon_{kij}\omega_i\kappa_j = 0 \,. \tag{10.22}$$

Local components of linear and angular velocity fulfills the equation:

$$\frac{\partial v_k}{\partial s} = \varepsilon_{ki1}\omega_i - \varepsilon_{kij}\kappa_i v_j \,. \tag{10.23}$$

Conditions for local components of linear acceleration may be expressed as follows:

$$\frac{\partial \mathbf{v}}{\partial t} = \frac{\partial v_k}{\partial t}\hat{\mathbf{e}}_k + v_k \frac{\partial \hat{\mathbf{e}}_k}{\partial t} = \left(\frac{\partial v_k}{\partial t} + \varepsilon_{kij}\omega_i v_j \right)\hat{\mathbf{e}}_k \,, \tag{10.24}$$

$$\frac{\partial^2 v_k}{\partial s \partial t} = \varepsilon_{ki1}\frac{\partial \omega_i}{\partial t} - \varepsilon_{kij}\left(\kappa_i \frac{\partial v_j}{\partial t} + v_j \frac{\partial \omega_i}{\partial s} \right) + v_i\left(\omega_i\kappa_k - \omega_k\kappa_i\right) . \tag{10.25}$$

10.2.2 Equations of Translational and Rotational Motion

Since the analyzed fibrous particle may be deformable, and therefore its various parts can move with different velocities (which are constrained, however, by the kinematic relationships derived previously), equations of motion have to be formulated locally, for the subsequent elements of the particle. Thus, let us take into consideration a very thin "slice" of the particle between two neighboring cross sections perpendicular to the local direction of vector $\hat{\mathbf{e}}_1$, which are located at arbitrarily taken distances s and $s + \Delta s$ from one chosen end of particle, where $\Delta s << L$, Figure 10.3. Upon these two cross sections act the resultant forces (internal forces from the point of view

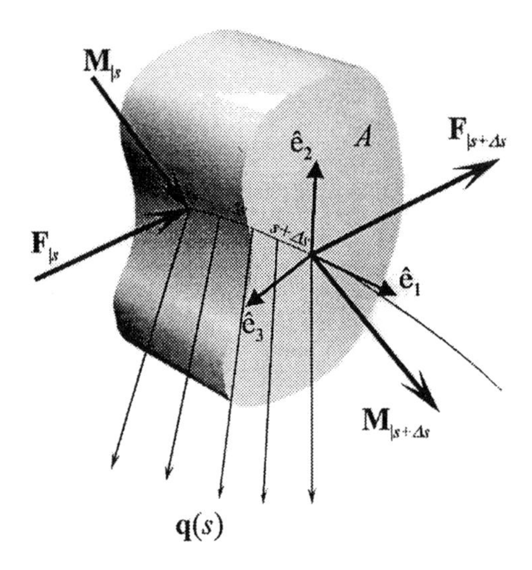

Figure 10.3 A slice of fibrous particle — definition of symbols for derivation of equations of motion.

of the entire particle, but external forces from the point of view of the slice analyzed): $\mathbf{F}_{|s+\Delta s}$ upon the front face, and $\mathbf{F}_{|s}$ upon the rear face. External force per unit length of the particle, which includes body force and viscous drag force per unit length, is denoted by vector **q**, and particle cross-sectional area by A. According to Newton's second law, the rate of change of linear momentum of the volume analyzed equals the total force acting on it; for the slice of the volume $\Delta V = A\Delta s$ we have

$$\iiint_{\Delta V} (\rho \mathbf{v}) dV = \int_{s}^{s+\Delta s} \mathbf{q} ds + \mathbf{F}_{|s+\Delta s} - \mathbf{F}_{|s} , \tag{10.26}$$

where ρ is density of particle material. Because

$$\iiint_{\Delta V} \frac{\partial}{\partial t}(\rho \mathbf{v}) dV = \int_{s}^{s+\Delta s} \left[\iint_{A} \frac{\partial}{\partial t}(\rho \mathbf{v}) dA \right] ds , \tag{10.27}$$

and

$$\mathbf{F}_{|s+\Delta s} - \mathbf{F}_{|s} = \int_{s}^{s+\Delta s} \left(\frac{\partial \mathbf{F}}{\partial s} \right) ds , \tag{10.28}$$

so that Equation 10.26 may be rearranged as follows:

$$\int_{s}^{s+\Delta s} \left[\iint_{A} \frac{\partial}{\partial t}(\rho \mathbf{v}) dA - \mathbf{q} - \frac{\partial \mathbf{F}}{\partial \mathbf{s}} \right] ds = 0 . \tag{10.29}$$

Since the volume of balance has been chosen arbitrarily, the above equation is fulfilled if the integrand vanishes identically at any point s along the particle major axis, that is when

$$\iint_{A} \frac{\partial}{\partial t}(\rho \mathbf{v}) dA = \mathbf{q} + \frac{\partial \mathbf{F}}{\partial \mathbf{s}} . \tag{10.30}$$

For a thin inextensible particle of constant density and cross-section area, we get the general vector form of equation of translational motion:

$$A\rho \frac{\partial \mathbf{v}}{\partial t} = \mathbf{q} + \frac{\partial \mathbf{F}}{\partial s} . \tag{10.31}$$

It should be noted that in Equations 10.26 through 10.30 vector **v** denotes velocity of any material point at a chosen cross section, while in Equation 10.31 and in other formulae it denotes local velocity of the mass center of a particle cross section (i.e., velocity of a material point lying on the major axis of the particle). Equation 10.31 can be directly rewritten for the Cartesian components of vectors:

$$A\rho \frac{\partial v_{x_k}}{\partial t} = \frac{\partial F_{x_k}}{\partial s} + q_{x_k} . \tag{10.32}$$

We can also write Equation 10.31 in terms of the vector components referring to the triad $\hat{\mathbf{e}}_1$, $\hat{\mathbf{e}}_2$, $\hat{\mathbf{e}}_3$. For any vector $\mathbf{w} = w_i\hat{\mathbf{e}}_i$ we can express its local partial derivatives as follows:

$$\frac{\partial \mathbf{w}}{\partial s} = \frac{\partial w_k}{\partial s}\hat{\mathbf{e}}_k + w_k\frac{\partial \hat{\mathbf{e}}_k}{\partial s} = \left(\frac{\partial w_k}{\partial s} + \varepsilon_{kij}\kappa_i w_j\right)\hat{\mathbf{e}}_k\,, \tag{10.33}$$

$$\frac{\partial \mathbf{w}}{\partial t} = \frac{\partial w_k}{\partial t}\hat{\mathbf{e}}_k + w_k\frac{\partial \hat{\mathbf{e}}_k}{\partial t} = \left(\frac{\partial w_k}{\partial t} + \varepsilon_{kij}\omega_i w_j\right)\hat{\mathbf{e}}_k\,, \tag{10.34}$$

where relationships 10.14 and 10.18 were used in rearrangements of the right-hand sides. Hence, expressing the partial derivatives in Equation 10.31 by means of formulae 10.33 and 10.34 we obtain local formulation of equations of translational motion in the local basis $\hat{\mathbf{e}}_i$:

$$A\rho\left(\frac{\partial v_k}{\partial t} + \varepsilon_{kij}\omega_i v_j\right) = \frac{\partial F_k}{\partial s} + \varepsilon_{kij}\kappa_i F_j + q_k\,. \tag{10.35}$$

Since a deformable body in three-dimensional space has six degrees of freedom, three equations of translational motion of the form 10.32 or 10.35 (for any cross section of the particle) do not give the complete description of motion. The remaining three relationships are the equations of rotational motion. They may be obtained by comparing the rate of change of moment of linear momentum with the total moment of forces about the chosen point that acts on the volume of balance. Referring to our slice, let us denote the resultant internal moments at the bounding cross sections: $\mathbf{M}_{|s+\Delta s}$ at the front face, and $\mathbf{M}_{|s}$ at the rear one. Taking the mass center of the rear cross section as the point of reference, we find that the resultant force $\mathbf{F}_{|s+\Delta s}$ has about this point moment $(\hat{\mathbf{e}}_1\Delta s) \times \mathbf{F}_{|s+\Delta s}$. Denoting tensor of moments of inertia of the particle cross section by $\mathbf{I}$, we have the balance:

$$\int_s^{s+\Delta s}\frac{\partial}{\partial t}(\mathbf{I}\cdot\boldsymbol{\omega})ds = \mathbf{M}_{|s+\Delta s} - \mathbf{M}_{|s} + \left[\int_s^{s+\Delta s}\hat{\mathbf{e}}_{1_{|s}}ds\right]\times\mathbf{F}_{|s+\Delta s}\,. \tag{10.36}$$

Replacing the finite difference $\mathbf{M}_{|s+\Delta s} - \mathbf{M}_{|s}$ by the appropriate definite integral of the derivative $\partial\mathbf{M}/\partial s$, similarly as in Equation 10.28, then rearranging Equation 10.36 to obtain one integral, and equating the integrand to zero, we obtain

$$\frac{\partial}{\partial t}(\mathbf{I}\cdot\omega) = \frac{\partial \mathbf{M}}{\partial s} + \hat{\mathbf{e}}_1\times\mathbf{F}\,. \tag{10.37}$$

Referring to the local triad $\hat{\mathbf{e}}_i$ being the versors of particle main axes, we can conclude that for inextensible body of constant density, the tensor $\mathbf{I}$ has diagonal form ($I_{ij} = 0$ when $i \neq j$) and the main moments of inertia of the particle cross section about these axes are constant in time. Thus, Equation 10.37 may be rewritten in terms of local components as follows:

$$I_{km}\frac{\partial \omega_m}{\partial t} = \frac{\partial M_k}{\partial s} + \varepsilon_{kij}\left(\kappa_i M_j - I_{jm}\omega_i\omega_m\right) - \varepsilon_{ki1}F_i\,, \tag{10.38}$$

where Equations 10.33 and 10.34 were used in rearrangements. For circular cross section of the diameter d, the nonzero elements of $\mathbf{I}$ are $I_{22} = I_{33} = \pi\rho d^4/64$, $I_{11} = I_{22} + I_{33} = \pi\rho d^4/32$. The set of equations derived above is formally (from the mathematical point of view only) a complete description of particle motion (if initial and boundary conditions, constitutive equations for particle material, and description of external forces are given). Such formulation is, however, very inconvenient for numerical simulations (see below).

10.2.3 Boundary Value Problems for Momentary Distribution of Internal Forces and Moments

When analyzing filtration of fibrous aerosol particles, or more generally, their deposition in a system considered, we are interested in computing trajectories of particles. They can be determined by integration in time of equations of motion in the form 10.32 or 10.35 and 10.38 accounting simultaneously the kinematic constrains 10.2 to 7 or 10.22 to 25. Unfortunately, equations of motion contain all unknown vectors, that is, linear and angular accelerations and momentary distributions of internal forces and moments, while the kinematic conditions do not contain them explicitly. In other words, instantaneous distributions of internal forces, $\mathbf{F}(s)$, and moments, $\mathbf{M}(s)$, which have to be known to calculate accelerations, must have the required form to keep the kinematic equations fulfilled. It is obvious that any direct iterative scheme for determination of $\mathbf{F}(s)$ and $\mathbf{M}(s)$ based only on the equations given above would be numerically very ineffective. Fortunately, it is possible to separate variables — accelerations, internal forces, and moments — and compute them independently, without iteration. Solution of this key problem is given below.

To eliminate linear acceleration let us differentiate with respect to s Equation 10.35; the result is

$$\frac{\partial^2 v_k}{\partial s \partial t} + \varepsilon_{kij}\left(\omega_i \frac{\partial v_j}{\partial s} + v_j \frac{\partial \omega_i}{\partial s}\right) = \frac{\partial P_k}{\partial s}, \tag{10.39}$$

where, for simplicity, we denoted by P_k the following expression:

$$P_k = \frac{1}{A\rho}\left(\frac{\partial F_k}{\partial s} + \varepsilon_{kij}\kappa_i F_j + q_k\right). \tag{10.40}$$

The second-order mixed derivative $\partial^2 v_k/\partial s \partial t$ in Equation 10.39 can be expressed by means of Equation 10.25. Rearrangement of the result obtained by means of Equation 10.22, 10.23, and 10.35 leads to the formula which does not contain linear acceleration:

$$\varepsilon_{ki1}\frac{\partial \omega_i}{\partial t} + \omega_1\omega_k - \delta_{k1}\omega_i\omega_i = \frac{\partial P_k}{\partial s} + \varepsilon_{kij}\kappa_i P_j\,. \tag{10.41}$$

Unknown angular acceleration can be then eliminated from the above expression using Equation 10.38. Taking into account diagonal form of $\mathbf{I}$ in the basis $\hat{\mathbf{e}}_i$ (i.e., $I_{ik} = I_{ij}\delta_{jk} = I_{kk}$), the local components of angular acceleration can be written as

$$\frac{\partial \omega_i}{\partial t} = I_{rs}^{-1}\delta_{is}\left(\frac{\partial M_r}{\partial s} + \varepsilon_{rmn}\kappa_m M_n - \varepsilon_{rm1}F_m - \varepsilon_{rmn}I_{np}\omega_m\omega_p\right) \tag{10.42}$$

Inserting formulae 10.40 and 10.42 into Equation 10.41, we obtain after some transformations using δ-ε identities the set of three linear boundary value problems describing momentary distribution of internal forces and moments along inextensible fibrous aerosol particle:

$$\frac{\partial^2 F_k}{\partial s^2} + 2\varepsilon_{kij}\kappa_i \frac{\partial F_j}{\partial s} + \varepsilon_{kij} F_j \frac{\partial \kappa_i}{\partial s} + \kappa_i\kappa_k F_i - \kappa_i\kappa_i F_k + A\rho I_{ri}^{-1}\varepsilon_{ki1}\varepsilon_{rm1}F_m$$
$$= -\frac{\partial q_k}{\partial s} - \varepsilon_{kij}\kappa_i q_j + A\rho\left(\omega_1\omega_k - \delta_{k1}\omega_i\omega_i\right) + \frac{A\rho}{I_{ri}}\varepsilon_{ki1}\left(\frac{\partial M_r}{\partial s} + \varepsilon_{rmn}\kappa_m M_n - \varepsilon_{rmn} I_{np}\omega_m\omega_p\right). \tag{10.43}$$

The detailed form of these equations in local coordinates is the following:

$$\frac{\partial^2 F_1}{\partial s^2} - F_1\left(\kappa_2^2 + \kappa_3^2\right) - 2\kappa_3 \frac{\partial F_2}{\partial s} + 2\kappa_2 \frac{\partial F_3}{\partial s} - \frac{\partial \kappa_3}{\partial s} F_2 + \frac{\partial \kappa_2}{\partial s} F_3 + F_2\kappa_1\kappa_2 + F_3\kappa_1\kappa_3$$
$$= -\frac{\partial q_1}{\partial s} - \kappa_2 q_3 + \kappa_3 q_2 - A\rho\left(\omega_2^2 + \omega_3^2\right), \tag{10.44}$$

$$2\kappa_3 \frac{\partial F_1}{\partial s} + F_1\left(\kappa_1\kappa_2 + \frac{\partial \kappa_3}{\partial s}\right) + \frac{\partial^2 F_2}{\partial s^2} - F_2\left(\kappa_1^2 + \kappa_3^2 + \frac{A\rho}{I_{33}}\right) - 2\kappa_1 \frac{\partial F_3}{\partial s} + F_3\left(\kappa_1\kappa_3 - \frac{\partial \kappa_1}{\partial s}\right)$$
$$= \frac{A\rho}{I_{33}}\left[\frac{\partial M_3}{\partial s} + \kappa_1 M_2 - \kappa_2 M_1 - \omega_1\omega_2\left(I_{22} - I_{11}\right)\right] + A\rho\omega_1\omega_2 - \kappa_3 q_1 - \frac{\partial q_2}{\partial s} + \kappa_1 q_3, \tag{10.45}$$

$$-2\kappa_2 \frac{\partial F_1}{\partial s} + F_1\left(\kappa_1\kappa_3 - \frac{\partial \kappa_2}{\partial s}\right) + \frac{\partial^2 F_3}{\partial s^2} - F_3\left(\kappa_1^2 + \kappa_2^2 + \frac{A\rho}{I_{22}}\right) + 2\kappa_1 \frac{\partial F_2}{\partial s} + F_2\left(\kappa_2\kappa_3 + \frac{\partial \kappa_1}{\partial s}\right)$$
$$= \frac{A\rho}{I_{22}}\left[-\frac{\partial M_2}{\partial s} + \kappa_1 M_3 - \kappa_3 M_1 + \omega_1\omega_3\left(I_{11} - I_{33}\right)\right] + A\rho\omega_1\omega_3 + \kappa_2 q_1 - \kappa_1 q_2 - \frac{\partial q_3}{\partial s}. \tag{10.46}$$

For a given moment of time, if we know the momentary position and shape of the particle characterized by the vectors $\mathbf{x}$ and κ, local velocities $\mathbf{v}$ and ω, and external forces $\mathbf{q}$, the three equations above contain six unknowns: three components of $\mathbf{F}$ and three of $\mathbf{M}$. Thus, to solve these equations three additional relationships have to be formulated, for example, constitutive equations for internal moments along the particle.

10.2.4 Constitutive Equations for Elastic Fibrous Particles — Discussion of General Equations for Limiting Mechanical Models

Description of fibrous aerosol particle motion has been as yet formulated completely for two limiting, from the mechanical point of view, cases. The first one is the simplest model of a rigid body preserving its shape independently of external forces applied. For such a body, description of motion is well known from classical mechanics. The second limiting case is a model of perfectly flexible body having no stiffness, i.e., a model of a filament or threadlike particle. Description of motion for such a body was presented by Podgórski and Gradoń (1990). It should be noted that the governing equations for both limiting cases do not contain any material properties of the particle (formally, the rigid body can be regarded as a body having infinite elastic modulus, while for a perfectly flexible one the value of this modulus is zero). The aim of this contribution is to present a general description of a deformable fibrous aerosol particle motion valid for any value of elastic modulus, including these two limiting cases as well as a complete spectrum of intermediate values.

For an elastic fibrous aerosol particle the constitutive equations for internal moments can be written in a general form as

$$\mathbf{M} = \mathbf{S} \cdot (\kappa - \kappa^{(eq)}), \tag{10.47}$$

where **S** is tensor of particle stiffness, and $\kappa^{(eq)}$ is local vector of equilibrium curvature of particle at rest (when no external forces and moments are present). We will later assume that the particle in equilibrium is straight, that is, $\kappa^{(eq)} = 0$. Thus, referring to the local triad $\hat{\mathbf{e}}_i$, we have

$$M_i = S_{ij}\kappa_j \tag{10.48}$$

In the local basis $\hat{\mathbf{e}}_i$ the tensor of particle stiffness has diagonal form ($S_{ij} = 0$ for $i \neq j$) and the nonzero components are: $S_{11} = GI_{11}/\rho$, $S_{22} = EI_{22}/\rho$, $S_{33} = EI_{33}/\rho$, where E is Young's elastic modulus, G is the shear elastic modulus, $G = {}^1/_2E/(1+\nu)$, and ν is Poisson's ratio ($\nu = {}^1/_2$ for an incompressible body, and then $G = E/3$). Eliminating M_i from Equation 10.44 through 10.46 by means of the above formulae, we get general equations of momentary distribution of internal forces along an elastic fibrous particle:

$$\begin{aligned}&\frac{\partial^2 F_1}{\partial s^2} - F_1\left(\kappa_2^2+\kappa_3^2\right) - 2\kappa_3\frac{\partial F_2}{\partial s} + 2\kappa_2\frac{\partial F_3}{\partial s} - \frac{\partial \kappa_3}{\partial s}F_2 + \frac{\partial \kappa_2}{\partial s}F_3 + F_2\kappa_1\kappa_2 + F_3\kappa_1\kappa_3 \\ &= -\frac{\partial q_1}{\partial s} - \kappa_2 q_3 + \kappa_3 q_2 - A\rho\left(\omega_2^2+\omega_3^2\right),\end{aligned} \tag{10.49}$$

$$\begin{aligned}&2\kappa_3\frac{\partial F_1}{\partial s} + F_1\left(\kappa_1\kappa_2 + \frac{\partial \kappa_3}{\partial s}\right) + \frac{\partial^2 F_2}{\partial s^2} - F_2\left(\kappa_1^2+\kappa_3^2+\frac{A\rho}{I_{33}}\right) - 2\kappa_1\frac{\partial F_3}{\partial s} + F_3\left(\kappa_1\kappa_3 - \frac{\partial \kappa_1}{\partial s}\right) \\ &= AE\left[\frac{\partial \kappa_3}{\partial s} + \kappa_1\kappa_2\frac{\left(I_{22}-GI_{11}/E\right)}{I_{33}}\right] + A\rho\omega_1\omega_2\left(\frac{I_{11}-I_{22}+I_{33}}{I_{33}}\right) - \kappa_3 q_1 - \frac{\partial q_2}{\partial s} + \kappa_1 q_3,\end{aligned} \tag{10.50}$$

$$\begin{aligned}&-2\kappa_2\frac{\partial F_1}{\partial s} + F_1\left(\kappa_1\kappa_3 - \frac{\partial \kappa_2}{\partial s}\right) + \frac{\partial^2 F_3}{\partial s^2} - F_3\left(\kappa_2^2+\kappa_3^2+\frac{A\rho}{I_{22}}\right) + 2\kappa_1\frac{\partial F_2}{\partial s} + F_2\left(\kappa_2\kappa_3 + \frac{\partial \kappa_1}{\partial s}\right) \\ &= AE\left[-\frac{\partial \kappa_2}{\partial s} + \kappa_1\kappa_3\frac{\left(I_{33}-GI_{11}/E\right)}{I_{22}}\right] + A\rho\omega_1\omega_3\left(\frac{I_{11}+I_{22}-I_{33}}{I_{22}}\right) + \kappa_2 q_1 - \kappa_1 q_2 - \frac{\partial q_3}{\partial s}.\end{aligned} \tag{10.51}$$

Boundary conditions for a "free" particle moving in fluid for Equation 10.44 through 10.46 and 10.49 through 10.50 have the form:

$$s = 0,\ s = L:\ F_i = M_i = \kappa_i = 0. \tag{10.52}$$

The general equations given above are simplified for the mentioned limiting cases. For perfectly flexible particle ($E = 0$), we have $M_1 = M_2 = M_3 = 0$, and $F_2 = F_3 = 0$ — only axial tensile, F_1, preserving inextensibility may be nonzero, so that in Cartesian coordinates $F_{xk} = F_1(\partial x_k)/(\partial s)$. Thus, the formulae describing distribution of internal forces reduces in the $\hat{\mathbf{e}}_i$ basis to the single equation for F_1:

$$\frac{\partial^2 F_1}{\partial s^2} - F_1\left(\kappa_2^2+\kappa_3^2\right) = -\frac{\partial q_1}{\partial s} + \kappa_3 q_2 - \kappa_2 q_3 - A\rho\left(\omega_2^2+\omega_3^2\right). \tag{10.53}$$

For the second limiting case of a stiff, straight fibrous particle, constitutive equations of the form 10.48 become undetermined ($E \to \infty$ but $\kappa_\iota = 0$). From Equation 10.22 it follows that $\partial\omega_i/\partial s = 0$, and from Equation 10.23 that $\partial v_k/\partial s = \varepsilon_{ki1}\omega_i$. Differentiating Equation 10.38 with respect to s we obtain three equations relating M_i and F_i:

$$\frac{\partial^2 M_k}{\partial s^2} = \varepsilon_{ki1}\frac{\partial F_i}{\partial s} \,. \tag{10.54}$$

Equations 10.44 through 10.46 for stiff, straight fiber are reduced to

$$\frac{\partial^2 F_1}{\partial s^2} = -\frac{\partial q_1}{\partial s} - A\rho\left(\omega_2^2 + \omega_3^2\right), \tag{10.55}$$

$$\frac{\partial^2 F_2}{\partial s^2} - \frac{A\rho}{I_{33}}F_2 = -\frac{\partial q_2}{\partial s} + \frac{A\rho}{I_{33}}\left[\frac{\partial M_3}{\partial s} + \omega_1\omega_2\left(I_{11} - I_{22} + I_{33}\right)\right], \tag{10.56}$$

$$\frac{\partial^2 F_3}{\partial s^2} - \frac{A\rho}{I_{22}}F_3 = -\frac{\partial q_3}{\partial s} + \frac{A\rho}{I_{22}}\left[-\frac{\partial M_2}{\partial s} + \omega_1\omega_2\left(I_{11} + I_{22} - I_{33}\right)\right]. \tag{10.57}$$

Hence, we have a set of six boundary value problems, Equations 10.54 through 10.57 for six unknown components of internal forces and moments inside the rigid fiber. The boundary conditions are given by Equation 10.52.

10.2.5 Working Form of General Equations — Example for the Two-Dimensional Case

The general description presented above allows us to calculate the fibrous particle trajectory in moving fluid. A numerical procedure could be the following: having instantaneous position $\mathbf{x}(s)$ and velocity $\mathbf{v}(s)$ of particle elements we can calculate the vectors $\kappa(s)$ and $\omega(s)$ from Equations 10.15 through 10.17 and 10.19 through 10.21, external forces $\mathbf{q}(s)$ (to be discussed later), then solving boundary value problems 10.49 through 10.51 we determine distributions $\mathbf{F}(s)$ and $\mathbf{M}(s)$. Inserting these results into equations of motion, we can calculate local acceleration of the particle elements, and after the time step of integration the new values of $\mathbf{x}(s)$ and $\mathbf{v}(s)$ are determined, and the procedure can be continued in the same way. Probably the simplest numerical implementation of the general theory is the case when the equations of motion are integrated in fixed Cartesian coordinates (i.e., we prefer to calculate v_{xi} rather than v_i), while internal forces and moments are determined in local basis (i.e., $\mathbf{F}_i$ and $\mathbf{M}_i$). Below we present without derivation and for simplicity for the two-dimensional case such a mixed, working formulation.

For the two-dimensional problem, when the particle major axis lies in the x_1–x_2 plane we can define:

$$\hat{\mathbf{e}}_1 = \frac{\partial x_1}{\partial s}\hat{\mathbf{i}}_{x_1} + \frac{\partial x_2}{\partial s}\hat{\mathbf{i}}_{x_2}\,, \tag{10.58}$$

$$\hat{\mathbf{e}}_2 = -\frac{\partial x_2}{\partial s}\hat{\mathbf{i}}_{x_1} + \frac{\partial x_1}{\partial s}\hat{\mathbf{i}}_{x_2}\,, \tag{10.59}$$

$$v_{x_1} = \frac{\partial x_1}{\partial s} v_1 - \frac{\partial x_2}{\partial s} v_2 , \tag{10.60}$$

$$v_{x_2} = \frac{\partial x_2}{\partial s} v_1 + \frac{\partial x_1}{\partial s} v_2 . \tag{10.61}$$

Vectors of curvature and angular velocity have only one nonzero component ($\kappa_1 = \kappa_2 = \omega_1 = \omega_2 = 0$):

$$\kappa_3 = \frac{\partial x_1}{\partial s}\frac{\partial^2 x_2}{\partial s^2} - \frac{\partial x_2}{\partial s}\frac{\partial^2 x_1}{\partial s^2} , \tag{10.62}$$

$$\omega_3 = \frac{\partial x_1}{\partial s}\frac{\partial v_{x_2}}{\partial s} - \frac{\partial x_2}{\partial s}\frac{\partial v_{x_1}}{\partial s} . \tag{10.63}$$

Equations of translational motion may be written for Cartesian components of linear acceleration as follows:

$$\frac{\partial v_{x_1}}{\partial t} = \frac{1}{A\rho}\left(q_{x_1} + \frac{\partial x_1}{\partial s}\frac{\partial F_1}{\partial s} + F_1 \frac{\partial^2 x_1}{\partial s^2} - \frac{\partial x_2}{\partial s}\frac{\partial F_2}{\partial s} - F_2 \frac{\partial^2 x_2}{\partial s^2} \right), \tag{10.64}$$

$$\frac{\partial v_{x_2}}{\partial t} = \frac{1}{A\rho}\left(q_{x_2} + \frac{\partial x_2}{\partial s}\frac{\partial F_1}{\partial s} + F_1 \frac{\partial^2 x_2}{\partial s^2} + \frac{\partial x_1}{\partial s}\frac{\partial F_2}{\partial s} + F_2 \frac{\partial^2 x_1}{\partial s^2} \right). \tag{10.65}$$

The local components of internal forces (F_1, F_2) along the particle can be found at any moment by solving the set of boundary value problems:

$$\frac{\partial^2 F_1}{\partial s^2} - F_1\left[\left(\frac{\partial^2 x_1}{\partial s^2}\right)^2 + \left(\frac{\partial^2 x_2}{\partial s^2}\right)^2\right] - 2\left(\frac{\partial x_1}{\partial s}\frac{\partial^2 x_2}{\partial s^2} - \frac{\partial x_2}{\partial s}\frac{\partial^2 x_1}{\partial s^2}\right)\frac{\partial F_2}{\partial s} - \left(\frac{\partial x_1}{\partial s}\frac{\partial^3 x_2}{\partial s^3} - \frac{\partial x_2}{\partial s}\frac{\partial^3 x_1}{\partial s^3}\right)F_2$$
$$= -A\rho\left[\left(\frac{\partial v_{x_1}}{\partial s}\right)^2 + \left(\frac{\partial v_{x_2}}{\partial s}\right)^2\right] - \frac{\partial q_{x_1}}{\partial s}\frac{\partial x_1}{\partial s} - \frac{\partial q_{x_2}}{\partial s}\frac{\partial x_2}{\partial s} , \tag{10.66}$$

$$\frac{\partial^2 F_2}{\partial s^2} - F_1\left[\left(\frac{\partial^2 x_2}{\partial s^2}\right)^2 + \left(\frac{\partial^2 x_1}{\partial s^2}\right)^2 + \frac{A\rho}{I_{33}}\right] - 2\left(\frac{\partial x_1}{\partial s}\frac{\partial^2 x_2}{\partial s^2} - \frac{\partial x_2}{\partial s}\frac{\partial^2 x_1}{\partial s^2}\right)\frac{\partial F_1}{\partial s} - \left(\frac{\partial x_1}{\partial s}\frac{\partial^3 x_2}{\partial s^3} - \frac{\partial x_2}{\partial s}\frac{\partial^3 x_1}{\partial s^3}\right)F_1$$
$$= \frac{\partial q_{x_1}}{\partial s}\frac{\partial x_2}{\partial s} - \frac{\partial q_{x_2}}{\partial s}\frac{\partial x_1}{\partial s} + \frac{A\rho}{I_{33}}\frac{\partial M_3}{\partial s} . \tag{10.67}$$

The third equation supplementing the set 10.66 and 10.67 is the constitutive equation for bending moment; for elastic particle of circular cross section of diameter d we have

$$M_3 = \frac{E\pi d^4}{64}\left(\frac{\partial x_1}{\partial s}\frac{\partial^2 x_2}{\partial s^2} - \frac{\partial x_2}{\partial s}\frac{\partial^2 x_1}{\partial s^2}\right). \tag{10.68}$$

The set of Equations 10.66 through 10.68 is reduced to the single equation for perfectly flexible particle ($E = F_2 = M_3 = 0$):

$$\frac{\partial^2 F_1}{\partial s^2} - F_1\left[\left(\frac{\partial^2 x_1}{\partial s^2}\right)^2 + \left(\frac{\partial^2 x_2}{\partial s^2}\right)^2\right] = -A\rho\left[\left(\frac{\partial v_{x_1}}{\partial s}\right)^2 + \left(\frac{\partial v_{x_2}}{\partial s}\right)^2\right] - \frac{\partial q_{x_1}}{\partial s}\frac{\partial x_1}{\partial s} - \frac{\partial q_{x_2}}{\partial s}\frac{\partial x_2}{\partial s}, \tag{10.69}$$

and for stiff particle it is simplified to

$$\frac{\partial^2 F_1}{\partial s^2} = -A\rho\left[\left(\frac{\partial v_{x_1}}{\partial s}\right)^2 + \left(\frac{\partial v_{x_2}}{\partial s}\right)^2\right] - \frac{\partial q_{x_1}}{\partial s}\frac{\partial x_1}{\partial s} - \frac{\partial q_{x_2}}{\partial s}\frac{\partial x_2}{\partial s}, \tag{10.70}$$

$$\frac{\partial^2 F_2}{\partial s^2} - \frac{A\rho}{I_{33}}F_2 = \frac{\partial q_{x_1}}{\partial s}\frac{\partial x_2}{\partial s} - \frac{\partial q_{x_2}}{\partial s}\frac{\partial x_1}{\partial s} + \frac{A\rho}{I_{33}}\frac{\partial M_3}{\partial s}, \tag{10.71}$$

$$\frac{\partial F_2}{\partial s} + \frac{\partial^2 M_3}{\partial s^2} = 0\,. \tag{10.72}$$

Similarly, the simplified equations of motion (10.64 and 10.65) for the limiting mechanical models are as follows:

- For perfectly flexible fiber:

$$\frac{\partial v_{x_1}}{\partial t} = \frac{1}{A\rho}\left(q_{x_1} + \frac{\partial x_1}{\partial s}\frac{\partial F_1}{\partial s} + F_1\frac{\partial^2 x_1}{\partial s^2}\right), \tag{10.73}$$

$$\frac{\partial v_{x_2}}{\partial t} = \frac{1}{A\rho}\left(q_{x_2} + \frac{\partial x_2}{\partial s}\frac{\partial F_1}{\partial s} + F_1\frac{\partial^2 x_2}{\partial s^2}\right). \tag{10.74}$$

- For straight, stiff fiber:

$$\frac{\partial v_{x_1}}{\partial t} = \frac{1}{A\rho}\left(q_{x_1} + \frac{\partial x_1}{\partial s}\frac{\partial F_1}{\partial s} - \frac{\partial x_2}{\partial s}\frac{\partial F_2}{\partial s}\right), \tag{10.75}$$

$$\frac{\partial v_{x_2}}{\partial t} = \frac{1}{A\rho}\left(q_{x_2} + \frac{\partial x_2}{\partial s}\frac{\partial F_1}{\partial s} + \frac{\partial x_1}{\partial s}\frac{\partial F_2}{\partial s}\right). \tag{10.76}$$

Description of the fibrous deformable aerosol particle motion becomes complete when external forces acting on it are specified. This topic is briefly discussed below.

10.2.6 Description of External Forces Acting on a Slender Fibrous Particle

To complete description of fibrous particle motion the vector of external forces per unit length of the fiber, $\mathbf{q}(s,t)$, must be determined. Among external forces we will consider three of the most common: viscous drag, gravitational force, and coulombic force acting on a charged particle in an electrostatic field. Let us denote: $\mathbf{f}(s,t)$ = vector of drag force per unit length of the fiber (to be discussed later); $\Delta\rho = \rho - \rho_f$, where ρ and ρ_f is particle and fluid density, respectively; $\hat{\mathbf{g}}$ = unit vector of orientation of gravity field with respect to fixed Cartesian coordinates x_1, x_2; $\mathbf{E}^{(el)}(s,t)$ = vector of electrostatic field intensity; $C^{(g)} \approx 9.81$ m/s^2, acceleration due to gravity; $\lambda(s,t)$ – linear density of electric charge of particle. In general, $\mathbf{q}(s,t)$ may be therefore written as

$$\mathbf{q} = \mathbf{f} + A\Delta\rho C^{(g)}\hat{\mathbf{g}} + \lambda\mathbf{E}^{(el)}. \tag{10.77}$$

The drag force $\mathbf{f}$ for an elongated particle in Stokes flow can be estimated from the slender body theory (Batchelor, 1970; Cox, 1970; Tilbet, 1970; Chwang and Wu, 1974; 1975; Keller and Rubinow, 1976; Johnson, 1980). Such estimates can be obtained by the method of matching the asymptotic inner and outer expansions or by the method of superposing of singularities (Stokeslets, doublets, rotlets, etc.) on the fiber centerline. The simplest result of the slender body theory has well-known form:

$$\mathbf{f}(s) = \frac{2\pi\mu}{\ln 2\beta}\left[2\hat{\mathbf{I}} - \frac{\partial\mathbf{x}(s)}{\partial s}\frac{\partial\mathbf{x}(s)}{\partial s}\right]\cdot\left[\mathbf{u}(s) - \mathbf{v}(s)\right], \tag{10.78}$$

where μ = fluid viscosity; $\mathbf{v}(s)$ = vector of local particle velocity; $\mathbf{u}(s)$ = vector of undisturbed (in the absence of particle) fluid velocity at the same points on fiber main axis as for $\mathbf{v}(s)$, that is, at $\mathbf{x}(s)$; and $\hat{\mathbf{I}}$ = unit (identity) matrix. This leading-order solution can be regarded as the first-order approximation of expansion of $\mathbf{f}$ in the series of $1/\ln 2\beta$, with neglected term $1/(\ln 2\beta)^2$ and higher powers being therefore limited to large values of β. The closed form of the second-order approximation, accurate to the term $1/(\ln 2\beta)^2$ given by Cox (1970) can be written in Cartesian coordinates as follows:

$$\begin{aligned}\mathbf{f}(s) = {} & \frac{2\pi\mu}{\ln 2\beta}\left[2\hat{\mathbf{I}} - \frac{\partial\mathbf{x}(s)}{\partial s}\frac{\partial\mathbf{x}(s)}{\partial s}\right]\cdot\left[\mathbf{u}(s) - \mathbf{v}(s)\right] - \frac{\pi\mu}{2(\ln 2\beta)^2}\left[2\hat{\mathbf{I}} - 3\frac{\partial\mathbf{x}(s)}{\partial s}\frac{\partial\mathbf{x}(s)}{\partial s}\right]\cdot\left[\mathbf{u}(s) - \mathbf{v}(s)\right] \\ & - \lim_{\varepsilon\to 0}\left\{\frac{2\pi\mu}{(\ln 2\beta)^2}\left[\mathbf{J}(s,\varepsilon) + \left(\mathbf{u}(s) - \mathbf{v}(s)\right)\ln 2\varepsilon\right]\cdot\left[2\hat{\mathbf{I}} - \frac{\partial\mathbf{x}(s)}{\partial s}\frac{\partial\mathbf{x}(s)}{\partial s}\right]\right\},\end{aligned} \tag{10.79}$$

where ε = small parameter, $\varepsilon << L$, and vector $\mathbf{J}$ is given by sum of two integrals, which limits of integration depend on ε:

$$\mathbf{J}(s,\varepsilon) = \frac{1}{2}\left[\int_0^{s-\varepsilon} \int_{s+\varepsilon}^{l}\right]\left\{\left[\frac{\hat{\mathbf{I}}}{|\mathbf{x}(s) - \mathbf{x}(z)|} + \frac{[\mathbf{x}(s) - \mathbf{x}(z)][\mathbf{x}(s) - \mathbf{x}(z)]}{|\mathbf{x}(s) - \mathbf{x}(z)|^3}\right]\cdot\left[\hat{\mathbf{I}} - \frac{1}{2}\frac{\partial\mathbf{x}(z)}{\partial z}\frac{\partial\mathbf{x}(z)}{\partial z}\right]\cdot\left[\mathbf{u}(z) - \mathbf{v}(z)\right]\right\}dz. \tag{10.80}$$

An even more accurate description with error of the order $1/\beta^2$ was given by Johnson (1980). However, in contrast to explicit Cox's approximation, this formulation has implicit form, in which formula for local value $\mathbf{f}(s)$ is expressed by an integral of unknown distribution $\mathbf{f}(s)$ along the entire particle. In the local basis $\hat{\mathbf{e}}_1$, $\hat{\mathbf{e}}_2$, components f_1, f_2 of $\mathbf{f}(s)$ may be calculated as follows:

$$f_i(s) = \frac{8\pi\mu}{H_i}\left[u_i(s) - v_i(s)\right] - \frac{1}{H_i}\int_0^L K_i(s,z)dz, \tag{10.81}$$

where $u_i(s)$, $v_i(s)$, $i = 1,2$, are components of fluid and particle velocity in local coordinates at point $\mathbf{x}(s)$ on fiber main axis located at distance s from a chosen end of the fiber, H_i has components: tangential, $H_1 = 4(\ln 2\beta - {}^1/_2)$, and normal (and binormal, for the three-dimensional case, too), $H_2 = 2(\ln 2\beta + {}^1/_2)$, and components of vector $K_i(s,z)$ in Equation 10.81 are given by

$$K_i = \frac{f_i(z)}{|\mathbf{R}(s,z)|} + \frac{\mathbf{f}(z)\cdot\mathbf{R}(s,z)}{|\mathbf{R}(s,z)|^3}R_i(s,z) + \frac{D_i f_i(z)}{|s-z|} \quad \text{(no sum on } i\text{)}, \tag{10.82}$$

in which z, $0 \le z \le L$, is an integration variable in Equation 10.81, $\mathbf{R}(s,z)$ is a vector between points $\mathbf{x}(s)$ and $\mathbf{x}(z)$, both lying on the body centerline, $\mathbf{R}(s,z) = \mathbf{x}(s) - \mathbf{x}(z)$, $R_i(s,z)$ denotes components of $\mathbf{R}(s,z)$ in local coordinates in the basis $\hat{\mathbf{e}}_1$, $\hat{\mathbf{e}}_2$, placed at point $\mathbf{x}(z)$, and D_i has components: $D_1 = 2$, $D_2 = 1$ (and $D_3 = 1$ for three-dimensional case). Equation 10.81 together with 10.82 can be solved, for example, iteratively. Taking $K_i = 0$ as a zeroth approximation, we obtain from Equation 10.81 the first iterate of $\mathbf{f}$ in the basis $\hat{\mathbf{e}}_1$, $\hat{\mathbf{e}}_2$:

$$f_1^{(1)}(s) = \frac{2\pi\mu}{\ln 2\beta - \frac{1}{2}}\left[u_1(s) - v_1(s)\right], \tag{10.83}$$

$$f_2^{(1)}(s) = \frac{4\pi\mu}{\ln 2\beta + \frac{1}{2}}\left[u_2(s) - v_2(s)\right]. \tag{10.84}$$

Then, this first approximation of $\mathbf{f}$ can be inserted into Equation 10.82, and after evaluation of the integral in Equation 10.81 the second approximation $\mathbf{f}^{(2)}(s)$ is obtained and the iteration procedure can be continued in the same way. Let us note that the result of the first iterate given by 10.83 and 10.84 may be rewritten for the two-dimensional problem in terms of Cartesian components f_{x1}, f_{x2} of $\mathbf{f}$ as

$$\mathbf{f}(s) = \frac{2\pi\mu}{\ln 2\beta - \frac{1}{2}}\left[2N\hat{\mathbf{I}} - (2N-1)\frac{\partial\mathbf{x}(s)}{\partial s}\frac{\partial\mathbf{x}(s)}{\partial s}\right]\cdot\left[\mathbf{u}(s) - \mathbf{v}(s)\right], \tag{10.85}$$

wherein $\mathbf{u}$ and $\mathbf{v}$ have Cartesian components u_{xi} and v_{xi} and N is defined as $N = {}^1/_2 H_1/H_2 = (\ln 2\beta - {}^1/_2)/(\ln 2\beta + {}^1/_2)$. Thus, for sufficiently slender particle, when $\ln 2\beta \gg {}^1/_2$, N tends to 1, and Equation 10.85 becomes identical with Equation 10.78. The iterative procedure of solving Equation 10.81 and 10.82 described above is usually quickly converging and only few (two to three) iterations are necessary to achieve satisfactory accuracy. An alternative approach is to solve numerically the system of integral equations 10.81 and 10.82 representing integrals on the right-hand side by means of composed quadratures for some number of subsections the fiber was divided for. In this way we get the system of linear algebraic equations that can be easily solved using standard procedures.

10.3 MODELING OF DEPOSITION OF FIBROUS PARTICLES IN A FIBROUS FILTER — EXAMPLES OF CALCULATIONS

The general theory of fibrous, deformable aerosol particles mechanics presented above can be used for theoretical studies of motion and deposition of such particles in various systems with an arbitrary pattern of carrier gas flow, provided that the flow around a fiber has Stokesian (creeping) character. Below we present some examples of calculations for deposition of flexible and stiff fibrous particles, as well as for spherical ones, on a cylindrical collector placed transversely to the direction of gas flow, which represents a single filtering element of a fibrous filter. The modeling of air filtration in a fibrous filter is a very difficult problem because of a complex, more or less random structure of the filter. Since the main topic of this chapter is the comparison of fibrous and spherical particle behavior rather than the general theory of filtration, we will use the simplest, two-dimensional model of the filter with a well-defined geometry, assuming a uniform arrangement of identical collectors on the hexagonal grid (Figure 10.4a). The structure of the filter is then divided into the sequence of parallel layers, and each layer into the set of rectangular elementary cells of

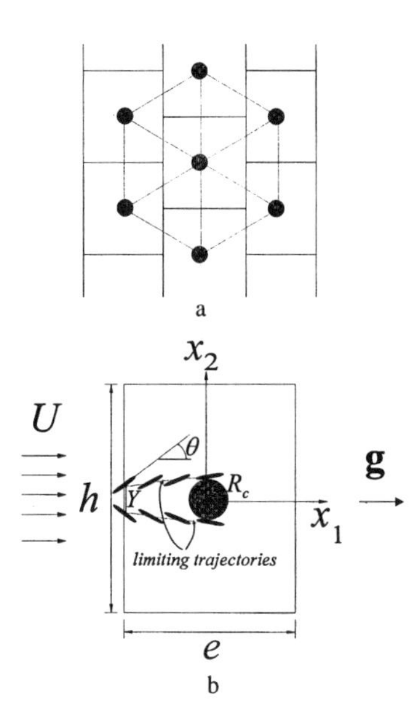

Figure 10.4 Model of a fibrous filter structure and definition of the single-cell deposition efficiency. (a) Staggered model of the filter (hexagonal grid). (b) Definition of single-cell deposition efficiency: $E = Y/h$.

height h and width e, with a single collector in the center. The cell dimensions can be calculated from the macroscopic porosity (void fraction) of the filter, ε, and the collector radius, R_c, as

$$\frac{h}{R_c} = \sqrt{\frac{2\pi\sqrt{3}}{3(1-\varepsilon)}}, \quad \frac{e}{R_c} = \sqrt{\frac{\pi\sqrt{3}}{2(1-\varepsilon)}}. \tag{10.86}$$

Modern fibrous filters are usually very porous ($\varepsilon \geq 0.95$), and hence the distances between neighboring fibers are much greater then their radii ($h >> R_c$). It enables us to neglect in a first approximation the influence of the other collectors when computing the fluid flow field around a cylinder considered. To describe the gas flow structure in the vicinity of the collector we used the exact analytical solution of the Oseen problem for the flow past a single circular cylinder (for details see Podgórski, 1993; Podgórski et al., 1995). Based on the concept of the described unit cell, an elementary measure of the filter performance can be a single cell deposition efficiency, E, defined as

$$E = \frac{Y}{h}, \tag{10.87}$$

where Y is a distance between the starting points of the upper and lower limiting trajectory (Figure 10.4b). The limiting trajectories divide the cell inlet into two parts: all particles whose mass centers enter the cell between the beginnings of these limiting trajectories will be deposited on the collector in the cell (we may say that the probability of deposition for these particles is 1); while other particles entering the cell beyond this region will pass over the collector (the deposition probability is 0). Such definition of the cell efficiency is unique for spherical particles, if there exist only the deterministic mechanisms of collection (i.e., interception, inertial impaction, sedimentation, and, eventually, electrostatic attraction), i.e., if the deposition due to random Brownian motion is negligible.

For submicron particles, the random motion becomes important and the method of evaluation of the deposition efficiency must be modified. Two different approaches are possible: (1) to determine separately the deposition efficiency due to deterministic mechanisms based on the concept of limiting trajectory as described previously, i.e., neglecting Brownian motion, and the deposition efficiency due to Brownian motion only, solving the diffusion–convection equation for the population of particles, and finally to calculate the total deposition efficiency assuming that the deterministic and stochastic mechanisms are independent; (2) to calculate the probability of deposition for various starting points at the cell inlet solving repeatedly the stochastic equation of motion with a random force term (Langevin equation), and then to determine an expected value of the deposition efficiency in the cell by averaging particular probabilities of deposition for various starting points.

In this chapter we consider larger particles only, limiting our interest to the deterministic mechanisms of deposition. For nonspherical particles, however, the course (pattern) of limiting trajectories is also dependent on the initial orientation of the particle, which for an initially straight fiber can be defined as an angle of inclination, θ, of its major axis with respect to the direction of gas flow, Figure 10.4b. By solving numerically equations of motion for a fibrous particle given in the previous section, the trajectories around a collector were determined; then the limiting trajectories were found iteratively and the single-cell deposition efficiency was calculated according to Equation 10.87 for various initial orientations θ. The set of constant parameters for the calculations presented below was the following: collector radius, $R_c = 20$ μm; filter porosity, $\varepsilon = 0.95$; particle density, $\rho = 1500$ kg/m^3; fluid (air) properties, $\rho = 1.2$ kg/m^3; $\mu = 1.83 \times 10^{-5}$ Pa s. Figure 10.5 presents an example of the relation between the single-cell efficiency and the angle of initial orientation θ for a flexible and a stiff fiber having length $L = 20$ μm and slenderness $\beta = 10$, and for gas flow velocity $U = 0.5$ m/s. For the comparison, the value of E for spherical particles of the same volume and density as the fibers is also shown. It can be seen that for both flexible and stiff fibers E is a complex function of θ. This complex shape is a result of two effects which are strongly dependent on the initial fiber orientation. The main mechanisms of deposition for these conditions are interception and inertial impaction.

It is obvious that an elongated fiber approaching the collector perpendicularly oriented to the direction of gas flow has more chances to be captured than the same fiber with parallel orientation. However, the spatial orientation of the fiber in the vicinity of the collector depends on the whole "history" of the particle motion, also on its initial orientation at the cell border. This orientation affects both translation (drag force depends on the fiber orientation; roughly, it is two times greater

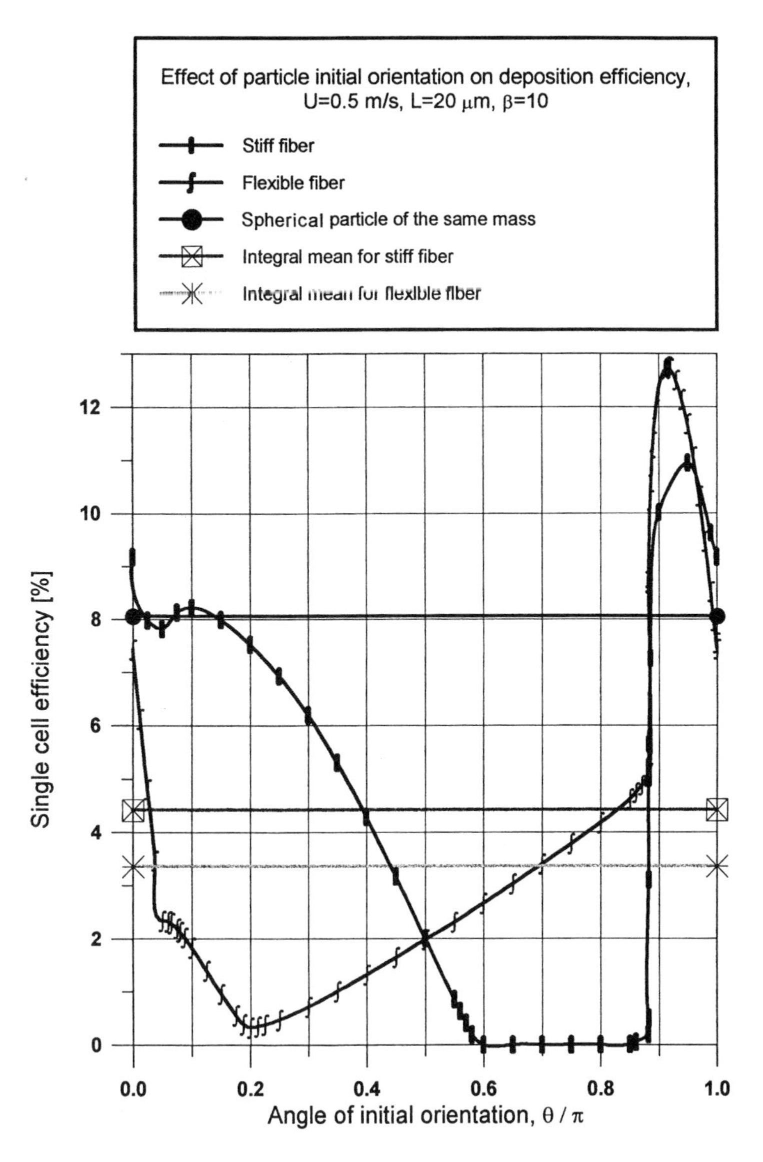

Figure 10.5 Relation of the single-cell deposition efficiency on the angle of initial orientation for stiff and flexible fibrous particle.

for perpendicular than for the parallel one, thereby the residence time in the cell depends on θ) and rotation of the fiber (which causes continuous change of orientation, leading to the alignment of the fiber major axis to the local direction of fluid streamlines). The actually observed tendency of a fibrous particle to follow fluid streamlines results in an unexpected decrease of the deposition efficiency for elongated fibrous particles in comparison to the compact, spherical particles of the same volume — most fibers move around the collector peripherally rather than transversely; this phenomenon lowers the capture probability because of their very small thickness (diameter). It is important to note that for the majority of initial orientations the deposition efficiency for both flexible and stiff fibers is usually smaller than for a spherical particle of the same mass. After determination of the function $E(\theta)$ one could calculate the average single-cell efficiency for a fibrous particle as

$$E_{av} = \int_0^{2\pi} E(\theta)p(\theta)d\theta\,, \tag{10.88}$$

where $p(\theta)$ is the probability density function that the fiber at the cell entrance is inclined at the angle θ to the gas flow direction. Determination of the function $p(\theta)$ is necessary to obtain a real measure of the filter performance but it requires further studies. Seemingly, it depends on the filter structure and porosity. If the filter structure is completely random and the porosity not extremely high, one could expect that all initial orientations are equally probable, so it might be assumed that $p(\theta)$ is constant and equals $1/2\pi$. On the other hand, for a filter with more regular structure and very high porosity, $p(\theta)$ should possibly have a very narrow shape of distribution with a maximum near $\theta = 0$. This conclusion can be implied from the observation that most fibers which are not captured become almost straight with their main axes parallel to the direction of gas flow (what corresponds to $\theta = 0$) at sufficiently far distance from the collector. Taking into account the criterion of random initial orientation, meaning that the average cell efficiency for a fibrous particle is simply the integral mean of the particular efficiencies averaged uniformly over all possible angles 0 to 2π, we can find from Figure 10.5 that both stiff and flexible fibers show a significantly lower average efficiency than the spherical particle of the same volume, and the deformable fiber is more penetrable through a filter than the identical rigid fiber. However, taking into account the second criterion, the conclusions will be different. Assuming, as an extreme case, that all the fibers far from the collector travel exactly along the fluid streamlines, and that in a very porous filter gas streamlines at the cell border are almost straight and coincide with the direction of gas flow, we have the probability distribution function in the limiting form: $p(\theta) = \{1, \text{for } \theta = 0; 0, \text{for } \theta \neq 0\}$. In this case the average efficiencies (equal to $E(\theta = 0)$) for all kinds of analyzed particles are comparable, the highest one for a stiff fiber, slightly lower for a spherical particle, and the lowest one for a flexible fiber. Since, as we mentioned before, the problem of correct determination of the function $p(\theta)$ has not been resolved as yet, we will limit our further analysis to some arbitrarily chosen angles θ. The three subsequent plots in Figure 10.6 show the influence of the gas velocity in a wide range 0.01 to 1 m/s on the single-cell collection efficiency of flexible and stiff fibers of length $L = 20$ μm for initial orientations $\theta = 0$, $\pi/4$, and $\pi/2$. For the comparison, the relationship $E(U)$ for spherical particle of the same volume and density as the fibers is also shown. We can observe that for high gas velocities, when the inertial mechanism of deposition prevails, the efficiencies for all kinds of particles are similar, but at intermediate and low flow rates, the differences become significant, depending on the angle of initial orientation. For parallel initial orientation ($\theta = 0$) and $U < 0.15$ m/s, Figure 10.6a, the lowest efficiency has the flexible fiber, the intermediate value the spherical particle, and the highest, the stiff fiber. This case, $\theta = 0$, emphasizes the importance of mechanical properties (stiffness). For $\theta = 0$ the moment of external forces acting on the entire fiber is too weak to initiate the rotation significant enough to cause specific spatial configuration which enables the particle to pass over the collector. A completely flexible particle, however, easily changes its shape even under action of very small external forces, which yields a significantly lower deposition efficiency. On the other hand, for transverse initial orientation, $\theta = \pi/2$ (Figure 10.6c), rotation of both stiff and flexible fibers is sufficiently strong to cause alignment of the major axis to the local direction of gas flow. It results finally in similar for both kinds of fibers, and significantly lower than for spherical particle, deposition efficiencies. For intermediate initial orientation, $\theta = \pi/4$ (Figure 10.6b), the situation is entirely different; for spherical particles and flexible fiber we can observe a minimum of deposition efficiency as a function of gas velocity, while for a stiff fiber there is generally a permanent increase of E with U. The nature of the observed minimum of the function $E(U)$ is such that for low gas velocities, when sedimentation and direct interception are the predominating mechanisms of deposition, an increase in gas flow rate causes a decrease in deposition efficiency, since for these mechanisms the drop of the residence time in the cell is unfavorable, while for higher gas velocities the efficiency for predominating then mechanism of

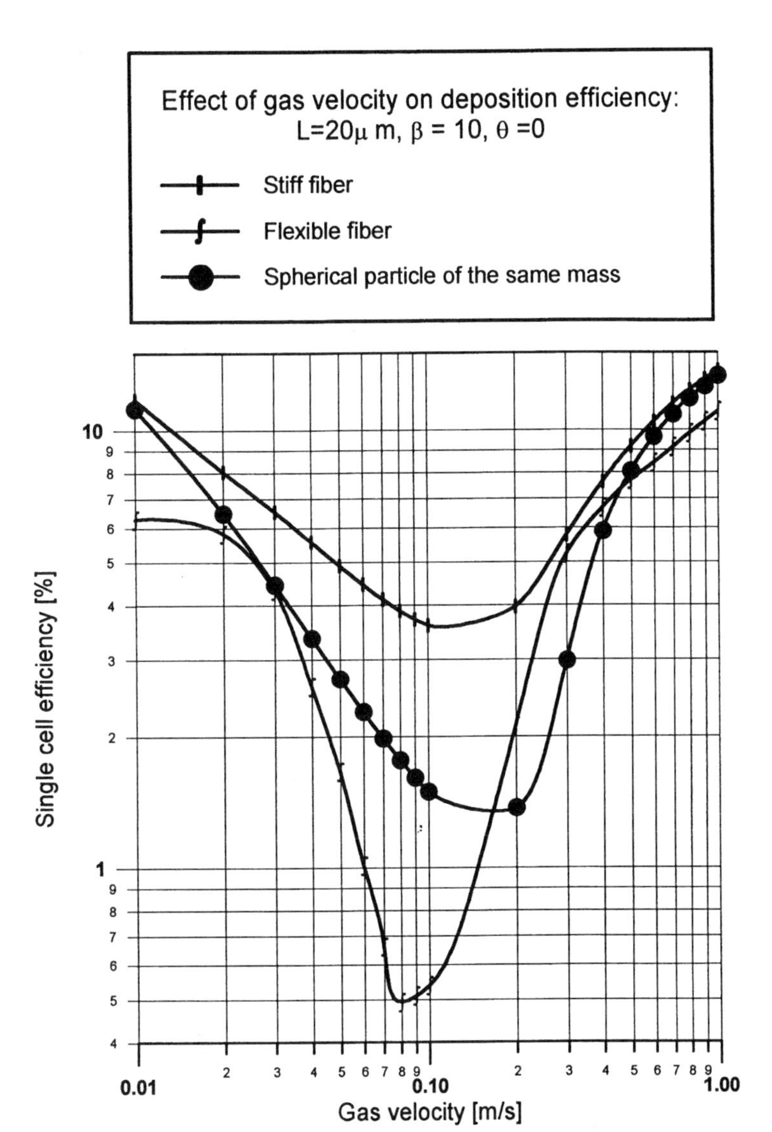

Figure 10.6 Influence of gas velocity on the single-cell deposition efficiency for various initial orientations of the fibrous particle: (a) $\theta = 0$; (b) $\theta = \pi/4$; (c) $\theta = \pi/2$.

inertial impaction increases obviously with an increase of the flow rate. This simple interpretation is somewhat complicated by the influence of gas velocity and initial orientation on the rotation, which is a very important phenomenon in the motion of an elongated fiber.

The next parameter taken into account in the analysis is the mass of the particle. Figure 10.7 shows two examples of computation results for low gas velocity, $U = 0.02$ m/s, and for the high one, $U = 0.5$ m/s, for $\beta = 10$ and $\theta = \pi/2$. The size of the fibrous particle is presented in terms of the diameter of a sphere having the same volume, d_{eq}. Since the diffusional mechanism of deposition (important for submicron particles) is not considered here, we observe an increase in the collection efficiency with the increase of the particle size, as this is favorable for all other mechanisms (sedimentation, interception, impaction). For low gas velocities, e.g., for $U = 0.02$ m/s (Figure 10.7da), the relationships $E(d_{eq})$ are represented in log–log plots by almost straight lines for all spherical, fibrous stiff, and fibrous flexible particles. For higher gas velocities, e.g., for $U = 0.5$ m/s (Figure 10.7b), when inertial effects in rotation are more important, the shapes of the curves $E(d_{eq})$

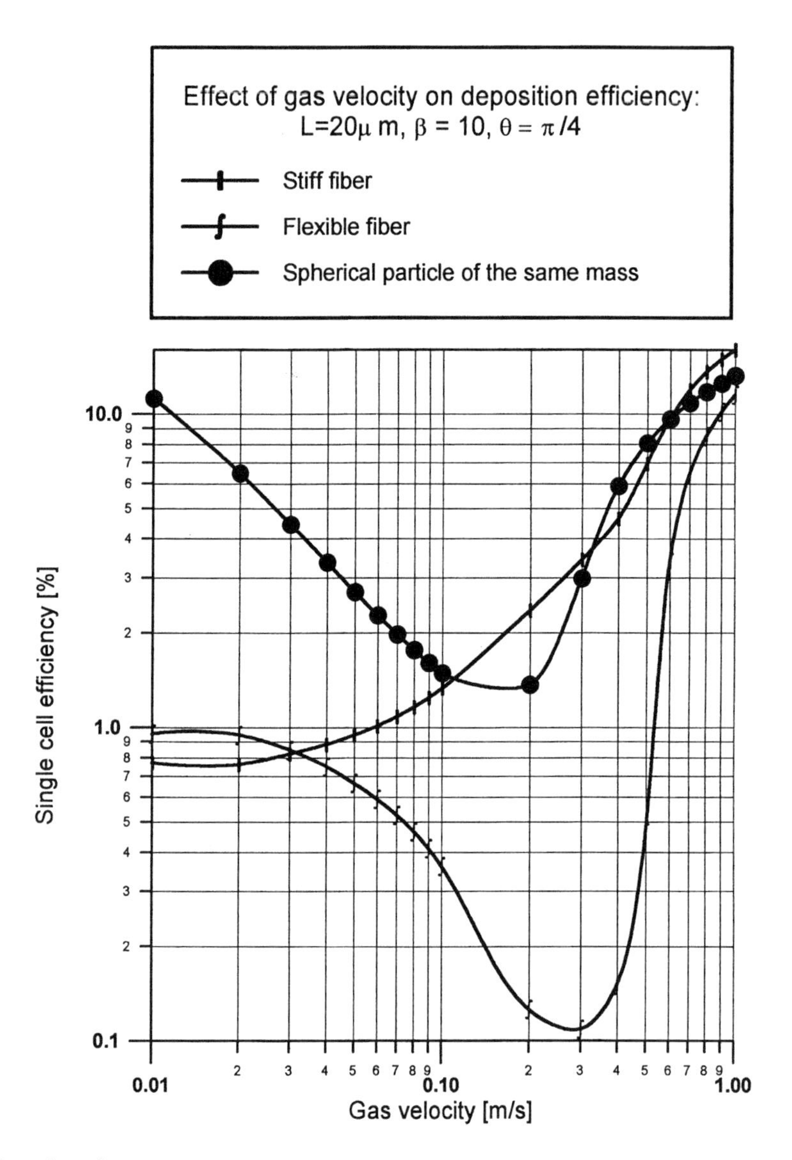

Figure 10.6 (continued)

become sigmoidal. Nevertheless, the general conclusion arising from the comparative analysis of the deposition efficiency as a function of the particle size is similar to that one, which may be drawn considering the influence of gas velocity: in a wide range of variation of particle size and gas flow rate, the collection efficiency for spherical particles is usually higher than this for stiff fibers of the same volume, and the lowest efficiency is achieved by flexible fibers.

All the results presented above concerned the particle of constant aspect ratio $\beta = 10$. To characterize the size of a nonspherical particle we have to know two dimensions, length and diameter. Analyzing the effect of the slenderness ratio β of the fibrous particle of fixed volume on the deposition efficiency for various initial orientations we have found that for both stiff and flexible fibers the function $E(\beta)$ shows usually a minimum in the range $5 < \beta < 15$, depending on θ and U.

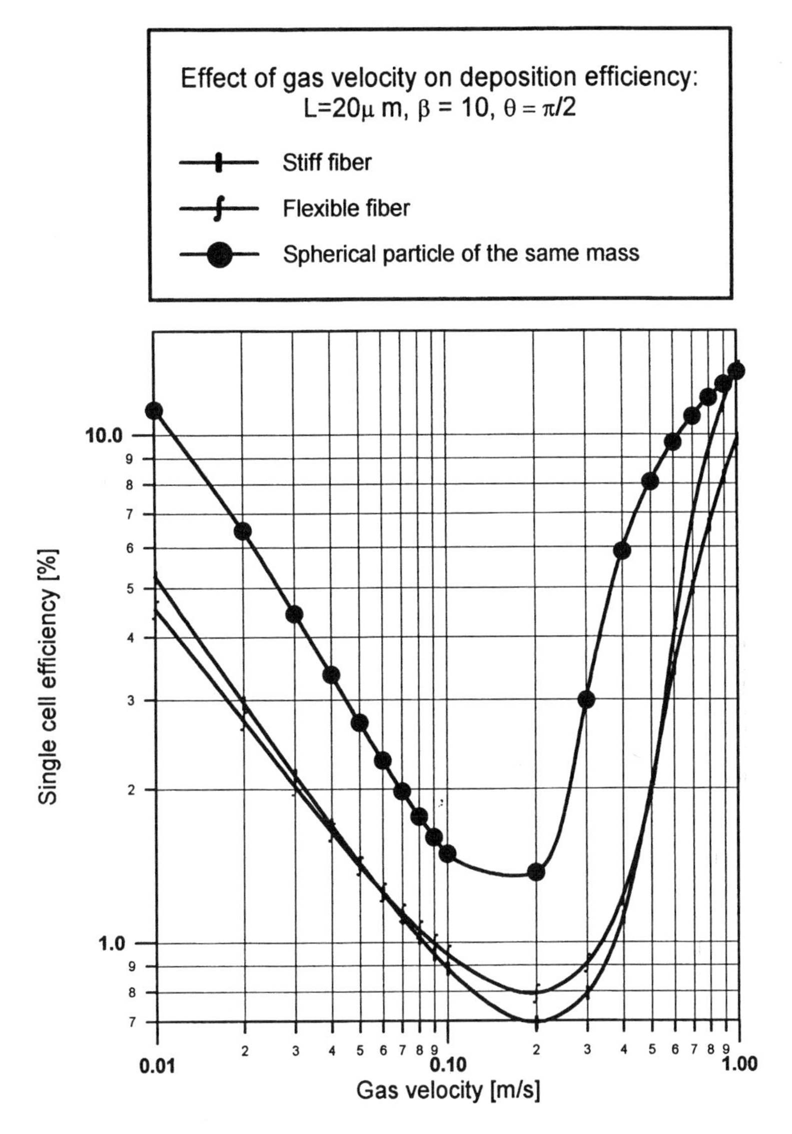

Figure 10.6 (continued)

10.4 SUMMARY

The complete set of equations describing deterministic motion of a deformable fibrous aerosol particle in a moving fluid was presented. This general formulation is valid for any stiffness of a particle, filling the gap in existing, and forming a bridge between the two limiting cases of completely stiff and perfectly flexible fiber, for which the description was already known. The theory can be used for analysis of a fibrous aerosol particle behavior in various systems; in this chapter, as an illustration, some examples of simulations for a simple model of filtration in a fibrous filter were shown. It was found that rotation and shape deformation (the latter for a flexible particle) are very important phenomena which lead to the alignment of the fiber major axis to the local direction of fluid streamlines, causing in most cases an unexpected drop of the deposition efficiency in comparison with that for equivalent spherical particles. Because aerodynamic behavior of a fibrous particle (drag

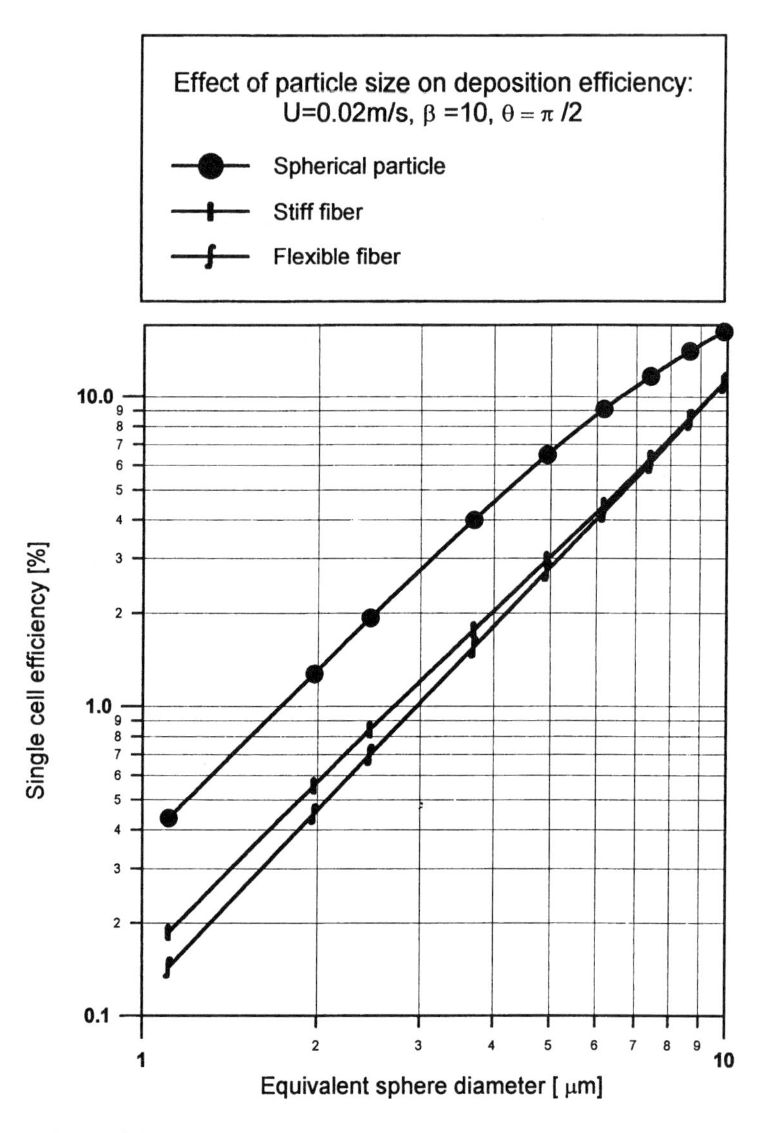

Figure 10.7 Dependence of the single-cell deposition efficiency on particle size for two gas velocities: (a) U = 0.02 m/s; (b) U = 0.5 m/s.

force, external moment of forces) depends on the momentary configuration of the particle in the gas flow field, and since this orientation varies in the filter continuously (due to rotation and, eventually, deformation), the shapes of the efficiency curves as a function of various parameters differ significantly for fibrous and spherical particles. Therefore, one should not expect that the filter performance for fibers removal could be predicted only on the basis of the well-known data for spherical particles. It was also found that the mechanical properties of fibrous particles are of importance and influence the deposition efficiency — usually, flexible fibers are worse filtrate than stiff ones. Numerical analysis of deformable fibrous aerosol behavior is very complicated, but seems to be necessary, taking into account the dangerous health effects and still unresolved problems with experimental techniques (generation of monodisperse testing fibers and measuring their sizes — length and diameter — and concentration). Finally, we have to point out that any transfer of data valid for spherical particles is impossible; fibers and spherical particles take part in different phenomena during their motion and these phenomena imply different mechanisms of deposition.

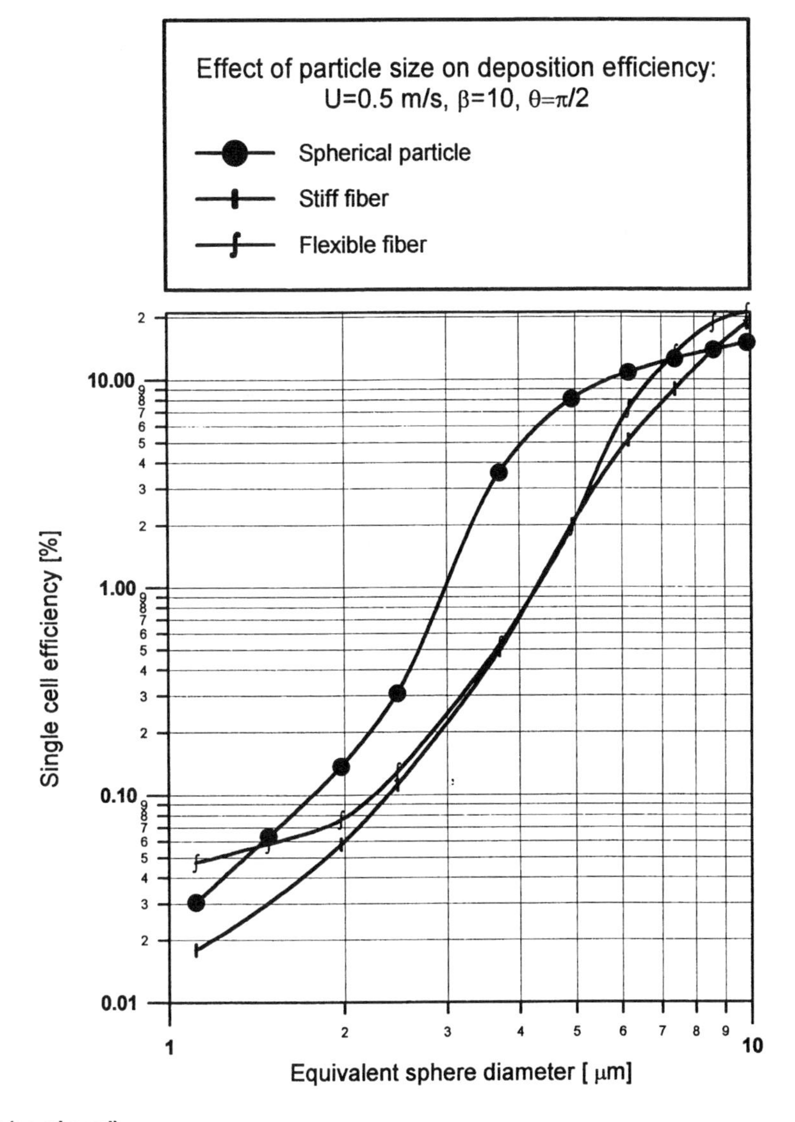

Figure 10.7 (continued)

REFERENCES

B. Asgharian, L. Gradoń, and C. P. Yu, *Aerosol Sci. Technol.*, 9, 113, 1988.
G. K. Batchelor, *J. Fluid Mech.*, 44, 419, 1970.
O. Bernstein and M. Chapiro, *J. Aerosol Sci.*, 25, 113, 1994.
A. T. Chwang, T. Y.-T. Wu, *J. Fluid Mech.*, 63, 607, 1974.
A. T. Chwang, T. Y.-T. Wu, *J. Fluid Mech.*, 67, 787, 1975.
R. G. Cox, *J. Fluid Mech.*, 44, 791, 1970.
N. A. Fuchs, *The Mechanics of Aerosols*, Pergamon, New York, 1964.
J. Gallily and E. M. Krushkal, *J. Colloid Interface Sci.*, 99, 141, 1984.
J. W. Gentry, K. R. Spurny, and S. A. Soluen, *J. Aerosol Sci.*, 19, 141, 1988.
L. Gradoń and A. Podgórski, *Chem, Eng. Sci.*, 45, 3435, 1990.
L. Gradoń and A. Podgórski, and W. Pilaciński, *Chem, Eng. Sci.*, 43, 1253, 1988.
L. Gradoń, A. Podgórski, and P. Grzybowski, *J. Aerosol Sci.*, 20, 971, 1989.

W. D. Griffith, *J. Aerosol Sci.,* 19, 703, 1988.
P. Grzybowski and L. Gradoń, *Chem. Eng. Sci.,* 47, 1453, 1992.
E. J. Hinch, *J. Fluid Mech.,* 74, 317, 1976.
R. E. Johnson, *J. Fluid Mech.,* 99, 411, 1980.
J. B. Keller and S. I. Rubinow, *J. Fluid Mech.,* 75, 705, 1976.
T. L. Odgen and W. H. Walton, *Ann. Occup. Hyg.,* 18, 157, 1975.
A. Podgórski, *J. Aerosol Sci.,* 24, S277, 1993.
A. Podgórski, *Abstracts of Conf. on Aerosol Sci. & Technol.,* Bombay, IASTA, 1994a, 79.
A. Podgórski, *J. Aerosol Sci.,* 25, S193, 1994b.
A. Podgórski and L. Gradoń, *J. Aerosol Sci.,* 21, 957, 1990.
A. Podgórski, L. Gradoń, and P. Grzybowski, *Chem. Eng. J.,* 58, 109, 1995.
M. R. Shamberger, J. E. Peters, and K. H. Leong, *J. Aerosol Sci.,* 21, 539, 1990.
M. Shapiro and M. Goldenberg, *J. Aerosol Sci.,* 24, 65, 1993.
J. P. K. Tillett, *J. Fluid Mech.,* 44, 401, 1970.
V. Timbrell, *Ann. N.Y. Acad. Sci.,* 132, 225, 1965.
C. C. Wang, J. R. Pao, and J. W. Gentry, *J. Aerosol Sci.,* 19, 508, 1988.

Chapter 11

Nature, Stability, and Effectiveness of Electric Charge in Filters

Richard C. Brown

CONTENTS

11.1 INTRODUCTION

Filter material carrying an electric charge is widely used in particulate respirator capsules and disposable filtering face pieces, but it also has a place in vacuum cleaners, small-scale air cleaners, air-conditioning units, and filters on a larger scale. The advantage of the material is that it functions by means of an attractive force between airborne particles and filter fibers, which augments filtration efficiency without any penalty of increased airflow resistance.

0-87371-830-5/98/$0.00+$.50

The greater part of the mass of most aerosols consists of electrically charged particles, but all particles respond to an electric field by means of induced dipole forces. A permanent charge on a particle is stable even if the material of the particle is conducting, because the particle is an isolated system and the charge cannot leak away. Fugitive charge on filter fibers is relatively common, but stable charge requires materials that are very good insulators.

11.2 ELECTRICALLY CHARGED MATERIALS, OLD AND NEW

Permanently charged filter materials may be classified by the means in which the material becomes electrically charged: triboelectric charging, corona charging, and charging by induction. A further process, the freezing in of polarization charge, has been demonstrated experimentally.

11.2.1 Tribolectrically Charged Material

The first recognized electrostatic filter, the resin–wool or Hansen filter (Hansen, 1931) was used for a number of years before its mechanism of action was understood. It consists of particles of resin mixed with fibers of wool, both elements being clearly visible in the electron micrograph, shown in Figure 11.1a. When the raw materials are made into a fleece with a microscopic structure suitable for a filter, by the familiar textile process of carding, the vigorous contact that takes place between the two components causes them to exchange charge so that the resin becomes negatively charged and the wool positively (Walton, 1942; Feltham, 1979).

Resin is an extremely good insulator, and its low conductivity is sufficient to ensure that the charge on the filter material is stable. Wool can be considered to be a conductor, but it is important to realize that the wool will not conduct away the charge on the resin because most of is charge does not come into physical contact with the wool.

The wool used in this type of material usually consists of short fibers, which are unsuitable for textile manufacture. Although the electrical properties of the wool are independent of the length of the fibers, the structure of material made from short fibers tends to be irregular, with regions of high fiber density, called neps, within a lower-density matrix. This results in nonuniformity of flow, which reduces the filtration efficiency.

Exploitation of triboelectricity has led to the development of a further material, simpler in structure and taking advantage of the highly insulating nature of the synthetic fibers that have come into widespread use since the development of the resin–wood filter. The material is made from a binary mixture of polymer fibers (Smith et al., 1988), which become charged during carding.

Two yarns of dissimilar polymers rubbed together will usually exchange charge in a consistent way so that one species develops a positive charge and the other a negative. In most instances the charge is low and unstable. Sometimes it decays before a measurement can be made, so that the sign of charge developed must be inferred from the charge on the companion fiber. Measurements of this sort can be used to produce a triboelectric series like that shown in Table 11.1, defined so that fibers of a type high in the series develop a positive charge when rubbed against those below them (Hersch, 1975). Different observers do not quite agree on the series, probably because generic names are used to describe classes of polymers, the members of which are not chemically identical; but wool is the most electropositive fiber, nylons are slightly less so, followed by cellulose-based polymers and polyolefins in that order, with aromatic and halogenated fibers at the negative end of the series. Although the triboelectric series gives an indication of the sign of the charge developed by each fiber type of a pair, it gives no indication of the level of charge, nor of its stability, two properties critical to filter performance. In fact, the mixed-fiber filter now in use consists of a polypropylene fiber, which imparts charge stability, like the resin, and a chlorinated polymer, which exchanges charge particularly well with the polypropylene. The chlorinated fiber, like wool, is not a particularly good insulator. An electron micrograph of this material is shown in Figure 11.1b.

Figure 11.1 Electron micrographs of electrically charged filter material. (a) Resin wool; (b) mixed-fiber material; (c) fibrillated electret material; (d) electret material charged after carding; (e) solvent-spun material; (f) melt-spun material. (© Crown copyright; by permission of Her Majesty's Stationery Office.)

The material is made from normal crimped textile fibers, which give it a uniform structure when produced in the form of a fleece. It can be strengthened by needling, a process in which barbed needles are punched through the fabric, causing entanglement of the fibers, and compacting

Table 11.1 Tribolectric Sereies of Textile Yarns

Positive
Wool
Hercosett wool
Nylon 66
Nylon 6
Silk
Regenerated Cellulose
Cotton
Polyvinyl alcohol
Chlorinated wool
Cellulose triacetate
Calcium alginate
Acrylic
Cellulose diacetate
Polytetrafluorethylene
Polyethylene
Polypropylene
Poly (ethylene terephthalate)
Poly (butylene terephthalate)
Modacrylic
Chorofiber
Negative

the material, increasing its mechanical strength. The chemical nature of the fibers is such that the material can be fabricated by means of thermal or ultrasonic welding (Brown, 1993a).

11.2.2 Corona-Charged Material

A point electrode at a high potential emits ions of its own sign, and these will drift under the influence of the electric field to a collecting surface of lower potential. The surfaces of a thin sheet of polymer can, in this way, be oppositely charged, and in such a form the material is often termed an *electret*. Strictly speaking, this term is incorrect; true electrets are polymers with a frozen-in electric polarization, their form being analogous with that of a magnet, just as the word *electret* is analogous with the word *magnet*. Polypropylene forms a good electret in the loose sense of the word, and if a sheet of this polymer is stretched it suffers considerable molecular realignment, becoming strong along the direction of stretching and weak in the direction perpendicular to it. This allows the sheet to be split into straight fibers with a line-dipole configuration of charge, which can be carded and made into a filter (Turnhout et al., 1980). An electron micrograph of the material is shown in Figure 11.1c, from which the rectangular cross section of the fibers is apparent.

The corona-charging process can be applied to a filter material as well as to the precursor of the filter fibers. An electron micrograph of an example (Baumgartner and Loeffler, 1987) is shown in Figure 11.1d. Alternatively, the dipolar charge can be imparted by freezing in a polarization charge after the material is felted (Shimokobe et al., 1985). If the latter is carried out the material will be a true electret, consisting of polarized fibers. The dipole moments of fibers polarized when in the form of complete materials will be aligned, unlike those of the material produced from split polarized fibers. The exact configuration of charge on materials charged by corona after production is less certain.

11.2.3 Material Charged by Induction

Filter fibers may be charged in a process similar to that used in the production of electrically charged sprays. A conductor placed in an electric field will develop a surface charge that will reduce its own internal field to zero. If the conductor is isolated, the charge developed will be dipolar in form, but if it is attached to another body and subsequently detached, a unipolar charge may be imparted. The charged body can then move under the influence of the electric field that induced the charge. If the conductor is a liquid, electrostatic spraying will take place.

In electrostatic spraying the process of detachment occurs simultaneously with the process of charging (Taylor, 1969), and this can be conceived as the extrusion of a filament of charged liquid which breaks up into droplets under the influence of surface tension (Rayleigh, 1892). If, however, this liquid is replaced by a polymer solution or melt, its high viscosity will delay the breakup into droplets, and under appropriate conditions the polymer filament may solidify into a fiber before this can occur. When such a fiber eventually becomes detached, it may be deposited on the collecting electrode. An accumulation of such fibers will produce a structure rather like an air-laid felt. It is important that the conductivity of the material in its liquid form is sufficiently high for real charge to be induced, and that the conductivity in the solid form is sufficiently low for the charge to be stable.

One effect of the spraying is the production of very fine fibers. Conventional fiber-spinning rarely produces fibers with diameters much less than 20 μm, whereas electrostatically extruded fibers may have diameters as low as 2 or 3 μm. Figure 11.1e shows fibers produced by electrostatic extrusion of a solution of polycarbonate (Schmidt, 1980; Weghmann, 1982) and Figure 11.1f fibers produced by electrostatic extrusion of a melt of polypropylene (Trouihet, 1981).

The principal means whereby electric charge can be imparted have been outlined above, but new electrically charged materials are constantly appearing, and in many instances the manufacturers are either unable or unwilling to describe the exact charging method. The development of some charge on materials is relatively common, although some effort has to be expended if the charge is to be sufficiently high to be useful. In the case of the mixed-fiber triboelectrically charged material, a 50/50 mixture gives the best performance, but either polymer, carded by itself, will act as an electrostatic filter, though of low, and in one case short-lived, efficiency. Other attempts to produce charged materials have included mixing a fine-fibered material into a matrix of coarser fibers, although the structure of such a mixture is far from ideal. In addition, very fine air-blown materials frequently develop an electric charge without the use of any high-voltage source. The advantage of electric charge on a material is clear, but if the charge is short lived, it may lead to a misplaced confidence about filter performance which a greater understanding of filtration would have prevented.

11.3 DIRECT EXPERIMENTAL OBSERVATION OF ELECTROSTATIC EFFECTS

The relative importance of the various electrostatic capture processes can be demonstrated by controlling the charge both on the filter and on monodisperse particles in a challenge aerosol. The former charge can be removed by a large dose of ionizing radiation, details of which will be given later in the chapter, and the latter can be controlled by an aerosol neutralizer. The results of such an exercise using aerosols of precisely known size and charge (either singly charged or neutral) and filters of a common structure, and either charged or uncharged are shown in Figure 11.2, (Trottier and Brown, 1990) in the form of the quality factor (Brown, 1993b), which is proportional to single-fiber efficiency, against particle size.

When both the fiber and the particles are neutral, the efficiency resulting from interception and diffusional deposition effects has a minimum at about 0.25 μm. If the particles are charged, image effects increase the size of the most-penetrating particle, since a single fundamental charge produces the same image force whatever the particle size, but the mobility of small particles is larger. When the fiber alone is charged, polarization forces, which increase rapidly with particle size, augment the capture of larger particles and shift the efficiency minimum to a smaller particle size. When both the fiber and the particle are charged, long-range coulomb forces are dominant except at large

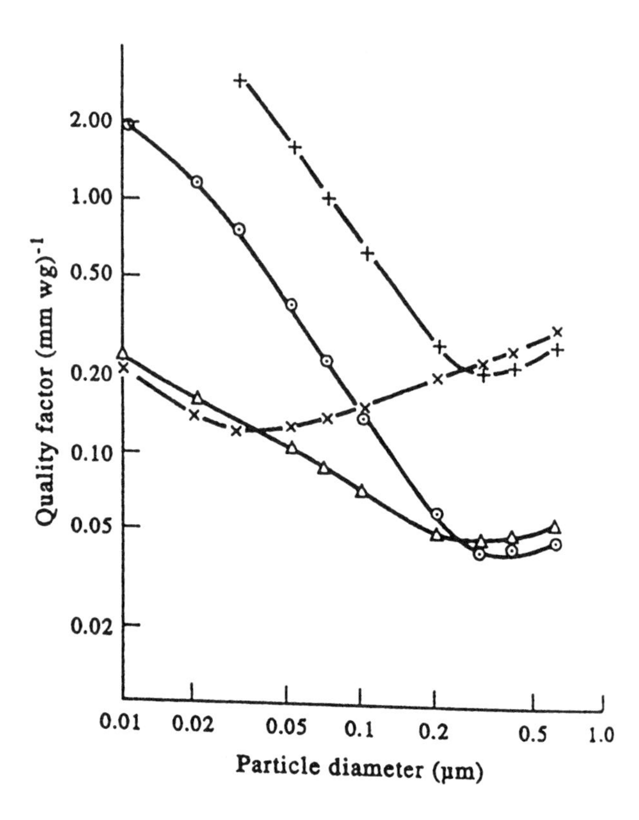

Figure 11.2 Quality factors for monodisperse aerosols and structurally identical filters. △ Filter neutral aerosol neutral; ⊙ filter neutral aerosol charged; x filter charged aerosol neutral; + filter charged aerosol charged. (© Crown copyright; by permission of Her Majesty's Stationery Office.)

particle size, where the polarization forces take over, Broadly similar effects have been observed by other workers (Baumgartner and Loeffler, 1987, 1988; Lathrache and Fissan, 1989).

11.4 CONFIGURATION OF CHARGE ON FIBERS AND CAPTURE EFFICIENCY

In early work on the theory of capture of particles by charged fibers the simplest assumption about the configuration of electric charge on the filter fibers was usually made, that the charge was distributed uniformly. However, details of the methods by which the fibers become charged have suggested that this is not the case, and more direct experimental evidence for nonuniform charge is given later in the chapter.

The simplest model of triboelectric exchange would suggest that the mixed-fiber filter consists of equal numbers of positively and negatively charged fibers according to the prediction of a triboelectric series. However, during the carding of the filter a fiber will contact not only chemically different fibers but also chemically similar ones, along with the metal components of the carding equipment itself. These contacts can easily give rise to charge of the "wrong" sign. The configuration of charge on the other triboelectrically charged filter, the resin–wood filter, is by its very nature complicated, varying not only along the fiber length but also around the fiber circumference.

By simple reasoning the electret filter fibers might be assumed to have a line-dipole configuration, but the amount of fiber–fiber contact that takes place after the line-dipole charge has been imparted is comparable with that which gives the material above all of the charge that it holds, and so it is unlikely that the dipole configuration is preserved exactly in the complete filter.

Electrically extruded fibers would be expected to carry charge of the same sign as that of the electrode from which they arise (Hayati et al., 1987), while they are in contact with it, but there are other processes by which melt- or solvent-spun materials may develop charge, since fine-spun material of this sort will normally carry charge even when no electrodes are involved in its manufacture (Wente, 1956).

This leads to the conclusion that the geometry of electric charge on permanently charged material may be very different from uniform, and so this must be accounted for in any credible theory of filter behavior. However, a moderating influence on the effect of charge configuration is that an electric field tends to be less variable than the charge producing it, because the electric field at any point depends on the electric charge at every point.

The electrostatic potential, V, outside a charged fiber can be expressed as the solution of Laplace's equation

$$\nabla^2 V = 0 \tag{11.1}$$

If the approximation is made that the fiber is an infinitely long circular cylinder, Laplace's equation can be written in cylindrical polar coordinates.

$$\frac{\partial^2 V}{\partial r^2} + \frac{1}{r}\frac{\partial V}{\partial r} + \frac{1}{r^2}\frac{\partial^2 V}{\partial \theta^2} + \frac{\partial^2 V}{\partial z^2} = 0 \tag{11.2}$$

Equation 11.2 can be solved by separation of variables. The variation of potential with z and θ is expressed by simple differential equations with sine and cosine solutions of the form $\cos(m\theta)$ and $\cos(nz)$. These two examples give rise to the following equation for V as a function of r

$$r^2\frac{d^2V}{dr^2} + r\frac{dV}{dr} - \left(n^2r^2 + m^2\right)V = 0 \tag{11.3}$$

The solution of Equation 11.3 is $K_m\,(nr)$, a modified Bessel function of the second kind, the value of which diminishes rapidly with increasing r when n and m are large (Abramovitz and Stegun, 1968). In general, a distribution of charge that is spatially rapidly varying is associated with a short-range electric field.

Much of filtration theory deals with two-dimensional representation of filters. Such models are less satisfactory representations of electrostatic behavior than of purely mechanical behavior because whereas flow may vary along the length of a fiber it will not suffer the rapid change in sign that is possible for electric charge. However, two-dimensional theory does illustrate the salient features of particle capture by electric forces.

Particle capture by electric forces, like mechanical capture, can be studied in terms of dimensionless parameters, derived from particle dynamics. The motion of a particle in response to an electric field, its drift velocity, U_d, is the product of the electric field and the electrical mobility of the particle μ_e.

$$\underline{U}_d = \frac{neCn}{3\pi\eta d_p}\underline{E} \tag{11.4}$$

where the particle, of effective diameter d_p, carries n fundamental charges, e; Cn is the Cunningham slip correction factor and η is the coefficient of viscosity of the air. The radial and tangential components of electric field due to a surface charge on a fiber, that varies sinusoidally with angle as $\cos(n\theta)$ are (Brown, 1993b).

$$E_r = \frac{\sigma R^{n+1} \cos(n\theta)}{\varepsilon_0 \left(1 + D_f\right) r^{n+1}} \tag{11.5}$$

$$E_\theta = \frac{\sigma R^{n+1} \sin(n\theta)}{\varepsilon_0 \left(1 + D_f\right) r^{n+1}} \tag{11.6}$$

where D_f is the dielectric constant of the material of the fiber, R its radius, and ε_0 the permittivity of free space. If the equation of motion of a particle under the influence of airflow and electric forces is reduced to dimensionless coordinates, a dimensionless parameter, relating the properties of the fiber and the particle to the filtration efficiency, emerges. This parameter closely resembles the quotient of the drift velocity and the convective velocity, and for capture of electrically charged particles by a line dipole fiber, it is

$$N_{\sigma q} = \frac{\sigma q}{3\pi\eta\left(1 + D_f\right) U \varepsilon_0 d_p} \tag{11.7}$$

For the capture of electrically neutral particles of dielectric constant, D_p, it is

$$N_{\sigma 0} = \frac{2}{3}\left(\frac{D_p + 1}{D_p + 2}\right) \frac{\sigma^2 d_p^2}{\varepsilon_0 \left(1 + D_f\right)^2 d_f \eta U} \tag{11.8}$$

Similar parameters emerge for other charge configurations, except that the fiber dielectric constant does not appear when the charge on the fiber is uniform, since in this situation there is no electric field within the fiber.

The functional dependence of single-fiber capture efficiency on the appropriate parameter will be weaker when the range of force is shorter. Calculated values for the variation of single-fiber efficiency with dimensionless parameter are shown in Figure 11.3 for various values of n, in the situations where the force on the particle is a central force. In the literature the functional form of this variation has been approximated by power laws, valid for limited ranges of the dimensionless parameter (Stenhouse, 1974; Brown, 1981).

This means that the variation of single-fiber efficiency as a function of dimensionless parameter gives qualitative information on the range of the relevant force law and, therefore, the charge configuration. Of course, experimental results give an average, but the data of Lathrache and Fissan (1989) and of Fjeld and Owens (1988), summarized in Figure 11.4, reveal a variation that is weaker than linear and which could, therefore, be attributed to a nonuniform charge configuration.

The fundamental importance of charge configuration can be illustrated by two extremes. The first of these is the situation in which an entire filter material is uniformly charged. This will result in a high field outside the material, which will not contribute to filtration, along with an internal field that is much smaller, in fact, zero at points of symmetry. Such a field will cause material-handling problems while contributing little to filtration performance. The other extreme is electric charge that has very rapid spatial variation, such as the charge held by the ions in an ionic crystal. Although the level of charge may be very high indeed, the spatial variation of the charge is so rapid that the electric field will not persist further than atomic dimensions from the surface, whereas to be effective in filtration electric fields need to extend over distances of the order of a micrometer at least.

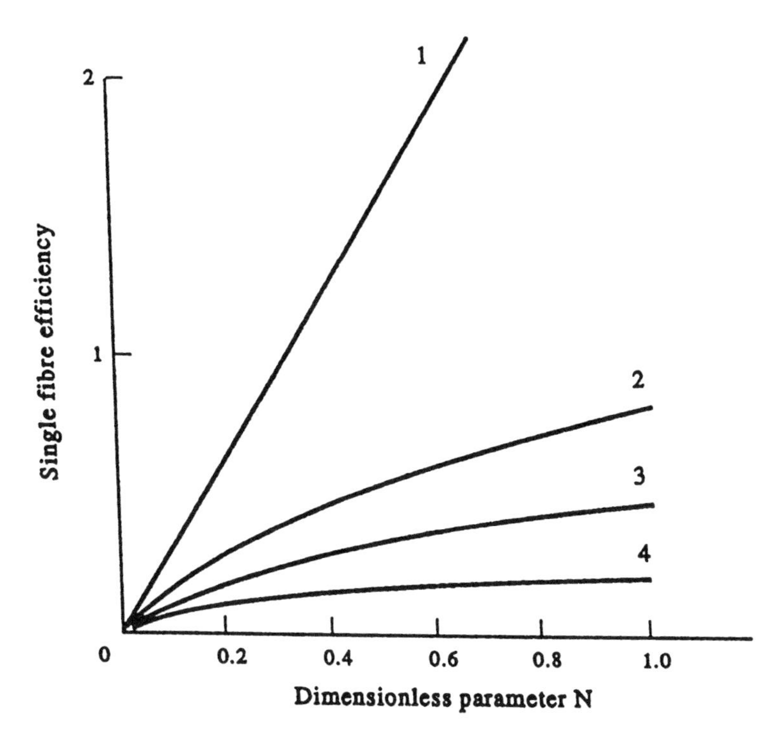

Figure 11.3 Single-fiber efficiency as a function of dimensionless parameter. a) Coulomb forces, uniformly charged fiber; b) polarization forces, uniformly charged fiber; c) polarization forces, line dipole charged fiber; d) polarization forces, $m = 5$ pole charged fiber. (© Crown copyright; by permission of Her Majesty's Stationery Office.)

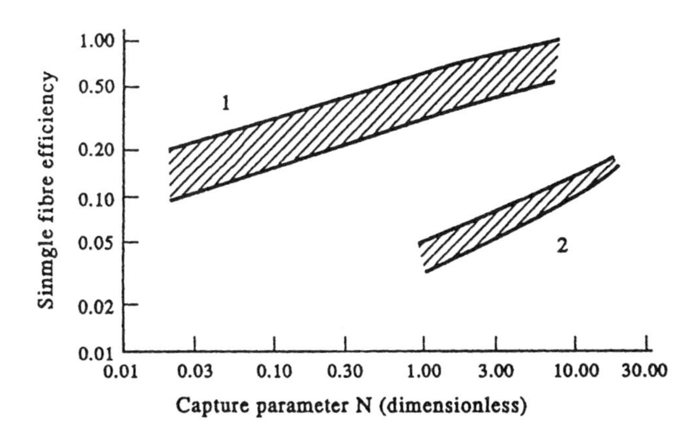

Figure 11.4 Measured values of single-fiber efficiency as a function of dimensionless parameter, for electrically charged fibers. (1, with permission of Springer-Verlag, Heidelberg; 2, © IEEE 1988.)

11.5 MEASUREMENT OF MACROSCOPIC CHARGE ON A FILTER

The algebraic sum of the electric charge on a filter can be easily measured with a Faraday ice pail connected to an electrometer. All of the materials described above show a very low level of charge in such experiments (Brown, 1979), behavior that could be anticipated from simple physical examination of the materials, for they show no obvious sign of being electrically charged. In textile vocabulary they would be described as "dead." This does not mean that they contain no electric charge; only that any charge that they do contain consists of positive and negative charge almost completely in balance.

In materials that contain a conducting component, this condition can be understood from simple energy considerations. A conductor will develop whatever charge is necessary to reduce the electrostatic energy of the system to a minimum; and the configuration that achieves this takes the form of charges that are the electrical images of that on the insulating component, charges of approximately equal magnitude and opposite sign. In materials that contain no good conductor, any excess of charge of one sign will tend, over a period of time, to be reduced by atmospheric ions since, as stated above, the charge will produce a long-range field extending into the region surrounding the filter.

Electrostatically spun materials usually contain just one component, but the filters are often faced with scrims of relatively conducting material. When the scrims are intact the material is "dead," but when they are removed the material will readily develop charge on contact with other bodies.

11.6 MEASUREMENT OF MICROSCOPIC CHARGE ON A FILTER

The ice pail measurements described above do not measure the component of charge that actually contributes to filtration. Particle capture depends on both the absolute magnitude of the positive and the negative charge on the material, and the configuration of the charge on the fibers. No method of measurement has yet been devised which gives a complete picture of the charge.

11.6.1 Measurement of Microscopic Charge by Ionizing Radiation

The strength of an X-ray beam is normally quantified in terms of the number of ions that it produces. A radiation dose of 1 R results from a beam that produces 3.33×10^{-4} C of charge of each sign, in the form of ions, in 1 m^3 of air at 0°C and 760 mmHg (Attix and Roesch, 1968). The ionization energy of the molecules in air is of the order of 30 eV, which is sufficient to enable the ions to neutralize the charge on filter fibers. If incremental doses of radiation are given to a filter, and the penetration of an aerosol through the filter is measured as a function of integrated dose, the charge on the filter can be inferred (Walton, 1942; Brown, 1979).

The free-air value for the ionization produced by a beam cannot be used directly in charge measurement because the rate at which ions are produced by the X-ray photons in the two-phase system of filter fibers and interstitial air is different from the free-air value, even though some 95% of the volume of the filter may be air. The X-ray photon is much more likely to interact with the material of the fiber than it is with the air, and if it does so it will liberate a high-energy photoelectron, which then produces a large number of low-energy ions in the interstitial air of the filter. This two-stage process can be accounted for by mans of a correction to the free-air ionization, based on modified cavity theory (Waker and Brown, 1988), and typical correction factors for common materials are of the order of two.

Figure 11.5 is a plot of minus the logarithm of the penetration of monodisperse aerosols (a function closely related to the single-fiber efficiency) against integrated radiation dose (Thorpe and Brown, 1991). At a high cumulative dose the function falls to a level that does not change with further irradiation, and which is related to the purely mechanical efficiency of the filter. The efficiency is easy to read from the results, but the end point is difficult to identify with precision. The irradiation test is useful as a qualitative test for whether or not the action of a filter has an electrostatic component. A significant increase in aerosol penetration as a result of irradiation is proof positive of electric charge. Ionizing radiation is the only agent that can be relied upon to remove electric charge without other complications.

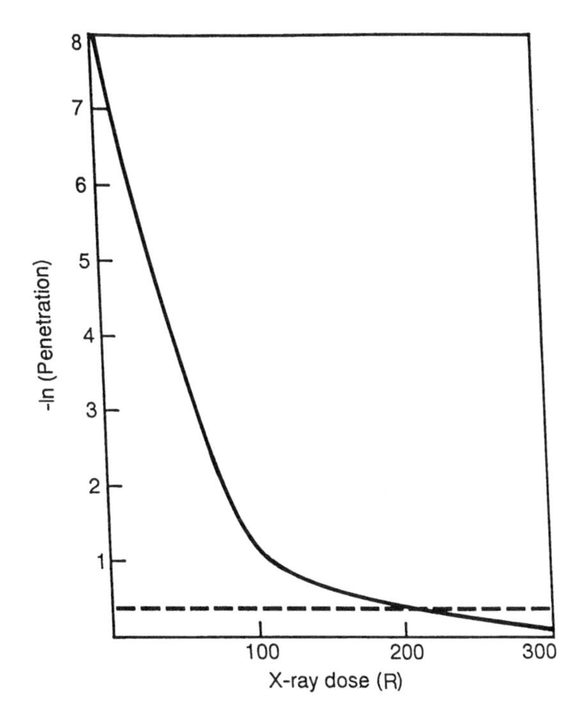

Figure 11.5 –ln (Penetration) as a function of cumulative radiation dose for electrically charged filter material. (© Crown copyright; by permission of Her Majesty's Stationery Office.)

In spite of the shortcomings outlined above, X-ray measurements shed some light on the configuration of charge on the material. In some instances the measured charge is greater than the normal breakdown charge for a uniformly charged cylinder, which itself exceeds the uniform field value of 3×10^6 Vm^{-1} (Harper, 1967) because of the short-range field. This points directly to an even shorter range of field and, therefore, to nonuniformity of charge on the fiber (Brown, 1979).

A further clue to the charge on configuration is given by the shape of the curve in Figure 11.5, which is concave upward, in contrast with the theoretical curves in Figure 11.3. Mechanical effects will occur only in the region of the curve below the dotted line (Pich et al., 1987; Thorpe and Brown, 1991) and so the phenomenon is caused purely by electric forces. There is no unique explanation, but one possibility is that the electric field becomes shorter in range as irradiation proceeds. If the charge configuration is complicated, regions of relatively uniform charge, since they will attract ions from large distances, would be selectively destroyed by the ionizing radiation, leaving a charge distribution on the fiber which becomes successively more variable as charge is removed. In such a situation, a description of the charge in terms of fractal geometry may be useful, although no rigorous theory exists.

11.6.2 Measurement of Microscopic Fiber Charge by Scanning

An alternative method of charge analysis is the examination of single fibers removed from the material (Baumgartner et al., 1985). A fiber is suspended vertically, by means of a small weight attached to its end, and a pair of electrodes producing a high field is moved down its length. Charge on a segment of fiber will cause it to be attracted to the oppositely charged electrode as it passes, and the deflection can be related to the charge on the fiber.

This method gives a microscopic picture of the variation of the electric charge along the length of the fiber. A modification enables the line-dipole moment of the charge to be measured. A significant drawback, however, is that the charge on a fiber will probably be altered by contact with

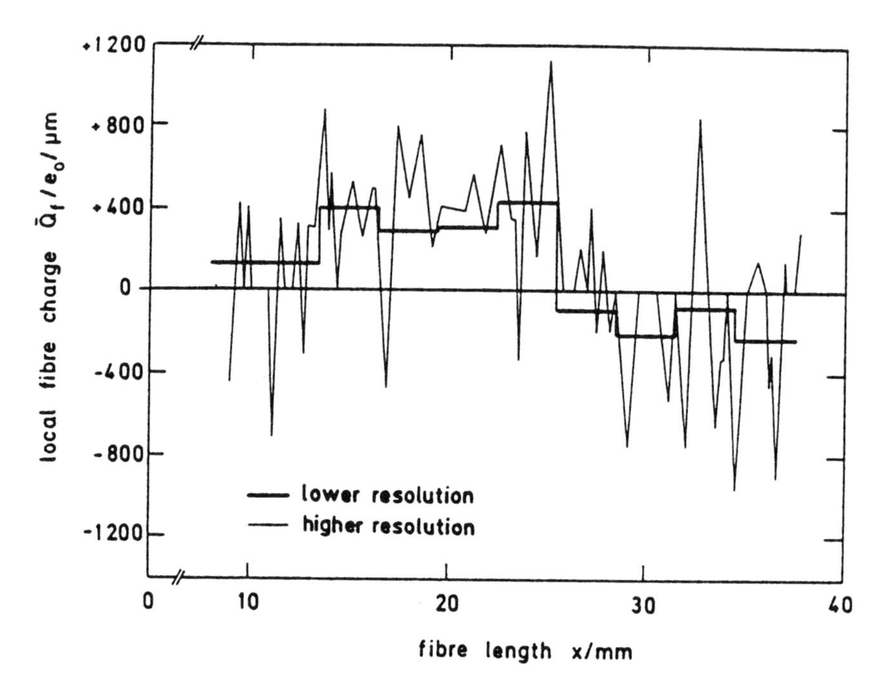

Figure 11.6 Voltage scan of fiber from electrically charged filter material. (© IEEE 1986.)

other fibers as the fiber under study is withdrawn from the material. The two graphs in Figure 11.6 show results obtained with probes of different spatial resolution. Whether or not measurements of this sort point toward a description of fiber charge in terms of fractal geometry would require more detailed measurements with yet finer probes.

11.7 INTRINSIC STABILITY OF ELECTRIC CHARGE

The electrically charged state is one of high energy and is, therefore, potentially unstable. This influences the shelf life of filters and the conditions in which they can be used. Previous sections have shown that fibers are likely to hold substantial amounts of spatially fixed positive and negative charge. The charge must lie in bound states, such that an excitation energy is required if it is to become mobile. If all of the charge were held in bound states with a binding energy Φ, then rate theory would predict that the lifetime of the charge would be given by τ where

$$\tau \sim \nu \exp\left(-\frac{\phi}{k_B T}\right) \tag{11.9}$$

ν is a characteristic frequency of the system, k_B is Boltzmann's constant, and T is the absolute temperature. The excitation energy would be thermal energy, and the charge loss would proceed exponentially with a time constant that became rapidly smaller as the temperature increased.

In practice, if the performance of a filter is measured immediately after the material is made and the measurement is repeated at intervals, a relatively rapid initial deterioration is observed, after which the performance becomes much more stable (private communication with several manufacturers). Moreover, if a filter is heated to any constant elevated temperature, it suffers a loss of efficiency in a relatively short period of time, after which its behavior stabilizes. If its temperature

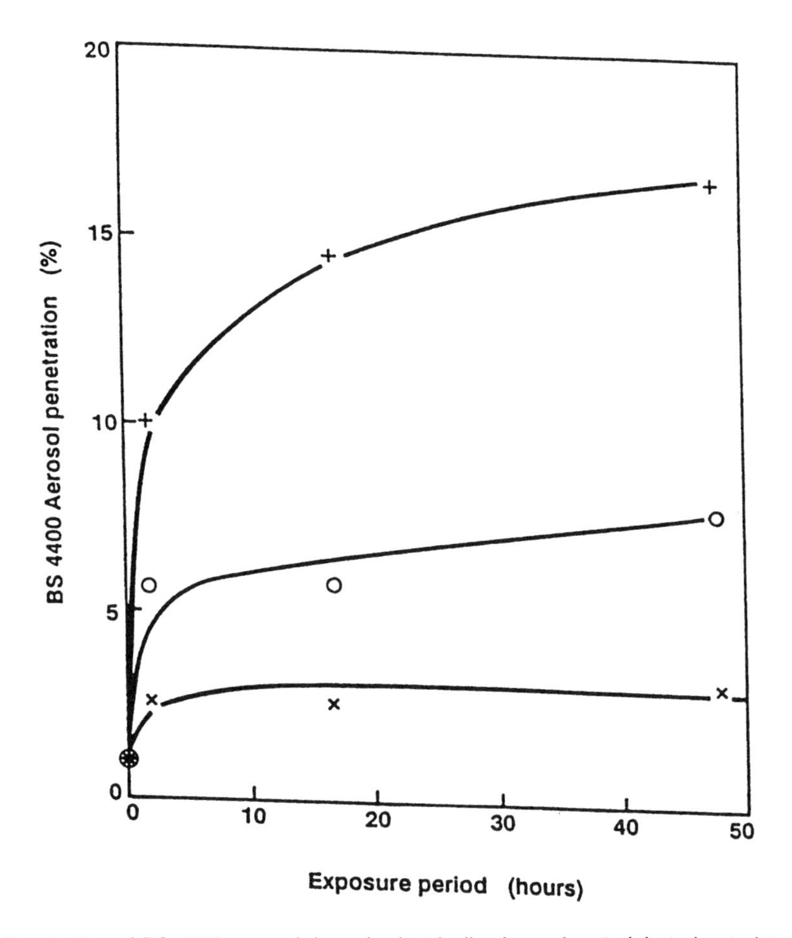

Figure 11.7 Penetration of BS 4400 aerosol through electrically charged material at elevated temperature: x, 90°C; ○, 100°C; +, 110°C. (© Crown copyright; by permission of Her Majesty's Stationery Office.)

is raised further, the pattern of behavior is repeated and the filter again stabilizes, but at a higher level of aerosol penetration than before. The stable level is temperature dependent, as shown in Figure 11.7 (Brown, 1989).

The explanation of this behavior is that the bound states in which the charge is held not have a single binding energy, but a range. At any temperature, charge that is held in relatively shallow levels is lost and that bound more tightly is retained, but the binding energy at which the behavior alters increases with increasing temperature. This follows the pattern of thermally stimulated discharge of electrets (Turnhout, 1980).

Even this description is incomplete, because the binding energy itself may alter with temperature. Humidity also has an effect on filter performance, and materials with a high moisture regain tend to be more likely to lose charge (Ackley, 1982). Surface conductivity and the conductivity of interfaces are particularly affected by adsorbed moisture, but the pattern is not easy to describe. When materials are humidity sensitive, the effect of elevated temperature and humidity together tends to be rather more serious than the effect of either separately.

The behavior of filters indicates that there may be some justification in annealing the charge on the filter, i.e., raising the temperature above normal for a period so that unstable charge is lost. The efficiency of the material is reduced by this process, but confidence can be increased over the stability of the remaining charge. Even without this, the charge stability of electrically charged materials tends to be good, unless they are subjected to extreme conditions.

11.8 EFFECT OF DUST LOAD ON ELECTRICALLY CHARGED FILTERS

The effect of loading electrically charged filters with dust is twofold. The dust will tend to form into structured deposits, dendrites, within the material just as it does in material that acts by purely mechanical means. Though the electric forces may affect the form, development, and stability of their structure, the general effect of the dendrites will be to increase both the filtration efficiency and the airflow resistance. The second effect, a reduction in filtration efficiency caused by interaction with the electric charge on the filter, with no effect on the resistance, has no analogy in mechanical filters. The behavior on loading depends on the properties of both the aerosol and the filter.

Figure 11.8 illustrates the behavior of a particular filer loaded intermittently with a sodium chloride aerosol. During the initial loading period the penetration falls and the pressure drop rises, behavior characteristic of simple clogging. However, when the filter is allowed to rest, the penetration reaches a level above the initial value and the pressure drop falls. Further loading cycles elicit basically similar behavior. During the resting phase, it is likely that the structured deposit of aerosol is collapsing into a form that offers less resistance to air or to airborne particles and that the increase in penetration is caused by screening of the fiber charge.

In more tightly packed filters both the pressure drop increase and the filtration efficiency increase are relatively better preserved during resting, so that the filter efficiency may never drop below its initial value, and the filter may show signs of marked clogging. In loosely packed filters the charge loss is the dominant process, and the dendrite formation much less so. In these instances, there may be a continuous loss of filtration performance on loading, without any evidence of clogging.

11.8.1 Semiempirical Theory of Charge Loss

A significant understanding of the process of charge loss can come about from the application of simple theory to experimental results. Filtration efficiency varies approximately exponentially with filter thickness; and the simplest linear approximation is to assume that the effect of aerosol loading in reducing the effectiveness of the electrical charge of the filter is independent of where the aerosol is deposited within the filter. It is then straightforward to show that a mass per unit area M_L of the loading aerosol results in a penetration $P_T(M_L)$ of the test aerosol which obeys the simple equation

$$\ln\left[\frac{P_T(M_L)}{P_T(0)}\right] = \beta_{TL} M_L \tag{11.10}$$

β_{TL} is an index that quantifies the extent to which a particular aerosol causes performance loss of a particular filter material. Where experiments can be carried out accurately, the exponential increase of penetration with M_L, predicted by Equation 11.10 is observed.

The β-values for a number of different aerosols and two filter materials of relatively similar structure are compared in Figure 11.9 (Brown et al., 1988), which indicates that the variation in β among different materials exposed to the same aerosol is relatively small, whereas the variation among samples of the same material exposed to different aerosols is considerable. In fact, the relationship between these β-values for the two materials is almost linear. This simple relationship can be explained if its is assumed that the aerosol acts by screening the fiber charge (Kanaoka et al., 1984). The screening charge would be expected to be approximately proportional to the charge being screened, and so the fractional reduction in single-fiber efficiency would be more or less independent of the charge. If, on the other hand, the process were one of charge neutralization, the effect would be more marked in the case of filter materials that held a lower level of charge. The former, and not the latter, is a good description of what is observed. Moreover, measurements of the level of charge on industrial aerosols show that typical masses captured do not hold sufficient charge to neutralize electrostatic filters (Johnston et al., 1985).

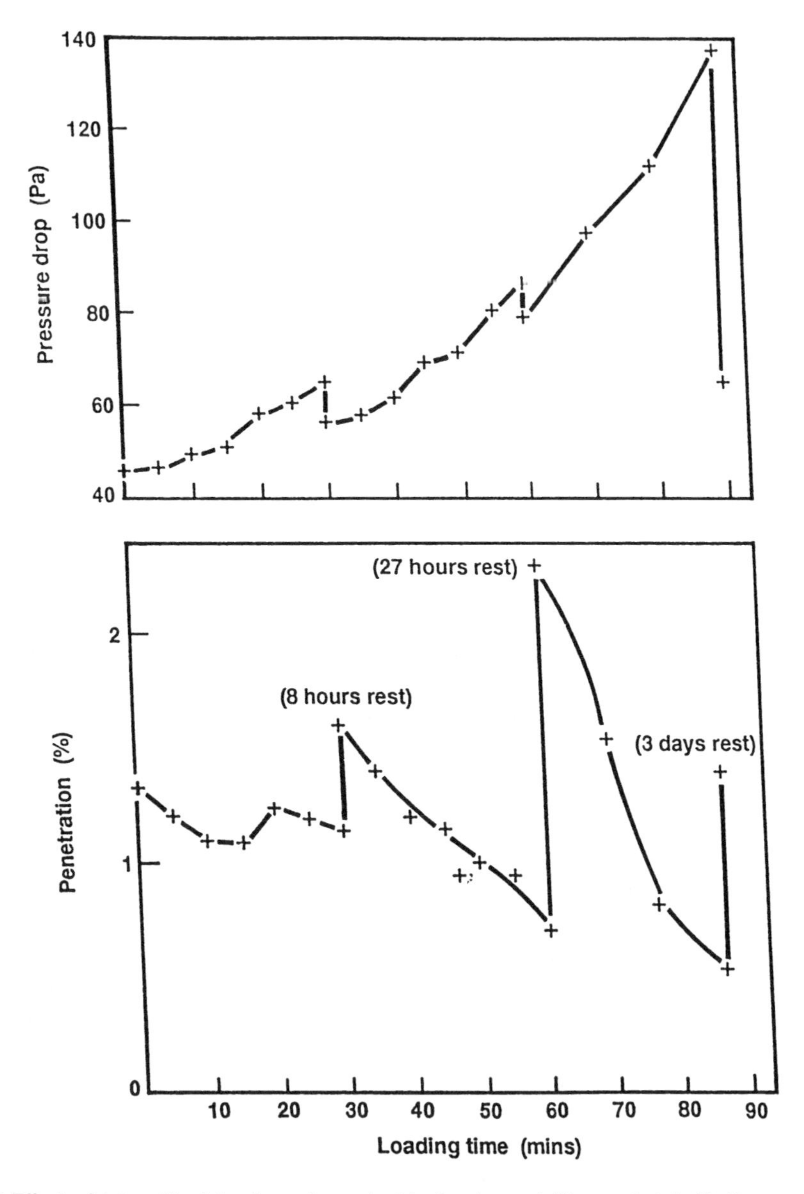

Figure 11.8 Effect of intermittent loading of an electrically charged filter material. (© Crown copyright; by permission of Her Majesty's Stationery Office.)

Situations in which this pattern is not followed indicate other mechanisms of charge loss. The β-value for polycarbonate material exposed to coal tar fume at a coke oven is higher than those of other materials, and this can be attributed to degradation by solvent constituents of the fume. When exposed to solvents in the laboratory, this material, which by necessity is made from a soluble polymer, suffers a much higher level of degradation than do materials made from insoluble polymers.

The practical implication of loss of efficiency with load should not exaggerated, because respirator filters can be disposed of after a relatively short period of use. The only other type of material capable of achieving the same high efficiency is fine fibered mechanical material; and this brings the problem of high airflow resistance, which can only be overcome by pleating, and the problem of detachment of fibers (Howie et al., 1986) posing a respiratory hazard of their own, which can be eliminated only by encapsulation of the filters.

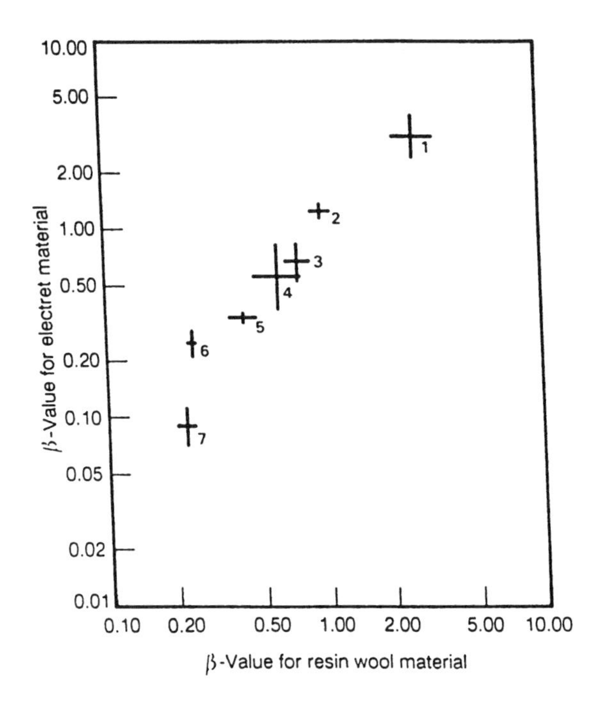

Figure 11.9 The β-values for resin–wool and electret material exposed to industrial aerosols. (© Crown copyright; by permission of Her Majesty's Stationery Office.)

11.8.2 Exposure of Electrically Charged Materials to Liquid Sprays

Filter material in personal protective devices may be used while other dust control systems are operative. One particularly useful means of preventing or reducing dust generation is the application of a spray of water, surfactant, or encapsulant to the dust-producing material, and it is important that the spray does not reduce the effectiveness of the filter. Vaughan et al. (1995) exposed resin–wool filters to the level of spray that they would be likely to receive in practice during a simulation of wet stripping of asbestos, and in no instance was a filter adversely affected. The same result was observed when spraying was carried out with penetrant or surfactant material. Moreover, electrostatic filters collected after actual use during wet spraying invariably gave adequate performance unless they had suffered mechanical damage. When filters are deliberately degraded, degradation is irreversible, and it can be concluded, therefore, that the level of performance shown by filters after use represents the worst performance that they have given. If this is acceptable, the filters are performing adequately throughout.

11.9 USE OF ELECTRICALLY CHARGED MATERIAL IN LARGER-SCALE FILTERS

The very high filtration efficiency and low airflow resistance of electrostatic filters makes them ideal for respirator filters, which are required to remove a substantial fraction of respirable dust from breathed air. Their use in larger-scale units carries the same benefit and the same problem. In the case of recirculation, ever an appealing option in an energy-conscious environment, their high efficiency makes them good candidates for achieving the filtration efficiency necessary to clean air to an acceptable quality for recirculation, and their low resistance makes them ideal for free-standing units with power limitations. It is necessary, though, that both properties are maintained throughout the period of use. Convincing data can be obtained if their efficiency is measured over a period of time using the dust with which they will be challenged in practice. A specification for use requires that both the challenge concentration and the level of dust that will be acceptable in the filtered air are known.

A specimen feasibility study of this type has been carried out for wood dust (unpublished work). The operation parameters of particular wood sanders have been observed on-site and duplicated in an automated laboratory simulation (Thorpe and Brown, 1994). These sanders produce dust with a mass median diameter of approximately 8 μm at a mean local concentration of 25 mg m^{-3}. The MEL (maximum exposure limit) for hardwood dust is 5 mg m^{-3} (Health and Safety Executive, 1995). Criteria for recirculation are being revised, but in the past an acceptable level of dust in recirculated air might have been set at one tenth the MEL, requiring a filtration efficiency of 98%. During wood sanding in a booth it is necessary that the booth be ventilated with air at a velocity of about 0.75 m s^{-1} in order to ensure that the dust does not enter the operator's breathing zone. If a filter is used in the form of a panel at the rear of the booth, its area could be comparable with that at the booth entrance and so its filtration velocity would equal the above.

An investigation carried out with two samples of electrically charged material of area weights 125 and 250 g m^{-2}. Each filter was followed by a high efficiency backup filter, and measurement of the increases in mass of the filters enabled the efficiency to be calculated and the dust concentration confirmed. Mean efficiencies are shown in Figure 11.10, which indicates that a mean dust penetration of 2%, a filtration efficiency of 98%, is maintained by the thicker material for a period of 8 h of continual use and, therefore, for a longer period of intermittent use.

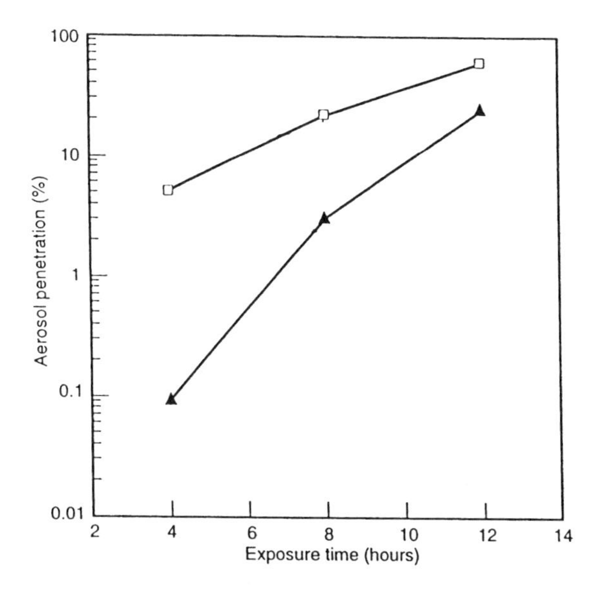

Figure 11.10 Cumulative penetration of wood-sanding dust through electrically charged filter. (© Crown copyright; by permission of Her Majesty's Stationery Office.)

A maintenance procedure would be needed, either in the form of replacing the filter after a set period or of installing a set-speed roll-up system. Cleanable filters, collecting dust in the form of a cake, could be used in an air cleaner of this sort, although power requirements would be significantly larger and there would remain the problem of cleaning the filter and disposing of the dust. The electrically charged filter, because of its open structure, would capture dust at least partly in depth and so it could be rolled up and disposed of along with the dust that it held.

The method of assessment above could be used to investigate the usefulness of such filters against other dusts. A measure of fine tuning would be possible by increasing the degree of compaction of the filter material. As described above this would serve to preserve filtration efficiency, but at the expense of a greater increase in pressure drop.

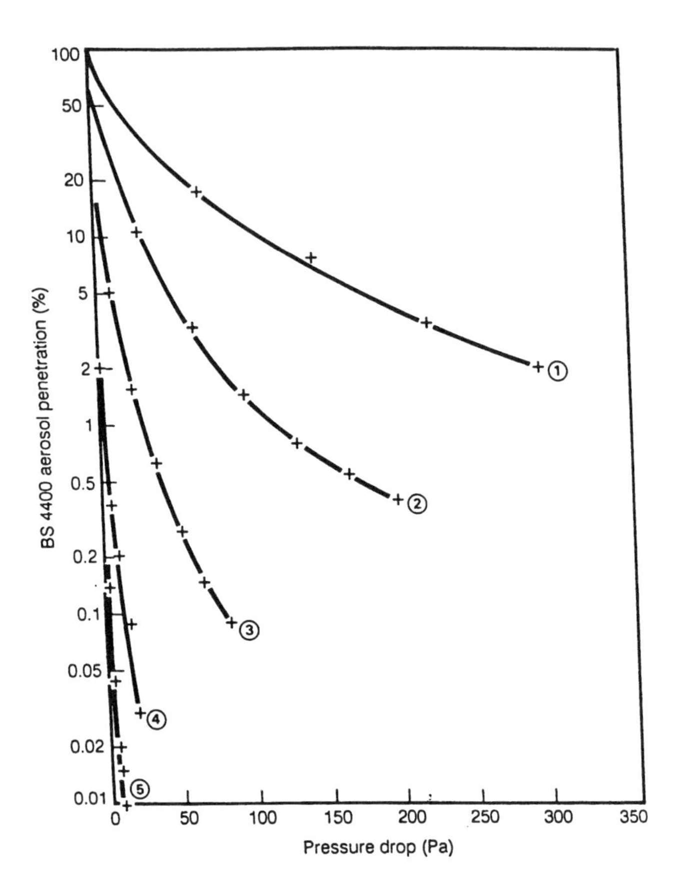

Figure 11.11 Penetration of BS 4400 sodium chloride through electrically charged filter as a function of pressure drop at various face velocities: (1) –1.36 m s^{-1}; (2) 0.68 m s^{-1}; (3) 0.29 m s^{-1}; (4) –0.084 m s^{-1}; (5) 0.030 m s^{-1}. (© Crown copyright; by permission of Her Majesty's Stationery Office.)

Large-scale filtration units treat large volumes of air. This tends to result in relatively high filtration velocities, but the benefits of reducing the velocity as far as possible (by, for example, pleating the filter) must not be ignored. Figure 11.11 shows the penetration of a standard aerosol though electrically charged filters of a common type, of various thicknesses, each graph corresponding to different velocities (Brown, 1988). An attempt to increase filtration efficiency by increasing filter thickness brings the penalty of an increase in pressure drop, and a situation of diminishing returns, characteristic of filtration of polydisperse aerosols; on the other hand, an increase in filter area will reduce the filtration velocity, increasing filtration efficiency by giving the electric forces longer to act, and will reduce the pressure drop and energy dissipation as well.

11.10 FLEXIBLE FILTER WITH AN EXTERNAL ELECTRIC FIELD

In static units an alternative means of employing electric forces in filters, the use of external fields, is possible. A dielectric fiber will be polarized in an applied field and the resultant dipole moment will augment the efficiency of the filter. A review of the process is given elsewhere (Brown, 1993b), and the single-fiber efficiency, ϕ, for capture of neutral particles by a fiber in a field E is

$$\phi_{p0} = \frac{1}{3}\left(\frac{D_p - 1}{D_p + 2}\right)\left(\frac{D_f - 1}{D_f + 1}\right)\frac{d_p^2 \varepsilon_0 E^2}{d_f \eta U} \tag{11.11}$$

Equation 11.11 indicates that filtration efficiency increases as the applied field is increased and as the fiber diameter is decreased, fine fibers being more efficient in almost all situations of filtration. The other changes that would increase filtration efficiency are increase in dielectric constant of the fibers and decrease in filtration velocity, the latter being held in common with all other electric capture effects. The expression for the capture of charged particles is more complicated than that in Equation 11.11, but it indicates that the same properties of the filter augment filtration efficiency. A recent investigation (Brown et al., 1994) has shown that the choice of fiber can be instrumental in improving the filter by an increase in dielectric constant and a decrease in filtration velocity.

The effect of dielectric constant follows from the results of tests on filters made from a variety of textile fibers but with the same basic structure, using a standard aerosol and a common electric field. The results of such an exercise are as shown in Table 11.2, where the nominal dielectric constant of the materials of the fibers is also given. The pattern is as expected, though the agreement is not quantitative. However, the values of dielectric constant are obtained from standard sources (Morton and Hearle, 1962; Kaye and Laby, 1966) and may be the results of measurements made with materials that are of the same type but not of identical chemical composition with the fibers tested.

Table 11.2 Electrical Properties of Dielectric Textile Fibers

Fiber	Dielectric Constant	Resistivity (Ω-cm)	Moisture Regain at 65% RH[d]
Wool	5.5[a]	10^8–10^9 [c]	16–18
Cotton	18.0	10^7 [c]	8
Hercosett wool	—	—	10
Nylon 66	4.0[a]	10^9–10^{12} [c]	4
Modacrylic	4.2[a]	10^8–10^9 [c]	1
PET	3.3–4.5[a]	$>10^{14}$ [b]	0.5
Polypropylene	2.6[b]	$>10^{16}$ [b]	0.1
PTFE	2.1[b]	10^{18} [b]	>0.1

[a] Measured at 65% RM (Morton and Hearle, 1962.
[b] Kaye and Laby 1986.
[c] Resistance of 1-cm-long specimen of 1g in weight (Morton and Hearle, 1962).
[d] Fractional increase in weight of specimen in equilibrium with air at RH 65% relative to dry specimen (Morton and Hearle, 1962).

The volume of air per unit time requiring treatment cannot be altered, and so the reduction of filtration velocity may only be brought about by an increase in filter area. This is achieved in mechanical materials and material with a permanent electric charge by means of pleating, which requires the material to be flexible. Most experiments on material with an external electric field have involved rigid metal electrodes, but flexible electrodes can be produced in the form of a conducting fiber layer which will bend with the dielectric part of the filter supporting it. In order to function as electrodes, the fibers must have a conductivity that is sufficiently high to allow the entire surface of the filter to develop a high electric potential (i.e., insulating fibers would not work) but sufficiently low to prevent shorting out of the filter by accidental contact of two fibers of opposite polarity. If fine metal or metal-coated fibers are used, shorting out is common. If, however, semiconducting fibers are used, shorting out does not usually occur. A possible explanation for this is that the fibers burn out as a result of the energy dissipated when they pass as current, maintaining the electrical integrity of the filter.

A relatively sparse electrode is suitable; in fact, mixing 5% of the conducting fibers with 95% of insulating fibers produces an electrode with a typical resistance of 50 to 300 Ω between two points on its surface, which is a perfectly satisfactory conductor. The performance of the filter is shown in Figure 11.12. A more dilute mixture than this does not function well as an electrode, and the reason probably lies with the number of fiber–fiber contacts within the material, such contacts

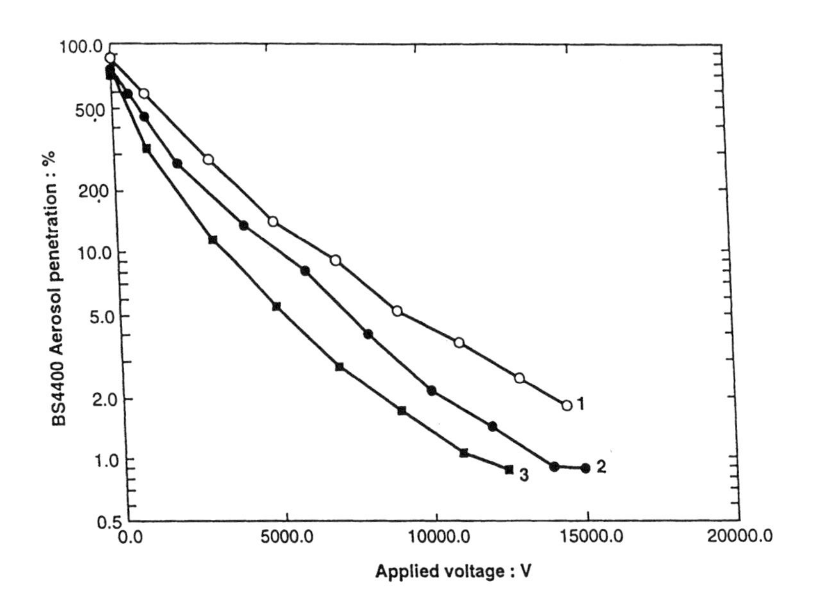

Figure 11.22 Effect of air velocity on the filtration efficiency of material acting by means of an external electric field: (1) 0.45 m s^{-1}; (2) 0.19 m s^{-1}; (3) –0.10 m s^{-1}.(© Crown copyright; by permission of Her Majesty's Stationery Office.)

being necessary to ensure electrical continuity. A 5% mixture of relatively conducting fibers is also found to be satisfactory in controlling the problem of static electricity in bag filters. The advantages of the electrically augmented filter outlined above, but the problem of flammability of the material or of the dust that it collects has yet to be resolved.

11.11 CONCLUSIONS

New electrically charged materials are continually appearing, and part of this chapter has been devoted to a particular material which comprises standard textile fibers, which develop an electric charge during the normal textile fabrication process. Existing technology is exploited to ensure that the structure of the material is open and uniform, in that crimped textile fibers of optimal length are used in is construction. The chemical nature of the material has the further advantage of allowing easy fabrication of filtering devices.

The configuration of electric charge on materials is as important as its magnitude, and both direct and indirect measurements show that the charge tends to be spatially variable. The charge is stable at ambient temperatures, and its pattern of loss at elevated temperatures indicates that it is retained in states with a spread of binding energies.

Loading with aerosols causes both clogging and screening of charge, and it is the latter process which determines the lifetime of the filters. Respirator filters will normally be disposed of before serious degradation has occurred; and the exposure of filters to liquid sprays used in normal dust control has no serious effect.

The materials show promise for use on a larger scale in recirculating systems, and it is important that their useful lifetime is not exceeded. Since they capture much of the dust in depth, by being disposable themselves they may facilitate the disposal of the dust that they capture.

Materials that act by means of an electric field can be produced with both the electrodes and the dielectric filter material in the form of textile fibers. The filter proper is best made from material with a high dielectric constant, and the material of the electrodes should be of fibers of semiconducting material. In this form they could be pleated or manipulated in similar ways to permanently charged materials.

REFERENCES

Abramovitz, M. and Stegun, I. A. (1968) *Handbook of Mathematical Functions,* Dover, New York.

Ackley, M. V. (1982) Degradation of electrostatic filters at elevated temperature and humidity, paper presented at 3rd World Filtration Congress, Filtration Society, London.

Attix, F. H. and Roesch, W. C. (1968) *Radiation Dosimetry,* Academic Press, New York.

Baumgartner, H. and Loeffler, F. (1987) Capture of submicrometer aerosol particles by electrostatically charged fibers — an experimental and theoretical study, paper presented at 3rd International Conference on Electrostatic Precipitation, Abano/Venice, Italy.

Baumgartner, H. and Loeffler, F. (1988) Abscheidung submikroskopisher Partikeln mit Tiefenfiltern aus elektrisch geladen Fasern (Elektretfilter), *Staub Reinhalt. Luft,* 48, 131–138 (in German).

Baumgartner, H., Loeffler, F., and Umhauer, H. (1985) Deep bed electret filters. The determination of single fiber charge states and collection efficiencies, *IEEE Trans.,* E1-21, 477–486.

Brown, R. C. (1979) Electrical effects in dust filters, in *Proceedings of the 2nd World Filtration Congress,* Filtration Society, London.

Brown, R. C. (1981) The capture of dust particles in filters by line dipole charged fibers, *J. Aerosol Sci.,* 12(4), 349–356.

Brown, R. C. (1988) An account of depth-filtration in air cleaning with particular reference to electrically charged filter material, in *Proceedings of 2nd Int. Symp. on Ventilation and Contamination Control,* London, 303–312.

Brown, R. C. (1989) Modern concepts of air filtration applied to dust respirators, *Ann. Occup. Hyg.,* 13(4), 615–644.

Brown, R. C. (1993a) A novel electrically charged filter material, *Mining Eng.,* 153(382), 4–5.

Brown, R. C. (1993b) *Air Filtration: An Integrated Approach to the Theory and Applications of Fibrous Filters,* Pergamon, Oxford.

Brown, R. C., Wake, D., Gray, R., Blackford, D. B., and Bostock, G. J. (1988) Effect of industrial aerosols on the performance of electrically charged filter material, *Ann. Occup. Hyg.,* 312, 271–294.

Brown, R. C., Wake, D., and Smith, P. A. (1994) An electrically augmented filter made from conducting and dielectric fibers, *J. Electrostatics,* 33, 393–412.

Feltham, J. (1979) The Hansen filter, *Filtration Separation,* 16, 370–372.

Fjeld, R. A. and Owens, T. M. (1988) The effect of particle charge on penetration in an electret filter, *IEEE Trans. Ind. Appl.,* 24, 725–731.

Hansen, N. L. (1931) Method for the Manufacture of Smoke Filters or Collector Filters, British Patent BP 384052.

Harper, W. R. (1967) *Contact and Frictional Electrification,* Oxford University Press, Oxford.

Hayati, I., Bailey, A. I., and Tadros, T. F. (1987) Investigation into the mechanisms of electrohydrodynamic spraying of liquids, *J. Coll. Int. Sci.,* 117(1), 205–211.

Health and Safety Executive (1995) EH 40/95 Occupational Exposure Limits.

Hersch, S. P. (1995) Resistivity and static behavior of textile fibers, in *Surface Characteristics of Fibers and Textile Surfaces,* M. J. Schick, Ed., Marcel Dekker, New York.

Howie, R. M., Addison, J., Cherrie, J., Robertson, A., and Dodgson, J. (1986) Fibre release from filtering facepiece respirators, *Ann. Occup. Hyg.,* 30(1), 131–133.

Johnston, A. M., Vincent, J. H., and Jones, A. D. (1985) Measurement of electric charge for workplace aerosols, *Ann. Occup. Hyg.,* 29(2), 271–284.

Kanaoka, C., Emi. H., and Ishiguro, T. (1984) Time dependency of collection performance of electret filters, in *Proc. 1st AAAR Conference 1984,* B. Y. H. Liu et al., Ed., Elsevier, New York, 614–616.

Kaye, G. W. C. and Laby, T. H. (1966) *Tables of Chemical and Physical Constants,* Longman, London.

Lathrache, R. and Fissan, H. (1989) Untersuchungen zum Abscheideverhalten der Elektret-Filter Teil 2: Bewertung der Filtrationseigenschaften, *Staub-Reinhalt. Luft,* 49, 365–370.

Morton, W. E. and Hearle, T. W. S. (1962) *Physical Properties of Textile Fibres,* Butterworth, Manchester, U.K.

Pich, J., Emi, H., and Kanaoka, C. (1987) Coulombic mechanism in electret filters, *J. Aerosol Sci.,* 18(1), 29–35.

Rayleigh, J. W. S. (1892) On the instability of a cyhlinder of viscous liquid under capillary force, *Phil. Mag.,* 34, 145–154.

Schmidt, K. (1980) Manufacture and use of felt pads made from extremely fine fibres for filtering purposes, *Melliand Textilber.,* 61, 495–497.

Shimokobe, I., Izumi, K., and Inoue, M. (1985) Electrostatic polarizing of non-woven polypropylene sheets, in *Proceedings of the Annual Meeting of the Institute of Electrostatics,* Japan, Oct. 19–20, 1985 (in Japanese).

Smith, P. A., East, G. C., Brown, R. C., and Wake, D. (1988) Generation of triboelectric charge in textile fibre mixtures and their use as air filters, *J. Electrostatics,* 21, 81–98.

Stenhouse, J. K. T. (1974) The influence of electrostatic forces in fibrous filtration, *Filtr. Sep.,* 25–26.

Taylor, G. T. (1969) Electrically driven jets, *Proc. R. Soc.,* A313, 453–475.

Thorpe, A. and Brown, R. C. (1991) A study of the electric charge in filters using monodisperse test aerosols, in *5th Aerosol Society Conference Proceedings,* 1991.

Thorpe, A. and Brown, R. C. (1994) Measurements of the effectiveness of dust extraction systems of hand sanders used on wood, *Ann. Occup. Hyg.,* 38(3), 279–302.

Trottier, R. A. and Brown, R. C. (1990) The effect of aerosol charge and filter charge on the filtration efficiency of submicrometre aerosols, *J. Aerosol. Sci.,* 21, S689–S692.

Trouilhet, Y. (1981) *Advances in Web Forming,* EDANA, Brussels.

Turnhout, J. van (1980) Thermally stimulated discharge of electrets, in *Electrets,* G. M. Sessler, Ed., Springer-Verlag, Berlin.

Turnhout, J. van, Adamse, J. W. C., and Hoeneveld, W. J. (1980) Electret filters for high efficiency air cleaning, *J. Electrostatics,* 8, 369–379.

Vaughan, N. P., Brown, R. C., and Evans, P. G. (1995) The effects of asbestos wet-stripping agents on filters used in powered respirators, 40(5), 539–553.

Waker, A. J. and Brown, R. C. (1988) Application of cavity theory to the discharge of electrostatic dust filters by X-rays, *Appl. Radiat. Isot.,* 39(7), 677–684, *Int. J. Radiat. Appl. Instrum. Part A.*

Walton, W. H. (1942) The Electrical Characteristics of Resin-Impregnated Filters, CDE Porton Report No. 236.

Weghmann, A. (1982) Production of electrostatic spun synthetic microfibre nonwovens and applications in filtration, in *Proceedings of the 3rd World Filtration Congress,* Filtration Society, London.

Wente, V. A. (1956) Superfine thermoplastic fibres, *Ind. Eng. Chem.,* 48, 1342–1346.

CHAPTER 12

Separation of Airborne Dust in Deep-Bed Filtration

Gabriel I. Tardos

CONTENTS

12.1 INTRODUCTION

Barrier filters are porous beds of either long fibers or large, mostly spherical granules that serve as collectors for dust carried by the gas flowing through the device. Small airborne dust particles or aerosols deposit upon impact on the collectors and are thus separated from the flow. The filtration process in fibrous filters was extensively studied in the pioneering work of Fuchs and co-workers (Kirsch and Fuchs, 1967; Stechkina and Fuchs, 1968; Fuchs 1973) starting in the early 1960s while extensive reviews can be found in Pich (1966) and more recently in Zhao et al. (1991). Granular filters were also extensively studied by Tardos and co-workers (Tardos et al., 1978; Gutfinger and Tardos, 1979; Tardos, 1994) and Tien who compiled a monograph on the subject (Tien, 1989). The theory and practice of fibrous and granular filters has long evolved along parallel paths, and the present account is an attempt to find a common line between them and to present the two processes together in a unified approach.

The separation of airborne dust in these kinds of filters takes place in either a "cake" or a "non-cake" (deep-bed) filtration mode depending on the region in the filter in which particle deposition actually occurs. During cake filtration, initially deposited dust layers serve as collection media for subsequent filtration, and the filter bed only serves as a support for the separated dust. The main mechanism of particle separation is sieving: incoming dust particles are retained on the already deposited dust. This results in a significant increase in thickness of the deposited layer as filtration

0-87371-830-5/98/$0.00+$.50

proceeds and is usually accompanied by a large increase in pressure drop. This in turn causes a compression of the deposited layer and, hence, results in a higher filtration efficiency as more and more dust accumulates on the surface of the cake. The efficiency of the filter in cake filtration is overwhelmingly a function of the pore size of the deposited layer and increases dramatically with pressure drop. If the size of the dust particle is larger than the pore size, dust is filtered and the efficiency is very high (practically 100%). However, if the dust is smaller than the open pore size, a cake is not formed and deep-bed filtration takes place.

During non-cake or "deep-bed" filtration, dust particles are captured on each and every one of the granules or fibers (collectors) of the filter. As filtration progresses, deposits of dust slowly fill the interstices of the porous bed starting with the contact points between collectors, without drastically altering the geometry of the filter or the pressure drop through the bed. The filtration in this case is overwhelmingly influenced by the size of dust particles and by the thickness of the filter in the direction of the flow.

The considerations presented in this section pertain only to the case of deep-bed (non-cake) filtration in fibrous and packed or moving granular filters, i.e., to those cases where dust is collected inside the filter on distinct collectors and/or particles deposit on each other *without significantly altering the geometry of the filter as the dust collection proceeds.* While deep-bed filtration is common in both fibrous and granular filters, there is, however, a significant difference in both the geometry and the packing density of collectors in the two kinds of devices. In a fibrous filter, collectors are, as the name implies, long, usually cylindrical textile, paper, metal, or ceramic fibers woven together and kept in position in the dusty gas flow by metal frames. The porosity or void fraction of this kind of filter is usually of the order $\varepsilon = 0.9$ or higher, although ceramic filters may have lower porosity. The size of the fibers is also quite small, from micron size to tens of microns in diameter. Granular beds, on the other hand, comprise more or less spherical particles of diameters in the range of fractions of millimeters or as large as several millimeters, held in place in the gas stream by screens or louvers. The porosity of these filter beds is of the order of $\varepsilon = 0.5$ to 0.55 so that the capacity of dust retention of these beds is much lower compared with fibrous filters.

For purposes of this chapter, the "collector" is either a fiber or a granule of equivalent diameter $d_f = 2a$ while the captured dust particle will be assumed to be mostly spherical with an equivalent diameter $d_p = 2r_p$. The porosity of the filer, be it granular or fibrous, will be denoted $\varepsilon = 1 - \beta$ where β is the solid fraction. Furthermore, fibrous filters are viewed as a loosely packed assemblage of single cylinders. Even though the fibers are oriented randomly in the filter, the bed is treated from a theoretical point of view as if every fiber is normal to the gas flow. In a granular filter, collectors or individual grains are assumed to be entirely surrounded by fluid and thus interparticle contact points are neglected. Although this is a quite simplified view of the filter and the process of filtration and is consistent with assuming that each collector is "independent" in the flow, it proves to be quite useful for both efficiency and pressure drop calculations.

12.2 TOTAL-BED EFFICIENCY

The efficiency with which dust is collected in a filter, η, can simply be calculated from the concentration of airborne dust entering n_{in}, and leaving the filter n_{out} as

$$\eta = 1 - n_{in}/n_{out} \tag{12.1}$$

Extensive studies of deep-bed filtration in both granular and fabric filters (Kirsch and Fuchs, 1967; Gal et al., 1985) have revealed that the total efficiency is an exponential function of the filter thickness and this can be expressed by the equation:

$$\eta = 1 - \exp[-K_1(1 - \varepsilon)(L/2a)E] \tag{12.2}$$

where $L/2a$ is the number of collector layers in the filter, ε is the relative void volume, and E is the so-called *single-collector efficiency.* The quantity E is defined as the ratio of the number of all airborne (dust) particles captured by a single collector in the bed to the total number of dust particles flowing toward it through its projected upstream area. The coefficient K_1 can be taken $1.27/\varepsilon$ for a fibrous bed (Flagan and Seinfeld, 1988) and either 1.5 (D'Ottavio and Goren, 1983), $1.5/\varepsilon$ (Paretski et al., 1971) or 1.875 (Schmidt et al., 1978) for a granular bed. The implicit assumption in Equation 12.2 is that all collectors act as if they were independent within the filter and hence experience similar filtration phenomena. Equation 12.2 can be used in a predictive way provided the single-collector efficiency E can be calculated from first principles.

A somewhat different but in principle equivalent way of computing the total-bed efficiency is to use the concept of the unit cell efficiency, e, (Tien, 1989) so that:

$$\eta = 1 - [1 - e]^n \tag{12.3}$$

The quantity e is defined for a granular bed as the ratio of the number of airborne dust particles captured by a collector to the total number of dust particles flowing toward it in a square duct of cross-sectional area l^2, where the length l is given by

$$l = 2[\pi/6(1 - \varepsilon)]^{1/3}a \tag{12.4}$$

For the case of the fibrous bed, the equivalent of Equation 12.4 becomes

$$l = [\pi/4(1 - \varepsilon)]^{1/2}d_f \tag{12.4'}$$

and the duct area is $l \times 1$. The quantity n is the number of layers of unit cells in the filter $n = L/l$. Comparing Equations 12.2, 12.3, and 12.4, the ratio of the single-collector and the unit cell efficiencies is given by

$$e/E = 1.2(1 - \varepsilon)^{2/3} \tag{12.5}$$

for a granular bed, taking $K_1 = 1.5$ (Tien, 1989) while the correlation for a fibrous filter is (Zhao et al., 1991):

$$e/E = 1.12(1 - \varepsilon)^{1/2}/\varepsilon \tag{12.5'}$$

While the definition of the single-collector efficiency, E, is somewhat arbitrary and its value can exceed unity in some cases (this may be difficult to justify on pure mechanistic grounds), the unit cell efficiency, e, has a clear physical meaning. For a detailed discussion of the different efficiencies and their definitions, the reader is directed to the exhaustive monograph on granular filtration by Tien (1989). The next section is dedicated to ways of calculating the single-collector efficiency E and hence allow the prediction of the total efficiency, η.

12.3 COLLECTION MECHANISMS IN DEEP-BED FILTRATION

Collection of small airborne dust by collectors in a deep-bed filter is due to external forces which cause the dust to deviate from the fluid streamlines and thereby to impact and stick to the collector. The forces which are most frequently associated with filtration are inertia, diffusion, gravity, and electrical effects. While inertial and gravitational forces are characteristic of large dust particles of the order of microns and tens of microns, diffusion becomes important only for very find particles in the submicron range; electrical forces, if present, are effective in the whole range of particle sizes. For relatively small particles and in the absence of electrostatics, the so-called interception becomes important. This is a purely geometric "mechanism" and is due to the finite size of the dust particles; i.e., even if the particles follow the fluid streamlines exactly some streamlines will approach the collector to a distance smaller than the radius, r_p, of the dust particle, as can be seen in Figure 12.1, thereby causing deposition.

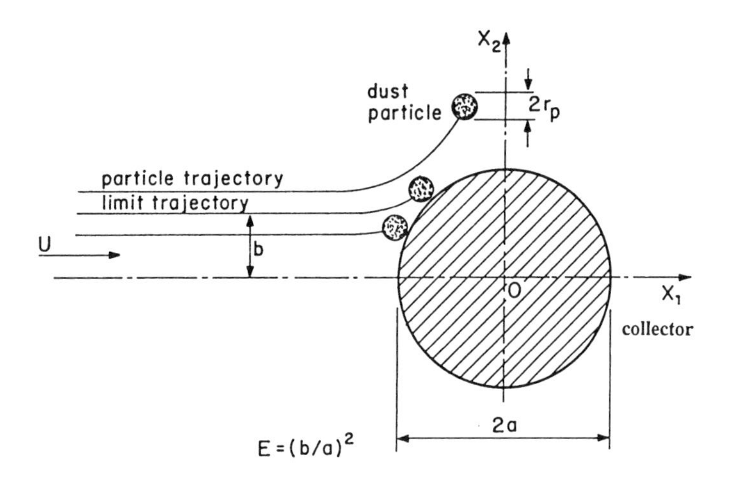

Figure 12.1 Schematic representation of dust deposition on a collector.

Table 12.1 presents a summary of the important mechanisms which cause deposition in a granular bed; each of the mechanisms is governed by a characteristic dimensionless number which is defined in the second column of the table. Since the most important electrical effects are caused by charges present on the particle and the collector or by an external electrical field, only the appropriate dimensionless numbers are given in the table. Table 12.2 is a summary of expressions for the correction factor $g(\varepsilon)$ which appears in the equations describing interceptional and diffusional efficiencies, while Table 12.3 contains theoretical and experimental relations to calculate the efficiency due to inertial effects. As seen in Table 12.1, the Reynolds number $Re_o = 2aU_o/\nu$ enters explicitly only in the expression of the inertial deposition; one has to note, however, that expressions for interception and diffusion are different for low and high Reynolds number flows, as shown in Table 12.1.

Table 12.4 presents a summary of single-collector (fiber) efficiencies for deposition in a fibrous bed and, as seen, the similarity with results given in Table 12.1 for granular beds is quite obvious. Again, the different dimensionless numbers characterizing the different mechanisms are defined in the table.

While it is quite simple to predict filtration efficiencies if only one mechanisms is active by using the expressions in Table 12.1 and 12.4, in reality a combination of effects almost always exists. A general practice in this case is to add the predicted values for each individual mechanism by using the equation:

$$E = 1 - (1 - E_R)(1 - E_D)(1 - E_G)(1 - E_I)(1 - E_{el}) \tag{12.6}$$

which, if all efficiencies are small compared with unity, simply becomes

$$E \cong E_R + E_D + E_G + E_I + E_{el} \tag{12.7}$$

Table 12.1 Collection Mechanisms in Granular Beds

Mechanism	Characteristic Dimensionless Number	Equation	Remarks
Interception	$R_p = r_p/a$; interception parameter	$E_R \cong 1.5\, g^3(\varepsilon) R_p^2$ $E_R \cong (3/\varepsilon) R_p$	$Re_o < 1$[a] $Re_o < 30$
Diffusion	$Pe = 2aU_o/D_B$; Peclet number	$E_D = 4g(\varepsilon)Pe^{2/3}$ $E_D = 4.52/(\varepsilon Pe)^{1/2}$	$Re_o < 1$ $Re_o < 30$[b]
Gravity	$Ga = ag/U_o^2$; Galileo number	$E_G \cong GaSt$	Independent of flow to a first approximation
Inertia	$St = 2C\rho_p U_o r_p^2/9\mu a$; Stokes number	$E_I = 2St'^{3.9}/(4.34^{-6} + St'^{3.9})$; $0.1 < St' < 0.03$	$St' = St[1 + 1.75Re_o/150(1-\varepsilon)]$[c]
Electrical effects	Electrical number		
	$K_c = Q_c Q_p/24\pi^2 \varepsilon_f r_p a^2 \mu U_o$	$E_{el} = 4K_c$	Coulombic force only
	$K_{ex} = CQ_p E_o/6\pi r_p \mu U_o$	$E_{el} = 4K_{ex}/(1 + K_{ex})$	External electric field only

[a] Geometric effect.
[b] See expressions for $g(\varepsilon)$ in Table 12.2.
[c] See other expressions in Table 12.3.

Table 12.2 Values of Correction Factor $g(\varepsilon)$

Author	$g(\varepsilon)$	Range
Pfeffer, 1964	$\{2[1-(1-\varepsilon)^{5/3}]/[2-3(1-\varepsilon)^{1/3}+3(1-\varepsilon)^{5/3}-2(1-\varepsilon)^2]\}^{1/3}$	$Re_o < 0.01$; $Pe \geq 1000$
Tardos et al., 1976	$\{\varepsilon/[2-\varepsilon-(9/5)(1-\varepsilon)^{1/3}-(1/5)(1-\varepsilon)^2]\}^{1/3}$	$Re_o < 0.01$; $Pe \geq 1000$
Sirkar, 1974, 1975	$\{[2+1.5(1-\varepsilon)+1.5[8(1-\varepsilon)-3(1-\varepsilon)^2]^{1/2}]/\varepsilon[2-3(1-\varepsilon)]\}^{1/3}$	$Re_o < 1$; $\varepsilon > 0.33$; $Pe \geq 1000$
Tardos et al., 1978	$1.31/\varepsilon$	$0.3 < \varepsilon < 0.7$; $Re_o < 0.01$; $Pe \geq 1000$
Tan et al., 1975	$1.1/\varepsilon$	$Re_o < 1$; $0.35 < \varepsilon < 0.7$
Wilson and Geankoplis, 1966	$1.09/\varepsilon$	$Re_o < 10$; $0.35 < \varepsilon < 0.7$
Thoenes and Kramers, 1958	$1.448/\varepsilon$	$Re_o < 10$; $\varepsilon = 0.746$
Karabellas et al., 1971	$1.19/\varepsilon$	$Re_o < 10$; $\varepsilon = 0.26$
Sorensen and Stewart, 1974, IV	$1.104/\varepsilon$ $1.17/\varepsilon$	$\varepsilon = 0.476$; $\varepsilon = 0.26$

The assumption behind Equation 12.6 is that different mechanisms act independently; this was demonstrated to be true for the case of diffusion, interception, and inertia (Gutfinger and Tardos, 1979), interception and gravity (Tardos and Pfeffer, 1980), and interception, diffusion, gravity, and weak electric effects (Pfeffer et al., 1981). Strong electric effects due to coulombic attraction and strong external electric fields (see Table 12.1) can not be combined with inertial effects and have to be considered separately (Nielsen and Hill, 1976; 1981; Tardos and Sbaddon, 1984). To improve precision some authors (Pich, 1966; Tien, 1989; Zhao et al., 1991) introduced different weighing coefficients in Equation 12.7 for those conditions under which single-collector efficiencies are not simply additive, but these equations are complicated and cumbersome to use and are not reproduced here. It is useful to remember that in using Equation 12.7, all efficiencies have to be small; in all other cases Equation 12.6 should be employed.

Table 12.3 Empirical Correlations for Single-Sphere Efficiency Due to Inertia Effects

Author	E	Range
Paretsky et al., 1971	$2 \times St^{1.13}$	St < 0.01
Meisen and Mathur, 1974	$0.00075 + 2.6 \times St$	St < 0.01
Doganoglu, 1975, 1978	$2.89 \times St$	$d_c \leq 100$ μm
	$0.0583 \times Re\ St$	$d_c \leq 600$ μm
Thambimuthu et al., 1978	$10^5 \times St^3$	0.001 < St < 0.01
Schmidt et al., 1978	$3.75 \times St$	St < 0.05
Goren, 1978	$1270 \times St^{9/4}$	0.001 < St < 0.02
Pendse and Tien, 1982[a]	$(1 + 0.04Re)[St\}$	—
D'Ottavio and Goren, 1983[b]	$St_{eff}^{3.55}/(1.67 + St_{eff}^{3.55})$	$0.33 < \varepsilon < 0.38$
Gal et al., 1985[c]	$2St'^{3.9}/(4.3 \times 10^{-6} + St'^{3.9})$	0.01 < St' < 0.02

[a] Interception neglected.
[b] $St_{eff} = f(Re,\varepsilon)St$; where the function $f(Re,\varepsilon)$ is given by $f(Re,\varepsilon) = (1 - \beta^{5/3})/(1 - 1.5\beta^{1/3} + 1.5\beta^{5/3} - \beta^2) + 1.14Re_o^{1/2}/\varepsilon^{2/3}$ where $\beta = 1 - \varepsilon$.
[c] $St' = St[1 + 1.75Re_o/150(1 - \varepsilon)]$.

Table 12.4 Collection Mechanisms in Fibrous Filters

Mechanism	Author(s)	Equation	Notations	Remarks
Interception	Ranz and Wang in Tan and Liang, 1984	$E_R = 1 + R_p - (1 + R_p)^{-1}$	$R_p = d_p/d_f$	Potential flow
	Lee and Liu, 1982	$E_R = (1 - \beta)R_p^2/K(1 + R_p)$	$K = 3$	$Re_o > 100$
Diffusion	Lee and Liu, 1982	$E_D = 2.6[(1 - \beta)/K]^{1/3}Pe^{-2/3}$	$K \approx 3$	
	Loeffler, 1971	$E_D = La^{1/3}Pe^{-2/3}$	$La = 2 - \ln Re_o$	$Re_o > 1$
	Loeffler, 1971	$E_D = K/Pe^{1/2}$	$K = 3$	$Re_o > 100$
Gravity	Tardos and Pfeffer, 1980	$E_G \approx GaSt$	$Ga = ag/U_o^2$	Independent of flow
Inertia	Landahl and Hermann in Pich, 1966	$E_I = St^3/(St^3 + 0.77\ St^2 + 0.22)$	$St = C\rho_p d_p^2 U_o/18\ \mu d t_f$	—
	Subramanyam in Tan, and Liang, 1984	$E_I = St/(St + 1.5)$	—	$Re_o > 100$ Potential flow
Electrical separation (charged particles and external electric field)	Kao et al., 1987 Zhao et al., 1991	$E_{el} = 3\pi(1 - \beta)K_{ex}/400\beta$ —	$K_{ex} = \omega_p E_o/U_o$ $\omega_p = CQ_p/3\pi\mu d_p$	Randomly arranged fibers

To complete the picture of collection of small airborne dust by a collector in a deep filter bed, the phenomenon of bounce-off has to be mentioned. It was observed by many researchers that at relatively high gas velocities and/or large particle sizes, while inertial effects ensure that dust particles collide with collectors following their tortuous way through the filer, the dust is in fact not collected and instead bounces off upon contact and is, in the end, not retained by the filter. This behavior results in a reduced efficiency at particle Stokes numbers larger than about St > 0.01. Tien (1989) introduces the coefficient of adhesion probability γ given by

$$\gamma = 0.00318\ St^{-1.248} \tag{12.8}$$

to account for this effect. For practical calculations, the efficiency E obtained from Equation 12.6 has to be multiplied by the factor γ if the Stokes number exceeds the value St = 0.01 even if the deposition is overwhelmingly influenced by electrostatic effects.

12.4 PRESSURE DROP IN DEEP-BED FILTERS

For pressure drop calculations in granular beds one can use with some confidence the well-known Ergun correlation (Ergun, 1952; Macdonald et al., 1979) which in dimensionless form is given as

$$f_o[\varepsilon^3/(1-\varepsilon)] = 180(1-\varepsilon)/\mathrm{Re}_0 + 1.8 \quad (12.9)$$

The actual pressure drop per unit thickness of filter is then evaluated from the equation:

$$\Delta p/L = f_0 \rho U_0^2/2a \quad (12.10)$$

where the Reynolds number is expressed as $\mathrm{Re}_o = 2aU_o\rho/\mu$. L is the thickness of the filter in the direction of the flow, a is some average granule radius, and U_o is the superficial gas velocity in the filter.

The pressure drop across clean beds of fibers can be expressed as (Kirsch and Fuchs, 1967; Van Osdell, 1988):

$$\Delta p/L = 4(1-\varepsilon)F/\pi d_f^2 \quad (12.11)$$

where F is the drag force per unit length of fiber and is listed for low and high Reynolds numbers in Table 12.5.

12.5 INFLUENCE OF DUST LOADING ON EFFICIENCY AND PRESSURE DROP

The theoretical considerations presented so far for both granular and fibrous filters only apply under conditions where the collector bed is relatively clean, i.e., during the initial filtration stage where dust particles deposit directly on the collector. For low concentrations of dust and for small aerosols or for very low flow rates in the filter, this relatively "clean state" covers a large period of the active filtration process before the filter media is replaced and/or cleaned. For high dust loadings, however, one has to consider the regime when the clean collector surface becomes covered with dust and subsequent collection occurs on top of the already deposited layer. The considerations that follow cover this regime, during which the filter elements, fibers (cylinders) and granules, maintain more or less their geometry but both their size and the overall porosity of the filter changes: as dust deposits the collector diameter increases while the void fraction decreases accordingly. Some geometry changes are inevitable under these conditions since some dust will deposit at the intersection of two fibers or at the contact point between two granules. *The assumption in the following calculations is that such changes in geometry are negligible* and while the initial collector size and porosity are $d_f = 2a$ and $\varepsilon = 1 - \beta$, the characteristic parameters of the "dust-loaded" filter become $d_f' = 2a'$ and $\varepsilon' = 1 - \beta'$ so that $d_f' > d_f$ and $\varepsilon' < \varepsilon$. Clearly, as large dust deposits are formed in the filter, the above simplified model will fail in that under these conditions deposits on different collectors will "grow" and touch each other, the bed geometry will change drastically, and the filtration process will switch over to cake filtration as explained earlier.

The purpose of the following considerations is to relate the dust-laden filter collector size d_f' and void fraction ε' to the total amount of dust collected in the filter. Following Zhao et al. (1991) and denoting L_f as the total length of fibers per unit volume of filter bed, the solid fraction of the clean filter is

$$\beta = \pi d_f^2 L_f/\pi \quad (12.12)$$

Table 12.5 Drag Force Exerted on Unit Length of Fiber

Authors	Drag Force	Expressions for Dimensionless Drag Force	Remarks
Sangani and Acrivos, 1982	$F = \mu U_o F^*$	$F^* = \dfrac{4\pi}{-0.5 \ln\beta - 0.738 + \beta - 0.887\beta^2 + 2.038\beta^2}$	Square arrangement; $\beta << 1$
		$F^* = \dfrac{9\pi}{2.\sqrt{2}}\left[1 - \left(\dfrac{\beta}{\beta_{max}}\right)^{-1/2}\right]^{-5/2}$	Square arrangement; $\beta_{max} - \beta << 1$; $\beta_{max} = \pi/4$
		$F^* = \dfrac{4\pi}{-0.5 \ln\beta - 0.745 + \beta - 0.25\beta^2}$	Staggered arrangement; $\beta << 1$
		$F^* \approx \dfrac{27\pi}{4.\sqrt{2}}\left[1 - \left(\dfrac{\beta}{\beta_{max}}\right)^{-1/2}\right]^{-5/2}$	Staggered arrangement; $\beta_{max} - \beta << 1$; $\beta_{max} = \pi/4$
Keller, 1964	$F = \mu U_o F^*$	$F^* = \dfrac{9\pi}{2.\sqrt{2}}\left[1 - 2(\beta/\pi)^{-1/2}\right]^{-5/2}$	Square arrangement
Kuwabara, 1959	$F = \mu U_o F^*$	$F^* = \dfrac{4\pi}{-0.5 \ln\beta - 0.75 + \beta - \beta^2/4}$	Staggered arrangement
Happel, 1959	$F = \mu U_o F^*$	$F^* = \dfrac{4\pi}{-0.5 \ln\beta - 0.5 + 0.5\beta^2/(1 + \beta^2)}$	Staggered arrangement
Hasimoto, 1959	$F = \mu U_o F^*$	$F^* = \dfrac{4\pi}{-0.5 \ln(\beta/\pi) - 1.3105 + \beta}$	Square arrangement
Tamada and Fujikawa, 1957	$F = \mu U_o F^*$	$F^* = \dfrac{4\pi}{-0.5 \ln(\beta/\pi) - 1.33 + \pi\beta/3}$	Square arrangement
Zhao et al., 1991	$F = \dfrac{1}{2}\rho_f U_0^2 d_f F^*$	$F^* = 224.8\left(\dfrac{4\beta}{\pi}\right)^{2.9}\left[1 - \left(\dfrac{4\beta}{\pi}\right)^{-1/2}\right]^{-2}$	Randomly arranged fibrous bed $Re_o > 10$

and the total length of all fibers in the filter is

$$L_b = L_f L A_b = 4\beta A_b / \pi d_f^2 \tag{12.13}$$

Here L is the bed thickness, and A_b is the area of the filter. Assuming that the dust-laden fiber diameter is d_f', the total weight of dust in the filter is

$$W_d = D_1 W_f = \pi\left(d_f'^2 - d_f^2\right) L_b \rho_p / 4 \tag{12.14}$$

Combining the above equations, one finds that d_f' is proportional to d_f:

$$d_f' = K_d d_f \tag{12.15}$$

where K_d is the dust-loading factor given by

$$K_d = [1 + W_d/\rho_p \beta A_b]^{1/2} = [1 + D_1 \rho_c/\rho_p]^{1/2} \tag{12.16}$$

where D_l is the *loading* defined as the ratio of the weight of deposited dust, W_d, to the initial weight of the clean filter, W_f, as shown in Equation 12.14. The correlation between the void fraction of the loaded and the clean filter is seen from Equation 12.13 to be

$$\beta' = K_d^2 \beta \tag{12.17}$$

For the case of the granular filter one can easily show that $a' = K_d a$ and that the equivalent of the dust-loading factor in Equation 12.16 becomes

$$K_d = [1 + W_d/\rho_p \beta L A_d]^{1/3} = [1 + D_l \, \rho_c/\rho_p]^{1/3} \tag{12.16'}$$

while the correlation of the bed porosities is

$$\beta' = K_d^3 \beta \tag{12.17'}$$

The advantage of the above analysis is that all equations to predict collection efficiencies and pressure drop developed for clean filters can now be employed for the dust-laden filter if the collector diameter $d_f = 2a$ and the void fraction $\varepsilon = 1 - \beta$ are replaced by d_f' and ε'. As explained above, it is implicitly assumed here that as dust deposition takes place, the collector diameters increase gradually but that deposits on different collectors do not touch. It is also assumed that as dust deposition occurs the surface properties of the collector do not change drastically due to the deposits. This last assumption can in some cases be relaxed by taking dust surface properties (such as electric constants, etc.) instead of collector properties for "dust-loaded" filters.

12.6 EXPERIMENTAL VERIFICATION

A schematic representation of the experimental apparatus to test a filter bed is depicted in Figure 12.2. One can see the particle generator which introduces dust particles or aerosols into the clean gas stream and the "particle analyzers" connected up- and downstream of the filter bed to measure aerosol concentrations. In the case of electrically enhanced filters, a wire mesh electrode is added to the top of the bed where the electric field is applied and a radioactive source, used to neutralize the generated aerosols (dust particles), is followed by a particle charger as shown in the figure. The complexity of the setup is required by the need to control the dusty gas flow, the particle and collector electric charge (or lack of it), and the gas humidity. A very careful procedure is also required to measure the average charge on the aerosols as they emerge from the corona charger; the system depicted in the figure, comprised of a Faraday cage, electrometer, and the Royco analyzer, accomplishes this according to the method described in Tardos et al. (1984). Filtration experiments usually require the generation of a dilute stream of test aerosols such as latex particles of known size which are subsequently pumped through the filter at known flow rate, and the concentration *in* and *out* of the bed is carefully measured. These experiments are repeated with a whole range of specially manufactured test dust or aerosols of different size and sometimes composition and electrical properties. To control electric charges, the test particles are first neutralized and then electrically charged to the appropriate level before entering the bed. Experiments are performed at different gas flow rates and at different electric fields if electric effects are present.

Figures 12.3 and 12.4 show measured (Tardos et al., 1979; Tardos and Snaddon, 1984) and calculated filter efficiencies using the equations given in Table 12.1. Dust particles used in these experiments are of the latex aerosol type which are commonly used in industry to test filters as mentioned above. Figure 12.3 shows filtration efficiencies, E, as a function of gas superficial velocity

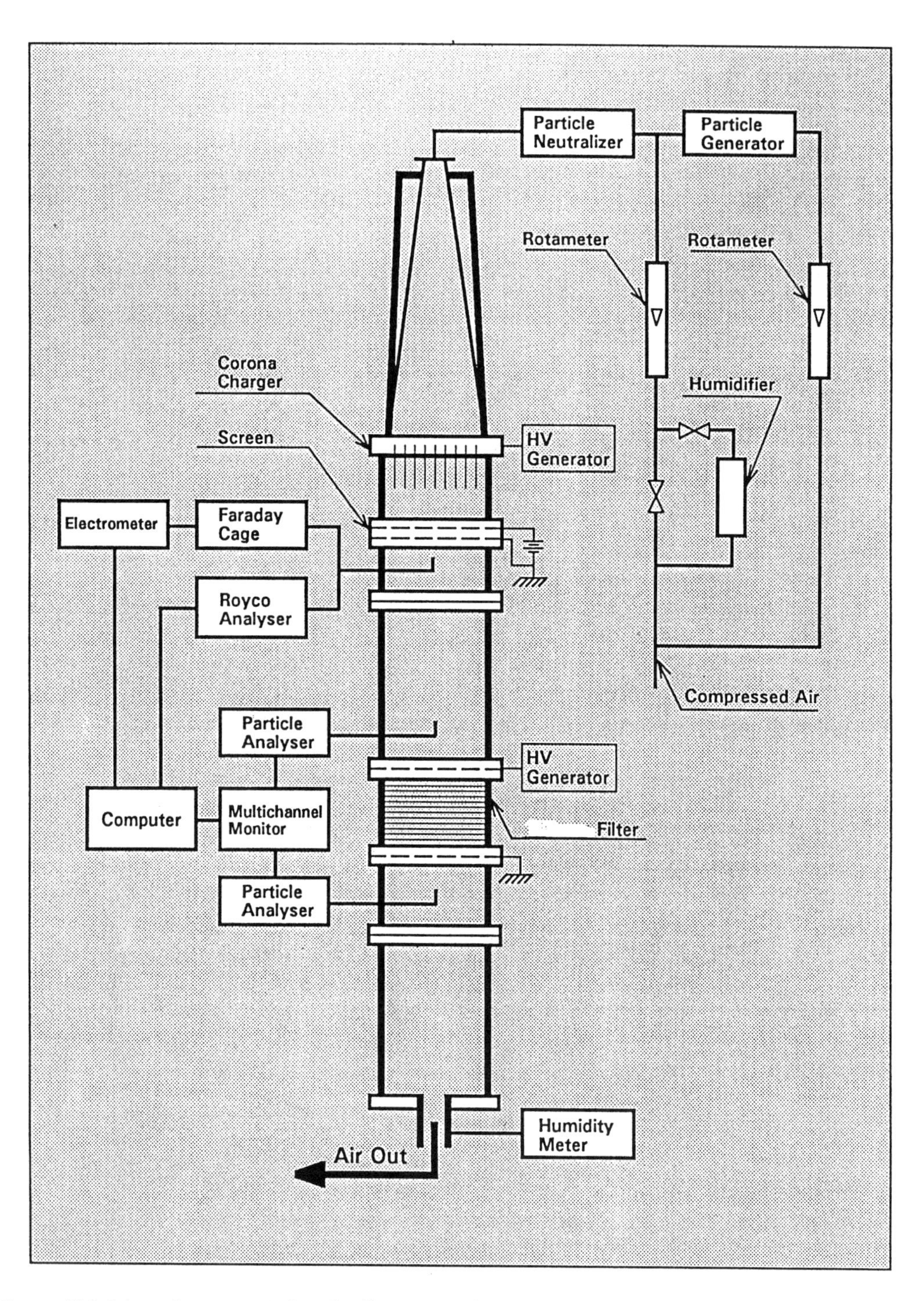

Figure 12.2 Schematic representation of a filter test section.

in a sand bed of grain average size 450 μm. Calculated values are for large Reynolds numbers (upper line in the figure) and very low Reynolds numbers (viscous flow) using the equations of Table 12.1 for diffusion, interception, gravity, and inertia. The experimental values of the Reynolds number are depicted with arrows on the lower side of the figure. As seen, the data follow the calculations as expected: for Reynolds numbers below about $Re_o = 3$, the data fit viscous flow

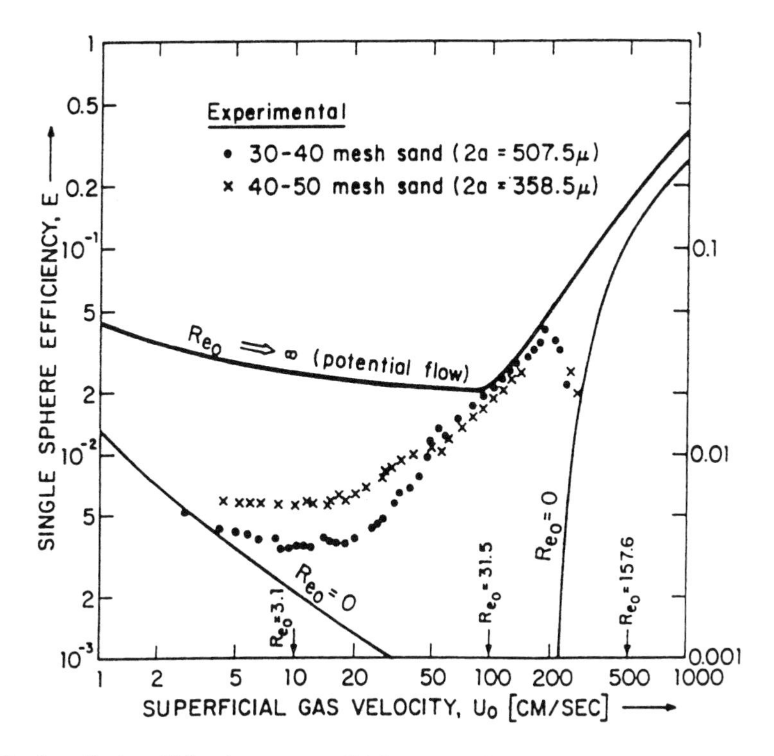

Figure 12.3 Single collector efficiencies vs. superficial gas velocity. Filtration of 1.1 microns in diameter latex aerosols. Theoretical values computed for: bed porosity $\varepsilon = 0.4$, granule diameter $2a = 0.45$ mm and particle density $\rho_p = 2$ g/cm^3.

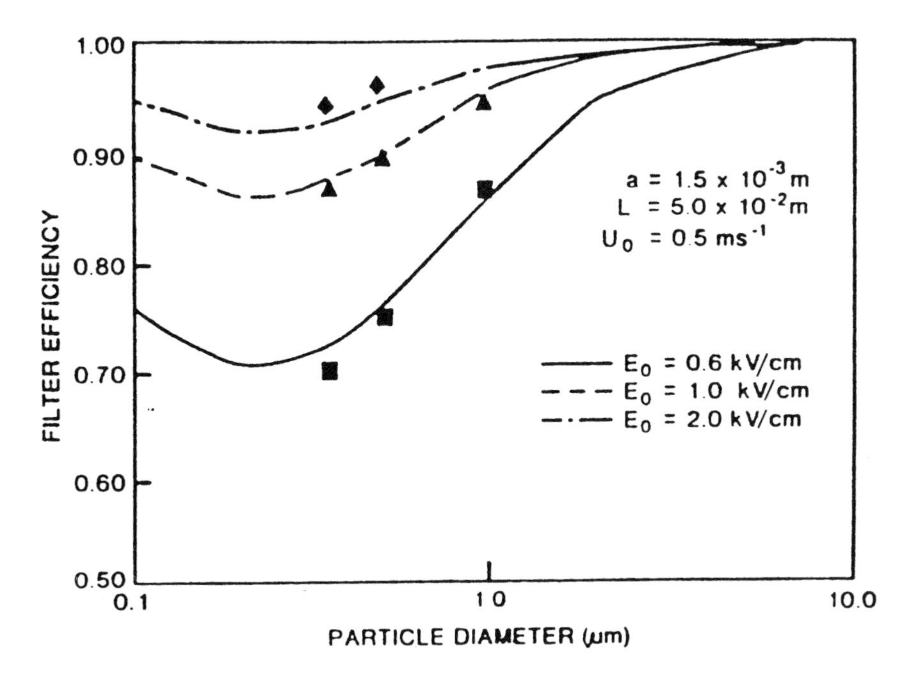

Figure 12.4 Total filter efficiencies vs. particle diameter. Comparison of model predictions with the "clean filter" theory ($U_0 = 0.5$ m/sec).

calculations well, while for values of the Reynolds number of the order or $Re_o = 30$ and higher, the measured data follow the calculations for potential flow. One can clearly see the effect of bounce-off at superficial gas velocities larger than about 2 m/s. One has to note here that the data presented above are an exceedingly exaggerated case in which the limits of the theoretical calculations are being checked. Granular filters are usually operated at gas velocities of the order of 2 to 30 cm/s where it is clearly seen that calculated values fall quite close to the measured ones.

Figure 12.4 shows results for an electrically enhanced granular filter operated with an external electric field. The shape of the efficiency curves (total efficiencies η in this case) are typical for this type of filter: efficiencies are high for small dust particles below 0.1 μm and large dust particles above 1 μm in diameter and are lower between these two limits. Increasing the applied electric field results in a significant improvement in efficiency even at the high gas velocity of $U_o = 0.5$ m/s as shown in the figure.

Figures 12.5 and 12.6 show results obtained from testing fibrous filters with liquid (dioctyl phtalate, DOP) and solid aerosols such as redispersed fly ash and alumina dust, by increasing the dust loading, D_l, from zero (clean filter) to about a value of 3 and measuring the efficiency and the pressure drop (Zhao et al., 1991). Figure 12.5 shows total bed filtration efficiencies vs. dust loading

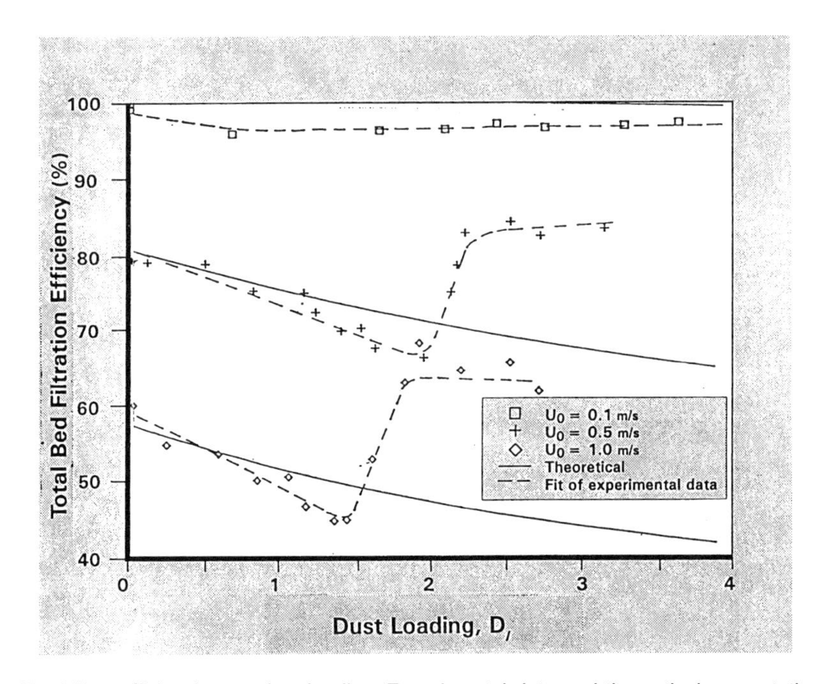

Figure 12.5 Total filter efficiencies vs. dust loading. Experimental data and theoretical computations using the "relatively clean filter" model.

for three different gas velocities of 0.1, 0.5, and 1 m/s. Theoretical results were calculated for all three cases by using Equations 12.2 and 12.6 and values obtained from Table 12.4 for both the clean and the dust-laden filter. The decrease in efficiency predicted theoretically and measured experimentally as dust loading increases is due to the increase in the size of the collectors $d_p' > d_f$ which is not compensated by the decrease in porosity $\varepsilon' < \varepsilon$. While the agreement between measured and calculated efficiencies is quite good for low values of the loading of up to about 1.5, one can see a sharp increase in efficiency not predicted by theory, above this value. The significance of this result is that it was obtained for both liquid (very sticky) and solid dust and thus seems to indicate that the theory breaks down at some critical loading. The critical value of the dust loading D_l was demonstrated by Zhao et al. (1991) to depend strongly on the gas velocity in the filter and also on the nature of the dust as shown in Figure 12.6. The same general kind of behavior was found by

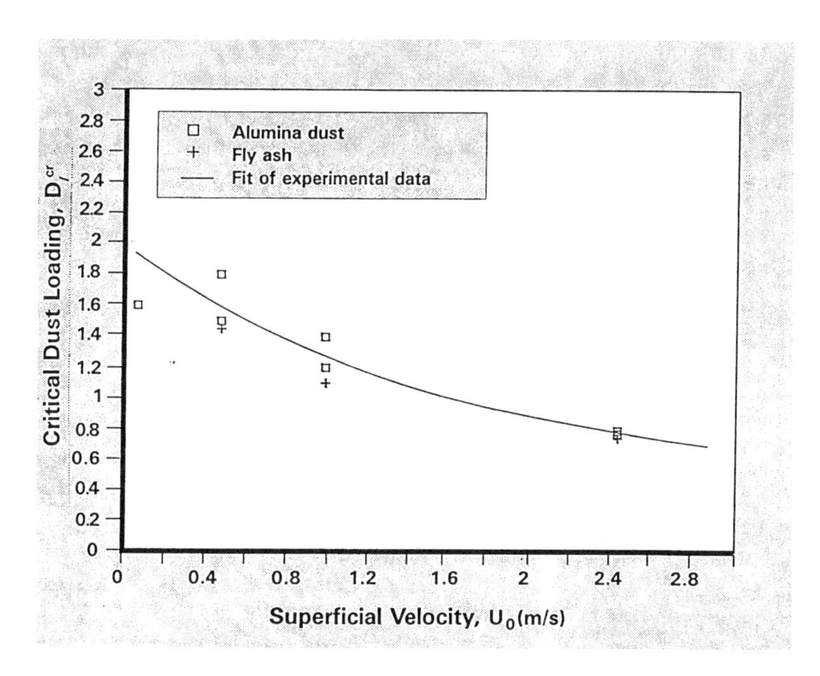

Figure 12.6 Measured *critical dust loading* vs. superficial gas velocity for filtration of solid dust particles.

measuring the pressure drop in the filter, i.e., slow increase below the critical loading and large increase above it. The physical picture of capture and dust layer growth on the collector appears to occur in two phases: an initial one without major geometry modifications where the collector diameter increases monotonously and a subsequent one when deposits essentially change both the geometry and the porosity. The theory presented earlier seems to be applicable to the first phase but breaks down for the second. The two phases are separated by the critical dust loading below which the theory holds and above which it usually underpredicts the efficiency. While it is true that the pressure drop follows the same trend, it was shown that the critical loading for pressure drop is somewhat lower as compared with that measured for efficiencies; the explanation for this result defies easy explanation. One has to note that dust loadings of $D_l = 2–3$, as measured during the experiments described above are quite high for deep beds and hence under normal conditions one would clean the filter before such large amounts of dust are actually collected.

12.7 CONCLUDING REMARKS

A theoretical approach was outlined to predict both the single-collector efficiency E and the total-bed efficiency η of a deep-bed filter from first principles. The method is based on the assumption that on the average each granule in the bed or fiber in the filter plays the role of a collector and that, overall, the effects of all collectors can be integrated to yield an exponential decay in concentration along the filter. The filtration process is then divided into individual mechanisms by which airborne dust deviates from the fluid streamlines and can, at least in principle, collide with a collector and stick to it. Some experimental evidence is given to show that this model is realistic and that carefully measured efficiencies can in fact be predicted theoretically with some degree of confidence.

The above approach was extended to "relatively clean" filters by assuming that initial dust deposits do not alter drastically the geometry of the filter and that deposits only increase the collector diameter and decrease the bed porosity. Careful experiments showed that the above assumption is valid up to a critical dust loading above which the theory fails. Values for the critical dust loading were obtained experimentally for both solid dust and liquid aerosols and seem to depend mostly on the gas velocity; it is proposed that filters be cleaned before the critical dust loading is reached.

Dust filtration in a cake clearly does not fit the above model and hence the reader is referred to the appropriate literature (Donovan, 1985; Flagan and Seinfeld, 1988) for further information. Fortunately, deep-bed filtration is almost always the important mode of separation of small particles while cake filtration becomes important for larger particles and when the filter becomes overwhelmingly clogged with dust. Because of pressure drop considerations, the operation of deep-bed filters in the clogged regime is not economically and technically attractive.

NOMENCLATURE

a	collector radius
a'	modified collector radius due to dust loading
A_b	area of filter bed (cm^2)
C	Cunningham correction factor
D_B	$KT/6\pi\mu r_p C$; dust particle diffusion coefficient
d_c	$2a$; collector diameter
D_ι	loading defined in Equation 12.14
d_f	$2a$; fiber (collector) diameter (cm)
d_f'	$2a'$; modified fiber (collector) diameter due to dust loading (cm)
d_p	$2r_p$; particle diameter (cm)
e	unit cell efficiency
E	single-collector efficiency
E_D	single-collector efficiency due to diffusion
E_{el}	single-collector efficiency due to electrical effect
E_G	single-collector efficiency due to gravity
E_I	single-collector efficiency due to inertia
E_0	applied electric field
E_R	sisngle-collector efficiency due to interception
F	force per unit length of fiber (dyn/cm)
F^*	dimensionless force per unit length of fiber
f_0	dimensionless pressure drop
g	acceleration of gravity
Ga	ag/U_0^2; Galileo number
$g(\varepsilon)$	Porosity dependent function given in Table 12.2
K	Boltzmann constant
K_c	dimensionless electric number due to external electric fields due to coulombic force
K_{ex}	dimensionless electric number
K_d	dust-loading factor
l	unit cell size
L	filter bed height
La	Lamb hydrodynamic factor
L_b	total length of fibers in filter bed (cm)
L_f	total length of fibers per unit volume of filter (cm)
n_{in}	inlet aerosol concentration
n_{out}	outlet aerosol concentration
n	L/l; number of unit cell layers
Pe	$2aU_d/D_B$; Peclet number
Q_c	collector electric charge

Q_p	particle (dust) electric charge
r_p	dust particle radius
R_p	r_p/a; interception parameter
Re_o	$2aU_o/\nu$; Reynolds number
St	$2C\rho_p U_o r_p^2/9\mu a$; Stokes number
T	absolute gas temperature
U_o	superficial flow velocity
Δ_p	pressure drop through the packed bed
W_d	weight of dust deposited (g)
W_f	weight of clean fibrous bed (g)

Greek Letters

ε	bed porosity
ε_f	fiber dielectric constant
ε'	modified bed porosity
ρ_p	dust particle density
ρ_f	fiber density
ρ_c	collector density
μ	gas viscosity
ν	μ/ρ; gas kinematic viscosity
η	total filtration efficiency
ρ	gas density
γ	adhesion probability coefficient
β	$1 - \varepsilon$; bed solid fraction
β'	$1 - \varepsilon'$; modified bed solid fraction

Subscripts

p	particle
c	collector
f	fiber
d	dust

REFERENCES

Doganoglu, Y., Ph.D. dissertation, McGill University, Montreal, 1975.

Doganoglu, Y. et al., Removal of fine particulates from gases in fluidized beds, *Trans. Inst. Chem. Eng.*, 56, 293, 1978.

Donovan, R. P., *Fabric Filtration for Combustion Sources,* Marcel Dekker, New York, 1985.

D'Ottavio, T. and Goren, L. S., Aerosol capture in granular beds in the impaction dominated regime, *Aerosol Sci. Technol.,* 2, 91, 1983.

Ergun, S., Fluid flow through packed columns, *Chem. Eng. Prog.,* 48, 89–94, 1952.

Flagan, R. C. and Seinfeld, J. H., *Fundamentals of Air Pollution Engineering,* Prentice-Hall, Englewood Cliffs, NJ, 1988.

Fuchs, N., *Mechanics of Aerosols,* Pergamon Press, Oxford, 1973.

Gal, E., Tardos, G. I., and Pfeffer, R., Inertial effects in granular bed filtration, *AIChE J.,* 31, 1093, 1985.

Goren, L. S., Aerosol filtration by granular beds., *EPA Symp. in Transfer and Utilization of Particulate Control Technology, Rpt. EPA-600-7-79-044,* 1978.

Gutfinger, C. and Tardos, G. I., Theoretical and experimental investigation of granular bed dust filters, *Atmos. Environ.,* 13(6), 853, 1979.

Happel, J., *AIChE J.,* 5, 174, 1959.
Henry, F. and Airman, T., *J. Aerosol Sci.,* 12, 137, 1981.
Henry, F. and Airman, T., *Particulate Sci. Technol.,* 11(2), 139, 1983.
Inculet, I. E. and Castle, G. S. P., *ASHRAE J.,* March, 47–52, 1971.
Kao, J. N., Tardos, G. I., and Pfeffer, R., *IEEE-IAS Trans.,* 23(3), 464–473, 1987.
Karabellas, A. J., Wegner, T. H., and Hanratty, T. J., Use of asymptotic relations to correlate mass transfer data in packed beds, *Chem. Eng. Sci.,* 26, 1581, 1971.
Keller, J., *J. Fluid Mech.,* 18, 96, 1964.
Kirsch, A. A. and Fuchs, A. A., *Ann. Occup. Hyg.,* 10, 23–30, 1967.
Kuwabara, S., The forces experienced by randomly distributed parallel circular cylinders or spheres in a viscous fluid at small Reynolds number, *J. Phys. Soc. Jpn,* 14, 527, 1959.
Lee, K. W. and Liu, B. Y. H., *Aerosol Sci. Technol.,* 1(2), 147–166, 1982.
Loeffler, R., Collection of particles by fiber filters, in *Air Polution Control, Part I,* W. Stauss, Ed., Wiley-Interscience, Academic Press, New York, 1971.
Macdonald, I. F., El Sayed, M. S., Mow, K., and Dullien, F. A. L., Flow through porous media — the Ergun equation revisited, *Ind. Eng. Chem. Fundam.,* 18, 199–208, 1979.
Meisen, A. and Mathur, K. B., Multi-phase flow systems, *Inst. Chem. Engs. Symp. Ser.,* 38, Paper K3, 1974.
Neilsen, K. A. and Hill, J. C., *I&EC Fund.,* 15, 149, 1976.
Neilsen, K. A. and Hill, J. C., *Chem. Eng. Commun.,* 12, 1/1, 1981.
Paretsky, L. C. et al., Panel bed filter for simultaneous removal of fly ash and sulfur dioxide, *J. APCA,* 21, 204, 1971.
Paytakes, A. C., Tien, C., and Turian, R. M., *AIChE J.,* 28, 677–696, 1973.
Pendse, H. and Tien, C., General correlation of the initial collection efficiency of granular filter beds, *AIChE J.,* 28(4), 677, 1982.
Pfeffer, R., *I&EC Fund.,* 3, 380, 1964.
Pfeffer, R., Tardos, G. I., and Pismen, L., Capture of aerosols on a sphere in the presence of weak electrostatic forces, *I&EC Fund.,* 20, 1981.
Pich, J., Theory of aerosol filtration by fibrous and membrane filters, in *Aerosol Science,* C. N. Davies, Academic Press, London, 1966.
Sangani, A. S. and Acrivos, A., *Int. J. Multiphase Flow,* 8(3), 193–206, 1982.
Schmidt, E. W. et al., Filtration of aerosols in a granular bed., *J. APCA,* 28(2), 143, 1978.
Sirkar, K. K., Creeping flow mass transfer to a single active sphere in a random spherical inactive particle cloud at high Schmidt numbers, *Chem. Eng. Sci.,* 29, 863, 1974.
Sirkar, K. K., Transport in packed beds at intermediate Reynolds numbers, *Ind. Eng. Chem. Fundam.,* 14, 73, 1975.
Sorenson, J. P. and Stewart, W. E., Computation of forced convection in slow flow through ducts and packed beds — I, II, III, IV, *Chem. Eng. Sci.,* 29, 819, 1974.
Stechkina, I. B. and Fuchs, N. A., Studies on fibrous aerosol filters, I, II, II, *Ann. Occup. Hyg.,* 9, 59–64; 10, 23–30, 1967; 11, 299–304, 1968.
Stenhouse, J. I. T. and Harrop, J. A., Particle capture mechanisms in fibrous filters, *Filtr. Sep.,* 41, 112–123, 1971.
Tan, A. Y. and Liang, F.-A., *Technology of Industrial Ventiliation and Dust Collection,* Chinese Architecture Publisher, Beijing, P. R., China, Chap. 11, 1984.
Tan, A. Y., Prasher, B. D., and Guin, J. A., Mass transfer in non-uniform packing, *AIChE J.,* 21(2) 396, 1975.
Tardos, G. I., *Granular bed filters,* in *Handbook of Powder Science and Technology,* 2nd ed., M. E. Fayed and L. Otten, Eds., Van Nostrand Reinholt, New York, 1997.
Tardos, G. I. and Pfeffer, R., Interceptional and gravitational deposition of inertialess particles on a single sphere in a granular bed, *AIChE J.,* 26(4) 698, 1980.
Tardos, G. I. and Snaddon, R. W. L., Separation of charged aerosols in granular beds with imposed electric fields, in *AIChE Symposium Series,* T. Knolton, Ed., 235, 80, 60, 1984.
Tardos, G. I., Abuaf, N., and Gutfinger, C., Dust deposition in granular bed filters — theories and experiments, *JAPCA,* 28(4), 354–363, 1978.
Tardos, G. I., Gutfinger, C., and Abuaf, N., High Peclet number mass transfer to a sphere in a fixed or fluidized bed, *AIChE J.,* 22, 1146–1149, 1976.
Tardos, G. I., Gutfinger, C., and Pfeffer, R., *J. Coll. Int. Sci.,* 71(3), 616, 1979.

Tardos, G. I. and Snaddon, R. W. L., and Dietz, P. W., *Trans. IEEE Ind. Appl.* IA-20, 6, November/December, 1984.

Thambimuthu, K. V. et al., Symp. Deposition and Filtration of Particles from Gases and Liquids, *Soc. Chem. Ind.,* London, 107, 1978.

Thoenes, D. and Kramers, H., Mass transfer from a sphere in various regular packings to a flowing fluid, *Chem. Eng. Sci.,* 8, 271, 1958.

Tien, C., *Granular Filtration of Aerosols and Hydrosols,* Butterworths, Boston, 1989.

Van Osdell, D. W., in *Proceedings of the International Technical Conference on Filtration and Separtation,* Ocean City, MD, 1988, 173–180.

Wilson, E. J. and Geankoplis, C. J., Liquid mass transfer at very low Reynolds numbers in packed beds, *Ind. Eng. Chem. Fundam.,* 5, 9, 1966.

Zhao, Z. M., Tardos, G. I., and Pfeffer, R., Separation of airborne dust in electrostatically enhanced fibrous filters, *Chem. Eng. Commun.,* 108, 307–332, 1991.

CHAPTER 13

Quality Assurance of Glass Fiber Filter Media: Control of Structure Inhomogeneities

W. Mölter and H. Fissan

CONTENTS

0-87371-830-5/98/$0.00+$.50

13.1 INTRODUCTION

Many technological developments and innovations have been advanced in the field of microelectronics and micromechanics. In these fields, particle-free and clean process environments — so-called clean rooms — are required.

The quality of the installed filter media plays a role in determining the maximum air quality inside a clean room. The filter media used in clean rooms are nonwovens consisting of polydisperse glass fibers with fiber diameters between 0.1 and 10 μm. These media are produced in a sheet thickness from 0.4 to 1 mm, with fiber volume fractions (packing density) of about 5%. In other words, only 5% of the medium volume is occupied by the fibers.

The skill as well as the difficulty of the manufacturing process is the arrangement of the fibers in a way that a homogeneous medium is produced. Due to the polydisperse fiber diameters and different fiber length, it is very difficult to arrange the fibers in a random but comparable manner all over the filter medium with fibers touching only a few others. In addition, a medium is called homogeneous if the sheet thickness keeps constant, as well.

In real nonwovens, however, local fluctuations of the fiber matrix (see Figures 13.1 and 13.2) and variations of the sheet thickness are not preventable. Unwanted inhomogeneities of the structure occur which deteriorate the quality of the filter medium.

During the production process, incomplete fiber dispersion and the occurrence of fluffs lead to structure inhomogeneities. The manufacturer is able to reduce such inhomogeneities via empirical control but the inhomogeneities are not completely avoidable. The structure inhomogeneities deteriorate the quality of filter media. There are local positions inside the medium showing lower fiber volume fractions and positions showing higher fiber volume fractions. Because the fluid takes the path of smallest resistance, those local positions are squeezed with higher and lower flow velocities, respectively. Thus, the inhomogeneities lead to a fluctuation of the local flow velocity. Because the diffusional capture mechanism of the particles is dominating, high flow velocities lead to a decrease of the local filter efficiency. In addition, the particle flux onto that weak position is increased and the flux to the regions with higher packing densities is decreased. Thus, the positions with higher packing densities and therefore with higher efficiencies are out of process (more or less) and do not compensate the weak efficiency of the lower packed regions. These nonlinear effects deteriorate the quality of the filter medium.

At the moment, these deteriorations of quality are compensated for by the manufacturer by increasing the sheet thickness of the medium. From the perspective of the quality warranty, this is a proper way and also the most economical way. But if it is possible to control the occurrence of structure inhomogeneities while still in the production line, less-expensive manufacturing of filter media could be established.

For this purpose, measuring procedures must be developed and installed to control the fluctuations of the structure during the manufacturing process. Recent developments of quasi-on-line measuring techniques for the detection of the local area mass and the local sheet thickness of a filter medium are presented in Adam (1995). The local area mass is determined using the absorption of β-radiation or using the different permittivity of glass and air. The local sheet thickness is measured by means of a laser triangulation principle. The advantages of these measurement techniques are the external and nondestroyable application. But the local sheet thickness as well as the local area mass are integral parameters averaged over the area of the measuring position, and the area mass is additionally averaged over the sheet thickness. Thus, the real structure inside

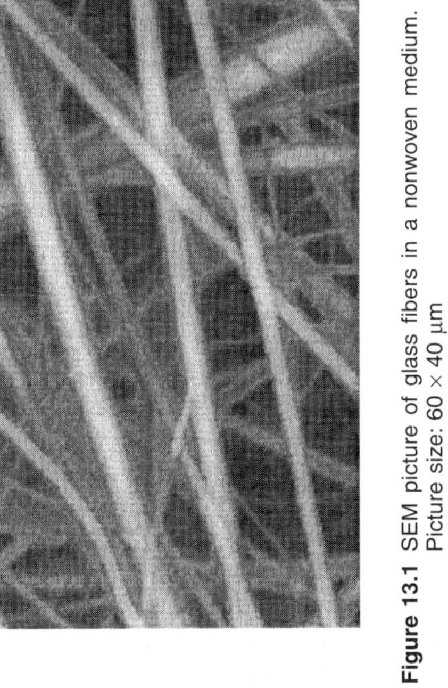

Figure 13.1 SEM picture of glass fibers in a nonwoven medium. Picture size: 60 × 40 μm

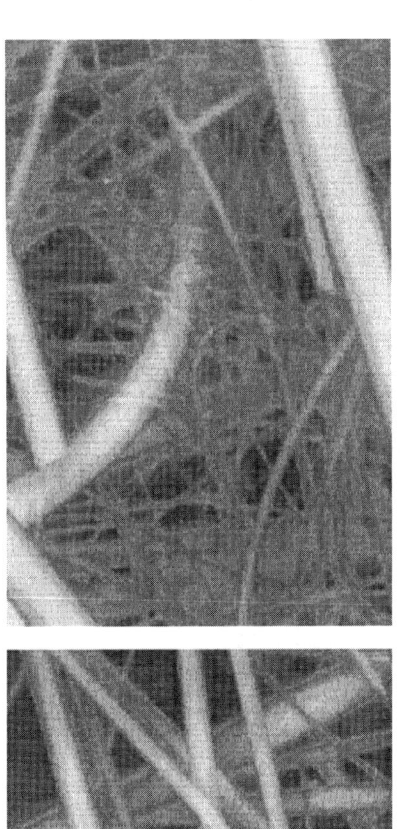

Figure 13.2 SEM picture of another position of the same medium. Picture size: 60 × 40 μm

the filter medium is not measured and no information is given to show the arrangement and the local internal frequency of the fibers with different fiber diameters.

But the filtration properties of glass fiber filter media depend strongly on the real internal structure. Thus, the relevance of the integral and averaged structure parameters for media quality has to be proved and guaranteed. In particular, the proof is important because the integrating measuring procedures give no image of the fiber diameter distribution which influences tremendously the filtration properties and the quality of the filter media.

13.2 THE QUALITY OF GLASS FIBER FILTER MEDIA

The quality of glass fiber filter media is guaranteed currently by the manufacturer by off-line controlling of some randomly selected 100 cm^2 area. The controlling procedure covers two filtration properties: pressure drop and particle capture efficiency. In this way the specified efficiency is ensured, and in addition it is shown that the accompanying pressure drop does not exceed the limit.

High efficiency or low penetration coupled with low pressure drop characterizes a high-quality air filter. This is valid in all fields of filtration, but it depends on the application whether the initial emission of particles, the mechanical stability, or the processing properties of the medium must be considered as quality features, as well.

Based on the pressure drop and penetration, a quality factor Q is defined and different definitions are possible. In the literature (Lathrache, 1987) Q is defined via the negative logarithm of the penetration P divided by the pressure drop Δp (see Equation 13.1).

$$Q = \frac{\ln P}{\Delta p} \tag{13.1}$$

Because this definition is not dimensionless, it is not very useful. Equation 13.2 shows the more common dimensionless definition, where the dimensionless single-fiber efficiency η is divided by the dimensionless resistance coefficient c_D. The resistance coefficient describes the flow resistance of a fiber in the fluid.

$$Q = \frac{\eta}{c_D} = \frac{\rho U_0}{2(1-\alpha)} \frac{-\ln P}{\Delta p} \tag{13.2}$$

In both equations the pressure drop and the penetration determine the quality of the filter media and can be measured via the procedures mentioned above. But the measurement of the filtration properties of a probe of 100 cm^2 delivers the actual value of the quality, which is influenced by the deterioration caused by the structure inhomogeneities. This actual value of Q is always smaller than the ideal quality that would be reached if the given fiber blend could be arranged homogeneously.

The local structure inhomogeneities lead to local fluctuations of the filtration properties. The stronger the inhomogeneities, the stronger the fluctuations and the stronger the difference between actual value and ideal value of the medium quality. Thus, the amount of the fluctuation of the local filtration properties quantifies the deterioration of Q.

If local data of the filtration properties can be measured, both, the actual quality as well as the quality deterioration — caused by the inhomogeneities — are given. A measuring procedure to evaluate the local filtration properties is presented in the following section.

13.3 MEASURING PROCEDURE FOR EVALUATION OF THE LOCAL FILTRATION PROPERTIES

For the investigation, a commercially available glass fiber filter medium was chosen with the technical data as shown below

Name	GFS Mikroglasfaservlies 5772/300 HEPA Quality H 13
Efficiency	99.97% at MPPS; face velocity 2.5 cm/s
Permeability of air	18.6 l/m² s at 100 Pa

This medium was chosen to apply the developed measuring procedure to materials actually in commercial use. The next section discusses the procedure and the results of the measurement of the local filtration properties.

13.3.1 Experimental Setup

The measurement of the local filtration properties permeability (responsible for pressure drop) and penetration were done with one common experimental setup, which is shown schematically in Figure 13.3 (see also Mölter and Fissan, 1994).

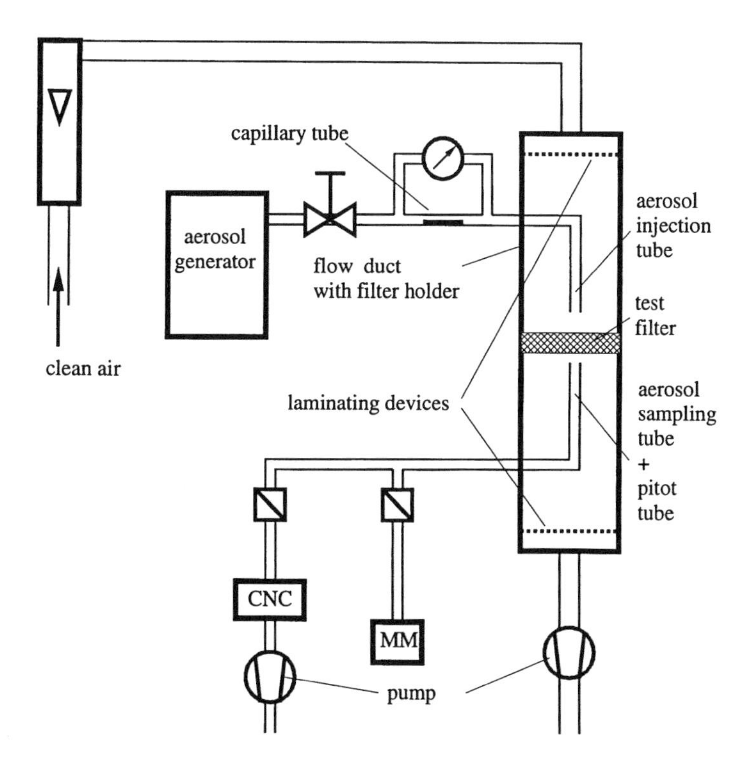

Figure 13.3 Schematic diagram of the experimental setup.

The velocity of the main flow is adjusted by means of the clean air supply. Two laminating devices homogenize the flow. The test filter is positioned in between the laminating devices in a section of well-defined flow conditions.

To determine the local permeability the local flow velocity downstream of the filter is measured by means of a pitot tube and a micromanometer (MM). The measurement of the penetration is done by defined injection (via capillary tube) of test aerosol (particle size 0.15 μm, monodisperse) and simultaneous measurement of the penetrating particle flow via the particle counter CNC. A reconstruction of the setup is not necessary because the probe, which is installed downstream works twice, first as a pitot tube and second as an aerosol sampling tube.

13.3.2 Procedure for Measurement of the Local Flow Velocity

The local fluctuation of the structure leads to local fluctuation of the permeability k in a filter medium and these different permeabilities lead to a fluctuation of the local flow velocity U. The local flow velocities are directly correlated to the structure-dependent permeability. The correlation depends on the fluid used and the correlation parameter is the overall pressure drop Δp. In the field of HEPA filtration, the dependency of pressure drop, permeability, and flow velocity is given by Darcy's law (Equation 13.3; see also Pich, 1985).

$$\Delta p = \frac{\mu}{k} UH \tag{13.3}$$

In Equation 13.3, H describes the sheet thickness of the medium and μ the viscosity of the fluid. As mentioned, the local flow velocity is measured via a kinetic pressure tube consisting of a pitot tube switched to a micromanometer. The kinetic pressure tube is installed downstream of the filter. When installing the tube, one has to consider that the fluctuation of the velocity decreases very rapidly with increasing distance SA (see Figure 13.4) from the filter surface. In the literature, (Kirsch and Stechkina, 1978) this distance of relaxation is estimated at several millimeters. The distance in the experimental setup used here is adjusted to 1 mm.

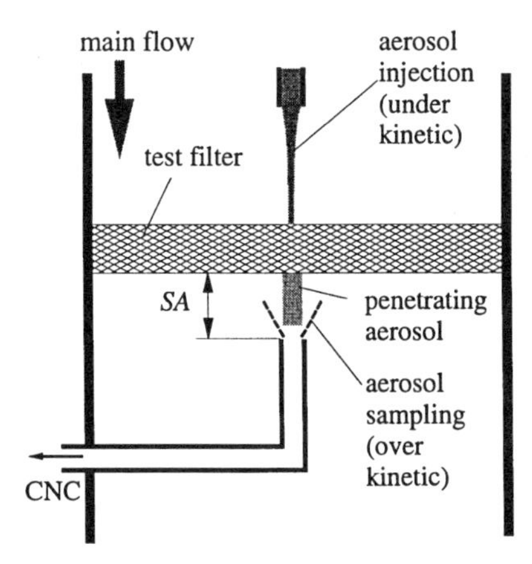

Figure 13.4 Positioning of the various tubes.

In normal application, the fluid flows through the medium at a velocity of about 1 cm/s. The kinetic pressure of such an air velocity amounts to 60 μPa and is not measurable by means of the installed micromanometer. So, the flow velocity was increased to 20 cm/s where the kinetic pressure is expected to be an amount of about 24 mPa which is measurable with sufficient accuracy. Regarding the filtration properties of the medium, the increase of flow velocity is allowed because the linear relationship between flow velocity and medium pressure drop covers this range, too.

13.3.3 Procedure of Measurement of the Local Penetration

To measure the local penetration, the face velocity is reduced to a more realistic value of about 5 cm/s. At this velocity the average penetration differs from that which is expected at a usually adjusted face velocity of about 1 cm/s, but the matter of interest, i.e., the fluctuation of the penetration, is reliably measurable. The reasons for the increased velocity are the transport losses inside the injection tube and the consequent problems of adjustment of such a low aerosol flow rate which would be necessary at face velocities of about 1 cm/s.

The measurement of the penetration is based on the upstream seeding of the airflow with monosized particles combined with the downstream sampling and counting of the penetrated particles. Two different strategies of local measurement are possible.

1. Total upstream challenging of the whole filter area and local sampling of the penetrated particles, where exact isokinetic conditions are strongly required in the sampling system.
2. Local upstream seeding by defined local injection of particles and downstream sampling of all particles that penetrate the filter.

In this experiment the option 2 was chosen for the following reasons:

In the case of a total upstream challenge, the local position has to be defined using exact isokinetic sampling conditions downstream of the filter. Due to the expected local fluctuation of the flow velocity, that is not possible.

In case of the local injection, only the position under measurement is seeded with particles and an unwanted contamination of the neighboring areas is prevented.

The definition of the local position of measurement is given by the position of the aerosol injection tube. The local resolution is defined by the diameter of the injection tube, and, in addition, the resolution can be partially increased by the adjustment of the aerosol flow through the tube. The underkinetic processing of the injection tube focuses and stabilizes the aerosol flow. By adjusting the aerosol flow through the 2-mm tube to a velocity of about 3.5 cm/s, the aerosol jet is focused to a diameter of about 1.5 mm (see Figure 13.4).

To measure the number of penetrated particles, all the particles in the flow downstream of the filter must be sampled. Hence, the free-test-filter area (80 mm in ∅) and the demanded face velocity (5 cm/s) determine the total flow (15 1/min) through the whole duct. But there is no instrument on the market with such a sampling flow rate; thus, to sample the whole penetrated particle flux (spread by the filter) an overkinetic sampling device is required and the used CNC (sampling flow rate, 1.6 l/min) fulfills this requirement.

13.3.4 Measurement of Local Permeability and Penetration

As mentioned, the measurement of the local permeability is done via measurement of the local flow velocity. The local penetration is measured by counting the penetrating particles. The local flow velocities U_j at a specified position j is calculated from the kinetic pressure δp_j and the fluid density ρ applying Equation 13.4.

$$U_j = \sqrt{\frac{2\delta p_j}{\rho}} \tag{13.4}$$

$$P_j = \frac{\dot{n}_j}{\dot{n}_0} \qquad (13.5)$$

The local penetration P_j is given in Equation 13.5 where $\dot{n}_j$ describes the penetrated particle flow and $\dot{n}_0$ describes the flow of the locally injected particles. Ten separate positions were investigated and the results are as follows.

The average local penetration was found to be 0.00051 with a relative standard deviation of about 21%. The average local flow velocity was found to be 18.8 cm/s with a relative standard deviation of about 3%.

The average value of the local penetration was compared with the result from a second experiment, where the penetration through a filtration area of about 100 cm^2 was measured under the same operating conditions and was found to be 0.00053. Thus, from the local measurement of the penetration, the average value describes the actual value of penetration, and the fluctuation could be used to estimate the deviation from the ideal value of quality where the fluctuations are assumed to be zero.

The quantitative assessment of the deviation from the ideal quality caused by the fluctuations is not possible because ideal filter media cannot be produced even under laboratory conditions. Thus, ideal quality is not quantifiable. Nevertheless, media showing lower fluctuations in filtration properties are structured more homogeneously and the quality is closer to the optimum.

As shown, the measurement of the local filtration properties delivers the correct means of the filtration properties, but the assessment of the effect of the inhomogeneities is still incomplete. In addition, the measuring procedure is time-consuming because, to test the medium, the filtration process must be simulated, i.e., all the local positions must be seeded with aerosol particles and the penetrating particles must be counted. Thus, the measurement of a moving medium (e.g., in the paper machine) is nearly impossible, and due to the time expense the off-line application of the measuring procedure is not very practicable either.

To determine the local properties, measuring procedures must be applied that deliver useful local data, and quasi-on-line and time-saving procedures are also necessary. Since the filtration properties depend directly on the structural setup of the filter media, it is possible to investigate the local structure instead of local penetration and local permeability, if some preconditions are observed.

The demand on on-line ability allows only the measurement of structure parameters that are measurable not touching and from the outside. Such external parameters are the local sheet thickness and the local area mass, which are averaged over the local position area and, additionally, over the depth of the media.

From these on-line procedures, no information can be obtained on the real internal distribution of the fibers which determines essentially the filtration properties. But the time-consuming simulation of the filtration process to measure the local filtration properties becomes unnecessary. On the other hand, the usefulness of those external parameters must be tested and verified (see below).

In the following section the procedures given in the literature to measure the local external structure parameters sheet thickness and area mass will be introduced.

13.4 ON-LINE MEASUREMENT OF THE LOCAL EXTERNAL STRUCTURE PARAMETERS

External procedures to quantify the fluctuation of local structure parameters are the optical measurement of the local sheet thickness and the capacitative measurement of the local area mass.

In industrial applications a capacitative method is already used to monitor the constancy of the area mass of filter media with a very rough local resolution. But only a part of the total medium area is monitored, and the local resolution is too low for the assessment of the fluctuations specified here.

On the other hand, quasi-on-line procedures for the measurement of the local external parameters with satisfying resolution of about 1 to 2 mm in diameter d_{loc} (see Figure 13.5) are described in the literature (Adams, 1995).

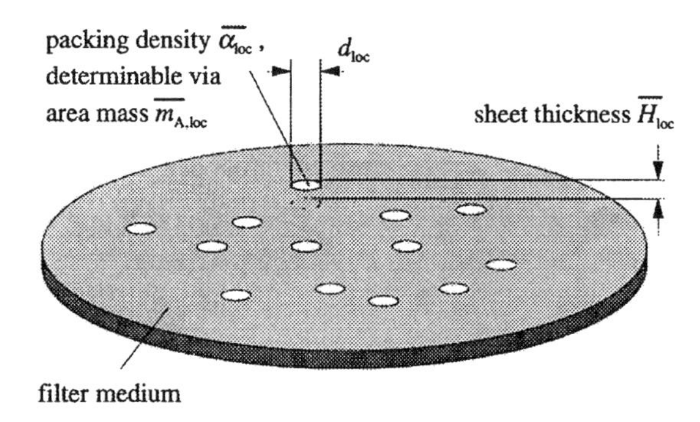

Figure 13.5 Various local positions in a filter medium.

These procedures deliver the local sheet thickness $\overline{H}_{loc}$ and the local area mass $\overline{m}_{A\,loc}$, both averaged over position and additionally over filter depth.

$$\overline{\alpha}_{loc} = \frac{\overline{m}_{A\,loc}}{\rho_f \overline{H}_{loc}} \tag{13.6}$$

Via Equation 13.6 the more important local parameter packing density $\overline{\alpha}_{loc}$ can be evaluated, and $\overline{\alpha}_{loc}$ is an averaged structure parameter as well (ρ_f: specific density of fiber material). Yet, the fiber diameter or the fiber diameter distribution of the investigated medium remains unknown.

13.4.1 Determination of the Local Sheet Thickness

An optical measurement procedure is used for determination of the local sheet thickness (Adam, 1995), which is based on the principle of optical laser triangulation. The surface of the medium sheet is locally illuminated by a laser beam spot and the diffusive reflection of the light is used. A part of the diffuse reflected light is focused to a CCD line detector by means of a diaphragm and the position of the focus at the detector determines the distance to the sheet surface. In a second simultaneously driven sensor on the opposite surface of the medium, the local sheet thickness can be measured with a thickness resolution of about 0.01 mm and a local resolution of about 0.32 mm^2.

13.4.2 Determination of the Local Area Mass

Two different procedures, i.e., a radiometric and a capacitative method, can be applied to evaluate the local area mass (Adam, 1995). The experimental setup for the radiometric measurement consists of a radioactive source (β-radiation, ^{85}Kr) and an appropriate sensitive device (Geiger–Müller counter). The attenuation of the radiation by an obstacle correlates to the area mass of the obstacle. The correlation is based on an exponential relation between the area mass and the radiation power with an without the obstacle. But, due to the stochastic nature of the radioactive source, the power of radiation fluctuates, which influences the accuracy of measurement.

The experimental setup for the capacitative measurement consists of a parallel capacitor of defined geometric size with the filter medium to be investigated in between. The procedure is based on the change of the amount of dielectric material (glass fibers) caused by the fluctuation of the local area mass when the medium is moved between the two plates of the capacitor. The dielectric constant of glass and air differs by a factor of 5; thus, the change of the amount of glass mass leads to a change of the capacity and the local resolution is determined by the area of the capacitor.

The sensitivity of the device is limited by the relatively low response of the capacity to the change of medium position. The dielectric constant of glass and air differs clearly, but only a very small part (5%) of the filter medium volume is occupied by the glass material. The response of the capacity to a specific change of the area mass of about 1 g/m^2 comes out to only 0.45 fF. Nevertheless, regarding the short time of measurement the capacitative method shows clear advantages compared with the radiometric method.

Thus, the mentioned methods are useful for measuring the amount and the local fluctuations of the external, averaged structure parameters and if some preconditions (listed in the next chapter) are fulfilled these local external structure parameters are useful for assessing the local fluctuation of the filtration properties and the quality level of the investigated filter medium.

13.5 PRECONDITIONS FOR APPLICATIONS OF THE EXTERNAL STRUCTURE PARAMETERS IN QUALITY ASSURANCE

If the quality of filter media is to be described by means of averaged, external structure parameters, a homogeneous distribution of the fibers inside the media must be assumed and, for that, three essentials must be fulfilled.

1. The fiber diameter distribution of all local positions of measurement must be nearly constant, because even small variations of the distribution lead to immense variation in filtration parameters even if local sheet thickness and local packing density keep constant.
2. The fluctuation of the sheet thickness within each local position must be relatively small. Otherwise, the filtration properties of the real medium compared with the medium described via averaged structure data are quite different.
3. For the same reason the internal fluctuation of the packing density within each position must be relatively small as well.

To apply the external measuring procedures these three preconditions must be fulfilled and they have to be proved to show the relevance of the procedures for quality assurance. The required basic information on the real internal structure can only be provided by detailed analysis. For that reason an appropriate procedure for analyzing the structure of glass fiber filter media is described in the next section and the results of measurement are presented and discussed.

13.6 INVESTIGATION OF THE QUALITY RELEVANCE OF THE AVERAGED STRUCTURE DATA VIA DETAILED STRUCTURE ANALYSIS

In the following section the employed procedure of analysis is described in detail. The main emphasis is the evaluation of the fractional packing density of the fibers (i.e., packing density vs. fiber diameter), the fiber angle distribution, and the internal fluctuation of the structure parameters.

For example, two local positions of the medium were investigated where the selection is based on the results of the measurement of the local filtration properties (see above). The positions No. 3 and No. 5 were selected; position 3, because local flow velocity and local penetration were equal to the average values, and position 5 because the local penetration at this positions differs clearly from the average.

For structure analysis the medium is potted in resin, sectioned along the selected local positions, and prepared for investigation via a scanning electron microscope (SEM). Figure 13.6 shows the sectioning of the positions in principle, and the selected positions are assigned by means of some neighboring markings.

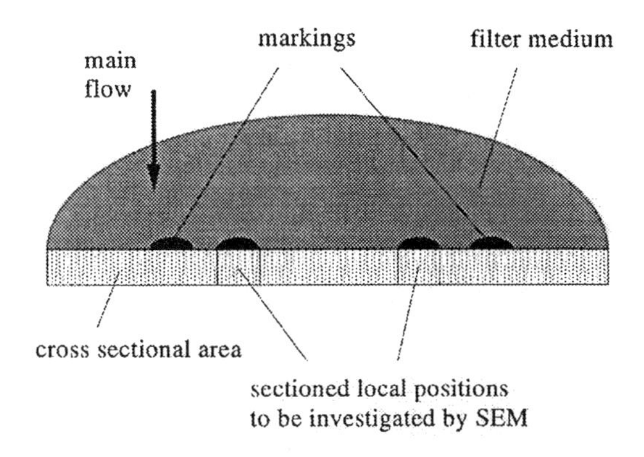

Figure 13.6 Sectioning of the local positions to be investigated.

The principle of measurement is based on the evaluation of the fractions of areas of different components (i.e., resin and glass). From the amount of different components the composition of the filter medium at the specified position can be assessed. To get the real internal distribution of the glass fibers the measured structure data must be related to the internal location unambiguously, even in the depth of the medium. But due to the necessary magnification, only small parts of the cross section are detected per single micrograph and thus the cross sections of the two positions have to be divided in 345 separate but touching single micrographs each. Figure 13.7 shows the arrangement of the single micrographs where they are numbered using the indexes m and n.

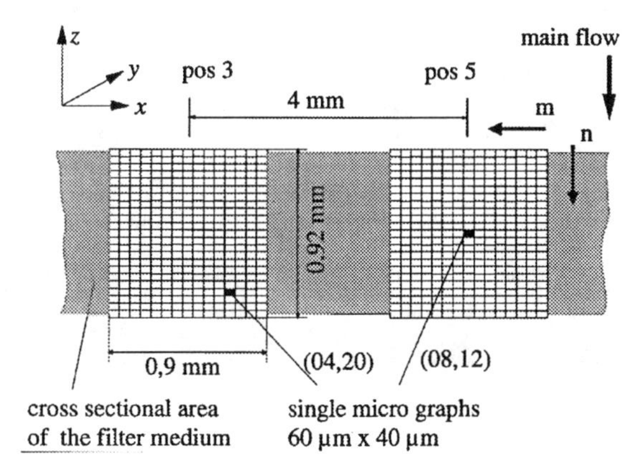

Figure 13.7 Internal arrangement of touching single micrographs.

The cross-sectional area per position is divided in 15 columns and the columns in 23 rows covering the depth of the filter medium. The structure data are obtained from all the micrographs as follows.

13.6.1 Preparation of the Samples for Microscopy

The cross-sectional area is prepared by sectioning the potted medium along the markings followed by grinding and polishing up to the desired level. Before the medium is potted, the fiber must be fixed with some glue vapor to prevent altering the structure (Vaughn and Brown, 1992; Schweers and Löffler, 1993). The final step of polishing is done using 60 nm diamond grinds. A magnification of 2000 for the SEM was chosen. Due to the analysis of the backscatter electrons, a very sharp contrast between glass and resin is given and in case of unknown subjects an analysis of the elements via EDX could be applied.

Figures 18.8 and 13.9 show two electron micrographs taken in this way. The magnification of 2000 was necessary to evaluate even thin fibers of about 0.1 μm, on the one hand, and, on the other hand, to get enough fibers per single graph for statistical reasons. The area of each graph amounts to 60 by 40 μm.

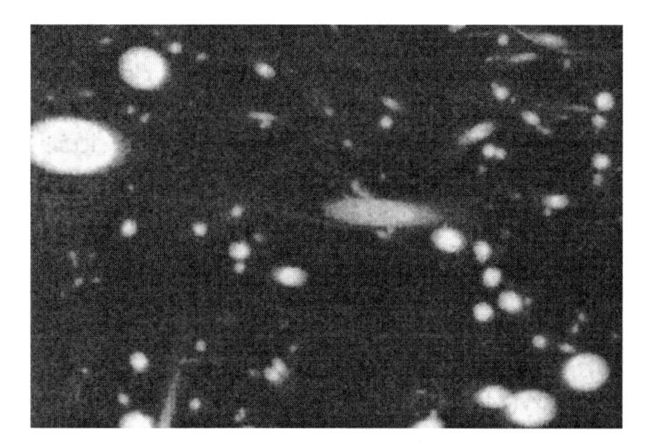

Figure 13.8 Real single micrograph (08.12 of position 5).

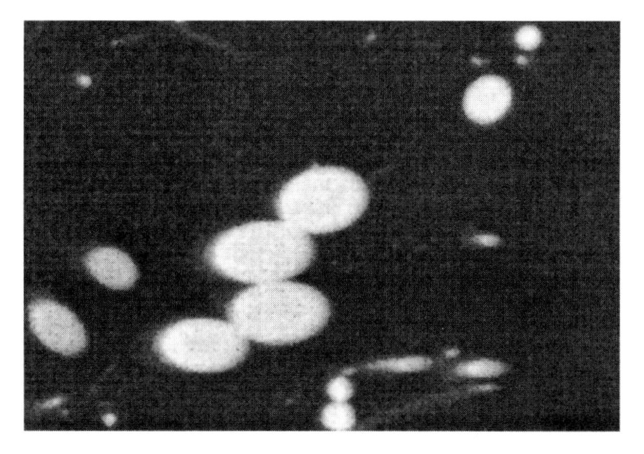

Figure 13.9 Real single micrograph (04.20 of position 3).

13.6.2 Data Analysis

The cross sections of cylindrical fibers are elliptical. The contents of information in the micrographs is the cross-sectional area of each fiber, the aspect ration, the coordinates, and the orientation of the ellipses. From the orientation, the shape and the size all the structure parameters mentioned above can be evaluated.

The evaluation is carried out by an automated image analysis which delivers the long axis and the area of each elliptical element, as well as their orientation and the coordinates related to each single graph. Touching fibers can be separated by the operator by manipulation of the binarized picture.

(1) Evaluation of the Fractional Packing Density

The most important structure parameter, i.e., the packing density of the fibers (fiber volume related to the medium volume), is given by the sum of the cross-sectional areas of all fibers related to the area of the single micrograph.

The packing density obtained in the described way is a local packing density which describes only the properties at the position of measurement. It is assumed that the properties found by the two-dimensional analysis on the cross sections are representative for the real three-dimensional arrangement of the fibers at the related positions.

The fiber diameter distribution of glass fiber filter media is polydisperse. Thus, the evaluation of the fractional packing density is the main priority. To do that, the real diameter of each fiber in the graphs must be measured at first, where the short axis of the ellipses represents the real diameter which can be calculated from the cross-sectional area and the long axis of each fiber. The summation of all cross-sectional areas of fibers belonging to each specified fiber diameter class determines the fractional packing density. A complete description of the procedure of evaluation of the fractional packing density is given by Mölter, 1995.

(2) Evaluation of the Internal Fluctuation of the Structure Parameters

To assess the degree of fulfillment of the preconditions mentioned above (low fluctuation in position-internal sheet thickness and packing density), the measured internal packing density per position can be used. The fluctuation in the height of the 15 columns and the fluctuation in the internal packing density can be obtained from the packing densities of all 345 single micrographs per position.

Regarding the fluctuation of the internal packing densities, one has to consider that two independent phenomena are responsible — first, the relevant fluctuation, i.e., the fluctuation caused by the fluffs produced during production, and, second, a certain basic fluctuation, caused by the random nature of the fiber arrangement, which can be described via a Poisson distribution of the fibers in an idealized (homogeneous) nonwoven.

The measured fluctuation contains both the relevant and the random fluctuation. Thus, to filter out the relevant fluctuation, the random fluctuation must be known. In the case of some basic assumptions concerning the fiber arrangement (e.g., perpendicular impinging angle and random azimuth angle, see below), the extent of fluctuation can be calculated. The relative standard deviation for internal packing density in the case of random distribution of fibers depends on the size of the single micrographs Δl_x and Δl_z and is given in Equation 13.7 (further details are given in Mölter, 1995).

$$\sigma_{\bar{\alpha}} = \sqrt{\sum_i \left[\frac{d_{f,i}^2}{\Delta l_x \Delta l_z} \frac{\pi}{4} \, 1.27 \left(0.5 \ln \frac{\Delta l_x}{d_{f,i}} + 1 \right) \bar{\alpha}_i \right]} \tag{13.7}$$

This relative standard deviation belongs to the systematic fluctuation, and the fluctuations caused by the fluffs interfere. Thus, to obtain the fluctuations caused by those fluffs, the difference between the measured standard deviation and the standard deviation calculated via Equation 13.7 must be extracted.

(3) Determination of the Impinging Angle and the Azimuth Angle

To calculate the filtration properties as a function of the structure parameters (see below), the method of choice is to apply given filtration models. The selection is based on the arrangement of the fibers, which was used to develop the model. The arrangement itself is described via the distributions of both, the inclination angle θ and the azimuth angle φ related to the main flow direction. The inclination angle is equal to the angle of impinging. Figure 13.10 shows the definition of the two basic angles, where the filter surface is parallel to the x-y plane and the main flow direction is from the top to the bottom ($-z$ direction).

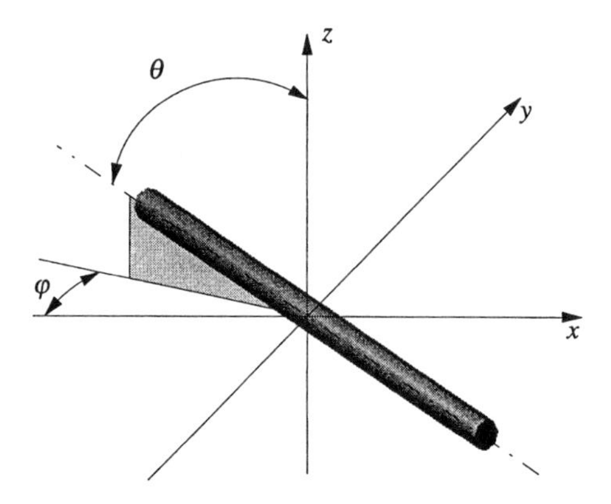

Figure 13.10 The definition of the orientation of a fiber.

As mentioned, the potted media are sectioned perpendicular to the filter surface, i.e., the x-z plane in Figure 13.10. Thus, the normals of the filter surface and the cross-sectional area are perpendicular to each other.

In the micrographs the orientation of the sectioned fibers is given and can be evaluated via the image analysis software. The azimuth angle in the cross-sectional area φ^* is given directly and the inclination angle in the cross-sectional area θ^* can be calculated from the long and the short axis of the elliptically shaped fiber sections. Figure 13.11 shows the definition of the relevant parameters.

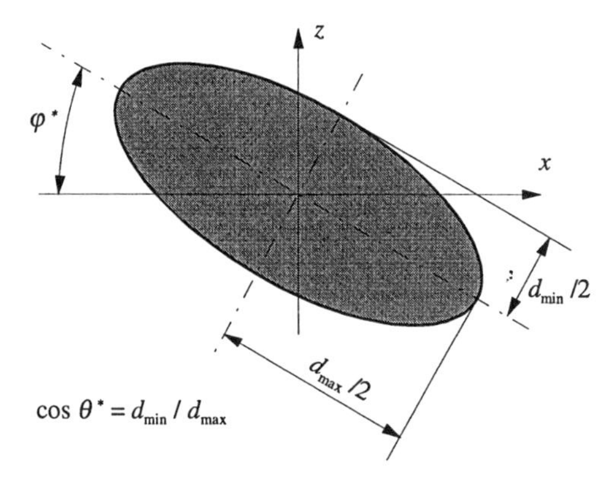

Figure 13.11 The definition of the angles related to the cross-sectional area.

There is a clear relationship between the angles θ^* and φ^* in the cross-sectional area and the angles θ and φ in the plane of the filter medium, and it is given via the eulerian angle definition (Bronstein-Semendjajew, 1976) (see Mölter, 1995 for details).

13.6.3 Results of Measurement and Discussion

The investigation of the local structure was done at the two selected positions of the test medium. By using an electron microscope 345 single micrographs per position were taken. The cross-sectional area was split into 23 × 15 single pictures, in main flow and in perpendicular direction, respectively. Thus, an area of about 0.92 × 0.90 mm per position was analyzed. This covers both the total thickness of the medium and the necessary local resolution (size of fluffs) (Adam, 1995). The average number of fibers per single micrograph was found to be approximately 70 at position 5 and approximately 50 at position 3.

(1) The Local Fractional Packing Density

The fractional packing densities of all the 690 single micrographs were obtained, where the range of fiber diameter was split into 64 classes between 0.06 and 10 μm. As an example, the Figures 13.12 and 13.13 show the results for the micrographs 08.12 of position 5 and 04.20 of position 3, respectively.

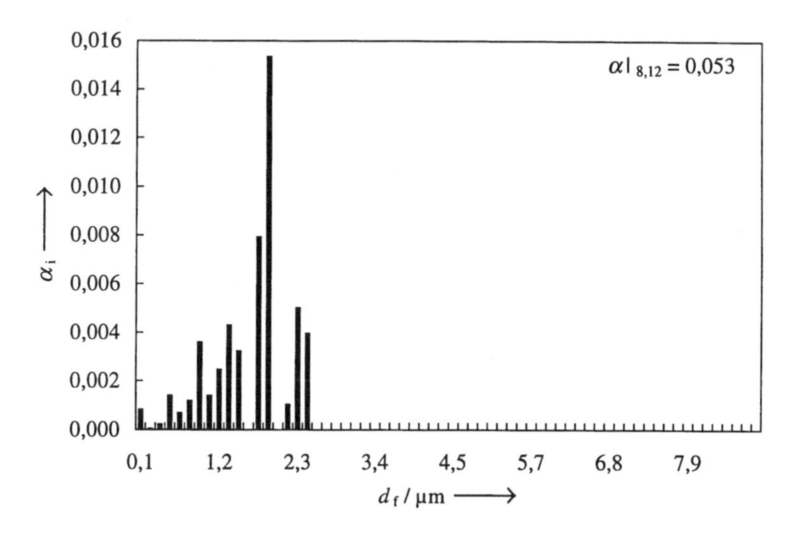

Figure 13.12 Fractional packing density of the micrograph 08.12 of position 5.

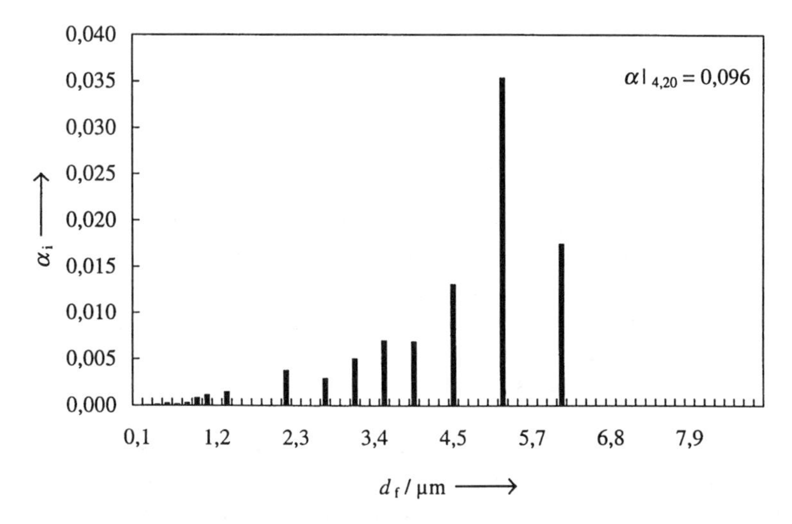

Figure 13.13 Fractional packing density of the micrograph 04.20 of position 3.

The large amount of data can be reduced depending on the goals of the investigation. For example, the two-dimensional distribution of the internal packing densities per micrograph gives an impression of the internal fluctuation within each position. Figure 13.14 shows the frequency distribution of the packing densities of position 5. The distribution of position 3 (not shown in the figure) shows nearly the same fluctuation, but the average packing density of position 5 is definitely higher. The average packing densities of positions 3 and 5 were found to be 0.0405 and 0.0558, respectively, and the respective standard deviations amount to 0.0277 and 0.0335. In comparison, the average packing density of the investigated medium in total was found in good agreement to be 0.0483.

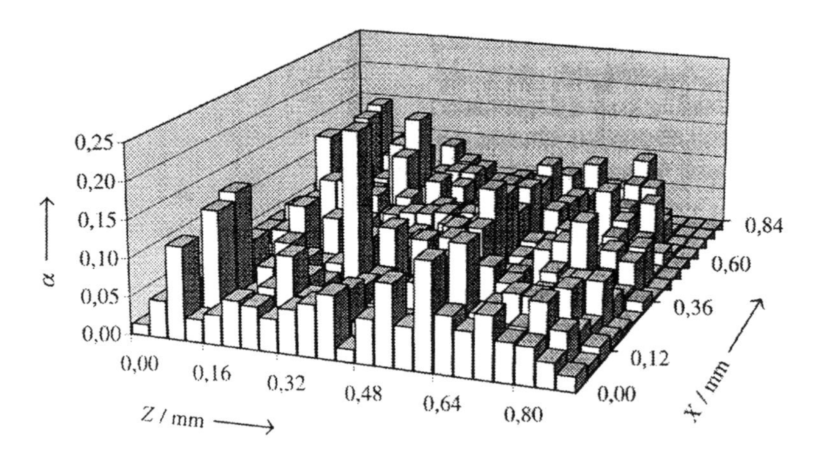

Figure 13.14 Frequency distribution of the packing density in the *x-z* plane of position 5.

A further possibility of reducing the large amount of data is provided by averaging the fractional packing densities of all the single micrographs per position where the splitting in diameter classes is retained. This leads to the most important structure parameter, the local fiber diameter distribution based on fiber mass.

Figure 13.15 shows the local fractional packing density of position 5. The comparison with those of position 3, shown in Figure 13.16, leads to the conclusion that the local fractional packing densities of the two investigated positions are, even in the case of different total packing densities, nearly constant. Thus (if statistical problems are neglected), a constant fiber diameter distribution all over the medium can be assumed.

(2) The Internal Fluctuation in the Structure Parameters

Table 13.1 shows the average values as well as the relative standard deviations of the local structure parameters sheet thickness and packing density. The internal fluctuation in the local sheet thickness of positions 3 and 5 were found to be relatively small, i.e., 3.7 and 6.4%, respectively.

The measured relative standard deviations of the local packing density were found to be 68% at position 3 and 60% at position 5. These experimentally obtained standard deviations include the relevant fluctuations due to the fluffs, as well as the fluctuations due to Poisson statistics of a homogeneous medium.

The latter were calculated via Equation 13.7 and are listed in Table 13.1, as well, and the differences between measured and calculated relative standard deviations are relatively small. Thus, the fluctuations in the packing density caused by the fluffs are also relatively small.

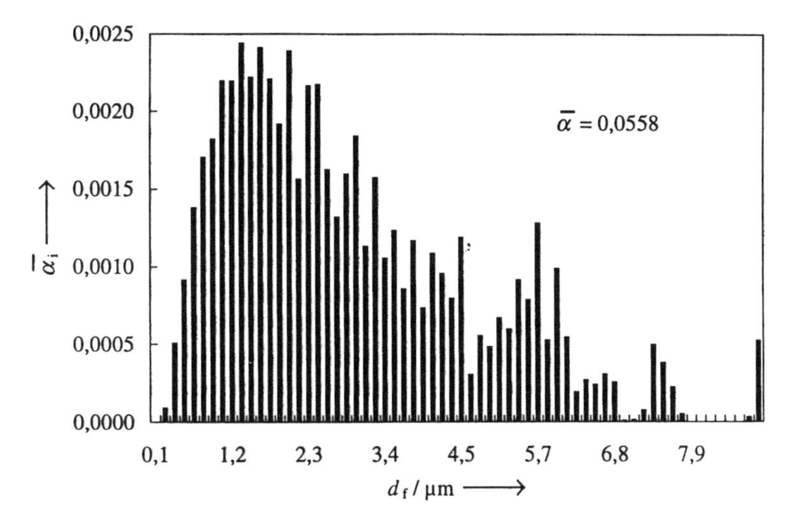

Figure 13.15 Averaged fractional packing density of position 5.

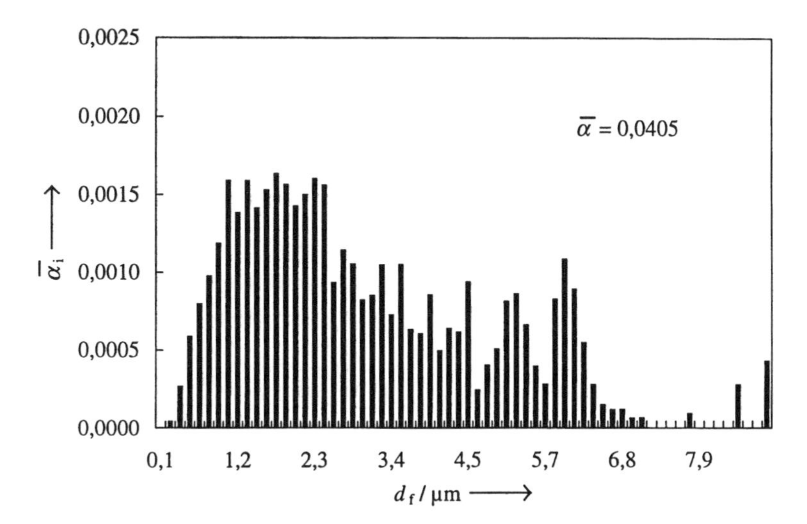

Figure 13.16 Averaged fractional packing density of position 3.

Table 13.1 Average Values and Relative Standard Deviation of the Local Structure Parameters

	Position 3	Position 5
Sheet thickness $\bar{H}_{loc}$	0.837 mm	0.825 mm
$\sigma_{H,rel}$	3.7%	6.4%
Packing density $\bar{\alpha}_{loc}$	0.0405	0.0558
$\sigma_{\alpha,rel}$ (measured)	68%	60%
$\sigma_{\alpha,rel}$ (according to Equation 13.7)	55%	46%

(3) The Angle Distributions

The results of the determination of the fiber angle distributions at the positions 3 and 5 are presented in Figures 13.17 and 13.18. Figure 13.17 shows the distribution of the impinging angles and Figure 13.18 the distribution of the azimuth angles.

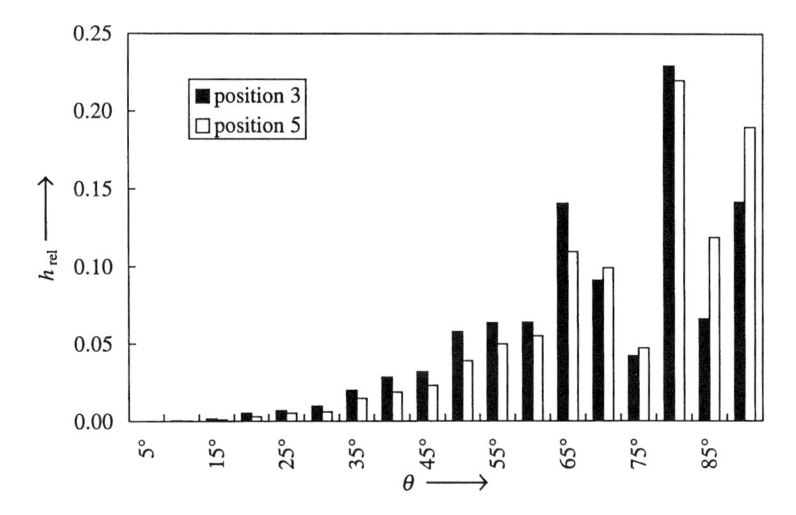

Figure 13.17 Frequency distribution of the impinging angles of positions 3 and 5.

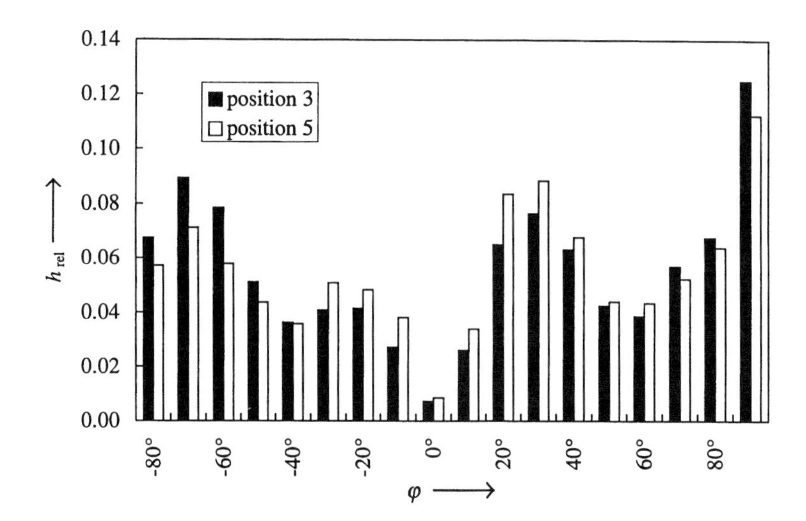

Figure 13.18 Frequency distribution of the azimuth angles of the positions 3 and 5.

From Figure 13.17 it can be seen clearly that most of the fibers are impinged at an angle between 60° and 90°. The minima at 70° and 85° are systematic errors and caused by the 15° resolution of the image analysis concerning the angle φ^*.

Regarding the azimuth angle distribution (Figure 13.18) there is a minimum at an angle of 0°. This minimum is caused by the fact that fibers of 0° azimuth are sectioned parallel to their length. They appear as very long-shaped ellipses with a high probability of crossing the edge of the micrographs, so the determination of these angles is not possible.

The maximum at the angle of 90° is caused by the very thin fibers which cannot be resolved as ellipses. So they are detected as circles, and circle-shaped elements are classified to have 90° azimuth angle. When these systematic errors are recognized, the azimuth angle distribution can be regarded as random. Thus, the result of the measurement of the angle distributions of the investigated medium leads to the conclusion that most of the fibers are impinged perpendicular and there is no other preferred fiber orientation.

13.6.4 Summary of the Structure Analysis with Respect to Quality Relevance of External Structure Data

The question of whether or not, in the case of the investigated medium, the external structure data are significant, i.e., whether or not they fulfill the above-mentioned preconditions, has been the motivation for analysis of the structure in detail. The results can be summarized as follows:

1. The demand of a constant fiber diameter distribution is fulfilled at the two investigated positions. The fractional packing density is nearly the same even in the case of differing averaging packing densities. A correlation coefficient of about 0.92 was found which denotes a highly significant correlation (Taylor, 1988).
2. The precondition to have only small fluctuations in the sheet thickness per position is fulfilled. The relative standard deviations were found to be 3.7 and 6.4% at the positions 3 and 5, respectively.
3. The precondition to have only small fluctuations in the internal packing density per position is fulfilled, as well. The relevant relative standard deviations caused by the fluffs are relatively small compared with the relative standard deviation caused by random arrangement of fibers in an idealized medium. The differences between measured and calculated standard deviations amount to 13 and 14 points of percent at the positions 3 and 5, respectively.

Thus, in case of the investigated medium it could be shown that the external procedures for the measurement of the averaged local structure data (sheet thickness and packing density) are relevant concerning the quality of filter media.

But this procedure of structure analysis is a destructive and time-consuming method. Therefore, it is not usable for on-line measurements and thus not applicable for quality assurance. The proof of the quality relevance of the external, on-line measuring procedures must be furnished in another way and the next section presents a concept developed for this purpose.

13.7 INVESTIGATION OF THE QUALITY RELEVANCE OF THE AVERAGED STRUCTURE DATA VIA ADDITIONAL MEASUREMENT OF THE LOCAL FILTRATION PROPERTIES

13.7.1 The Filtration Properties as Functions of the Averaged Structure Data

In the literature different idealized arrangements of fibers are considered in various filtration models. A survey of current models is given in Brown (1993) and Lücke (1944). The analysis of the fiber arrangement of the investigated medium showed that the majority of the fibers is impinged perpendicularly and no other dominating orientation was found. Thus, the FMF model (fan model filter) of Kirsch and Stechkina (1978) should be used for further calculations because this model was developed under the assumption of this idealized fiber arrangement.

In the case of real glass fiber filter media with polydisperse fiber diameter distributions, the developed equations have to be extended by summing up the contributions of all fiber diameter classes i to the permeability (pressure drop) and efficiency of the filter medium. This leads to the following Equations 13.8 and 13.9:

$$\Delta p = \frac{4\mu U_0 \overline{H}}{\pi} \sum_i \frac{F_i^* \overline{\alpha}_i}{d_{f,i}^2} \tag{13.8}$$

$$P = \exp\left(-\frac{4\overline{H}}{\pi(1-\overline{\alpha})} \sum_i \frac{\eta_i \overline{\alpha}_i}{d_{f,i}} \right) \tag{13.9}$$

In Equations 13.8 and 13.9 the internal fluctuation is neglected by using the average sheet thickness $\overline{H}$ and the average packing density $\overline{\alpha}$. The goal is a description of the filtration properties through external structure data only.

The permissibility of this neglect is investigated later on when the concept of validation is discussed. (The consideration of the internal fluctuation is described in Mölter, 1994, and will not be discussed here.) The application of Equations 13.8 and 13.9 is based on accurate data of the fiber diameter distribution, yet the procedure of measurement is very tedious and time-consuming. For that reason a so-called effective fiber diameter will be introduced in the next section which takes the contribution of the fiber diameter distribution into account.

13.7.2 Effective Fiber Diameter

The effective fiber diameter d_f^* is defined here as the overall effective diameter (compare the concept of the equivalent diameter in aerosol technology). Thus, d_f^* is the diameter of monodisperse fibers, which show the same filtration properties as the real polydisperse fibers.

The fiber diameter–dependent parameters in Equations 13.8 and 13.9 must be modified as follows:

$$\sum \frac{F^*(d_{f,i}) \cdot \overline{\alpha}_i}{d_{f,i}^2} \Rightarrow \frac{F^*(d_f^*) \cdot \overline{\alpha}}{d_f^{*2}}$$

$$\sum \frac{\eta(d_{f,i}) \cdot \overline{\alpha}_i}{d_{f,i}} \Rightarrow \frac{\eta(d_f^*) \cdot \overline{\alpha}}{d_f^*}$$

These equations are valid for media which are not influenced by local fluctuations of the external structure parameters. But in the case of real media, i.e., local inhomogeneous media, the local structure parameters must be taken into account. The local inhomogeneities cause fluctuation in the local permeability, which leads to fluctuation in the local flow velocity describe the effect of local fluctuation in external structure parameters on the local filtration properties.

$$U_{\text{loc}} = \frac{\Delta p \pi}{4\mu \overline{H}_{\text{loc}}} \frac{d_{f,\text{loc}}^{*2}}{\text{F}^*\left(d_{f,\text{loc}}^*\right)\overline{\alpha}_{\text{loc}}} \tag{13.10}$$

$$P_{\text{loc}} = \exp\left(-\frac{4\overline{H}_{\text{loc}}}{\pi\left(1-\overline{\alpha}_{\text{loc}}\right)} \frac{\eta\left(d_{f,\text{loc}}^*\right) \cdot \overline{\alpha}_{\text{loc}}}{d_{f,\text{loc}}^*} \right) \tag{13.11}$$

In Equations 13.10 and 13.11 local flow velocity and local penetration are shown, on the one hand, as functions of the external local structure parameters sheet thickness and packing density and, on the other, as functions of the local effective fiber diameter $d^*_{f,\text{loc}}$, which takes into account the contribution of the fiber diameter distribution as well as any influence exerted by internal structure fluctuation at the pertinent local positions.

Here a procedure was introduced to measure the local filtration properties — local flow velocity and local penetration. Moreover, a procedure to measure the external local structure parameters is also given in the literature, but to apply these methods in quality assurance the external parameters must be validated. A validation model, which is based on the effective fiber diameter, is presented in the following.

13.7.3 Validation Model

This model is based on the application of Equations 13.10 and 13.11, which may be formulated as follows:

$$U_{\text{loc}} = f\left(\overline{\alpha}_{\text{loc}}, \overline{H}_{\text{loc}}, d^*_{f,\text{loc}}\right) \tag{13.12}$$

$$P_{\text{loc}} = f\left(\overline{\alpha}_{\text{loc}}, \overline{H}_{\text{loc}}, d^*_{f,\text{loc}}\right) \tag{13.13}$$

If it can be shown that the local effective fiber diameter remains constant throughout the filter medium, then and only then are local filtration properties exclusively a function of local external structure parameters. In this case, Equations 13.12 and 13.13 can be reduced to

$$U_{\text{loc}} = f\left(\overline{\alpha}_{\text{loc}}, \overline{H}_{\text{loc}}\right)$$

$$P_{\text{loc}} = f\left(\overline{\alpha}_{\text{loc}}, \overline{H}_{\text{loc}}\right)$$

Hence, to validate external structure parameters effective fiber diameter must be determined and examined for constancy.

Using Equations 13.10 and 13.11, local effective fiber diameter can be calculated iteratively if the local structure parameters sheet thickness and packing density as well as the accompanying local filtration properties are known.

But measuring local filtration properties is more time-consuming than measuring local structure parameters. For that reason the following procedure for applying the model is recommended:

1. Measurement of the local external structure parameters of the entire area of investigation;
2. Measurement of the local filtration properties at several positions;
3. Calculation of the effective fiber diameters at those positions and checking for constancy.

Figure 13.19 makes this procedure clearer. Note that the procedure has to be applied only once for each sort of medium characterized by the same fiber blend and the same adjustment of the paper mill.

Based on the determination of $d^*_{f,\text{loc}}$, the quality relevance of the external structure parameters can be assessed. If $d^*_{f,\text{loc}}$ is not constant, the external parameters provide quantitative data on the local fluctuation in the fiber mass but not in the filtration properties. In this case, the local filtration properties can only be estimated using additional information usually based on experience.

However, if the effective fiber diameter is constant, it has been proved that local measurement of the average sheet thickness and the average packing density is sufficient to assess fluctuation in the filtration properties and, finally, to assess the grade of deterioration of filter medium quality caused by the structure inhomogeneities.

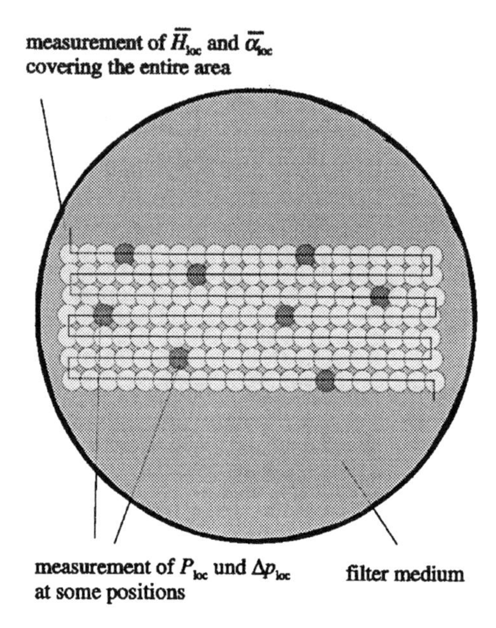

Figure 13.19 The measurement and validation of the local external structure parameters.

13.8 CONCLUSION

The assessment and quantification of structural inhomogeneities are an important part of quality assurance for filter media for high-efficiency air filtration. Structure inhomogeneities occur during the manufacturing process, caused by incomplete fiber dispersion and fluffs. They lead to local fluctuations in filtration properties, which causes (due to their nonlinearity) a deterioration in the quality of the filter media.

To assess the deterioration of the quality a measurement procedure was introduced to determine the local fluctuations in the filtration properties permeability and penetration, which are responsible for the quality of the filter medium. The average values of the local filtration properties describe the actual quality, and the standard deviations describe the deterioration cause by the inhomogeneities. However, these methods are not very practicable in quality assurance; in other words, on-line methods have to be applied.

In the literature, on-line methods were presented which can be used as an alternative to determine local fluctuations in the external structure parameters sheet thickness and packing density. These integral data are averaged over the area of the local position and the depth of the filter medium. For that reason they describe the local filtration parameters only if a homogeneous distribution of the fibers at the relevant local position can be assumed. The external structure parameters have relevance for quality only if this precondition is fulfilled.

In practice, the quality relevance must be shown only once for each type of media and is continuously valid if fiber blend and manufacturing parameters are kept constant.

In the case of the media which were investigated here the mentioned preconditions have been tested in detail by means of an extended procedure of structure analysis. The procedure is based on the image analysis of electron micrographs of the cross-sectional areas of potted media and delivers the essential internal structure parameters, such as fiber diameter distribution, fiber angle distribution, and the geometric distribution inside the medium.

The method was applied to two different local positions of the investigated medium, and from the results it can be seen that in case of the investigated medium the fiber diameter distribution (averaged over sheet thickness) keeps nearly constant even if the local packing density differs considerably. Further, it is shown that the internal fluctuation of the sheet thickness and packing density (caused by fluffs) is relatively small within the investigated positions.

But for practical reasons, the application of this detailed analysis method to quality assurance is not usable. So a concept has been developed to test the quality relevance of the external structure data by means of time-saving methods that are applicable from outside the medium exclusively. The concept is based on the effective fiber diameter, and relevance exists if this effective diameter keeps constant and independent of local positions of the filter medium. The overall concept covers, first, the extensive local measurement of the external structure parameters and, second, the validation of these parameters by means of the determination of the local effective fiber diameter. The check of the effective diameter is possible by additional measurement of the filtration properties at some pertinent local positions and the required relevance of the on-line procedures to measure the external structure parameters can be monitored.

The time-consuming and tedious measurement of the real internal structure and of the fiber diameter distribution is no longer necessary in case of a constant effective fiber diameter. The assurance of quality is possible in practice even when quality deterioration by inhomogeneities is taken into account.

NOMENCLATURE

C_D	resistance coefficient
d	diameter of local position
d_f	fiber diameter, m
d_f^*	effective fiber diameter, m
d_{max}	long axis of ellipse, m
d_{min}	short axis of ellipse, m
d_p	particle diameter, m
k	permeability, m^2
$\overline{m_A}$	average area mass, kg/m^2
$\dot{n}$	particle flow, 1/s
$\dot{n}_0$	upstream particle flow, 1/s
F*	dimensionless resistance force
H	sheet thickness, m
$\overline{H}$	average sheet thickness, m
P	penetration
Q	quality factor
U_0	face velocity, m/s
α	packing density
$\overline{\alpha}$	average packing density
η	single-fiber efficiency
φ	angle of azimuth related to medium area, °
φ^*	angle of azimuth related to cross-sectional area, °
θ	angle of inclination related to medium area, °
θ^*	angle of inclination related to cross-sectional area, °
μ	viscosity of fluid, Pa s
ρ	density of fluid, kg/m^3
ρ_f	density of fiber material, kg/m^3
σ_α	standard deviation of packing density
σ_H	standard deviation of sheet thickness
Δl_x	length of micrograph border in x-direction, m
Δl_z	length of micrograph border in z-direction, m

Δ_p pressure drop, Pa
δp kinetic pressure, Pa

Indexes

i class of fiber diameter
j position number
loc local position
m indication of position in z-direction
n indication of position in x-direction
x,y,z Cartesians

REFERENCES

Adam, R., Kennzeichnung inhomogener Filter für die Schwebstoffiltration, Ph.D. thesis, Technische Universität Dresden, 1995.

Bronstein, I. and Semendjajew, K., *Taschenbuch der Mathematik,* 16th ed., Harry Deutsch, Zürich, 1976.

Brown, R. C., *Air Filtration, An Integrated Approach to the Theory and Application of Fibrous Filters,* Pergamon Press, Oxford, 1993.

Kirsch, A. A. and Stechkina, I. B., The theory of aerosol filtration with fibrous filters, in *Fundamentals of Aerosol Science,* D. T., Ed., Wiley, New York 1978, 165–256.

Lücke, T. Ph.D. thesis, Filtrationseigenschaften inhomogener Filtermedien für die Schwebstoffiltration, Technische Universität Dresden, 1994.

Lathrache, R., Einführung in die Filtertheorie, in *Seminar: Praxisorientierte Überprüfung von Faserfiltern,* Palas, Karlsruhe, 1987.

Mölter, W., Inhomogenitäten in Schwebstoffiltermedien und ihre Auswirkungen auf die Filtrationseigenschaften, Ph.D. thesis, Gerhard-Mercator-Universität, Duisburg, 1995.

Mölter, W. and Fissan, H., Zur Qualitätssicherung von HEPA/ULPA — Medien durch Prüfung lokaler Eigenschaften, in *Aerosol Technologie Seminar,* Fa. Palas, Karlsruhe, 1994.

Pein, J., Glasfaserpapiere für Luftfilter, in *Seminar: Luftfilter im industriellen Einsatz,* Haus der Technik, Essen, 7.4, 1992.

Pich, J., Permeabilities of paper filters, *Staub Reinhalt. Luft,* 45, 222–224, 1985.

Schweers, E. and Löffler, F., Analyse der Struktur technischer Tiefenfilter, *Staub Reinhalt. Luft,* 53, 101–107, 1993.

Taylor, J. R., *Fehleranalyse,* VCH Verlagsgesellschaft, Weinheim, 1988.

Vaughan, N. P. and Brown, R. C., Measurement of filter structure, *J. Aerosol Sci.,* 23(Suppl. 1), S741–S744, 1992.

CHAPTER 14

The Lung as a Filter for Inhaled Aerosols — Particle Deposition Patterns in Human Airway Bifurcations

Werner Hofmann, Imre Balásházy, and Thomas Heistracher

CONTENTS

14.1 INTRODUCTION

Health effects of inhaled particulate matter in the human lung depend on the total number or mass of particles deposited in bronchial and alveolar airways during a full breathing cycle (ICRP, 1994). A characteristic feature of most toxicologic or therapeutic effects of inhaled environmental or medical aerosols in the human lung is their apparent site selectivity (Schlesinger and Lippmann, 1978). For example, spontaneously occurring and radiation-induced lung cancers have been observed primarily in segmental and subsegmental bronchial airways (Spencer, 1985; Yamamoto et al., 1987), possibly originating at bronchial airway bifurcations (Hofmann et al., 1991). This site selectivity of biological response in the lungs is caused by the superposition of heterogeneous deposition patterns with heterogeneous target cell distributions. Thus, any assessment of biological response to inhaled particles has to be based on local deposition patterns in bronchial and alveolar airways rather than on total deposition (Martonen and Hofmann, 1986; Martonen, 1991).

While total and regional (i.e., extrathoracic, bronchial, and alveolar) deposition in human test subjects can be studied by various experimental methods (Stahlhofen et al., 1989; ICRP, 1994), it is at present not possible to measure particle deposition in individual airways *in vivo*. Although some limited experimental information about local deposition can be derived from measurements in bronchial airway casts (Schlesinger and Lippmann, 1978; Cohen and Asgharian, 1990), upper airway models (Martonen et al., 1987; Martonen, 1991), and single bifurcation models (Kim and

0-87371-830-5/98/$0.00+$.50

Iglesias, 1989; Kim et al., 1989), deposition in individual airways can only be estimated by mathematical modeling. To model local deposition of inhaled particles in the human lung, the respiratory tract is treated as a series of filters through which aerosol particles pass during inspiration and expiration. This approach requires information about the deposition probability in each filter as a function of particle parameters and flow rate.

In the present study, the human airway filter system is represented by a serial succession of Y-shaped airway bifurcations, each consisting of a parent branch and two daughter branches. The deposition probabilities of inhaled particles in such an anatomic filter are obviously affected by the nature of the fluid motion in which they are entrained. For idealized regular airway geometries, such as straight or bent cylindrical tubes, and idealized flow profiles, such as uniform or parabolic laminar flow, the equations of particle motion and deposition by a specific physical mechanism can be solved analytically (Ingham, 1975; Balásházy et al., 1990; 1991). However, the prediction of particle deposition patterns in an anatomically realistic airway bifurcation geometry under physiologically realistic flow conditions, while accounting for the simultaneous action of different physical deposition mechanisms, requires the application of numerical techniques (Gradon and Orlicki, 1990; Lee and Goo, 1992; Balásházy and Hofmann, 1993a; Asgharian and Anjilvel, 1994; Li and Ahmadi, 1995).

The simulations of particle deposition patterns in bronchial airway bifurcations presented here are based on our recently developed numerical model for the calculation of airflow velocities and particle trajectories in three-dimensional bifurcation models (Hofmann and Balásházy, 1991; Balásházy and Hofmann, 1993a; 1995; Balásházy, 1994; Hofmann et al., 1995; Balásházy et al., 1996). First, airflow is computed by solving the Navier–Stokes equations with finite-difference or finite-volume methods. Next, trajectories of aerosol particles entrained in the airstream under the simultaneous effects of Brownian motion, inertial impaction, gravitational setting, and interception are simulated by Monte Carlo methods. Finally, particle deposition patterns are determined by the intersection of individual particle trajectories with the surrounding walls of the bifurcation.

Deposition in a bifurcation filter depends on the bifurcation geometry, airflow rate, and particle size. Thus, theoretical simulations of inspiratory and expiratory particle deposition patterns are presented here for different bifurcation geometries, airflow rates, and particle sizes. These predictions are used (1) to determine the deposition efficiency in such anatomic filters, (2) to illustrate the inhomogeneity of particle deposition within airway bifurcations, and (3) to quantitate this inhomogeneity by defining local deposition densities and enhancement factors.

14.2 AIRFLOW PATTERNS IN BRONCHIAL BIFURCATIONS

Upon inspiration and expiration, particles are transported by air through the human airway system. Hence, it is necessary to know the fluid dynamics patterns within the airways of the human lung to understand the factors affecting the motion of inhaled particles. Airflow patterns within the lung are determined by a complex interplay of structural features (i.e., individual airway dimensions and the spatial configuration of the branching network) and ventilatory conditions (i.e., tidal volumes and breathing frequencies), as demonstrated in several experimental (Pedley et al., 1977; Chang and Masry, 1982; Isabey and Chang, 1982) and theoretical (Balásházy and Hofmann; 1993a; Martonen et al., 1993; 1994; Hofmann et al., 1996) studies.

In the three-dimensional airway bifurcation model proposed by Balásházy and Hofmann (1993a), the parent and daughter branches are represented by straight cylindrical tubes, joined together by a central transition zone. Depending on the width of this transition zone, two different bifurcation models have been defined: an idealized "narrow" bifurcation model (generally used in deposition experiments) and a biologically more realistic "wide" bifurcation model. In both bifurcation models, the carina (i.e., the dividing spur) is modeled by a sharp wedge. To approximate the real shape of bronchial airway bifurcations, Heistracher and Hofmann (1995) constructed a "physiologically realistic bifurcation" (PRB) model, in which the parent and daughter airways are connected by a smooth transition zone and the carinal ridge may adopt the different shapes observed in anatomic studies.

For simulations of airflow patterns in the airway bifurcations, the fluid dynamics equations can only be solved by numerical methods. For example, the results of our flow simulations presented in subsequent figures are based on the solution of the three-dimensional Navier–Stokes equations by the finite-volume program package FIRE® (Heistracher and Hofmann, 1995). The resulting three-dimensional flow field is characterized by the velocity vectors at the grid points of the three-dimensional computational mesh. For visualization, the three-dimensional flow is commonly represented by the primary (i.e., axial) and secondary (i.e., radial) velocity profiles.

The effects of bifurcation geometry, flow rate, and breathing mode on the primary and secondary flow profiles are illustrated in Figures 14.1 through 14.4. The diameters and lengths of the parent and daughter branches of the bifurcation models used here correspond to the airway generation 3–4 junction in Weibel's (1963) symmetric Model A (*note:* the trachea is denoted as generation 0) Primary flow profiles are displayed by isolines in the plane of the bifurcation and secondary flow profiles are represented by velocity vector plots in defined cut-planes. To quantitate their magnitude, the secondary motions are characterized by a secondary motion intensity factor (SMIF), which is defined as the ratio of the maximum secondary velocity in a given cut-plane to the average velocity in the parent branch (Heistracher and Hofmann, 1995).

Primary and secondary flow patterns in the symmetric narrow bifurcation model with a 35° branching angle are shown in Figure 14.1 for inspiratory parabolic inlet flow. The tracheal flow rate of 60 l min^{-1} corresponds to light physical exercise breathing conditions. The distinct asymmetry in the primary flow profile downstream of the central bifurcation zone as well as the strong secondary motions directed toward the carinal ridge suggest that particles will be deposited preferentially at the carinal ridge and at the inner parts of the daughter branches.

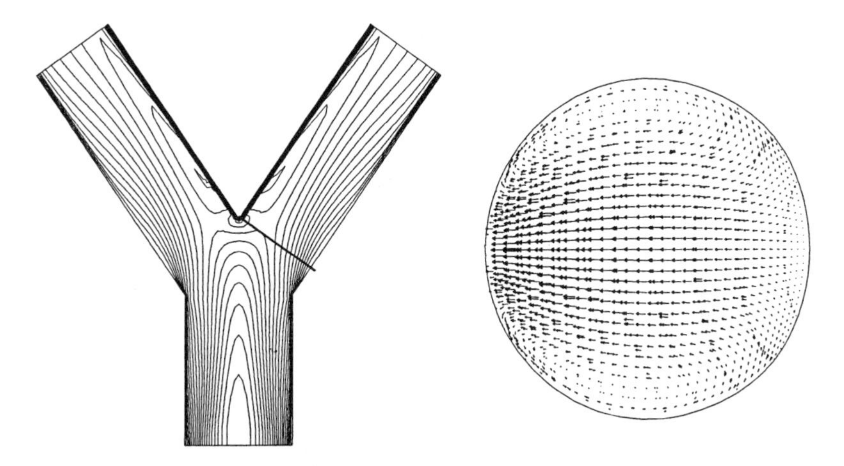

Figure 14.1 Inspiratory airflow patterns in a symmetric narrow bifurcation, with a branching angle of 35°, for a tracheal flow rate of 60 l min^{-1} with a parabolic inlet velocity profile. Left panel: isolevel representation of the airflow velocity in the plane of the bifurcation (20 isolines with $v_{max,\ inlet}$ = 10.2 m s^{-1}). Right panel: secondary velocity vector plot in a cut-plane at the carinal ridge (SMIF = 0.453).

Corresponding simulations for a symmetric PRB model are displayed in Figure 14.2. Thus, the effect of bifurcation geometry (i.e., narrow vs. physiologically realistic) on airflow patterns can be studied by comparing Figures 14.1 and 14.2. Regarding primary flow profiles, distinct discontinuities of the narrow bifurcation cause a marked decrease in air velocity near the outer walls. In

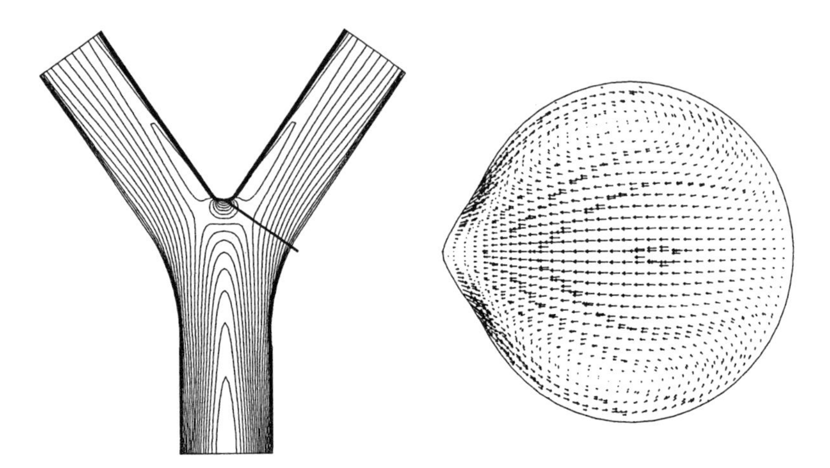

Figure 14.2 Inspiratory airflow patterns in a symmetric PRB, with a branching angle of 35°, for a tracheal flow rate of 60 l min^{-1} with a parabolic inlet velocity profile. Left panel: isolevel representation of the airflow velocity in the plane of the bifurcation (20 isolines with $v_{max,\,inlet}$ = 10.2 m s^{-1}). Right panel: secondary velocity vector plot in a cut-plane at the carinal ridge (SMIF = 0.476).

contrast, the smoother geometry of the PRB model produces a less abrupt change in axial velocity. Regarding secondary flows, both models show strong secondary motions toward the carinal ridge of comparable magnitude (i.e., SMIF values are similar), although secondary velocity vectors are pointing *away* from this site in the PRB model. Both observations of primary and secondary flow patterns suggest that the commonly used narrow bifurcation models may overestimate carinal deposition of particles for inspiratory flow conditions.

In Figure 14.3, primary and secondary flow patterns are computed in the same PRB model as used in Figure 14.2, but for a flow rate of 10 l min^{-1} (characteristic of resting breathing conditions). Thus, differences in related flow patterns can be attributed exclusively to differences in flow rates. Both the asymmetry of the primary flow and the magnitude of the secondary motions are reduced at the low flow rate as compared with the high flow rate (Figure 14.2).

In order to investigate the effects of the breathing mode (i.e., inspiration vs. expiration) on flow patterns, primary and secondary flows for expiratory breathing conditions in the PRB model are plotted in Figure 14.4 for a tracheal flow rate of 60 l min^{-1}. Comparison with the corresponding calculations displayed in Figure 14.2 indicates a relatively uniform profile for axial flow in the parent branch on exhalation (Figure 14.4). Additionally, the characteristic secondary flow pattern with four symmetric vortices on expiration (Figure 14.4) is completely different from the nearly unidirectional flow pattern during inspiration (Figure 14.2), suggesting that deposition upon expiration is enhanced at the bottom and top parts of the parent branch (see Figure 14.13), whereas deposition upon inspiration is enhanced at the carinal ridge.

14.3 PARTICLE DEPOSITION EFFICIENCIES IN BRONCHIAL BIFURCATIONS

In order to determine the deposition of a particle entrained in the airstream, the velocity of that particle has to be related to the velocities of the flow field at the grid points of the computational mesh. In our numerical particle deposition model (Balásházy and Hofmann, 1993a), the trajectory of a particle, whose position at the inlet cross section is randomly selected, through the whole bifurcation is computed for prespecified time intervals. If the trajectory intersects the surface of the bifurcation, then the coordinates of this intersection point define the spatial location of the deposition event.

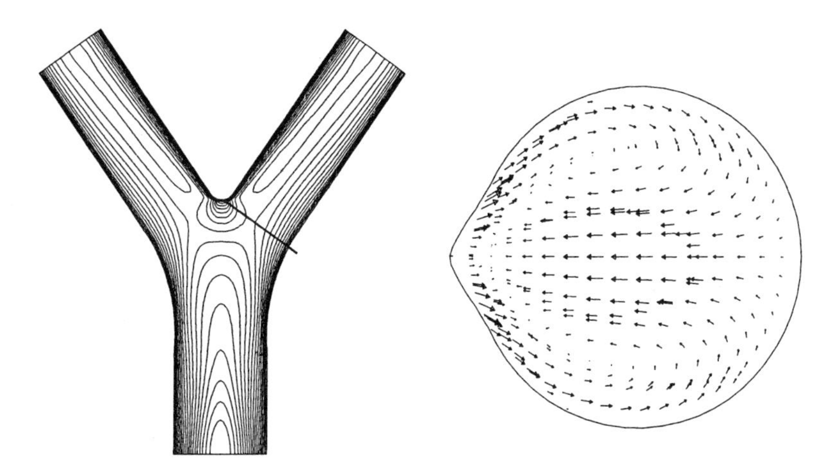

Figure 14.3 Inspiratory airflow patterns in a symmetric PRB, with a branching angle of 35°, for a tracheal flow rate of 10 l min^{-1} with a parabolic inlet velocity profile. Left panel: isolevel representation of the airflow velocity in the plane of the bifurcation (20 isolines with $v_{max,\ inlet}$ = 1.7 m s^{-1}). Right panel: secondary velocity vector plot in a cut-plane at the carinal ridge (SMIF = 0.223).

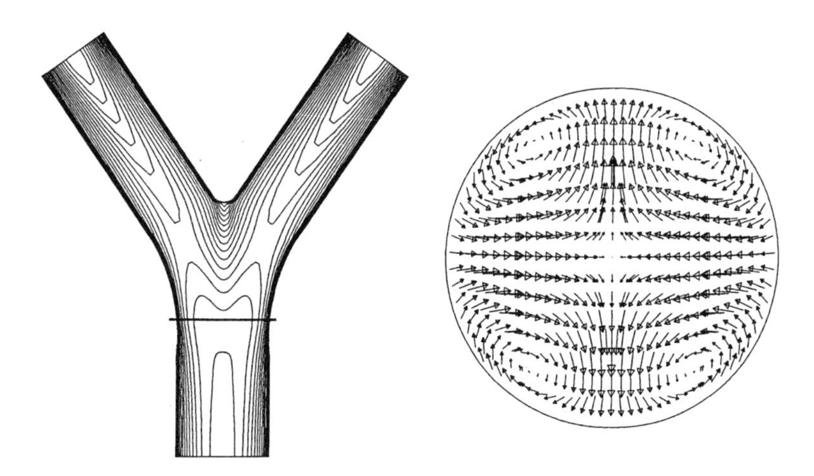

Figure 14.4 Expiratory airflow patterns in a symmetric PRB, with a branching angle of 35°, for a tracheal flow rate of 60 l min^{-1} with a parabolic inlet velocity profile. Left panel: isolevel representation of the airflow velocity in the plane of the bifurcation (20 isolines with $v_{max,\ inlet}$ = 7.9 m s^{-1}). Right panel: secondary velocity vector plot in a cut-plane at the distal end of the parent branch (SMIF = 0.149).

The first characteristic feature of particle deposition in an airway bifurcation filter is the deposition efficiency, i.e., the number of particles deposited in a bifurcation normalized to the number of particles entering that bifurcation. In our numerical deposition model, the deposition efficiency can be obtained by integrating over all deposition events in a given bifurcation regardless of their specific location. Predicted deposition efficiencies in the symmetric narrow and wide airway bifurcations are plotted in Figure 14.5 as a function of the Stokes number and compared with the experimental data of Kim and Iglesias (1989) and Johnston et al. (1977) (*note:* the Stokes number is defined here as St = $u\tau/R$, where u is the particle velocity, τ the particle relaxation time, and R the airway radius). To facilitate comparison with the experimental data, the bifurcation geometry used in the experiments of Kim and Iglesias was chosen in these simulations. The geometry corresponds approximately to the diameters and lengths of airway generations 3–4 in Weibel's

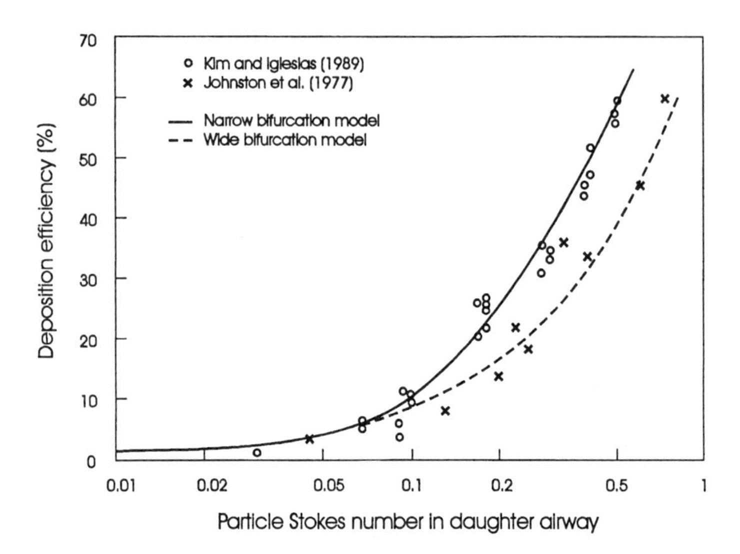

Figure 14.5 Comparison of computed inspiratory deposition efficiencies with experimental data in bronchial bifurcation models, plotted as a function of the Stokes number. Results are presented for narrow and wide symmetric bifurcation models with branching angles of 35° under parabolic inlet flow conditions.

(1963) Model A. If parabolic inlet flow in a narrow bifurcation is assumed, then excellent agreement is obtained between experiment and theory. In contrast, the smaller deposition efficiencies at higher Stokes numbers in the Johnston et al. (1977) experiments can be approximated best by assuming parabolic inlet flow conditions in a wide bifurcation model. This suggests that the observed differences in deposition efficiencies between the two experimental efforts may be attributed to differences in the geometry of the central transition zone of the bifurcation.

We now shall compare the numerical model predictions to results (1) obtained by the widely applied analytical deposition model of Yeh and Schum (1980), in which deposition equations were derived for idealized flow conditions in straight and bent tubes and deposition mechanisms were considered as acting independently, and (2) computed by our previous analytical bifurcation model (Balásházy et al., 1991), in which idealized flow profiles were corrected for physiologically more realistic flow conditions upon inspiration and sedimentation and impaction were treated simultaneously. Deposition efficiencies for 10 μm unit density particles are illustrated in Figure 14.6 for different flow conditions in asymmetric upper and lower airway bifurcations. Diameters, lengths, branching angles, and gravity angles of the two wide bifurcation models of large bronchi (generations 4–5) and terminal bronchioles (generations 15–16) represent average values of a stochastic morphometric model of the human lung (Koblinger and Hofmann, 1985; Hofmann et al., 1996) (*note:* here the trachea is denoted as generation 1). The two tracheal flow rates of 20 and 120 l min^{-1} correspond to low and heavy physical activity breathing conditions.

A comparison of the simulations for airway generations 4–5 and 15–16 under low and high inspiratory flow conditions indicates that the numerical bifurcation model consistently predicts the highest deposition efficiencies, while the analytical tube model always predicts the smallest values. The primary reason for the systematic underprediction of the deposition efficiency by the analytical tube model is the assumption of idealized constant flow profiles in straight or bent tubes during the passage of the airstream through the bifurcation, thereby neglecting the formation of asymmetric flow profiles and the existence of secondary flows (Balásházy and Hofmann, 1993a). While both analytical tube and bifurcation models predict similar deposition efficiencies for the lower flow rate, there is fair agreement between the numerical and the analytical bifurcation models in the case of the high flow rate. This can be explained by the applied asymmetric flow correction in the

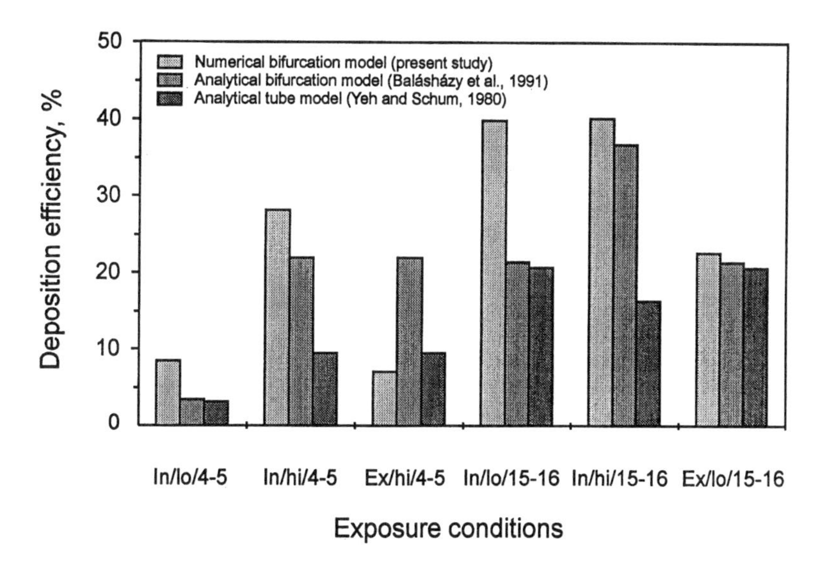

Figure 14.6 Computed deposition efficiencies for 10 μm unit density particles in upper (generations 4–5) and lower (generations 15–16) bronchial airway bifurcations (Koblinger and Hofmann, 1985) for inspiration and expiration. Numerical predictions for these asymmetric wide bifurcation models are compared with the results obtained by two different analytical models. The exposure conditions are denoted by in (inspiration), ex (expiration), lo (low tracheal flow rate, 20 l min^{-1}), and hi (high tracheal flow rate, 120 l min^{-1}).

analytical bifurcation model, which accounts, in an empirical fashion, for the asymmetric flow profiles in the daughter branches. Since flow asymmetries and secondary flows increase in magnitude with increasing flow rate, the agreement in deposition efficiency with the numerical model is best at the higher flow rates.

The situation is quite different for expiratory flow conditions. Here, the numerical model predictions are very similar to those obtained by the analytical tube model. The reason for the overestimation of the deposition efficiency by our analytical bifurcation model (Balásházy et al., 1991) in the case of the high flow rate in the upper airway bifurcation is that the flow correction, which specifically refers to inspiratory flow asymmetries, simply does not apply to expiratory flow conditions. Because straight and bent tube models do not distinguish between inspiratory and expiratory flow, the same analytical deposition efficiencies were obtained for inhalation and exhalation under identical geometry and flow conditions.

Even in a symmetrically dividing bifurcation, partial constriction or total blockage of one of the daughter airways can produce significant asymmetries in flow division. As a result, local deposition patterns in a diseased lung may be quite different from those in a healthy lung (Balásházy and Hofmann, 1994; 1995). Such a difference in deposition efficiencies and localized deposition patterns may have important implications for risk assessment and therapeutic application of inhaled aerosols in humans with lung diseases.

The significance of a totally blocked airway branch on inspiratory particle deposition in bronchial bifurcations is shown in Figure 14.7 for unit density particles, ranging in size from 10 nm to 10 μm, in the symmetric narrow bifurcation model used by Kim and Iglesias (1989) at a flow rate of 8 l min^{-1} in the parent branch with a parabolic inlet flow profile. While there is practically no deposition in the size range from 0.1 to 1 μm for a symmetrically dividing airflow, the deposition efficiency is approximately 25% when one of the daughter airways is completely blocked. Thus, a blocked airway in a bronchial bifurcation can significantly increase the deposition efficiency for all particle sizes.

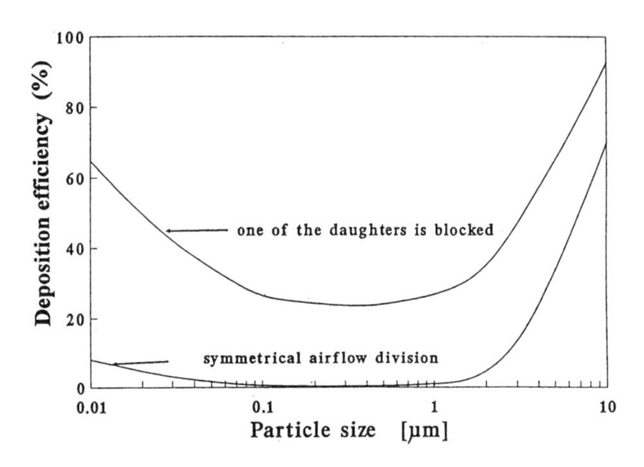

Figure 14.7 The effect of airway blockage on inspiratory particle deposition as a function of particle size in a symmetric narrow bifurcation with a branching angle of 30°. The inspiratory flow rate in the parent branch is 8 l min^{-1}.

14.4 PARTICLE DEPOSITION PATTERNS IN BRONCHIAL BIFURCATIONS

The heterogeneity of the particle deposition pattern in a given bifurcation is the second characteristic feature of an airway bifurcation filter in the human lung. For the etiology of health effects following the inhalation of particulate matter, such as bronchial carcinomas, local enhancements of particle deposition may even be more relevant than the deposition efficiency, which is conceptually equivalent to the assumption of uniform deposition. Graphical illustrations of the inhomogeneity of the particle deposition pattern in airway bifurcations are shown in Figures 14.8 through 14.13, and local enhancement of deposition is quantified by deposition density plots and deposition enhancement factors (Figures 14.14 and 14.15).

Projections of spatial deposition patterns of 10 μm unit density particles are illustrated in Figure 14.8 for inspiratory flow conditions. The linear dimensions of this symmetric narrow bifurcation model correspond to airway generations 3–4 in Weibel's (1963) Model A (Kim and Iglesias, 1989). The flow rate in the parent branch is assumed to be 7.5 l min^{-1}, which is equivalent to a tracheal flow rate of 60 l min^{-1}. A characteristic feature of the distribution of deposition sites is the high degree of inhomogeneity: (1) a distinct hot spot is formed at the carinal ridge due to inertial impaction and (2) enhanced deposition is observed along the inner walls of the daughter branches as a result of the asymmetric flow profile (see Figure 14.1). The computed deposition efficiency for this narrow bifurcation is 41.9%.

Corresponding simulations for 10 μm unit density particles in a physiologically realistic version of the above bifurcation model under identical breathing conditions are displayed in Figure 14.9. Comparison of both figures indicates that deposition patterns are quite similar for the two geometries and that the deposition efficiency in the PRB model is only slightly higher (45.2%) than that for the narrow one (41.9%).

Spatial deposition patterns of 10,000 trajectories of 10 nm particles randomly selected at the inlet of the parent branch according to the flow profile are plotted in Figure 14.10. The geometry of the airway bifurcation (PRB model) and the breathing conditions (tracheal flow rate of 60 l min^{-1}) are identical to those used for the 10 μm particles (Figure 14.9), thereby allowing comparison of particle deposition patterns caused by different physical deposition mechanisms, i.e., Brownian motion (10 nm) vs. inertial impaction and gravitational settling (10 μm). Although Brownian

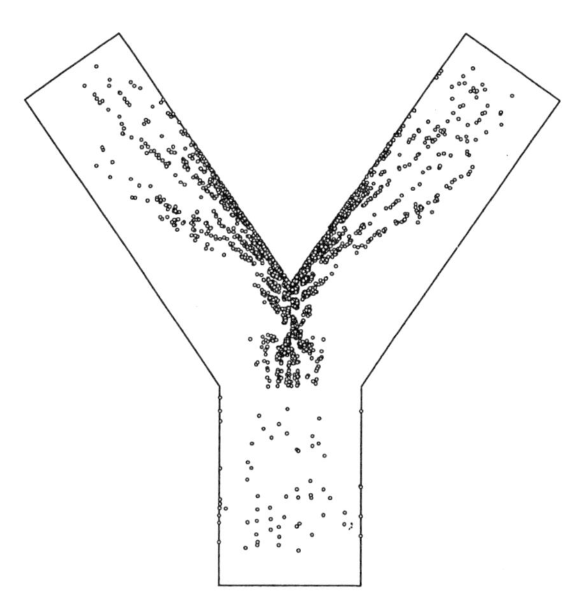

Figure 14.8 Spatial deposition patterns of 10,000 randomly selected 10 μm unit density particles in a symmetric narrow bifurcation model, with a branching angle of 35°, for parabolic inspiratory inlet flow with a tracheal flow rate of 60 l min^{-1}.

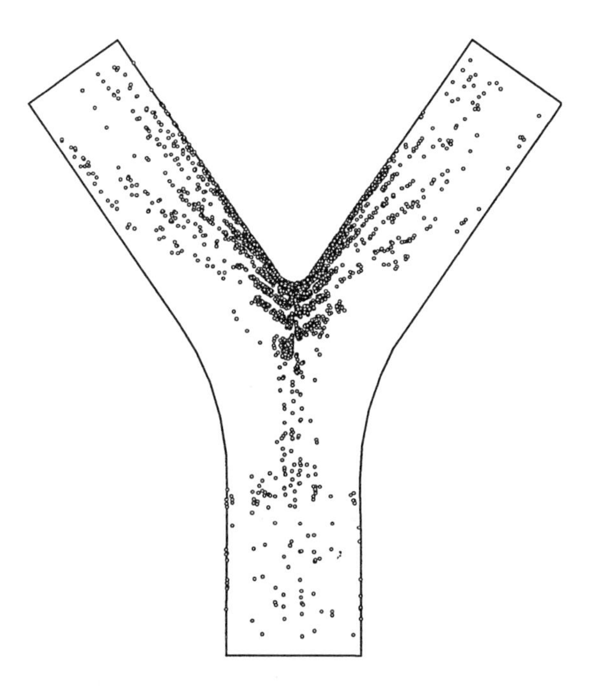

Figure 14.9 Spatial deposition patterns of 10,000 randomly selected 10 μm unit density particles in a symmetric PRB model, with a branching angle of 35°, for parabolic inspiratory inlet flow with a tracheal flow rate of 60 l min^{-1}.

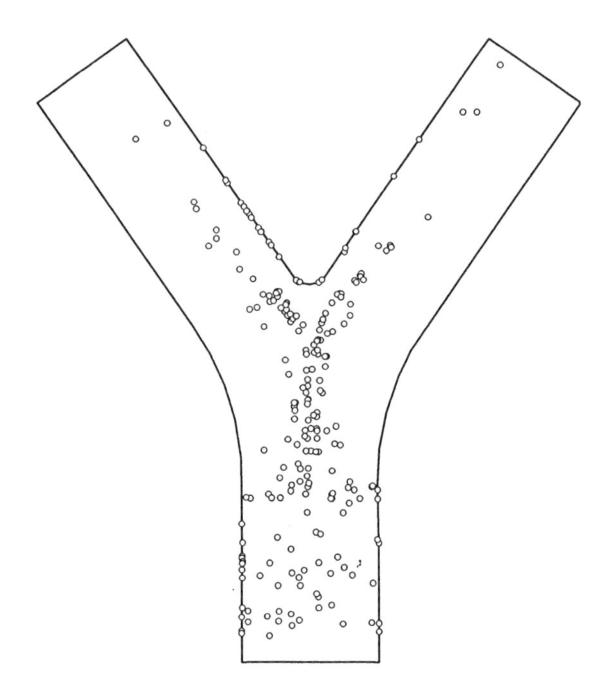

Figure 14.10 Spatial deposition patterns of 10,000 randomly selected 10 nm unit density particles in a symmetric PRB model, with a branching angle of 35°, for parabolic inspiratory inlet flow with a tracheal flow rate of 60 l min^{-1}.

diffusion is uniform with respect to the spatial motion (except close to the walls), the highly asymmetric three-dimensional flow pattern causes a distinct hot spot at the dividing spur and enhanced deposition at the inner sides of the daughter airways. Even for 10,000 particles, the random walk of small particles produces slightly different deposition patterns in the two daughter airways. The deposition efficiency for the 10 nm particles is 2.3%.

To examine the effect of asymmetry in daughter airway diameter and branching angle, an idealized narrow bifurcation geometry was selected where the diameter of the major daughter is equal to that of the parent airway and twice as large as that of the minor daughter airway, while the branching angle of the major daughter branch is only half that of the minor daughter branch. Particle deposition patterns of 10 nm particles in this asymmetric bifurcation are shown in Figure 14.11 for 10,000 inhaled particles, of which 701 are deposited in the bifurcation (i.e., the deposition efficiency is 7.0%). The inspiratory flow rate is again 8 l min^{-1} in the parent airway. Consistent with the earlier observations, preferential deposition occurs at the carinal ridge. In addition, deposition is significantly higher in the larger daughter airway than in the smaller daughter branch (*note:* the number of particles entering the major daughter branch is also higher). The total deposition efficiency for the whole bifurcation, however, is similar to the corresponding value for a symmetrically dividing airway bifurcation under the same inhalation conditions.

The deposition sites of 10 μm unit density particles in asymmetric wide bifurcation models representing bronchial airway generations 4–5 and 15–16 of the stochastic lung model (Koblinger and Hofmann, 1985) (see Figure 14.6) are plotted in Figure 14.12 for both the 20 and 120 l min^{-1} tracheal flow rates. The four panels illustrate the effect of flow rate on particle deposition patterns due to differences in (1) the location of the bifurcation relative to the trachea and (2) breathing maneuvers. In the generation 4–5 bifurcation model, inspired particles produce an intense hot spot at the carinal ridge for both flow rates. The higher deposition efficiency at the higher flow rate is caused by the increased effectiveness of the inertial impaction mechanism. On the other hand, the effect of gravity is practically negligible for both inhalation conditions. The situation is quite

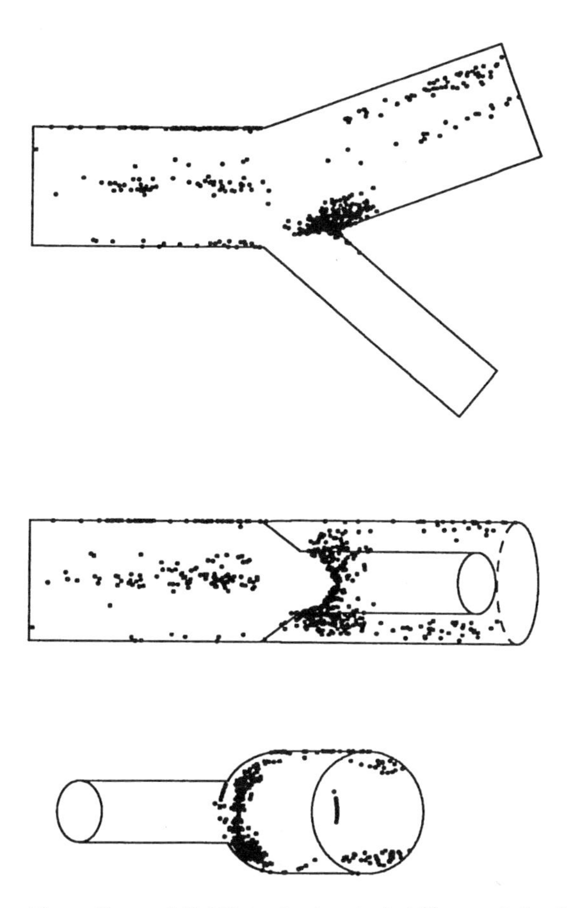

Figure 14.11 Spatial deposition patterns of 10,000 randomly selected 10 nm unit density particles in a symmetric narrow bifurcation model for parabolic inspiratory inlet flow rate of 8 l min^{-1} in the parent branch.

different in the generation 15–16 bifurcation. Here, at the low flow rate, there is practically no hot spot, but the deposition efficiency is about five times higher than that in the upper airways for the same tracheal flow rate. The side view of the bifurcation effectively illustrates that gravitational settling is the dominant deposition mode at the lower flow rates encountered in the peripheral airways. At the higher flow rate, however, the additional effect of impaction produces a hot spot in the vicinity of the carinal ridge.

Due to the asymmetric branching in the upper and lower bronchial bifurcation models, the ratio of the number of 10 μm particles deposited in the major daughter branch to that in the minor daughter branch is about 2 for the high flow rate and about 1.5 for the low flow rate in both the upper and lower bronchial bifurcation models. In other words, more particles are deposited in the larger daughter airway, consistent with the results presented in Figure 14.11.

Computed deposition sites for a parabolic inlet flow at the two daughter branches in a narrow symmetric bifurcation filter are presented in Figure 14.13 for both 10 μm and 10 nm unit density particles under expiratory flow conditions (the flow rate in the parent branch is 8 l min^{-1}). Bifurcation geometry and flow parameters are identical to those used in the experimental studies of Kim et al. (1989). In contrast to the simulations of inspiratory particle deposition patterns (Figures 14.8 through 14.12), narrow, elongated, and very distinct hot spots are formed downstream of the central bifurcation zone on the top and bottom parts of the parent airway. The locations of these hot spots are consistent with the sites of high secondary motions toward the surface of the bifurcation (see Figure 14.4). Thus, inspiration and expiration under identical geometry and flow conditions cause very different localized deposition patterns.

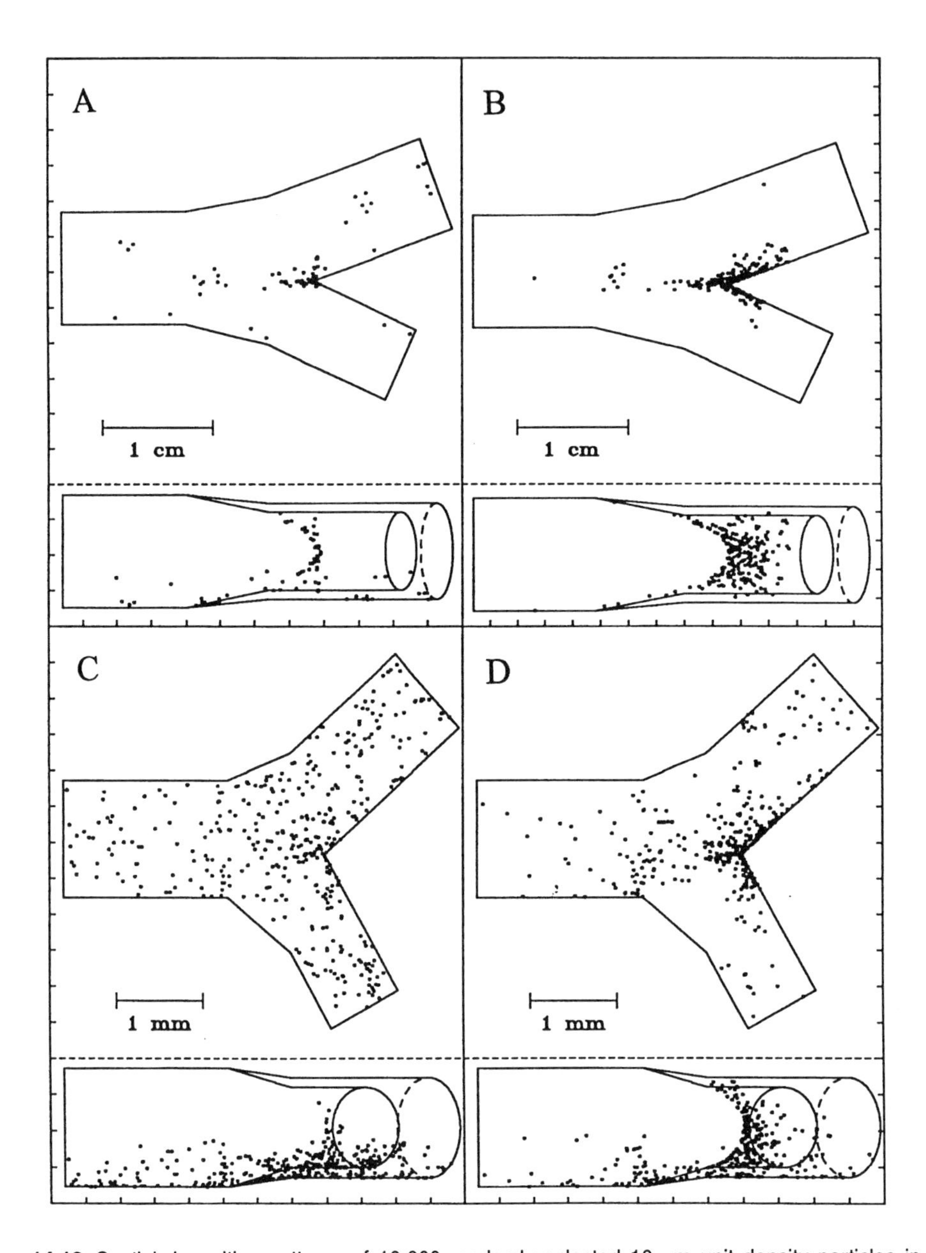

Figure 14.12 Spatial deposition patterns of 10,000 randomly selected 10 μm unit density particles in upper (generations 4–5) (panels A and B) and lower (generations 15–16) (panels C and D) bronchial airway bifurcations (Koblinger and Hofmann, 1985) at low (20 l min^{-1}) (panels A and C) and high (120 l min^{-1}) (panels B and D) tracheal flow rates during inhalation. The axes of the parent and daughter branches of these asymmetric wide bifurcation models lie in one horizontal plane.

To quantify the localized inhomogeneities of the particle deposition patterns in the various bifurcation models illustrated in Figures 14.8 through 14.13, deposition density plots are constructed and deposition enhancement factors are defined. For the calculation of deposition density plots upon inspiration, all deposition events in a daughter airway segment directly at the carina, i.e., where the deposition hot spots are found, are projected onto a cut-plane at the carinal ridge. In the present chapter, the length of the cylindrical segment is assumed to be half of the airway diameter. Consistent with expiratory deposition patterns, the whole parent airway is selected in the case of expiration and deposition events are projected onto a cut-plane at the end of the

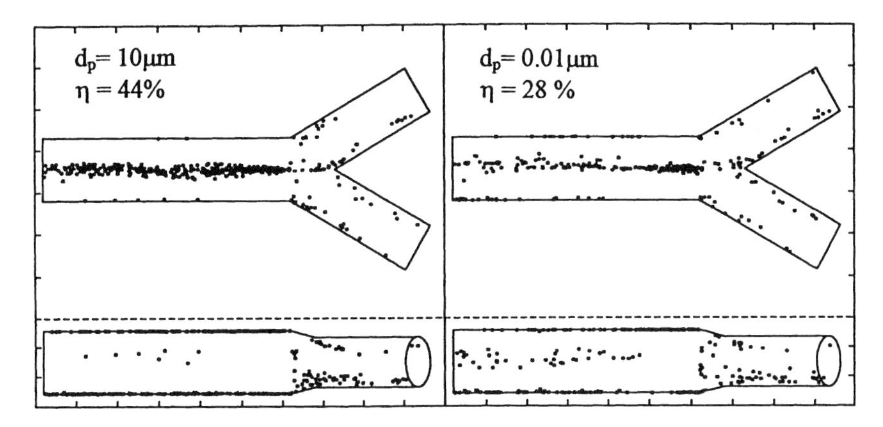

Figure 14.13 Spatial deposition patterns of 10,000 randomly selected 10 μm and 10 nm unit density particles in a narrow symmetric bifurcation model under expiratory flow conditions, assuming a parabolic inlet flow profile at both daughter branches with a flow rate of 8 l min^{-1} in the parent branch.

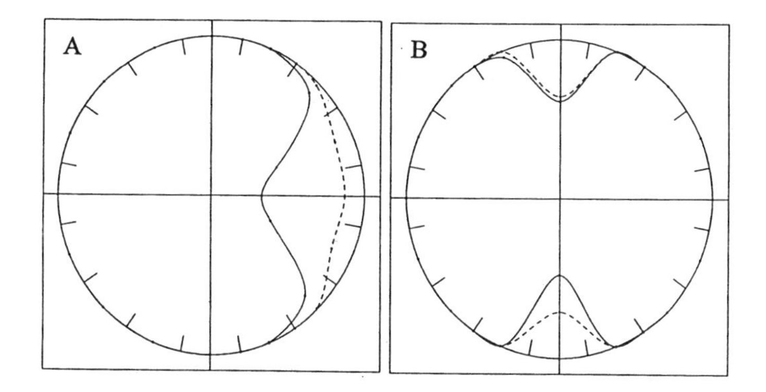

Figure 14.14 Deposition density plots for 10 μm (solid line) and 10 nm (broken line) unit density particles in a symmetric wide bifurcation model (generations 3–4) (Weibel, 1963), with a branching angle of 35°, under inspiratory (panel A) and expiratory (panel B) flow conditions with a tracheal flow rate of 120 l min^{-1}. Calculations are based on 1000 simulations of particle trajectories.

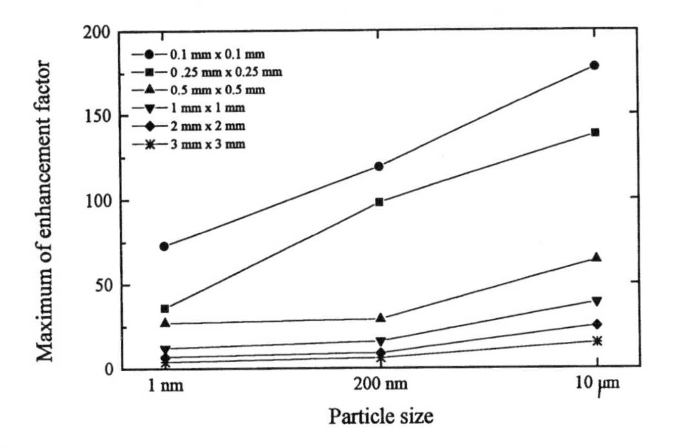

Figure 14.15 Maximum deposition enhancement factors for three different particle sizes in a symmetric physiologically realistic bronchial bifurcation model (generations 3–4) (Weibel, 1963) at a flow rate of 7.5 l min^{-1} in the parent branch (corresponding to 60 l min^{-1} in the trachea) for parabolic inspiratory inlet flow. The size of the scanning surface elements ranges from 0.1 × 0.1 to 3 × 3 mm.

central transition zone. In each segment, the cross section is divided into 16 equal sectors. The deposition densities are then normalized to a specified maximum number of deposition events; i.e., the center point corresponds to this defined number (100%) and the perimeter to no deposition at all (0%).

Deposition density plots for 10 μm and 10 nm unit density particles in a wide symmetric bronchial bifurcation (generation 3–4 junction in Weibel's, 1963, Model A) at a tracheal flow rate of 120 l min^{-1} are displayed in Figure 14.14 for inspiratory and expiratory flow conditions. On inspiration, a deposition maximum is formed at the inner side of the daughter airways downstream of the carinal ridge, which is consistent with the strong secondary motions directed toward the carina (Figures 14.1 through 14.3). On expiration, deposition is significantly enhanced at the top and bottom of the parent airway, which is again consistent with the corresponding secondary motions pointing at these locations (Figure 14.4). For both breathing modes, maximum deposition densities are higher for large particles (due to inertia) than for ultrafine particles (due to convective diffusion). The conformity between secondary flow patterns and related deposition density plots indicates a strong correlation between local fluid dynamics and particle deposition (Heistracher et al., 1995).

Deposition enhancement factors are defined here as the ratio of the number of particles deposited in a surface element relative to the number deposited in the whole bifurcation, divided by the surface area of the surface element relative to the total surface area of the bifurcation. For the calculation of deposition enhancement factors, a defined square surface element scans the whole surface of the finite volume mesh and computes the maximum and the distribution of the enhancement factors.

For example, the maximum deposition enhancement factors for 1 nm, 200 nm, and 10 μm unit density aerosols in a symmetric physiologically realistic bronchial bifurcation model (generations 3–4 in Weibel's, 1963, Model A) are illustrated in Figure 14.15 for inspiratory flow conditions. The flow rate of 7.5 l min^{-1} in the parent branch corresponds to flow rate of 60 l min^{-1} in the trachea. The size of the nonoverlapping scanning surface elements ranges from 0.1×0.1 to 3×3 mm. Maximum deposition enhancement factors increase with rising particle size, consistent with the more-localized deposition patterns due to inertial impaction relative to convective diffusion. Furthermore, the increase of the maximum enhancement factor with decreasing size of the scanning surface element indicates that the finer the scanning mesh, the better the spatial resolution of localized deposits. For the smallest scanning element of 0.1×0.1 mm, which is equivalent to an area of about 10×10 epithelial cells, maximum enhancement factors are 73 (1 nm), 119 (200 nm), and 178 (10 μm). This suggests that localized deposits may be two orders of magnitude higher than the average deposition density obtained by dividing the total deposition by the total surface area of the bifurcation.

14.5 IMPLICATIONS FOR RISK ASSESSMENT

For the inhalation of airborne particulate matter the human respiratory tract can be considered as a sequence of airway bifurcation filters (ICRP, 1994). Since each airway bifurcation gives rise to two smaller bifurcations, the number of bifurcation filters increases and the flow rate of the inspired air decreases on the way from the trachea to the alveolar sacs. At each level, deposition in airway bifurcations reduces the number of particles reaching the downstream bifurcations (i.e., deposition in upper bronchial airways reduces deposition in peripheral airways). More importantly, however, deposition in a given bifurcation determines the resulting biological response at that site. As a first approximation (i.e., assuming equal sensitivity), health effects depend on the number or mass of particles deposited in a given bifurcation filter, which are proportional to the deposition efficiency (ICRP, 1994). The suspected preferential occurrence of bronchial tumors at airway branching sites (Hofmann et al., 1991) suggests, however, that the deposition enhancement factor may be a dosimetric quantity as relevant as the deposition efficiency (Martonen and Hofmann, 1986; Hofmann et al., 1990).

Current models of particle deposition generally assume idealized flow profiles, most commonly a fully developed laminar flow, to obtain analytical expressions for particle deposition in straight and bent cylindrical airways (e.g., Ingham, 1975; Balásházy et al., 1990). These assumptions, however, neglect flow irregularities introduced by the division of flow in a bifurcation, such as asymmetries of the primary velocity profiles and the formation of secondary flows. Both the degree of asymmetry of the primary flow and the magnitude of secondary flows determine the resulting deposition efficiency and the localized distribution of deposited particles. Therefore, it is not surprising that inspiratory deposition efficiencies obtained by the currently used analytical models are consistently smaller than those predicted by the present numerical model.

Current models of particle deposition generally assume identical flow profiles on inspiration and expiration, thereby using the same analytical deposition equations for both breathing modes (e.g., Yeh and Schum, 1980; Koblinger and Hofmann, 1990). However, the observed differences in primary and secondary flow patterns for inspiratory and expiratory flow conditions lead to significant differences in local deposition patterns. Indeed, our simulations have demonstrated that the currently used analytical equations for deposition in straight and bent tubes may significantly underestimate deposition efficiencies in bifurcations for inspiration, but may provide reasonable estimates for expiration.

Current models of particle deposition generally assume a lung morphometry based on the generation numbering scheme, in which all airways in a given airway generation have identical linear dimensions (e.g., Weibel, 1963; Yeh and Schum, 1980). Consequently, all airway filters in that specific generation have identical deposition efficiencies. In reality, however, airway bifurcations exhibit a significant variability in linear airway dimensions in the same generation, branching preferentially in an asymmetric manner (Koblinger and Hofmann, 1985). Moreover, a large variability in central transition zone geometries and carinal ridge shapes has been observed (Horsfield et al., 1971; Hammersley and Olson, 1992; Martonen et al., 1994). The results of the numerical simulations presented here have shown that deposition efficiencies and distributions of deposition sites are different for (1) symmetrically and asymmetrically branching airway bifurcations and (2) for varying shapes of the central transition zone as modeled here by narrow, wide, and physiologically realistic bifurcation geometries. As a consequence, the biological variability in lung morphometry and physiology causes a distribution of deposition efficiencies even in a given airway generation. In the special case of a blocked daughter airway, the deposition efficiency may dramatically increase relative to an undisturbed symmetric flow division (Balásházy and Hofmann, 1995).

Current models of particle deposition based on analytical formulations can provide information about deposition efficiencies and, thus, fractional deposition, but not about inhomogeneities of the associated deposition patterns on a microscopic scale (e.g., Yeh and Schum, 1980; Koblinger and Hofmann, 1990). In contrast, the particle trajectory approach utilized in our numerical model allows the prediction of the three-dimensional deposition pattern. Two methods have been presented here to quantify the nonuniformity of the deposition patterns: (1) the construction of deposition density plots and (2) the definition of deposition enhancement factors. Localized enhancements of deposition may have important implications for risk assessment of inhaled particulate matter (Hofmann et al., 1990). Indeed, preliminary studies on deposition enhancement factors for a 0.1×0.1 mm scanning surface element indicates that local cellular doses may be two orders of magnitude higher than average cellular doses.

The present findings may be incorporated into currently used deposition models for risk assessment purposes by (1) deriving semiempirical equations for deposition efficiencies from the numerical simulations as explicit functions of the bifurcation geometry to account for structural variability and (2) computing deposition enhancement factors to account for deposition inhomogeneity.

REFERENCES

Asgharian, B. and Anjilvel, S. (1994), *J. Aerosol Sci.,* 25, 711–721.
Balásházy, I. (1994), *J. Comput. Phys.,* 110, 11–22.
Balásházy, I. and Hofmann, W. (1993a), *J. Aerosol Sci.,* 24, 745–772.
Balásházy, I. and Hofmann, W. (1993b), *J. Aerosol Sci.,* 24, 773–786.
Balásházy, I. and Hofmann, W. (1994), *J. Aerosol Sci.,* 25, Suppl. 1, S483–S484.
Balásházy, I. and Hofmann, W. (1995), *J. Aerosol Sci.,* 26, 273–292.
Balásházy, I., Hofmann, W., and Martonen, T. B. (1990), *Aerosol Sci. Technol.,* 13, 308–321.
Balásházy, I., Hofmann, W., and Martonen, T. B. (1991), *J. Aerosol Sci.,* 22, 15–30.
Balásházy, I., Heistracher, T., and Hofmann, W. (1996), *J. Aerosol Med.,* 9, 287–301.
Chang, H. K. and El Masry, O. A. (1982), *Respir. Physiol.,* 49, 75–95.
Cohen, B. S. and Asgharian, B. (1990), *J. Aerosol Sci.,* 21, 789–797.
Gradoń, L. and Orlicki, D. (1990), *J. Aerosol Sci.,* 21, 3–19.
Hammersley, J. R. and Olsen, D. E. (1992), *J. Appl. Physiol.,* 72, 2402–2414.
Heistracher, T. and Hofmann, W. (1995), *J. Aerosol Sci.,* 26, 497–509.
Heistracher, T., Balásházy, I., and Hofmann, W. (1995), *J. Aerosol Sci.,* 26, Suppl. 1, S615–S616.
Hofmann, W. and Balásházy, I. (1991), *Radiat. Prot. Dosim.,* 38, 57–63.
Hofmann, W., Martonen, T. B., and Ménache, M. G. (1990), *Radiat. Prot. Dosim.,* 30, 345–259.
Hofmann, W., Crawford-Brown, D. J., Ménache, M. G., and Martonen, T. B. (1991), *Radiat. Prot. Dosim.,* 38, 91–97.
Hofmann, W., Balásházy, I., and Koblinger, L. (1995), *J. Aerosol Sci.,* 26, 1161–1168.
Hofmann, W., Balásházy, I., Heistracher, T., and Koblinger, L. (1996), *Aerosol Sci. Technol.,* 25, 305–327.
Horsfield, K., Dart, G., Olson, D. E., Filley, G. F., and Cumming, G. (1971), *J. Appl. Physiol.,* 31, 207–217.
Ingham, D. B. (1975), *J. Aerosol Sci.,* 6, 125–132.
International Commission on Radiological Protection (ICRP) (1994), *Human Respiratory Tract Model for Radiological Protection, ICRP Publication 66,* Pergamon Press, Oxford.
Isabey, D. and Chang, H. K. (1982), *Respir. Physiol.,* 49, 97–113.
Johnston, J. R., Isles, K. D., and Muir, D. C. F. (1977), in *Inhaled Particles IV,* Walton, W. H., Ed., Pergamon Press, Oxford, 61–73.
Kim, C. S. and Iglesias, A. J. (1989), *J. Aerosol Med.,* 2, 1–14.
Kim, C. S., Iglesias, A. J., and Garcia, L. (1989), *J. Aerosol Med.,* 2, 15–27.
Koblinger, L. and Hofmann, W. (1985), *Phys. Med. Biol.,* 30, 541–556.
Koblinger, L. and Hofmann, W. (1990), *J. Aerosol Sci.,* 21, 661–674.
Lee, J. W. and Goo, J. H. (1992), *J. Aerosol Med.,* 5, 131–154.
Li, A. and Ahmadi, G. (1995), *Aerosol Sci. Technol.,* 23, 201–223.
Martonen, T. B. (1991), *J. Aerosol Med.,* 4, 25–40.
Martonen, T. B. and Hofmann, W. (1986), *Radiat. Prot. Dosim.,* 15, 225–232.
Martonen, T. B., Hofmann, W., and Lowe, J. E. (1987), *Health Phys.,* 52, 213–217.
Martonen, T. B., Zhang, Z., and Lessmann, R. C. (1993), *Aerosol Sci. Technol.,* 19, 133–156.
Martonen, T. B., Yang, Y., and Xue, Z. (1994), *Aerosol Sci. Technol.,* 21, 119–136.
Pedley, T. J., Schroter, R. C., and Sudlow, M. F. (1977), in *Bioengineering Aspects of the Lung,* West, J. B., Ed., Marcel Dekker, New York, 163–265.
Schlesinger, R. B. and Lippmann, M. (1978). *Environ. Res.,* 15, 424–431.
Spencer, H. (1985), *Pathology of the Lung,* Pergamon Press, Oxford.
Stahlhofen, W., Rudolf, G., and James, A. C. (1989), *J. Aerosol Med.,* 2, 285–308.

Weibel, E. R. (1963), *Morphometry of the Human Lung,* Springer, Berlin.
Yamamoto, T., Kopecky, K., Fujikura, T., Tokuoka, S., Monzen, T., Nishimori, I., Nakashima, E., and Kato, H. (1987), *J. Radiat. Res. Tokyo,* 28, 156–171.
Yeh, H. C. and Schum, G. M. (1980), *Bull. Math. Biol.,* 42, 461–480.

CHAPTER 15

Effect of Deposition on Aerosol Filtration

Chi Tien

CONTENTS

15.1 INTRODUCTION

For emission control and air pollution abatement, fibrous and granular filters are often applied. The operating principles of both types of filters are the same: a gas–solid suspension to be treated is passed through a filter (either fibrous or granular). As the suspended particles pass through the filter, some (or all) of them are transported to the surface of the collecting elements (either fibers or granules) because of the various forces acting on them and become deposited, thus removing these particles from the gas phase.

For a filter in operation, the removal and retention of the suspended particles from the gas phase causes the filter medium to undergo a continuing change in its porosity, structure, and possibly surface characteristics. As a result, the pressure gradient–flow rate relationship of the medium at

0-87371-830-5/98/$0.00+$.50

a given time may differ significantly from the initial relationship when the media is clean. This is shown by the widely recognized fact that the pressure drop required to maintain a constant gas throughput of a filter increases as the extent of deposition increases.

The change of the medium structure due to deposition also causes a change of the ability of the medium to capture particles passing through the medium. It is well known that as a medium becomes clogged, its collection efficiency increases accordingly. For filters operating under constant throughput conditions; both the pressure drop and collection efficiency increases with time.

In spite of abundant experimental evidence on the effect of deposition on aerosol filtration, systematic study of the deposition effect did not begin until recent years. Further, most of the studies are concerned with the effect on the collection efficiency while some attempts have been made to relate the change of medium permeability with the extent of deposition. These developments by and large have not reached the stage for practical applications.

The purpose of the present chapter is to provide a summary and review of studies concerning the effect of deposition on aerosol filtration reported in the literature on a systematic and consistent basis. Equal weight is placed on analysis and experimental investigations. The use of some of these results in practical applications, such as predicting the transient behavior of aerosol filtration, is also given.

15.2 EARLIER INVESTIGATIONS

Davies (1973) considered the effect of deposition on fibrous filtration in two aspects. First, the pore distribution of a clogged medium became more uniform as compared with that of its initial distribution, as deposition tends to take place preferentially in large pores. Second, deposited particles, as a rule, do not distribute themselves evenly over fibers, but build up to form chain aggregates (or dendrites) which act as additional particle collectors, thus causing an increase in collection efficiency. That particle deposition leads to the growth of chain aggregates was confirmed experimentally by several investigators (Watson, 1946; Leers, 1957; Billings, 1966).

The effect of deposition, as stated earlier, will lead to a decrease of the medium permeability. Based on data collected from various types of filters, Davies (1970) found that the rise in flow resistance of a filter during service against a constant aerosol with solid particles may be expressed as

$$w/w_o = e^{\alpha t} \tag{15.1}$$

where w is the flow resistance and w_o is the value of w initially; α is a constant and t, the time. This expression holds true for filters with fairly narrow pore size distributions. Since the pressure drop required to maintain a given flow rate is directly proportional to the flow resistance, w/w_o therefore is the same as $\Delta P/\Delta P_o$ for filters operating under constant gas throughput conditions. ΔP and ΔP_o are the pressure drop and the initial pressure drop, respectively.

A phenomenological expression relating the change in pressure drop with the extent of deposition in fibrous filters was proposed by Juda and Chrosciel (1970) as

$$\frac{\Delta P}{\Delta P_o} \approx \frac{\ln \varepsilon_f + K}{\ln\left(\varepsilon_f + \sigma\right) + K} \tag{15.2}$$

where ΔP and ΔP_o are the pressure drop and the initial pressure drop, ε_f is the fiber packing density (i.e., volume fraction occupied by fibers), and σ, the volume of deposited particles per unit filter volume (or the specific deposit). K is an empirical constant.

Filter performance is commonly given by its overall collection efficiency, E, defined as

$$E = \frac{C_{\text{in}} - C_{\text{eff}}}{C_{\text{in}}} \tag{15.3}$$

where C_{in} and C_{eff} are influent and effluent particle concentration. Alternatively, it may be described by penetration, P, defined as

$$P = \frac{C_{\text{eff}}}{C_{\text{in}}} = 1 - E \tag{15.4}$$

The single-fiber experimental data of Billings (1966) yield the following expressions on the decrease in penetration with deposition:

$$P = P_o \exp(-\delta N_A) \tag{15.5}$$

where P_o is the initial value of P. N_A is number of deposited particles per unit fiber surface area in number of particles per square centimeter. δ is an empirical constant ranging from 0.9×10^{-9} to 6.9×10^{-9} cm^2 per deposited particles.

The data of Leers (1957) and Davies (1970) on the improvement of the penetration may be related to the corresponding data on pressure drop increase (or increase in flow resistance) by the expression

$$P/P_o = \gamma^{-\left(\sqrt{w/w_o}-1\right)} \tag{15.6}$$

where γ is a dimensionless constant. The magnitude of γ can be used to classify the performance of fibrous filters. As a guide, for poor filters, $\ln \gamma \simeq 10$. On the other hand, for good filters $\ln \gamma = 1$.

By combining Equations 15.6 with 15.3, the transient behavior of fibrous filtration may be described as

$$P/P_o = \exp[-(e\alpha^{t/2} - 1)\ln \gamma] \tag{15.7}$$

All the above expressions are empirical and based on limited experimental data. The extent to which these expressions may be used for prediction is questionable. More importantly, both the total collection efficiency (or penetration) and the pressure drop across a filter are macroscopic quantities. However, since deposition within a filter is not uniform, it is important to have information which relates the changes in local pressure gradient (or permeability) and local unit collector (or single-collector) efficiency with the extent of local deposition. The expressions shown above do not give this kind of information.

15.3 A DENDRITE GROWTH MODEL OF AEROSOL FILTRATION

As stated before, Billings (1966) observed the formation of dendrite-like particle aggregates in his single-fiber deposition experiments. Subsequent model filter studies by Ushiki and Tien (1985) and Yoshida and Tien (1985) confirmed Billings's observations. For a dendrite growth process, Payatakes and Tien (1976) presented a set of kinetic equations. This idea was further extended by Payatakes and co-workers (Payatakes, 1976a,b; Payatakes et al. 1977; Payatakes and Gradon, 1980; Payatakes and Okuyama, 1982) for predicting the transient behavior of fibrous filtration. In the following, we present a brief account of the work of Payatakes and Tien (1976) which forms the basis of the aforementioned investigations.

15.3.1 Representation of Particle Dendrite

A particle dendrite formed in aerosol filtration is depicted in Figure 15.1a. In formulating their model, Payatakes and Tien (1976) proposed the use of an ideal dendrite configuration as an approximation of the actual geometry. An ideal dendrite is assumed to have a number of layers with a uniform thickness of $2a_p$, where a_p is the particle radius. Furthermore, for any given particle of a dendrite, it is placed entirely within a layer. A depiction of the ideal dendrite configuration as compared with an actual one shown in Figure 15.1a is given in Figure 15.1b.

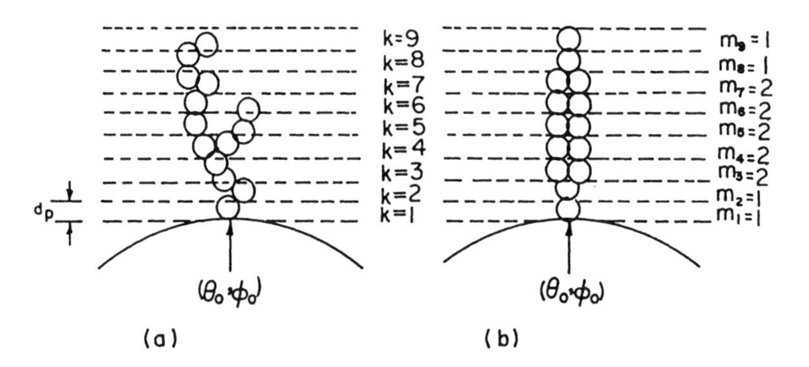

Figure 15.1 Approximation of (a) an actual dendrite by (b) an idealized dendrite.

With the use of the ideal dendrite configuration, a dendrite may be described by a set of numbers $[m_k]k = 1,2,\ldots$ where m_k is the number of particles in the kth of a dendrite. The kinetics of dendrite growth can then be represented by a set of differential equations with $[m_k]k = 1,2,\ldots$ as the dependent variables and t, time, the independent variable for a particular location on a collector.

15.3.2 Kinetics of Dendrite Growth

Payatakes and Tien (1976) formulated their kinetic equations in the following manner. First, the value m_k $(\mathbf{x},t)$ is considered as the expected number of particles of a dendrite formed at a particular location on the surface of the collector with $\mathbf{x}$ being the position vector. On this basis, m_k is real but not necessarily an integer. Furthermore, it is assumed that

$$m(\mathbf{x},t) = 1 \tag{15.8}$$

This assumption arises from the fact that during the initial stage of deposition, particles become attached to a collector in a discrete way as shown in Billings's experiments (Billings, 1966). Accordingly, it is reasonable to assume that a dendrite originates from one deposited particle.

As time passes, for layers other than the first one, we may anticipate that $m_k k \neq 1$ increases with time. However, it is also reasonable to assume that this increase does not go on indefinitely, but is limited by the number of particles present in the preceding layer. The following assumption is therefore used.

$$m_k \leq a m_{k-1} \tag{15.9}$$

where a is a constant greater than unity. In the work of Payatakes and Tien (1976), a is taken to be 2. The kinetic equations for dendrite growth are assumed to be

$$\frac{dm_1}{dt} = 0 \tag{15.10a}$$

$$\frac{dm_k}{dt} = \alpha\phi_k m_{k-1}\left[1 - \frac{m_k}{am_{k-1}}\right] k \neq 1 \tag{15.10b}$$

$$m_1 = 1 \quad t = 0 \tag{15.11a}$$

$$m_k = 0 \quad t = 0 \tag{15.11b}$$

Equation 15.10a results directly from Equation 15.8. The rationale used in formulating Equation 15.10b is as follows. The rate of the increase of particles in the kth layer results directly from collisions between approaching particles and those in the $(k - 1)$th layer. Therefore, dm_k/dt can be expected to be proportional to m_{k-1}. Also by Equation 15.9, it is clear that am_{k-1} represents the maximum number of particles of the kth layer and m_k is the actual number of particles present in the kth layer at a given time. Accordingly, $1 - (m_k/am_{k-1})$ is the fraction of unfilled space in the kth layer and to which the dm_k/dt is proportional. Based on this interpretation, $\alpha\phi_k$ is simply the rate constants of the kth layer particle growth.

The solutions to Equation 15.10a and 15.10b subject to the initial conditions of Equations 15.11a and b are

$$m_1 = 1 \tag{15.12a}$$

$$m_2 = a\left[1 - \exp\left(-\frac{\alpha\phi_1}{a}t\right)\right] \tag{15.12b}$$

$$m_k = a^{k-1}\left[\sum_{j=1}^{k=1} \sum_{1 \le l \le k-1} \frac{\phi_l}{\phi_l - \phi_j}\left(-\frac{\alpha\phi_j}{a}t\right)\right] \quad k = 3, 4, \ldots \tag{15.12c}$$

From Equation 15.12c, the following identity relationship must be obeyed

$$\sum_{j=1}^{k-1} \underset{1 \le l \le k}{\pi} \frac{\alpha\phi_l}{\alpha\phi_l - \alpha\phi_j} = 1 \tag{15.13}$$

15.3.3 Application of Dendrite Growth Results

The dendrite growth results shown above describe the growth of a single dendrite. To apply the results to aerosol filtration, two functions describing the spatial and temporal distributions of dendrite growth $N(\mathbf{x},\tau)$ or $\chi(\mathbf{x},\tau,t)$ must be known or assumed. First, over a differential area of a collector, dA, at times, τ, the number of dendrites over the area is

$$N(\mathbf{x},\tau)dA$$

Furthermore, to account for the fact that dendrites were formed at different times, let the fraction of the dendrite with age between t and $t + dt$ be

$$\chi(\mathbf{x},\tau,t)dt$$

The two density functions $N(\mathbf{x},\tau)$ and $\chi(\mathbf{x},\tau,t)$ obey the conditions

$$\int_{A_c} N(\mathbf{x},\tau)dA = A_c\,\tilde{N}/t \tag{15.14a}$$

$$\int_0^\tau \chi(\mathbf{x},\tau)dt = 1 \tag{15.14b}$$

where A_c is the collector surface area and $\tilde{N}$ is the dendrite density function at τ. To quantify the increase in collection efficiency, one may write

$$\eta/\eta_o = 1 + \frac{1}{I_o}\left[\int_A \left\{\int_0^\tau \sum_k^M \frac{dm_k}{dt}\chi(\mathbf{x},\tau,t)dt\right\} N(\mathbf{x},\tau)dA_c\right] \tag{15.15}$$

where M is the number of layers of particle dendrite present at positions $\mathbf{x}$. I_o is the particle flux when the collector is free of deposited particles.

To apply the results given above for predicting filtration performance, the value of the rate constant, $\alpha\phi_k$ (see Equation 15.10b) and the two density function expressions, $N(\mathbf{x},\tau)$ and $\chi(\mathbf{x},\tau,t)$ are required. In principle, the rate constant may be determined from trajectory calculations as pointed out by Tien (1989). For the density function, Payatakes and Gradoń (1980) and Payatakes and Okuyama (1980) assumed them, based on a dimensional argument, to be of certain forms. The validity of this assumption, however, remains to be tested.

15.4 STOCHASTIC SIMULATION OF AEROSOL FILTRATION

A direct and conceptually simple way of studying aerosol filtration is to recreate the process through computer experiments. In computer experiments, trajectories of every approaching particle are determined and monitored to determine whether or not they make contact with the collector and become deposited and, if so, at what position. The simulation results then provide information of deposition rates (or collection efficiency), the morphology of particle deposits, and its evolution.

15.4.1 Basic Concept

In the original formulation of the stimulation method, Tien et al. (1977) and Wang et al. (1977) pointed out the two basic features important to simulating aerosol filtration: the finiteness of particle size and the manner in which suspended particles are distributed throughout the gas phase. The finiteness of particle size leads to the so-called shadow effect: namely, once a particle becomes deposited on a particular location on the surface of a collector, the area immediately adjacent to the deposited particle (i.e., shadow area) becomes inaccessible to subsequent approaching particles for deposition. The magnitude of the shadow area is largely a function of the size of the deposited particle, although it is also dependent on the fluid velocity.

Shadow areas give rise to two consequences. First, since no deposition takes place in shadow areas, it means that particle deposition occurs at discrete locations. Furthermore, if shadow areas created by deposition are substantial, deposits cannot be in the form of smooth coatings.

The other consequence is the formation of dendritelike particle aggregates as observed before (Billings, 1966; Barot et al., 1980). Since no deposition within shadow areas is possible, particles that would have been deposited in shadow areas had there been no deposition now attach themselves to deposited particles, leading to the growth of particle dendrites.

The second feature arises from the manner with which suspended particles are distributed spatially and temporarily. Although particles present in a suspension may be considered as uniform macroscopically, their locations are not ordered. Thus, over a specific area (of the order of the collector), the positions on this area through which approaching particles pass are randomly distributed. Furthermore, the approaching particles pass through the area one at a time. Tien et al. (1977) termed this behavior random and singular.

The characterization of particle distribution as random is valid for dilute aerosol suspensions. The singular assumption is valid if particles obey the Poisson distribution (Tien, 1989). Thus, in actual simulation, the initial positions of each particle considered are randomly assigned. The number of the total particles considered, M, and the corresponding time t are related by the expression

$$M = U_\infty C_\infty S(t) \tag{15.16a}$$

where S is the area of the surface through which the approaching aerosols originate U_∞ the approach velocity, and C_∞ the particulate concentration of the main flow.

If the porous media model used for characterizing the filter media permits the inlet cross section of the collector to be the control area, then

$$m = \int_0^t SU_\infty C_\infty \eta \cdot dt \tag{15.16b}$$

where η is the unit collector efficiency. From the above two expressions, one has

$$\eta = \frac{dm}{dt} \tag{15.16c}$$

Equation 15.16c provides the basis of evaluating the collector efficiency from simulation results.

15.4.2 Principle of Simulation

In principle, simulation of particle deposition may be carried out in the following steps:

1. Selection of a model characterizing the filter medium; specifically, the geometry of collecting elements and flow fields around the elements.
2. Selection of a control surface through which particles move toward the collection.
3. Assignment of the initial positions of approaching particles on the control surface.
4. Determination of trajectories of approaching particles.
5. Determination of the outcome of approaching particles; namely, whether an approaching particle will make contact with filer elements or not. It it does, will contact lead to deposition?
6. From the knowledge of the outcome of the approaching particles the extent of particle deposition, the collection efficiency, the deposit morphology, and its evolution can be determined as functions of time. This information, in turn, provides the basis of predicting filter performance.

15.4.3 Simulation Results

Theoretically speaking, stochastic simulation presents a rigorous and direct method for predicting filter performance (including both particle collection efficiency and pressure drop increase). In practice, stochastic simulation can only be made approximately because of a lack of certain information including:

1. With significant deposition, the geometry of filter elements of a filter becomes irregular and difficult to characterize. Thus, it is difficult, if not impossible, to consider the change of fluid field around filter elements in a heavily loaded filter.
2. Present theoretical knowledge does not allow a quantitative prediction of the outcome of the impaction of a particle upon a surface. Theories such as those advanced by Dahneke (1971; 1972; 1973) are often difficult to apply because of the unavailability of the numerical values of the material constants present in his theory.

In an earlier simulation work, Tsiang et al. (1982) considered the deposition of particles on single fibers. In carrying out their simulations, the effect of deposited particles on the flow field around the fiber was ignored. In considering the adhesion of impacting particles, the adhesion energy was arbitrarily set to be ten times the value given by the classical Bradley–Hamaker expression (in other words, the adhesion energy was used as a fitting parameter). The simulation results were found in good agreement with experiments as shown in Figure 15.2. More recently, Jung and Tien (1993) presented extensive simulation results of aerosol filtration in granular media. The main features of Jung and Tien's work involved the use of an empirical correction factor, the capture probability, to account for the main complicating factors mentioned before, namely, the change in flow field due to particle deposition, hydrodynamic interaction, and the possible bounce-off of impacting particles. By definition, the capture probability may be considered as the probability of an approaching particle to be collected upon making contact with the collector (or previously deposited particles) on the assumption that there are no changes in flow field due to particle deposits and no hydrodynamic interaction between the particle and the collector (including deposited particles).

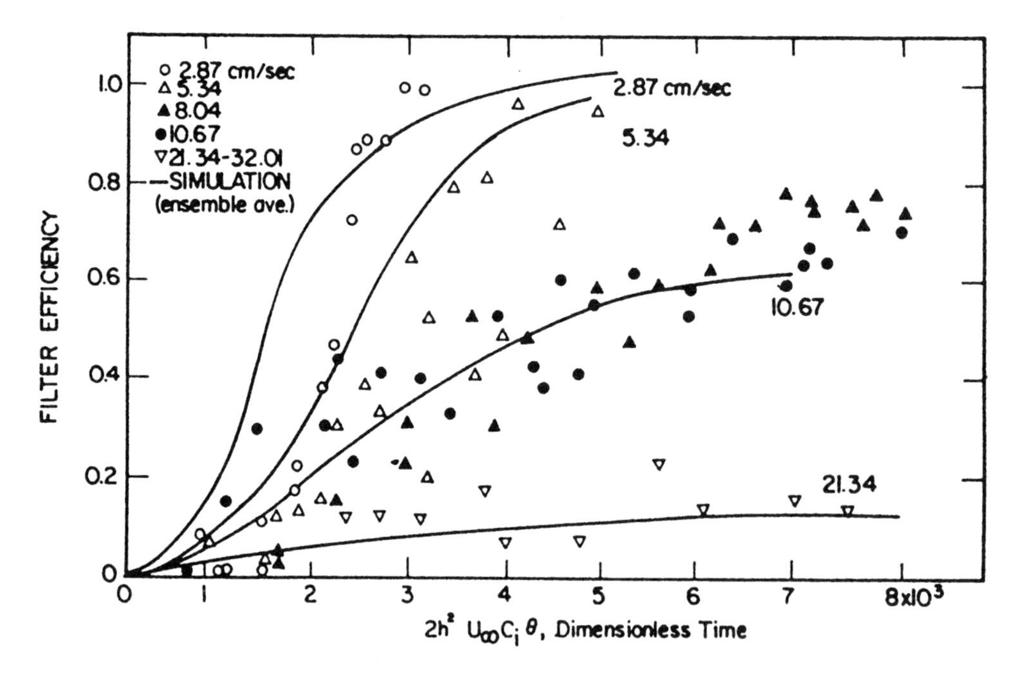

Figure 15.2 Comparisons between simulation and experiments. (From Tsiang, R. C. et al., *Chem. Eng. Sci.*, 37, 1661–1673, 1982. With permission.)

The determination of the capture probability was made empirically. For a given simulation, an arbitrary value was assumed for the capture probability and the simulation results obtained were then compared with experiments. By trial and error, the capture probability is found to be a function of the Stokes number, N_{St} at the interception, N_R (see Table 15.1 for the results).

Table 15.1 Values of Capture Probabilities at Different Conditions

N_{st}	N_R	Capture Probability
3.924×10^{-3}	4.198×10^{-3}	0.29
5.491×10^{-3}	4.198×10^{-3}	0.21
7.223×10^{-3}	4.198×10^{-3}	0.23
1.958×10^{-3}	2.095×10^{-3}	0.15
3.605×10^{-3}	2.095×10^{-3}	0.15
5.504×10^{-3}	2.095×10^{-3}	0.18
7.547×10^{-3}	2.095×10^{-3}	0.20
6.219×10^{-3}	3.850×10^{-3}	0.14

Source: Jung, Y. and Tien, C., *Aerosol Sci. Technol.*, 18, 418–440, 1993. With permission.

As expected, simulation results obtained by ignoring the change in flow field and possible bounce-off of impacting particles tend to grossly overestimate the extent of particle deposition. By using the capture probability correction factor, good agreement between simulations and experiments can be obtained as seen from Figure 15.3.

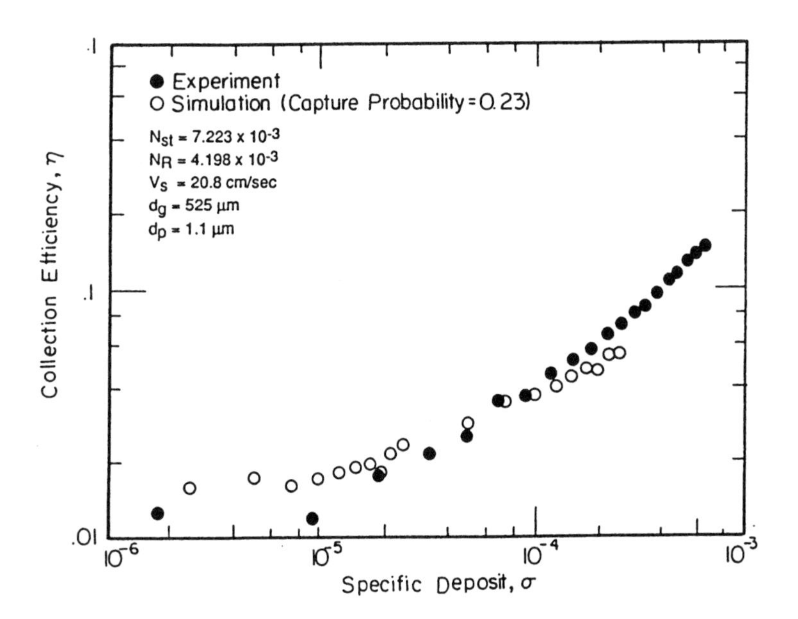

Figure 15.3 Comparisons between simulation and experiments using the capture probability correction factor. (From Jung, Y. and Tien, C., *Aerosol Sci. Technol.*, 18, 418–440, 1993. With permission.)

15.5 EXPERIMENTAL INVESTIGATIONS

15.5.1 Experiments with Fibrous Filters

A number of earlier investigators including those mentioned previously (see Section 15.1) have observed experimentally the change of the overall collection efficiency with deposition in fibrous filters. However, none of these results was analyzed or interpreted in a way which yields direct relationships between the local collection efficiency (in terms of unit collector or single-collector efficiency) with the extent of local deposition. Such a study was made only recently by Emi et al. (1982). These investigators determined the single-fiber collection efficiency and pressure drop of fibrous model filters as a function of mass deposit with aerosols of diameter of 1.1 μm. The model filters were composed of layers of wire screens (with wire diameter ranging from 25 to 52.7 μm and fiber distance, center to center, from 79.5 to 132 μm). A typical set of the data reported by Emi et al. is shown in Figure 15.4.

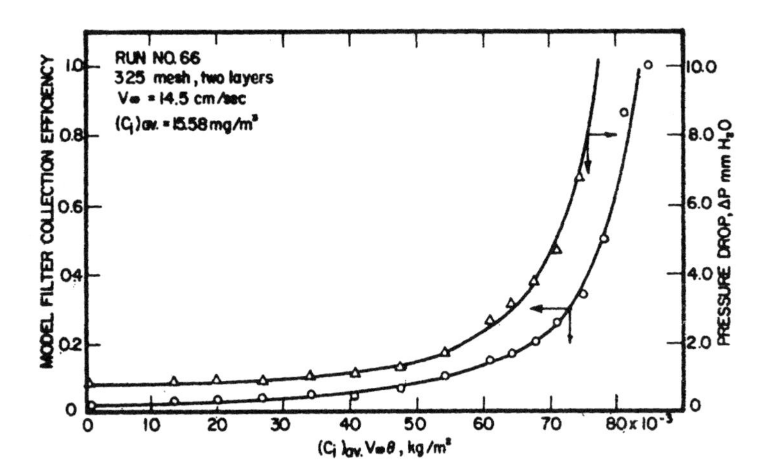

Figure 15.4 A typical set of model filter data collected by Emi et al. (From Emi, H. et al., *AIChE J.*, 28, 397–405, 1982. With permission.)

To relate the change of the single-fiber efficiency with the extent of deposition, first an expression relating the single-collector efficiency and the overall collection efficiency (which can be determined experimentally) was established. As shown by Tsiang (1980), the collection efficiency of a single wire screen, w_f, and the single-fiber efficiency of the wires constituting the screen, n_f, is

$$1 - E_f = \left[1 - n_f \frac{a_c}{h}\right]^2 \tag{15.17}$$

where a_c is the radius of the wire and h is the half distance (center to center) between two adjacent wires.

Consider two model filters composed with k layers and $(k = 1)$ layers of identical wire screens. Let $(E)_k$ and $(E)_{k-1}$ denote the respective total collection efficiencies. From Equation 15.17 and the definition of collection efficiency, one has

$$\frac{C_k}{C_{k-1}} = \left[1 - (\eta_f)_k \frac{a_c}{h}\right]^2 \tag{15.18}$$

$$C_k/C_{\text{in}} = 1 - (E)_k \tag{15.19}$$

$$C_{k-1}/C_{\text{in}} = 1 - (E)_{k-1} \tag{15.20}$$

where $(\eta_f)_h$ is the single-fiber efficiency of the kth layer wires. C_{k-1} and C_{k1} are the aerosol concentrations of gas stream leaving the $(k - 1)$th and kth layer, respectively. C_{in} is the influent aerosol concentration. Combining Equations 15.18, 15.19, and 15.20, one has

$$\left(n_f\right)_k = \frac{h}{a_c}\left[1 - \sqrt{\frac{1-(E)_k}{1-(E)_{k-1}}}\right] \tag{15.21}$$

The extent of deposition of the kth layer can be defined by its mass deposit, m_k (in kg m^{-2}), namely, the amount of deposited particles per unit surface. By definition, m_k can be written as

$$m_k = \int_0^t \left(C_{k-1} - C_k\right) U_\infty d\theta$$

or

$$m_k = \int_0^{c_{\text{in}}U_\infty t} \left[(E)_k - (E)_{k-1}\right] d\left(C_{\text{in}} U_\infty \theta\right) \tag{15.22}$$

where t is time, U_∞, the superficial gas velocity. The change of η_f vs. m can be readily obtained experimentally. A set of data reported by Emi et al. is shown in Figure 15.5.

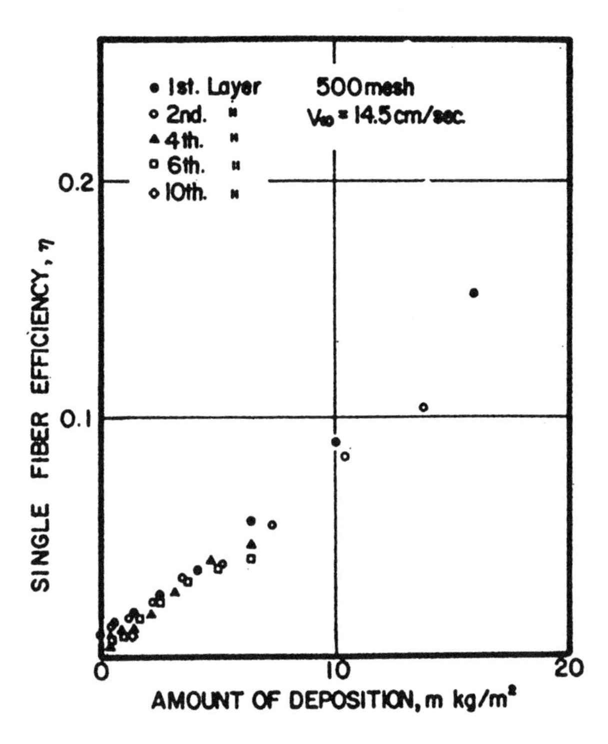

Figure 15.5 Single-fiber efficiency as a function of the amount of deposit obtained by Emi et al. (From Emi, H. et al., *AIChE J.*, 28, 397–405, 1982. With permission.)

The following empirical expression relating η_f and m was proposed by Emi et al. (1982):

$$\frac{\eta_f}{\left(\eta_f\right)_o} = 1 + \left(\frac{m}{m_o}\right)^b \tag{15.23}$$

$$m_o = A(\eta_f)_o U_\infty^{0.25} \tag{15.24}$$

where $(\eta_f)_o$ is the initial single-fiber efficiency. m_o is the mass deposit at which the increase in (η_f) over its initial value is 100%. m_o can be estimated from Equation 15.24 in which U_∞ is expressed in cm s^{-1} and m_o is in kg m^{-2}. The values of the two empirical constants, A and b are

Screen Size	A	b
200	2.7×10^{-3}	1.15
325	1.5×10^{-3}	1.23
500	1.1×10^{-3}	1.34

Figure 15.6 gives a comparison between the correlation of Equation 15.23 and experimental data.

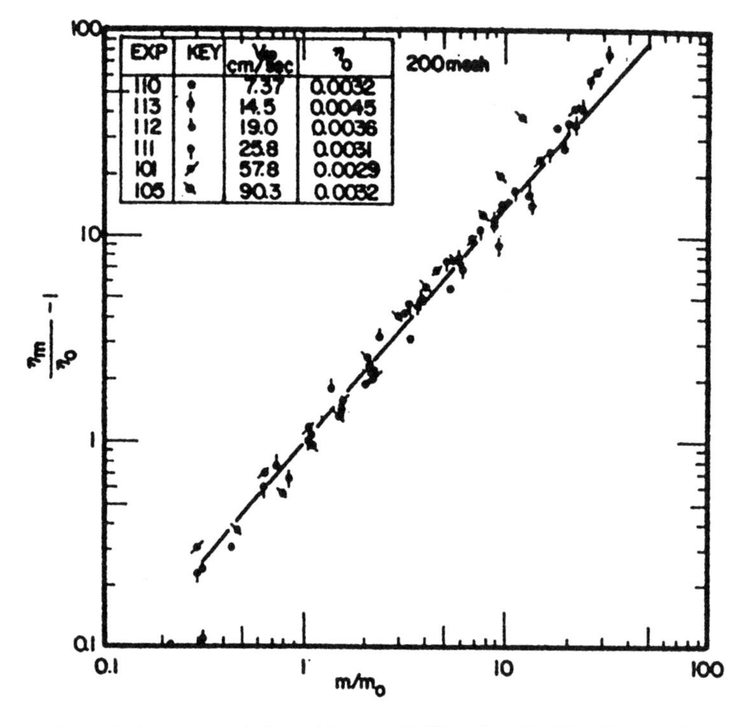

Figure 15.6 Comparisons between correlation of Emi et al. (Equation 15.23) with experiments. (From Emi, H. et al., *AIChE J.*, 28, 397–405, 1982. With permission.)

Similar correlation on pressure drop was also attempted and not successful. The scattering of the pressure drop data is shown in Figure 15.7.

15.5.2 Granular Filtration of Monodispersed Aerosols

As shown in the preceding section, the relationship between the single-fiber efficiency and the extent of deposition in model filters was determined from data taken from model filters composed of k layers of wire screen and $(k-1)$ layers of wire screen. This approach, however, is not applicable to granular filters since it is difficult, if not impossible, to add one layer of granules to or substract one layer of granules from a filter.

Walata et al. (1986) proposed a procedure for extracting information on the effect of deposition on filter performance from effluent concentration and pressure drop data. For a granular filter of

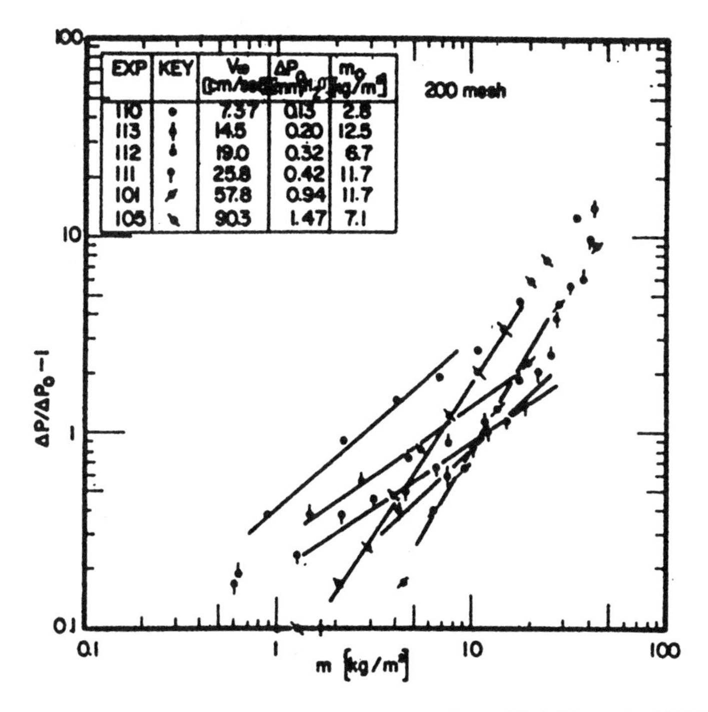

Figure 15.7 Pressure drop increase due to deposition in model filters (Emi, H. et al., *AIChE J.*, 28, 397–405, 1982. With permission.)

length L, if deposition is assumed to be uniform, the extent of deposition expressed as an average specific deposit, $\bar{\sigma}$ may be expressed as

$$\bar{\sigma} = \frac{1}{L}\int_o^t \left(C_{\text{in}} - C_{\text{eff}}\right)dt \tag{15.25}$$

The average pressure gradient, $\left(\overline{dp/dz}\right)$, is

$$\left(\overline{dp/dz}\right) = \Delta P/L \tag{15.26}$$

where ΔP is the pressure drop.

The unit collector efficiency under the unform deposition assumption, $\bar{\eta}$, can be expressed as (Takahashi et al., 1986)

$$\bar{\eta} = 1 - \left(\frac{C_{\text{eff}}}{C_{\text{in}}}\right)^{1/N} \tag{15.27}$$

$$N = L/l \tag{15.28}$$

$$l = \left[\frac{\pi}{6\left(1-\varepsilon_0\right)}\right]^{1/3} d_g \tag{15.29}$$

where l is known as the length of periodicity (i.e., the axial distance corresponding to one unit collector). d_g and ε_0 can then filter grain diameter and the initial porosity of filter media.

From Equations 15.25, 15.26, and 15.27, and the effluent concentration and pressure drop data for one set of conditions, one may establish empirical correlations between $\overline{\eta}$ and $\overline{\sigma}$ and $\left(\overline{dp/dz}\right)$ and $\overline{\sigma}$ as

$$\frac{\overline{\eta}}{\eta_o} = 1 + \overline{\alpha}_1 \overline{\sigma}^{\overline{\alpha}_2} \tag{15.30}$$

$$\frac{\overline{(\partial P/\partial z)}}{\partial P/\partial z_o} = 1 + \overline{\beta}_1 \overline{\sigma}^{\overline{\beta}_2} \tag{15.31}$$

where η_o and $(\partial P/\partial z)_o$ are the initial values of $\overline{\eta}$ and $\left(\partial \overline{P}/\partial z\right)$. The empirical constants $\overline{\alpha}_1$ and $\overline{\alpha}_2$ (or $\overline{\beta}_1$ and $\overline{\beta}_2$), in general, depend upon the length of the filter. By evaluating these constants from effluent concentration and pressure drop data obtained using filters of different height, one may obtain, by extrapolation, the values of these constants at zero height, thus establishing the relationships between the increase in unit collector efficiency (or pressure drop) with the extent of deposition expressed in σ.

The procedure of Walata et al. was later used by Takahasi et al. to establish empirical correlations for estimating the changes in η and $(\partial P/\partial z)$ (or permeability which is inversely proportional to pressure graduent) due to deposition. However, this is a rather tedious and time-consuming task as it requires extensive data collection using filters of different heights.

More recently, Jung and Tien (1991), based on simulation results, found that if filter height is sufficiently shallow; or $L \leq 10l$, the results obtained by using the uniform deposition assumption and those obtained by applying the extrapolation procedure of Walata et al. become identical. Based on data collected by Jung and Tien (1991), the following correlations were proposed.

$$\frac{\eta}{\eta_o} = 1 + \alpha_1 \sigma^{\alpha_2} \tag{15.32}$$

$$\alpha_2 = 0.4416 N_{\mathrm{st}}^{-0.3469} N_R^{0.2397} \tag{15.33a}$$

$$\alpha_1 = 0.09545 N_{\mathrm{St}}^{-1.478} N_R^{0.4322} 10^{3\alpha_2} \tag{15.33b}$$

$$\frac{(\partial P/\partial z)}{(\partial P/\partial z)_o} = 1 + \beta_1 \sigma^{\beta_o} \tag{15.34}$$

$$\beta_2 = 3.5134 N_{\mathrm{St}}^{-0.0925} N_R^{0.2748} \tag{15.35a}$$

$$\beta_1 = 0.3484 N_{\mathrm{St}}^{-1.199} N_R^{0.8568} 10^{3\beta_2} \tag{15.35b}$$

where the Stokes and interception parameters, N_{st} and N_R, are

$$N_{\mathrm{St}} = \left(\rho_p d_p^2 U_\infty C_s\right) / \left(9 \mu d_g\right) \tag{15.36a}$$

$$N_R = d_p/d_g \tag{15.36b}$$

where d_p and d_g are the aerosol and filter grain diameters, respectively, ρ_p is the particle density, μ, the gas viscosity, U_∞, the gas superficial velocity, and C_s, the Cunningham correction factor.

15.5.3 Granular Filtration of Polydispersed Aerosols

The results presented in the preceding section are those of monodispersed aerosols. In actual applications, particles to be removed from gas streams are likely to cover a range of sizes. Thus, in predicting the performance of filters, the values of the unit collector efficiencies of aerosols of different sizes are required (or the so-called fractional efficiencies). In determining the changes in the fractional efficiencies due to deposition, one must take into account of the fact that the deposited particles are polydispersed.

The effect of the size of deposited particles on collector efficiency was studied experimentally by Jung and Tien (1993) and Wu and Tien (1995). The experiment devised by these investigators may be described as follows. A shallow experimental filter was first subjected to a monodispersed aerosol flow (diameter d_{pj}) for a specified period of time followed by a different type of monodispersed aerosol flow (diameter d_{pi}) for another specified period of time. The procedure was then repeated. Although both types of aerosols were deposited within the filter, by adjusting the respective periods of exposure and the aerosol concentrations, it was possible to create a situation in which the dominant amount of deposited particles was of one type (say, d_{pj}). Thus, the unit collector efficiency of particles of diameter d_{pi} in a filter with deposited particles of diameter d_{pj} or $\eta(i/j)$ may be determined.

A set of experimental results obtained by Jung and Tien is shown in Figure 15.8. In Figure 15.8a, the unit collector efficiency of aerosols of diameter 1.1 μm in filters with deposited particles of diameters 1.1, 2.02, and 3.09 μm is shown. It is clear that for a fixed amount of deposition (i.e., constant σ), the collection efficiency decreases as the size of the deposited particles increases. The same conclusion was also reached for the two other cases.

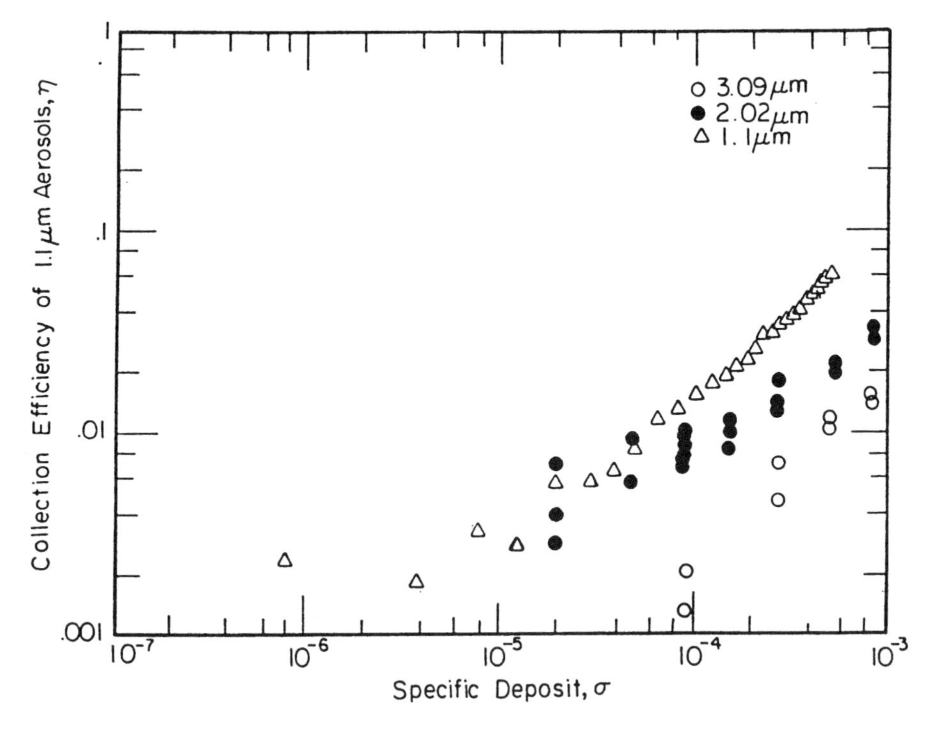

Figure 15.8 Effect of deposited particle size on the increase of the unit collector efficiency reported by Jung and Tien (1992), u_s = 11.3 cm S^{-1}, d_g = 525 μm, L = 0.42 cm.

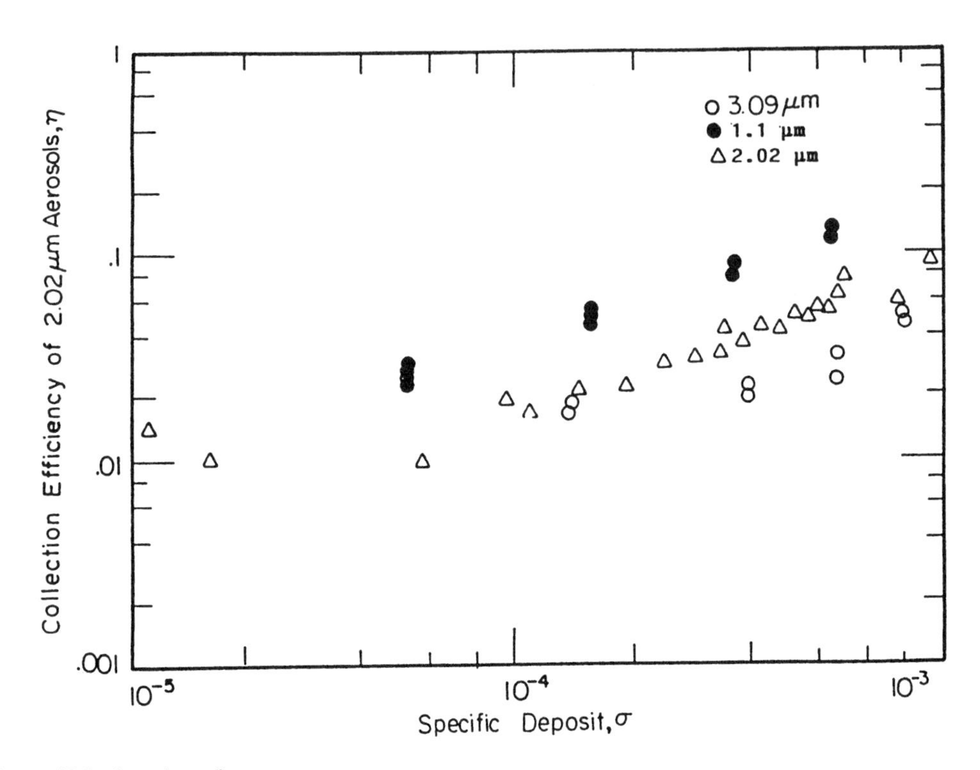

Figure 15.8 (continued)

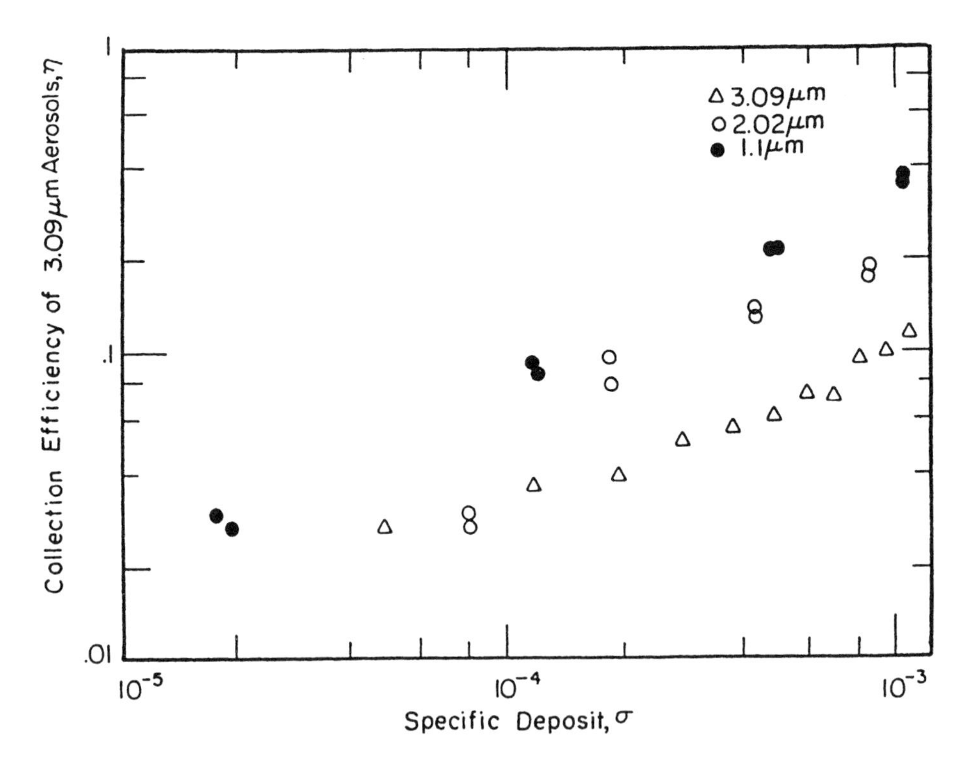

Figure 15.8 (continued)

The experimental results demonstrate that

1. $$\eta(i/j) < \eta(i/i) \quad \text{if} \quad d_{pj} > d_{pi} \tag{15.37}$$

2. The experimental results can be approximated by the power law expression of the type of Equation 15.32 or

$$\eta(i/j) = 1 + \alpha_1(i/j)\sigma^{\alpha 2(i/j)} \tag{15.38}$$

where the symbol (i/j), denotes a variable of particles of diameter d_{pi} in a filter with deposited particles of diameter d_{pj}.

In polydispersed aerosol filtration, the deposited particles are of different sizes. Furthermore, the composition of the deposited particles can be expected to vary with both temporally and spatially within the filter. Jung and Tien (1993) hypothesized that for polydispersed aerosols of n different sizes, the increase in the collection efficiency of the particles of the ith type (i.e., with diameter d_{pi}) may be approximated to be the sum of the contributions due to deposits of each of the n type of particle with their volume fractions as the weighing factors, or

$$\left(\frac{\eta}{\eta_o}\right)_i = 1 + \sum_{j=1}^{n} \alpha_1(i/j)\sigma^{\alpha_2(i/j)} w_j \tag{15.39}$$

$$w_j = \sigma_j/\sigma \tag{15.40}$$

$$\sigma = \sum_{i=1}^{\infty} \sigma_j \tag{15.41}$$

where σ_j is the specific deposit of the jth type particles and σ, the total specific deposit. w_j is the fraction of the jth particle of the deposits. $\alpha_1(i/j)$ and $\alpha_2(i/j)$ are the empirical constants of Equation 15.38.

The validity of Equation 15.39 was confirmed experimentally. In Figure 15.9, we present the experimental data obtained by Jung and Tien using a tridispersed aerosol (with d_p = 1.1, 2.02, and 3.09 μm). The collection efficiency of 1.1 μm particles is shown as a function of the total specific deposit. The experimental values agree with predictions based on Equation 15.39. Also included in this figure are the results of the collection efficiency for the monodispersed case. The effect of smaller deposited particles is shown rather convincingly.

With the validation of Equation 15.39, predicting the performance of polydispersed granular filtration becomes possible if the empirical constants of $\alpha_1(i/j)$ and $\alpha_2(i/j)$ (of Equation 15.38) are known. In a recent study, Wu and Tien (1995) attempted correlations of the constants. The correlations are

$$\alpha_1(i/j) = \alpha_1(i/i) \tag{15.42}$$

$$\frac{\alpha_2(i/j)}{\alpha_2(i/i)} = 1.0324\left(d_{pi}/d_{pi}\right)^{0.1339} N_{\text{St}}^{-0.0815} N_R^{0.0787} \text{ for } d_g = 250\ \mu m \tag{15.43a}$$

$$= 0.5167\left(d_{pi}/d_{pi}\right)^{0.2046} N_{\text{St}}^{0.0498} N_R^{-0.1642} \text{ for } d_g = 525\ \mu m \tag{15.43b}$$

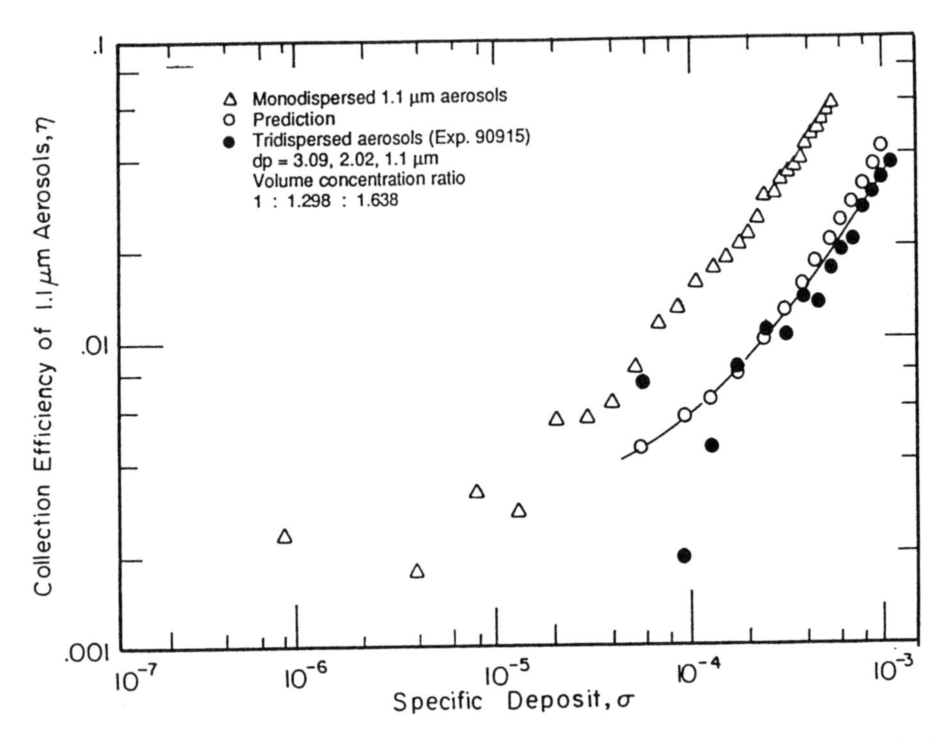

Figure 15.9 Comparisons between experiments and predictions based on Equation 15.39. Data obtained using tridispersed aerosols at u_s = 11.3 cm S^{-1}, d_g = 525 μm, L = 0.42 cm.

In other words, the constants are related to their counterparts of the monodispersed case; namely, $\alpha_1(i/i)$ and $\alpha_2(i/i)$ are given by Equations 15.33a and b.

15.6 PREDICTION OF GRANULAR FILTRATION OF POLYDISPERSED AEROSOLS

With the correlation of $\alpha_1(i/j)$ and $\alpha_2(i/j)$ available, it becomes possible to predict the granular filtration of polydispersed aerosol filtration. In the following section, we outline the algorithm proposed by Wu and Tien (1995) recently.

The macroscopic conservation equations and the filtration rate equation for polydispersed aerosols are (Tien, 1989)

$$U_\infty + \frac{\partial C_s}{\partial C} + \frac{\partial \sigma_i}{\partial \theta} = 0 \tag{15.44a}$$

$$\frac{\partial \sigma_i}{\partial \theta} = U_\infty \lambda_i C_i \quad i = 1, 2, \ldots\ n \tag{15.44b}$$

where z is the axial distance of a filter and θ, the corrected time, is defined as

$$\theta = t - \int_o^t \frac{\varepsilon dz}{U_\infty} \tag{15.45*}$$

and C_i is the concentration of ith type of aerosols of the gas phase. σ is the specific deposit of the ith type aerosol and U_∞, the superficial velocity. The filter coefficient, λ_i, can be related to the unit collector efficiency as

$$\lambda_i = \eta_i / l \tag{15.46}$$

where l is the length of periodicity defined before. The unit collector efficiency η_i can be found from Equation 15.39 and is a function of $\sigma_i's$ and known and with $\alpha_1(i/j)$ and $\alpha_2(i/j)$ estimated from Equations 15.42 and 15.43, the local instantaneous η_i can be determined readily.

The initial and boundary conditions are

$$C_i = (C_i)_{\text{in}}, \quad i = 1,2 \ldots n \quad \theta > 0, \quad z = 0 \tag{15.47a}$$

$$C_i = (C_w)_\infty, \quad i = 1,2 \ldots n \quad z > 0, \quad \theta = 0 \tag{15.47b}$$

$$\sigma_i = (\sigma_i)_o$$

where $(C_i)_o$ and $(\sigma_i)_o$ are the initial values of C_i and σ_i within the filter (for a clear filter, they are zero) and $(C_i)_{\text{in}}$ is the influent value of C_i.

The effluent concentration, $(C_{\text{in}})_{\text{eff}}$ can be obtained from the numerical solution of Equations 15.44a and b with the initial and boundary conditions of Equation 15.47a and b. The dependent variables are σ_i and C_i. If a filter is viewed as an assembly of unit collectors, C_i at the exit of a collector can be found from the inlet value of C_i and the knowledge of η_i which, in turn, requires information of the extent of deposition, i.e., σ_i $i = 1,2,\ldots n$. A comparison between experiments and predictions made by Wu and Tien is shown in Figure 15.10.

ACKNOWLEDGMENT

The study was conducted with partial support from a research grant awarded to the Environmental Technology Enterprise, National University of Singapore from the National Science and Technology Board, Republic of Singapore.

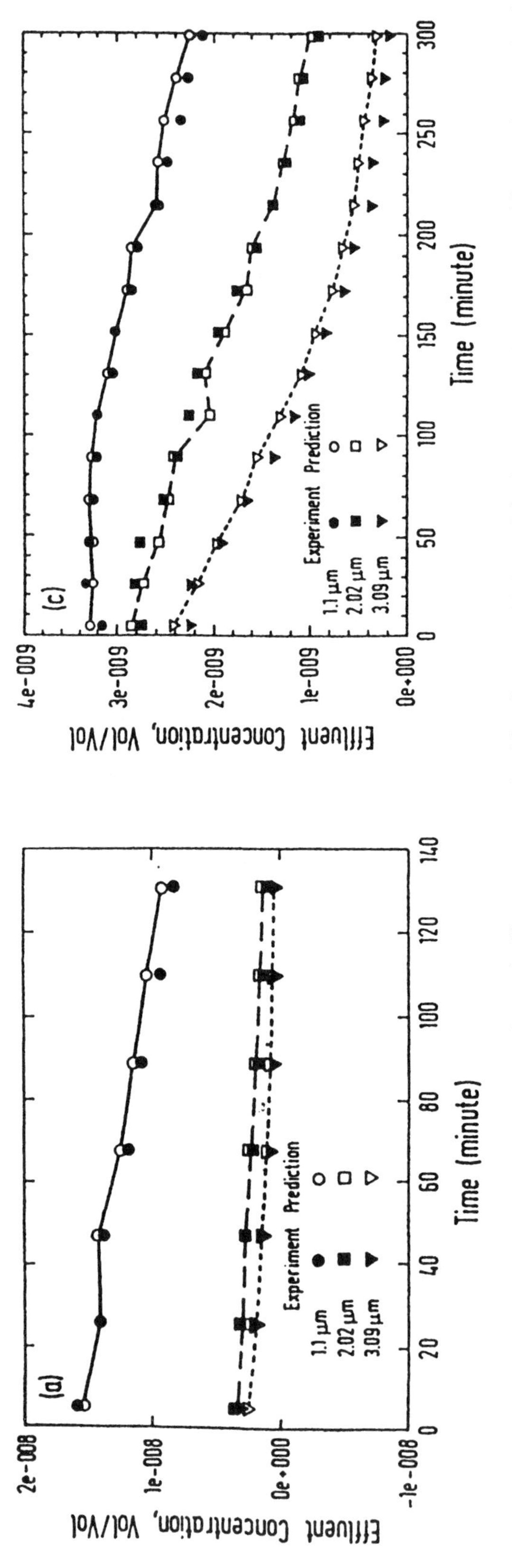

Figure 15.10 Comparisons between predicted and experimental effluent concentration histories. u_s = 11.3 cm S^{-1}, d_g = 525 µm, L = 0.42 cm, tridispersed aerosols of diameter 1.1, 2.05, and 3.09 µm. Initial number concentration ratio 170.5:6.1:1. (From Wu, X. and Tien, C., *Separation Technol.*, 5, 63–75, 1995. With permission.)

REFERENCES

Barot, D. T., Tien, C., and Wang, C. S. (1980). *AIChE J.,* 26, 289–290.

Billings, C. E. (1966). Effect of Particle Accumulation in Aerosol Filtration, Ph.D. dissertation, California Institute osf Technology, Pasedena.

Dahneke, B. (1971). *J. Colloid Interface Sci.,* 34, 342–353.

Dahneke, B. (1972). *J. Colloid Interface Sci.,* 40, 1–13.

Dahneke, B. (1973). *J. Colloid Interface Sci.,* 45, 584–590.

Davies, C. N. (1970). *J. Aerosol Sci.,* 1, 35–49.

Davies, C. N. (1973). *Air Filtration,* Academic Press, London.

Emi, H., Wang, C. S., and Tien, C. (1982). *AIChE J.,* 28, 397–405.

Juda, J. and Chrosciel, S. (1970). *Staub Reinhalt. Luft,* 30, 196–198.

Jung, Y. and Tien, C. (1993). *Aerosol Sci. Technol.,* 18, 418–440.

Jung, Y. and Tien, C. (1991). *J. Aerosol Sci.,* 22, 187–200.

Jung, Y. and Tien, C. (1992). *J. Aerosol Sci.,* 23, 525–537.

Leers, R. (1957). *Staub,* 50, 402–417.

Payatakes, A. C. (1976a). *Filtr. Sep.,* 13, 602–607.

Payatakes, A. C. (1976b). *Powder Technol.,* 14, 267–278.

Payatakes, A. C. and Gradon´, L. (1980). *Chem. Eng. Sci.,* 35, 1083–1096.

Payatakes, A. C. and Okuyama, K. (1982). In *Proceedings of International Symposium on Powder Technology '81,* The Society of Powder Technology (Japan), Kyoto, 501–516.

Payatakes, A. C. and Tien, C. (1976). *J. Aerosol Sci.,* 7, 85–100.

Takahashi, T., Walata, S. A., and Tien, C. (1986). *AIChE J.,* 32, 684–690.

Tien, C. (1989). *Granular Filtration of Aerosols and Hydrosols,* Butterworths, Boston.

Tien, C., Wang, C. S., and Barot, D. T. (1977). *Science,* 196, 983–985.

Tsiang, R. C. (1980). Mechanics of Particle Deposition in Model Filters Composed of an Array of Parallel Fibers, Ph.D., dissertation, Syracuse University, Syracuse, NY.

Tsiang, R. C., Wang, C. S., and Tien, C. (1982). *Chem. Eng. Sci.,* 37, 1661–1673.

Ushiki, K. and Tien c. (1985). *AIChE Sym. Ser. No. 241,* 80, 137–148.

Walata, S. A., Takahashi, T., and Tien, C. (1986). *Aerosol Sci. Technol.,* 5, 23–37.

Wang, C. S., Beizaie, M., and Tien, C. (1977). *AIChE J.,* 23, 879–889.

Watson, J. H. L. (1946). *J. Appl. Phys.,* 17, 121–127.

Wu, X. and Tien, C. (1995). *Sep. Technol.,* 5, 63–75.

Yoshida, H. and Tien, C. (1985). *Aerosol Sci. Technol.,* 4, 365–381.

CHAPTER 16

Performance of An Air Filter at Dust-Loaded Condition

Chikao Kanaoka

CONTENTS

Abstract — The performance of a dust-loaded air filter is reviewed, such as (1) The collection process of particles and the morphology of particle accumulates on a single fiber, including experimental observation and computer simulation by various collection mechanisms; (2) collection efficiency and pressure drop of a dust-loaded filter, (3) prediction of filter performance with dust load; (4) improvement of filter service life.

16.1 INTRODUCTION

Aerosol filtration by a fibrous air filter is a widely adopted and highly efficient method to separate particles from the gas stream. As particles deposit on a fiber inside a filter and/or on

0-87371-830-5/98/$0.00+$.50

previously captured particles, they form complicated accumulates, which lead to a marked increase in collection efficiency and pressure drop. Since filters are operated under dust-loaded conditions and since the mass of dust load in a filter changes the performances of air filters, in this chapter the following topics will be reviewed:

Collection of particles and morphology of particle accumulates on a single fiber;
Collection efficiency and pressure drop of a dust-loaded filter;
Prediction of filter performance under dust load;
Improvement of filter service life.

16.2 COLLECTION OF PARTICLES AND MORPHOLOGY OF ACCUMULATES ON A SINGLE FIBER

16.2.1 Observation of Deposition Pattern

When a fiber is clean, particles are collected directly on it, but once a particle is collected, the flow pattern changes, and this enhances the collection of particles because of the existence of captured particles. This can be clearly seen from Figure 16.1, which shows the time changes of accumulates of lead particles (d_p = 1 μm) on a tungsten wire (d_f = 10 μm) at a velocity of v = 50 cm/s (Stk = 3.5). At 1.5 min, few particles are collected on the fiber surface but the number of particles increases enormously from 1.5 to 10 min. Furthermore, the average distribution does not change much with time; i.e., the maximum deposition appears around the front stagnation and then decreases gradually to the side edge of the fiber. Figure 16.2 shows the deposition pattern by diffusion. Particles are captured not only on the front but also on the back side of the fiber and particles form tall and coarse accumulates.

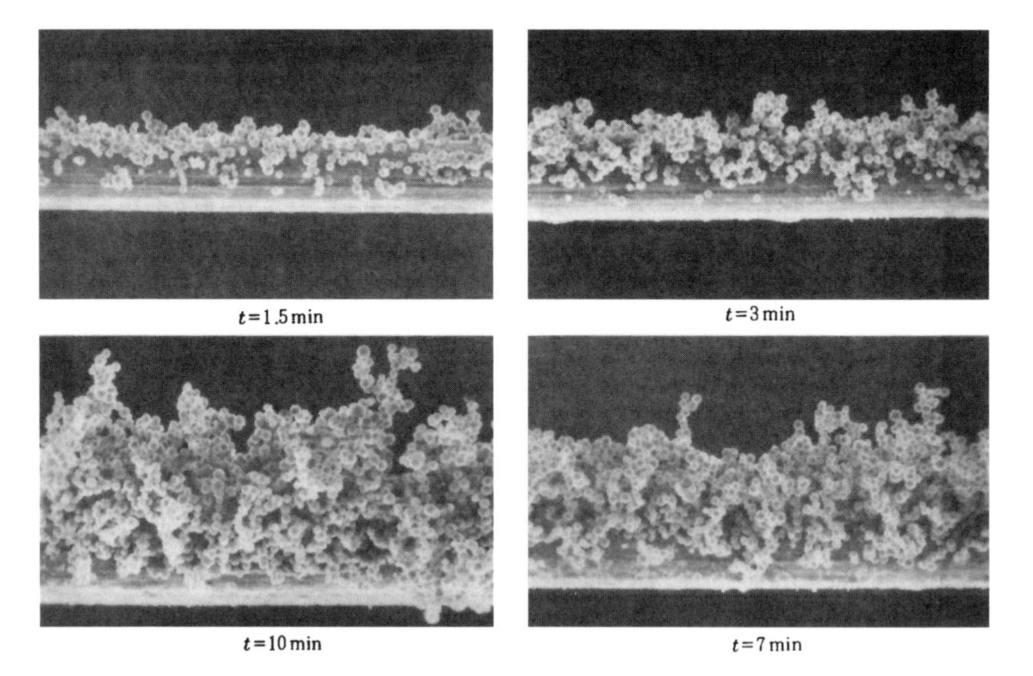

Figure 16.1 Time dependency of particle agglomerates on a tungsten wire d_f = 10 μm, d_p = 1 μm, v = 50 cm/s, ρ_p = 11.34 g/cm^3, Stk = 3.5, R = 0.1.

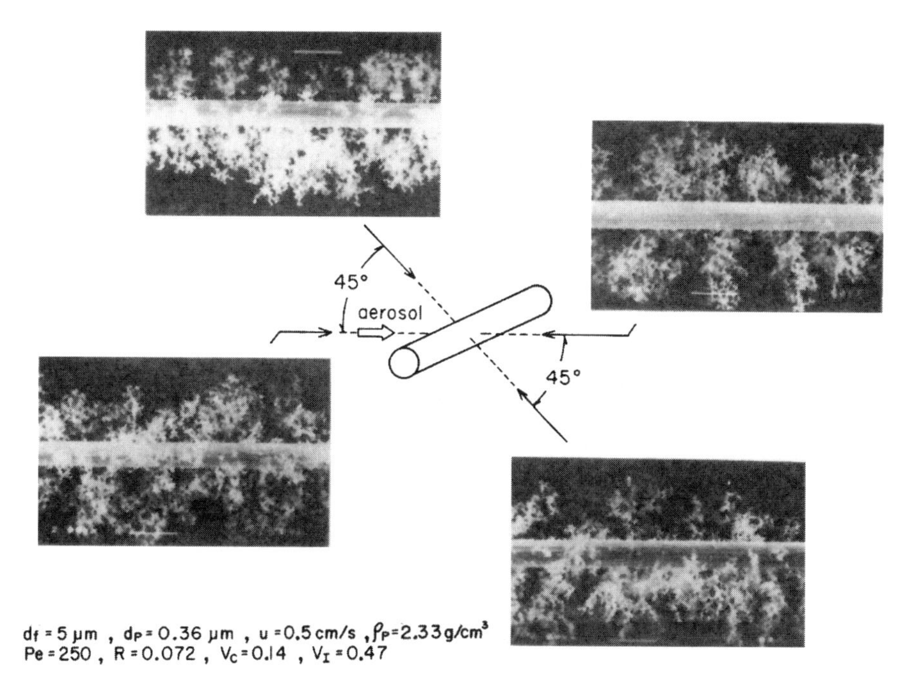

Figure 16.2 Particle agglomerates formed by diffusional effect.

Based on the experimental observation, the general features of the deposition pattern are summarized in Figure 16.3.

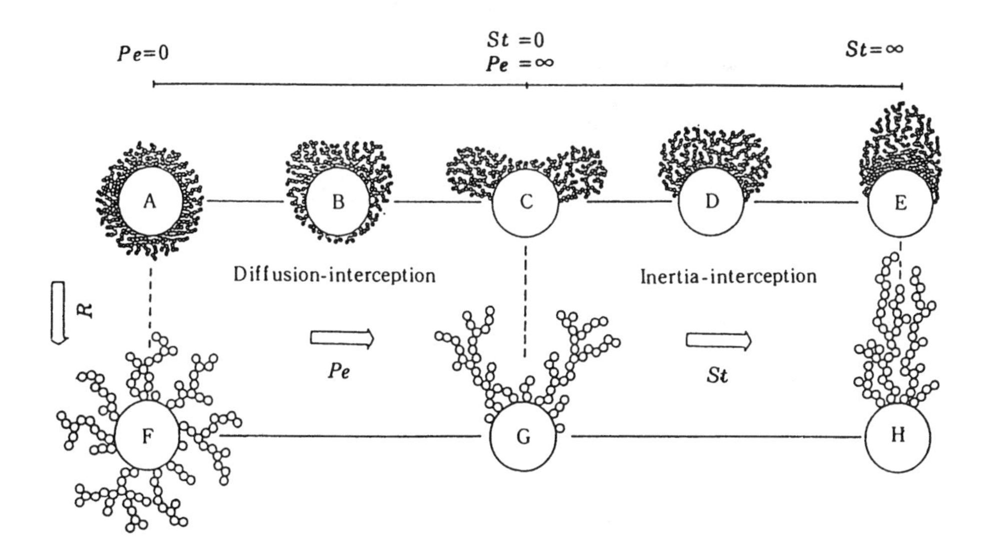

Figure 16.3 Shape of particle accumulates by the change of collection mechanism.

16.2.2 Deposition Pattern of Charged Particles

When particles and/or a fiber are electrically charged, the deposition pattern is different from mechanical collection because of electrostatic effects. Figure 16.4 shows the particles on an electret fiber. When uncharged particles are collected, they attach all around the fiber and form chainlike agglomerates and then they become irregular and complicated because of the weak electrostatic

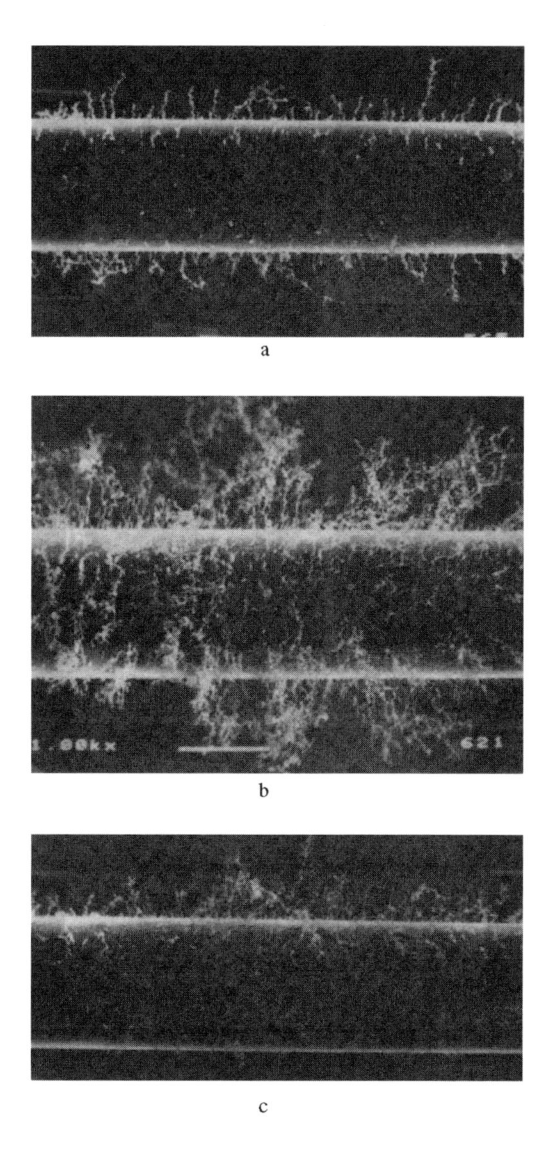

a

b

c

Figure 16.4 Particle agglomerates formed by electrostatic effect; $d_f = 30$ μm, $d_p = 0.39$ μm, $v = 15$ cm/s, $\rho_p = 2.33$ g/cm³, Stk = 0.015, Pe = 5 × 104, $R = 0.013$. (a) Uncharged particle $K_{In} = 0.004$, $V_c = 4.8 \times 10^{-4}$; (b) uncharged particle $K_{In} = 0.004$, $V_c = 1.5 \times 10^{-3}$; (c) charged particle $K_C = 0.016$, $V_c = 4.9 \times 10^{-4}$.

effect. For charged particles, the shape is similar to the former but agglomerates concentrate in a limited area of opposite polarity to the particles.

16.2.3 Simulation of Collection Process

Since the collection process is random and the flow field around it changes by every collection of a particle, it has to be calculated to simulate the process accurately for every collection of particles. However, the calculation of flow field around such an irregular fiber is almost impossible. There are two main approaches to simulate the particle agglomerates on a fiber, namely, the deterministic model[11-13] and the stochastic model.[1-10,14-16] The former is capable of describing the average growth of dendrites but does not express the random nature. In contrast, the latter can handle any random process but may take a long time to get meaningful results. The author has

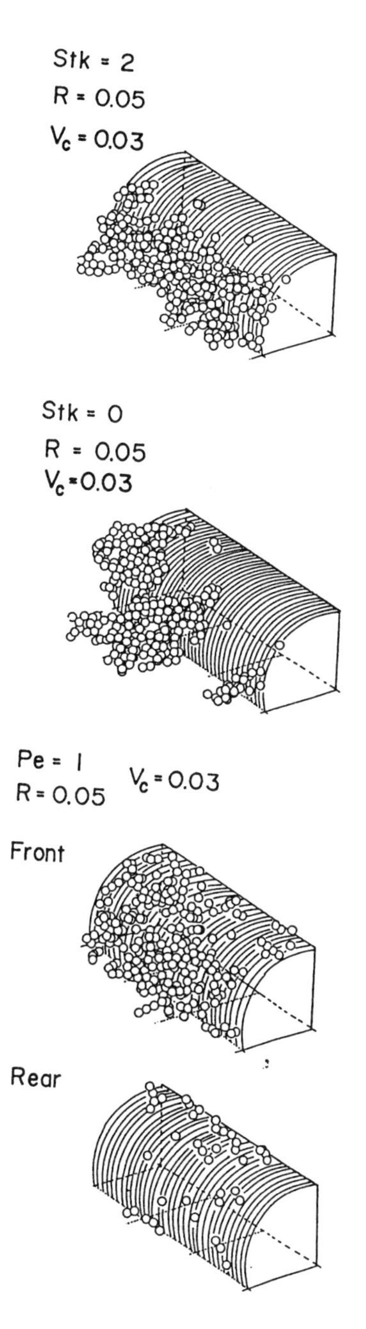

Figure 16.5 Simulated agglomerate.

developed a three-dimensional simulation algorithm. Figure 16.5 shows typical simulation accumulates by diffusion and inertial mechanisms.

16.3 COLLECTION EFFICIENCY AND PRESSURE DROP

16.3.1 Collection Efficiency

Collection efficiency of a clean fiber is defined by the ratio of inlet height of the limiting particle trajectory to the fiber diameter. This is also applied even for a dust-loaded fiber, although its apparent

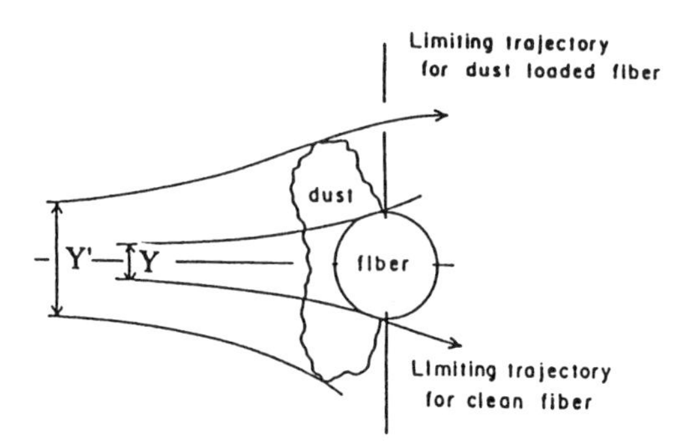

Figure 16.6 Definition of single-fiber collection efficiency.

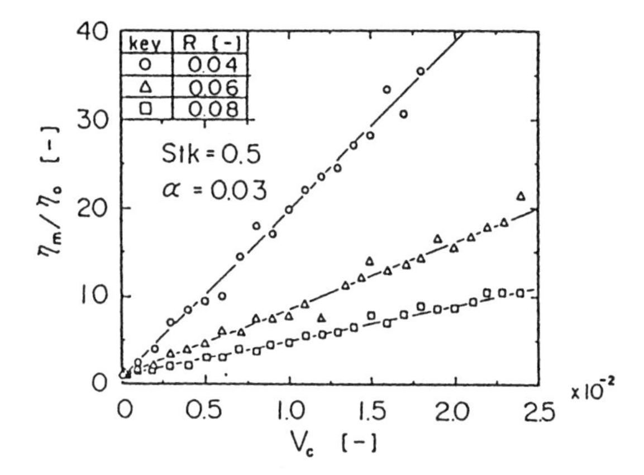

Figure 16.7 Experimentally obtained normalized single-fiber collection efficiency.

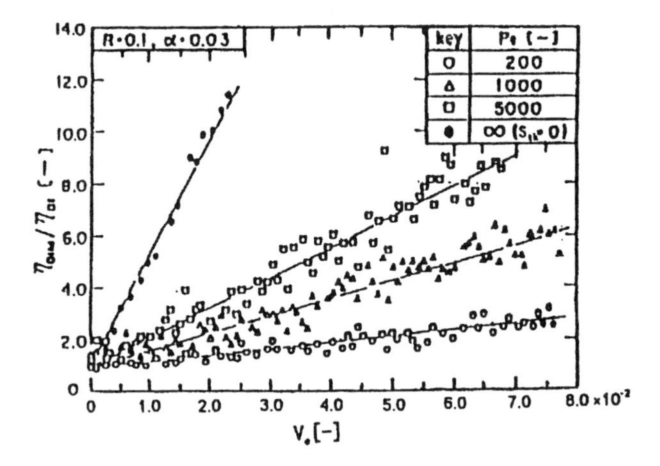

Figure 16.8 Simulated normalized single-fiber collection efficiency.

shape and size changes with time as shown in Figure 16.6. Figures 16.7 and 16.8 show the experimental and simulated efficiencies. Normalized efficiencies increase almost linearly with dust load so that it can be approximated by the following equation.

$$\eta_{\alpha m}/\eta_{\alpha 0} = \Lambda V_c = 1 + \lambda m \tag{16.1}$$

Here, Λ and λ are collection efficiency–raising factors and are plotted against collection efficiency of a clean fiber $\eta_{\alpha 0}$ in Figure 16.9. Both simulated and experimental Λ decrease with $\eta_{\alpha 0}$ and with Stk and R, but experimental λ is smaller than the prediction because of the difference in the microstructures of model and actual filters, and reentrainment of captured particles from fibers might occur in experiments but it is not taken into account in the simulation.

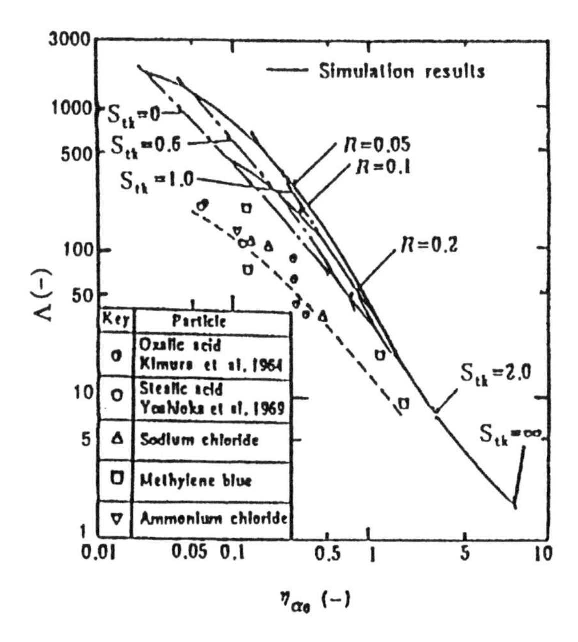

Figure 16.9 Relation between collection efficiency–raising factor and collection efficiency of a clean fiber.

16.3.2 Pressure Drop

When an air filter is not loaded too much, its pressure drop is related to the drag force acting on the fibers in it as

$$-\Delta P_m A = l_f \int_0^L F_m A\, dx \tag{16.2}$$

$$F_m = C_{Dm} d_{fm} \rho_f \, v^2/2 \tag{16.3}$$

here, l_f is total fiber length in a unit filter volume and is defined as

$$l_f = 4\alpha/\pi d_f^2 \tag{16.4}$$

Since C_{Dm} and d_{fm} change with dust load,

$$F_m = \left(C_{Dm}/C_{D0}\right)\cdot\left(d_{fm}/d_{f0}\right)\cdot C_{D0} d_{f0}\cdot \rho_f \, v^2/2 = \left(C_{Dm}/C_{D0}\right)\cdot\left(d_{fm}/d_{f0}\right)\cdot F_0 \tag{16.5}$$

$$-\Delta P_m = -\Delta P_0 \int_0^L \left(C_{Dm}/C_{D0}\right)\cdot\left(d_{fm}/d_{f0}\right)\cdot dx/L \tag{16.6}$$

In the above equation, C_{Dm} and d_{fm} are not constant but change with dust load and filtration condition, etc. so that pressure drop can be evaluated if both values are given by some measurable parameters. However, as mentioned before, captured particles do not accumulate uniformly on a fiber; they have to be determined experimentally. From the experiment, d_{fm}/d_f is approximated by the following equations:

$$\frac{d_{fm}}{d_f} = \begin{cases} 1 + aV_c & (V_c \le 0.05) \\ \sqrt{bV_c + c} & (0.05 \le V_c) \end{cases} \tag{16.7}$$

here, a, b, c stand for the experimental coefficients.

C_{Dm}/C_D also has to be determined through measurement of pressure drop of a thin layer of a model fan filter.

16.4 PREDICTION OF FILTER PERFORMANCE WITH DUST LOAD

When an air filter is composed of uniformly packed fibers with the same diameter, its efficiency is estimated by Equation 16.8, but cannot be used at dust-loaded condition.

$$E = 1 - \exp\left(-\frac{4}{\pi} \cdot \frac{\alpha}{1-\alpha} \cdot \frac{L}{d_f} \eta\alpha \right) \tag{16.8}$$

Under dust-loaded conditions, filter efficiency E_m and dust load can be estimated by solving Equations 16.9 and 16.10 with Equation 16.1 and initial and boundary conditions.

$$\frac{\partial C}{\partial x} = -\frac{4}{\pi} \cdot \frac{\alpha}{1-\alpha} \frac{\eta_{\alpha m}}{d_f} C \tag{16.9}$$

$$\frac{\partial C}{\partial x} = \frac{1}{v} \cdot \frac{\partial m}{\partial t} \tag{16.10}$$

Collection efficiency

$$E_m = 1 - \frac{C_o}{C_i} = 1 - \frac{\exp(-\lambda B C_i v t)}{\exp(-\lambda B C_i v t) + \exp(BL) - 1} \tag{16.11}$$

Dust load

$$m = -\frac{1}{\lambda} \cdot \frac{\exp(-\lambda B C_i v t) - 1}{\exp(-\lambda B C_i v t) + \exp(Bx) - 1}$$

$$B = -\frac{4}{\pi} \cdot \frac{\alpha}{1-\alpha} \frac{\eta_{\alpha 0}}{d_f} \tag{16.12}$$

Figures 16.10 and 16.11 show the estimated collection efficiency and pressure drop calculated from the above equations and the experimental results, in which methylene blue particles of 0.8 µm in diameter are collected at a filtration velocity of 1 m/s by a fan model filter that is composed of 25 model filter layers. Stainless steel wires of 30 µm in diameter are attached parallelly on each layer. As seen from the figure, the estimated efficiency and pressure drop agree very well with the experimental results.

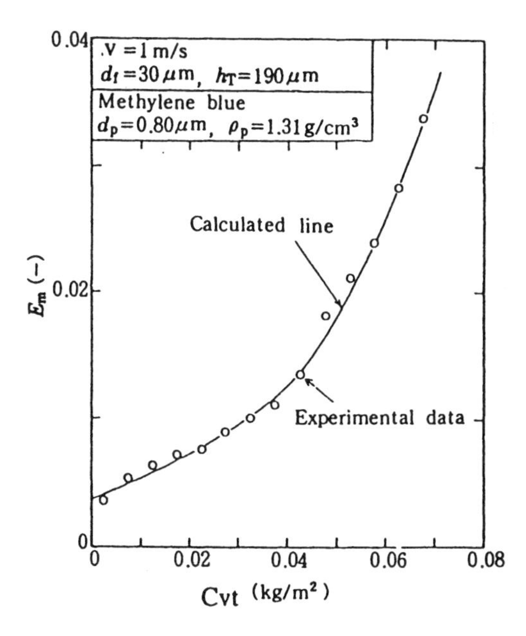

Figure 16.10 Comparison of estimated and experimental efficiency of a fan model filter.

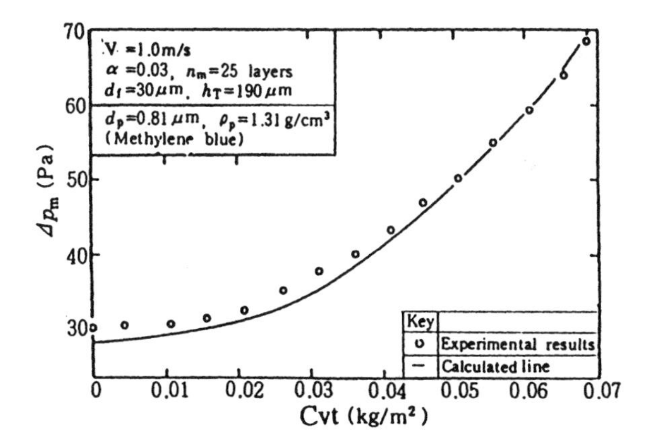

Figure 16.11 Comparison of estimated and experimental pressure drop of a fan model filter.

16.5 IMPROVEMENT OF FILTER SERVICE LIFE

Usually, an air filter is not equipped with a dust-cleaning system so that it is discarded when captured particles clog the air passage. Hence, it is necessary to design a filter with larger dust-holding capacity at the same pressure drop, i.e., a filter with a longer service life. Filter performance, of course, depends not only on filtration conditions and particle properties, but also on filter properties, such as fiber diameter, packing density, packing structure, and so on. This suggest that the performance changes by filter structure.

Here, the performance of two filters with the same packing density but with different distribution along thickness, shown in Figure 16.12, is compared. Major calculation conditions are listed in Table 16.1. Figures 16.13 and 16.14 show m and $\Delta P_m/\Delta_x$, respectively. When fibers are packed uniformly (filter A), both m and $\Delta P_m/\Delta_x$ have maximums at $x = 0$ at any Cvt and decrease very steeply inside the filter, while the filter with density distribution (filter B) the maximum appears at the outlet at $Cvt = 0$ and the maximum position of m shifts to the front side with Cvt. Since the drag of fiber is fairly large in this case, the maximum appears at the filter outlet and the same location of maximum m.

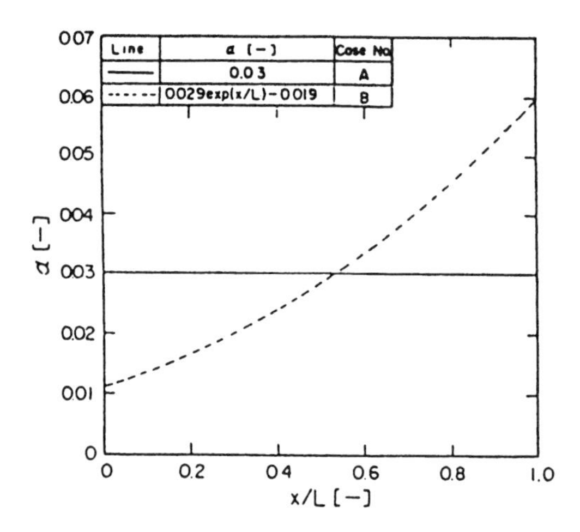

Figure 16.12 Fiber packing density distribution along thickness.

Table 16.1 Calculation Conditions

Fiber diameter d_f (μm)	30
Fiber packing density α(−)	0.03
Filter thickness L (cm)	5
Particle diameter d_p (μm)	1
Particle concentration C (mg/m³)	100
Filtration velocity v (m/s)	1

Figure 16.15 shows the time dependency of filter efficiency against total incoming particles per unit filtration area. Both efficiencies increase rapidly with Cvt but that of B raises more rapidly than that of A. Circles in the figure correspond the points of Cvt when pressure drop of the filter reaches two times the initial drop. Since an air filter is usually discarded when its pressure drop becomes double or triple the time of the initial, those Cvt values indicate the filter service life. In this sense, the service life of filter B is more than two times of filter A. In this calculation, all conditions other than the packing density distribution of fibers are set the same. This suggests that similar improvement can be expected by controlling or changing filter properties, such as fiber diameter distribution, combination of packing density distribution, and so on.

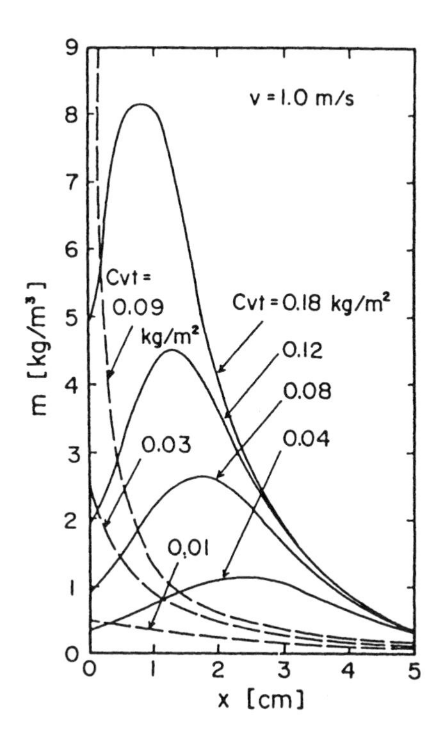

Figure 16.13 Distribution of dust inside a filter.

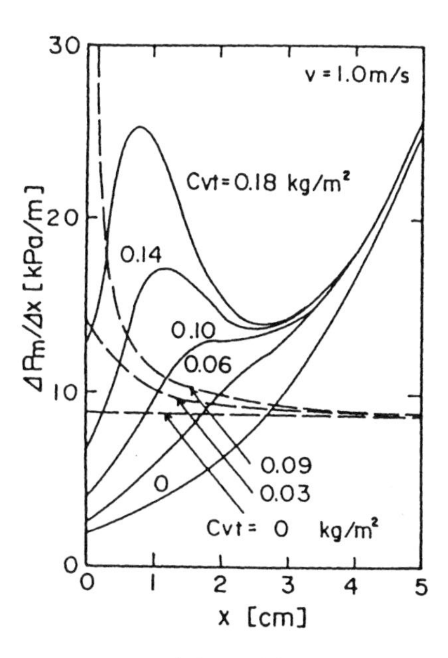

Figure 16.14 Distribution of pressure drop per unit filter thickness.

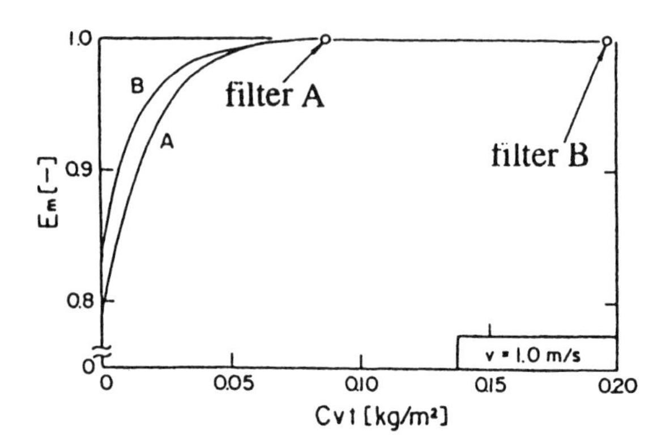

Figure 16.15 Time change of filter efficiency.

NOMENCLATURE

A	filtration area
a, b, c	constants used in Equation 16.10
C	particle concentration
C_c	Cunningham's correction factor
C_D	drag coefficient
D_{BM}	Brownian diffusion constant
d	fiber and particle diameters
E	filter efficiency
F	external force or fluid drag
L	filter thickness
l	total fiber length in a filter
m	dust load in a unit filter volume
ΔP	pressure drop
Pe	Peclet number (= $D_f u_0 / D_{BM}$
R	interception parameter (= d_p/d_f)
Stk	Stokes number (= $C_o \rho_p d_p^2 u_0 / 9\mu d_f$)
u	gas velocity
V_c	dimensionless volume of accumulated particles on a fiber (= $m\rho_p\alpha$)
v	particle velocity
x	location
α	packing density of fiber
η	single-fiber collection efficiency
Λ	dimensionless collection efficiency-raising factor (= $\lambda\rho_p\alpha$)
λ	collection efficiency–raising factor
μ	gas viscosity
ρ	air and particle densities

Subscript

f	fluid or fiber
m	dust loaded
p	particle
0	initial

REFERENCES

1. Bahrot, D. T., Tien, C. and Wang, C. S., *AIChE J.,* 26, 289, 1980.
2. Kanaoka, C., Emi, H., and Myojo, T., *Kagaku Kogaku Ronbunshu,* 5, 535, 1978.
3. Kanaoka, C., Emi, H., and Myojo, T., *J. Aerosol Sci.,* 11, 377, 1980.
4. Kanaoka, C., Emi, H., and Tanthapanichakoon, W., *AIChE J.,* 29, 895, 1983.
5. Kanaoka, C., Emi, H., and Hiragi, S., *Proc. of 2nd Int. Aerosol Conf.,* Berlin, 1986, 674.
6. Kanaoka, C., Emi, H., Hiragi, S., and Myojo, T., *Soc. Powder Tech. Jpn.,* 24, 74, 1987.
7. Kanaoka, C., Hiragi, S., and Yamada, K., *Kagaku Kogaku Ronbunshu,* 16, 252, 1990.
8. Kanaoka, C. and Hiragi, S., *J. Aerosol Sci.,* 21, 127, 1990.
9. Kanaoka, C. and Hiragi, S., *Proc. 12th Int. Symp. on Contamination Control,* 1994, 59.
10. Myojo, T., Kanaoka, C., and Emi, H., *J. Aerosol Sci.,* 15, 483, 1984.
11. Payatakes, A. C., *Filtr. Sep.,* 13, 602, 1976.
12. Payatakes, A. C., *AIChE J.,* 23, 192, 1977.
13. Payatakes, A. C. and Gradoń, *AIChE J.,* 26, 443, 1980.
14. Tanthapanichakoon, W. and Kanaoka, C., *Proc. 6th World Filtration Cong.,* 1993, 514.
15. Wang, C. S., Beizaie, M., and Tien, C., *AIChE J.,* 23, 879, 1977.
16. Wongtry, M., Tanthapanichakoon, W., and Kanaoka, C., *Adv. Powder Tech.,* 2, 11, 1991.

CHAPTER 17

Predicting Pressure Response Characteristics Across Particle-Loaded Filters

Vincent J. Novick and Jane F. Klassen

CONTENTS

17.1 BACKGROUND

The efficient design of filtration systems continues to be important in the control of air pollution for many applications. In general, the proper design of a filtration system requires the consideration of two parameters, the collection efficiency and particle mass–loading characteristics of the filter. The collection efficiency of the filter is the fraction of particles that are collected and do not penetrate the filter into the environment, compared to the total quantity of particles challenging the filter. The particle–loading characteristics describe how the pressure difference, ΔP, increases with the quantity of particles collected on the filter. The mass-loading characteristics of the filter are particularly important in situations where there are large quantities of particles that need to be removed from an airstream before the air can be discharged to the environment. Specific applications include radioactive or other toxic particle control situations where it is not viable to select a regenerative filtration technique due to particle resuspension increasing the possibility of particle migration into the environment.

The need to predict the relationship between mass loading and ΔP is apparent from the number of investigations found in the literature. For example, mass loading of fly ash on baghouse filters from coal-fired power plants has been extensively studied (Durham and Harrington, 1971; Airman and Helfritch, 1977). There has also been significant research done in the nuclear industry on sodium fire aerosols loading high-efficiency particulate air (HEPA) filters (Gunn and Eaton, 1976;

0-87371-830-5/98/$0.00+$.50

McCormack et al., 1978; Jordan et al., 1981; Pratt and Green, 1987). However, most of this research was conducted on a specific type of aerosol and a specific filtration method without regard to factors such as particle size, shape, particle density, velocity, or humidity effects. The specific nature of most of the studies makes it difficult to generalize the results to different aerosols or different filters, or even using the same aerosols and filters but varying the conditions of particle collection.

The focus of this chapter will be to examine various models relating the mass loading and the increase in ΔP across a filter, with emphasis on research that can be applied to a variety of filtration applications, allowing either the ΔP or mass loading to be predicted. There are two approaches that can be taken to establish a relationship between ΔP and mass loading. The first approach is to model physically the relationship between ΔP and mass loading and then determine what factors are necessary to make the model fit the data. The other approach is to develop an empirical equation correlated from a large quantity of data measuring mass loading as a function of ΔP. This chapter will present some of the models developed by various researchers that describe the relationship between ΔP and aerosol mass loading, with particular attention to the advantages and disadvantages of each model.

This chapter will focus primarily on HEPA filters because HEPA filters typically are nonregenerative and, therefore, most susceptible to being overloaded or blinded by the accumulation of particles. A HEPA filter is defined as any filter that demonstrates a collection efficiency of better than 99.97% for 0.3-μm-diameter particles. Most research on the mass loading of HEPA filters deals with solid particles at relatively low humidities and at typical HEPA filtration velocities of about 3 cm/s.

The chapter will also examine the applicability of the models to other filters, including prefilters, which are used in conjunction with a HEPA filter in many situations. For completeness, liquid models will be briefly addressed. Finally, this chapter will examine the research on effects of humidity on the mass-loading relationship with ΔP.

17.2 FILTRATION MODELS — SOLIDS

A number of different solid particle–gas filtration models can be found in the literature. Four of these models have been selected for examination in this section. Three of the models start with principles and develop models that describe the mass loading as a function of pressure drop. The other is a correlation-type model where a number of experiments are used to develop an equation that can predict mass loading as a function of pressure drop.

One of the earliest filtration models developed to describe the total pressure drop across a filter describes the total pressure drop across a filter as the sum of the pressure drop across the clean filter plus the pressure drop across the filter cake due to particle loading.

$$\Delta P = \Delta P_0 + \Delta P_{cake} \tag{17.1}$$

This model assumes that the filter and filter cake are both rigid, porous beds, superimposed on each other, and that the flow through them is laminar. This leads to each component of ΔP being described by D'Arcy's law.

$$\Delta P = K_1 V + K_2 VM/A \tag{17.2}$$

where $\Delta P_0 = K_1 V$ and $\Delta P_{cake} = K_2 VM/A$.

The values for K_1 and K_2 are the unknown parameters in this model, since the velocity can be calculated from the total volumetric gas flow rate and the total surface area available for filtration. K_1 depends solely on the properties of the individual filter, and can easily be measured for a particular filter from the slope of ΔP as a function of velocity curve. K_2 can be modeled by considering the pressure drop across the filter for a cake comprising isolated spheres of porosity (ε) approaching 1. K_2 can be determined using Stokes law,

$$\Delta P_{\text{cake}} = 3\pi\mu V d_p N/C \tag{17.3}$$

and the mass of particles collected,

$$W = N\rho(\pi/6)d_p^3 \tag{17.4}$$

and Equation 17.2, to yield

$$K_{2\text{-Stokes}} = 18\mu/\rho d_p^2 C \tag{17.5}$$

This equation is valid for aerosol spheres that are of uniform distribution and far enough apart that the gas flow around a particle is not affected by a neighboring particle. In actuality, the particles do touch and the gas streamlines around one particle are affected by neighboring particles. This causes a pressure drop increase greater than that expected from Stokes law. Therefore, a resistance factor, R, is used to account for the difference.

$$K_2 = RK_{2\text{-Stokes}} \tag{17.6}$$

The resistance factor, R, is always greater than 1, increasing as the porosity decreases and approaches 1 as the porosity increases.

In order to determine R, Kozeny and Carman (Carman, 1956) assumed that gas flows through capillaries with a surface area equal to that of the particles making up the cake. The capillary volume is set equal to the cake void volume. Then, R can be expressed as

$$R = 2K_{\text{KC}}(1 - \varepsilon)/\varepsilon^3 \tag{17.7}$$

where K_{KC} is defined as the Kozeny–Carman constant and is equal to 4.8 for spheres and equal to 5.0 for irregular shapes.

A review of fabric filtration by Leith and Allen (1986), concludes that Equation 17.7 should not be used for situations where $\varepsilon > 0.7$. However, this condition is not universally accepted. Japuntich et al. (1994) calculates values for ε between 0.87 and 0.89 using an expression derived from Equations 17.2 through 17.7 and compares these results with values for ε measured by Vold (1959) and Kirsch and Lahtin (1975) that ranged from 0.83 to 0.875.

One potential factor that may account for the differing conclusions is that the model requires an empirical factor K_{KC}, which accounts for particle shape but not other parameters. In an attempt to avoid the need for an empirical constant, K_{KC}, Rudnick and First (1978) developed the following equation for R:

$$R = [3 + 2(1 - \varepsilon)^{5/3}]/[3 - 4.5(1 - \varepsilon)^{5/3} - 3(1 - \varepsilon)^2] \tag{17.8}$$

However, while all of the parameters are based on physical properties, the functional dependence of the cake porosity is still unknown. The porosity can be determined from experimental measurements. This process of directly measuring the porosity is rather difficult and is usually calculated from the slope of a pressure drop vs. mass-loading data. Clearly, there is a wide range of reported values for the cake porosity. It should also be noted that for every 1% change in ε, the calculated ΔP to mass-loading ratio changes by 10%. Therefore, this model is of limited use in predicting mass loading as a function of ΔP, unless sufficient data have already been taken for the specific aerosol and filter system combination such that an accurate value exists for the cake porosity. It is also limited to solid particles and high-efficiency filters that allow a particle cake to form rapidly on the filter surface.

Bergman et al., (1979) developed an alternative model that accounts for particle collection throughout a porous filter based on an equation developed by Davies (1973). Bergman's model further differs from the Kozeny–Carman model by not requiring that a filter cake form on the filter surface. Instead, the particles trapped in the filter are treated as new fibers. Assuming the filter fibers are of uniform diameter and the challenge particles are also of uniform diameter, the following equation was developed:

$$\Delta P = 64\mu Vh[(\alpha_f/d_f^2) + (\alpha_p/d_p^2)]^{1/2}\ [(\alpha_f/d_f) + (\alpha_p/d_p)] \tag{17.9}$$

This model is considered general and describes the behavior of a variety of particle filters including prefilters and HEPA filters. Equation 17.9 can be simplified when considering HEPA filters since there is a large difference between the high fiber volume fraction and the small fiber size. Combining a Taylor series expansion with the assumption that the volume density of the fibers is greater than the particle volume density and the fiber effective collection diameter is less than the particle structure effective collection diameter, the following expression can be derived.

$$\Delta P - \Delta P_0 = 64\mu Vh(\alpha_f^{1/2}/d_f)\ (\alpha_p/d_p) \tag{17.10}$$

Bergman used this model to accurately fit data of sodium chloride aerosols collected at air velocities between 12 and 66 cm/s for mass loadings of up to 2 kg resulting in final pressure drops up to 800 Pa. The challenge in using this model is to determine the volume fraction of trapped particles, α_p, and the effective particle diameter, d_p. For HEPA filters, d_p requires a knowledge of the fraction of particle mass penetrating into the filter compared with the quantity that forms a cake on the surface. In addition, the effective particle diameter d_p is not necessarily the challenge aerosol particle diameter, but rather the diameter of the chainlike structures that are formed by the particles collected in the filter volume. Typically, d_p and α_p are adjusted to fit the data collected for a given aerosol filtration system. Therefore, it is difficult to use this model for predictive purposes.

The third model is as modification of the Bergman model. Instead of treating the collected particles as being uniformly distributed throughout the filter, Letourneau et al. (1989; 1991; 1993) treats the filter as a series of disks of a certain thickness. In this case, the particle volume density, α_p, then becomes a function of the thickness, and the pressure drop can be described by Equation 17.11,

$$\Delta P = \int_0^z 64\mu Vh\left[\left(\alpha_f/d_f^2\right)+\left(\alpha_p(x)/d_p^2\right)^{1/2}\right]\left[\left(\alpha_f/d_f\right)+\left(\alpha_p(x)/d_p\right)\right]dx \tag{17.11}$$

where $\alpha_p(x)$ is defined as

$$\alpha_p(x) = m_s\ \{(ke^{-kx})/[\rho(1 - e^{-kx})]\} \tag{17.12}$$

In order to prove the assumptions of this model, an adhesive tape peeling method was used to determine the penetration of particles in the filter. The authors contend that the results of the peeling method show that ΔP is not affected by the particle cake formed on the surface of the filter. This model was used to model accurately data sets for fluorescein particles of 0.15 μm mass median diameter collected at filtration velocities ranging from 3 to 50 cm/s through HEPA filters. The maximum mass collected was about 7 g/m² at a flow velocity of 10 cm/s resulting in a ΔP of about 800 Pa.

The model, while extremely accurate in describing the history of a filter that has been loaded, appears to be limited in its ability to predict loadings on untested filters under different aerosol collection conditions. As in the Bergman model, the equations require values for α_p, d_p, and the new coefficient k. These values can only be obtained from previously loaded filters for a given set of collection conditions since they depend on factors such as the particle size, shape, density, and filtration velocity. However, Letourneau does state that the factor k does not appear to be a function of particle velocity.

The fourth model attempts to avoid the need to determine factors, such as the cake porosity, by accumulating a large quantity of data on the mass loading of filters as a function of aerosol particle size, density, and shape and then correlating the data to an equation that can be applied to a wide variety of aerosol filter system designs. This approach is less satisfying from a scientific point of view, but may be more useful for predicting the relationship between ΔP and aerosol mass loading for untested filter systems. Novick et al. (1992) developed on empirical equation from experiments performed using three different aerosol materials that were chosen based on their variety of particle shape and density characteristics. The materials were sodium chloride, ammonium chloride, and aluminum oxide. The flow rate through the HEPA filter was held constant during the tests. For each material and particle size, the mass loading was plotted as a function of ΔP. Typical loading curves are shown in Figure 17.1 as a function of the aerosol particle diameter. As expected from the models and other experimental reports, the mass loading increases as the mass median diameter (MMD) of the aerosol is increased for a given ΔP.

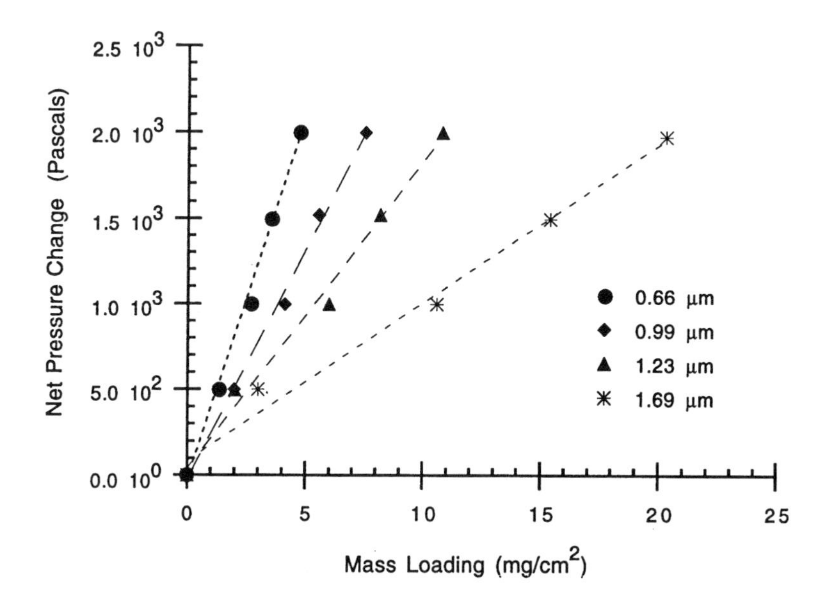

Figure 17.1 Typical HEPA filter loading response to NACl aerosols. (From Novick, V. J. et al., *J. Aerosol Sci.*, 23(6), 657–665, 1992. With permission.)

Bergman et al. (1979) showed that such a set of loading curves for a single material with different particle diameters could be normalized to a single curve by plotting the mass collected as a function of the increase in pressure divided by the particle diameter. However, Novick et al. (1992) found that this simple normalization was insufficient when considering particles of different densities. Instead, the slope of each line was determined for a given data set of pressure change as a function of mass loading. This value was then set equal to the specific resistance, K_2, times the gas filtration velocity, V, as in the simple model described by Equation 17.1. Each K_2 data point obtained from a graph of the mass loading as a function of ΔP was plotted as a function of the inverse particle diameter as shown in Figure 17.2. The particle diameter was chosen to be the MMD of the distribution because it represents the mass loading on the filter and is generally an easily measurable value.

The least-squares linear curve fit to the data when combined with the initial ΔP_0 of the clean filter results in Equation 17.13.

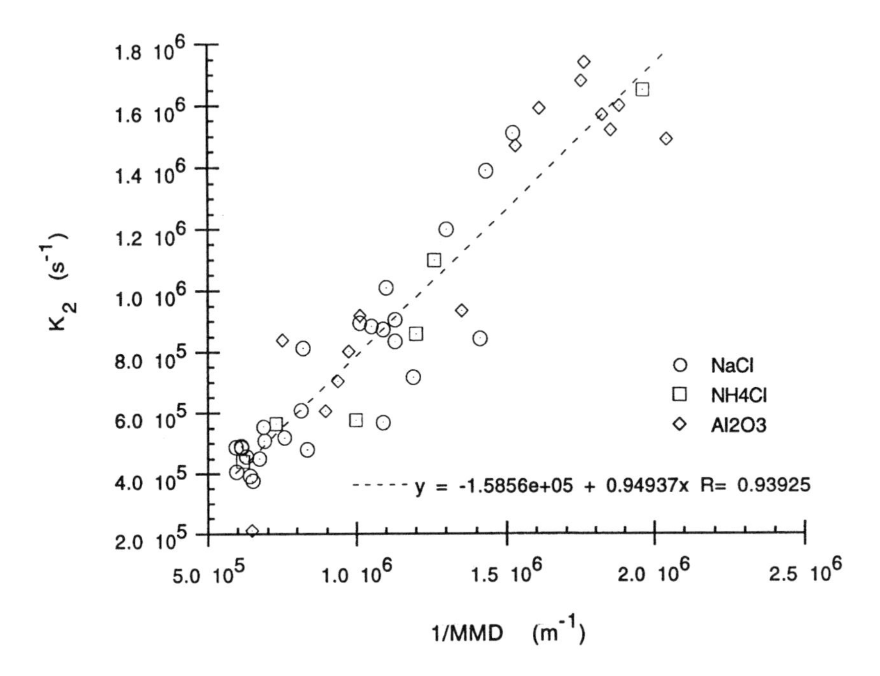

Figure 17.2 Specific resistance (K_2) as a function of particle diameter for different materials. (From Novick, V. J. et al., *J. Aerosol Sci.,* 23(6), 657–665, 1992. With permission.)

$$\Delta P_H - \Delta P_0 = [(0.963/MMD) - 1.64 \times 10^5]VM/A \tag{17.13}$$

It should be noted that the particle shape factor had no significant effect on the correlation. This is consistent with Kozeny and Carmen's work which observed only a 4% difference in value for the specific resistance between spherical and irregular particles.

The data presented in Figure 17.2 can also be plotted using a $1/\text{MMD}^2$ relationship and also provides an equally good curve fit. In both cases, the fit factor is better than 0.9. The *1/MMD* correlation was selected because it provided a significantly better fit in the large-particle region where the MMD was greater than 1 μm.

For this particular model, it is required that the initial clean ΔP across the filter, the gas velocity through the filter media, and the filtration area be known. Then, if the MMD of the particle that will be challenging the filter can be estimated or measured, a prediction can be made for the relationship between mass loading and ΔP.

The HEPA correlation has been successfully used in a number of areas. The correlation was used to predict mass loadings on HEPA-type filters ranging in size from 12.7 mm diameter (1.27×10^{-4} m^2 filtration area) to 6 × 6 in. pleated HEPA filter (3.85 m^2 filtration area). The correlation was also used to predict the mass loading on pleated HEPA filters placed downstream of a prefilter by calculating the MMD diameter of the aerosol penetrating the prefilter and challenging the HEPA filter (Novick et al., 1993). This diameter was calculated using data on the initial particle size distribution challenging the prefilter and the prefilter efficiency as a function of particle size. The calculated MMD of the particles challenging the HEPA filter was used in conjunction with Equation 17.13 to predict the mass loading as a function of ΔP. Table 17.1 presents the measured values and calculated results for comparison.

In order to be useful, a correlation must also compare favorably to data reported by other researchers. Letourneau's data (Letourneau et al., 1991) determines the specific resistance K_2, for 0.15 μm fluorescein particles at filtration velocities of 5 cm/s to be 6.3×10^6, while the specific resistance, predicted by extrapolation from Equation 17.13, is 6.2×10^6. Also, the correlation was used to predict the specific resistance for respirator mask filters. Hinds and Kadrichu (1994) present

Table 17.1 Calculated HEPA Filter Mass Loadings Based on Equation 17.13 vs. Experimentally Measured Mass Loadings Following a Prefilter

Challenge MMD	Penetrating MMD	Prefilter Efficiency	Calculated HEPA Mass	Measured HEPA Mass	Difference, %
1.69	0.71	0.98	0.55	0.65	15
1.60	0.71	0.98	1.27	4.95	74
1.58	0.71	0.98	1.46	1.75	17
1.58	0.71	0.98	2.44	6.25	61
1.34	0.71	0.97	2.49	2.40	–4
0.70	0.41	0.85	2.05	2.60	21
0.68	0.41	0.84	5.78	6.90	16
0.61	0.41	0.81	4.67	4.65	0
0.61	0.41	0.81	3.11	3.60	14
0.50	0.33	0.74	1.62	1.95	17
0.51	0.33	0.74	2.80	3.35	16
0.42	0.33	0.66	6.22	7.25	14
0.52	0.33	0.75	4.83	5.65	15
			Average Difference		21
			Standard Deviation		22

mass-loading data for two types of filters using AC Fine Test Dust (MMAD = 2.8 μm). Assuming the dust has the same density as silica ($\rho = 2.6$), the MMD of the test dust is 1.74 μm. The dual cartridge filter was measured to have a clean collection efficiency of 99.5% for the test dust. This is a less-efficient filter compared with measured efficiencies of 99.97% at 0.15 μm and 99.999% at 0.4 μm for HEPA filters (Novick et al., 1991), but is still considered an efficient filter. The experimentally determined average specific resistance for the dual cartridge filter was 3.6×10^5. The value determined from Equation 17.13 is 3.8×10^5. Therefore, the correlation appears to be applicable to any high efficiency filter, such that a particle cake is rapidly formed on its surface.

The second filter tested by Hinds and Kadrichu was a disposable filter with a measured efficiency of only 95% for the 1.7 μm particles. This efficiency is typical of prefilter efficiencies measured by Novick et al. (1991) and of various filter papers tested by Japuntich et al. (1994). These filters can also produce a linear ΔP response with mass loading, but the slope changes as the particle filtration mechanism proceeds from fiber filtration to cake filtration. The measured specific resistance for this filter, using the same AL fine test dust, was 5.0×10^4. However, the predicted value of K_2 is still 3.8×10^5. Since prefilters and other low-efficiency filters collect a significant fraction of particles in the depths of the filter material rather than on the surface as a cake, it is not surprising that the correlation cannot accurately predict the specific resistance for the disposable filter.

17.3 FILTRATION MODELS — LIQUIDS

The previous models considered only the filtration of solid aerosol particles. Much less work has been reported for liquid droplet collection. A model developed by Liew and Condor (1985) describes the ratio of the pressure drop for a liquid-coated filter compared with a dry clean filter:

$$\Delta P_{wet}/\Delta P_0 = A_1[d_f/\alpha_f h)^{0.561} (\text{At} \cos \phi/Q\mu)^{0.477}] \qquad (17.14)$$

This model only describes the pressure difference across a filter which is in equilibrium between mass collected by filtration and mass removed by drainage. The model is certainly limited to prefilters for this reason, because only prefilters are designed to have sufficient porosity to allow liquid to drain. The model does not describe how the pressure difference increases as the filter is loaded to the equilibrium point. This model contains two factors that tend to further limit its usefulness. First, the effective fiber diameter can become an uncertain quantity as the filter fibers become coated, especially with liquid aerosols with high surface tensions. Second, the contact angle of the droplet with respect to the fiber is generally unknown.

Some experimental results of liquid aerosol collection were reported by Novick et al. (1993), for both prefilters and HEPA filters. The prefilters were found to be properly designed such that the liquid drained from the filter resulting in no measurable increase in the pressure drop regardless of the total mass of liquid collected. However, the HEPA filters, as expected, did not drain and the increase in pressure drop as a function of liquid mass loading was exponential. These results are presented in Figure 17.3. Di-ethylene glycol (MMAD = 0.71) and dioctyl phthalate (MMAD = 1.53) were used as challenge aerosols due to their low vapor pressures.

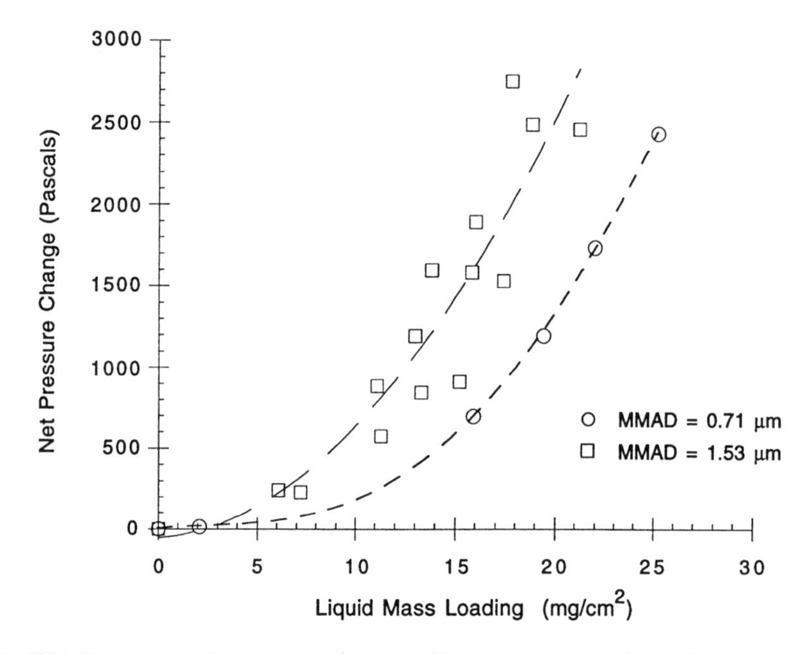

Figure 17.3 HEPA filter pressure increase as a function of liquid mass loading for di-ethylene glycol and dioctyl phthalate aerosols penetrating a prefilter. (From Novick, V. J. et al., in *22nd DOE/NRC Nuclear Air Cleaning Conference Proceedings,* 1993. With permission.)

The results indicate little change in the rate of pressure increase with mass loading for different aerosol particle droplet sizes. The difference in the two curves is attributed to differences in material properties between the two challenge liquids. Different liquids will have different surface tension and contact angles and should be expected to load a HEPA filter in a similar, nonlinear manner, but with a different functional relationship between mass loading and ΔP. Significantly more work needs to be done before predictions can be made for liquid particles challenging filtration systems.

Novick et al. (1993) also reported experiments on solid–liquid combined aerosols on HEPA filters. The data are not considered useful for other applications due to the functional dependence on the liquid surface tension, contact angles, and the solid-to-liquid ratio affecting the wicking of the liquid into the HEPA filter itself. Clearly, the pure solid models and pure liquid models cannot be used to predict the pressure response when loading solid–liquid combined aerosols. In general, the solid and liquid combined aerosol has the capability of blinding, plugging, or raising the ΔP to much higher levels for a given mass loading than either the pure solid or the pure liquid case.

17.4 HUMIDITY EFFECTS

While the research involving pure liquid or liquid–solid mixed aerosol collection is limited, a number of investigators have been interested in the more common problem of quantifying humidity effects on filter loading. Durham and Harrington (1971) observed a decrease in ΔP as relative humidity (RH) increased or fly ash particles but did not note any change for cement dust, pulverized limestone, for amorphous silica. However, Ariman and Helfritch (1977) noted a pressure drop decrease with increased humidity for not only fly ash, but also ground silica dust and asphalt rock dust.

Gupta et al. (1993) performed an extensive laboratory study to determine the effects of humidity on mass loading and ΔP characteristics of HEPA filters as a function of particle hygroscopicity and size, in an attempt to resolve the conflicts in trends noted by earlier researchers. The tests were performed using two materials, one hygroscopic and one nonhygroscopic, and particle sizes of 0.5 and 1.0 μm (MMD). Relative humidities ranging from 1% to 100% RH for nonhygroscopic material and between 35% and 90% RH for hygroscopic material were tested. A constant filtration velocity of 3.00 cm/s was used for all tests.

The results of these tests showed that for nonhygroscopic particles of aluminum oxide, there was a decrease in the specific resistance, k^2, with increasing humidity for both particle size distributions tested. However, this decrease was evident for the 0.5 μm particles only above 90% RH, while it was evident over the whole range of RH for the 1.0 μm particles. The results were reproducible, but unexplained.

The hygroscopic particles of sodium chloride also showed a decrease in k^2 for both particle sizes below 75% RH (deliquescent point for NaCl). However, above 75% RH, the mass-loading capacity of the filter is greatly reduced because of the change in phase of the aerosol particles. It was also noted that the higher the humidity of the hygroscopic particles, the more mass that could be loaded onto the HEPA filter. This was explained by considering the wicking phenomenon, in that particles closer to the deliquescent point are much more solid and would have a tendency to reside solely on the surface of the filter. At these humidity levels, the liquid viscosity would be insufficient to allow the filter to wick the liquid away from the front surface of the filter. At higher humidities, the hygroscopic particle would be more fluid and, hence, allow the liquid to be wicked into the depth of HEPA filter volume. At sufficiently high humidities, the mass-loading curve would be expected to parallel that obtained for a pure liquid collected on a HEPA filter.

While reproducible and explainable, an equation describing this behavior does not exist. Furthermore, correlation of the results into an empirical equation is not possible due to the limited data set.

17.5 CONCLUSIONS

The work presented in this chapter allows explanations for most of the data reported describing both linear and nonlinear increases in the pressure drop across a filter as a function of mass loading. Linear relationships are obtained for solid particles collected at relatively low filtration velocities (< 5 cm/s). Nonlinear curves are obtained when solid particles are collected at significantly higher velocities or when a wet aerosol is collected. A wet aerosol can be a pure liquid, a solid and liquid mixed aerosol, or a hygroscopic aerosol at high humidities. Hybrid curves, those that are initially linear but then become nonlinear, can usually be explained by a change in the collection parameters. For example, collecting solid particles at low humidities with a pleated filter initially produces a linear curve, which can become nonlinear when the valleys of the pleats begin to fill with particles such that the total surface area available for filtration is reduced as a function of time. It is also possible to have the humidity increase or the aerosol change from solid to wet during the collection period.

While it is relatively easy to explain the general behavior of the pressure drop across a filter as it is loaded with particles, it is much more difficult to predict a value for ΔP at a specified mass loading. Therefore, the design of filtration systems based on total mass or pressure limitations remains mostly guesswork unless the specific characteristics of the system have been previously determined or the system will be operated in a regime that allows the application of one of the models discussed.

For solid particle loading on any filter, if sufficient information is known about the particles and the manner in which they collect inside of a filter, i.e., the diameter and porosity of the structures formed by the collected particles, then the Letourneau (1993) and Bergman (1979) models would be the best choice for predicting relationships between ΔP and mass loading. If this information is not available and cannot be obtained, the models are of limited usefulness in predicting the pressure response as a function of mass loading.

For reasonably high-efficiency-type filters, such that the particle cake forms on the filter surface (the model defined by Equation 17.1), the method using the correlations developed by Novick et al. (1992) allows a fairly accurate (better than ± 25%) prediction of the relationship between the pressure increase and the particle mass loading (i.e., the specific resistance). This method allows predictions of the pressure change as a function of mass loading based solely on the knowledge of the MMD of the challenge aerosol.

A significant amount of work still remains in order to develop models and equations that can predict, *a priori*, mass loading as a function of ΔP for liquids, hygroscopic aerosols with or without humidity, and aerosol containing solids mixed with liquids.

NOMENCLATURE

A	surface filtration area
A_1	Liew and Condor correlation coefficient
C	Cunningham slip correction factor
d_f	fiber diameter
d_p	particle diameter
h	depth of filter material
k	mean penetration factor in the filter medium
K_1	filter constant depending on clean filter parameters
K_2	specific cake resistance
m_s	surface mass of aerosol collected on the filter
M	aerosol mass collected on filter
MMAD	mass median aerodynamic particle diameter
MMD	mass median particle diameter
N	number of spheres per unit area
ϕ	contact angle of droplet with respect to surface of the fiber
ΔP	total pressure difference
ΔP_0	pressure drop across clean filter
ΔP_{cake}	pressure drop due to particulate cake
ΔP_H	pressure drop across a HEPA filter
Q	volumetric gas flow rate
R	resistance factor (d_p/d_f)
t	liquid surface tension
V	filter media velocity
W	mass of collected particles

α_f	filter solidity or packing (volume) density
α_p	particulate volume density
ε	cake porosity
ρ	particle density
μ	gas viscosity

REFERENCES

Airman, T. and Helfritch, D. J. (1977), *Filtr. Sep.,* 14, 127–130.

Bergman, W. et al. (1979), in *15th DOE Nuclear Air Cleaning Conference Proceedings*, CONF-760822.

Carman, P. C. (1956), *Flow of Gases through Porous Media,* Academic Press, New York.

Davies, C. N. (1973). *Air Filtration,* Academic Press, New York.

Durham, J. F. and Harrington, R. E. (1971), *Filtr. Sep.,* July/August, 389–398.

First, M. W. and Rudnick, S. N. (1981), in *16th DOE Nuclear Air Cleaning Conference Proceedings,* CONF-801038.

Gunn, C. A. and Eaton, D. M. (1976), in *14th ERDA Air Cleaning Conference Proceedings.*

Gupta, A., Novick, V. J., Biswas, P., Monson, P. R. (1993), *Aerosol Sci. Technol.,* 19(1), 94–107.

Hinds, W. C. (1982), *Aerosol Technology — Properties, Behavior and Measurement of Airborne Particles,* John Wiley & Sons, New York.

Hinds, W. C. and Kadrichu, N. P. (1994), *Applied Occupational and Environmental Hygiene,* 9, 10, 700–706.

Japuntich, D. A., Stenhouse, J. I. T., and Liu, B. Y. H., (1994), *J. Aerosol Sci.,* 25(2), 385–393.

Jordan, F., Alexas, A., and Lindner, W. (1981), in *16th DOE Air Cleaning Conference Proceedings.*

Kirsch, A. A. and Lahtin, U. B. (1975), *J. Colloid Interface Sci.,* 52, 270.

Leith, D. and Allen, R. W. K., (1986), *Progress in Filtration and Separation 4,* Elsevier, New York.

Letourneau, P. et al. (1989), in *20th DOE/NRC Nuclear Air Cleaning Conference Proceedings*, CONF-880822.

Letourneau, P. et al. (1991), in *21st DOE/NRC Nuclear Air Cleaning Conference Proceedings*, CONF-900813.

Letourneau, P. et al. (1993), in *22nd DOE/NRC Nuclear Air Cleaning Conference Proceedings*, CONF-9020823.

Liew, T. P. and Condor, J. R. (1985), *J. Aerosol Sci.,* 16(6), 497–509.

McCormack, J. D. et al. (1978), in *15th ERDA Air Cleaning Conference Proceedings*, CONF 760822.

Novick, V. J., Higgins, P. J., Dierkschiede, B., Abrahamson, C., and Richardson, W. B. (1991), in *21st DOE/NRC Nuclear Air Cleaning Conference Proceedings*, CONF-900813.

Novick, V. J., Monson, P. R., and Ellison, P. E. (1992), *J. Aerosol Sci.,* 23(6), 657–665.

Novick, V. J., Klassen, J. F., Monson, P. R., and Long, T. A. (1993), in *22nd DOE/NRC Nuclear Air Cleaning Conference Proceedings*, CONF-9020823.

Pratt, R. P. and Green, B. L. (1987), in *19th DOE/NRC Nuclear Air Cleaning Conference Proceedings,* CONF-860820.

Rudnick, S. N. and First, M. W. (1978), in *3rd Symposium on Fabric Filters for Particulate Collection Proceedings,* EPA-600/7-789-087.

Vold, M. J. (1959), *J. Colloid Interface Sci.,* 14, 168.

CHAPTER 18

Fundamentals of the Compression Behavior of Dust Filter Cakes

Wilhelm Höflinger

CONTENTS

18.1 GENERAL DATA

Filtration resistance is a main parameter determining pressure drop and energy consumption in dust baghouse filtration, whereby the specific cake resistance of the separated dust cake is dependent on the manner in which the porous structure of the cake built up during cake formation. The dust cake itself can be seen as a network of particle agglomerates or dendrites which are very sensitive to compression by the gas flowing through the cake. Knowledge about the mechanism and reasons causing compression of the cake is still very poor. Cake compression resulting from a fluid flow through a cake does not only occur in solid–gas filtration, but also occurs in solid–liquid filtration, which has a much longer research tradition. Comparison of these two analogous filter cakes can help to improve understanding of the mechanism inside the cake during buildup.

Liquid filter cakes are mostly built in a two-phase system (solid and liquid phase) in comparison with dust filter cakes which are built sometimes in a three-phase system (solid, liquid, and gaseous) and introduces more influencing parameters and creates therefore a more complex situation. Generally, the tendency to compress is dependent on the nonequilibrium between the pressure drop or compression pressure and the resistance forces of the cake material (interparticulate forces and stiffness of the particles) resisting compression.

In liquid filter cakes the applied pressure drops are usually high (0.5 to 100 bar, due to the relatively high viscosity of liquids), whereas in dust filter cakes pressure drops are only about 0.003 to 0.03 bar. Adhesion forces between particles are, however, approximately of the same magnitude

0-87371-830-5/98/$0.00+$.50

in both cases or even higher in dust filter cakes; therefore, slight changes in adhesion forces, which can be neglected in liquid filter cakes, may become significant for dust filter cakes. This is even more the case if a three–phase system exists, allowing even greater adhesion forces due to possible liquid bridges between particles. Therefore, the number of influences governing the compression behavior of a filter cake is increased for gas filtration and focuses mainly on the adhesion properties between particles. In liquid filter cakes the main influence on specific cake filtration resistance α is the pressure drop Δp, which is demonstrated by some simple model equations (e.g., Equation 18.1),[1,2] where n is a measure of the compressibility.

$$\alpha = \alpha \Delta p^n \quad \text{or} \quad \alpha_m = \Delta p^n + \alpha_0 \tag{18.1}$$

These equations are of very practical use in designing liquid filtration processes. In dust cake filtration the influence of the adhesion forces is dominating and complex, so no comparable models are available. There is, however, an empirical relation (Equation 18.2),[3] which relates the specific cake resistance K_2 to the filtration velocity v and indicates that there must be some kind of compression. The specific cake resistance K_2 is determined from pressure drop curves for different constant filtration velocities v:

$$(K_2)_2 = (K_2)_1(v_2/v_1)^n \tag{18.2}$$

An application of this equation to the design of dust filtration processes is not possible, due to the large number of influencing parameters as mentioned above, thereby limiting the transferability of results gained from test runs. The determination of the specific cake resistance K_2 from the pressure drop curves is demonstrated in Figure 18.1.[4]

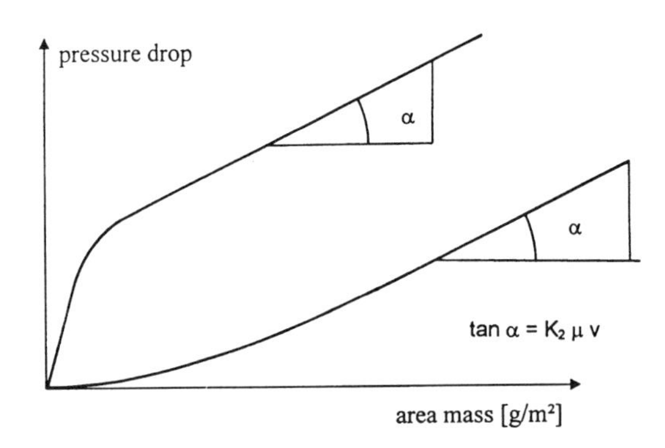

Figure 18.1 Pressure drop curves for different filter media and determination of the specific cake resistance K_2.

The value K_2 is then determined for a limited range of Δt toward the end of the curve. Sometimes this portion of the curve is not linear and demonstrates a slight progressive increase or sudden "jumps," indicating that the cake is compressible and the calculated K_2 values can be only seen as an average value over the time period Δt. A more-detailed characterization of the specific cake resistance and its relation to compression requires still more research work.

The shape of the pressure drop curve depends on the type of filter medium used.[5] If the cake is built up on needle felt, a low pressure drop curve results. After an initial pressure drop increase, which corresponds to the clogging of the needle felt, a nearly linear increase can be seen. The same result is obtained if a coarse metal mesh is used to build up a precoat dust cake (see Figure 18.2). The higher pressure drop curve is achieved if a filter fabric is used, as particles are initially separated within the meshes and the higher fluid velocity through the voidage of these particles produces the high pressure drop increase initially. After this initial clogging range, the velocity of the fluid inside the growing cake becomes lower, resulting in a slower rate of increase of the pressure drop. As can be seen in Figure 18.1, the initial filtration time period is very important in determining whether the pressure drop curve proceeds on a high or a low level. This initial filtration period must be excluded when determining K_2.

18.2 EXPERIMENTAL INVESTIGATION OF COMPRESSION BEHAVIOR

Experimental investigation of compression behavior also includes structure analysis of a dust cake. Analyzing the structure or the porosity of a dust cake is rather difficult and involves considerable expenditure.[23] A simpler method to get some insight into the structure variation in a dust cake is to use a very regular filter medium and to measure the cake height at different filtration times using a microscope.[6] In that case, diatomaceous test dust was filtered on a coarse metal mesh. Use of a coarse metal mesh as filter medium has the advantage that measurement of the cake height was easier, due to the mechanically stable metal mesh structure. In addition to cake height measurement, the pressure drop, the turbitity of the clean gas, and the remaining open filter area were determined. The results of these experiments are shown in Figure 18.2.

It can be seen that after the closure of every mesh opening by dust material, the height of the cake layer continues to increase strongly, but later at a lower rate. This indicates that at least two different layers of the cake were formed. One layer is a loose, very porous layer with a dentritic structure, corresponding to the rapid height increase initially and which always forms the upper part of the cake layer, where freshly filtered dust is deposited. A compressed layer develops beyond this dentritic layer due to the collapse of the dentritic structures near the surface as dust is deposited onto the surface. This compressed layer behaves more or less incompressibly or will compress further only at larger pressure differences. These results have also been reported by Reference 7. Further it seems to be logical to assume that there must be an intermediate layer between these two layers in which the collapsing of the dentrites takes place. Hence, there appears to be at least three different layers of a dust cake present.

18.3 COMPUTER SIMULATION OF CAKE BUILDUP

A computer simulation program was written in order to get more insight into the compression mechanism. A detailed description of this program and the model structures can be found in References 8 through 10. The model structure is based on the filtration of spherical particles (in a two-dimensional manner) on a horizontal line. The position of the particles over the line before filtration is chosen randomly. The pressure drop over the cake is then calculated using the Carman–Kozeny model equation:

$$\alpha_j = K_0(6/d)^2(1 - \varepsilon_j)^2/\varepsilon_j^3 \tag{18.3}$$

where K_0 is the Carman–Kozeny constant and d is an average particle diameter of the dust.

The porosity ε_j necessary for calculation is derived from the geometry of the spherical particles in the simulation. A so-called withstandable stress f_j can also be calculated from the geometry, which in comparison to the pressure drop Δp_{kj} is a measure for the compressibility of the cake. This withstandable stress can be determined throughout the cake height and is equivalent to the pressure drop required to start particle movement. It is assumed that two kinds of force interactions act between the particles, the friction angle φ and a maximum adhesion force Z_{max} (Figure 18.3). A force balance over a partial cake layer j results then in the withstandable stress f_j for this partial layer. The simulation results for different friction angles and different maximum adhesion forces

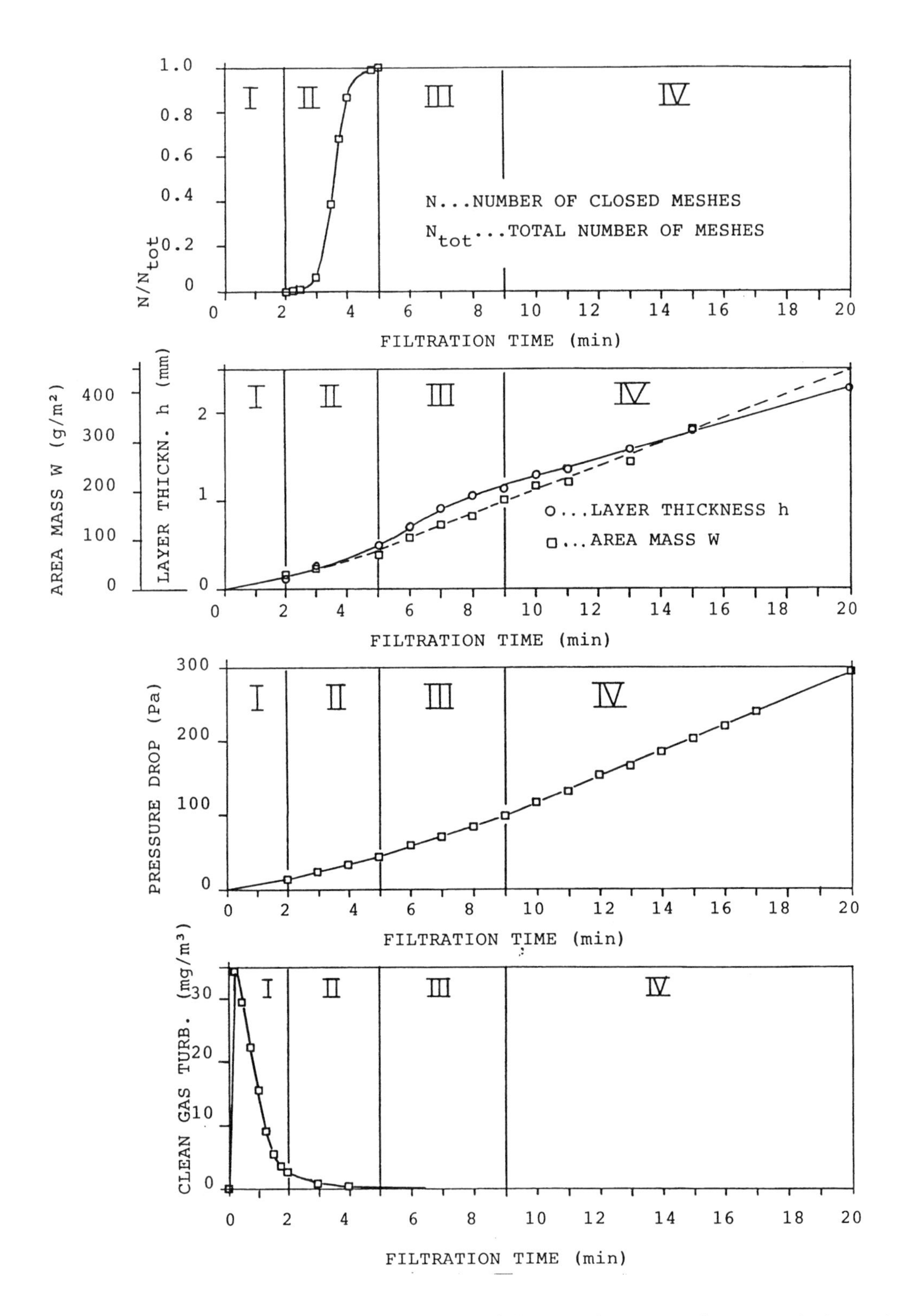

Figure 18.2 Buildup of a dust cake (diatomaceous earth Celite 545, $d_{p50,3}$ = 32 μm) on a metal wire mesh (square meshing, mesh size: 200 μm, wire diameter: 125 μm), filtration velocity: 4.49 m/min, solid content: 3.2 g/m³, gas: ambient air.

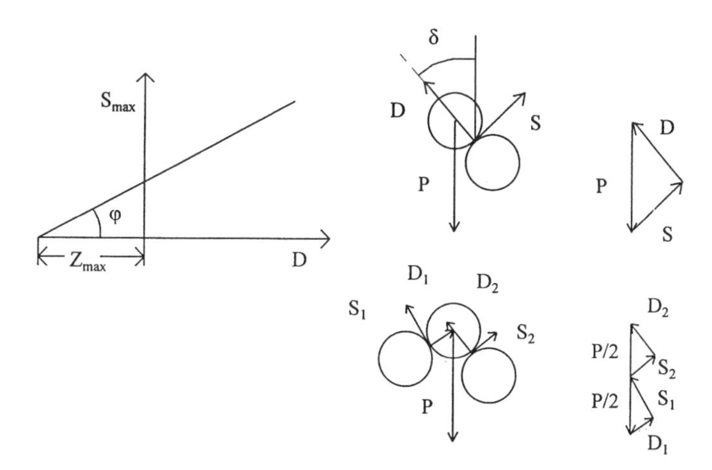

Figure 18.3 Forces at particle contact (Mohr Coulomb), two and three particle contact. S, S_{max}, S_1, S_2: shear forces between two particles; D, D_1, D_2: pressure forces between two particles; P: load force for one particle; Z_{max}: maximum adhesion force; φ: friction angle; δ: angle between two particles.

are shown in Figures 18.4 and 18.5 (d = 15 μm, K_0 = 10, $\mu = 1.85 * 10^{-5}$ Pa s, ρ = 2000 kg/m³, and the filtration velocity u = 0.05 m/s).

The ability of the cake to compress can be seen by comparing the curve of the pressure drop $\Delta p_{k,j}$ with the curve of the withstandable stress f_j. In Figure 18.4 the friction angle is held constant at 20°, and Z_{max} is varied. For low values of Z_{max} (e.g., 10^{-100} N), there will be very small maximum shear forces and the particles can easily move in the cake even if the compression pressure is low. Thus, the recently filtered particles will immediately move and become tightly fixed. The strength in this layer increases greatly, and the compression pressure cannot reach the maximum strength of the cake. This results in a rather homogeneously compressed cake, which behaves incompressibly at increased pressure loads. If Z_{max} is higher (e.g., $Z_{max} = 10^{-9}$ N), two cake layers can be observed: a compressed layer and an upper layer, in which the withstandable stress occasionally approaches a value near the compression pressure. In this layer, particle movement will occur if the compression pressure increases further. For Z_{max} values between $5 * 10^{-9}$ and 10^{-8} N, compression occurs locally in positions randomly distributed throughout the layer nearest to the horizontal line representing the filter medium, where the withstandable stress or the strength is near the compression pressure, and the cake is therefore compressible. This results in the lower layer being compressed than the upper layer.

For $Z_{max} = 3 * 10^{-8}$ to $4 * 10^{-8}$ N, the whole cake is very compressible, because the strength is approximately equal to the compression pressure (or slightly above). This high grade of compressibility can also be seen in the porosity ε_i over the cake thickness. For Z_{max} values of about 10^{-7} N, the resulting maximum shear forces are already so high that again two different layers can be recognized. In this case the strength in the upper layer is greater than the compression pressure, which is hardly visible in the figure. This layer did not compress at all and shows a constant high porosity. Deeper in the cake, the porosity begins to decrease, and at the same time the strength is nearly equal to the compression pressure, which indicates that the layer below is compressible. At high Z_{max} values ($5 * 10^{-7}$ N), the strength is always above the compression pressure, and the very porous cake behaves incompressibly. In Figure 18.5 Z_{max} is held constant and φ is varied. As in Figure 18.4, a partition of the cake into three different layers is seen. The influence of φ on the cake buildup and the compression behavior is much greater then the influence of Z_{max}. If $\varphi = 0°$, the cake almost completely collapses. The value of φ determines the nonreversibility of the particle movement. If a particle is in contact with another particle at an angle δ which is less than φ, it is not able to move any further and is considered irreversible. This results in the building of stable

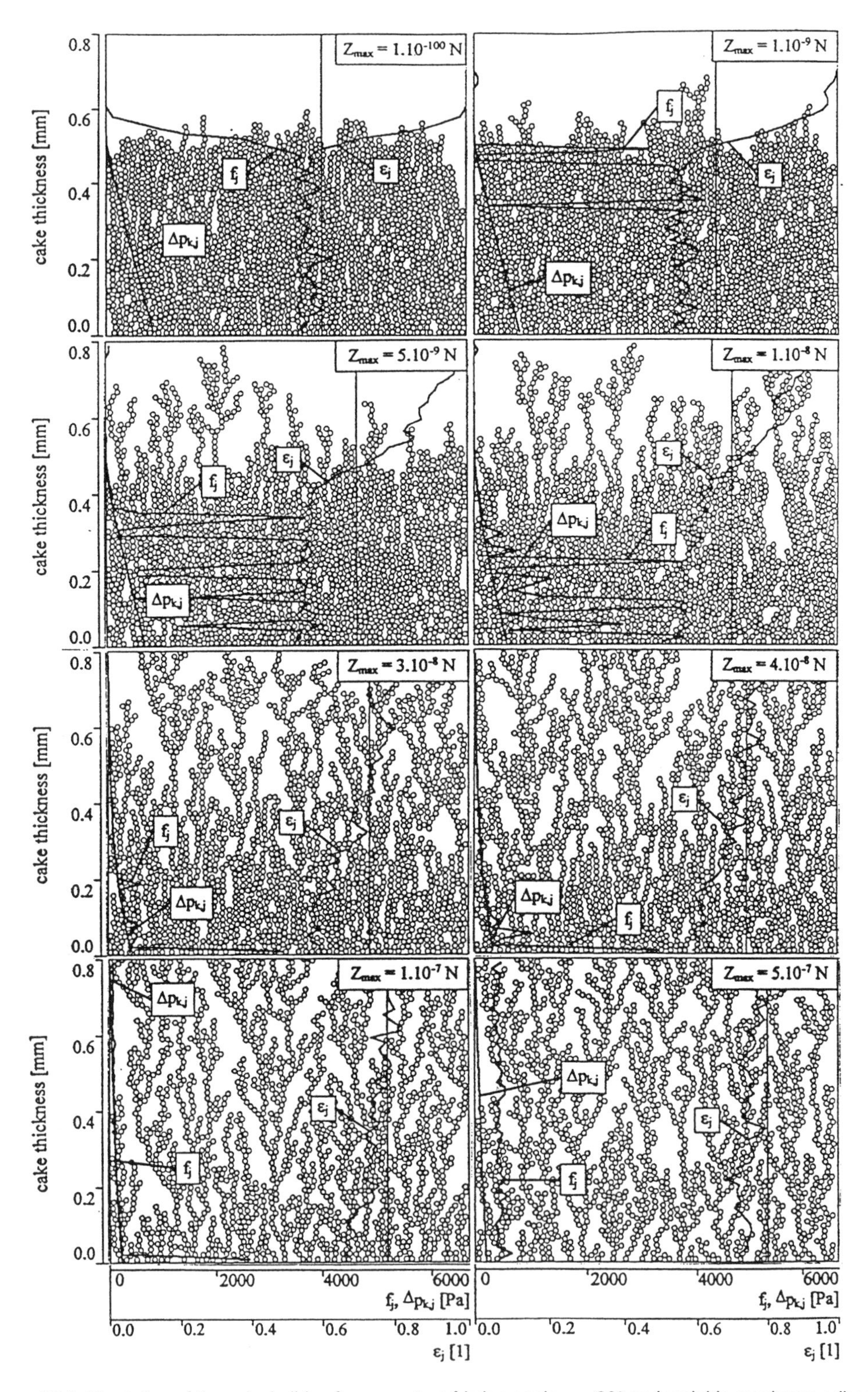

Figure 18.4 Simulation of the cake buildup for a constant friction angle $\varphi = 20°$ and variable maximum adhesion forces Z_{max}.

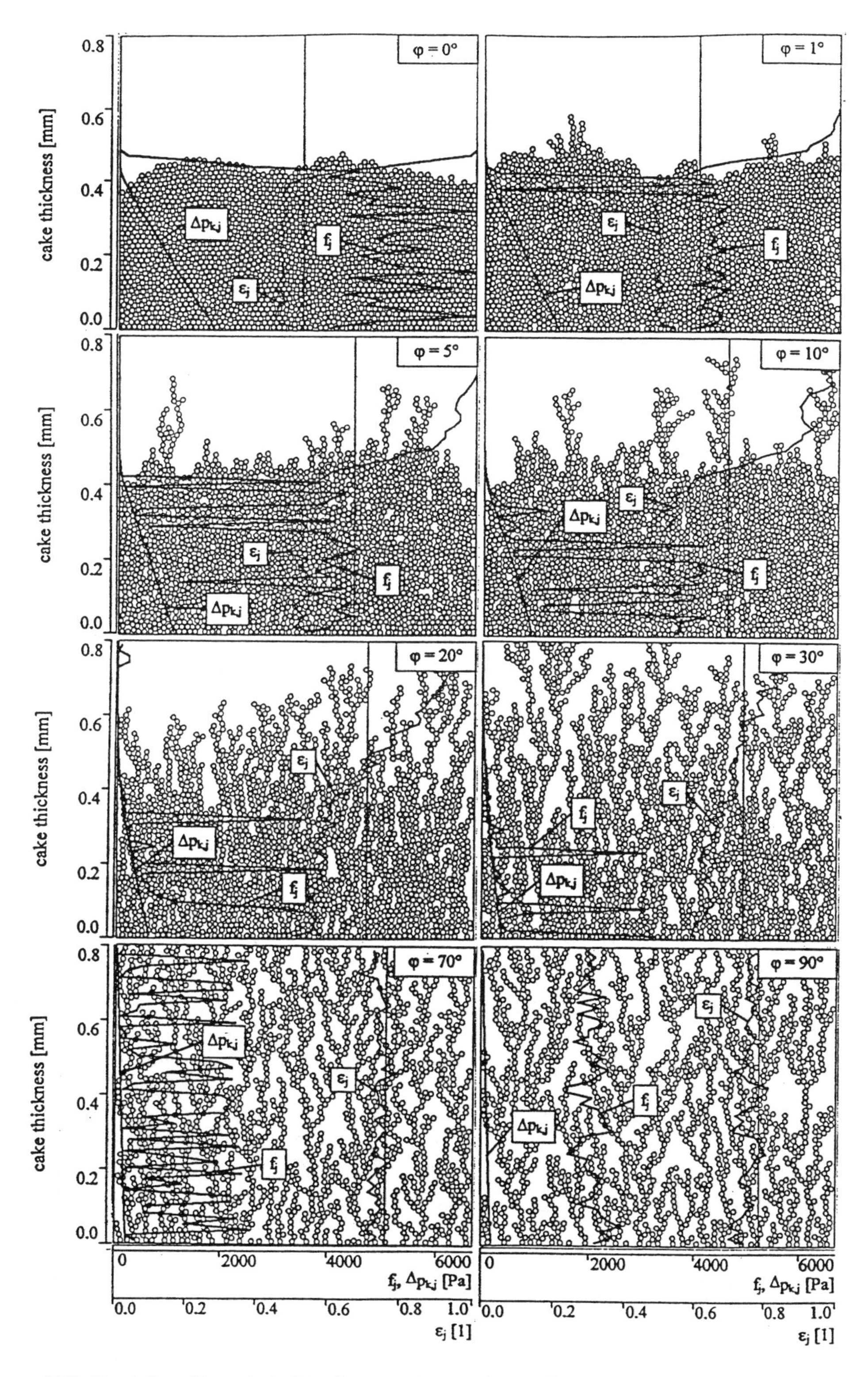

Figure 18.5 Simulation of the cake buildup for a constant maximum adhesion force $Z_{max} = 1 * 10^{-8}$ N and variable friction angles φ.

bridges and holes inside the cake, which are resistant to compression. These holes will increase in number and size as φ increases.

In summary, the cake was found in general to consist of at least three layers of different structures, as already recognized from the experimental investigations described above. The thickness of each layer in the cake or whether the cake consists of only one or two layers depends on the values of the friction angle φ and the maximum adhesion force Z_{max} and of the whole cake thickness.

Figure 18.6 shows the structure of a three-layer cake. The filtered particles initially form a loose layer on the filter medium, with a low compression pressure $\Delta p_{k,j}$ so that the strength parameter f_j in this layer was not exceeded. This layer would be stiff and incompressible. If the compression pressure reaches the strength of the cake, the cake begins to compress and the strength increases. At first, the strength increases only slightly, because only very few particles start to move. In this portion of the layer the curve representing the strength runs parallel, but slightly above, the curve representing the compression pressure. With increasing compression pressure, more and more particles are forced to move. It is possible that the number of particles in a partial layer increases greatly, since particles can change partial layers when they move, and hence the strength can reach a value which lies above the maximum compression pressures. This partial layer will then appear incompressible. If in all partial layers below this partial layer the strength is also high, this part of the cake will behave as an incompressible cake. In summary, there exists a loose upper layer, a strong but compressible middle layer, and a compressed underlayer which cannot be further compressed without destroying the particles.

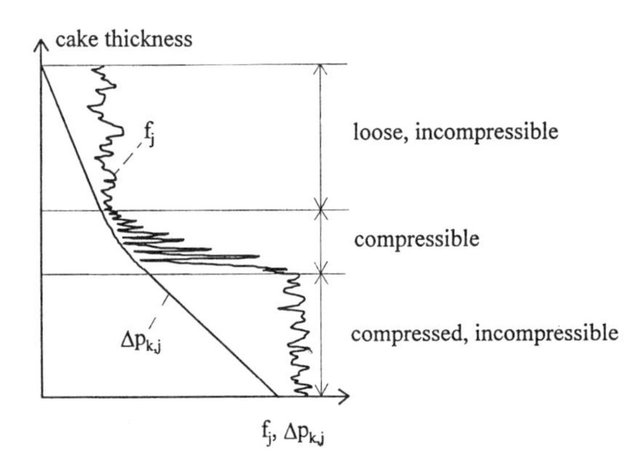

Figure 18.6 Three-layer structure in a dust filter cake.

For the stimulation examples in Figures 18.4 and 18.5, the total pressure drop over the area mass is calculated and shown in Figure 18.7 and 18.8. Detailed information about calculation formulas and data can be found in References 8 and 10. It can be seen that the shape of the pressure drop curve is dependent on the type and number of layers in the cake. For example, in Figure 18.7 the highest value of Z_{max} ($5 * 10^{-7}$ N) produces a single homogeneous and very porous cake, which is incompressible. This results in a very low and linear pressure drop increase. If Z_{max} is lower (3 to $4 * 10^{-8}$ N, for example) the main part of the cake forms a compressible layer, as can also be seen in Figure 18.4, so the pressure drop curve has a steeper slope after a short initial linear increase. The starting point of the portion of the curve with a steeper slope moves more to lower area masses, as Z_{max} becomes smaller. At small Z_{max} values only a gradually increasing slope is visible at the beginning, and thereafter the pressure drop increase is again linear, but with a steeper gradient. This is a result of the existence of only a small compressible layer just below the cake surface. Most of the cake consists of a single, compressed layer, which is responsible for the large linear

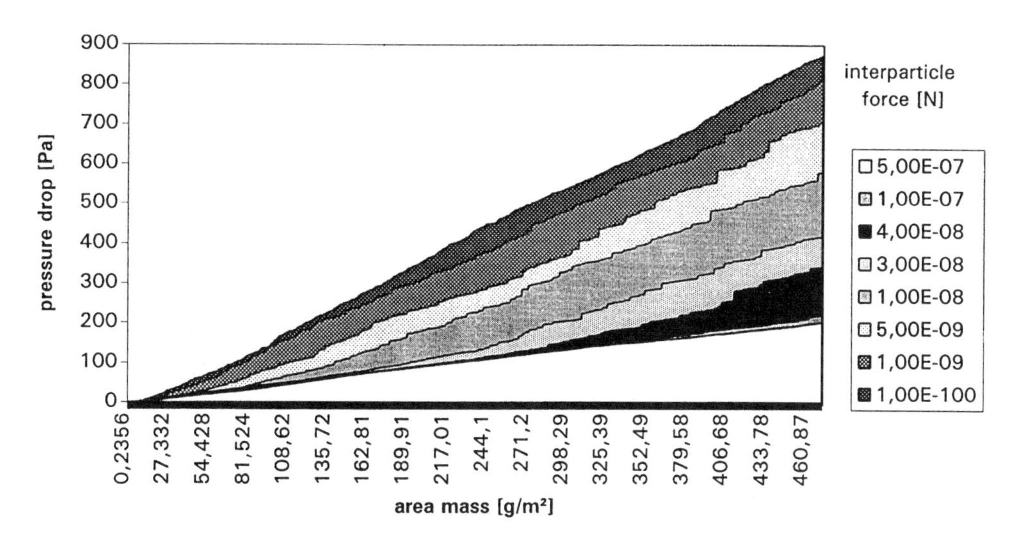

Figure 18.7 Pressure drop over the filtered area mass for different maximum adhesion forces at $\varphi = 20°$.

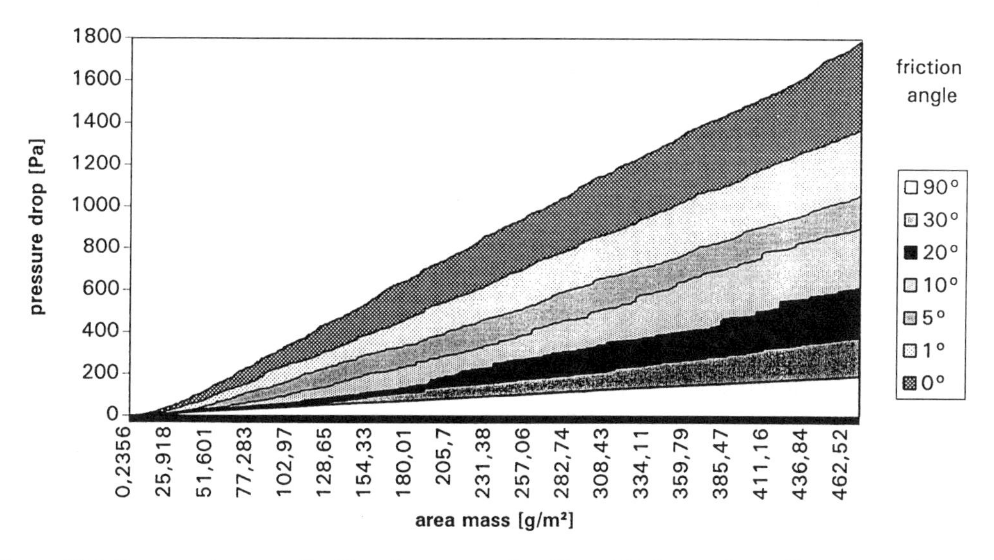

Figure 18.8 Pressure drop over the filtered area mass for different friction angles at $Z_{max} = 1 * 10^{-8}$ N.

pressure drop increase just after the initial steeper slope. The curves in Figure 18.8, in which the maximum adhesion force is held constant and the friction angle is varied, show analogous results.

The simulation results in Figures 18.7 and 18.8 demonstrate that different adhesion and shear forces between particles produce different increases in pressure drop during filtration, which in turn results in different cake resistances. The pressure drop curve is only linear for large and small adhesion or shear forces. The gradient is then high for small forces and low for large forces. For average forces the cake behaves compressibly, and the pressure drop curve shows an increasing slope. Pressure drop jumps are also visible in these curves, as is observed in reality.

As can be seen from these investigations, the influence of the interparticulate forces is very important considering compression and pressure drop curves during the cake buildup. Summarizing, the resistance of a filter cake and its relation to compression can be described by the Carman–Kozeny constant K_0 and two compression parameters, the maximum adhesion force Z_{max} and the friction angle φ.

18.4 INFLUENCE OF RELATIVE HUMIDITY ON CAKE COMPRESSION

Unfortunately, knowledge of the adhesion forces and their influence on cake compression is still very incomplete. Generally, there are three types of adhesion forces between dust particles,[11] electrostatic forces, van der Waal forces, and forces from adsorption layers on particles or liquid bridges resulting from sorption or condensation mechanisms due to the existence of a humid gas. Several experimental investigations of electrostatic forces are known, which show that precharging the dust particles before filtering on a filter medium produces a considerable reduction in the pressure drop, for example, see References 12 through 14. This is due to the fact that the particle charge is polarized on the particle and creates therefore a more porous filter cake with lower resistance. A general model relating the particle charge parameters to the cake resistance in the form of an equation is not known to the author.

Forces resulting from liquid bridges between particles or adsorption layers are dependent on the humidity of the gas. Humid layers on particles can also influence electrostatic and van der Waal forces. The humidity of the gas can therefore be seen as a major parameter influencing particle adhesion forces and compressibility of dust filter cakes. Scientific work on this influence is very rare. Initial investigations[15-17] report a reduction in cake resistance with increasing humidity. As mentioned above, humidity does not only influence directly adhesion forces by way of liquid bridges but also indirectly via other parameters, e.g., via the changing of the electrostatic and van der Waal forces. For that reason, it is interesting to investigate humidity in combination with other parameters. Rzepa[18] investigated, for example, the influence of humidity in combination with temperature and the CO_2 concentration of the gas on cake resistance of Kalkhydrat layers. Others[19,20] have investigated the influence of humidity in combination with the particle size on the specific cake resistance. The specific cake resistance was determined in an approximate way, as demonstrated in Figure 18.1. The results of these measurements, executed for a coarse and for a fine diatomaceous test dust at different relative humidities, are shown in Figure 18.9. It can be seen that at low relative humidities, the specific cake resistance for the fine dust is high compared with the coarse dust. This seems to be logical due to the size of the particles of the dust. With increasing humidities, the resistance fall is steeper for the fine dust than for the coarse dust. At high humidities, the filtration resistance of the fine dust falls below that of the coarse dust. The phenomenon requires explanation.

An explanation for this behavior can be given by considering the structure of the loose upper layer of the dust cake. This upper layer, as can be seen in Figure 18.4, is not homogeneous. There are ranges of clusters of particles with low porosity and ranges of large voidage inside this layer. These are due to the randomness of the particle filtration and also due to the agglomerate formation of particles, before the particles are filtered onto a cake, which was not considered for the simulation, but does in fact occur. The clusters or the agglomerates forming the upper layer of the cake have an internal strength which is important for further compression. This internal strength of the agglomerates can be estimated (Löffler and Raasch[21]):

$$\sigma = Hk(1 - \varepsilon)/(d^2\varepsilon) \tag{18.4}$$

where the particle size is d, the adhesion force between the particles of the agglomerate H, and the porosity ε. By assuming that at constant relative humidity the resulting adhesion forces H between the particles generated over liquid bridges are proportional to the particle diameter d,[22] then the strength of the agglomerates σ is inversely proportional to the particle diameter. On the other hand, at constant particle diameters high adhesion forces or a high strength of the agglomerate will be

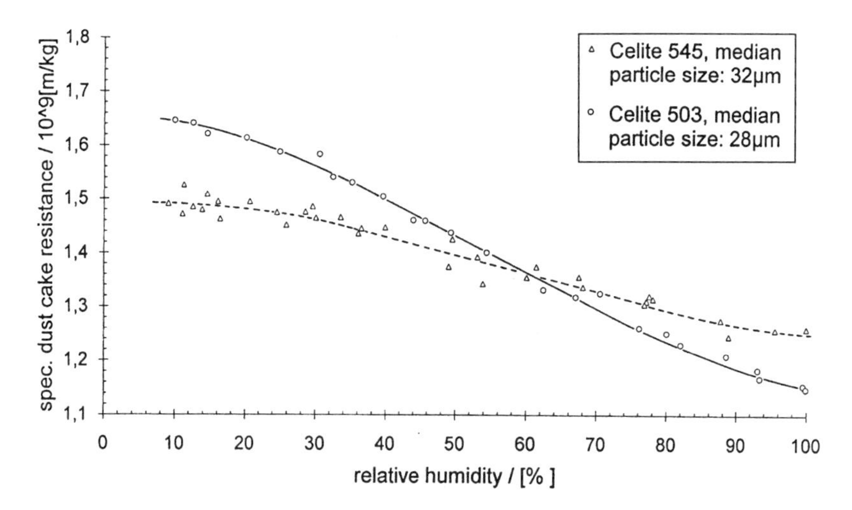

Figure 18.9 Specific dust cake resistance of a coarse and a fine diatomaceous dust for different relative humidities, filtration velocity: 0.055 m/s.

produced by larger numbers of liquid bridges between particles. This is assumed to be the case for high relative humidities. At low relative humidities very few liquid bridges will be generated and the strength of the agglomerate is therefore low.

The overall specific cake resistance will now be considered. The dominant factor is the specific resistance of the compressed layer in the cake, which in turn depends on how the agglomerates or the clusters break down into single particles. At low relative humidities the strength inside the agglomerate of the upper layer is relatively low and the agglomerates or clusters can easily break down into smaller pieces, especially into their single particles.

The specific cake resistance will be determined by the size of these single particles, which in turn produces a resistance which is higher for the fine dust than for the coarse dust. At high relative humidities very high strength will be generated in the agglomerate, especially for clusters or agglomerates consisting of small single particles, because of the large number of contact points. These clusters will not easily break down into their single particles, but will break down into smaller clusters or smaller agglomerate pieces which do not change the porosity and the specific agglomerate surface very much and therefore have a relatively low specific cake resistance.

Coarse dust at high relative humidities will now be considered. Those agglomerates in the upper layer consisting of large single particles have internal strengths which are not as high as for the fine dust at high relative humidities, because the number of contact points is fewer. These agglomerates can more easily break down into much smaller clusters or pieces compared to the fine dust agglomerates. Therefore, at high relative humidities the resulting cake resistance of the coarse dust is high and can even reach values above those of the fine dust cake, as can be seen from Figure 18.9.

18.5 CONCLUSION

The compression behavior of a dust filter cake during cake buildup was experimentally and theoretically analyzed. It was found that the cake consists of different partial layers. In order to approach the system theoretically, a computer simulation program was developed, in which spheric dust particles are filtered. In this computer model the compression behavior is described by a friction angle and a maximum adhesion force between particles.

The simulation results show the same layered structure of the cake as found experimentally and gives a theoretical explanation for its existence and development. The interparticulate forces, responsible for compression are affected very much by humidity, and so humidity of the gas can be seen as a main parameter determining the dust cake resistance and should be of central concern in future research. Some results concerning the influence of relative humidity and dust particle size on specific dust cake resistance are discussed in this chapter.

REFERENCES

1. Autorenkollektiv, *Lehrbuch der Chemischen Verfahrenstechnik,* VEB Deutscher Verlag für Grundstoffindustrie, Leipzig, 1983.
2. Orr, C., *Filtration, Principles and Practices,* Marcel Dekker, New York, 1977.
3. Dennis, R. and Klemm, H. A., Modelling Coal Fly Ash Filtration with Glass Fabrics, EPA-600/7-78-087, NTIS, Springfield, VA, 1978.
4. Löffler, F., Dietrich, H., and Flatt, W., Staubabscheidung mit Schlauchfiltern und Taschenfiltern, Vieweg Verlag, Braunschweig, 1984.
5. Wayne, T. D., Paul, J., and Kenneth, E. N., The generation and evaluation of fabric filter performance curves from pilot plant data, *Filtr. Sep.,* Nov./Dec., 555–560, 1976.
6. Höflinger, W. and Hackl, A., Coarse metal mesh for dust precoat filtration, in *Proceedings PARTEC,* March, 1992, 309.
7. Schmidt, E. and Löffler, F., Verification of dust cake compression by means of structure analysis and the effect on filtration performance, *Proceedings of the 6th World Filtration Congress,* Nagoya, Japan, May, 1993, 54–59.
8. Stöcklmayer, Ch., Computersimulation des Kompressionsvorganges beim Aufbau eines Staubfilterkuchens, Dipl. Arbeit TU, Wien, 1993.
9. Höflinger, W., Stöcklmayer, Ch., and Hackl, A., Model calculations of the compression behavior of dust filter cakes, in *Proceedings FILTECH Conference,* Karlsruhe, Oct. 1993.
10. Höflinger, W., Stöcklmayer, Ch., and Hackl, A., Model calculations of the compression behavior of dust filter cakes, *Filtr. Sep.,* Dec., 804–811, 1994.
11. Löffler, F., *Staubabscheiden,* Georg Thieme Verlag, Stuttgart, 1988.
12. Helfritch, D. J., Performance of an electrostatically aided fabric filter, *Chem. Eng. Prog.,* 73, 54–57, 1977.
13. Lamb, G. E. R. and Constanza, P. A., A low energy electrified filtered system, *Filtr. Sep.,* 17, 319–322, 1980.
14. Greiner, G. P. et al., Electrostatic simulation of fabric filters, *J. Air Pollut. Control Assoc.,* 31, 1125–1130, 1981.
15. Durham, J. F. and Harrington, R. E., Influence of relative humidity on filtration resistance and efficiency of fabric dust filters, *Filtr. Sep.,* 7/8, 389–392/398, 1971.
16. Kempe, W. and Jugel, W., Untersuchungen zum Einfluß der Gasfeuchte bei der Gewebeentstaubung, *Luft Kaltetech.,* 2, 60–64, 1976.
17. Ariman, T. and Helfritch, D. J., How relative humidity cuts pressure drop in fabric filters, *Filtr. Sep.,* 3/4, 127–130, 1977.
18. Rzepa, M., Einfluß von wasserdampt und Kohlendioxid auf das Durchströmungsverhalten von Kalkhydratfilterschichten, *Fortschr. Ber.,* VDI, 3, 322, 1993.
19. Mausschitz, G., Feuchtigkeitsunfersuchungen bei der Staubabscheidung mit Precoatschichten, Dipl. Arbeit TU, Wien, 1991.
20. Höflinger, W. and Hackl. A., New developments and investigations in dust precoat filtration, in *Proceedings 6th World Filtration Congress,* Nagoya, Japan, May 1993, 439.
21. Löffler, F. and Raasch, J., Grundlagen der Mechanischen Verfahrenstechnik, Vieweg Verlag, Braunschweig, 1992.
22. Schubert, H., Kapillarität in porösen Feststoffsystemen, Springer-Verlag, Berlin, 1982.
23. Schmidt, E. and Löffler, F., Präparation von Staubkuchen, *Staub Reinhalt. Luft,* 49, 429–432, 1989.

CHAPTER 19

Filter Efficiency Modeling, Electrically Stimulated Agglomeration, and Separation of Bioaerosols — Some Recent Trends in Aerosol Research

Eberhard Schmidt, Andreas Gutsch, and Ralf Maus

CONTENTS

0-87371-830-5/98/$0.00+$.50

19.1 INTRODUCTION

This chapter reviews some fundamental projects in the field of aerosol science actually running at the Institut für Mechanische Verfahrenstechnik und Mechanik of Karlsruhe University. Initiated by the late Professor Friedrich Löffler, these quite different projects illustrate the broad field of his work and his interests. The first example deals with a computer code for the calculation of transient collection efficiencies for surface filters by single-particle tracing. The second one describes some aspects of the conditioning of nanosized solid particles by electrostatic effects. Finally, the third one treats the separation of airborne bacteria out of gases with fibrous filters. The authors try to continue the launched research projects just as Professor Löffler would have done.

19.2 CALCULATION OF TRANSIENT COLLECTION EFFICIENCIES FOR CAKE-FORMING FILTRATION

19.2.1 Background

Periodically regenerable cake-forming surface filters, in particular bag filters incorporating reinforced nonwoven filter media, are widely used in particulate control technology. The particles collected on the filter media surface build up a so-called dust cake. The operational characteristics of such filter apparatus, especially the time-dependent clean gas particle concentration, are strongly influenced by the structure of the dust cake formed. The structure itself depends among others on particle size distribution, particle charge, adhesive and cohesive properties, and pressure drop. Dust cake mass and therefore its flow resistance with respect to pressure drop increase with filtration time. Due to this, one can find an exponential decrease in particle penetration through the dust cake filter medium complex. This chapter presents a computer simulation with which the creation and growth of a dust cake can be modeled during the filtration via particle trajectory calculations. The influence on the gas flow field exerted by those particles which have already been deposited is taken into account in a macroscopic way. The input parameters include the size distribution of the raw gas particles, the separation characteristics of the clean filter medium, and the filter face velocity. The output parameters include the local dust cake porosities, the time-dependent pressure loss, and fractional separation efficiency trends, together with the clean gas particle concentration.

19.2.2 Conception of the Computer Simulation Program

(1) Description of the Fluid Flow

For the cake-forming dust collection concerned here, the filter medium is permeated by the particle-loaded fluid. In the case of an ideal filtration, the particles do not penetrate into the filter medium, but are deposited upon its surface. Here they form a so-called filter cake, itself serving as a filter for subsequent particles. Therefore, the collection efficiency improves tremendously with time. An exponential decrease in clean gas particle concentration can be measured during each filtration cycle. The continually growing particle layer also enhances the permeation resistance, hence demanding the periodic removal of the cake from the filter medium. The filter media are usually designed as tubes or bags. The presented simulation draws on a plane subsection, with a perpendicular approach flow.

In order to model the fluid flow, the surface region above the filter medium is subdivided into approximately 1000 equal-sized cubic elements (or cells), all of which are initially empty. From an adequate distance, the filter medium is then uniformly approached and permeated by the gas flow. The absolute pressure beneath the filter medium on the clean gas side is a fixed parameter (in this case, 1 bar). A further boundary condition which may be set, as required, is the filter medium permeation resistance, which is assumed to be constant. Following the start of the separation, the cells become more or less rapidly filled with particles, depending on the cell distance from the filter medium. Each element can now be allocated with a permeation resistance which depends on the number and size of the particles retained. The incorporation of all permeation resistances in the emerging system of connected equations allows the numerical calculation of the average perpendicular fluid velocity components for all six faces of each individual element. The three-dimensional velocity components at any location within the cells is then derived via interpolation. Moreover, one obtains the pressure within each element, which allows the calculation of the pressure loss across the whole particle layer.

(2) Description of the Particle Motion

The simulation of the dust cake formation is executed by means of particle trajectory calculations. In each instance, a single spherical particle of a specific size, density, and electrical charge is released with a given velocity at an adequate distance from the filter medium at the upper edge of the initial cell layer. The particle trajectory through the individual volume elements is then calculated step for step from each randomly defined start position in the x-y plane. In addition to the inertia force, the parameters for the equation of motion include the flow resistance force, gravity, electrical field forces, and a stochastic force. The latter allows the Brownian molecular motion to be taken into account, which is especially important for particles smaller than 1 μm. The result of the numerical solution of the formulated stochastic differential equation supplies the velocity and location of each individual particle in dependence of time.

(3) Layer Growth

For each time increment, the program interrogates whether or not an overlap exists between the moving particle and one which has already been collected. In the case of a positive result, the central coordinates of the corresponding initial collision are stored, a new particle is released, and the procedure restarted. For the face velocities used here (i.e., $v < 0.1$ m/s), particle rebound is improbable. Should a particle touch the plane filter surface, then it may permeate it with a given probability corresponding to the particle size. The collection probability is specified in accordance with separation efficiency trends which are typical for fibrous filters. Those particles which pass through the filter are then of no further significance for the rest of the simulation. In contrast, however, the local coordinates of the particles which are collected are stored before a new particle is initiated.

The data of the particles are set within defined limits via a random number generator before being released. Following a specific number of particles (in this case 500), the fluid flow field is calculated anew. The program is terminated when the preset number of particles ($10^4 \ldots 10^5$) have been released, or a specific layer thickness is attained (approximately 50 mean particle diameters).

19.2.3 Results of the Simulation Calculations

(1) Dust Cake Structure

A small section (2500 μm²) of a filter element with a small part of the simulated dust cake may be seen in Figure 19.1 from the side and the top, respectively. The position (not the permeability) of the filter medium is represented by an array of parallel cylinders. The spherical particles are between 6 and 10 μm in diameter. This representation has been derived with the aid of a computational ray-tracing process, using the file containing the diameters and local coordinates of the collected particles. By studying such calculated pictures, one can observe that the initial particles are solely retained by the filter medium. However, after a brief space of time, dendrite-shaped

Figure 19.1 Side and top view of a numerically simulated dust cake deposited on a filter medium (particle diameter between 6 and 10 μm).

structures begin to grow and form bridgelike complexes. Finally, the arriving particles are collected mainly by particles belonging to the dust cake; they no longer reach the filter medium.

The calculated average porosities of such filter cakes within thin layers parallel to the filter medium as functions of the distance from it correspond to those of real filter cases when no compression of the particle layer occurs (Schmidt and Löffler, 1991). An average porosity of approximately 85% has been determined in most cases. Furthermore, one can recognize the interfacial influence of the plane filter medium, together with that of the dendrite growth at the cake surface: a porosity minimum exists at a distance of half of the mean particle size from the surface of the filter medium; the porosity increases continually in the upper 50 μm of such cakes up to 100% (Schmidt, 1994).

(2) Pressure Loss

The pressure loss trends of three different particle fractions as functions of time are plotted in Figure 19.2 (particle size distribution equally spread between the given limits, filter face velocity v = 0.03 m/s, raw gas particle concentration c_{raw} = 10 g/m^3). Following a slight, but progressive initial pressure loss increase, the curves then possess a linear trend. This complies with the values measured with real, incompressible dust cakes (Schmidt, 1993). In the linear range, the slope of the three curves is almost proportional to the square of the respective mean particle size reciprocal. This was to be expected, as the porosities of the three dust cakes are almost identical.

(3) Particle Collection

Figure 19.3 demonstrates the given fractional collection efficiency of the clean filter medium, together with the calculated separation efficiencies which were established after different dusting durations. The curves drawn for the latter are curves of best fit. The separation efficiencies were calculated by dividing the number of particles collected within a specific time by the number of

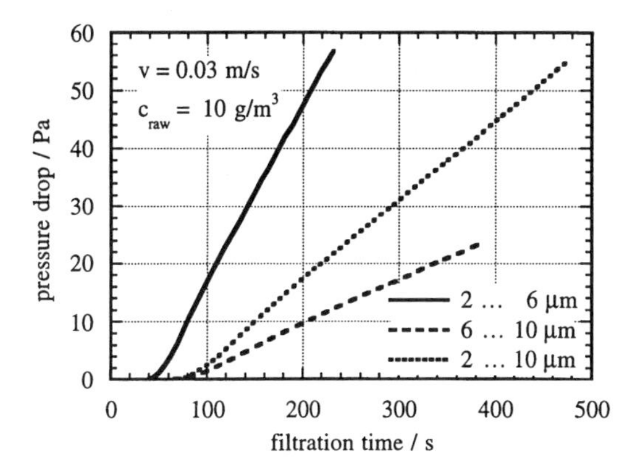

Figure 19.2 Pressure loss as a function of the filtration duration for three dust cakes with different particle size distributions.

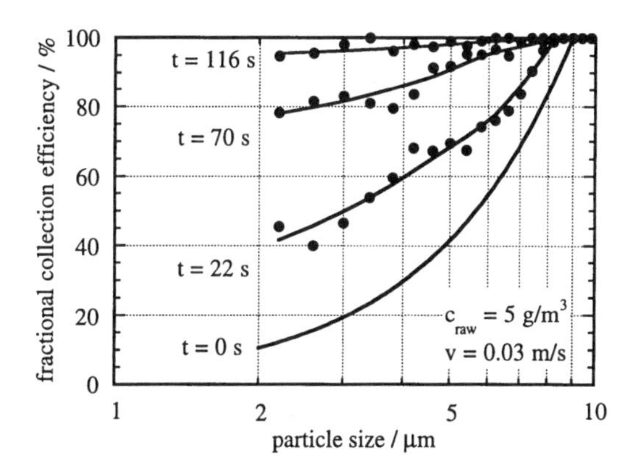

Figure 19.3 Fractional separation efficiency of the filter medium/dust cake complex for different periods or dust loading states (the dots represent calculated values).

particles released, for different size intervals. The result complies extremely well with experimentally determined separation efficiencies (Löffler, 1988).

Upon calculating the particle concentration within the clean gas for the same confining parameters, one receives the result plotted in Figure 19.4. Here, again, the calculated values are the plotted dots and the curve is an approximation function. The calculated time-dependent exponential concentration reduction may also be observed in reality. The relatively high emission values determined at the start of the simulation may be traced back to the unfavorable filter medium collection characteristics assumed.

19.2.4 Summary

Despite the main weakness of an unsatisfactory description of the flow field in the near vicinity of the collected particles, the simulation program issues almost genuine dust cake structures. The time-dependent pressure loss trends and those for the separation efficiency and clean gas particle concentrations are also correctly predicted. The influence exerted by different operational parameters on, for example, the transient fractional collection efficiency, may be explicitly investigated, which is often empirically not possible, due to the complex interrelationships concerned. Upon allocating adhesion forces to the individual contact points, then it should be possible to describe

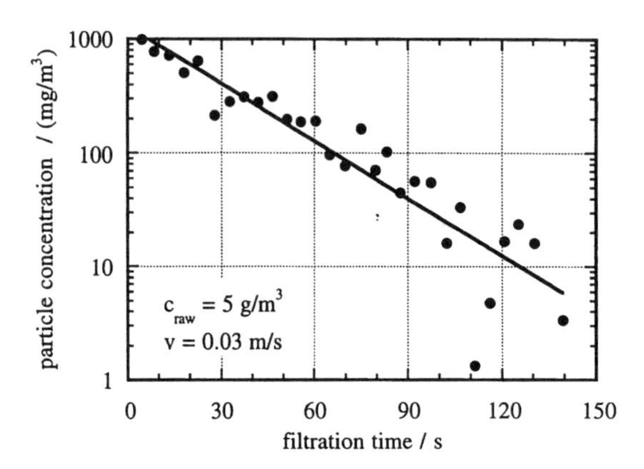

Figure 19.4 Time-dependent clean gas particle concentration (the dots represent calculated values).

both the compression processes (Schmidt, 1993) and the cake removal (e.g., as a result of a flow reversal), for real dust cakes. These aspects are the subject of current investigations.

19.3 BIPOLAR CONDITIONING OF NANOSIZED PARTICLES

19.3.1 Background

Conventional separators such as cyclones or wet scrubbers have low separation efficiencies in the submicron size range. Even high-efficiency cyclones are not useful for collection of particles below 1 μm (Mothes, 1982; Schmidt, 1993). Electrostatic precipitators are principally useful for separating of particles below 0.05 μm. However, the problem of efficient precipitation in the size range from approximately 0.1 to 1 μm remains to be solved (Riehle, 1992). Fabric filters, which have high-grade efficiency in all particle size ranges, show an increasing pressure drop and a decreasing cycle time when they are operated with submicron particles (Löffler et al., 1984). To avoid the above problems new techniques of aerosol conditioning are used to increase the particle size upstream to the separators. Thus, a higher separation efficiency can be achieved even with conventional separation techniques.

In general two different methods of aerosol conditioning have to be distinguished:

- Wet conditioning, where the particle growth takes places due to water condensation on the particle (Schabel et al., 1994).
- Dry conditioning, where the particle enlargement is based on agglomeration.

The agglomeration processes are subdivided into acoustic and electrical methods. Acoustic methods enlarge the particle mobility by using external sound fields. As a result of the increased particle mobility, a higher collision rate and so a size enlargement is obtained. Magill et al. (1989) showed that strong acoustic waves enlarge the size of soot particles by a factor of 10. Funcke and Frohn (1994) found the most distinctive enlargement by acoustic agglomeration of glycerin droplets of approximately 1 μm. However, the method of acoustic conditioning of submicron particles is a power-consuming process. Therefore, large-scale application of acoustic conditioning has not been done yet.

Electrical conditioning processes enable the enhancement of the particle mobility as well as the attractive interparticle forces, such as coulomb or polarization forces. Hautanen et al. (1995) developed an electrical agglomerator, where charged aerosol particles are fed into an alternating electric field (AC-field). In response to the oscillation of charged particles in the AC-field, they measured a reduction of fine particles as a result of agglomeration between fine and large particles. Eliasson et al. (1987) obtained an enhanced agglomeration after mixing two oppositely charged fractions of limestone particles and glass beads. They increased the agglomeration rate through a repeated charging and mixing of the oppositely charged fractions. However, these methods were not expanded into the submicron particle size range. Wadenpohl (1994) investigated the electrical agglomeration of diesel soot particles. They found the soot particles to be enlarged by a factor of 10...1000 after separation and redispersion in a tubelike electrostatic precipitator.

This chapter presents a new method of bipolar charging of submicron aerosol particles. As a result of the presence of oppositely charged particles, enhanced agglomeration is obtained. Thus, the dominant conditioning mechanism is the agglomeration of oppositely charged particles due to attractive coulomb forces.

19.3.2 Setup for Measurement of the Charge Distribution

Figure 19.5 shows a schematic of the experimental setup for measurements of the charge distribution. The aerosol is generated according to the method of Scheibel and Porstendörfer (1983). A boat filled with high-purity NaCl is heated to temperatures between 600 and 750°C. At these temperatures sufficient sublimation of NaCl into the nitrogen carrier gas (flow rate 2 l/min) takes place. After cooling the carrier gas, particle formation starts due to homogeneous nucleation and heterogeneous condensation. The polydisperse aerosol is fed into the first differential mobility analyzer (TSI, DMA 3071), which selects particles of a specific electrical mobility. The "monodisperse" outlet of the DMA I (flow rate 2 l/min) is connected to the inlet of the charging unit. Thus, measurements of the charge distribution were conducted starting from a monomobile fraction of the originally polydisperse aerosol.

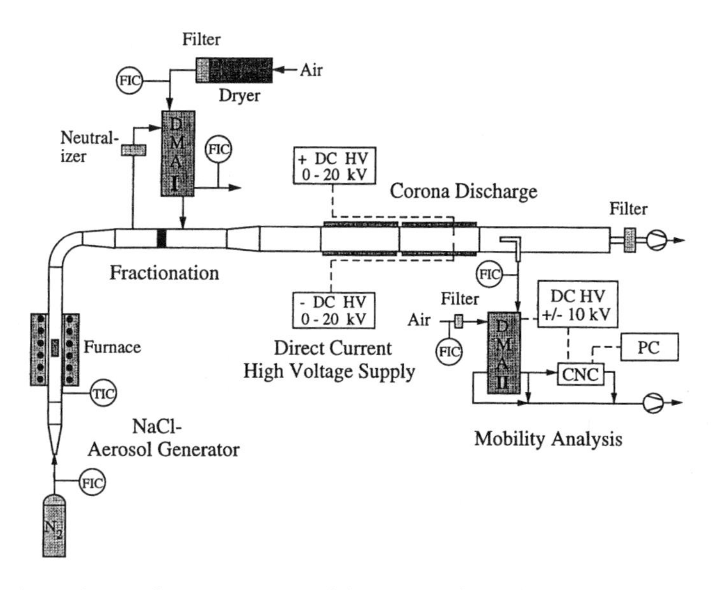

Figure 19.5 Experimental setup for measurement of the charge distribution.

Figure 19.6 shows a schematic of the charging unit. It consist of two stainless steel needle electrodes centered in a quadratic PVC-channel. Each electrode is separately connected to a direct-current high-voltage (DC HV) supply of opposite polarity. If the local electrical field strength on the electrode tips exceeds the corona-onset field strength, a discharge on both needles is obtained. With regard to the opposite potential of the electrodes, positive ions are emitted from one electrode and negative ions from the other. The opposite ions follow the electrical field lines that are

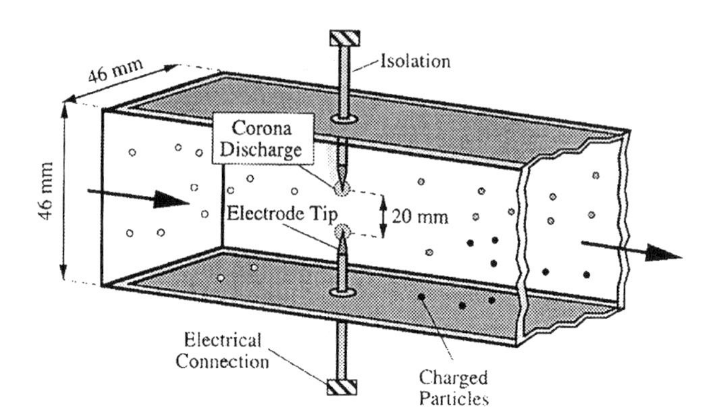

Figure 19.6 Schematic of the corona-charging unit (two needle electrodes of opposite electrical potential).

perpendicular to the direction of the aerosol flow. Thus, aerosol particles are charged bipolarly. The charged particles are directly fed into a scanning mobility particles sizer (TSI, SMPS 3934) without passing through a neutralizer prior to the SMPS. Therefore, the aerosol keeps its charge distribution.

Considering that initially a monomobile fraction of particles is fed into the charging unit, the mobility distribution measured with the SMPS enables the calculation of charge distribution of the monodisperse fraction of the NaCl aerosol. The mobility Z_p of a charged particle (diameter d_p) is related to the particle charge q_p as follows:

$$Z_p = \frac{q_p C}{3\pi\eta d_p} \tag{19.1}$$

where C is the Cunningham correction (Hinds, 1982), η the dynamic viscosisty, and d_p the particle diameter.

To determine the charge distribution of a bipolarly charged aerosol the mobility distribution is measured with positive and negative classifier voltage at the SMPS. By taking into account that all particles fed into the charging unit are preliminary positively charged, the total initial number concentration is measured by the SMPS without additional charging. Thus, the initial number concentration, as well as the number concentration of positively and negatively charged particles after passing through the charging unit, is determined.

19.3.3 Results

(1) Charge Distribution

Figure 19.7 shows the measured charge distribution of a monodisperse fraction of 0.12 μm as a function of the electrical potential of the discharge electrodes for a symmetric potential ratio (amount of positive potential of one discharge electrode equal to the amount of negative potential of the other discharge electrode; $U_+/|U_-| = 1$).

The primary monodisperse fractions exclusively contain positively charged particles. Thus, below the corona onset voltage, in our case about 4 kV, only positively charged particles are measured. Above the corona onset voltage, the number of single positively charged particles decreases and additional negatively charged particles are obtained. If the discharge voltage exceeds 6 kV, the fraction of multiple negatively charged particles increases whereas the fraction of positively charged particles decreases with increasing electrical potential. This indicates that a symmetric potential ratio results in an asymmetric charge distribution. However, with regard to the

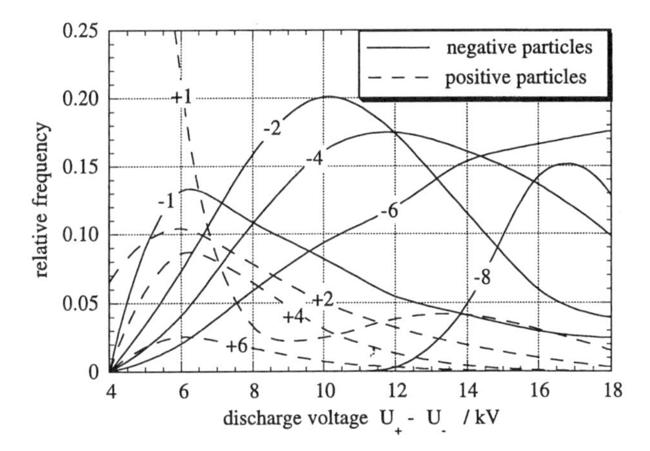

Figure 19.7 Charge distribution as a function of the discharge voltage of the electrodes for a symmetric potential ratio ($U_+/|U_-| = 1$).

enhanced agglomeration a symmetric charge distribution is required. It was found that for air as carrier gas (note that the originally pure nitrogen carrier gas is mixed with clean and dry sheath air in the first DMA) a potential ratio in the range between 1.3 and 1.5 results in a symmetric charge distribution, so that the number of positively charged particles equals the number of negatively charged particles. Table 19.1 presents the maximum charges per particle measured for symmetric potential ratio of three different particle sizes.

Table 19.1 Maximum Particle Charge after Bipolar Charging ($U_+/|U_-| = 1$)

Diameter	0.04 µm	0.06 µm	0.12 µm
Charge	$+2e_0/-3e_0$	$+2e_0/-3e_0$	$+6e_0/-8e_0$

Following Table 19.1 it can be seen that the maximum number of charges is not found to be proportional to the particle diameter as predicted by the diffusion charging theory (White, 1963).

The results shown in Figure 19.7 prove the developed electrode configuration to be able to charge aerosol particles bipolarly. Consequently, a corona discharge between needle electrodes of opposite electrical potential should increase the agglomeration rate of submicron aerosols.

(2) Agglomeration

The investigation of the electrical agglomeration was conducted using the same experimental setup as for the measurements of the charge distribution (see Figure 19.5). Instead of merely passing the polydisperse aerosol to the first DMA, it is directly fed into the charging and agglomeration unit. Thus, the total primary particle number concentration c_{prim} in the conditioning zone is of about 10^7 cm^{-3}.

Figure 19.8 shows a schematic of the agglomeration device. Each pair of discharge electrodes is followed by a residence time zone of 10 cm length. Since Gutsch and Löffler (1993) showed that the presence of external electrical fields reduces the agglomeration rate of oppositely charged particles no external electrical fields are applied at the residence time zone. With regard to the recombination of charges parallel to agglomeration of oppositely charged particles, a loss of attractive particle–particle forces occurs while the aerosol passes through the residence time zone. Therefore, the residence time zone is followed by an additional charging unit that is again followed by a residence time zone. The successive addition of the charging units and residence time zones leads to a multistage bipolar agglomeration process.

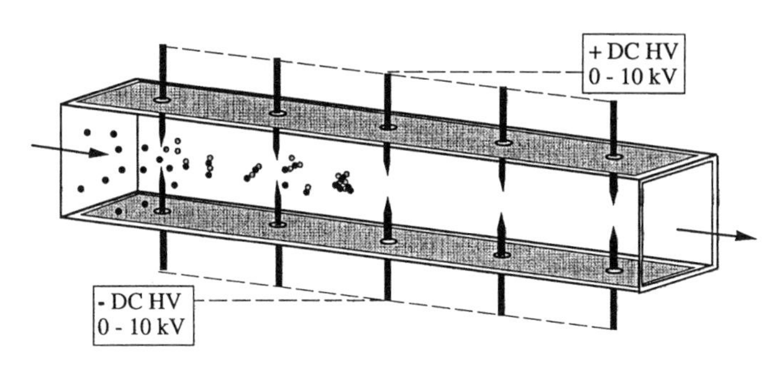

Figure 19.8 Schematic of the multistage charging and agglomeration device.

The extent of agglomeration is indicated by the comparison of SMPS measurements of the aerosol with and without bipolar charging. In any case the samples for the SMPS analysis are taken after passing the agglomeration device. Thus, the residence time between particle formation and analysis is the same for each experiment which guarantees that changes of the size distributions are almost exclusively related to the electrical agglomeration and not to other phenomena, e.g., Brownian coagulation.

Figure 19.9 shows the cumulative number distribution of the primary aerosol (solid line) and of one to fivefold bipolar charged aerosols (dotted lines). A successive shift in the size distribution is found when additional agglomeration units are operated. The primary particle size distribution covers a range from 0.024 to 0.22 μm and the final distribution corresponds to a decrease of the total number concentration from initially 10^7 to 3×10^6 cm^{-3}. The overall residence time of the aerosol in the agglomerator was about 50 s.

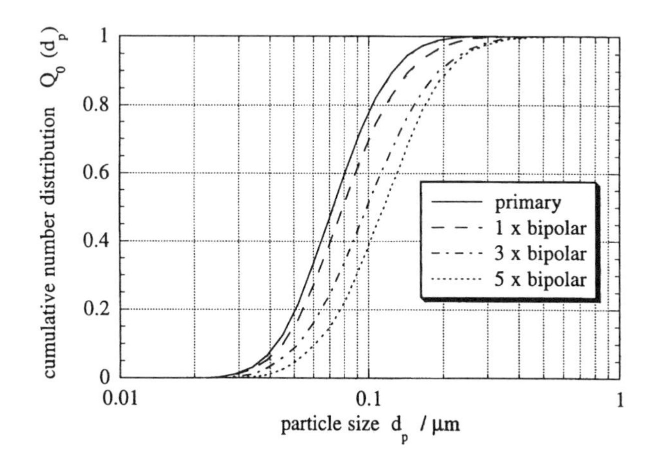

Figure 19.9 Cumulative number distribution for one- to fivefold bipolar agglomeration.

In addition to the analysis of the shift of size distribution, another method of analysis is of interest, especially in the case of homogeneous agglomeration where the transformation of the size distribution is related to a loss of particles in the lower size classes and an increase of particles in the upper size classes.

The concentration ratio ε is defined according to Equation 19.2:

$$\varepsilon = \frac{c_{\text{agg}}\left(d_p\right)}{c_{\text{prim}}\left(d_p\right)} \tag{19.2}$$

where $c_{agg}(d_p)$ is the number concentration of particles of diameter d_p after the agglomeration and $c_{prim}(d_p)$ is the number concentration of these particles before the agglomeration.

Values of ε smaller than 1 indicate a loss of particles. Values of ε larger than 1 show an excess of particles in a certain size class. It should be noticed that Equation 19.2 is restricted to the size range of the primary particle size distribution. Figure 19.10 shows the cumulative number distribution and the concentration ratio ε for a primary aerosol with a number median diameter of 0.017 μm. The concentration ratio indicates a significant loss of particles in the size range below 0.02 μm and an increasing number of particles in the size range above 0.03 μm. Furthermore, it can be seen that operating additional charging units leads to a larger loss in the lower size range and, in correspondence to that, to a larger increase in the upper size range. The same relationship is found for primary particles in the size range between 0.03 and 0.4 μm.

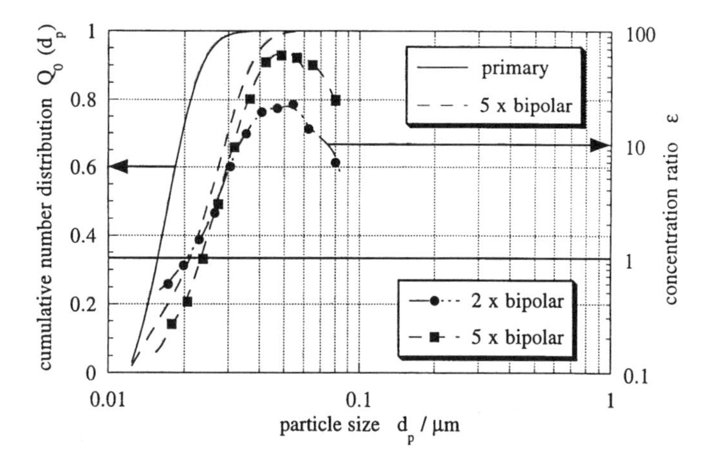

Figure 19.10 Cumulative number distribution and concentration ratio for bipolar agglomeration (primary median diameter = 0.017 μm).

Figure 19.11 presents the concentration ratio and the cumulative number distribution for multistage bipolar agglomeration of an aerosol of a primary number median diameter of 0.098 μm. According to the larger primary particles, the loss of particles now occurs below 0.1 μm, whereas a significant increase of particles is obtained in the range above 0.2 μm.

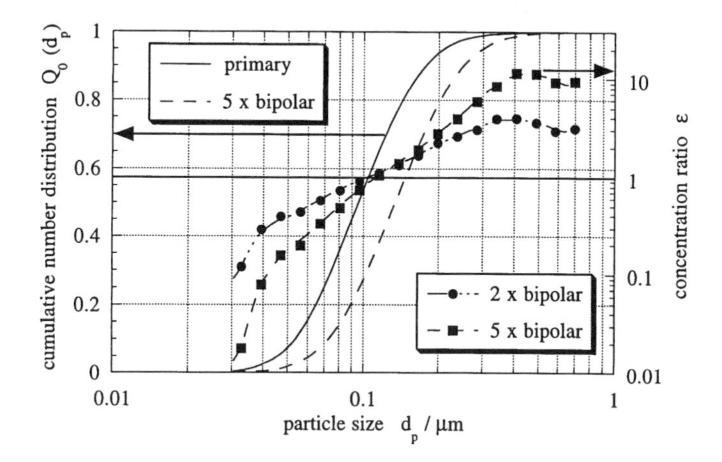

Figure 19.11 Cumulative number distribution and concentration ratio for bipolar agglomeration (primary median diameter = 0.098 μm).

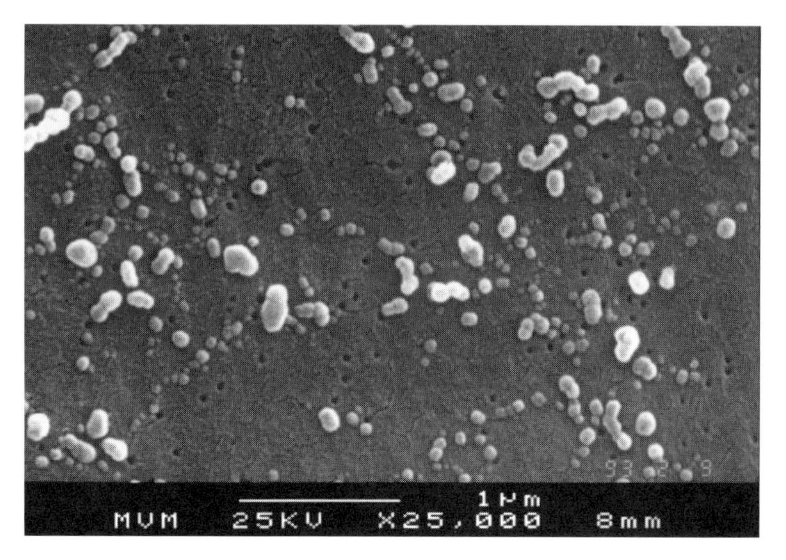

Figure 19.12 SEM of the primary NaCl particles.

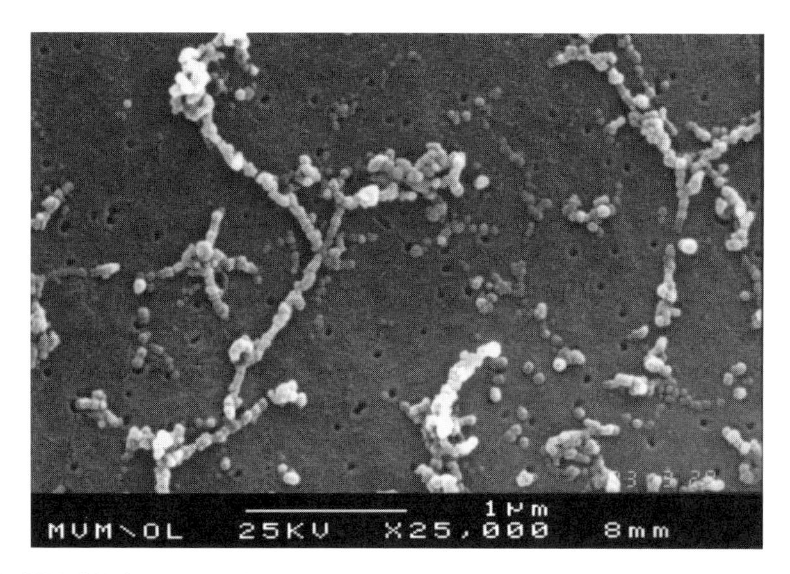

Figure 19.13 SEM of NaCl particles after fivefold bipolar agglomeration.

The evolution of the particle morphology caused by the electrical agglomeration can be seen from Figure 19.12 and 19.13. Figure 19.12 shows a scanning electron micrograph (SEM) of primary NaCl particles. Besides a few agglomerates of less than four primary particles, mainly single spherical particles are found. This is totally different after electrical agglomeration. Figure 19.13 is dominated by elongated chainlike agglomerates of more than 100 primary particles.

The chainlike morphology of the agglomerates is a result of the inhomogeous electrical fields in the vicinity of the charged aerosol particles. Theoretical investigations of Gutsch and Löffler (1993) showed that oppositely charged particles tend to grow in this chainlike manner.

Although a significant change of the particle morphology is obtained through the bipolar agglomeration, the corresponding shift of the size distributions is insufficiently small for a significant increase of the separation efficiency. However, it is found that the electrical agglomeration of submicron aerosols at least allows influencing of the particle morphology. Therefore, the results shown above may rather have a bigger impact on gas-phase particle synthesis than on fine particle separation processes.

Industrial aerosols, for example, combustion aerosol, have broader size distribution than the NaCl aerosol that was used here. The presence of large particles beside small particles allows an enhancement of the agglomeration rate between the small submicron particles and the large micron particles. Since particles in the range of about 5 to 10 μm can easily be separated with conventional separators, the scavenging of fine particles by large particles is a more useful application of the electrically enhanced agglomeration. A result of the electrical agglomeration between fine and coarse particles is presented in Figure 19.14. This SEM shows a latex sphere of 0.8 μm which was mixed with the primary NaCl aerosol and exposed to fivefold bipolar charging. The surface of the latex sphere is covered by small NaCl particles. Without bipolar charging the extent of agglomeration between the coarse latex particles and the small NaCl particles was extremely low. Thus, the deposits of NaCl particles at the latex sphere are a result of the electrically enhanced agglomeration caused by bipolar charging of the bimodal mixture.

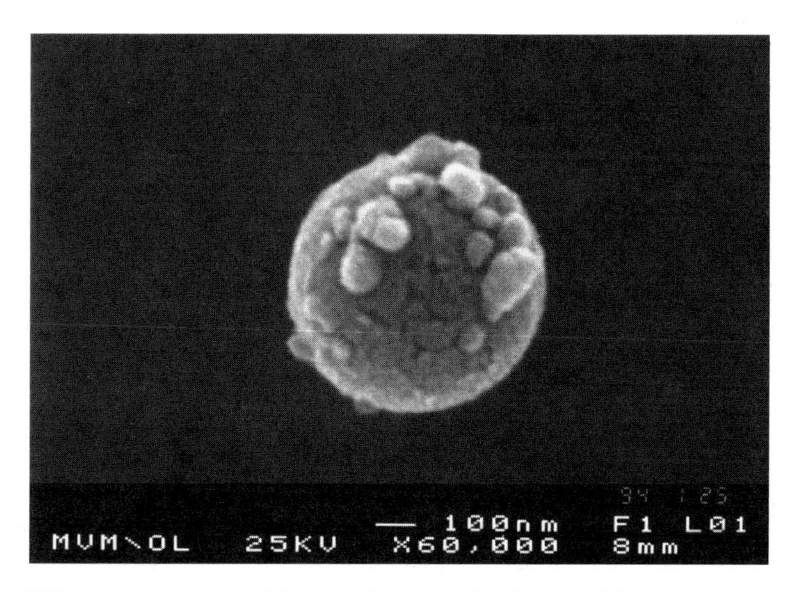

Figure 19.14 SEM of latex sphere (d_p = 0.8 μm) that is covered by NaCl particles due to bipolar agglomeration of the mixture.

Although the developed method of bipolar charging results in a random distribution of particle polarity, a significant increase of the agglomeration rate is obtained. A higher agglomeration rate could have been achieved if the coarse particles were charged with the opposite polarity than the fine particles. Further investigations will be done with respect to this demand.

19.3.4 Summary

A new method of bipolar charging of submicron aerosol particles is introduced. The particles are charged by corona discharge between two needle electrodes of opposite electrical potential. An increased particle charge is obtained with increasing corona discharge voltage as well as with increasing particle size. As a result of the bipolar charge, an enhanced agglomeration rate between oppositely charged particles is achieved. The extent of agglomeration is increased by repeated bipolar charging. The resulting particles are mostly chainlike, elongated agglomerates. In addition to the enhanced agglomeration of particles of nearly the same size, an increased agglomeration between fine and coarse particles can be achieved by bipolar charging of bimodal aerosols. This method of electrically enhanced scavenging of fine particles might be successfully applied for separation of submicron particles from gases after further developments.

19.4 FRACTIONAL EFFICIENCIES OF FIBROUS FILTERS FOR AIRBORNE BACTERIA

19.4.1 Introduction

Fibrous filters are widely used in gas- or air-handling systems to separate dust and airborne particulate matter. In common air-conditioning and ventilation systems fibrous filters collect not only dust particles but also biological particles, i.e., airborne bacteria, fungi, pollen, etc., which are also present in the ambient air. They usually occur in relatively low number concentrations ranging from 10 to 1000 viable particles per cubic meter of air. Only in the vicinity of an emission source the concentrations are considerably higher (Maroni et al., 1993; Sawyer et al., 1993; Reponen et al., 1994). But the values are still much lower than the typical concentrations observed for dust particles. However, airborne microorganisms and pollen, even in low number concentrations, can cause a variety of infectious and allergic diseases for human organisms (Crook, 1992; Deininger, 1993; Lacey and Dutkiewicz, 1994).

When biological particles containing allergenic or pathogenic substances are inhaled and separated in the respiratory tract, they can induce allergic symptoms (mucous membrane irritations, rhinitis, allergic asthma, alveolitis, etc.) or cause a range of different types of infections (plague, tuberculosis, etc.). In industrial processes, like food sterilization, procedures can be ineffective in the presence of airborne viable particles. Airborne bacteria and fungal spores can contaminate food products after pasteurization or sterilization. The contamination with pathogenic and spoilage microorganisms will reduce shelf-life and product safety (Kang and Frank, 1989). These are some reasons that in recent times researchers pay more attention to airborne biological particles and their reduction in various environments with the help of fibrous filters (Ohgke et al., 1993; Maus and Löffler, 1994; Brosseau et al., 1994).

Concerning the collection efficiency of fibrous filters, one has to consider that in comparison to mineral dust biological particles show different aerodynamic and adhesion behavior. The different particle density and shape influence the aerodynamic properties of the particles and thus the transport to the filter fibers. The adhesion properties of biological particles are different because the particle substance and the surface structure, which both influence the adhesion mechanisms, are different compared with mineral particles. These characteristics of biological particles will consequently lead to different collection behavior in fibrous filters, where transport and adhesion to the fibers determine the collection efficiency. Therefore, the collection efficiencies of filters determined with mineral dust cannot be assigned to biological particles in a simple way; instead they need to be verified experimentally.

In this chapter fractional efficiencies of fibrous filters for limestone dust and microbial aerosols are presented. The collection or fractional efficiency of a filter describes, in addition to the pressure drop, the performance of any air filter. The experimental setup operates with two optical particle counters for the *in situ* measurement of the particle flux and size before and behind a test filter. Optical particle counters were chosen for particle analysis because they yield many advantages for the measurement of biological particles. Usually, airborne bacteria are collected in samplers where no direct or short-term reading of the concentration is possible. The data can only be obtained after a time-consuming procedure (Willeke and Baron, 1993) in which the microorganisms have to be cultivated on agar plates. Instead, the *in situ* measurement in the raw and clean gas with optical particle counters allows the quick determination of the particle size-dependent collection efficiency of filters within a few minutes.

19.4.2 Methods

The experimental setup consists of an aerosol generation unit, a measurement unit, and a filter holder (Figure 19.15). As an aerosol generator for dispersing a highly concentrated suspension of bacteria serves a suspension nebulizer. Dust powder can be dry disseminated with the help of an aerosol generator working with a rotating brush (Zahradnicek and Löffler et al., 1976). Right behind the aerosol generator an aerosol neutralizer is installed to reduce the electrical charges of the particles, which they acquired during the dispersing process. The primary aerosol is then mixed with conditioned air to adjust the relative air humidity and assure that all droplets evaporate. The number concentrations in the aerosol range from 10^8 to 10^{10} m^{-3}. After the mixing zone, the aerosol is directed in a tubular flow channel (inner diameter = 50 mm) which contains the actual measuring unit with the optical particle counters and the filter holder.

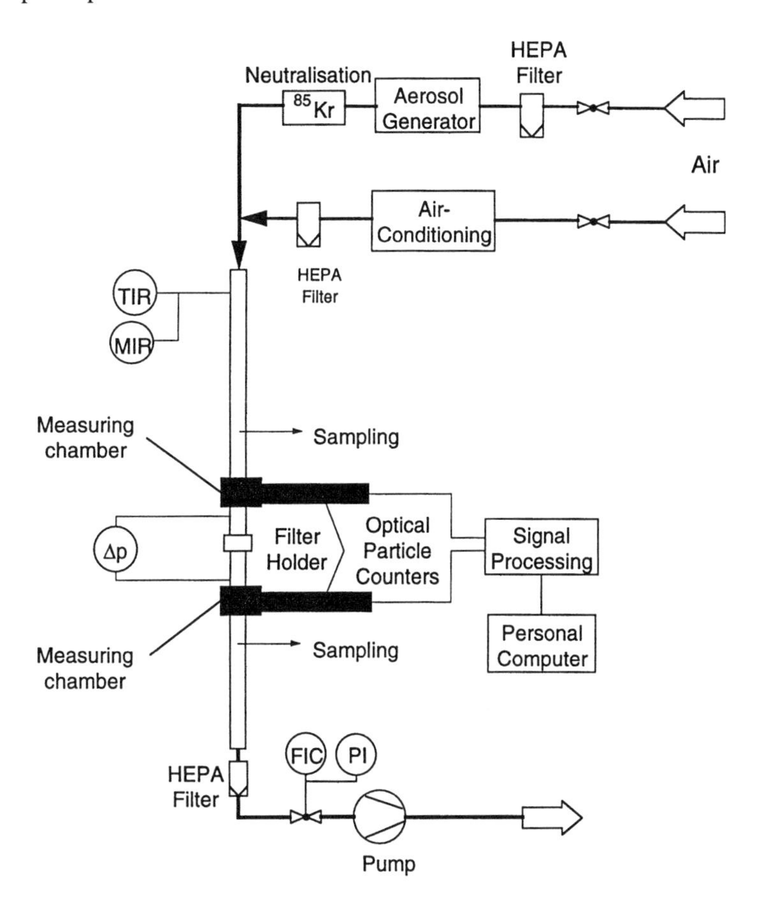

Figure 19.15 Schematic diagram of the experimental setup.

The measuring unit incorporates two optical particle counters for the *in situ* measurement of the particle flux and size within the given flow upstream and downstream of the test filter. The optical particle counters operate with a purely optically defined measuring volume and include a system for the elimination of the border zone error (Umhauer, 1983). Prior to this study the optical particle counters were calibrated with monodisperse fractions of polystyrene latex spheres.

The measurements take place within measuring chambers which are part of the flow channel (Figure 19.15) and which have the same inner diameter as the flow channel. By this arrangement the measurement of the particles can be done without disturbing the tubular aerosol flow. The design of the measuring unit is similar to that of Jodeit (1985) and Baumgartner and Löffler (1986).

In between the two measuring chambers, the filter holder is located. The dimensions of the filter holder and the test filter were chosen in the way that the occurrence of nonuniform aerosol flow through the test filter is minimized. Following the test filter and the second measuring chamber, the aerosol flows through a high-efficiency filter as a backup filter for collecting the remaining particles in the clean air. A vacuum pump behind the backup filter generates the airflow through the hole system.

(1) Filters

The investigated filters are rated according to the European standard EUROVENT 4/5 as filters from class EU4 to EU5 (fine and coarse dust filters). These filters are mainly used in air-conditioning and ventilation systems to guarantee a relatively low particle concentration in the ventilated areas. The classification is done by the measurement of the total collection efficiency for synthetic (EU1 to EU4) or atmospheric (EU5 to EU9) dust. Usually no information about the fractional efficiencies is given. Air filters with higher collection efficiencies as EU9 filters are classified as high-efficiency filters (HEPA or ULPA filters).

The test filters used in this investigation are made of polyester or polycarbonate fibers. The fiber diameters range from 5 to 50 μm depending on the filter media. The filters show porosities >98% and are 0.5 to 2.5 cm thick. Circular samples (diameter = 52 mm) of each filter medium were used as test filters. Additionally, filter media were investigated that include a layer of electrically charged fibers to enhance their collection efficiency. These filter media are called electret filters and are not classified according to EUROVENT 4/5 (see above). They are commonly used in those fields in which a high collection efficiency is required along with a low flow resistance (Brown, 1993), e.g., respirator filters, vacuum filters, automobile air filters, etc. They are scarcely used in air-conditioning or ventilation systems mainly because of their relatively high manufacturing costs.

(2) Bacteria

For the experiments, the bacteria species of *Micrococcus luteus* (DSM 20030), *Escherichia coli* K12 (DSM 498), and *Bacillus cereus* (wild-type strain) were chosen. These microorganisms are common types encountered in bioaerosols (Kang and Frank, 1989; Crook, 1992; Deninger, 1993; Sawyer et al., 1993; Maroni et al., 1993). The bacteria *M. Luteus* and *E. coli* were cultured for 48 h in Standard I nutrient broth at 30°C. The cells were then harvested and washed three times with deionized water using a centrifuge at 2600 × g. The final suspension of about 10^8 viable cells/ml was stored at 4°C until use. Bacillus *cereus* was grown on Standard I nutrient agar at 30°C for 10 days. The colonies were harvested from the cultures with sterile deionized water +0.1% TWEEN. The suspensions were then centrifuged at 2600 × g, washed three times with deionized water, and subsequently used for the experiments. It was verified by microscopical observations that the suspensions contain mainly vegetative cells of *B. cereus.*

19.4.3 Results and Discussion

(1) Calibration

Because the optical properties of the airborne bacteria are not known, the size distributions obtained with the optical particle counters have to be verified with alternative methods. A modified six-stage impactor (Andersen Viable Sampler) was used to determine the size distribution of the viable bacteria in the aerosol. Also, the size distribution of all bacteria in the aerosol was obtained by collection of the particles on membrane filters (pore size = 0.2 μm) and subsequent image analysis of the electron micrographs (Maus et al., 1993).

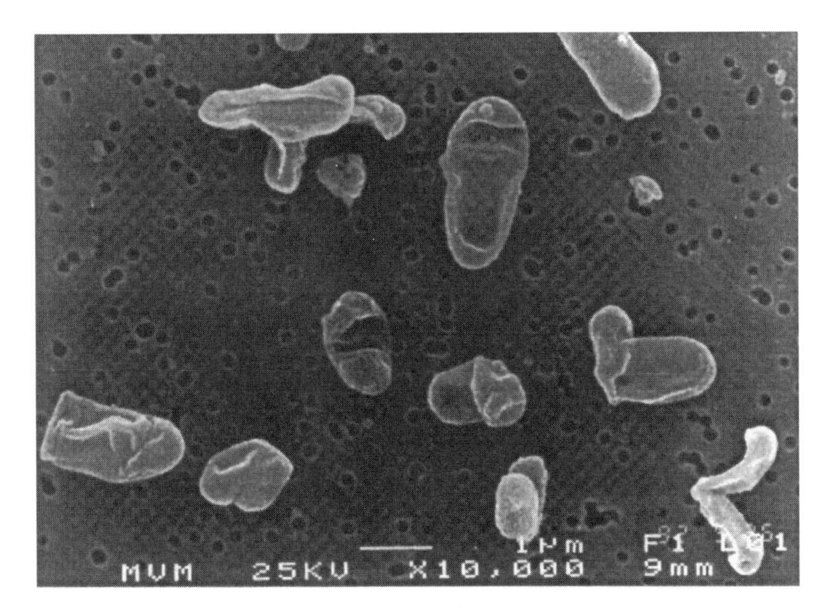

Figure 19.16 Airborne *E. coli* bacteria collected on a membrane filter (pore size 0.2 μm).

Figure 19.16 displays a micrograph of a membrane filter with sampled airborne *E. coli* cells. These micrographs were used to determine the size distribution of the bacterial aerosol by means of image analysis. Because of their exposure to air and the preparation technique for the SEM, the cells on the membrane filter are dehydrated. But still, the typical rod shape of the *E. Coli* cells can be recognized.

In Figure 19.17 the cumulative number distribution of airborne *E. coli* bacteria determined by different methods is plotted.

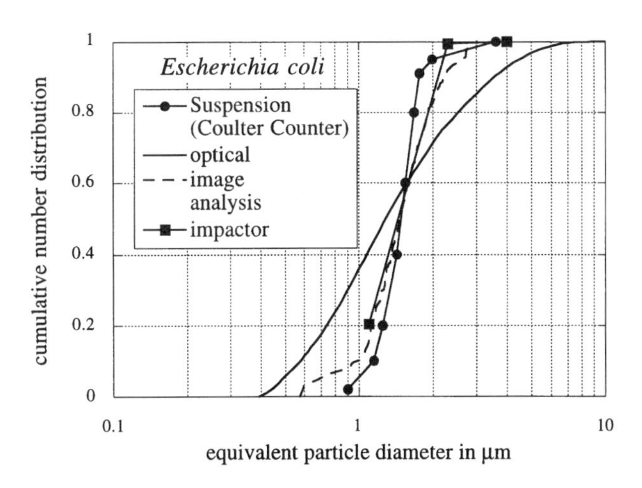

Figure 19.17 Size distribution of airborne *E. Coli* as determined by three different methods.

The size of the airborne bacterial cells is in the range from 0.6 to 3 μm. The curves for the different methods are close together and very similar in shape and position. The size distribution measured by the optical particle counters is slightly broader than the distributions determined by the alternative methods. This fact is due to the nonuniform optical properties of the bacterial cells which can vary within a given cell suspension. Similar results were received with the bacteria of *M. luteus*. With these findings no particular calibration of the optical particle counters were performed for the airborne bacteria, because the optical equivalent diameter resembles approximately the aerodynamic diameter of the bacterial cells. However, for *B. cereus* the optical particle counters measured a broader distribution than determined by image analysis and the count median diameters were significantly different from each other. In this case a special calibration was performed with the results of the image analysis.

(2) Fractional Efficiencies

The fractional efficiencies of the filter samples were determined at the initial stage of filtration. This implies that the loading of the filter fibers with particles did not affect the filtration efficiency in a significant manner. The concentrations in the raw gas stream lay in the range from 10^8 to 10^{10} m^{-3}. Each fractional efficiency curve displayed in the following plots resembles the mean of at least three to five single trial runs in which several thousands of particles were analyzed by the counters. The particle size range of the fractional efficiency curves is limited because of the relatively narrow size distribution of the bacteria (Figure 19.17). The efficiencies for the bacterial aerosols are compared with the results obtained with limestone dust. The correspondent experiments were conducted with the same experimental setup introduced in this chapter.

Figure 19.18 shows the fractional efficiencies of two medium-efficiency filters for microbial aerosols and for limestone dust. The efficiencies of the EU4 filter are lower than the efficiencies of the EU5 filter, which reflects the different filter class of both filters. Efficiencies of over 90% are only achieved for particles larger than 8 to 10 μm in size. For microbial aerosols the filters show similar efficiencies as for limestone particles. The differences between the curves are within the variations between single measurements and not significant.

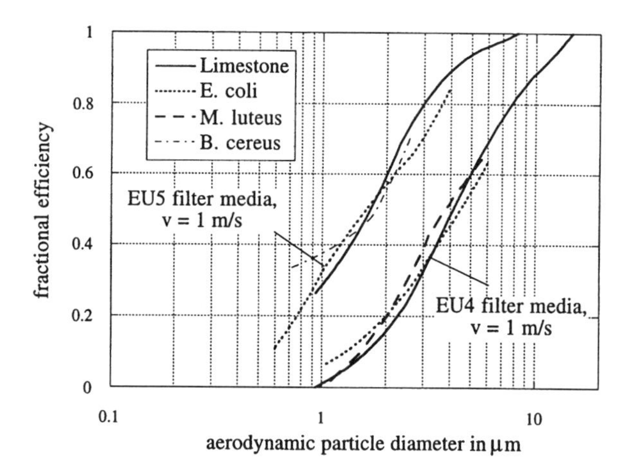

Figure 19.18 Fractional efficiencies of medium-efficiency filters (EU4 and EU5) for microbial aerosols and limestone dust at filter face velocity $v = 1$ m/s.

The transport of particles >0.5 μm to the filter fibers in the investigated filters is governed by inertia forces, and the aerodynamic particle diameter is an appropriate particle characteristic to describe the behavior of particles under the influence of inertia forces. The aerodynamic particle diameter is the diameter of the unit density sphere that has the same settling velocity as the considered particle. It can be seen that the separation behavior of the microbial and the dust (limestone) particles is equivalent under the applied conditions. Thus, with the results of collection efficiency measurements performed with dust the efficiency for airborne microorganisms can be predicted.

In Figure 19.19 the fractional efficiencies of a high-efficiency filter (electret fiber medium) for limestone dust and for different bacterial aerosols are shown. The efficiency of the electret fiber medium is significantly higher than the efficiency of the EU4 and EU5 filter (Figure 19.18). The

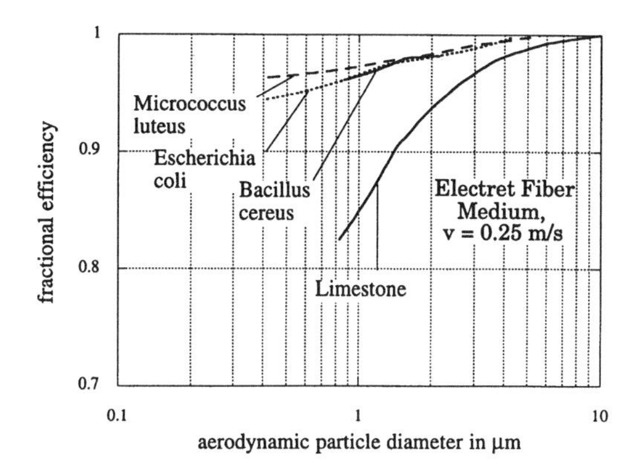

Figure 19.19 Fractional efficiencies of a high-efficiency filter (electret fiber medium) for bacterial aerosols and limestone dust.

cut size of this filter medium for limestone particles is <0.8 µm compared with a cut size of 1.6 and 4.0 µm of the medium-efficiency filters. In contrast to the EU4 and EU5 filter media the efficiency of the electret medium for airborne bacteria (*E. coli, M. luteus*) is significantly higher than for limestone dust.

The separation of the particles in the electret fiber medium is not only due to inertia forces but also to electrical forces (coulombic attraction, image force) because this filter media includes a layer of electrically charged fibers to enhance the efficiency of the filter. Especially the image force increases with increasing dielectric constant of the considered particles. Water, which is the main constituent of bacteria, has a dielectric constant of 80 while limestone has a dielectric constant of about 8. Thus, the bacteria will experience a higher attractive force toward the electrically charged fiber than the limestone particles. Because of this mechanism the collection of airborne bacteria in this filter medium is more efficient than the collection of limestone particles of the same aerodynamic particle size (Baumgartner, 1987). Thus, it is not possible to predict the collection efficiencies of electret filters for airborne bacteria when knowing the efficiencies for mineral dust.

The presented results show that the use of electret fiber media is advantageous in environments of high or hazardous microbial content because of their enhanced collection efficiency for airborne bacteria compared with dust particles.

REFERENCES

Baumgartner, H.-P. (1987), Elektretfaserschichten für die Aerosolfiltration — Untersuchungen zum Faserladungszustand und zur Abscheidecharakteristik, *VDI Fortschr. Ber.,* 3(146).

Baumgartner, H.-P. and Löffler, F. (1986), The collection performance of electret filters in the particle size range 10 nm to 10 µm, *J. Aerosol Sci.,* 17, 438–445.

Brosseau, L. M., Chen, S.-K., Vesley, D., and Vincent, J. H. (1994), System design and test method for measuring respirator filter efficiency using *Mycobacterium* aerosols, *J. Aerosol Sci.,* 25, 1567–1578.

Brown, R. C. (1993), *Air Filtration,* Pergamon Press, Oxford.

Crook, B. (1992), Exposure to airborne microorganisms in the industrial workplace, *J. Aerosol Sci.,* 23(Suppl. 1), 559–562.

Deininger, C. (1993), Pathogene Bakterien, Pilze und Viren am Arbeitsplatz, *Staub Reinhalt. Luft,* 293–299.

Eliasson, B., Egli, W., Ferguson, J. R., and Jodeit, H., (1987), Coagulation of bipolarly charged aerosols in a stack coagulator, *J. Aerosol Sci.,* 18, 869.

Eurovent (1988), EUROVENT-Dokument 4/5, Prüfung von Luftfiltern für die allgemeine Lufttechnik, Maschinenbau Verlag, Frankfurt, Germany.

Funcke, G. and Frohn, A. (1994), Experimental investigation of the influence of sound intensity on acoustic agglomeration, in *Proc. of the 4th Int. Aerosol Conference,* Los Angeles.

Gutsch, A. and Löffler, F. (1993a), Numerical simulation of the electrically induced aerosol agglomeration in the submicron range, in *Proc. 6th Int. Symp. on Agglomeration,* Nagoya, Japan, 46.

Gutsch, A. and Löffler, F. (1993b), Electrically induced aerosol agglomeration: formulation of the collision frequency and simulating agglomerate growth, *J. Aerosol Sci.,* 24, 505.

Hautanen, J., Kilpeläinen, M., Kauppinen, E. I., Jokiniemi, J., and Lethinen, K. (1995), Electrical agglomeration of aerosol particles in an alternating electric field, *Aerosol Sci. Technol.,* 22, 181.

Hinds, C. W. (1982), *Aerosol Technology,* Wiley & Sons, New York.

Jodeit, H. (1985), Untersuchung zur Partikelabscheidung in technischen Tiefenfiltern, *VDI Fortschr. Ber.,* 3, 108, VDI-Verlag, Düsseldorf.

Kang, Y.-J. and Frank, J. F. (1989), Biological aerosols: a review of airborne comtamination control and its measurement in dairy processing plants, *J. Food Protection,* 52, 512–524.

Lacey, J. and Dutkiewicz, J. (1994), Bioaerosols and occupational lung disease, *J. Aerosol Sci.,* 25, 1371–1404.

Löffler, F. (1988), *Staubabscheiden,* Thieme, Stuttgart, 208.

Löffler, F., Dietrich, H., and Flatt, W. (1984), *Staubabscheidung mit Schlauchfiltern und Taschenfiltern,* Vieweg, Braunschweig, Wiesbaden.

Madelin, T. M. (1994), Fungal aerosols: a review, *J. Aerosol Sci.,* 25, 1405–1412.

Magill, J., Pickering, S., Fourcaudot, S., Gallego-Juarez, J. A., Riera-Franco De Sarabia, E., and Rodriguez-Corral, G. (1989), Acoustic aerosol Scavenging, *J. Nucl. Mater.,* 166, 208.

Maroni, M., Bersani, M., Cavallo, D., Anversa, A., and Alcini, D. (1993), Microbial contamination in buildings: comparison between seasons and ventilation systems, *Proc. Indoor Air,* 4, 137–142.

Maus, R. and Löffler, F. (1994), Abscheidung von biologischen Partikeln (Bioaerosole) aus Luft in technischen Tiefenfiltern (Vorprojekt), Kernforschungszentrum, Karlsruhe, KfK-PEF 120.

Maus, R., Umhauer, H., and Löffler. F. (1993), Size analysis of biological particles by scattered light measurements, *J. Aerosol Sci.,* 24(Suppl. 1), 429–430.

Mothes, H. (1982), Bewegung und Abscheidung der Partikeln im Zyklon, Dissertation, Universität Karlsruhe (TH), Karlsruhe.

Ohgke, H., Senkspiel, K., and Beckert, J. (1993), Experimental evaluation of microbial growth and survival in air filtes, *Proc. Indoor Air,* 6, 521–526.

Reponen, T., Hyvärinen, A., Ruuskanen, J., Raunemaa, T., and Nevalainen, A. (1994), Comparison of concentrations and size distributions of fungal spores in buildings with and without mould problems, *J. Aerosol Sci.,* 25, 1595–1603.

Riehle, C. (1992), Bewegung und Abscheidung von Partikeln im Elektrofilter, Dissertation, Universität Karlsruhe (TH), Karlsruhe.

Sawyer, B., Elenbogen, K. C., Rao, K. C., O'Brien, P., Zenz, D. R., and Lue-Hing, C. (1993), Bacterial aerosol emission rates from municipal wastewater aeration tanks, *Appl. Environ. Microbiol.,* 59, 3183–3186.

Schabel, S., Heidenreich, S., Sachweh, B., Büttner, H., and Ebert, F. (1994), Separation of submicron particles supported by heterogenous condensation, *J. Aerosol Sci.,* 25, 459.

Scheibel, H. G. and Porstendörfer, J. (1983), Generation of monodisperse Ag- and NaCl aerosols with particle diameter between 2 and 300 nm, *J. Aerosol Sci.,* 14, 113.

Schmidt, E. (1993), Zur Kompression von auf Filtermedien abgeschiedenen Staubschichten, *Staub-Reinhalt. Luft.,* 53, 369–376.

Schmidt, E. (1994), Simulation des dreidimensionalen Aufbaus von Staubschichten mittels Partikelbahnrechnung am Beispiel der Oberflächenfiltration, *Chem. Ing. Tech.,* 66, 718, 720.

Schmidt, E. and Löffler, F. (1991), The analysis of dust cake structures, Part. Part. Syst. Charact., 8, 105–109.

Schmidt, M. (1993), Theoretische und experimentelle Untersuchungen zum Einfluß elektrostatischer Effekte auf die Naßentstaubung, Dissertation, Universität Karlsruhe (TH).

Umhauer, H. (1983), Particle size distribution analysis by scattered light measurements using an optically defined measuring volume, *J. Aerosol Sci.,* 14, 765–770.

Wadenpohl, C. (1994), Elektrostatisch unterstützte Abscheidung von Dieselrußpartikeln in Fliehkraftabscheidern, Dissertation, Universtität Karlsruhe (TH).

Wang, S. C. and Flagan, R. C. (1990), Scanning mobility spectrometer, *Aerosol Technol.,* 13, 230.

White, H. J. (1963), *Industrial Electrostatic Precipitation,* Addison-Wesley, Reading, MA.

Willeke, K. and Baron, P. (1993), *Aerosol Measurement,* Von Nostrand Reinhold, New York.

Zahradnicek, A. and Löffler, F. (1976), Eine neue Dosiervorrichtung zur Erzeugung von Aerosolen aus vorgegebenen feinkörnigen Feststoffen, *Staub Reinhalt. Luft.,* 36, 425–427.

CHAPTER 20

On the Filtration and Separation of Fibrous Aerosols

Kvetoslav R. Spurny

CONTENTS

20.1 INTRODUCTION

In recent years there has been considerable interest in asbestos and other natural and anthropogenic mineral fibers as health hazards in the working environment and in the indoor and outdoor atmospheres.[1] Fabric filters manufactured from fine organic and inorganic fibers are used to clean exhaust gases in the asbestos mining and manufacturing industries, as well as in industries manufacturing anthropogenic mineral fibers. To sample asbestos and other fibrous aerosols in workplaces, in buildings, and in ambient air for monitoring their concentration, membrane and Nuclepore filters (NPF) are used.

There is a need for theoretical and experimental knowledge concerning the separation of fibrous particles in technical and analytical filters. During the last two decades the problem of filtration and separation of fibrous aerosols has been investigated by several authors and has seen important progress. Nevertheless, there is still need for more theoretical and experimental knowledge.

0-87371-830-5/98/$0.00+$.50

20.2 SEPARATION THEORIES

The first description and evaluation of fiber separation and deposition in human airways ("lung filter"), based on the Timbrell observations, were published by Harris and Fraser in 1976.[2] Feigly[3] attempted to describe a simple mathematical model for deposition of asbestos fibers in a fibrous aerosol filter. He applied the classical filtration theories existing for the separation of isometric particles and modified them for the separation of fibrous particles.

The physical principles of the separation of isometric particles of filters are well known and mathematically described.[4,5] These models and theories are based on a combined action of several partial mechanisms, whose influence is in the first approximation a function of the aerodynamic diameter of spherical particles. In the case of fibrous aerosols, the individual particles are not spherical; they have a prolongated form and can be approximated, e.g., by a prolate ellipsoid. In order to use the classical aerosol filtration theories, mainly the equations, describing the collection of spherical particles by different physical mechanisms, a definition for the aerodynamic "diameter" of a fibrous particle with a length L_F and diameter D_F is needed.

The aerodynamic behavior of fibrous particles dispersed in the gas phase is in addition influenced by their spatial orientation in the gas phase. This fact is of crucial importance. The orientation of a fibrous particle in the gas flow can be described best by means of a probability orientation function.[7] Nevertheless, in practical solutions the aerodynamic diameter of fibrous particles can be approximately described as a function of the parameters L_F, D_F, their aspect ratio $\beta = L_F/D_F$, and, for special cases, that is, for the parallel ||, perpendicular $\perp$, and random orientation of the fibrous particles in the gas flow.[6] The relevant equations for these cases are

1. Parallel orientation:

$$D_{ae} \parallel = \frac{3}{2} D_F \sqrt{\frac{\sigma}{\sigma_0}\left[\ln(2\beta) - 0.5\right]} \tag{20.1}$$

2. Perpendicular orientation:

$$D_{ae} \perp = \frac{3}{4} D_F \sqrt{\frac{2\sigma}{\sigma_0}\left[\ln(2\beta) + 0.5\right]} \tag{20.2}$$

3. Random orientation:

$$D_{ae} = \frac{3}{2} D_F \sqrt{\frac{\dfrac{\sigma}{\sigma_0}}{\dfrac{0.358}{\ln(2\beta) - 0.5} + \dfrac{1.230}{\ln(2\beta) + 0.5}}} \tag{20.3}$$

Here are σ and σ_0 are the densities of particles and of the gas.

Also, the shape factors K of fibrous particles can be similarly described.

$$K_{\parallel} = \frac{4(\beta^2 - 1)}{3\left[\dfrac{2\beta^2 - 1}{\sqrt{\beta^2 - 1}}\ln\left(\beta + \sqrt{\beta^2 - 1}\right) - \beta\right]} \tag{20.4}$$

$$K_{\perp} = \frac{8(\beta^2 - 1)}{3\left[\frac{2\beta^2 - 3}{\sqrt{\beta^2 - 1}}\ln\left(\beta + \sqrt{\beta^2 - 1}\right) + \beta\right]} \tag{20.5}$$

See Figure 20.1.

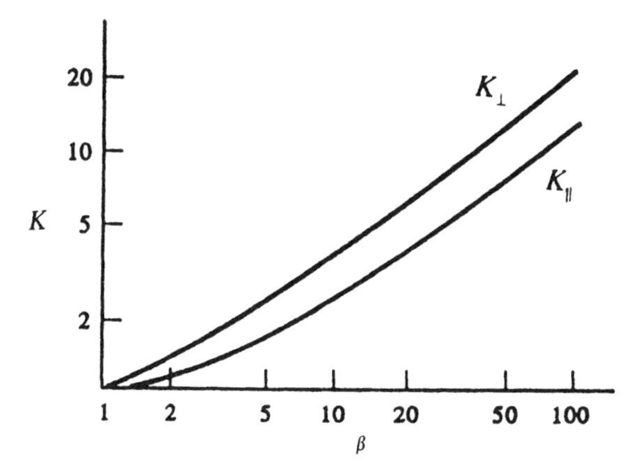

Figure 20.1 Dependencies of the particle shape factors K on the aspect ratio $\beta = L_F/D_F$ and on fiber orientation. (From Spurny, K., *Faserige Mineralstäube*, Springer-Verlag, Heidelberg, 1997. With permission.)

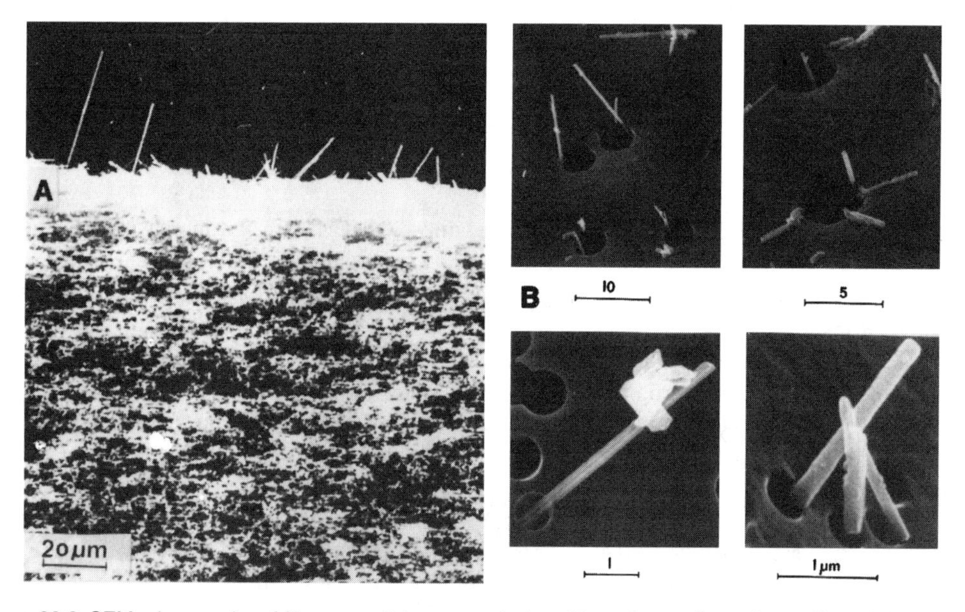

Figure 20.2 SEM micrographs of fibrous particles separated on the surfaces of membrane filters (A) and NPFs (B) showing the influence of spatial fiber orientation.

The practical importance of the orientation of fibrous particles, e.g., immediately before the contact with the surface of filter elements, is demonstrated and visualized well in Figure 20.2. Feigley used the classical filtration theory and the mentioned equations. Then, he described the partial collection efficiencies for three basic mechanisms: for the direct interception E_R, for the inertial separation–impaction E_I, and for the diffusion E_D. The total filter collection efficiency E was given by

$$E = E_D + E_{DR} + E_R + E_I \tag{20.6}$$

The Feigley model made the assumption that the aerodynamic behavior of fibrous particles can be approximated by that of small cylinders or rods. The motion of fibers relative to the surrounding fluid was assumed to be in the Stokes regime. Some results of Feigley's computations are shown in Figure 20.3. The filter efficiency E was plotted vs. the aerodynamic particle radius R_{ae} ($2R_{ae} = D_{ae}$) at five different aspect ratios (β = 10, 25, 63, 159, and 316: curves 1 to 5) and for a solidity (packing density) $\beta_p = 0.05$.

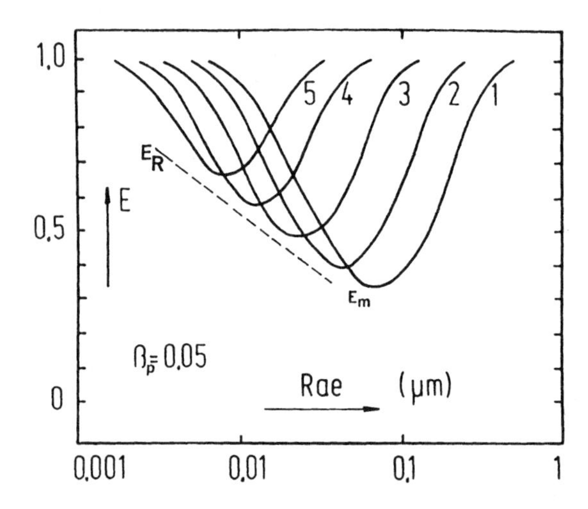

Figure 20.3 Computed collection efficiencies E for a fiber filter with a packing density $\beta_p = 0.05$ as a function of the aerodynamic particle radius R_{ae}. E_m means the minimum of the collection efficiency. (From Feigley, C. E., *Clean Air* (Australia), Nov., 65–71, 1975. With permission.)

The increasing E values with respect to increasing β, that is with the increasing fiber length L_F, demonstrate the importance of the partial separation mechanisms E_R (direct interception). Gradon et al.[7] tried to describe theoretically the motion and the deposition of fibrous particles on a single-fiber element inside a filter.

The definition of the collection efficiency of an aerosol particle for the partial mechanisms of the direct interception (E_R), inertia (E_I), and sedimentation can be reduced to the definition of the limiting trajectory of the particle (Figure 20.4a). In the case of a spherical particle or of a compact particle which can be approximated by a sphere of equivalent diameter a_p, the effect of particle rotation is neglected, because the position of its edge in relation to the surface of the cylinder (diameter a_f) of the filter element is independent of the form of motion. The situation differs in the case of a fibrous particle (Figure 20.4b). An analysis of the motion of the trajectory of the center of the particle mass reveals that the position of its edge in relation to the filter element depends not only on the component of the translatory motion but also on that of the rotary motion. This effect increases with the value of the aspect ration β. The calculations have shown that collection efficiency of fibrous particles is higher than of spherical particles with the same mass. It is the higher, the greater is the value of the aspect ratio β of the particle. The value of the total collection efficiency attains a minimum value for gas velocities in the range of 5 to 10 cm/s and depends again on the value of β.

In the following calculations[8,10] the total collection efficiency E was obtained in dependency on the parameter β and on the gas flow velocity U, which lay in the range between 0.01 and 1 m/s.

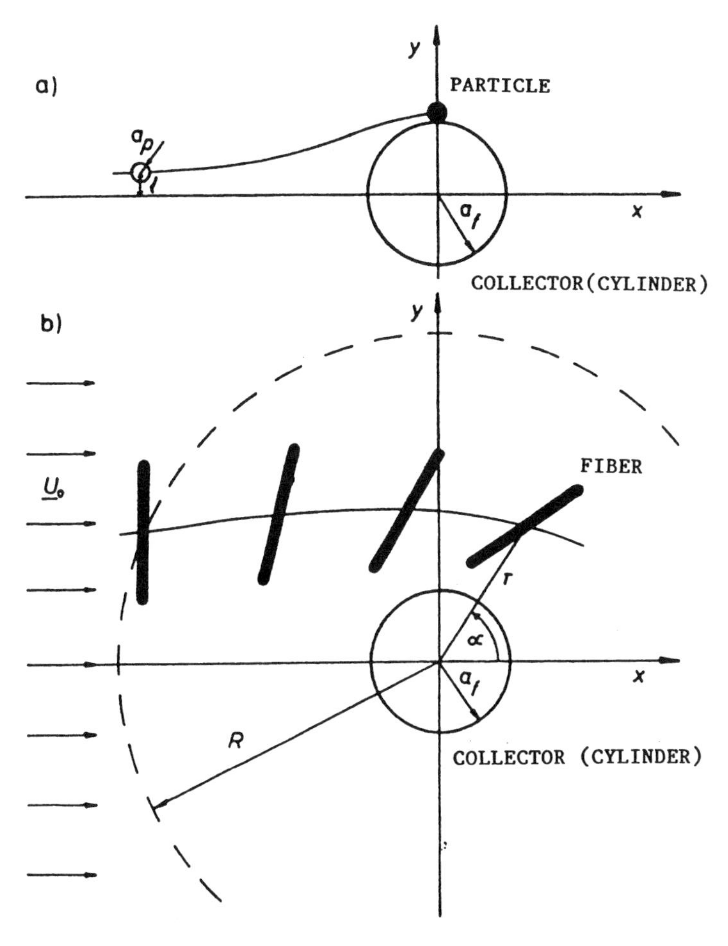

Figure 20.4 Spherical and fibrous particle limit trajectory. (From Gradon, L. et al., *Chem. Eng. Sci.,* 43, 1253–1259, 1988. With permission.)

The results are plotted in Figure 20.5. For longer and long fibers with β values between 10 and 100, the collection efficiencies were considerably increased.

20.2.1 Collection on Charged Collectors

The effect of electrostatic forces was investigated by Schamberger et al.[11] When the nonspherical particles/fibers are uncharged, the influence of the electrostatic torque can lead to either increases or decreases in collision efficiencies. Initial particle orientation is shown to be important for low collector charges but insignificant for high collector charges, when the electrostatic torque aligns the prolate particles parallel to the electric field. For the charged fibrous particle case, where the coulombic force is dominant, the collision efficiencies for particles having large aspect rations, β, are significantly lower than those for spherical particles (Figure 20.6).

Depending on the fiber and collector charge levels, both front and back capture modes were found. Image forces were shown to be insignificant for highly charged fibers. From the practical point of view, the electrostatic collection mechanism should be of great importance. The fibrous particles, e.g., asbestos fibers, as well as fibrous collectors, e.g., aerosol filters made from organic electret fibers, may have relatively high electric charge.

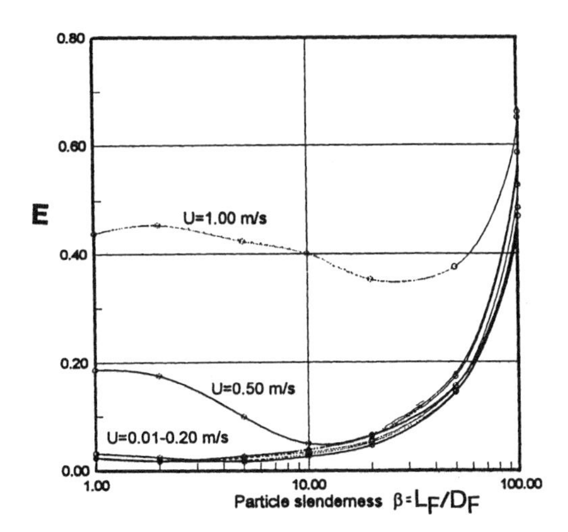

Figure 20.5 Filter collection efficiency for fibrous particles (*E*) as a function of the particle slenderness (aspect ratio) at different gas flow velocities. (From Grzybowski, P. and Gradon, L., *J. Aerosol Sci.*, 26(Suppl. 1), S725–S726, 1995. With permission.)

20.3 FIBROUS AEROSOLS AND FILTER TESTING

The methods for testing filter properties are developed and standardized well for the application of model aerosols consisting of spherical or isometric particles.[12] Methods for testing filters by means of fibrous aerosols are now also well developed. Nevertheless, they differ substantially from the first ones. These methods use suitable aerosol generators for the generation of finely dispersed fibrous test aerosols. By means of electron microscopical procedures — scanning electron microscope (SEM), transmission electron microscope (TEM), energy dispersive x-ray analysis (EDXA), selected area electron diffraction (SAED) — particle concentrations, sizes, forms and chemical compositions are measured at high resolution before and after exposure to the test filter. Penetration curves can be then plotted for the obtained data showing the penetration parameter P as a bivariate function of L_F, D_F and of the flow rates. Electron scanning microscopy can be used to investigate the collected particles on the surface and inside the tested filter, as well as to study structural changes of the filter material (Figure 20.7).

A vibrating bed aerosol generator (VBAG) is a very suitable piece of equipment for the generation of different types of fibrous aerosols (asbestos, anthropogenic fibers, carbon fibers, etc.). The generator is schematically shown in the Figure 20.8, and had already been published elsewhere.[13-16] By using the VBAG, dry homogenized fibrous material is aerosolized in the vibrating bed cylinder F. The air flow L through the fluidized bed and the membrane filter MF is controlled by the pressure regulator D, a manometer M_1, a rotameter R, and is dried by means of of silica gel column T. The vibration is produced by an electrodynamic system RC and SE, and the frequency and amplitude are measured by the vibration velocity detector system SA and mV. The concentration of the generated fibrous aerosol and the size distribution of the fibrous particles are governed by the physical conditions in the fluidized bed, especially the frequency and amplitude of the mechanical vibrations.

The generation of fibrous aerosols by the VBAG is limited for the dispersion of fibrous particles with diameters over 0.3 μm. For generating fibrous aerosols with fibrous particles thinner and much thinner than 0.3 μm, another generating system has to be used (Figure 20.9). Suspensions of very thin fibers (e.g., chrysotile asbestos) are dispersed in the generator DV. The liquid suspension of fibers in the container is treated ultrasonically (US) to prevent fiber agglomeration. The dispersed microdroplets containing almost one fiber are dried in the columns SG (silica gel) and C (charcoal).

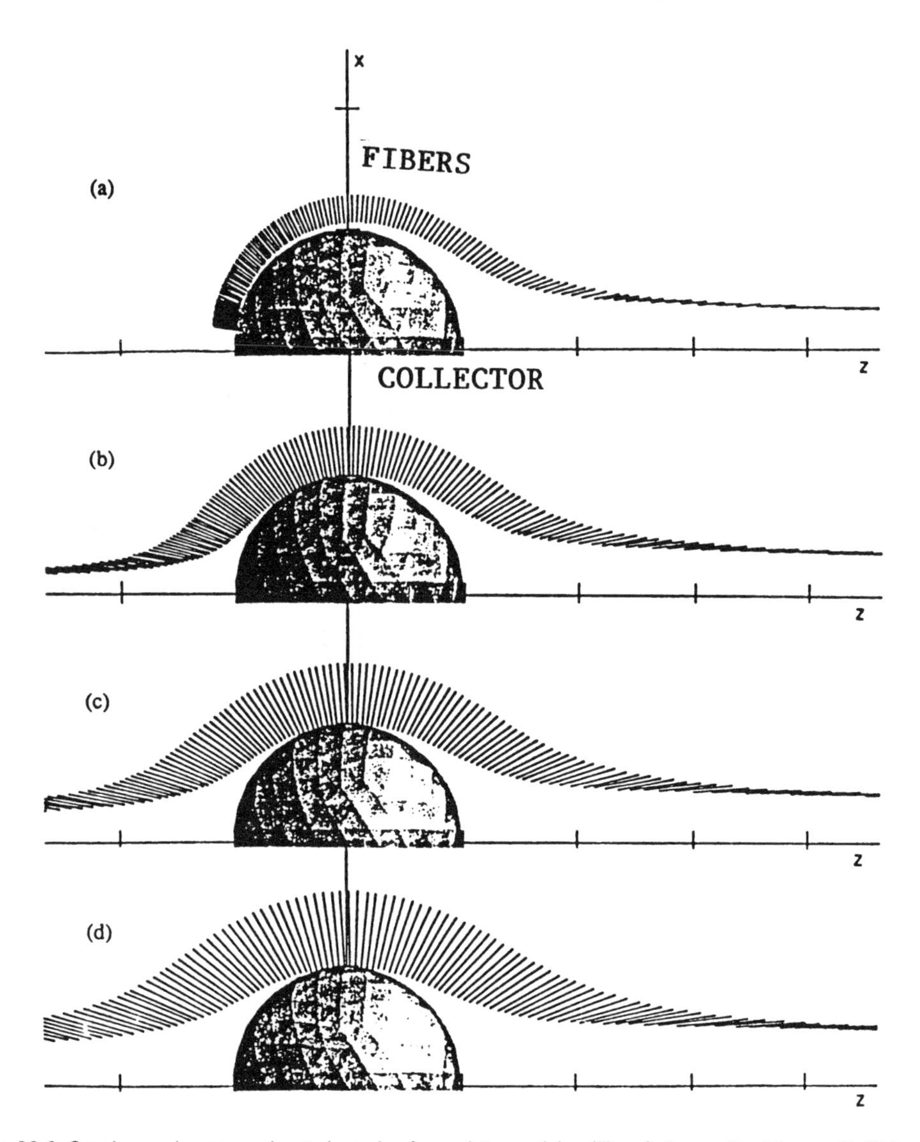

Figure 20.6 Grazing and near grazing trajectories for prolate particles (fibers) charged to 10 e⁻ and initial parallel orientation, collector was charged to 1.10⁻¹³ C. The computations were done for different aspect ratios β: 2 (a); 5 (b), 7 (c), and 10 (d). (From Schamberger, M. R. et al., *J. Aerosol Sci.*, 21, 359, 1990. With permission.)

The produced fibrous aerosol is discharged by a radioactive source (Kr-85). The resulting electrically neutral fibrous aerosol is then used for evaluation of the testing filter F.

AF designates absolute filters (NPFs) for particle sampling before and behind the tested filter F. The other parts of the equipment are flowmeters R_1, R_2, valve V, pump P, reservoir Re, and a hood Ho. The penetration through the tested filter P is evaluated by means of SEM procedures.

20.4 PENETRATION AND COLLECTION CHARACTERISTICS

By using the generating and testing system shown in Figure 20.9, size-selective penetration coefficients can be obtained. Figure 20.10 demonstrates how the testing procedure works. The fibrous aerosol is sampled by NPFs (N) before and behind the tested filter W; SEM micrographs

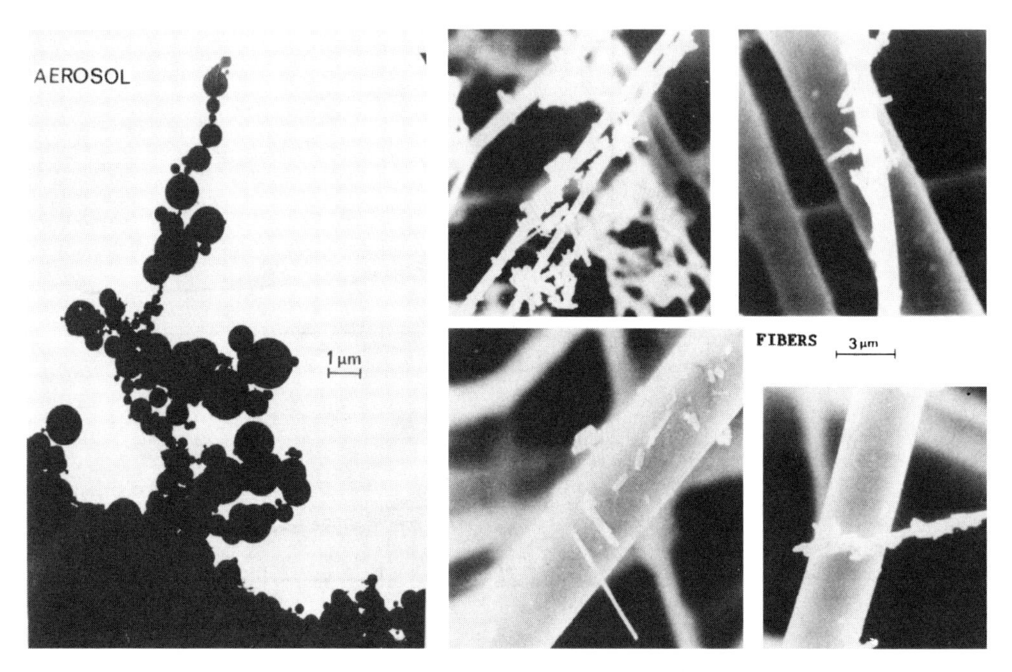

Figure 20.7 SEM micrographs showing the deposition of spherical (aerosol) and fibrous (fibers) particles in a fiber filter.

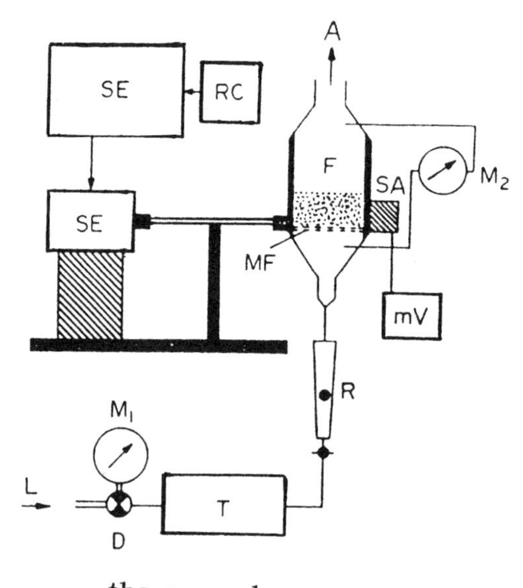

Figure 20.8 Schematic diagram of the generator for preparation of fibrous test aerosols.

show the upper (U) and the lower (L) sides. From the quantitative evaluation of the fibers (number and sizes) deposited on both NPFs, penetration P and collection efficiency E are determined ($P = 1 - E$). The measured penetrations could be plotted as a function of fiber sizes (fiber length LF and fiber diameter DF). An example of such measurement is shown in Figure 20.11.

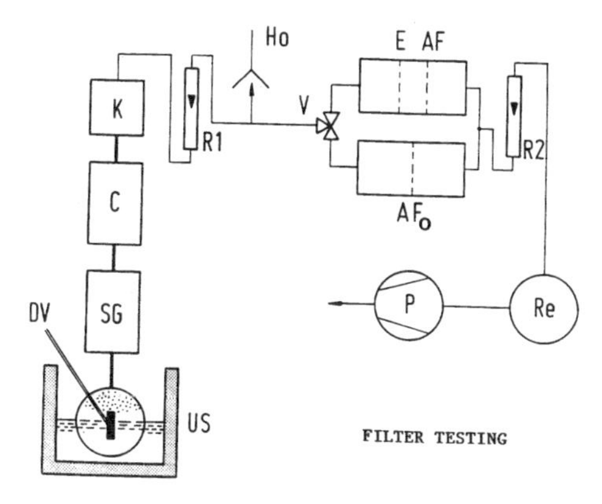

Figure 20.9 Schematic picture of the testing equipment. Description in the text.

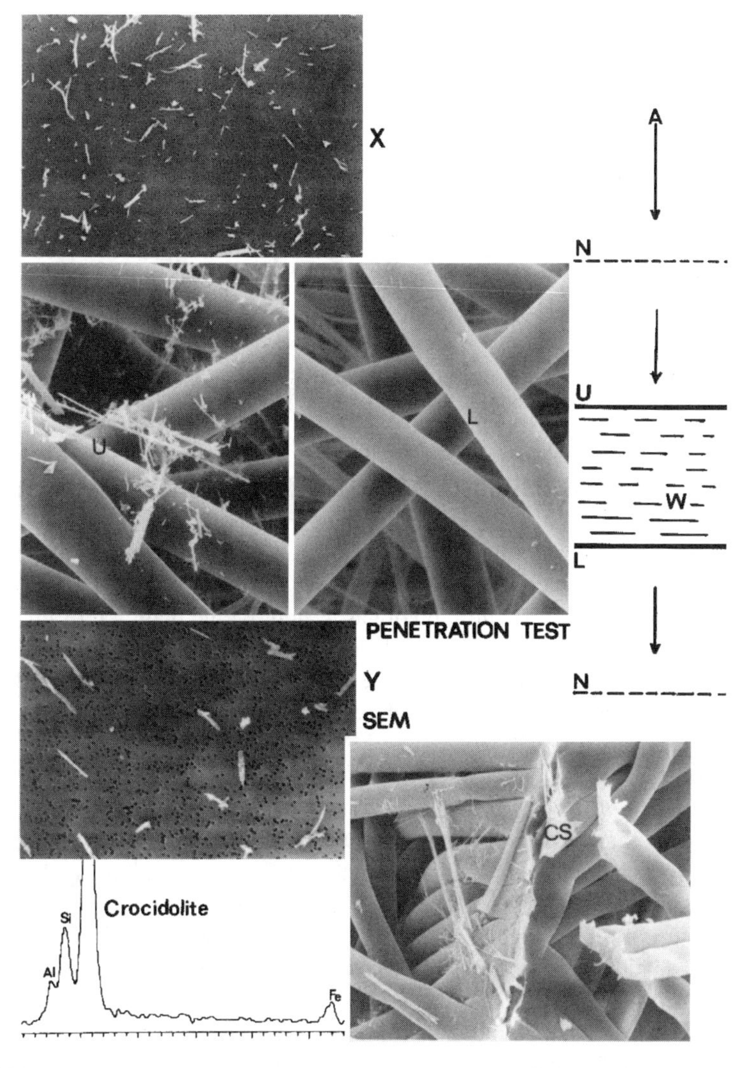

Figure 20.10 SEM micrographs of the material structure and asbestos fibers that penetrated.

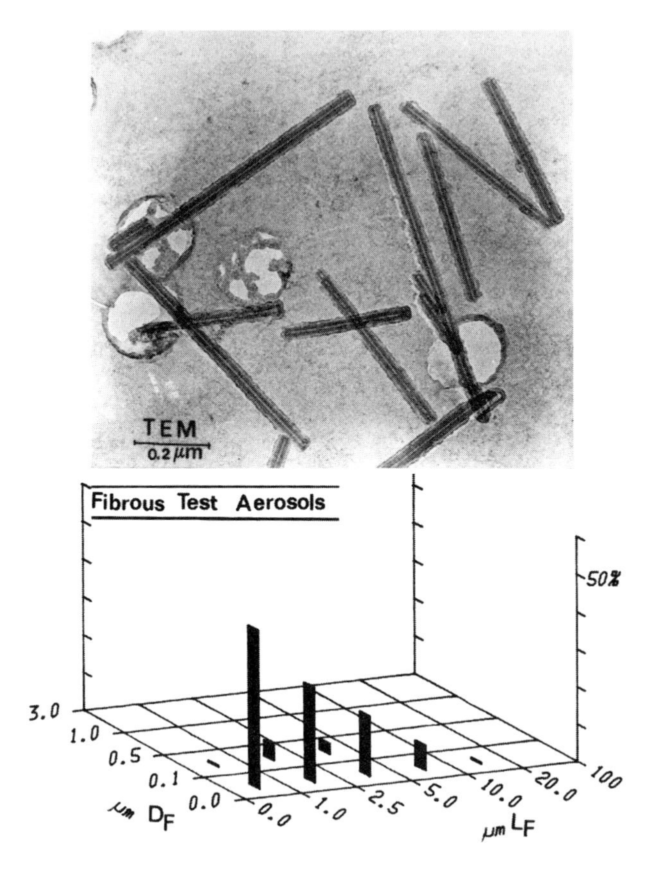

Figure 20.11 Characterization of the fine fibrous test aerosols: SEM micrographs of fibrous particles (TEM) and the bivariate size distribution (lower picture).

20.5 FILTRATION AND SAMPLING OF FIBROUS AEROSOLS IN PORE FILTERS

Pore filters, mainly membrane filters and NPFs, differ from the structure of fiber filters. They consist of porous foam or similar structures (membrane filters) or have cylindrical pores (NPFs). Both are routinely used in laboratory filtration of gases, liquids, and mainly for sampling particulate suspensions in gases and liquids. Because of their applications in the filtration and sampling of aerosols, knowledge of their penetration characteristics is of great importance. Both membrane filters and NPFs were tested by means of the above-mentioned procedures.

20.5.1 Membrane Filters

The classical membrane filter is a deep filter, with a thickness of about 100 μm. During the filtration process, fibrous particles are separated on the filter surface and inside the inner structure. These phenomena are demonstrated well in Figures 20.12 through 20.14. In Figure 20.12 the inner porous foam structure of a commercial membrane filter is shown. Figure 20.13A shows a cross section of a membrane filter with asbestos fibrous particles separated on the surface, while in Figure 20.13B asbestos fibers deposited inside the open porous structure can be seen. Collection and efficiency measurements on commercial membrane filters have shown that the separation of fibrous particles depends again in the first approximation of the fiber length L_F. This is demonstrated well in Figure 20.14.

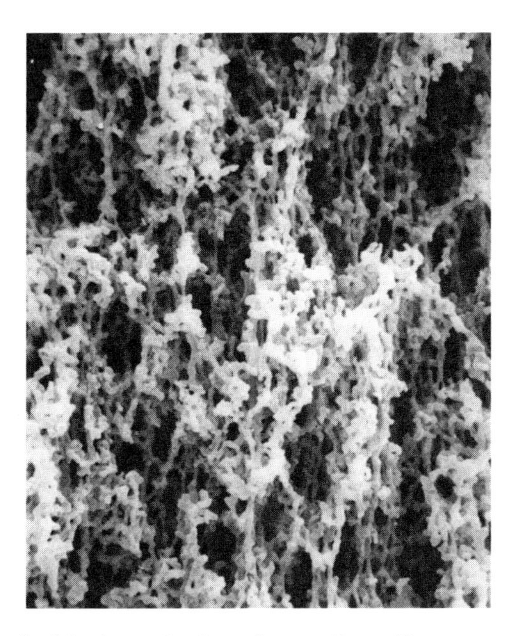

Figure 20.12 SEM micrograph of the inner structure of a membrane filter.

Membrane filters with pore sizes between 0.1 and 8 μm were tested for the separation of fine dispersed fibrous aerosols.[17,18] It could be concluded that membrane filters with pore sizes ≤0.8 μm are sufficiently efficient ($P < 0.5\%$) for collecting fibrous particles with fiber diameters over 0.03 μm. Such filters are therefore very convenient for sampling all existing kinds of fibrous aerosols in the workplace, in the indoor and outdoor atmospheres.[19]

20.5.2 Nuclepore Filters

NPFs are standardized for the sampling of fibrous aerosols.[18,19] Information about their collection efficiencies is therefore of basic importance. By using the mentioned procedure, collection efficiencies of NPFs were measured by means of fine dispersed asbestos aerosols, mainly chrysotile aerosol. The collection efficiencies were measured using two filters in series exposed to the flow of fibrous aerosol. The generated fibrous aerosol was drawn through the test filter by means of a vacuum pump and the flow rate was measured. The second "absolute" filter was placed in series to collect fibrous particles that had penetrated through the test filter.[17,18] The filter surface air velocities V_F were varied between 1 and 50 cm/s. SEM and TEM were used to count the fibers and to measure the fiber sizes collected on both filters. Penetration and collection efficiencies were then calculated. The measured collection efficiencies of different types of NPFs are presented in Table 20.1. The collection efficiencies were high enough for NPFs with pore sizes ≤0.8 μm. For sampling of fibrous aerosols, NPFs with a pore size of 0.4 μm (pore diameter = 2 R) seem to be optimal. Their collection efficiency is high enough and the pressure drop is still reasonably "low."

The collection efficiencies for fibrous aerosols in a filter battery consisting of six NPFs (pore size 5 μm) in series is shown in Figure 20.15. To collect a well-represented sample, three such filters in series had to be used. The collection efficiencies E were also measured as a function of the gas flow velocity V_F. Figure 20.16 shows an example of such measurements. Filters with small pore sizes do not change their collection efficiencies with changing flow rates. The parameter E of NPFs with larger pore sizes depends on the flow rate; when thin chrysotile fibrils are filtered, E decreases with increasing V_F. This phenomenon was observed also by testing fiber filters. The penetration of fibrous particles through pore filters is also size dependent, as can be seen well in

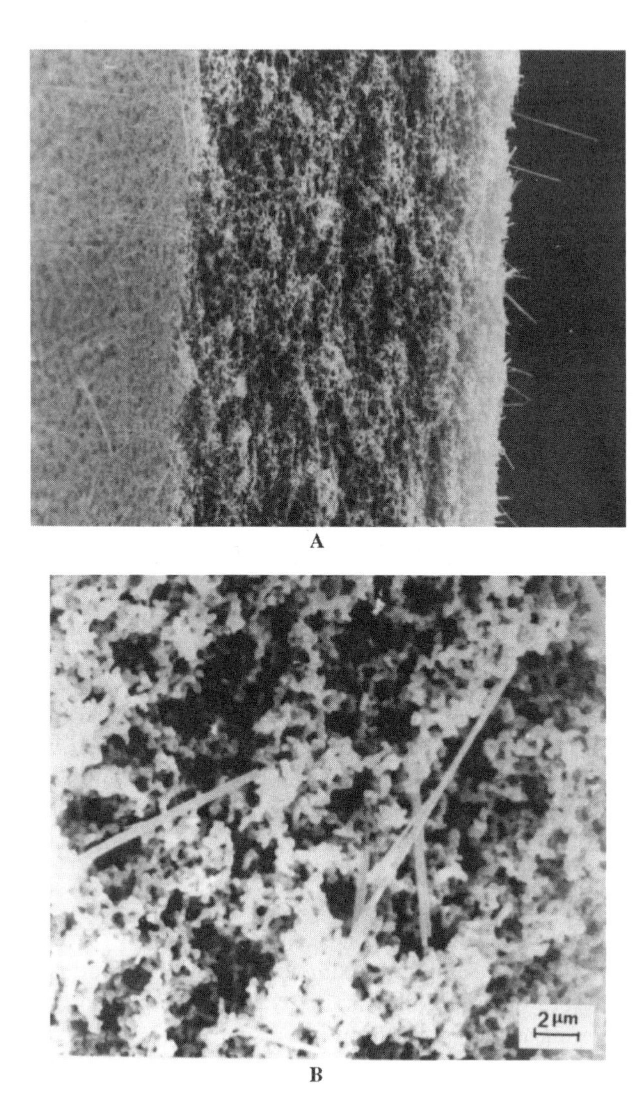

Figure 20.13 SEM micrographs showing fibers sampled on the surface and inside of a membrane filter.

Figure 20.17. The fibrous test aerosol contains fibers with lengths about 20 μm. The penetrated fibers are shorter or much shorter than 5 μm.

20.5.3 Fiber Alignment Mechanism

It has already been noted that the most important collection mechanism for separating fibrous particles in filters is the mechanism of direct interception. This was also evident in examinations by electron microscopical methods. The electron micrographs in Figure 20.18, show the importance of the particle shape and orientation. An important factor influencing the fiber penetration through the pores of an NPF is the spatial orientation of the single fibrous particles in the gas flow approaching the filter surface and pore opening. Fibers oriented in the direction of the flow can easily penetrate pores with a diameter which is only a little greater than the diameter of the fibrous particle. This can be seen well in Figure 20.19.

In further investigations the influence of the alignment on the penetration of fibrous particles was studied experimentally. The results are summarized in Figure 20.20. The collection efficiencies

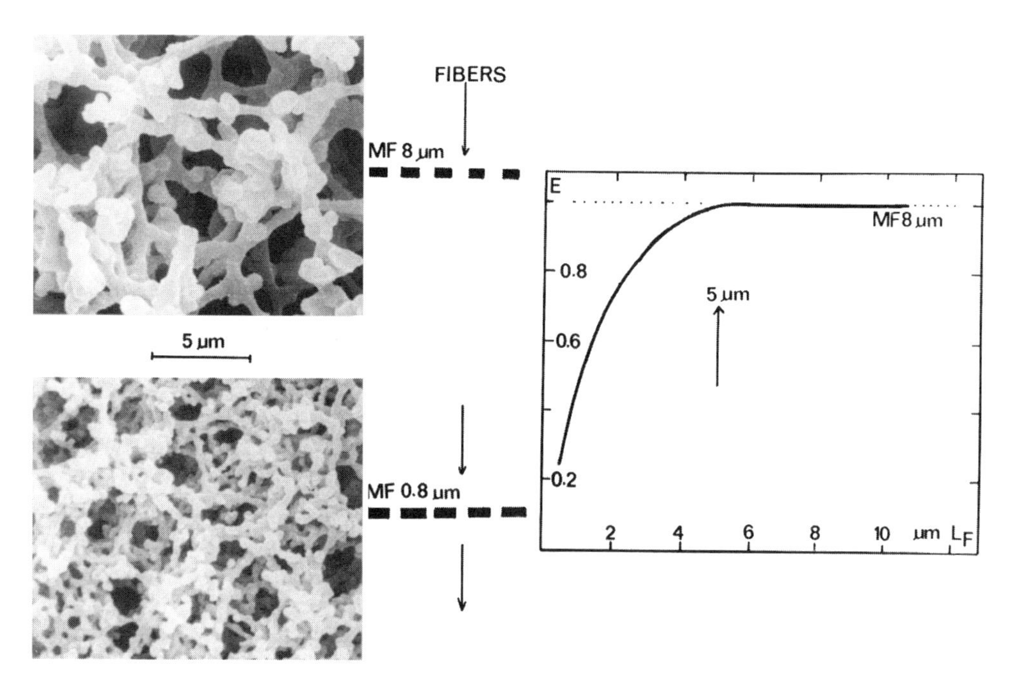

Figure 20.14 SEM micrographs of the inner structure and the separation characteristics of a membrane filter with pore size of 8 µm.

Table 20.1 Collection Efficiencies for Fibrous Aerosols (V_F = 3.5 cm/s)

NPFs (pore diameter, µm)	Mean Collection Efficiency (%)	Standard Deviation ± (%)
12	38.241	1.023
8	54.733	0.322
5	81.661	0.092
3	94.984	0.043
0.8	98.783	0.018
0.4	99.887	0.009

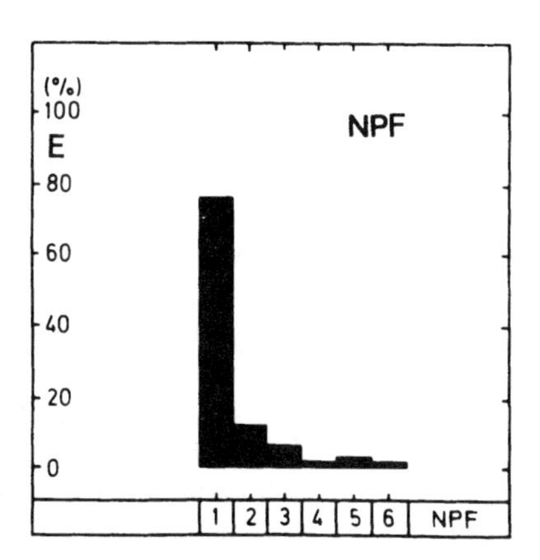

Figure 20.15 Fibrous particles deposited on six NPFs in series.

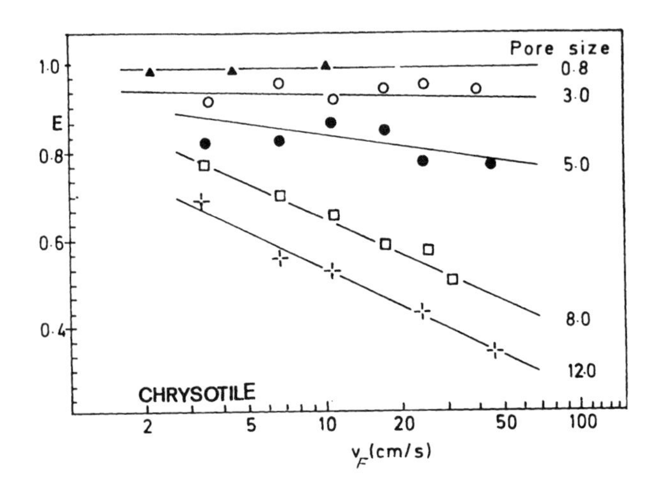

Figure 20.16 Collection efficiencies E of NPFs with pore sizes between 0.8 and 12 μm as a function of gas flow velocities v_F

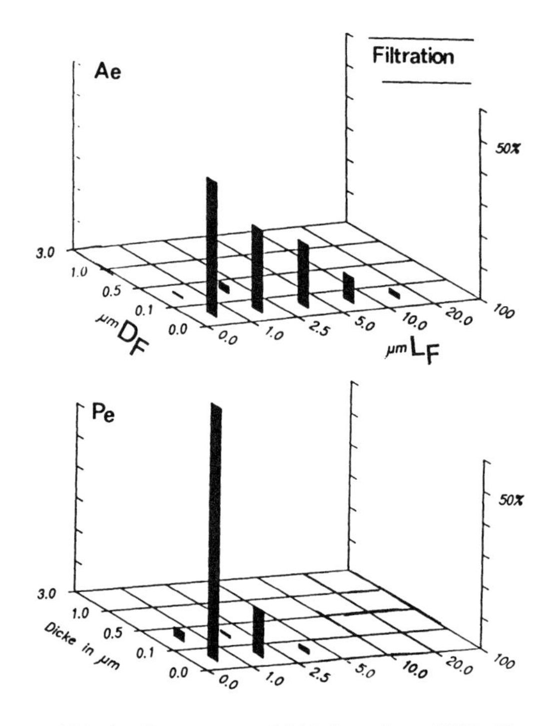

Figure 20.17 The penetration (P_e) of a fibrous aerosol (A_e) through an NPF with a pore diameter of 2 μm.

were plotted as a function of a dimensionless parameter, $G = (2R/L_R)$; R is the pore radius of the NPF used. The measurements were done for pore diameters ($2R$) between 0.6 and 3 μm and the computation values were plotted for pore diameters 1, 2, and 3 μm.[22]

Measured (points) and computed (lines)[22] E-values were correlated under the assumption of a free rotation of fibers within an NPF pore. This assumption overestimated fiber removal in the pores. On the other hand, the experimental values showed clearly that the fiber orientation and the fiber length are the most important parameters associated with the fiber separation in the pores. A better quantitative description of the fiber removal in an NPF could probably be achieved if we

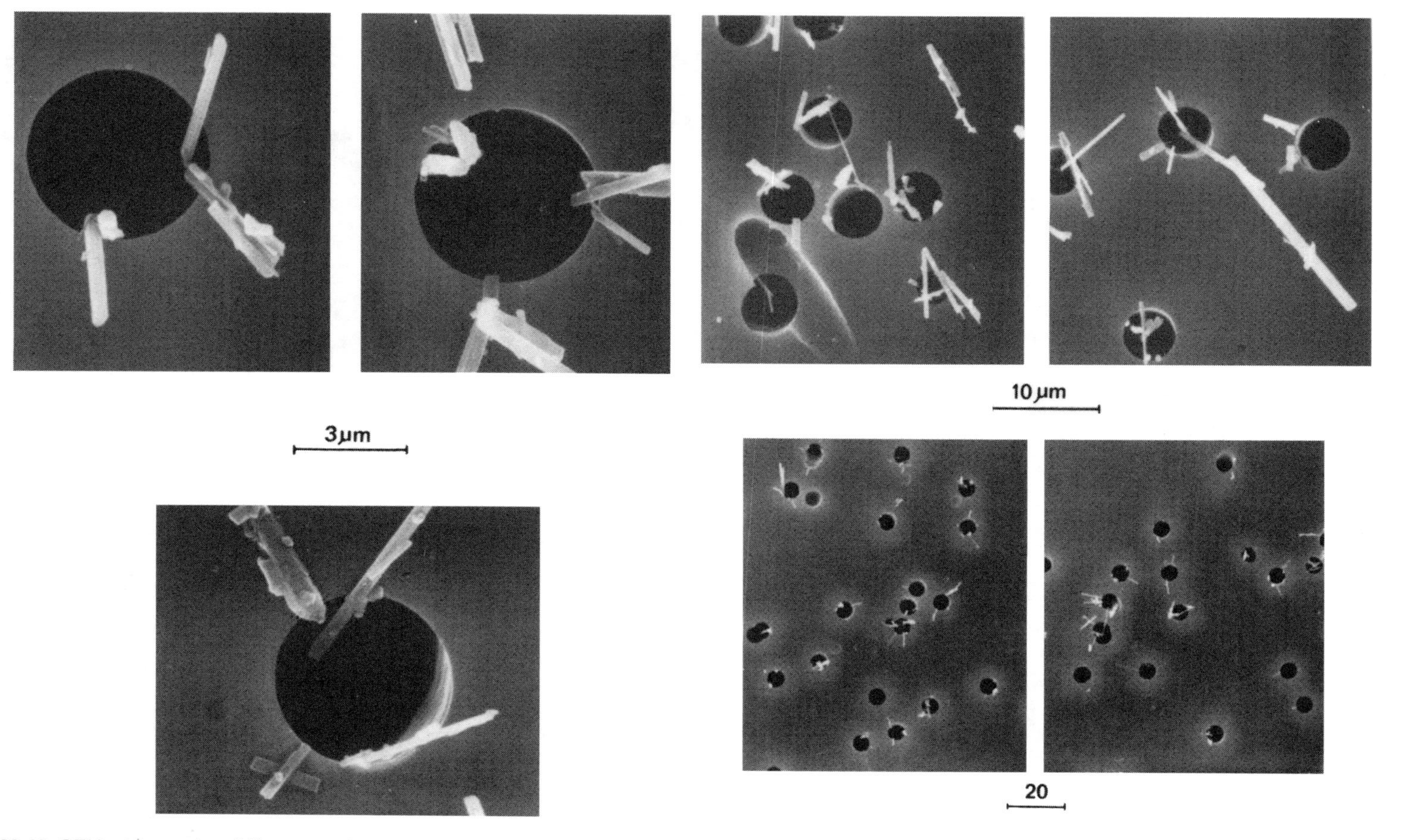

Figure 20.18 SEM micrographs of fibrous particles separated in the pores of NPFs.

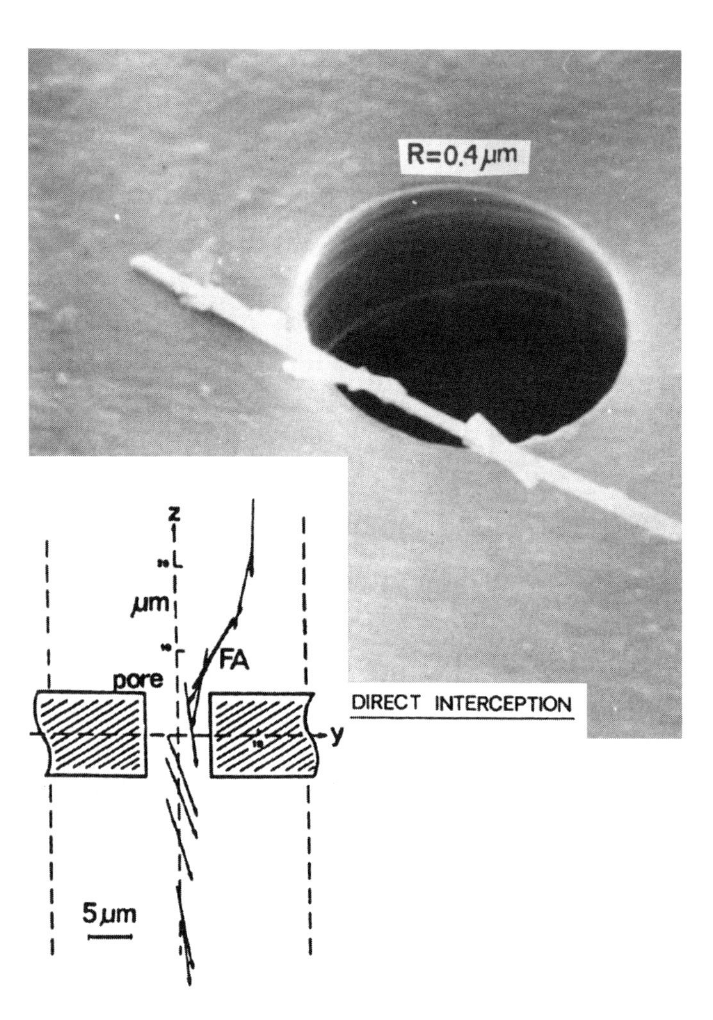

Figure 20.19 SEM micrograph of one NPF pore and a separated fibrous particle as well as a schematic picture of fibrous particle orientation in the gas flow. (From Spurny, K. R. in *Proc. 5th World Filtration Congress,* Nice, France, 1990, 547–556. With permission.)

knew the dependency between the alignment angle of the fibrous particle approaching the NPF pore at the gas flow velocity V_F and the pore size $2R$. If L_F is the fiber length, then the mechanism of direct interception could be described by the parameter N_R:

$$N_R = \frac{L_F \cdot O_P}{2R} \quad (0 < O_P < 1) \tag{20.7}$$

Here $N_R < 1$ and O_P is the orientation probability. A special case of direct interception capture is the sieving mechanism, when $N_R > 1$ (Figure 20.19). Obviously, for this case all fibers are captured by the NPF.

Gentry et al.[21] calculated the angular orientation of fibrous particles and estimate theoretically the collection efficiencies as a function of D_F and I_F (Figure 20.21). The theoretical efficiencies showed agreement with experiment when the fibers had a strong tendency to align. In other cases the collection efficiencies obtained were significantly higher than those calculated from relations developed for spherical particles. In all cases the Gentry model underestimated the interception effect.

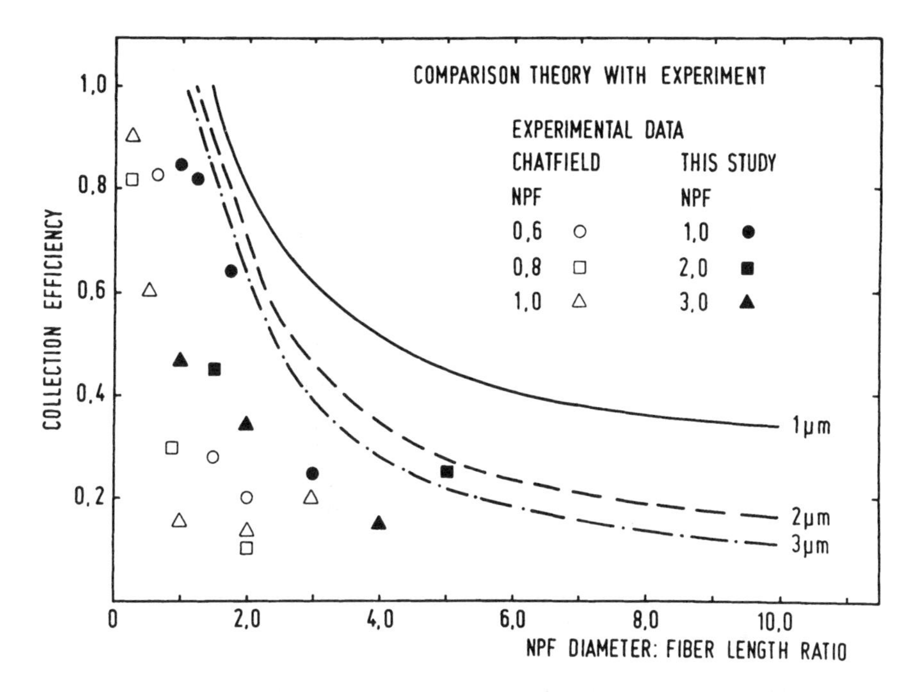

Figure 20.20 Comparisons of experiments with theory: collection efficiencies are plotted as function $2R/L_F$ (R is the radius of the NPF).

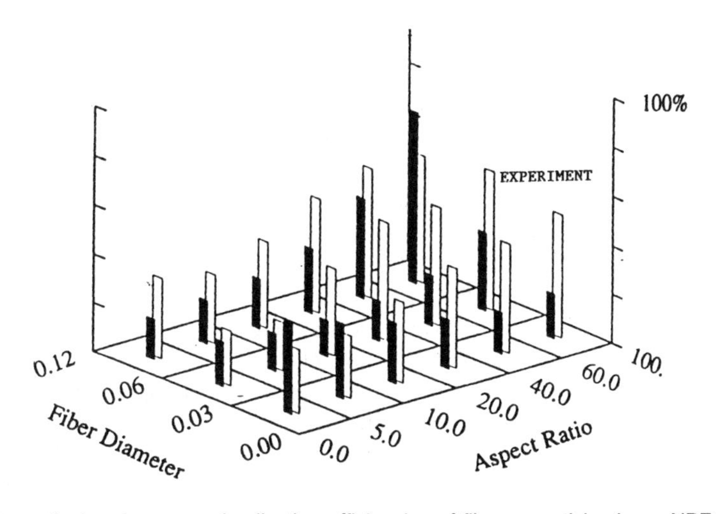

Figure 20.21 Theoretical and measured collection efficiencies of fibrous particles in an NPF. (From Gentry, J. J. et al., *J. Aerosol Sci.*, 21, 711–714, 1990. With permission.)

20.6 SEPARATION OF FIBROUS PARTICLES IN INERTIAL EQUIPMENT

The problem of the uncertainty of estimating the spatial orientation of fibrous particles in different gas flow types is also very evident in the case of different inertial samplers, like elutriators, cyclons, and impactors. Several measurements showed that a theoretical prediction of collection efficiencies for the sampling of fibrous particles, e.g., in cyclons, elutriators, impactors, is almost impossible. This fact is well demonstrated for the case of cascade impactors (Figure 20.22).[23] Asbestos dust was collected by means of an Andersen cascade impactor having seven impaction stages

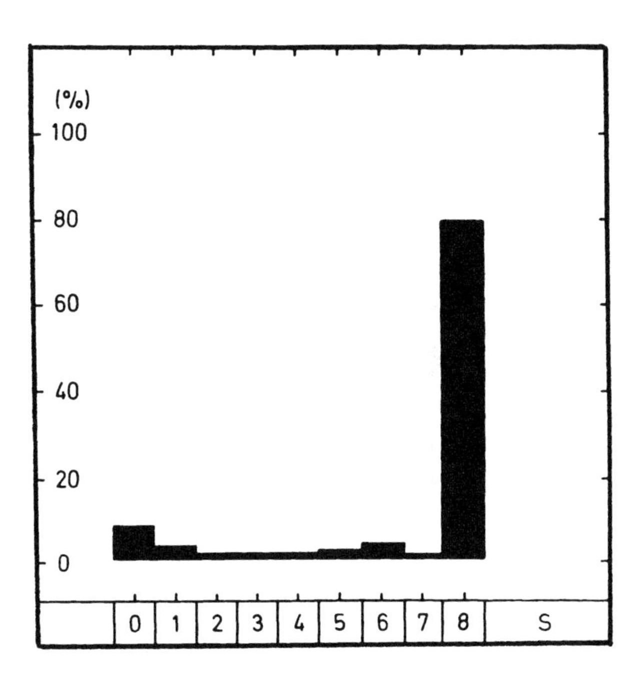

Figure 20.22 Separation of fibrous particles in a cascade impactor (Stages 0 to 7) and in the backing filter. (From Grzybowksi, P. and Gradon, L., *J. Aerosol Sci.*, 26(Suppl. 1), 5725–5726, 1995. With permission.)

and a backing high-efficiency membrane filter.[8] With respect to the estimated aerodynamic diameters of the sampled fibrous dust particles, more than 50% of the particles should be collected on the seven impaction stages. In reality, more than 80% of the particles penetrated all states of the impactor; that is, the orientation, mainly the alignment of the fibrous particles, was responsible for the penetration. Therefore the inertial dust samplers are not suitable for sampling fibrous dusts and aerosols.[23]

Asgharian et al.[36] have done the first calculations of the collection efficiency of fibrous particles in an impactor. Fibers in a flow field are subjected to a coupled motion of translation and rotation. The movement of a fiber is greatly influenced by its changing orientation in the flow field. By using mathematical formulations describing the mechanics of fibrous aerosols, previously developed by the authors, collection efficiency curves for the separation of fibrous particles in one-stage impactors were computed and plotted as functions of the particle aspect ratio $\beta = L_F/D_F$ (fiber length/fiber diameter) and the inertial parameters, Stokes number $\text{Stk} = \rho \cdot D_F^2 - \beta \cdot U/18\mu \cdot D$. ($\rho$ and μ are the density of the particle and the gas viscosity, U the gas flow velocity in the impaction nozzle with a diameter D). The calculated efficiency curves for the inertial separation of spherical and fibrous particles in one-stage impactors are shown in Figure 20.23. The efficiency curves for spherical and fibrous particles have qualitatively a similar shape and form. Nevertheless, the parameter β (Figure 20.23A) and therefore the fiber length L_F (Figure 20.23B) play an important role in the impaction of fibrous aerosols. It can be seen that fibrous particles are effectively separated by much higher Stk than spherical particles with an equivalent size. Separation of thin fibers ($D_F < 2$ µm) in nozzle impactors is very poor (Figure 20.23B). As the parameter β increases and the cutoff size diameter of the fibers decreases, the fiber collection efficiency curve for fibers moves to a higher Stokes number.

20.7 CONCLUSIONS

A full satisfactory theory for the filtration and separation of nonspherical aerosol particles, mainly fibers, is still not available. The previously published experimental as well as theoretical

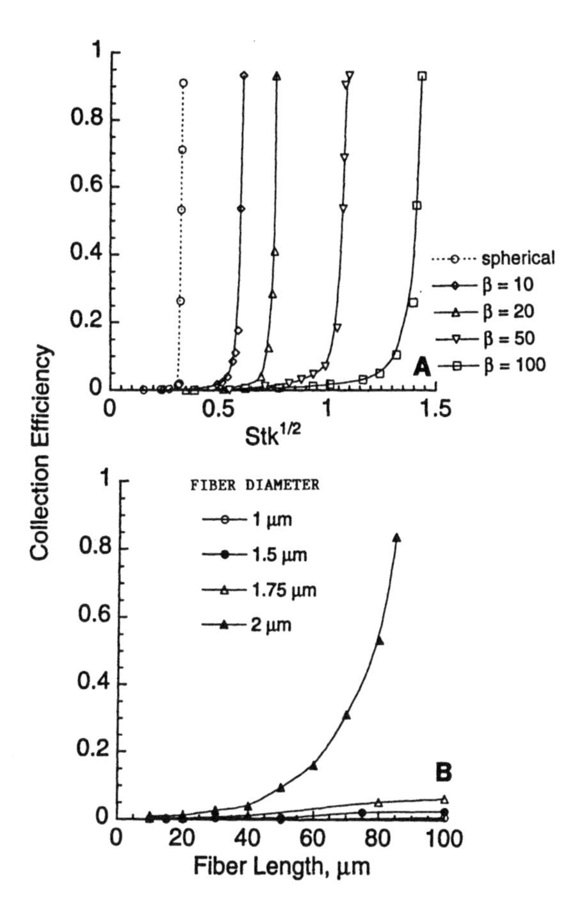

Figure 20.23 Calculated collection efficiency of spherical particles and fibers in a straight orifice geometry for a Reynolds number 1100 (A) and as a function of fiber length and diameter. (From Asgharian, B. et al., *J. Aerosol Sci.*, 28, 277–287 (1997). With permission.)

results have confirmed that significant differences exist between separation of spherical and fibrous particles.

1. While the spatial orientation or rotation of spherical and isometric particles can be neglected in all separation mechanisms, both these parameters play an important role in the case of the filtration and separation of fibers. Knowledge of an orientation probability function, e.g., the time- and flow rate–dependent fiber orientation outside and inside the filter, plays a crucial role in any quantitative description of the collection mechanisms for fibers.
2. The previously published experimental data offer several important pieces of information for practical applications. The direct interception dependent on the aspect ratio β, and so on the fiber length, is the most important separation mechanism for fibrous particles. The electric charge (charged elements and/or charged fibrous particles) influences the collection effectiveness considerably more significantly than in the case of the filtration of isometric particles. The collection efficiencies are higher in the case of fiber filtration than in the case of filtration of equivalent isometric particles. The total penetration of fibers as well as the maximal length of fibers, which can penetrate (MLP), increases with increasing gas flow rate. Very approximately it can be stated that any filter whose collection efficiency is sufficient for the separation of highly dispersed isometric aerosols can be even more effective for the filtration of fibrous aerosols.[24]
3. The differences in filter characterization for the separation of isometric and fibrous particles are well demonstrated in Figure 20.24. While the most-penetrated particle size (MPPS) is the most important parameter for the evaluation of filter quality in the case of filtration of isometric particles, in the case of the filtration of fibrous particles, knowledge of the MLP is decisive and most important.

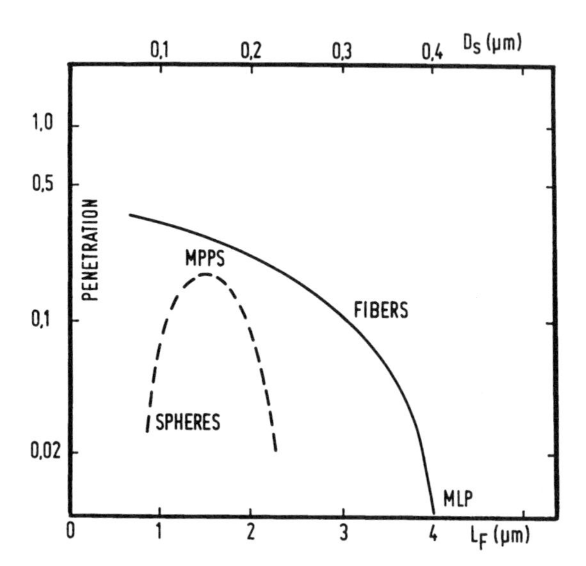

Figure 20.24 Differences in the characterization of filters in cases of the filtration of fibrous and spherical aerosols.

20.8 THE "LUNG FILTER"

Fibrous dusts and aerosols that can be present in the workplace, indoor and outdoor atmospheres, are inhaled by workers and by the general population and may produce harmful health effects, including lung cancer. Single fibrous particles enter into the inner structure of the airways and are there separated and deposited. The whole airway system is divided into the nasal, tracheobronchial, and pulmonary spaces. The tracheobronchial part forms a "tree" of tubings and capillaries with many bifurcations (Figure 20.25). The structure as well as the dimensions of the tracheobronchial tree, including the alveolar region, duct bifurcations, and intrathoracic airways, are already relatively well known and mathematically described.[25,26] The particle separation in such a lung filter is governed by the same mechanisms that act in mechanical filters and is satisfactorily described by mathematical models.[27] Nevertheless, to describe the separation of spherical particles is again easier than is the case for nonspherical, e.g., fibrous particles. Fibers are defined by three parameters — D_F, L_F, and β as well as by their spatial orientation in the airflow in the lung. In general, fibers in the airways are neither parallel to the gas flow nor randomly oriented. The dynamics of a fiber in a fluid flow is therefore more complex than of a spherical particle.

The existing theories describing the separation of fibrous particles in the lung filter consider the superposing action of several partial mechanisms, such as sedimentation, impaction, interception, and diffusion. The relative contribution of these partial mechanisms to the total separation of fibrous particles in airways depends upon the fiber size and flow rate as well as the dimension and location in the respiratory tract.[28] The best existing mathematical models describing the fiber

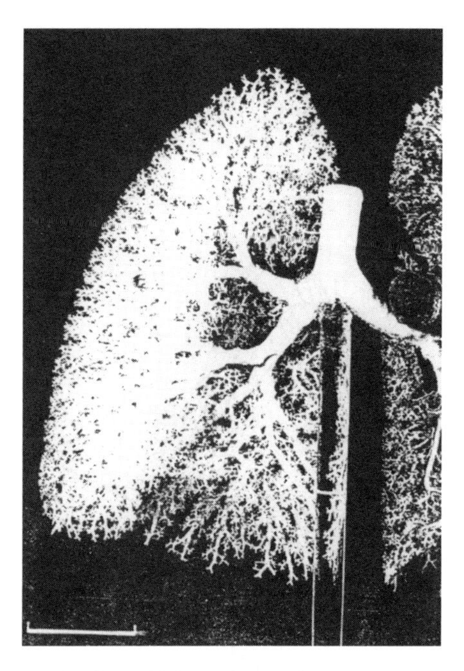

Figure 20.25 Airway cast of the tracheobronchial tree from a human.

separation in the airways were published only during the last decade.[29-32] These models have used, as usual, the classical models of the separation of spherical particles in the airways and modified them for fiber dimensions and orientations. By means of these models, computations characterizing the separation of fibrous particles in different parts of human as well as animal airways are certainly possible. Examples are shown in Figure 20.26.

The total separation of fibrous particles in the airways differs naturally from the separation characteristics of spherical particles (Figure 20.27) and depend again on the parameter of the aspect ratio β, and also on the fiber length L_F. Following the deposition curves in Figure 20.26, we can see that the maxima (M) of these curves for fibrous particles are shifted into the direction of finer particles and decrease with increasing values of β.[33]

Recently, Podgorski et al.[34] studied theoretically and experimentally the separation of fibrous aerosols in a granular bed filter. The existing theory of the capture of spherical aerosol particles was extended to fibrous aerosol filtration. The same method was also used for the estimation of fiber deposition in the human lung. The granular bed filter model was applied for simulation of human airways. Measurements have shown an approximate agreement of this model[35] with the mentioned Asgharian model.[28] The minimum alveolar deposition lay at the equivalent mass particle diameter of about 0.5 μm.

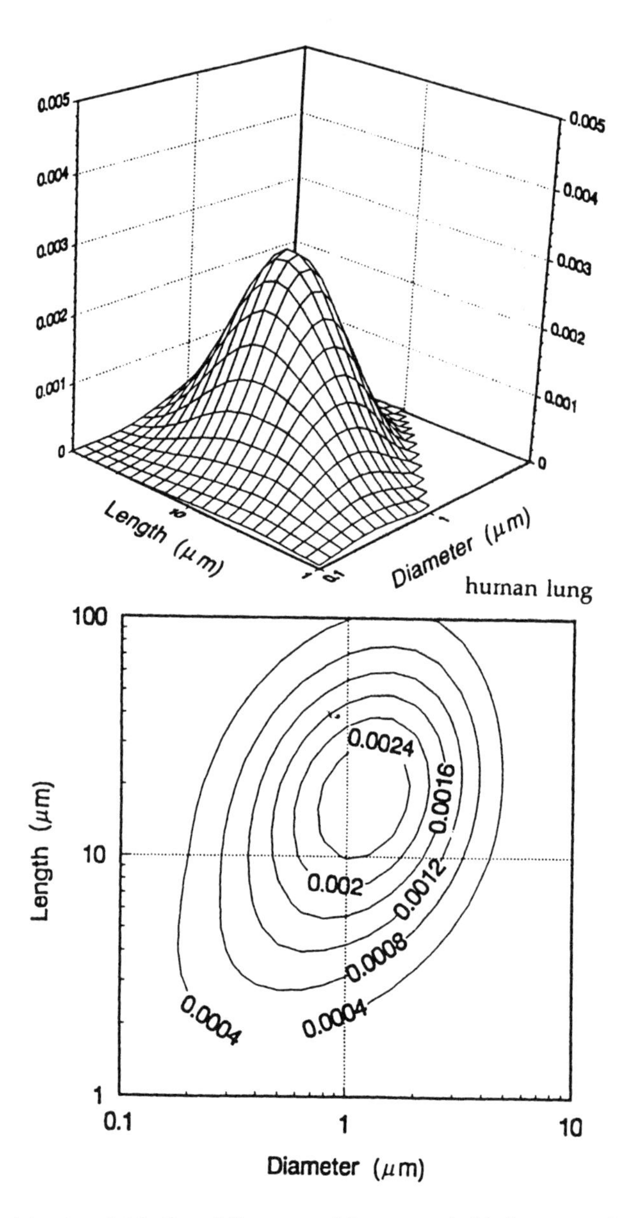

Figure 20.26 Bivariate size distribution of fibrous particles separated in human and rat lungs. (From Yu, C. P. et al., *J. Aerosol. Sci.*, 25, 407–417, 1994. With permission.)

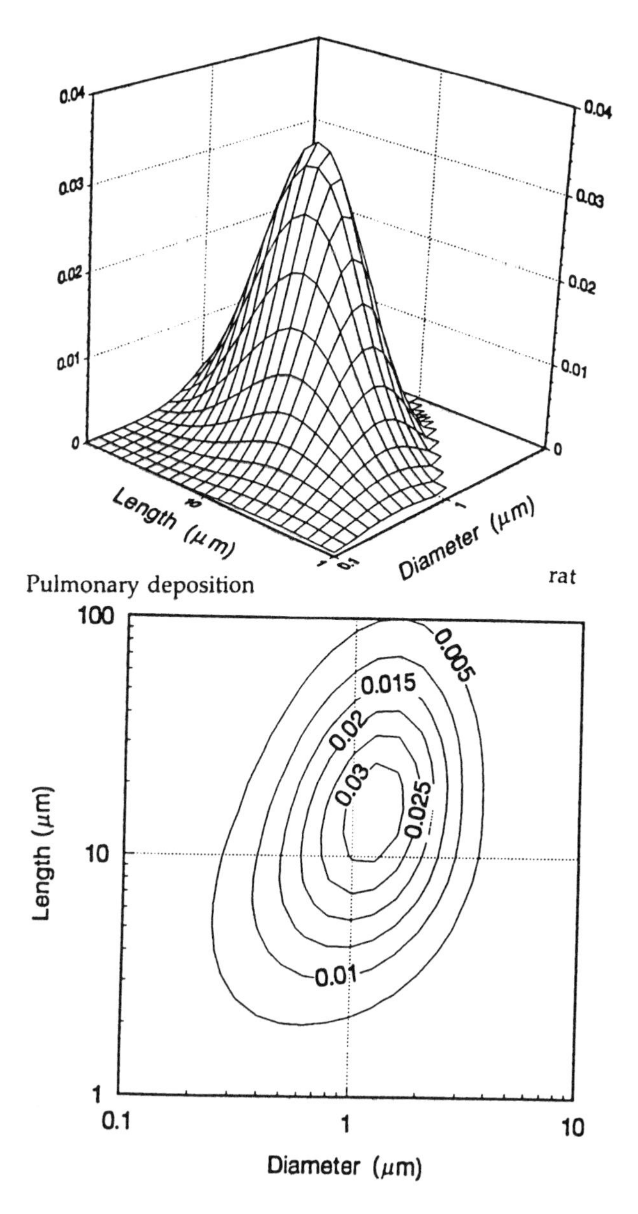

Figure 20.26 (continued)

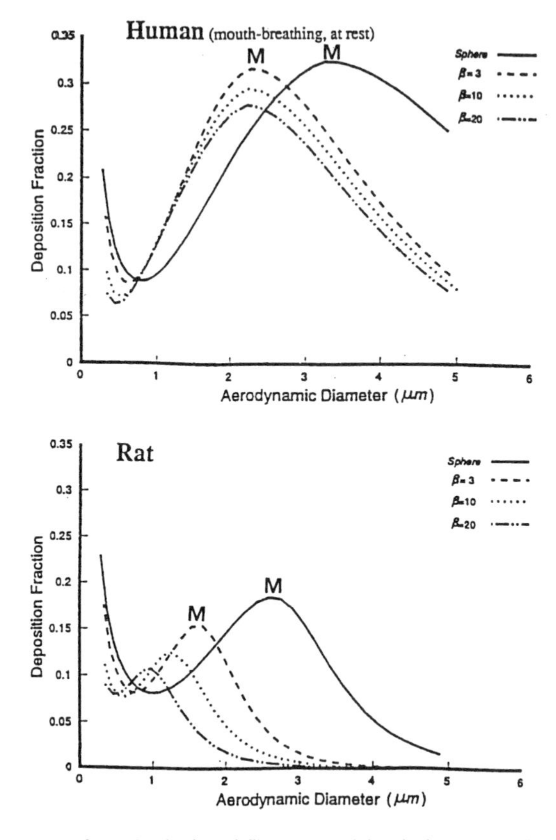

Figure 20.27 Deposition curves for spherical and fibrous particles in human and rat lung. (From Oberdörster, G., U.S. EPA Report, University of Rochester, 1–49, 1995.)

REFERENCES

1. Spurny, K. (1997) *Faserige Mineralstäube [Fibrous Mineral Dusts],* Springer-Verlag, Heidelberg.
2. Harris, R. L. and Fraser, A. A. (1976) *Am. Ind. Hyg. Assoc. J.,* 37, 73–89.
3. Feigley, C. E. (1975) *Clean Air* (Australia), Nov., 65–71.
4. Pich, J. (1987) Gas filtration theory, in *Filtration,* Matteson, M. J. and Orr, C., Eds., Chemical Industries 127, Marcel Dekker, New York, 1–152.
5. Brown, R. C. (1993) *Air Filtration,* Pergamon Press, Oxford, 1–152.
6. Gallily, I. (1984) Transport and mechanisms of non-spherical particles, in *Aerosols,* Liu, B. Y. H., Pui, D. H. Y., and Fissan, H. F., Eds., Elsevier, Amsterdam, 793–795.
7. Gradon, L., Grzybowski, P., and Pilaciski, W. (1988) *Chem. Eng. Sci.,* 43, 1253–1259.
8. Grzybowski, P. and Gradon, L. (1995) *J. Aerosol Sci.,* 26(Suppl. 1), S725–S726.
9. Podgorski, A. and Gradon, L. (1990) *J. Aerosol Sci.,* 21, 957–867.
10. Podgorski, A. (1995) *J. Aerosol Sci.,* 26(Suppl. 1), 651–652.
11. Schamberger, M. R., Peters, E. J., and Leong, K. H. (1990) *J. Aerosol Sci.,* 21, 539–554.
12. Dorman, R. G. and Ward, A. S. (1987) Filter evaluation and testing, in *Filtration,* Matteson, M. J. and Orr, C., Eds., Marcel Dekker, New York, 675–721.
13. Spurny, K. R. (1980) Fiber generation and length classification, in *Generation of Aerosols and Facilities for Exposure Experiments,* Willeke, K., Ed., Ann Arbor Sci. Inc., Ann Arbor, MI, 257–298.
14. Spurny, K. R. (1980) *Staub Reinhalt. Luft,* 41, 330–335.
15. Spurny, K. R., Schörmann, J., and Opiela, H. (1990) *Am. Ind. Hyg. Assoc. J.,* 51, 36–43.
16. Spurny, K. R. (1994) *Filtr. Sep.* (Germany), 8, 166–175.
17. Spurny, K. R., Weiss, G., and Opiela, H. (1986) *Staub Reinhalt. Luft,* 45, 106–111.

18. Spurny, K. R. (1986) *Sci. Total Environ.,* 52, 189–199.
19. Spurny, K. R. (1994) *Analyst,* 119, 41–51.
20. Gentry, J. W., Spurny, K. R., and Schörmann, J. (1989) *Aerosol Sci. Technol.,* 8, 184–195.
21. Gentry, J. W., Spurny, K. R., and Han, R. H. (1990) *J. Aerosol Sci.,* 21, 711–714.
22. Spurny, K. R. (1990) Asbestos dust filtration, in *Proc. 5th World Filtration Congress,* Nice, France, 547–556.
23. Spurny, K. R., Gentry, J. W., and Stöber, W. (1978) Sampling and analysis of fibrous aerosol particles, in *Fundamentals of Aerosol Science,* Shaw, D. T., Ed., J. Wiley, New York, 257–324.
24. Feigley, E. and Chen, H. C. (1992) *Am. Ind. Hyg. Assoc. J.,* 53, 767–772.
25. Yeh, H. C. and Harkeme, J. R. (1993) Gross morphometry of airways, in *Toxicology of the Lung,* Gardner, D. E., Crapo, J. D., and McClellan, R. O., Eds., Raven Press, New York, 55–79.
26. Heyder, J. (1986) Single-particle deposition in human airways, in *Physical and Chemical Characterization of Individual Airborne Particles*, Spurny, K. R., Ed., E. Horwood Ltd., Chichester, U.K., 72–85.
27. Stöber, W., McClellan, R. O., and Morrow, P. E. (1993) Approaches to modelling disposition of inhaled particles and fibers in the lung, in *Toxicology of the Lung,* Gardner, D. E., Crapo, J. D., and McClellan, R. O., Raven Press, New York, 527–601.
28. Asgharian, B. and Yu, C. P. (1988) *J. Aerosol. Med.,* 1, 37–50.
29. Asgharian, B. and Yu, C. P. (1989) *J. Aerosol Sci.,* 20, 355–366.
30. Asgharian, B. and Yu, C. P. (1989) *Aerosol Sci. Technol.,* 11, 80–88.
31. Yu, C. P. and Asgharian, B. (1993) Mathematical models of fiber deposition in the lung, in *Fiber Toxicology,* D. Warheit, Ed., Academic Press, San Diego, 73–98.
32. Yu, C. P., Zhang, L., Oberdörster, G., Mast, R. W., Glass, L. R., and Utell, M. J. (1994) *J. Aerosol Sci.,* 25, 407–417.
33. Oberdörster, G. (1995) Developing Test Guidelines for Respirable Fibrous Particles, Report U.S. EPA, University of Rochester, 1–49.
34. Podgorski, A., Zhou, Y., Bibo, H., and Marijnissen, J. (1996) *J. Aerosol Sci.,* 27(Suppl. 1), 479–480.
35. Zhou, Y., Podgorski, A., Marijnissen, J., Lemkowitz, S. M., and Bibo, H. B. (1996) *J. Aerosol Sci.,* 27(Suppl. 1), 491–492.
36. Asgharian, B., Zhang, L., and Fang, C. P. (1997) Theoretical calculation of the collection efficiency of spherical particles and fibers in an impactor, *J. Aerosol Sci.,* 28, 277–287.

Chapter 21

Carbon Fiber Filters with Aerosol Separation and Gas Adsorption Properties

Kvetoslav R. Spurny

CONTENTS

21.1 INTRODUCTION

During the last three decades, glass fibers and organic fibers with fiber diameters less or much less than 3 μm have been the most frequently used materials for the production of high-efficiency aerosol filters, e.g., for the production of HEPA and ULPA filters. In several industrial and environmental situations, as well as in analytical sampling procedures, air particulates (aerosols) and gaseous impurifications have to be removed from air or gases simultaneously. At present, separation batteries consisting of prefilters, charcoal filters (or other modern adsorbents), and "after" or "backing" aerosol filters are used as complex air-cleaning units. The same is valid for a simultaneous sampling of volatile substances, which exist in the atmosphere as vapors and aerosols at the same time.

By the end of the 1970s we had started investigations of aerosol filters made of coarse as well as of fine carbon fibers, which could be activated and used as good filters and adsorbers at the same time. The results have shown that such filters, activated carbon filters (ACFF), could be very useful for air cleaning as well as for air sampling.[1-3] The only disadvantage at that time was the relatively high production price of such filtration and adsorption materials in comparison with commercial glass fiber filters. In several publications during the two last decades, our preliminary scientific conclusions were confirmed and the production costs for carbon fibers and carbon fiber papers were substantially decreased.[4-26] Therefore, in our opinion, the time is coming when new types of air filters — e.g., ACFF — will complete the family of classical fibrous aerosol filters.

0-87371-830-5/98/$0.00+$.50

21.2 CARBON FIBERS AND CARBON FIBER FILTERS

The classical methods for the production of carbon fibers, reviewed elsewhere,[1,2] were substantially improved during the last decade. Carbon fibers produced now by heterogeneous pyrolysis of pitch or of other organic liquids and gases[4-16] are fine, dispersed and their production is also less expensive. They can be used for the production of carbon filter mats, felts, papers, and filters. They can also be activated well.

Activated carbon fibers (ACF) are microporous (pore sizes of about 10 Å with large specific surface area (SSA) ranging from 700 to 3000 m^2/g, and have random structures consisting of an assembly of micrographites with dimensions of approximately 20 × 20 Å. The electric resistivities are temperature dependent and lie on the order of 10^{-3} W · m, and the magnetic susceptibilities are relatively small and lie approximately between 10^{-7} and 10^{-6} emu/g.[5,6]

In our previous investigations,[1,2] we used carbon fiber felt produced in Germany by the SIGRI Company (Bavaria). It was manufactured from coarse polyacrylonitrile fibers. The mean fiber diameter was of about 8 μm, the felt thickness was between 2 and 10 mm, and the area density was 900 g/m^2. The SSA of the activated felt was about 800 m^2/g. The pressure drop Δ_P was of about 1600 Pa at the gas flow velocity of 1 cm/s. The collection efficiencies for aerosol particles of 0.3 μm in diameter lay in the range of 70%. The microstructure of this filtration felt is shown in Figure 21.1A, and in Figure 21.2 the effect of particle separation on a single-fiber element is shown. The TOYOBO Company in Japan[7] produces similar fiber felts and papers. They consist also of relatively coarse carbon fibers with a mean diameter of about 3 μm (Figure 21.3). The SSAs of these ACFs lie between 1000 and 1600 m^2/g, with micropore sizes between 5 and 10 Å. The speed of adsorption is 10 to 100 times higher than is the case in the application of conventional granular activated carbon. The area densities lie between 90 and 270 g/m^2, with apparent density of 0.045 g/cm^2, and the adsorption capacity for benzene is 30 to 58%. These carbon fibers are produced by using cellulosic fibers as the base material.[7] ACFF, of this production have already found useful applications in the fields of air cleaning, vapor adsorption, and air sampling.

Nipon KYNOL, Inc. produces ACFs with SSAs between 900 and 2500 m^2/g (Table 21.1). The carbon contained in these fibers is above 90%, and the N_2 and O_2 contaminations are less than 1%. The chemical surface analysis done by x-ray photoelectron spectroscopy (XPS) is shown in Figure 21.4. These ACFs have also shown very good adsorption properties for VOC (volatile organic compounds); see also Table 21.2. ACFs of Russian production are also available.[8] These ACFs are ionic chemiadsorbents. They are well resistant against temperatures up to 400°C and have SSAs as much as 1500 m^2/g. They have already found applications for gas adsorption land aerosol separation.

The carbon fibers produced can be coated by inorganic films for more resistance against temperature and acidic vapors. Very successful coating can be done, e.g., by SiC films. This film protects filters from oxidation to almost 1000°C.[10] In our previous investigations[1,2] carbon fibers with much smaller fiber diameters (in the range of 3 μm and thinner) were produced by a laboratory method and used for the preparation of fine porous carbon filters or in combination with coarser fibers (see also Figure 21.1B). The filtration properties of thin fibrous sheets were much better than the filtration properties of filter sheets made of coarse fibers. This is evident because the collection efficiency of a fibrous filter increases with the decreasing diameter of the single fibers forming the filter.

Recently, new types of very fine carbon fibers (VFCF) were described. They are also now commercially available.[11] They are manufactured by Applied Science, Inc. (Cedarville, OH). They are very thin, with fiber diameters ranging from 0.1 to 0.2 μm, and are longer than 100 μm. Their density is about 2.0 g/cm^3. In contrast to the straight morphology of the continuous carbon fibers, the VFCFs have a bent morphology resembling cotton wool. Such activated microfibrous material itself or in combinations with coarser ACF seems to be very promising for production of ACFFs with high collection efficiency for aerosols and adsorption capacity for vapors and gases.

Figure 21.1 Scanning aerosol micrographs showing the inner structure of aerosol filters made of coarse (A) and mixed (B) carbon fibers.

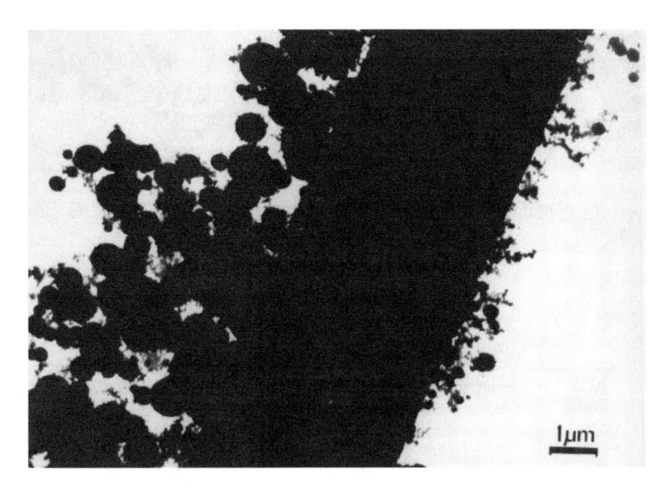

Figure 21.2 Electron micrograph of aerosol particles deposited on a single carbon fiber.

21.3 OTHER SOLUTIONS

The combined simultaneous separation of aerodispersed and gaseous pollutants and toxic substances from the gas phase can also be achieved by other physical and chemical procedures.[4] The collection efficiency of a classical fibrous filter can be improved by means of surface modification with chemicals, and the adsorption rate and capacity can be enhanced by different chemical impregnations.

Figure 21.3 Scanning electron micrographs of carbon filters produced by TOYOBO Company. Fiber diameter approximately 10 μm.

Table 21.1 SSA and Elemental Composition of ACF Samples

ACF	Surface Area[a]	C (wt%)	H (wt%)	O (wt%)
ACF-15	900	92.8	1.04	6.12
ACF-20	1610	95.4	0.68	3.92
ACF-25	2420	95.4	0.59	3.97

[a] BET SSA (m^2/g).

Source: Nipon KYNOL, Inc.

Since, for example, surface polarity of activated carbon is changed by treatment with certain chemicals, it may be possible for ACFs to retain surface charge and this charge could be utilized to enhance collection efficiency for aerosol particles. The collection efficiency of fibrous filters, including ACFFs, for fine particles can be further enhanced by means of electrical charging of filter fibers and/or aerosol particles. There is also the possibility of converting gaseous components into

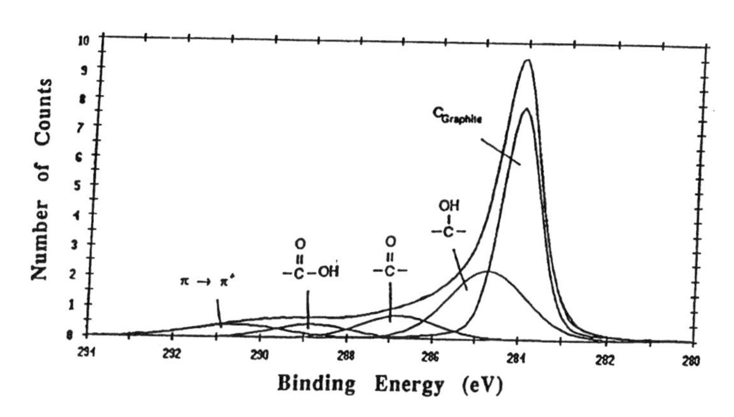

Figure 21.4 XPS analysis of the Nippon KYNOL ACF showing their surface composition. (From Nakayama, A. et al., *Bull. Chem. Soc. Jpn.*, 6l9, 333–339, 1996. With permission.)

Table 21.2 Effective Pore Volume for Selected VOCs

	Adsorbent (Volume Adsorbed cc/g)		
Adsorbate	**ACF-15**	**ACF-20**	**ACF-25**
Acetone	0.326	0.613	0.859
Cyclohexane	0.314	0.638	0.805
Benzene	0.323	0.653	0.849
Toluene	0.345	0.632	0.877
1,1,1-Trichloroethane	0.319	0.643	0.834
Mean pore volume	0.325	0.636	0.845

particles and then separating them by conventional high-efficiency filters. The applied corona discharge can be used to convert vapors (mainly the VOC) to particles.

These few examples show that classical aerosol filtration can achieve important improvements in the near future. This is also valid for the whole range of aerosol filtration research.[12]

21.4 APPLICATIONS

The ACFF, especially after further improvement, will have, in our opinion, large and important applications in the field of technological air and gas cleaning, as well as in the field of air sampling. The second task seems to be as important as the first one. Some successful applications have already been published.[19-26] ACFFs have been, e.g., tested for the adsorption of several basic air pollutants — NO, SO_2, NH_3, and CO_2 — and have been found to be very convenient (Figures 21.5 and 21.6). The adsorption of CO_2 at 273 and 298 K at subatmospheric pressure is a very suitable technique to characterize the narrow microporosity of the ACF.[19]

ACFFs have shown advantages in comparison with classical charcoal filters in the sampling of gaseous and aerodispersed air pollutants, including pesticides. Their application for the sampling of volatile air pollutants can be expected to be much broader in the future. ACFFs were, as mentioned, applied very successfully for atmospheric sampling of pesticides and fungicides. After the sampling on an ACFF, an extraction by a benzene–ethanol solution in combination with ultrasonic treatment follows. The concentrations are then determined by gas chromatography (GC) procedures.[25,26]

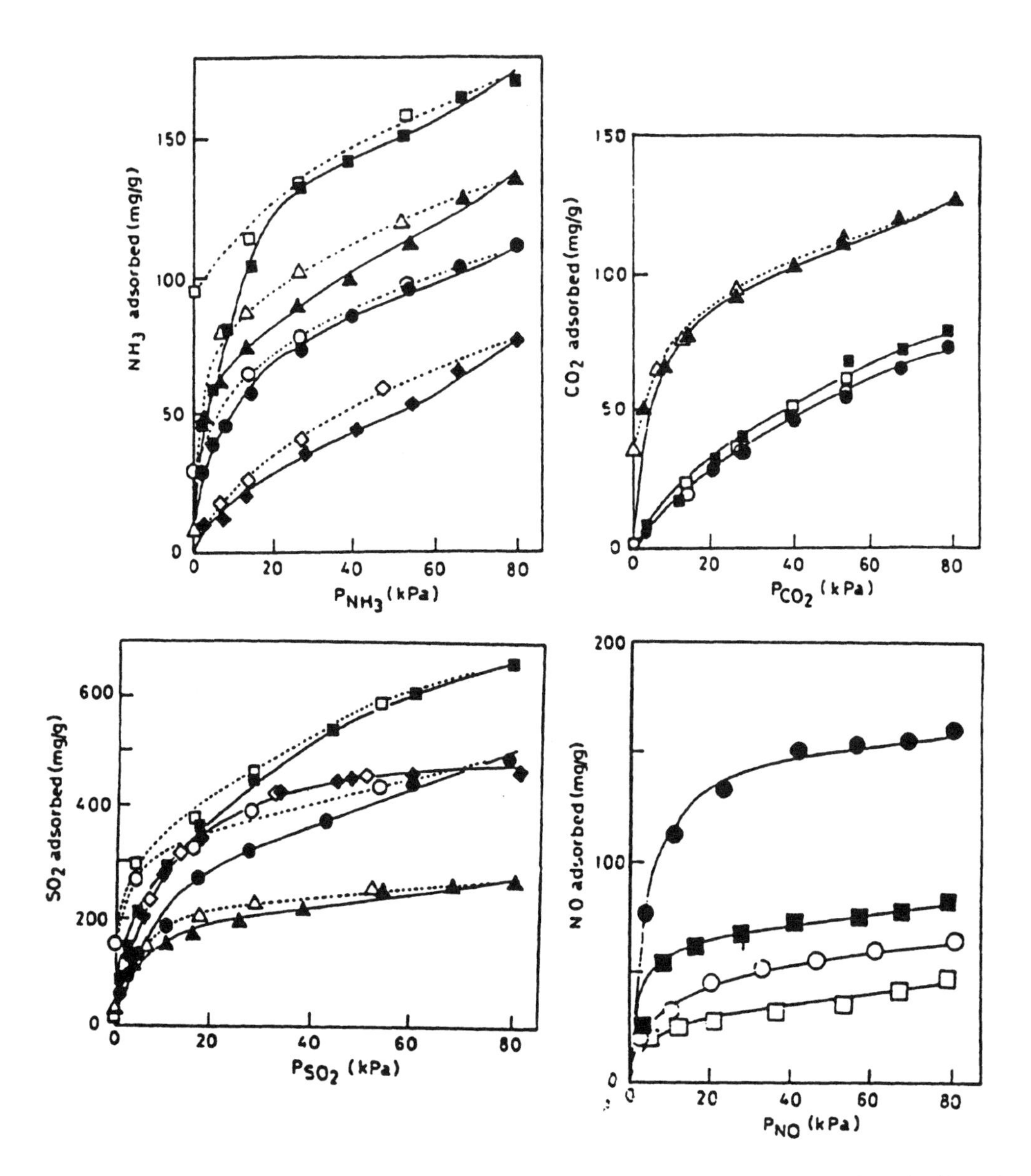

Figure 21.5 Adsorption isotherms for basic air pollutants — NH_3, NO, SO_2, and CO_2. (From Kaneoko, K. et al., *Colloid Polym. Sci.*, 265, 1018–1026, 1987. With permission.)

21.5 CONCLUSIONS

The previously published results, as well as the results of our earlier investigations,[1,2] have shown that ACFFs can be well manufactured and used in many practical technological and environmental applications. The improvement that must be made in the near future is the production of ACFFs containing fine dispersed carbon fibers, i.e., with fiber diameters less than 3 μm or thinner than 1 μm, as well. The technologies necessary for the production of such fine carbon fibers are already available.

Thin carbon fibers and the new carbon fiber filters exhibit properties recommending them for widespread applications in industrial and environmental technologies beyond the scope of aerosol filters made from glass and organic fibers. These new filters will have great advantages in industrial

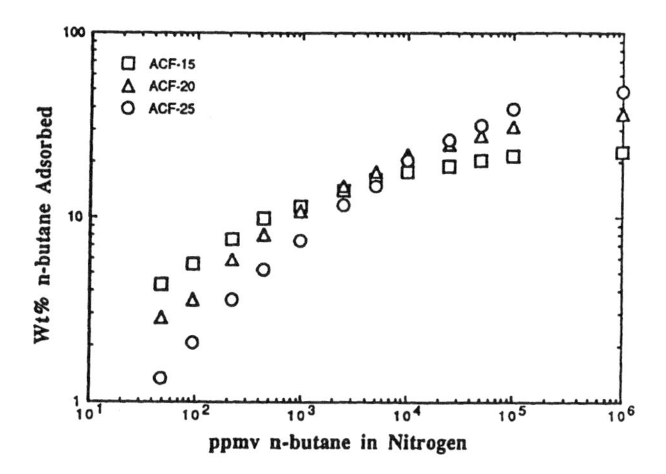

Figure 21.6 Adsorption isotherm of *n*-butane on ACFF. (From Kaneoko, K. et al., *Carbon,* 26, 327–332, 1988. With permission.)

filtration of inert or reducing gases at higher temperatures, in separation processes removing aerosol and gaseous impurities simultaneously, in filtration processes involving corrosive gases, etc. Further advantages can exist in applications as analytical filters for aerosol, vapor, and gas sampling.

REFERENCES

1. Spurny, K. R. (1979) Development in carbon fibre filters, in *The 2nd World Filtration Congress,* Filtration Society, London, 249–256.
2. Spurny, K. R. (1979) Zur Herstellung und zu den Eigenschaften von Kohlenstoffaser-Filtern, *Dräger Sonderhelft,* Mai, 66–81.
3. Spurny, K. R. (1991) La filtration des aerosols combinee avec l'adsorption des gaz"COFERA, 8e, *Journ. Etud. Aerosols* (Paris), 111–116.
4. Otani, Y., Emi, H., Mori, J., and Nioshino, H. (1991) *J. Aerosol Sci.,* 22, S793–794.
5. Enoki, T., Kobayashi, N., Nakayama, A., Suzuki, K., Hosokoshi, Y., Kinoshita, M., Endo, M., and Shindo, N. (1994) *Mater. Res. Soc. Symp. Proc.,* 349, 73–78.
6. Nakayama, A., Suzuki, K., Enoki, T., Koga, M., Endo, M., and Shindo, N. (1996) *Bull. Chem. Soc. Jpn.,* 619, 333–339.
7. Toyobo Comp. (1990) The Activated Carbon Fibers, Information Report, Osaka, Japan.
8. Nipon KYNOL, Inc. (1990) Carbon Fibers, Japan.
9. FRAM Comp. (1992) Activated Fibrous Carbon, Information, Brno, Czech Rep.
10. Kusakabe, K., Sea, B. K., Hayashi, J. I., Maeda, H., and Morooka, S. (1996) *Carbon,* 34, 179–185.
11. Hudnut, S. W. and Chung, D. D. L. (1995) *Carbon,* 33, 1627–1631.
12. Spurny, K. (1990) Aerosol filtration and separation, in *Aerosols,* Masuda, S. and Takahashi, K., Eds., Pergamon Press, Oxford, 34–38.
13. Kasaoka, S., Sakata, Y., Tanaka, E., and Naitoh, R. (1989) *Int. Chem. Eng.,* 219, 101–107.
14. Benissad, F., Gadelle, P., Coulon, M., and Bonnetain, L. (1988) *Carbon,* 26, 425–432.
15. Tibbetts, G. G. (1991) Aerosols and vapor-grown technology, *AAAR Annual Aerosol Conference.*
16. Ruland, W. (1990) Carbon fibers, *Adv. Mater.,* 2, 528–535.
17. Kaneoko, K., Ozeki, S., and Inouye, K. (1987) *Colloid Polym. Sci.,* 265, 1018–1026.
18. Kaneoko, K., Nakahigashi, Y., and Nagata, K. (1988) *Carbon,* 26, 327–332.
19. Cazorlaamoros, D., Alcanizmonge, J., and Linaressolano, A. (1996) *Langmuir,* 12, 2820–2824.
20. Balieu, E. (1989) *Ann. Occup. Hyg.,* 33, 181–195.
21. Balieu, E. (1990) *Ann. Occup. Hyg.,* 34, 1–11.
22. Peschke, J., Stray, H., and Oehme, M. (1988) *Fresenius J. Anal. Chem.,* 330, 581–587.
23. Niehaus, R., Scheulen, B., and Dürbeck, H. W. (1990) *Sci. Total Environ.,* 99, 163–172.

24. Henriks-Eckerman, M. L. (1990) *Chemosphere,* 21, 889–904.
25. Kawata, K., Moriyama, N., Kasahara, M., and Urushiyama, Y. (1990) *Bunseki Kagaku* (Japan), 39, 423–425.
26. Kawata, K., Moriyama, N., and Urushiyama, Y. (1990) *Bunseki Kagaku* (Japan), 39, 601–604.

CHAPTER 22

Pore Filters — Aerosol Filtration and Sampling

Kvetoslav R. Spurny

CONTENTS

22.1 INTRODUCTION

The analytical filtration of liquids has been known since ancient times.[1,2] The physicist Réaumur, in 1714, was the first to use ultrafilters in research. Membrane filters (MF) made of cellulose derivatives were prepared by Fick in 1896.[3] At the beginning of the 19th century Bechhold[4] studied pore size and its control when forming filters made by a colloid-chemical process, by evaporation of cellulose nitrate solutions. The colloid chemist Zsigmondy[5,6] developed the first technology for producing cellulose derivative MFs between 1915 and 1918. His method was later improved, e.g., by Elford[7] and Grabar.[8]

0-87371-830-5/98/$0.00+$.50

There are several books dealing with MFs. Chronologically, the first is by Jander and Zakowski[9] and covers mainly MF applications in chemical analysis. The second is a brochure by Tovarnickij and Glucharev[10] which focuses on the problem of MF preparation. And the third, by Daubner,[11] discusses primarily the methods of filter preparation and their applications in hydrobacteriology. This book includes about 500 references related to MFs and their applications.

22.2 CELLULOSE ESTER MFS

Cellulose ester MFs were produced in Germany beginning in 1927. This production was based mainly on the research done by Zsigmondy.[5,6,12-14] After World War II production of MFs started in the U.S.,[15] England, Russia, Czechoslovakia, etc. The largest producers of classical MFs are the Gelman and Millipore Filter Corp. in the U.S. and the Fa. Sartorius GmbH in Germany.

The classical MF is prepared by a colloid-chemical process in which concentrated colloidal solutions of polymers are allowed to solidify. A classical MF has three distinct structures (Figure 22.1): the upper surface structure, the inner structure, and the lower surface structure.[2] Its structure differs slightly from the capillary model of a pore filter in that its structure is less regular. Classical MFs, because they have void volume (porosity) in the range of 65 to 85%, allow relatively high flow rates. MFs have an average of 10^8 pores/cm^2 and are available in graded pore sizes from 10 μm to almost 25 Å. Cellulose ester MFs are stable in the dry state to temperatures up to 125°C in the presence of oxygen, and to 200°C in oxygen-free atmospheres.

22.3 METALLIC MF

Pore filters can be made also of thin metallic sheets. Walkenhorst[16,17] developed a method of preparing filters which consisted of a sheet of Al_2O_3 perforated by fine pores of very fine and uniform diameter near 0.025 μm. These filters could be used at temperatures up to 700°C. Incineration of dust samples to remove organic matter and carbon can thus be carried out on the filter used for sampling.

Other metallic filters made with a powder-metallurgical process had been described by Richards[18] and are known as SMF (silver membrane filters). In 1964 the Selas Corp. in the U.S. developed commercial production of SMFs made of pure silver with relatively uniform porosity (Figure 22.2A). One advantage of the SMF as well as of other metallic MFs over organic MFs lies in the ability to conduct electricity (specific resistance at 20°C is 1.6 μΩ cm), thereby eliminating electrostatic charging. SMs are thermally stable from –130 to +370°C in all atmospheres. An SMF is resistant to chemical attacks by all fluids that do not attack pure silver.

Desorbo and Cline[19] developed a complex method for the preparation of metallic MFs with very uniform submicrometer-size pores. Such MFs were produced by selectively etching the rod phase of directionally solidified eutectics of NiAl-Cr and NiAl-Mo and by a two-state replication process. The resulting filters had very uniform 0.5-μm-diameter holes (Figure 22.2B) with number densities as high as 8.5×10^7 cm^{-3}. Unfortunately, commercial production has not been realized because of high costs.

22.4 NUCLEPORE FILTERS

Since 1965 a new type of porous filter, called the Nuclepore filter (NPF), has come on the market.[20,21] In the period 1960 to 1965, Fleischer et al.[20] studied radiation damage in solids by exposing samples to high-energy radiation. The same method was used for producing small holes in certain materials. These studies of charged particle tracts led to the development of a new porous

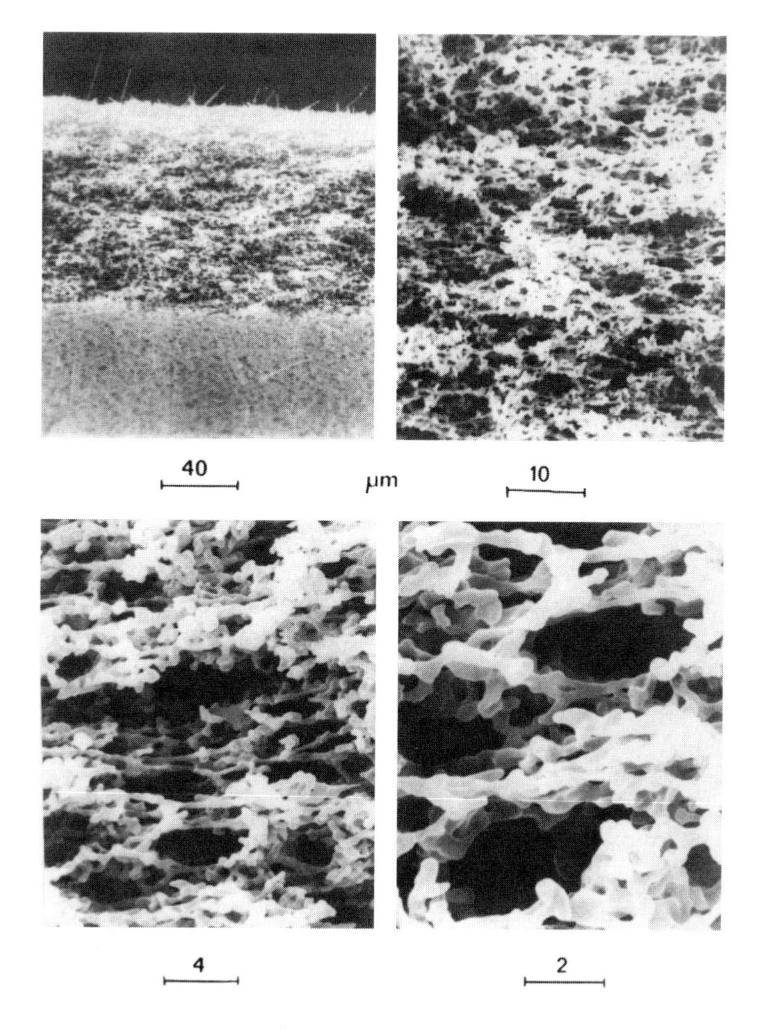

Figure 22.1 Scanning electron micrographs of the inside porous structure of a cellulose ester MF.

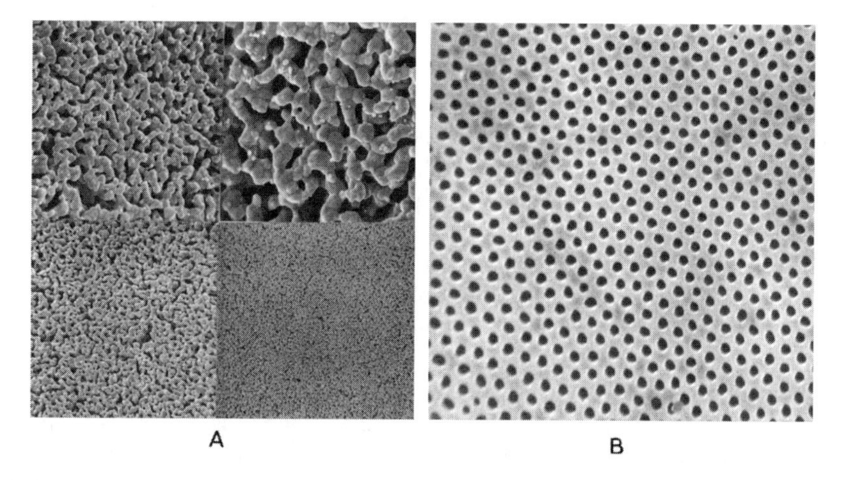

Figure 22.2 Scanning electron micrographs of the structure of two kinds of metallic MFs: silver membrane (A) and nickel alloy pore filter (B).

filter — the NPF. Following irradiation of a thin sheet of suitable material with a collimated beam of fission fragments, it was possible to produce a controlled number of tracks perpendicularly through the sheet of film (Figure 22.3). Irradiation may be accomplished by placing the sheet in an evacuated box with a natural uranium layer and then exposing the box in a nuclear reactor. Fission fragments from the thermal neutron capture of ^{235}U bombard the sheet.

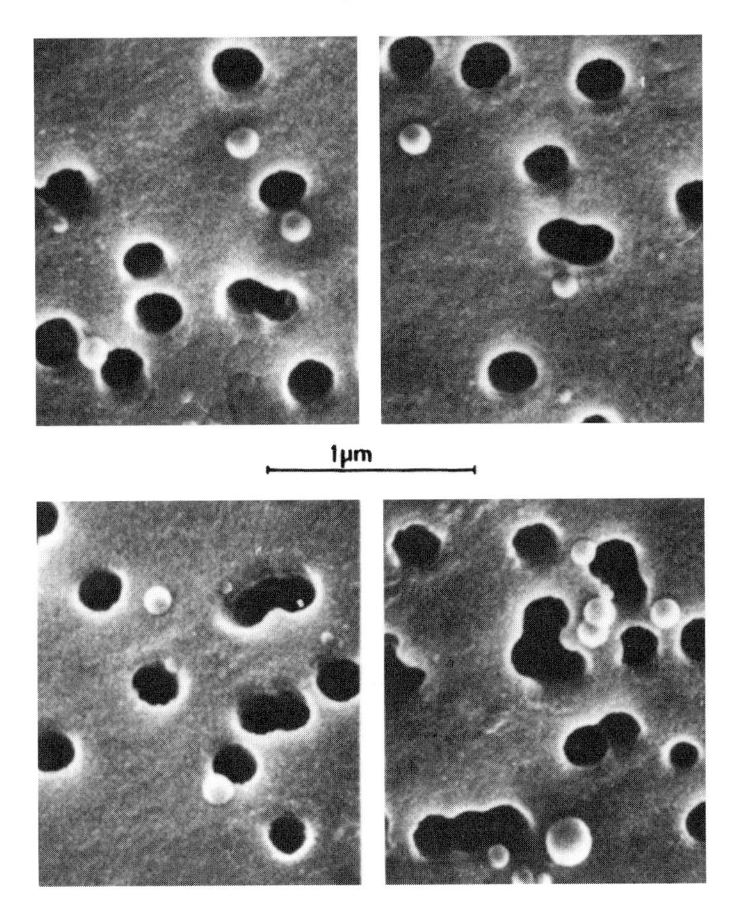

Figure 22.3 Scanning electron micrographs of the structure of an NPF with sampled latex particles.

The plastic sheets are then placed in an etching bath and fine, hollow channels (pores) are etched along the tracks of the fission fragments. Pore density (pores/cm^2) is controlled by the residence time in the irradiator. The tracks are not visible before the film is etched. The etching process (warm 10% NaOH solution) selectively dissolves the damaged material, leaving cylindrical straight capillaries or pores in the material. The pore size is controlled by the etching procedure. Longer etching times result in larger pores. Subsequent treatment can be used to render the membrane hydrophilic, partially hydrophobic, or totally hydrophobic.[22]

Commercial production of NPFs made of polycarbonate films started in the U.S. by General Electric in 1965 and was then continued by Nuclepore Corp. (Pleasanton, CA). In contrast to classical MFs (Figure 22.1), NPFs have a uniform structure (Figure 22.3) and relatively narrow pore-size distribution. The pores are round cylinders arranged perpendicularly (±30%) to the surface, with an even random distribution over it. A classical MF is a depth-type filter. Its "pore size" is only a hypothetical parameter indicating a relative comparison in size. NPFs are screenlike surface-acting filters. They are manufactured with a wide range of pore sizes from 0.015 to 12 µm. (Table 22.1). The pore density is closely related to the pore size. NPFs have an extremely smooth

Table 22.1 Nuclepore Membrane Specifications

Pore Size[a] (μm)	Nominal Pore Density (Pores/cm²)	Nominal Thickness[b] (μm)	Flow Rates[c] Air (1 · min⁻¹ · cm⁻²)	Flow Rates[c] Water (ml · min⁻¹ · cm⁻²)
12.0	10^5	6	65	2000
10.0	10^5	8	50	2000
8.0	10^5	10	45	2000
5.0	4×10^5	10	42	2000
3.0	2×10^6	10	42	2000
2.0	2×10^6	10	36	2000
1.0	2×10^7	10	36	300
0.8	3×10^7	10	36	300
0.6	3×10^7	10	16	80
0.4	10^8	10	16	80
0.2	3×10^8	10	6	26
0.1	3×10^8	5	2.2	8
0.08	6×10^8	5	2.2	8
0.05	6×10^8	5	0.25	0.45
0.03	6×10^8	5	0.25	0.03
0.015	10^9	5	0.002	0.002

[a] Maximum pore diameter (mean diameter about 10% smaller).
[b] Filter sizes: disks with diameter 13, 25, 37, 43, 47, 50, 62, 76, 82, 90, 142, 150, 298 mm (also rectangles, rolls, and tapes).
[c] At 25°C and 760 mmHg.

and flat surface and are therefore ideal for electron microscope and other analysis of aerosols and single particles.[23] NPFs with pore sizes of more than 0.2 μm are about 10 μm thick. Filters of smaller pore sizes are about 5 μm thick. NPFs have a low tare weight varying from 0.5 to 0.1 mg/cm². They are flexible and strong and cannot easily be ripped or torn. The average ash weight is less than 1 μg/cm². They give very little trace element contamination. They resist attack by most acids and organic solvents (except halogenated hydrocarbons). The best solvents for polycarbonate NPFs are chloroform, methylene chloride, pyridine, and strongly basic solutions, such as 6 M potassium hydroxide, sodium hydroxide, or ammonia solution. Their specific gravity ranges from 0.94 to 0.97. Their use at elevated temperatures is limited, but they remain stable in air up to 140°C. They have sufficient transparency for light microscopy and are birefringent with refractive indexes of 1.616 and 1.584.

The polycarbonate NPFs are the most-used analytical pore filters, besides the classical MFs. Nevertheless, other nuclear track filters have also been produced and described. The author[2] was the first to investigate and describe the physical and separation properties of polycarbonate NPFs[21] in the 1960s. Based on his personal communication with Russian aerosol and filtration scientists, production of NPFs made of other organic materials started also in Russia.[24,25] The Russian NPFs were designated as PTM (polymeric track membranes). Their development started in the Joint Institute for Nuclear Research in Dubna in the 1980s. Polymer films (thickness between 10 and 20 μm) of polyethylene terephthalate (PETP), of polyvinylidene fluoride (PVDF), as well as of polypropylene (PP), were used as standard polymer material. Xenon ion beams of high intensity from the cyclotrons (U-300 and U-400) were used for irradiation of these polymeric films. The chemical etching was done by alkali solutions. The PTMs produced have pore sizes in the range between 0.1 and 1 μm, with pore densities about 10^8 cm^{-2}. Fleischer et al.[20] had already described the preparation of nuclear membrane filters of mica. Nevertheless, a larger production was done much later in Germany. The mica NPFs (MNPF) were produced by irradiation of thin (50 μm) mica foils. These foils were irradiated in a special high-energy accelerator with xenon, lead, or uranium ions at specific energies up to 8.5 MeV/amu (doses of 10^3 to 10^6 cm^{-2}). The irradiated foils are then etched to the desired pore diameter with hydrofluoric acid. The length of a side of

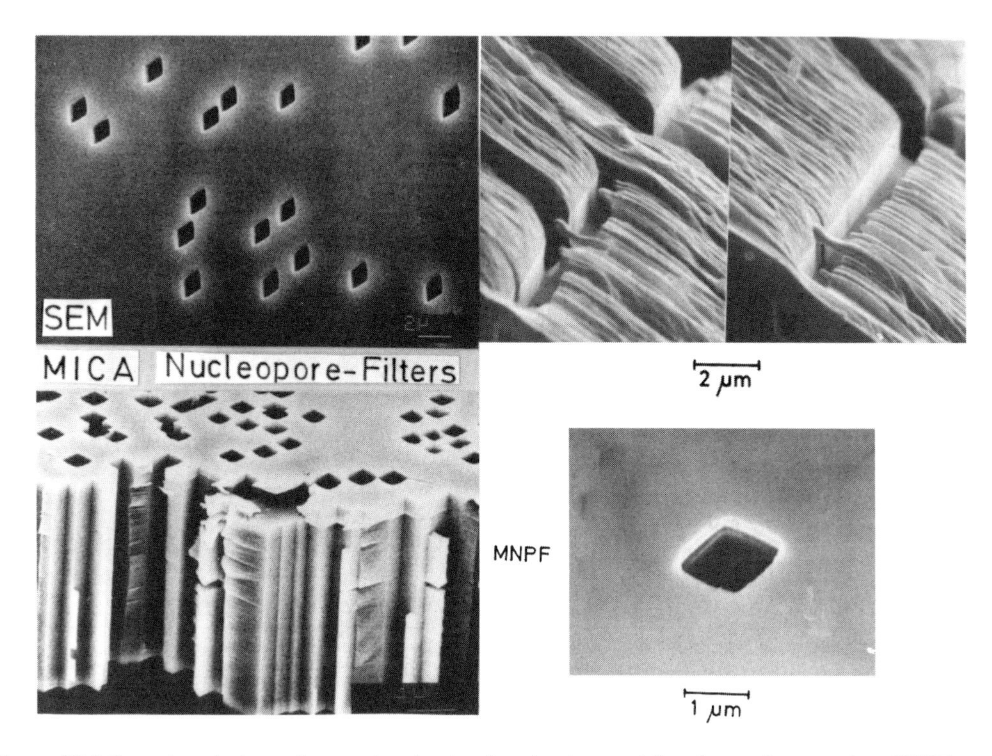

Figure 22.4 Scanning electron micrographs showing the structure and the shape of pores of an MNPF.

the rhomboidally shaped pores (Figures 22.4 and 22.5) depends linearly on the etching time, and has a Gaussian distribution. The pore size is constant throughout of the filter.[26]

MNPFs should be considered as complementary to classical MFs and to NPFs. They possess important resistance properties — to high temperatures (up to 600°C), organic solvents, and most acids. The use of MNPFs for filtration and sampling solid particles in gases and liquids under extreme conditions is therefore an important potential application. Furthermore, these filters are resistant to high levels of radioactivity and can be used in sampling and analyzing highly radioactive aerosols.

Similar track membranes were prepared by irradiating quartz crystals of approximately 100 μm thickness with heavy ions (e.g., by ^{238}U and by following etching with a KOH solution.[27,28]

22.5 FILTRATION PROPERTIES

By *filtration properties* is meant the hydrodynamics of the gas flow through the pores as well as the physical mechanisms by means of which the aerosol particles move to the filter and pore surfaces and are then captured, and also the clogging of the pores during filtration.

22.5.1 Gas Flow and Pressure Drop

For investigation of fluid flow through pore filters (MF, NPF), etc.), pressure drop, and filtration mechanisms, the capillary model is applied. This model, according to Guerout,[29] envisions a system of equidistant parallel, circular capillaries perpendicular to the filter surface having a length equal to the filter thickness L. The flow of gas through an NPF produces a pressure difference across the filter which is designated as the filter pressure drop Δp. The pressure drop is one of the important material parameters of any filter. the Δp for a capillary model of a pore filter can be described by

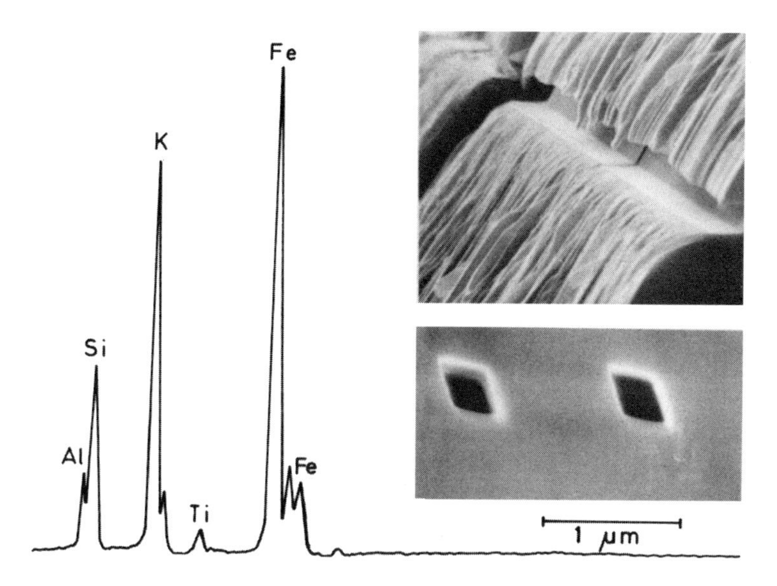

Figure 22.5 Scanning electron micrograph and elemental composition of an MNPF. (From Vater, P. et al., *Nucl. Instrum. Methods,* 173, 205–211, 1980. With permission.)

known hydrodynamic equations for gas and fluid flow through capillaries. These equations differ depending on the flow regime (classical, slip flow, free molecular).[30,31]

Nevertheless, experimental investigations have shown differences in experimental and calculated Δp values of NPF, with larger pore sizes.[32] An external pressure drop Δp_E was derived for the whole filter and an internal pressure drop Δp_I across a cylindrical pore was described by the Hagen–Poiseuille formula:

$$\Delta P_E = \frac{24 v \eta}{N D^3}$$
$$\Delta P_I = \frac{128 Q \eta L}{\pi D^4 (1 + 11.0 l/D)} = \left(\frac{16}{3}\frac{L}{\pi D}\right) \Delta P_E \left(\frac{1}{1 + 11.0 l/D}\right) \quad (22.1)$$

The total pressure drop, Δp_T, is obtained as an addition:

$$\Delta p_T = \Delta p_E + \Delta p_I \quad (22.2)$$

The full meaning of the symbols is given in the original paper.[32] A short designation is: Q is the flow rate per pore, η is the fluid viscosity, D is the pore diameter, v is the gas flow velocity, and N is the number of pores, cm^{-2}. A similar definition of Δp_E was also published by Kanaoka et al.[33]

22.5.2 Particle Separation

Apparently, the first investigation of the mechanisms of particle retention in pore filters, at that time of MEs, was by Duclaux and Errera.[34] Later, Cox and Hyde[35] investigated and physical factors involved in MF filtration. In these early papers attention was focused mainly on the sieve mechanism of particle retention. However, MFs, and pore filters generally, have the property of retaining particles much smaller than their pore size.

A pore filter is simulated mathematically as a system of straight parallel capillary tubes of a length equal to the thickness of the filter L, as mentioned. The pore radius of the capillaries is equal

Figure 22.6 Model of MF surface. Circular holes are filter pores; lined area is the impaction area of one filter pore.

to the mean pore radius of the real filter, as estimated by indirect methods.[2] The pores are arranged in a hexagonal lattice in the schematic model shown in Figure 22.6. The choice of hexagonal arrangement is arbitrary. Various mechanisms of filtration were considered.

(1) Diffusion

Fine particles diffuse to the walls of the capillaries as well as to the filter surface as they pass through the filter. They are captured then as fine particle deposition. The diffusion of fine particles to the walls inside the pore capillaries follows the well-known equations for collection efficiency n_D and penetration P_D, which have been described in several papers and reviews.[22,36-40]

$$n_D = 2.56 N_D^{2/3} - 1,2 N_D - 0.177 N_D^{4/3}$$

$$\begin{aligned} P_D = {} & 0.819 \exp(-3.657 N_D) \\ & + 0.097 \exp(-22.3 N_D) \\ & + 0.032 \exp(-57 N_D) \\ & + 0.0157 \exp(-106.7 N)_{D+\dots} \end{aligned} \tag{22.3}$$

n_D means a dimensionless parameter: $n_D = (DL)/(R^2 v)$. (D is the diffusion coefficient of particles, R is the pore radius, and v the gas flow velocity.)

In the Table 22.2 diffusion coefficients of fine aerosol particles are presented. The particle diffusion to the filter surface as well as the combination of both (pores and filter surface) were solved numerically in 1979.[41,42] Very recently, Fang and Cohen[72] tested the diffusion collection mechanism of NPFs with monodispersed aerosols in a range of 0.01 to 0.2 μm (particle diameter). The collection efficiencies of the NPFs were determined by comparing upstream and downstream concentrations at different filter face velocities.

It was found that particle diffusion deposition first decreases, then increases with increasing face velocity. There exists a face velocity at which there is minimal deposition. The results suggested that the submicrometer particles deposit on both the pore surface and planar surface by diffusion. At low face velocities, deposition on the pore surface, which is enhanced by decreasing face velocity,

Table 22.2 Diffusion and Sliding Coefficients for Small Aerosol Particles

Size of Microparticles D_p (μm)	Diffusion Coefficient D (cm²/s)	Cunningham Coefficient of Sliding Friction C
0.1	0.92×10^5	2.909
0.2	0.23×10^5	1.886
0.3	0.10×10^5	1.567
0.4	0.62×10^6	1.416
0.5	0.41×10^6	1.330
1.0	0.10×10^6	1.163

Source: Ovchinnikov, V. V. et al., *Aerosol Sci. Technol.*, 17, 159–168, 1992. With permission.

dominates, while deposition on the planar surface increases with increasing face velocity and is more important at high face velocities.

(2) Inertial Deposition

Larger particles are deposited on the front face of the filter due to their inertia, since the flow into each pore converges and impingement (impaction) of particles takes place around the orifice of each capillary. It was Pich[43,44] who calculated for the first time the efficiency of inertial deposition in NPFs. The inertial collection efficiency is a function of the Stokes number, Stk.

(3) Direct Interception and Sieving

Particles that contact the surface of the pore walls are captured as a result of mechanisms of direct interception, calculated first by Spurny.[21] The collection efficiency for direct interception is characterized by the parameter $N_R = r/R$, where r and R are the radius of the aerosol particle and the filter pore. When $N_R < 1$, the particle will be captured by the interception. If $N_R > 1$, the particle will be separated by a simple sieving effect.

(4) Gravitational Deposition

The problem of gravitational deposition of particles in a circular tube under laminar flow was investigated by Natanson.[46] The collection efficiency by gravitational deposition is proportional to the dimensionless parameter N_G:

$$N_G = (3v_sL)/(8rv) \qquad (22.4)$$

where v_s is the sedimentation velocity of the particle. Applications of this theory on MFs and NPFs were further developed by Spurny[47] and Pich.[37]

(5) Deposition by Electrostatic Forces

The first observations and measurements dealing with electrostatic charges on MFs were done in the 1950s.[48] MFs and NPFs are very good insulators and therefore the filter surface can get relatively high electric charges.[49] Figure 22.7 shows a scanning electron micrograph with the distribution of electric charges on the surface of an NPF. The white halos around the pores are a visualization of high electric charges on the edges of cylindrical pores. In Figure 22.8 electric field lines surrounding the charged NPF surface, including a pore and a polarized aerosol particle, are

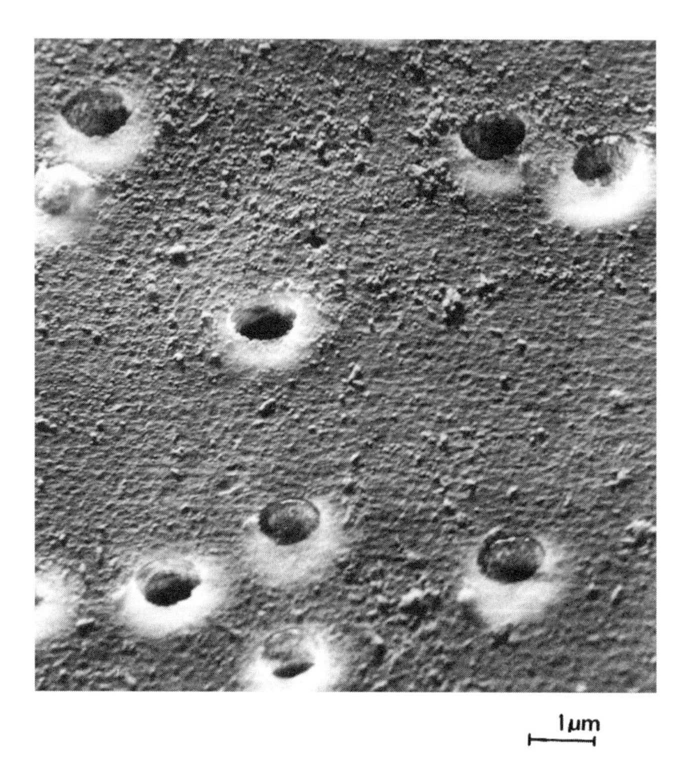

Figure 22.7 Scanning electron micrograph of an NPF surface showing the electric charge enhancement around the pores.

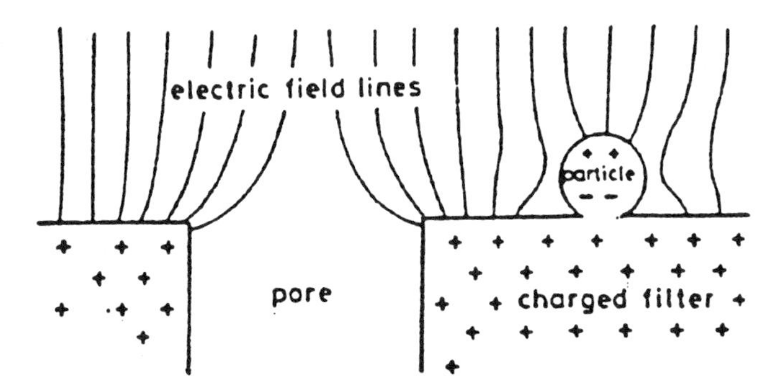

Figure 22.8 Proposed electric field surrounding the NPF pores. (From Romo-Kröger, C. M. and Morales, J. R., *Phys. Scripts,* 37, 270–273, 1988. With permission.)

shown.[50,51] The first theoretical description of the electrostatic deposition of aerosol particles in an NPF was published by Zebel.[52] His model takes into account the influence of electrostatic forces between charged aerosol particles and the pore charges. The influence of electrostatic forces is given by the dimensionless parameter E:

$$E = n \cdot e \cdot q \cdot B/R \cdot v \tag{22.5}$$

where n is the number of electric elementary charges e of the particle, B is the mobility of the particle, q the electric charge per unit length of the pore edge, R the pore radius, and v the flow velocity toward the filter.

The electric charge on filters and particles may influence the process of filtration in two ways: the particles may be attracted to (or repelled from) the filter surface from a greater distance or the charge makes the particle "stick" to the filter surface. For the first, the coulombic force may be more important, because it is of longer range than the bipolar force. For the effect of particles sticking on the surface, in turn, the dipolar interactions play a principal role. Particles that fall across the filter surface by inertial impaction or direct interception (or by influence of coulomb force) may be strongly fixed near the pore edges due to the intensive dipolar force. Dipolar force contributes to avoiding the detachment of particles. Microscopic analysis of particles on the NPF surface has demonstrated well that these particles are mainly trapped by the direct interception at the pore edges and the dipolar force sticking them. The most recent measurements have also shown that the dipolar force greatly exceeds the coulomb force near the pore edge.[51a] The electric charge on the surface of an NPF has a spatial distribution with the highest charge concentration near the edge of the pore (see also Figure 27.7).

(6) Combination of Partial Collection Mechanisms

As in the case of other aerosol filters, the total collection efficiency of pore filters is a function of all partial efficiencies and parameters. A first method for combining two mechanisms was proposed by Spurny[53] and Pich.[54] They considered only particle inertia and diffusion in the pores. By the combination of three mechanisms, three filters were supposed to be in series each operating by a single mechanism. These efficiencies, e.g., E_I (impaction), and F_D (diffusion), are, of course, strongly size dependent, each in a different way. The total efficiency (F_T) is, therefore,

$$E_T = E_I + E_R + E_D - E_I E_R - E_R E_D + E_I E_R E_D \tag{22.6}$$

The total efficiency of a pore filter gives a minimum as a function of particle size and filter flow velocity. Filtration efficiency minima were also found as the functions of gas pressure and gas temperature.[55] Using computer techniques, collection efficiency tables for pore filters were published.[56] They can be used to find the approximate collection efficiency of commercial MFs and NPFs and to determine the optimal gas flow for aerosol sampling.

Caroff et al.[57] have solved the total efficiency equation by computer for distributions of particle size and pore size. The results showed that the error resulting from neglect of the particle size distribution was small, but the pore size distribution had a pronounced effect, being larger than errors arising from neglect of particle interception, E_R. The minimum of the curve efficiency against particle size was shifted by making allowance for nonuniformity of size of particles and pores.

22.6 EXPERIMENTAL INVESTIGATIONS

Several aerosol physical and electron microscopical methods were used in order to investigate the basic separation mechanisms and verify the theoretical results. In such ways have applications of MFs and NPFs in aerosol sampling and analysis been enhanced and promoted. For such tests and measurements well-defined model aerosols were prepared and used.[58-60]

22.6.1 Collection Efficiency Measurements

Experimental studies on MFs and NPFs have mainly two purposes: to estimate the contribution and the validity of the most important partial particle deposition mechanisms and to find the maximum penetrated particle size, which is a parameter having great practical importance for choosing suitable MFs and NPFs for filtration or sampling of a real aerosol. Experimental model aerosols used for such studies can be mono- or polydisperse, solid or liquids, isometric or nonisometric, etc. As mentioned, we have used in our studies mainly quasi-mondispersed radioactive-labeled solid and liquid aerosols with approximately isometric particles, and monodispersed latex aerosols. The advantages of such model aerosols were the high detection sensitivity for the penetrated aerosol fraction, as well as the broad option of particle size ranges.

Scanning electron microscopy has been found to be a very useful tool for studying the configurations of deposited particles. Our studies[20,21,30-33] have for the first time demonstrated well the existence of different partial deposition mechanisms. By the filtration of highly dispersed aerosols, the existence of a diffusional deposition of very fine particles inside the pores and on the filter surface could be demonstrated well and confirmed (Figures 22.9 and 22.10). The existence of the most important partial deposition mechanisms is shown also in the Figure 22.11. Except the mentioned diffusional deposition (Figure 22.11A), the partial mechanisms of interception (Figure 22.11B) and impaction (Figure 22.11C) are also demonstrated well by means of deposited latex particles. When two or more partial separation mechanisms are acting together, the collection efficiency characteristics (the total efficiency E can be plotted as the function of the particle radius r) define the filter collection properties. In Figure 22.11D results of measured and theoretical E values for an NPF with pore size of 5 μm are shown. The existence of the collection efficiency minima for MFs and NPFs could be well confirmed. Similar measurements done in the mid 1970s by using monodispersed di-octyl phthalate (DOP) aerosols have shown that these efficiency minima, depending on the particle size, pore size, and gas flow velocity, lay in the submicronic particle size range.[61]

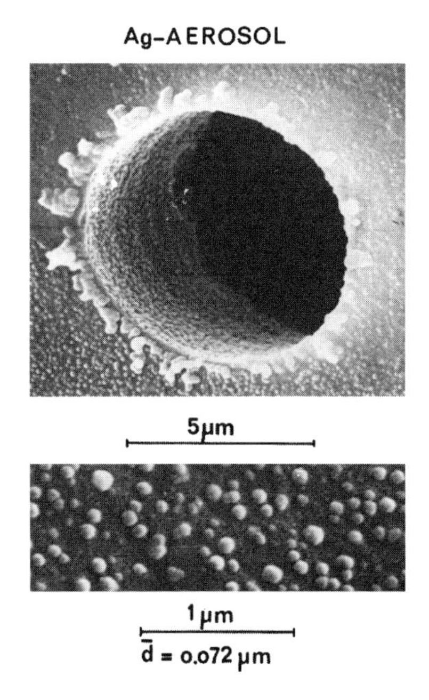

Figure 22.9 Scanning electron micrograph showing a pore of an NPF with small particles of a silver aerosol deposited inside the pore and on the filter surface.

22.6.2 Testing the Electrostatic Effects

Observations that MFs can get electrically charged had already been made by the early 1960s,[2] when these filters started to be used as the standardized method for measurement of dust and aerosol mass concentrations in workplaces and in ambient air. Considerable errors in gravimetric analysis of deposited material in the filter have been attributed to the presence of an electric charge on it. On the other hand, this charge may produce interactions between particles and filter (attraction and

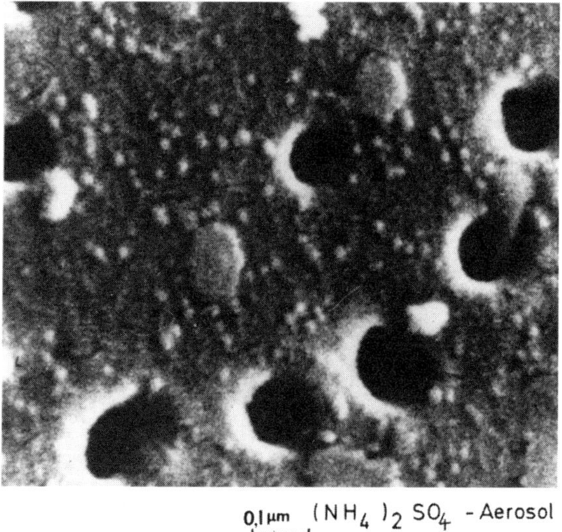

Figure 22.10 Scanning electron micrograph of an NPF surface with sampled particles of a highly dispersed atmospheric ammonium sulfate aerosol.

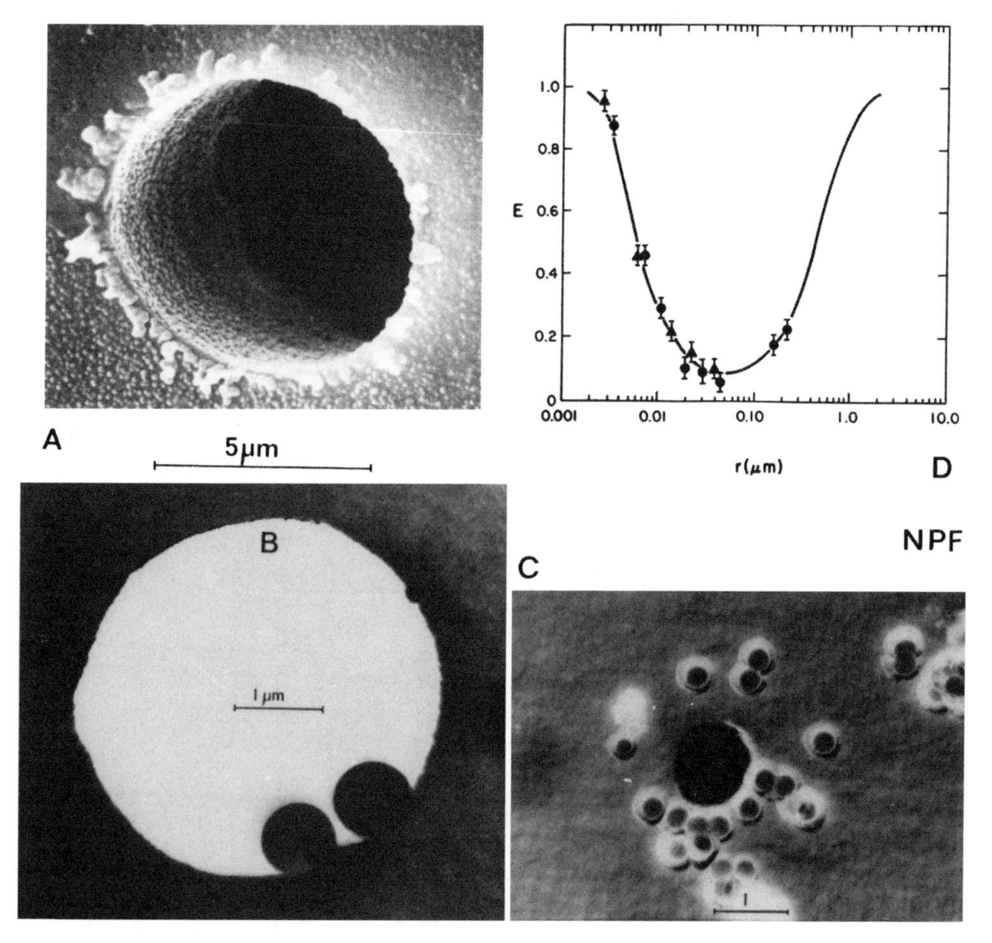

Figure 22.11 Picture demonstrating the main separation mechanisms for aerosol particles in an NPF. Description in the text.

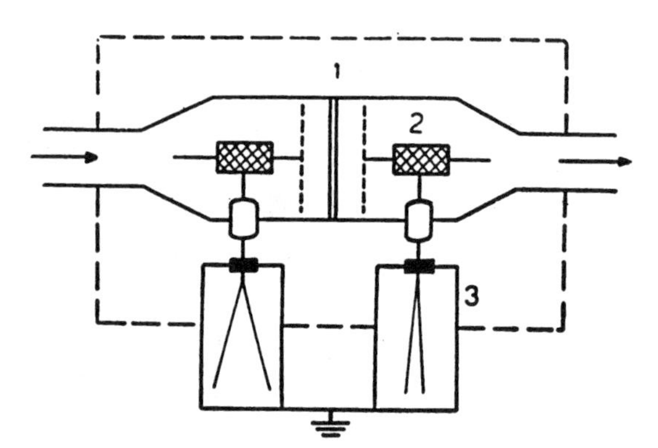

Figure 22.12 Schematic picture showing the apparatus for measuring electrostatic charges on MFs. (2) Filter, (2) electrodes, (3) electrometer.

repulsion coulomb forces, electric dipolar forces, etc.), altering the efficiency of performance. Later observations have shown that NPFs can also get relatively high electric charges.[62-64]

In our early studies[48,49] done on MFs, the original filter charges as well as their changes during the dust sampling were semiquantitively detected. By means of an induction method, the electric charges of the MF (1, in Figure 22.12) were measured by means of two metallic screen electrodes (2, in the figure) and two microscopical electrometers (3, in the figure). Examples of such measurements are shown in Figure 22.13. Under dry air conditions (Figure 22.13A) the electric charges on the MF increase with the sampling time t and the deposited particle mass, later reaching a steady state. At higher air humidities (Figure 22.13B) the filter charge decreases after reaching a maximum.

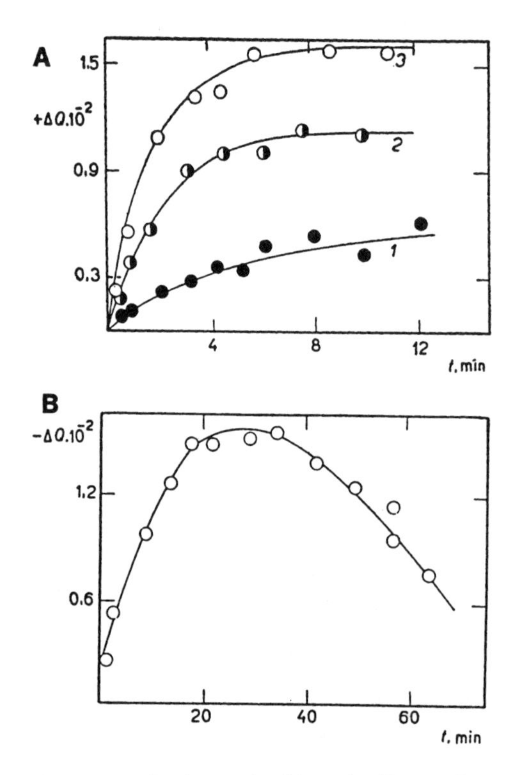

Figure 22.13 Measured electric charges (in elst. un./cm^2)* on the filter surfaces as a function of time (t) and flow rate, 1, 2, 3 (A). Filter charging at high air humidity (B). *elst. un. (electrostatic units). 1 E.S. cgs unit of electrostatic charge = 3.33 × 10^{-8} C (coulomb).

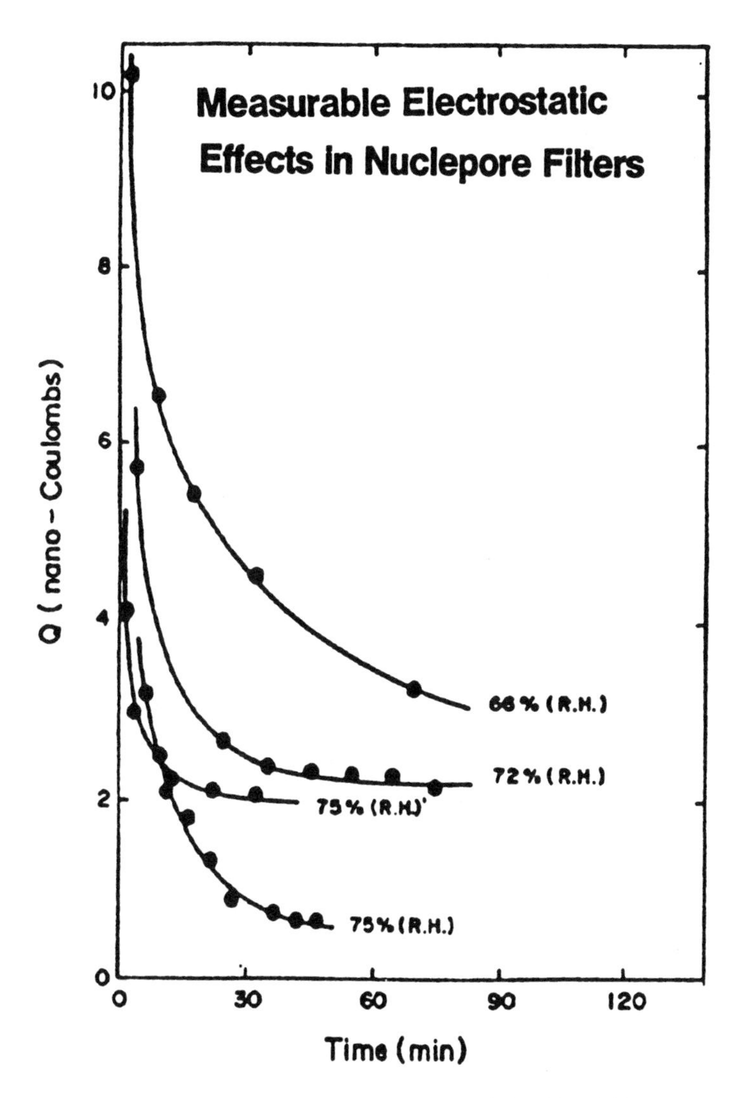

Figure 22.14 Electric charge of an NPF as a function of time and air humidity (RH).

Similar measurements on NPFs were realized at the end of the 1980s[50,51] and similar results were also obtained. The relatively high initial static charge Q of the NPF decreased during the filtration process, and furthermore this decrease depended strongly on the relative humidity of air (Figure 22.14). An important enhancement of the collection efficiency was measured when particles or filter or both were electrically charged (Figure 22.15).[65] Nevertheless, when gas flow velocities through the filter increase, the effect of electrical collection efficiency enhancements get smaller.

22.7 PORE CLOGGING

During filtration or sampling, particles are collected in the pores and on the filter surface. The collection efficiency, as well as the pressure drops, changes because the filter gets clogged and its inner structure and porosity change, too. Experimental results as well as theoretical descriptions were published in the 1970s.[66-69] The solid aerosol particles line the walls of the pores and decrease during the filtration time the effective pore size (Figure 22.16). The clogging characteristics depend

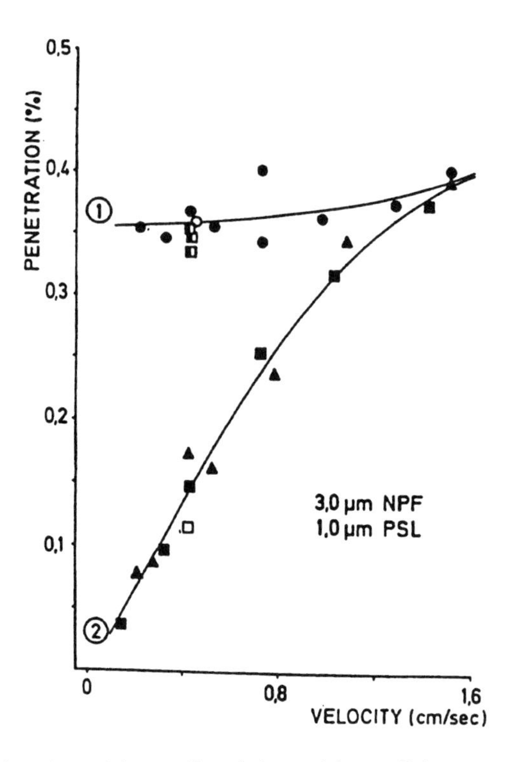

Figure 22.15 Penetration of uncharged (curve 1) and charged (curve 2) latex particles (PSL) through an NPF.

on the particle size, particle shape, gas flow velocity, etc. Resulting from the clogging process, filter pressure drop (ΔP) and the collection efficiency (E) characteristics have qualitatively very similar shapes. Both parameters increase during the filtration time in dependency on the deposited amount of particles and their size distribution. Examples of the ΔP clogging characteristics for solid and liquid particles are shown in Figure 22.17. By developing the "clogging theory" the equations describing the parameters ΔP and E of clean pore filters were modified for description of a time dependency, considering the number of deposited particles as well as the dominant separation mechanism.

22.8 APPLICATIONS IN AEROSOL ANALYTICS

Applications of pore filters — MF and NPF — in air sampling and aerosol analysis are very broad.[2,22,23,70,71] Both MFs and NPFs have high chemical purity and constant weight; they are well suited for gravimetric and chemical analysis. For aerosol sampling and analysis, both large- and fine-pore filters are useful. Large-pore filters are applied in size-selective sampling. Fine-pore filters, mainly with pore sizes between 0.2 and 0.8 µm, are useful for atmospheric and laboratory sampling. NPF, are well suited for determinations of both particle mass concentration and particle number concentration.

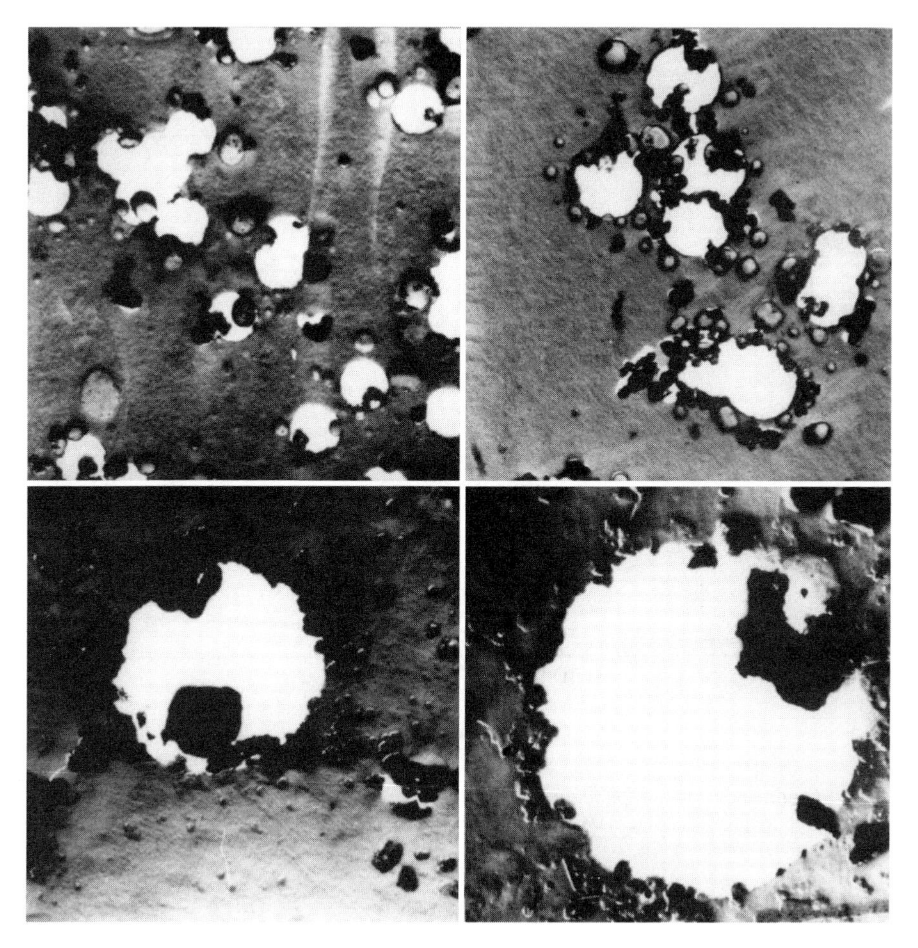

Figure 22.16 Scanning electron micrographs demonstrating the beginning phase of pore clogging with atmospheric dust particles.

Because of the transparency of NPFs, it is possible to count the number of particles on the filter surface by using light microscopy. For counting and measuring very small particles, scanning or transmission electron microscopy provides the best methods. Because of their relatively good chemical resistance and chemical purity, NPFs can be used very successfully for microchemical analysis. The microsample can be dissolved and analyzed for some or all components by standard methods (polarography, spectrophotometry, mass spectrometry, atomic-absorption spectrometry, etc.). Samples can be examined directly — i.e., without prior dissolution — by means of neutron-activation analysis (NAAA), X-ray fluorescence analysis, proton-induced X-ray emission (PIXE), X-ray diffraction analysis, infrared analysis, various microprobe techniques, etc. Last, but not least, they are excellent sampling supports for the modern analysis of individual airborne particles.[23]

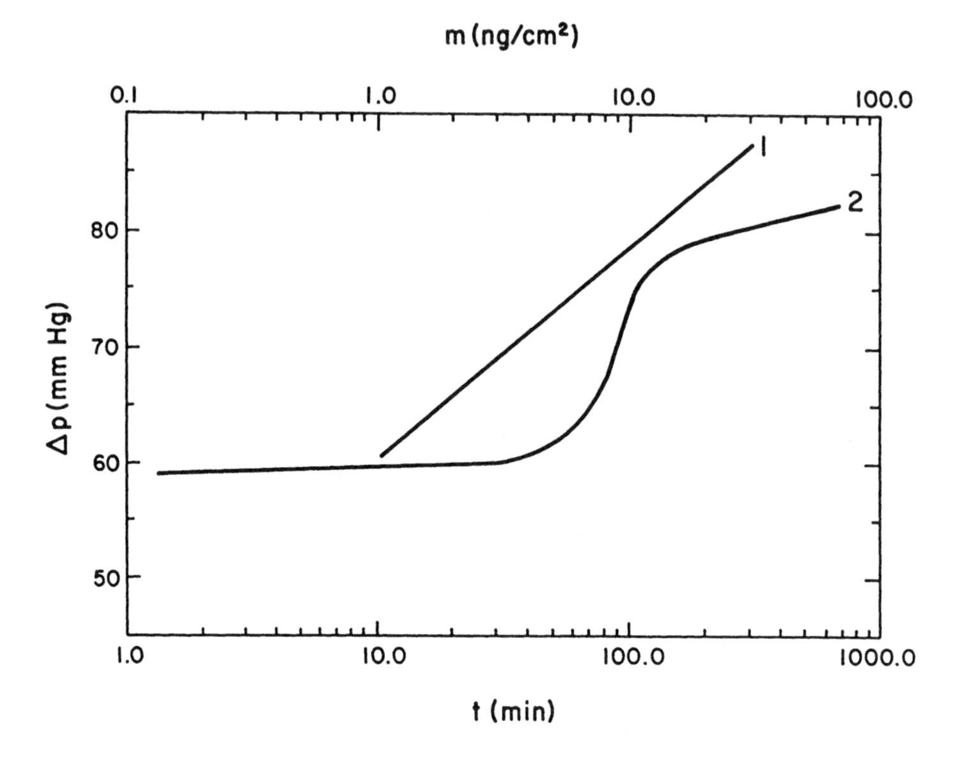

Figure 22.17 Experimental time dependency of pressure drop (Δp) for an NPF. Line 1: liquid aerosol of pyrophosphoric acid ($r = 0.02$ μm, $n_o = 3.3 \times 10^6$ cm^{-3}, $q = 7.5$ cm s^{-3}, $R_o = 1.0$ μm). Line 2: solid selenium aerosol ($r = 0.1$ μm, $n_o = 1.2 \times 10^4$ cm^{-3}, $q = 7.5$ cm s^{-1}, $R_o = 1.0$ μm).

REFERENCES

1. Gelman, C. (1965) *Analy. Chem.,* 37, 29A–37A.
2. Spurny, K. (1965–67) *Zbl. Aerosol Forschung,* 12, 369–407 1965; 13, 44–101, 1966; 13, 398–451, 1967).
3. Brown, N. (1915) *Biochem. J.,* 9, 320–327, 591, 597.
4. Bechhold, H. (1907) *Z. Phys. Chem.,* 60, 257–266.
5. Zsigmondy, R. and Bachmann, W. (1916) German Patents 329-060 (May 9) and 329-117 (August 22).
6. Zsigmondy, R. and Bachmann, W. (1918) *Zs. Anorg. Chem.,* 103, 119–129; *J. Soc. Chem. Ind.,* 37, 453A–458A.
7. Elford, W. J. (1930, 1931) *Proc. R. Soc. London.,* Ser. B, 105, 216–223, 1930; *J. Pathol. Bacteriol.,* 34, 505–511, 1931.
8. Grabar, P. (1935) *Bull. Soc. Chim. Biol.,* 17, 965–971, 1245–1249.
9. Jander, G. and Zakowski, J. (1929) *Membran-Cella und Ultrafilter,* Akad. Verlagsgesellschaft, Leipzig.
10. Tovarnickij, V. I. and Glucharev, G. P. (1951) *Ultrafilters and Ultrafiltration* (in Russian), Medgiz, Moscow.
11. Daubner, I. (1960) *Membrane Filters and Their Application in Hydrobacteriology* (in Slovak), Slovak Academy of Science, Bratislava.
12. Zsigmondy, R. (1919) *Z. Anal. Chem.,* 58, 241–246.
13. Zsigmondy, R. (1926) *Biochem. Zs.,* 171, 198–304.
14. Zsigmondy, R. (1926) *Angew. Chem.,* 398–404.
15. Lippmann, M. (1972) Filter media in air sampling, in *Air Sampling Instruments,* (Am. Conf. Govern. Ind. Hygienists, Cincinnati, OH.
16. Walkenhorst, W. (1970) *Staub Reinhalt. Luft,* 30, 246–249.

17. Walkenhorst, W. (1972) New sampling methods for particulate matter with finely porous filters, in T. T. Mercer, P. E. Morrow, and W. Stöber, Eds., *Assessment of Airborne Particles,* C.C. Thomas, Springfield, IL, 4195–518.
18. Richards, R. T., Donovan, D. T., and Hall, J. R. (1967) *Am. Ind. Hyg. Asoc. J.,* 28, 5190–598.
19. Desorbo, W. and Cline, H. E. (1970) *J. Appl. Phys.,* 41, 2099–2105.
20. Fleischer, R. L., Price, P. B., and Walker, R. M. (1965) *Science,* 149, 383–393.
21. Spurny, K. R., Lodge, J. P., Frank, E. R., and Sheesley, D. C. (1969) *Environ. Sci. Technol.,* 3, 453–468.
22. Spurny, K. R. and Gentry, J. W. (1979) *Powder Technol.,* 24, 129–142.
23. Spurny, K. R., Ed. (1986) *Physical and Chemical Characterization of Individual Airborne Particles,* Ellis Horwood, Chichester, U.K.
24. Apel, P. Yu. (1991) *Nucl. Tracts Radiat. Meas.,* 19, 29–34.
25. Apel, P. Yu. (1995) *Radiat. Meas.,* 25, 667–674, 717–720, 733–734.
26. Vater, P., Tress, G., Brandt, R., Genswürder, B., and Spor, R. (1980) *Nucl. Instrum. Methods,* 173, 205–211.
27. Holländer, W., Brandt, R., Pape, E., and Plachky, M. (1986) *J. Aerosol Sci.,* 17, 505–509.
28. Holländer, W., Dunkhorst, W., Brandt, R., and Vater, P. (1987) *J. Aerosol Sci.,* 18, 907–912.
29. Guerout, M. (1872) *C. R. Acad. Sci.,* 75, 1809–1814.
30. Spurny, K. (1977) *Staub Reinhalt. Luft,* 39, 328–334.
31. Pich, J. (1987) Gas filtration theory, in: M. J. Matteson and C. Orr, Eds., *Filtration Chem. Ind.,* 27, Marcel Dekker, New York, 1–132.
32. John W., Hering, S., Reischl, G., Sasaki, G., and Goren, S. (1983) *Atmos. Environ.,* 17, 115–119.
33. Kanaoka, C. H., Emi, H., and Aikura, T. (1979) *J. Aerosol Sci.,* 10, 29–35.
34. Duclaux, J. and Errera, J. (1924) *Rev. Gen. Colloides,* 2, 130–141.
35. Cox, H. and Hyde, R. R. (1932) *Am. J. Hyg.,* 16, 667–672.
36. Spurny, K. and Pich, J. (1963, 1965) *Collect. Czech. Chem. Commun.,* 28, 2886–2894, 1963; 30, 2276–2287, 1965.
37. Pich, J. H. (1966) Aerosol filtration, in C. N. Davies, Ed., *Aerosol Science,* Academic Press, London.
38. Spurny, K. R. and Lodge, J. P. (1968) *Staub Reinhalt. Luft,* 28, 179–188.
39. Spurny, K. R., Lodge, J. P., Sheesley, D. C., and Frank, E. R. (1969) *Environ. Sci. Technol.,* 3, 453–468.
40. Spurny, K. and Hrbek, J. (1969) *Staub Reinhalt. Luft,* 29, 70–74.
41. Manton, M. J. (1979) *Atmos. Environ.,* 13, 525–532.
42. Manton, M. J. (1979) *Atmos. Environ.,* 13, 1669–1674.
43. Pich, J. (1964) *Staub,* 24, 60–62.
44. Pich, J. (1964) *Coll. Czech. Chem. Commun.,* 29, 2223–2227.
45. Ovchinnikov, V. V., Belushkina, I. A., Akapiev, G. N., Shestakov, V. D., and Pilvinis, R. P. (1992) *Aerosol Sci. Technol.,* 17, 159–168.
46. Natanson, G. L. (1964) in N. A. Fuchs, *The Mechanics of Aerosols,* Pergamon Press, Oxford.
47. Spurny, K. (1970) Aerosol Filtration by Means of Analytical Pore Filters, Final Report, IAAE Contract RC 411, Vienna, 23–109.
48. Spurny, K. and Polydorova, M. (1961) *Coll. Czech. Chem. Commun.,* 26, 921–931.
49. Havlicek, V., Polydorova, M., and Spurny, K. (1961) *Coll. Czech. Chem. Commun.,* 26, 932–936.
50. Romo-Kröger, C. M. and Morales, J. R. (1988) *Phys. Scripta,* 37, 270–273.
51. Romo-Kröger, C. M. (1989) *J. Air Pollut. Control Assoc.,* 39, 1465–1466.
51a. Romo-Kröger, C. M. and Diaz, V. (1996) *J. Aerosol Sci.,* 27, 751–757.
52. Zebel, G. (1974) *J. Aerosol Sci. Technol.,* 5, 473–482.
53. Spurny, K. R. (1972) Aerosol filtration by means of analytical pore filers, in T. T. Mercer, P. E. Morrow, and W. Stöber, Eds., *Assessment of Airborne Particles,* C. C. Thomas, Springfield, IL, 54–80.
54. Pich, J. (1970) *J. Aerosol Sci.,* 1, 17–24.
55. Machacova, J., Hrbek, Y., Hampl, V., and Spurny, K. (1970) *Coll. Czech. Chem. Commun.,* 35, 2087–2099.
56. Spurny, K. R. and Lodge J. P. (1972) Collection Efficiency Tables for Membrane Filters Used in the Sampling and Analysis of Aerosols and Hydrosols, NCAR, Technical Note, Boulder, CO.
57. Caroff, M., Choudhary, K. R., and Gentry, J. W. (1973) *J. Aerosol Sci.,* 4, 93–102.
58. Spurny, K. R. and Hampl, V. (1965) *Coll. Czech. Chem. Commun.,* 30, 507–514.
59. Spurny, K. R. and Hampl, V. (1967) *Coll. Czech. Chem. Commun.,* 32, 4190–4196.

60. Spurny, K. R. and Lodge, J. P. (1968) *Atmos. Environ.,* 2, 429–440.
61. Liu, B. Y. H. and Lee, K. W. (1976) *Environ. Sci. Technol.,* 10, 345–350.
62. Engelbrecht, D. R., Cahill, T. A., and Feeney, P. J. (1980) *J. Air Pollut. Control Assoc.,* 30, 391–398.
63. Feeney, P. J., Cahill, T. A., Olivera, J., and Guidara, R. (1984) *J. Air Pollut. Control Assoc.,* 31, 376–381.
64. Fan, K. C., Leaseburge, C., Hyun, Y., and Gentry, J. W. (1978) *Atm. Environ.,* 12, 1797–1804.
65. Gentry, J. W., Spurny, K. R., Boose, C., and Schörman, J. (1985) *J. Aerosol Sci.,* 16, 379–383.
66. Spurny, K. R. and Madelaine, G. (1971) *Coll. Czech. Chem. Commun.,* 36, 2857–2862.
67. Spurny, K. R., Havlova, J., Lodge, J. P., Ackerman, E. R., Sheesley, D. C., and Wilder, B. (1974) *Environ. Sci. Technol.,* 8, 758–761.
68. Spurny, K. R., Havlova, J., Lodge, J. P., Ackerman, E. R., Sheesley, D. C., and Wilder, B. (1975) *Staub Reinhalt. Luft,* 35, 77–122.
69. Fan, K. C. and Gentry, J. W. (1978) *Environ. Sci. Technol.,* 12, 1289–1294.
70. Heidem, N. Z. (1981) *Atmos. Environ.,* 15, 891–904.
71. Matteson, M. J. (1987) Analytical applications of filtration, in M. J. Matteson and C. Orr, Eds., *Filtration,* Marcel Dekker, New York, 607–672.
72. Fang, C. P. and Cohen, S. (1996) Characteristics of particle diffusion deposition of microporous filters, in *15th National AAAR Aerosol Conference, Abstracts,* Orlando, FL, 129.

Part II

Aerosol Separation

CHAPTER 23

Inertial Impaction

Jozef S. Pastuszka

CONTENTS

23.1 INTRODUCTION

Most aerosol measurements are performed by actively drawing aerosols into a sensor or onto a sampling surface by means of a pump or other air mover. In aerosol sampling the particle trajectory must be separated from the air streamline. To achieve this, different physical forces are used, as illustrated in Figure 23.1. From the technical point of view, the sampling process is carried out mainly by partitioning off the aspirating aerosol flow by different kinds of collection substrates (when the inertial force is mainly used) or by applying external forces. If this partition medium is a suitable porous substrate, i.e., a filter, such a process of collection is called filtration (Figure 23.1a). Liquid impingment, Figure 23.1b, is a method in which particles are removed by impacting the aerosol particles into a liquid. Using a solid plate as a partition medium in a sampling device makes it possible to construct a so-called impactor, in which aerosol particles with sufficiently high inertia in a deflected airstream are impacted onto a collecting surface (Figure 23.1c).

Figure 23.1d illustrates a virtual impactor, which is also based on the inertial behavior of particles. Size separation of particles occurs when the small, low-inertia particles flow with the airstreams and the large, high-inertia particles cross the "virtual" impaction surface and are collected below this interface (Nevalainen et al., 1992). In a cyclone (Figure 23.1e) the direction of airflow is changed in a relatively long distance. In this case the particle separation is caused by a centrifugal force. It also uses the inertial behavior of the particle, but in a radial geometry. Particles may also be removed from the airstreams by externally applied forces such as electrical force or a thermal gradient perpendicular to the flow (Figure 23.1).

0-87371-830-5/98/$0.00+$.50

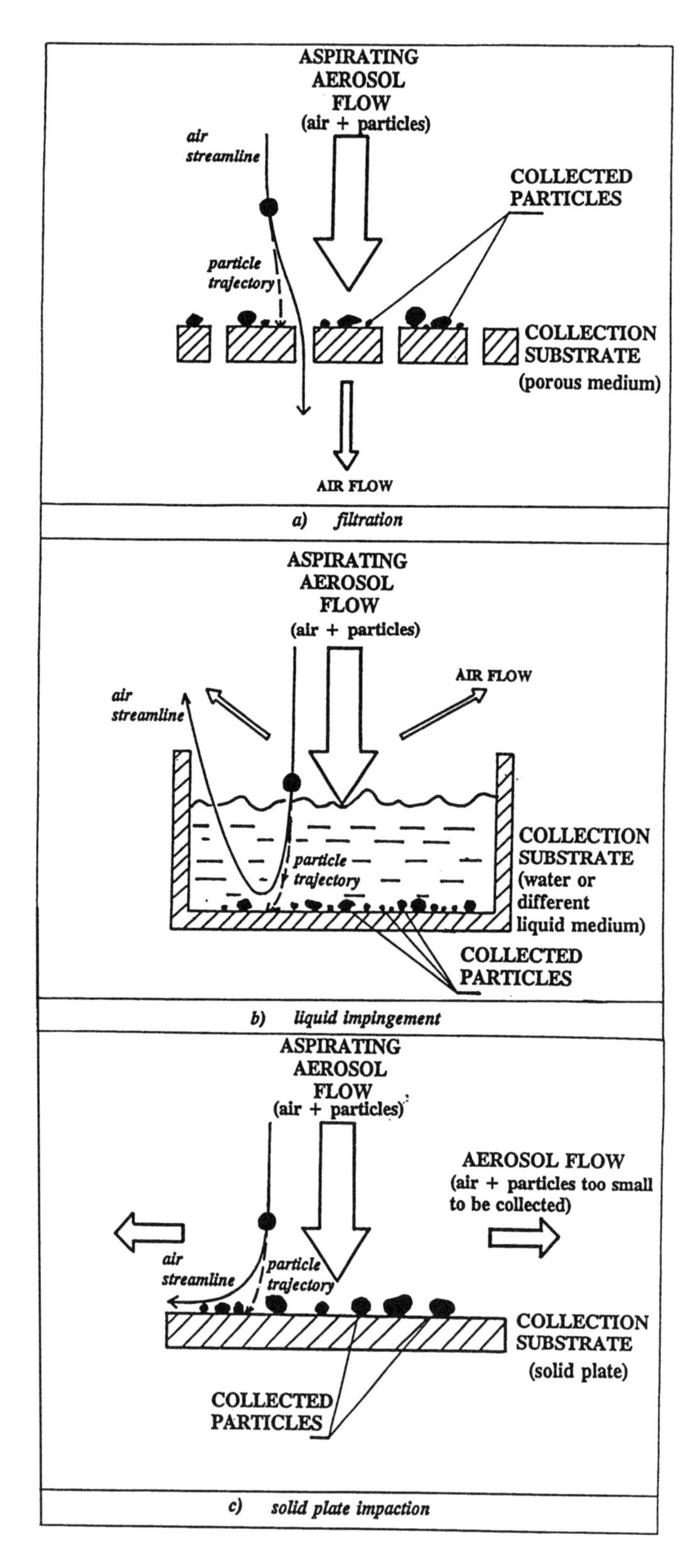

Figure 23.1 Methods of aerosol particle collection. (a) Filtration; (b) liquid impingement; (c) solid plate impaction; (d) virtual impaction; (e) centrifugal impaction; (f) external force.

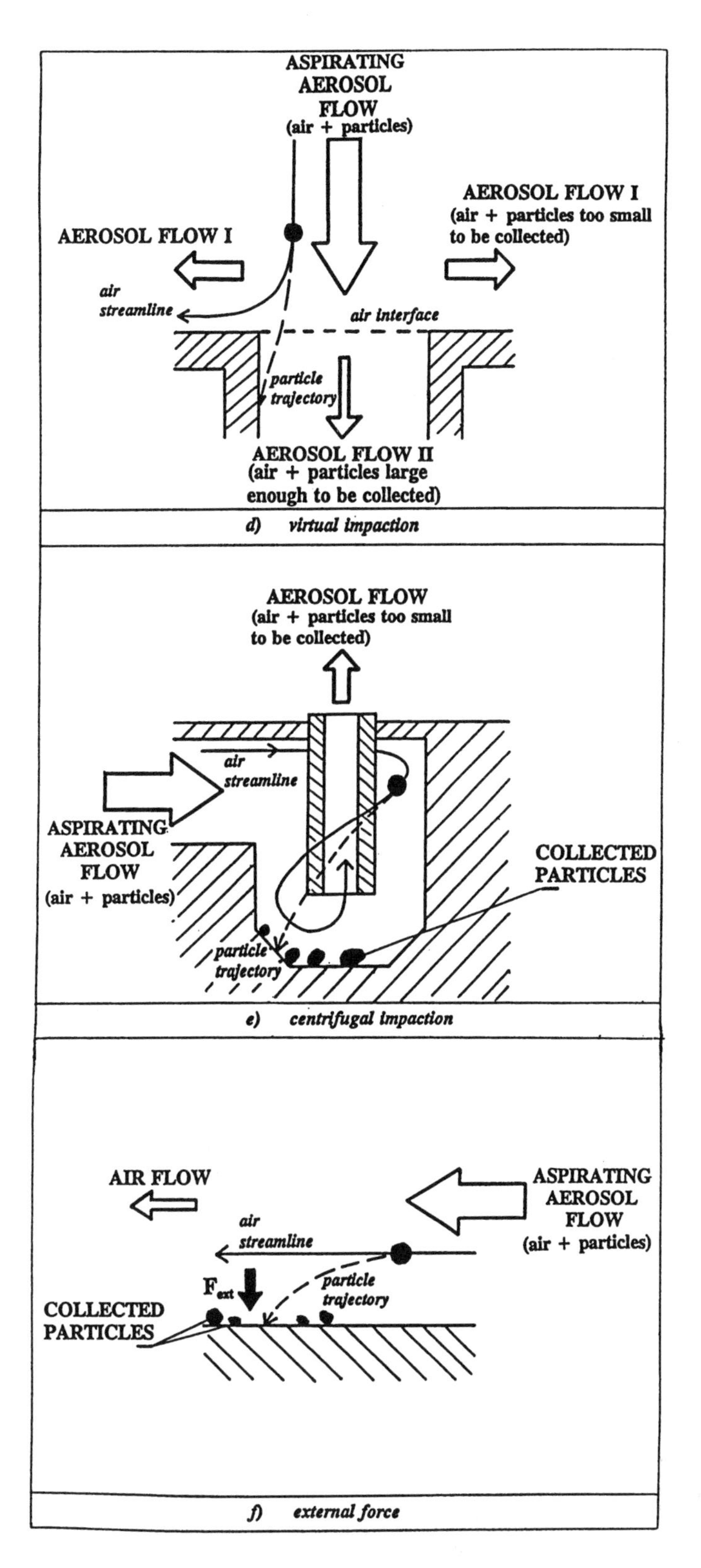

Figure 23.1 (continued)

Impactors and cyclones have become the most popular samplers for collection of airborne particles during the last decade. They are used as single, independent devices or in a combination with the filtration method for separating coarse fractions (for example, particles larger than 10 μm). For this reason, impaction will be discussed in more detail.

23.2 FORMULATION OF THE PROBLEM OF INERTIAL DEPOSITION

The basic mechanism for inertial deposition of particles on surfaces is a specific motion of aerosol particles in the vicinity of a stagnation point (impaction zone). In this region the directions of the streamlines change abruptly, and, because of inertia, a small, rigid particle traveling with the flow leaves its initial streamline and continues its movement on a new one.

Deceleration and collection of a particle in the impaction zone is determined by its stopping distance, S, which is defined as the distance the particle continues to travel in its original direction (until it stops if there are no external forces). Although in the impactor, near the collection plate, the airflow tries to move the inertial traveling particles according to the air streamlines, nevertheless, the concept of the stopping distance is very useful in the description of the inertial deposition process.

The stopping distance may be written as follows:

$$S = \int_0^t V(t)dt \tag{23.1}$$

where $V(t)$ is the particle velocity, changing with time from the initial velocity $V_0 = V(t = 0)$, usually equal to the incoming flow velocity, U.

It is known that, in the absence of external forces, the velocity of a particle traveling in still air with an initial velocity V_0 decreases according the equation (Hinds, 1982):

$$V(t) = V_0\ e^{-t/\tau} \tag{23.2}$$

where τ is the relaxation time and relates to the time required by a particle to adjust or "relax" itself to a new set of forces or conditions (Hinds, 1982).

Neglecting the gravitational force in an impaction process and using Equation 23.2 leads to

$$S = \int_0^t V_0 e^{-t/\tau} dt = V_0\left[-\tau e^{-t/\tau}\right]_0^t \tag{23.3}$$

which finally integrates to

$$S = V_0 \tau \tag{34.4}$$

because $\exp(-t/\tau)$ is equal to 1 and almost zero for $t = 0$ and for $t >> \tau$, respectively. The simple Equation 23.4 is sometimes used as a definition of stopping distance (for example, Baron and Willeke, 1993).

It should be noted that the relaxation time, τ, is the time which a particle needs to pass through the stopping distance if this travel could be continued with the constant velocity, V_0. In reality, however, a particle is traveling with decreasing velocity. Therefore, although τ is very useful for calculating S, the relaxation time is not the time that a particle will take to travel its entire stopping distance. Such time must be longer. Nevertheless, it can be shown (Hinds, 1982) that the particle travels 95% of that distance (S) very rapidly, in a time equal to 3τ, and during the time τ the particle can travel more than 60% of S.

To understand the relationship between the relaxation time and other particle parameters it will be necessary to make a little digression on the forces acting on the particle during its motion in gas. At first, it should be noted that the flow pattern in the pipe. Whether it is smooth or turbulent is governed by the ratio of the inertial force of the gas to the friction force of the gas moving over the surface. This ratio has come to be called the Reynolds number (Re) and can be expressed by the following, nondimensional quantity (Ower, 1949):

$$\mathrm{Re} = \frac{\rho_g U W}{\eta} \tag{23.5}$$

where U is the mean speed of flow, η is the coefficient of gas viscosity, ρ_g is the mass density of the gas, and W is the pipe diameter. For rectangular shape of the pipe, W should be doubled ($2W$ instead of W). It was found that laminar flow occurs when the flow Reynolds number is less than about 2000, while turbulent flow occurs for Re above 4000. In the intermediate range, the gas flow is sensitive to the previous history of the gas motion (Baron and Willeke, 1993). This is important in that if the gas velocity is increased into this intermediate range slowly, the flow may remain laminar. To further simplify the analysis, only occurrence of laminar flow during the inertial deposition is assumed.

It should be mentioned that apart from the flow Reynolds number discussed above, there is also the Reynolds number for particles which describes the gas flow around the particle. In such case, to calculate Re for a moving particle, in Equation 23.5 instead of W and U, the particle diameter, d, and the relative velocity between the particle and gas flow must be used, respectively. Since the difference between these velocities is generally small, and the particle dimension is also very small, the particle Reynolds number usually has a very small numerical value. In the next considerations of the inertial impaction only the flow Reynolds number is used.

Let us now consider the action of a drag force on a particle moving in the air. This force is net result of the distribution and magnitude of the local static pressure and shearing forces over its surface. In principle, these forces may be obtained from solutions of the Navier–Stokes equations of motion for the air flowing over the particle with relevant boundary conditions. More than 140 years ago Stokes found that for the spherical particle the drag on this particle depends only on gas viscosity, η, particle velocity, V, and particle diameter, d_p. This led to the following expression of the Stokes law:

$$F_{\mathrm{drag}} = \frac{3\pi\eta V d_p}{C_c} \tag{23.6}$$

C_c is the Cunningham correction factor. This factor is important for particles with diameter below 1 μm. Such particles are sufficiently small compared with the distance between molecules in the air so that the air no longer behaves like a continuous medium, but the particles "slip" between molecules, and the action of drag force on the particles is weaker than this equation without C_c would predict. Also for particles of other shapes the drag force may be different from the force calculated by use of Equation 23.6. It is important to note that only for the small velocity is the drag force proportional to the particle speed. The upper limit of precise applicability is set by the point at which the particle begins to generate a "wake" behind it, thus bringing into play additional aerodynamic forces (Lodge et al., 1981). Substituting Equation 23.2 in Equation 23.6, we have

$$F_{\mathrm{drag}} = \frac{3\pi\eta}{C_c} V_0 e^{-t/\tau} d_p \tag{23.7}$$

On the other hand, according to Newton's second law of the dynamics, it can be written

$$F_{\text{drag}} = -m\frac{dV}{dt} \tag{23.8}$$

In above equation a "minus" is used because the particle due to the drag force is decelerated. This deceleration can be calculated using Equation 23.2

$$\frac{dV}{dt} = V_0 \frac{d}{dt}\left(e^{-t/\tau}\right) = -\frac{V_0}{\tau} e^{-t/\tau} \tag{23.9}$$

The mass for a spherical particle can be easily obtained as follows:

$$m = \frac{1}{6}\rho_p \pi d_p^3 \tag{23.10}$$

Combining Equations 23.8, 23.9, and 23.10, we have

$$F_{\text{drag}} = \frac{\rho_p \pi}{6\tau} V_0 d_p^3 e^{-t/\tau} \tag{23.11}$$

so that from Equations 23.7 and 23.11

$$\tau = \frac{\rho_p d_p^2 C_c}{18\eta} \tag{23.12}$$

Hence, the stopping distance can be expressed as follows:

$$S = \frac{V_0 \rho_p d_p^2 C_c}{18\eta} \tag{23.13}$$

Values of S for an initial velocity of 5 m/s (typical for many impactors) are given in Table 23.1. From this table note that for very small particles, with diameter ranging from 0.01 to 0.1 μm, the stopping distance is only three to four times longer than these diameters, but then rapidly increases. For 1 μm particles, S is 18 times longer than d_p, while for 10 μm particles their stopping distance exceeds the particle diameter 100 times. Big particles, with radii of about 100 μm, have the stopping distance 1000 times longer than their size.

23.3 ANALYSIS OF PARTICLE MOTION IN THE IMPACTION ZONE

During the last decade a number of works dealing with deposition in stagnation flows have appeared. Most of these works considered diffusion or combined inertial–diffusional deposition (de la Mora and Rosner, 1981; Stratmann et al., 1988; Peters et al., 1989). In all these works it was assumed that a small stopping distance causes only small deviations of the particle trajectory from the fluid streamline. However, when a particle moves close to the stagnation streamline, there is a region in which the direction of the streamline changes abruptly. In this region, even a very small particle crosses many streamlines. Fichman with co-workers (1990) found recently, in a very

Table 23.1 Particle Stopping Distance for Unit Density Particles under Standard Conditions[a]

Particle Diameter d_p [μm]	Slip Correction Factor C_c[b]	Stopping Distance S [μm] for V_0 = 5 m/s
0.01	22.22	3.4×10^{-2}
0.1	2.867	0.44
1	1.164	18
10	1.016	1560
100	1.002	154×10^3 (154 mm)

[a] Physical constants used in preparation of this table are $\rho_p = 1$ g/cm³, $\eta = 1.810 \times 10^{-4}$ P $= 1.810 \times 10^{-4}$ g/cm s.

[b] From *Tables for Use in Aerosol Physics. Unit Density Spheres,* Edited by BGI, Inc., Waldham, MA, 1980.

general consideration, that the computation of the particle trajectories depends on the ratio of these stopping distances S to the radii of curvature of the fluid streamline R. They stated that for large and medium S/R ratios the trajectories exhibit boundary layer–type behavior, while for small ratios no such behavior is seen. Their result indicates that the chance to collect a particle on the impaction plate is high if the particle stopping distance is significantly higher than the radius of curvature of the air streamline (Figure 23.2). Unfortunately, it is very difficult to measure the radius R. It is

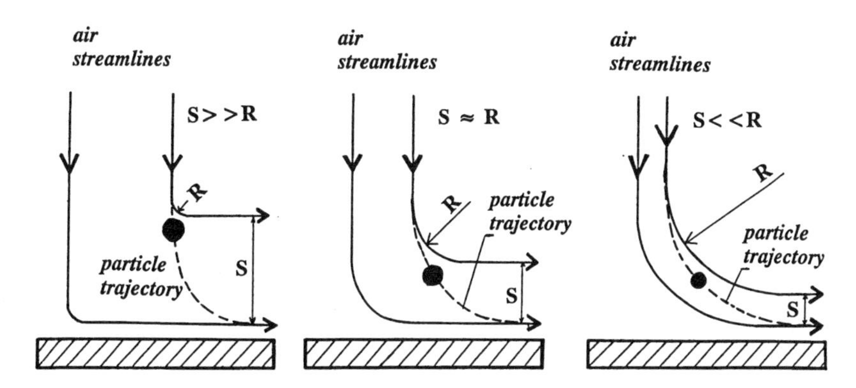

Figure 23.2 Streamlines and particle trajectories for various S/R ratios. (Adapted from Fichman, M. et al., *Aerosol Sci. Technol.,* 23, 281–296, 1990. With permission from Elsevier Science.)

much easier to estimate the impact distance, marked in Figure 23.3 as I, which is a function of R, and defines the region, where the free streamline has deflected from a straight line by 5 to 10% relative to the half-width of the throat (Marple et al., 1974a; Willeke and McFeters, 1975). Distance I results from considerations of the boundary conditions and fluid inertia terms in the Navier–Stokes equations. In a real impactor device, when the nozzle-to-plate distance, Y is sufficiently large (see Figure 23.3), it is found that (Willeke and McFeters, 1975).

$$\begin{aligned} I &= W \quad \text{for rectangular jet} \\ I &= W/2 \quad \text{for round jet} \end{aligned} \tag{23.14}$$

As it can be seen from the analysis of the Figure 23.3, the most important question is, how long the stopping distance must be for a given particle, with the diameter d_p and density ρ_p, to collect this particle by impaction (V_0 is equal to the jet velocity U)? Of course, the particle will be collected if $S > I$, but the stopping distance may be also shorter because in the region where the particle trajectory is located very close to the impaction plate other mechanisms, i.e., diffusion, gravitation,

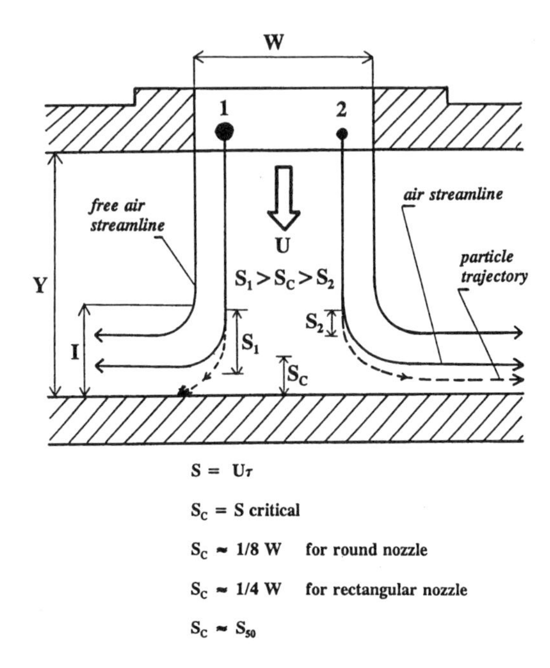

Figure 23.3 Mechanism of inertial impaction. (Adapted from Nevalainen, A. et al., *Atmos. Environ.*, 26A, 531–540, 1992. With permission from Elsevier Science.)

and electrostatics, are also effective and contribute to the deposition of a particle onto the collection plate. That means that the value of the critical stopping distance, S_c, must be known as a function of some geometric parameters of impactor, depending also on the starting position of the particle at the impactor entrance (if the particle starting trajectory is the same as such air streamline which next, in the impaction zone, will run very close to impaction plate, this particle can be collected even if its stopping distance is relatively short). Fortunately, according to the Willeke and McFeters's analysis (1975) of a number of theoretical and experimental works, mostly done by Marple and co-workers (Marple and Liu, 1974; Marple et al., 1974a and b), it can be concluded that, in practice, the starting position may be neglected and S_c is about $^1/_4 I$. Then, using Equation 23.14, the critical stopping distance is successively determined according to

$$\begin{aligned} S_c &\approx W/4 \quad \text{for rectangular nozzle} \\ S_c &\approx W/8 \quad \text{for round nozzle} \end{aligned} \tag{23.15}$$

By using Equation 23.13, the particle diameter, d_p, can be expressed in terms of S and V_0 as

$$d_p = \sqrt{S \frac{18\eta}{V_0 \rho_p C_c}} \tag{23.16}$$

If we include in Equation 23.16 a critical value of stopping distance, S_c, we will obtain the critical diameter, i.e., the separation point of the impactor. The ideal impactor should collect all particles larger than this separation size and none of the smaller one, which would correspond to the sharp ideal cutoff collection efficiency characteristics illustrated in Figure 23.4. The collection efficiency curve for an impactor (or impactor stage) gives the fraction of particles of a given size collected from the incident stream as a function of particle size. The efficiency curve of a typical, real impactor stage is also shown in Figure 23.4. As it can be seen, for this real efficiency curve, d_{50} is the value of the aerodynamic diameter for which 50% of the particle (with diameter equal to d_{50})

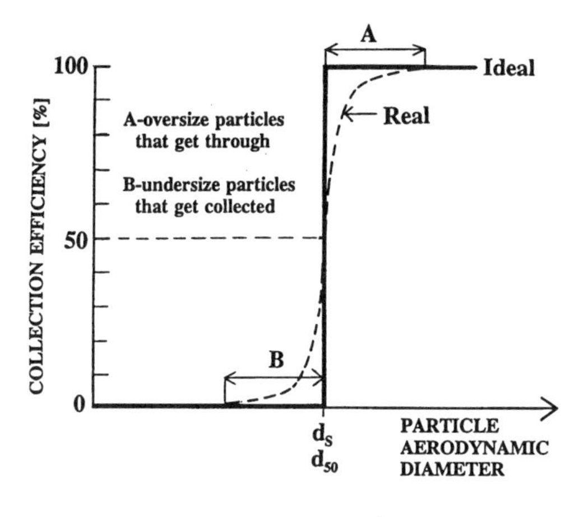

Figure 23.4 Particle collection efficiency curves for ideal and real impactors.

are collected and 50% pass through the impactor. However, it has long been recognized that the collection efficiency of almost all particles larger than d_p is about 100%. On the other hand, Figure 23.4 clearly shows that the range of undersize particles that get collected (with significant efficiency) is also very narrow. Thus, it can be reasonably assumed that the separation diameter is equal to d_{50} and, consequently, that $S_c = S_{50}$ (where S_{50} is the value of the stopping distance corresponding to d_{50}, the diameter of particles collected with 50% efficiency).

Since d_{50} is widely used as the effective impactor cutoff diameter, it would be useful to rewrite Equation 23.16 using relationships 23.15 and knowing that V_0 for a rectangular nozzle can be calculated as follows

$$V_0^{(\mathrm{rect})} = \frac{Q}{nWL} \tag{23.17a}$$

where Q is the flow rate, n is the number of holes, and L is the length of a nozzle, while for the round nozzle

$$V_0^{(\mathrm{round})} = \frac{4Q}{n\pi W^2} \tag{23.17b}$$

Therefore, substituting Equations 23.15 and 23.17 into Equations 23.16, we may write

$$d_{50}^{(\mathrm{rect})} = 3W\sqrt{\frac{\eta nL}{2\rho_p C_c Q}} \tag{23.18a}$$

and

$$d_{50}^{(\mathrm{round})} = \frac{3}{4}\sqrt{\frac{\pi\eta nW^3}{\rho_p C_c Q}} \tag{23.18b}$$

for rectangular and round nozzles, respectively.

As already stated (see Equation 23.15) the critical stopping distance is shorter for the round nozzle than for the rectangular one. Hence, it can be concluded that also the effective cutoff diameter should be smaller for the impactor with a round hole than for the impactor having a rectangular nozzle. A quantitative comparison of these two types of impactor can be obtained by calculating the ratio of d_{50} for these different nozzles.

$$\frac{d_{50}^{(\mathrm{round})}}{d_{50}^{(\mathrm{rect})}} = \frac{1}{2}\sqrt{\pi \frac{W}{L}} \tag{23.19}$$

Equation 23.19 expresses *explicitly* the fact that the impactor cutoff size, for a given aerosol, can be designed not only by selection of suitable nozzle size and flow rate but also by choosing the nozzle shape.

Designing the impactor for assumed or needed value of d_{50}, we should remember that some of the parameters in Equations 23.18a and b are also related to each other in a different way. First, since the flow in an impactor must be laminar, the Reynolds number (Equation 23.5) should be at least below 4000, which gives the limitation of the throat size depending on the flow rate, and which, therefore, influences the d_{50}. It is interesting that the importance of the Reynolds number is related also, and even more to the sharpness of cut, than to the cut-size of the impactor. Theoretical analysis has shown, and experiments have verified, that the efficiency curve will be sharpest if the Reynolds number is kept between 500 and 3000 (Marple et al., 1993).

Also the ratio of the nozzle-to-plate distance Y (see Figure 23.3) to the nozzle diameter, W, cannot be quite free. The recommended Y/W value should be greater than 1.0 for round impactors and 1.5 for rectangular impactors, and preferably not larger than 10 (Marple et al., 1993).

As it was shown, for a particle to be collected on the plate the stopping distance has to be sufficiently large. Hence, to collect very small particles by impaction the diameter of the nozzle must be small or the impactor must operate at a low pressure (Herring et al., 1979). If the pressure is so low that the jet Mach number (which is the ratio of jet velocity to sound velocity) is much greater than 1, then the impactor is said to operate in the hypersonic regime. Because of the high speed, hypersonic impactors have the capacity of sizing particles in the nanometer range (de la Mora et al., 1990). Opposite to the classical, subsonic impactor, it has been shown that the cutoff diameter, d_{50}, of a hypersonic impactor depends strongly on the nozzle-to-plate distance, Y (de la Mora, 1990; Olawoyin et al., 1995).

In this chapter we discussed only the inertial impaction of spherical particles and such non-spherical particles with shapes similar enough to spheres to be well described by the aerodynamic diameter without any additional parameters. However, the situation becomes more complicated when we try to analyze the impaction of fibrous particles. Such a particle, as a rule, is defined as a cylindrical particle with a length L_F and diameter d_F. Both these parameters can be combined in the aspect ratio parameter $\beta = L_F/d_F$. The aerodynamic particle diameter d_{ae} again depends strongly on the parameters d_F, β, and the orientation angle (Spurny, 1995). In the impaction zone where flow conditions are rapidly changing, the fiber mechanics are governed by initial orientation and flow relaxation time besides the usual parameters observed for spherical particles (Gallily et al., 1986; Baron, 1993). For instance, fibers with large rotational inertia (especially long fibers) may not orient completely or may be only approximately defined by the stopping distance S. Experimental measurements of fiber deposition in impactors have been carried out by Burke and Esmen (1978) and by Prodi and co-workers (1982).

There is one final aspect of this analysis which needs to be mentioned. In classical description of impaction, it is common practice to relate the particle diameter not to stopping distance but to the square root of the Stokes number, Stk, which is the ratio of the particle stopping distance to the radius, or half-width, of the impactor throat.

$$\mathrm{Stk} = \frac{S}{\frac{1}{2}W} \tag{23.20}$$

In such notation, the discussed condition for collection of a particle which can be expressed as $S > S_{c1}$, where the critical value of stopping distance is equal to S_{50}, now is following

$$\mathrm{Stk} > \mathrm{Stk}_{50}$$

or

$$\sqrt{\mathrm{Stk}} > \sqrt{\mathrm{Stk}_{50}} \tag{23.21}$$

and instead of Equation 23.15 there are the following relationships:

$$\begin{aligned} &\sqrt{\mathrm{Stk}_{50}} = 0.707 \quad \text{for rectangular nozzle} \\ &\sqrt{\mathrm{Stk}_{50}} = 0.5 \quad \text{for round nozzle} \end{aligned} \tag{23.22}$$

23.4 SOME PRACTICAL APPLICATIONS

Impactors are very useful samplers. They are widely used in atmospheric physics, industrial hygiene, and in different fields to measure the concentration of aerosol modes and to collect airborne particles for subsequent analysis.

If the aerosol is passed through several impaction stages connected in series, with each stage having a higher airflow velocity than the preceeding one, smaller and smaller particles will be impacted at each succeeding stage. In this way it is possible to separate the airborne particulates into distinct size ranges. Such a device is known as a cascade impactor (Figure 23.5).

The data obtained from the impactor measurements can be used to plot the mass size distribution of collected particles. Such information is very important and can produce a good picture of the real size dependence of the airborne particle concentration (see, for example, Horvath et al., 1989; Bouchertall, 1989; McGovern et al., 1994; Pastuszka et al., 1993; Pastuszka and Okada, 1995). Although such information is very often good enough for direct analysis, however, we should remember that the distribution of collected mass fractions, $M_i\,(d_p)$ (where "i" is a number of impactor stage) may differ from the size distribution function of particles suspended in the air $F\,(d_p)$. This difference exists because, generally speaking, it is impossible to perform exact measurements and, on the other hand, a knowledge of the behavior of the measurement instrument can be used to correct the data (Cooper, 1993). In order to obtain the real particle size distribution from the mass fractions deposited on the stages of the impactor, it is necessary to solve an inversion problem represented by the vector equation:

$$M_i(d_p) = \int F(d_p)E_i(d_p)dd_p \tag{23.23}$$

Hence, the analysis of obtained data should be extended and the goal of the second part of this analysis is to retrieve the particle size distribution function $F\,(d_p)$ from knowledge of the efficiency matrix $E_i\,(d_p)$ and the collected mass fractions $M_i\,(d_p)$. Several methods have been adapted and developed for solving the inversion problem (Twomey, 1975; Cooper and Spielman, 1976;

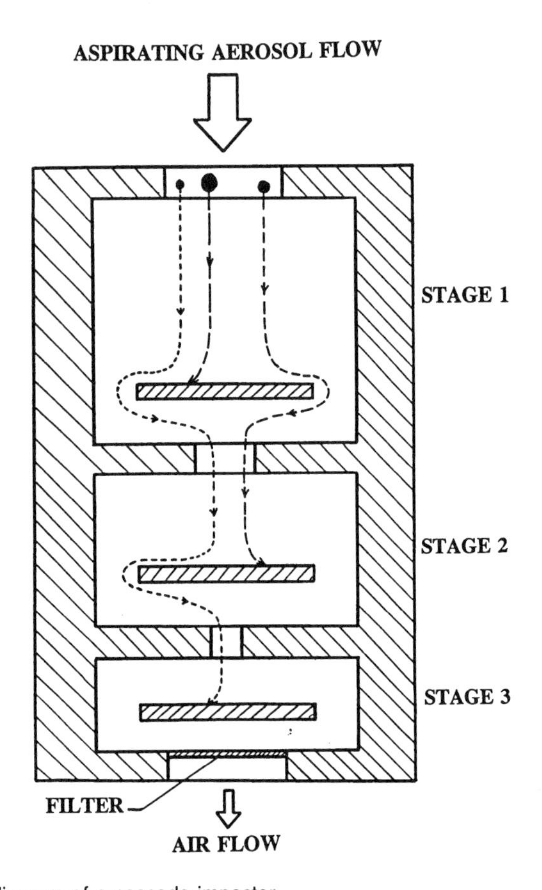

Figure 23.5 Schematic diagram of a cascade impactor.

Markowski, 1987; Wolfenbarger and Seinfeld, 1990; 1991; Diouri and Sanda, 1995). More details can be found in the review work by Cooper (1993).

Different problems of the size distribution data analysis are related to overlapping among the efficiency curves of individual stages. To eliminate the need for correction of the overlapping, it is recommended (Zhengyong et al., 1995) to give the measured cumulative efficiency for each stage or

$$E_i(d_p) = \frac{N_1(d_p) + N_2(d_p) + \cdots + N_i(d_p)}{N_1(d_p) + N_2(d_p) + \cdots + N_i(d_p) + \cdots + N_F(d_p)} \tag{23.24}$$

where $E_i\ (d_p)$ = the collection efficiency for a given particle aerodynamic diameter
$N_i\ (d_p)$ = the number or mass of particles collected at stage i
$N_F\ (d_p)$ = the number or mass of particles collected on the final filters

The common problem of the cascade impactors is the long sampling time, especially if we want to separate submicron particles into some modes. For this reason, some researchers have tried to construct new devices, still based on the cascade impactor but which could measure the size distribution automatically. Among a number of various models one of the most interesting seems to be the electrical low-pressure impactor made by Dekati Ltd., in Tampere, Finland (Keskinen et al., 1992; Tapper et al., 1995). This instrument covers a wide range of particle sizes from 30 nm up to 10 μm. The device consists of a particle charger and a conventional low-pressure impactor (see Figure 23.6). Aerosol is first sampled through a unipolar corona charger. The charged particles

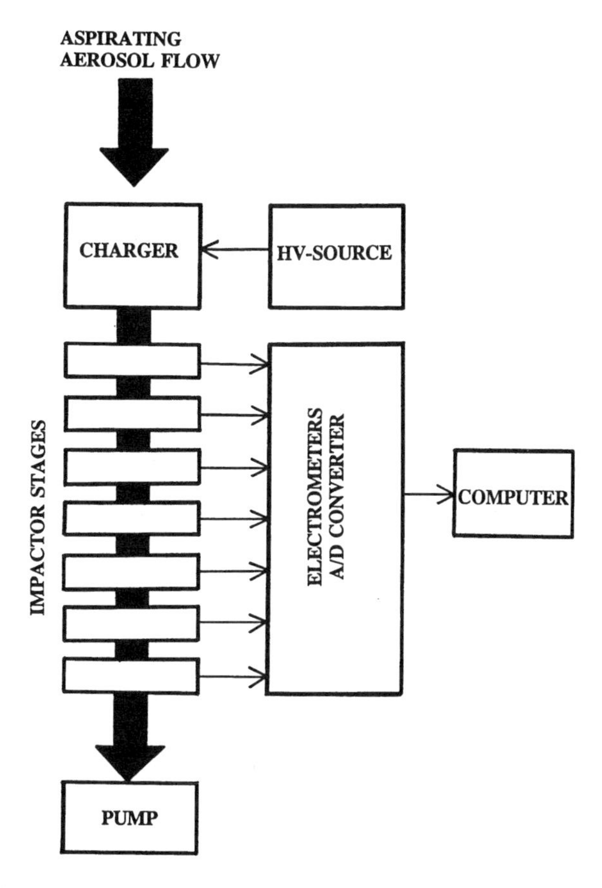

Figure 23.6 Electrical low-pressure impactor (Dekati model). Operating principle.

then pass into a low-pressure impactor with electrically isolated collection stages. The electric current carried by charged particles into each impactor stage is measured in real time by a sensitive multichannel electrometer. Measured current signals are converted to aerodynamic size distribution using particle size–dependent relations describing the properties of the charger and the impactor stages (Keskinen et al., 1992). The response time is less than 5 s.

Apart from cascade impactors, different inertial classifiers have also been developed. One of them is the inertial spectrometer shown in Figure 23.7. This instrument employs a filter instead of an impaction plate and introduces the particle through a nozzle, near one edge of the filter, with a volume of clean air on both sides. The position of the particles on the filter is a function of the aerodynamic diameter of the particles. Some spectrometers have the nozzle at the center of the circular filter (Prodi et al., 1988; Marple et al., 1993).

Inertial impaction is also the most widely used mechanism of particle removal in the currently used bioaerosol samplers. Readers are referred to Nevalainen and co-workers (1992; 1993) as well as to Juozaitis et al. (1994) for a discussion of bioaerosol collection characteristics and impactor design considerations.

23.5 SIZE DISTRIBUTION MEASUREMENTS AND CHEMICAL ANALYSIS

Because of the extreme diversity of chemical species that can become airborne, knowledge about the chemical composition of sized particles is needed in the chemistry of aerosols and in

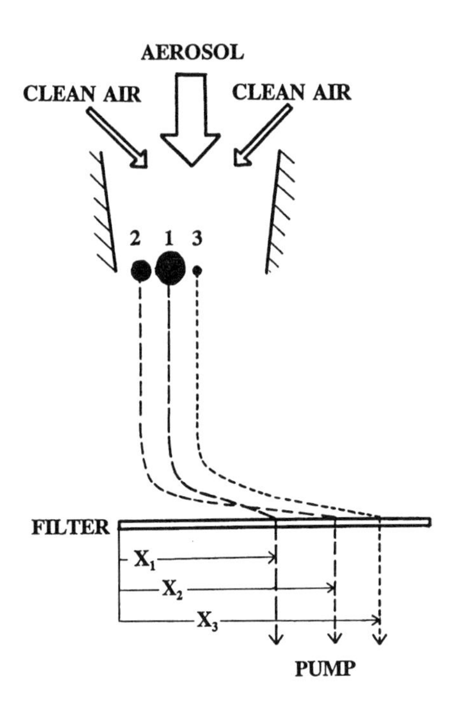

Figure 23.7 Schematic diagram of inertial spectrometer.

toxicology. Whether a particle, which can contain some toxic substances, is inhaled and retained in the lungs or not, primarily depends on the particle size.

Data on the size distributions of trace metals obtained using cascade impactors have been summarized by Milford and Davidson (1985) and Davidson and Osborn (1986). The investigation of current trace metal size distributions in the Los Angeles area was performed by Lyons et al. (1993). They observed that Pb and Mn distributions were multimodal with modes due to automotive emissions, particle growth, and suspended dust. Particle size distributions of Zn, Cu, and Ni showed significant amounts of these metals to be present in particles less that 1 μm in diameter, indicating anthropogenic sources, but suspended dust particles greater than 2 μm were also an important source of these metals.

One of the earliest investigations of size distribution of polycyclic aromatic hydrocarbons (PAHs) was carried out by Pierce and Katz (1975) at three urban and one rural sites in Canada using an Andersen impactor. Venkataraman et al. (1994) and Venkataraman and Friedlander (1994) measured size distributions of PAHs and elemental carbon (EC) in the Los Angeles air basin. They found the bimodal distributions of PAHs and EC with peaks in the 0.05 to 0.12 μm (mode I) and 0.5 to 0.1 μm (mode II) size ranges. Mode I was attributed to primary emissions from combustion sources, while mode II was attributed to the accumulation of secondary reaction products on primary aerosol particles.

Recently, increasing information has become available on the particle size distribution of emissions and, to some extent, the chemistry of emission as a function of particle size (e.g., Knapp and Bennett, 1990; Kauppinen and Pakkanen, 1990a and b).

From studies of the chemical composition of sized airborne particles outdoors and indoors, information about the infiltration of outdoor aerosol into the indoor environment can be obtained (e.g., Lewis and Zweidinger 1992; Górny et al., 1994). Such data also make it possible to estimate the contribution of indoor sources to total indoor pollution (Raiyani et al., 1993).

REFERENCES

Baron, P. A. (1993), Measurement of asbestos and other fibers, in *Aerosol Measurement,* Willeke, K. and Baron, P. A., Eds., Von Nostrand Reinhold, New York, 560–590.
Baron, P. A. and Willeke, K. (1993), Gas and particle motion, in *Aerosol Measurement,* Willeke, K. and Baron, P. A., Eds., Von Nostrand Reinhold, New York, 23–40.
Bouchertall, F. (1989), *Atmos. Environ.,* 23, 2241–2248.
Burke, W. A. and Esmen, N. A. (1978), *Am. Ind. Hyg. Assoc. J.,* 39, 400–405.
Cooper, D. W. (1993), Methods of size distribution data analysis and presentation, in *Aerosol Measurement,* Willeke, K. and Baron, P. A., Eds., Van Nostrand Reinhold, New York, 146–176.
Cooper, D. W. and Spielman, L. A. (1976), *Atmos. Environ.,* 10, 723–729.
Davidson, C. I. and Osborn, J. F. (1986), The sizes of airborne trace Metal containing particles in the Los Angeles area, in *Toxic Metals in the Atmosphere,* Nriagu, J. O. and Davidson, C. I., Eds., Wiley, New York, 355–390.
de la Mora, F. and Rosner, D. E. (1981), *Physiocochem. Hydrodyn.,* 2, 1–21.
de la Mora, F., Herring, S. V., Rao, N., and McMurry, P. H. (1990), *J. Aerosol Sci.,* 21, 169–187.
Diouri, M. and Sanda, I. S. (1995), *J. Aerosol Sci.,* 26, S763–S764.
Fichman, M., Pnueli, D., and Gutfinger, C. (1990), *Aerosol Sci. Technol.,* 13, 281–296.
Gallily, I., Schiby, D., Cohen, A. H., Holländer, W., Schless, D., and Stöber, W. (1986), *Aerosol Sci. Technol.,* 5, 267–286.
Górny, R. L., Pastuszka, J. S., and Jedrzejczak, A. K. (1994), Investigation of airborne particles and heavy metals concentration and deposition in Upper Silesia, Poland, indoors and outdoors, in *Toxicity, Hazard, Risk, Proc. 3rd Regional SECOTOX Meeting,* Balatonaliga, Hungary 1994, *Acta Biol. Debr. Oecol. Hung.,* 5, 69–77.
Herring, S. V., Friedlander, S. K., Collins, J. J., and Richards, L. W. (1979), *Environ. Sci. Technol.,* 3, 184–188.
Hinds, W. C. (1982), *Aerosol Technology,* John Wiley & Sons, New York.
Horvath, H., Habenreich, T. A., Kreiner, I., and Norek, C. (1989), *Sci. Total Environ.,* 83, 127–159.
Juozaitis, A., Willeke, K., Grinshpun, S. A., and Donelly, J. (1994), *Appl. Environ. Microbiol.,* 60, 861–870.
Kauppinen, E. I. and Pakkanen, T. U. (1990a), *Environ. Sci. Technol.,* 24, 1811–1818.
Kauppinen, E. I. and Pakkanen, T. U. (1990b), *Atmos. Environ.,* 24A, 423–429.
Keskinen, J., Pietarinen, K., and Lechtimäki, M. (1992), *J. Aerosol Sci.,* 23, 253–260.
Knapp, K. T. and Bennett, R. L. (1990), *Aerosol Sci. Technol.,* 12, 1067–1074.
Lewis, C. W. and Zweidisnger, R. B. (1992), *Atmos. Environ.,* 26A, 2179–2184.
Lodge, J. P., Jr., Waggoner, A. P., Klodt, D. T., and Crain, C. N. (1981), *Atmos. Environ,* 15, 431–482.
Lyons, J. M., Venkataraman, C., Main, H. H., and Friedlander, S. K. (1993), *Atmos. Environ.,* 27B, 237–249.
Markowski, G. R. (1987), *Aerosol Sci. Technol.,* 7, 127–141.
Marple, V. A., Liu, B. Y. H., and Whitby, K. T. (1974a), *J. Aerosol Sci.,* 5, 1–16.
Marple, V. A., Liu, B. Y. H., and Whitby, K. T. (1974b), *J. Fluids Eng.,* 96, 394–407.
Marple, V. A. and Liu, B. Y. H. (1974), *Environ. Sci Technol.,* 8, 648–654.
Marple, V. A., Rubow, K. L., and Olson, B. A. (1993), Inertial, gravitional, centrifugal, and thermal collection techniques, in *Aerosol Measurement,* Willeke, K. and Baron, P. A., Eds., Van Nostrand Reinhold, New York, 206–232.
McGovern, F. M., Krasenbrink, A., Jennings, S. G., Georgi, B., Spain, T. G., Below, M., and O'Connor, T. C. (1994), *Atmos. Environ.,* 28, 1311–1318.
Milfrod, J. B. and Davidson, C. I. (1985), *JAPCA,* 35, 1249–1260.
Nevalainen, A., Pastuszka, J., Liebhaber, F., and Willeke, K. (1992), *Atmos. Environ,* 26A, 531–540.
Nevalainen, A., Willeke, K., Liebhaber, F., Pastuszka, J., Burge, H., and Henningson, E. (1993), Bioaersol sampling, in *Aerosol Measurement,* Willeke, K. and Baron, P. A., Eds., Van Nostrand Reinhold, New York, 471–492.
Olawoyin, O. O., Raunemaa, T. M., and Hopke, P. K. (1995), *Aerosol Sci. Technol.,* 23, 121–130.
Ower, E. (1949), *The Measurement of Air Flow,* Chapman & Hall, London.
Pastuszka, J., Hlawiczka, S., and Willeke, K. (1993), *Atmos. Environ.,* 27B, 59–65.
Pastuszka, J. S. and Okada, K. (1995), *Sci. Total Environ.,* 175, 179–188.

Peters, M. H., Copper, D. W., and Miller, R. J. (1989), *J. Aerosol Sci.,* 20, 123–136.

Pierce, R. C. and Katz, M. (1975), *Environ. Sci, Technol.,* 9, 347–353.

Prodi, V., Zaiacomo, T. D., Hochrainer, D., and Spurny, K. (1982), *J. Aerosol Sci.,* 13, 49–58.

Prodi, V., Belosi, F., Mularoni, A., and Lucialli, P. (1988), *Am. Ind. Hyg. Assoc. J.,* 49, 75–80.

Raiyani, C. V., Shah, S. H., Desai, N. M., Venkaiah, K., Patel, J. S., Parikh, D. J., and Kashyap, S. K. (1993), *Atmos. Environ.,* 27A, 1643–1655.

Spurny, K. R. (1995), *J. Aerosol Sci.,* 26, S635–S636.

Stratmann, F., Fissan, H., and Peterson, T. W. (1988), *J. Environ. Sci.,* 31, 3l9–41.

Tapper, U., Mariamäki, M., Moisio, M., Kauppinen, E. I., and Keskinen, J. (1995), *J. Aerosol Sci.,* 26, S103–S104.

Twomey, S. (1975), *J. Comput. Phys.,* 18, 188–200.

Venkataraman, C., Lyons, J. M., and Friedlander, S. K. (1994), *Environ, Sci. Technol.,* 28, 555–562.

Venkataraman, C. and Friedlander, S. K. (1994), *Environ, Sci. Technol.,* 28, 563–572.

Willeke, K. and McFeters, J. J. (1975), *J. Colloid Interface Sci.,* 53, 121–127.

Wolfenbarger, J. K. and Seinfeld, J. H. (1990), *J. Aerosol Sci.,* 21, 227–247.

Wolfenbarger, J. K. and Seinfeld, J. H. (1991), *Aerosol Sci. Technol.,* 14, 348–357.

Zhengyong, L., Aiwu, K., and Cuiming, F. (1995), *Aerosol Sci. Technol.,* 23, 253–256.

Chapter 24

Development of Small-Cutpoint Virtual Impactors and Applications in Environmental Health

Constantinos Sioutas and Petros Koutrakis

CONTENTS

24.1 INTRODUCTION

Ambient aerosols are defined as suspensions of relatively stable solid or liquid particles in ambient air. Ambient particles range from 0.01 to 100 µm in diameter. The particle size range from 0.01 to 0.1 µm, known as the ultrafine mode (Hinds, 1982), contains the majority (in numbers) of the ambient particles. However, most of the ambient mass is distributed in two size modes: a coarse particle mode, centered around 10 to 20 µm diameter, and a fine or accumulation mode, centered around 0.2 to 0.8 µm diameter (Whitby et al., 1972). The bimodal distribution of atmospheric aerosols displays a "saddle" point in the size range 1 to 3 µm, which serves to distinguish the two modes: by convention, the coarse mode consists of particles larger than 2.5 µm in aerodynamic diameter, whereas the fine (or accumulation) mode consists of particles smaller than 2.5 µm in aerodynamic diameter.

The two modes have different chemical composition, sources, and lifetimes in the atmosphere. Particles in the coarse mode are produced by mechanical processes (grinding, erosion, and resuspension by the wind). They are relatively large, and settle out of the atmosphere within hours.

0-87371-830-5/98/$0.00+$.50

Particles in the accumulation mode are mostly anthropogenic in origin, they are generated through gas-to-particle conversion mechanisms including homogeneous and heterogeneous nucleation, and by condensation onto preexisting particles in the accumulation size mode. Because they are too small to settle out, particles of the accumulation mode have lifetimes in the atmosphere on the order of days (Hinds, 1982), and they can be transported over long distances. The major chemical constituents of fine particles are sulfate, nitrate, ammonium, organic and elemental carbon, as well as a variety of trace metals formed in combustion processes. Oxidation of the primary pollutants, sulfur dioxide and nitrogen oxides, emitted during the combustion of coal, oil, and other conventional fuels, results in the formation of sulfuric and nitric acids through homogeneous or heterogeneous atmospheric processes. Ammonium salts are formed through the neutralization of these acids by ambient ammonia. There is a variety of sources of trace metals found in fine ambient particles. They include coal, wood, and oil combustion, waste incineration, and metal mining and production. In addition to sulfate, nitrate, and metals, carbon-containing compounds also are found to be associated with inhalable particles. A variety of organic compounds including long-chain hydrocarbons, polycyclic aromatic hydrocarbons, and organics containing oxygen, nitrogen, or sulfur can be present at ng/m^3 to $\mu g/m^3$ levels. A host of viable species can also be present, such as fungi, bacteria, pollen, yeasts, and viruses. Moreover, high-molecular-weight hydrocarbons emitted from plants or petroleuum residues have been observed in urban environments.

Exposures to ambient particles, especially those of the fine mode, have recently received considerable attention as the result of findings from epidemiological studies, which showed associations between ambient particle concentrations and increased mortality rates. Lave and Seskin (1977) analyzed statistically the geographic differences in annual mortality and the corresponding pollution levels in 117 standard metropolitan areas (SMAs) from 1960. Their analysis demonstrated a significant positive association between mortality and total suspended particulates (TSP), as well as total suspended particulate sulfate. Evans et al. (1984) and Ozkaynak and Thurston (1987) revised and reanalyzed the data of Lave and Seskin, and confirmed the observed positive association. Ozkaynak and Thurston (1987) used 1980 U.S. vital statistics and available ambient pollution data for sulfate and fine, coarse, and total suspended particulate and effectively demonstrated the importance of including particle size, chemical composition, and source information in modeling health effects. Fine particles and sulfate concentrations were consistently and significantly associated with increased mortality rates. On the other hand, particle mass measurements that included coarse particles (inhalable, or total suspended mass) were often found to be nonsignificant predictors of total mortality. Recently, Schwartz and Dockery (1992) and Pope et al. (1992) analyzed associations between daily pollution levels and mortality in several cities in the U.S. Increased daily mortality with increased particle concentration was demonstrated. Kinney and Ozkaynak (1991) analyzed data on daily mortality in St. Louis, MO and Kingston, TN and found a strong association between mortality rates and exposures to fine particles. Similar associations, however, were not found for aerosol acidity, sulfur dioxide, ozone, or nitrogen dioxide concentrations. Ozkaynak (1993) reviewed the most recent epidemiological findings relating mortality to particle concentration.

Along with the effect of ambient particle concentration on mortality, field studies of adverse respiratory effects of particles conducted in a relatively high exposure community have shown significant associations between PM_{10} exposure and a number of morbidity outcomes, including hospital admissions for bronchitis and asthma, and longitudinal changes in peak flow rates, respiratory symptoms, and medication use (Pope et al., 1991; 1992). Thurston et al. (1992) observed an increase of about 2% in admissions for each 10 $\mu g/m^3$ increase in daily PM_{10} in metropolitan New York. Data from Boston and three other communities in Massachusetts showed a higher rate of pneumonia and influenza admissions (about 15%) for children under 15 years old associated with increases in daily PM_{10} levels by 10 $\mu g/m^3$ (Ozkaynak et al., 1990).

To better assess the impact of ambient particles on the environment and public health, fine particles need to be sampled separately from the surrounding gases. This is particularly important when chemical analysis of the particulate matter needs to be performed. Thus, the primary task of aerosol samplers is to separate the aerosol particles from the overwhelmingly larger mass of gases in the atmosphere.

24.2 PARTICLE/GAS SEPARATION METHODS

Three procedures are commonly used to separate particles from gases during sampling: filters, diffusion denuders, and inertial impactors. The simplest method for simultaneously sampling atmospheric acid particulate and gaseous pollutants employs two filters connected in series; the first filter retains the particulate phase and the second filter, which is coated with a suitable sorbent, retains the gas phase. Nevertheless, several studies in the literature reported sampling errors resulting in an underestimation or overestimation of the particulate phase. These errors, often known as "artifacts," are due to the following phenomena:

1. Gas-phase compounds, such as SO_2, HNO_3, and organic vapors, can be adsorbed on filter media, thus overestimating the particle concentration (Appel et al. 1983; 1984);
2. Volatilization of unstable ammonium particulate salts, such as NH_4NO_3 and NH_4Cl, forms NH_3 and NHO_3 or HCl (Koutrakis et al., 1992). The produced ammonia can diffuse to the collected acid sulfate particles and neutralize them. Typically, the produced acidic gases are collected by the coated filters of the filter pack that are placed downstream from the particle collection filter. Thus, the gas-phase concentrations contain the amounts that existed in the atmosphere during sampling plus the amounts that volatilized from the collected particles, resulting in an overestimation of the concentration of acidic gases and an underestimation of acidic particles;
3. The particle/gas phase partitioning of compounds distributed in both phases may shift during prolonged sampling, resulting in volatilization losses from the collected particles (Van Vaeck et al., 1984).

During the last decade, diffusion denuders have been used in a variety of atmospheric monitoring studies to collect gaseous atmospheric pollutants (Stevens et al., 1978; Forest et al., 1982; Possanzini et al., 1983; and Koutrakis et al., 1989). In these studies, glass or metallic hollow tubes are coated with an appropriate substance to collect the different gases selectively while allowing other gases and particles to penetrate. The particles are subsequently collected on filter media. By using this sampling technique, particles and gases are separated during sampling and sampling artifacts are minimized. Despite their wide use for inorganic gas collection (Stevens et al., 1978; Koutrakis et al., 1993), diffusion denuders have limitations in sampling organic compounds, since the nonpolar nature of some of these compounds (polycyclic aromatic hydrocarbons, polychlorinated biphenyls, and pesticides) complicates the choice of the appropriate coating. To avoid this problem, Krieger and Hites (1992) developed a denuder consisting of short sections of capillary gas chromatographic (GC) columns, thus making the GC stationary phase act as the denuder coating. However, this system has not been quantitatively characterized in terms of particle transmission efficiency. In addition, its breakthrough volume remains to be determined.

Conventional impactors can be used to separate atmospheric particles from gases. Particle separation can be efficiently performed for coarse particles (e.g., larger than 2.5 μm). However, since most of the mass in the accumulation mode is associated with particles in the size range 0.1 to 1.0 μm in diameter, impactors with very small cutpoints are required to separate the particulate from the gas phase efficiently. Low-pressure impactors have made sampling of particles as small as 50 nm possible (Hering et al., 1978; Berner et al., 1979). Nevertheless, the collected particles are exposed to low pressures (0.03 to 0.20 atm) and this can cause volatilization of the deposits into the airstream. In addition, aerodynamic cooling within the high-speed jet can cause vapor condensation and the particle size to increase (Biwas et al., 1987).

Another method to separate by impaction particles as small as 0.1 μm involves the use of micro-orifice impactors that reduce the impactor cutoff size by accelerating the aerosol to be sampled through small-diameter jet orifices (Marple et al., 1991). They operate at pressures closer to atmospheric (0.7 to 0.9 atm) and achieve cutpoints on the order of 0.1 μm by employing orifices 0.004 to 0.02 cm in diameter. Nevertheless, the flow rate in such impactors is limited to about 30 l/min, due to the very large number of micro-orifices required to achieve flow rates on the order of 200 to 300 l/min, often required to collect sufficient mass for analysis. Another problem that would need to be addressed is multiple jet interaction which could deteriorate the performance of the impactor (Fang et al., 1991), affecting both the 50% cutpoint and the internal losses.

To overcome all of the aforementioned problems in gas/particle separation, researchers at the Harvard School of Public Health developed virtual impactors operating at flow rates as high as 1000 l/min and with a small cutpoint (0.1 μm) to separate ambient fine particles from the surrounding gases (Sioutas et al., 1994a, b, and c).

24.3 SMALL-CUTPOINT VIRTUAL IMPACTORS

24.3.1 Design and Development

The physical principle of operation of a virtual impactor, shown in Figure 24.1, is similar to that of an inertial impactor; both methods use particulate inertia to separate particles from gases. A jet of particle-laden air is injected at a collection medium, which causes an abrupt deflection of the air streamlines. Particles larger than a certain size (the so-called cutpoint of the impactor) cross the air streamlines and, in the case of an inertial impactor, are collected on the medium, while particles smaller than a certain size follow the deflected streamlines. The main difference between an inertial and a virtual impactor is that in the latter, particles are injected at a collection probe rather than a collection medium. To separate larger particles continuously from the collection probe, a fraction of the total flow, referred to as the minor flow (typically 10 to 20% of the total flow), is allowed to pass through the probe, leaving particles larger than the cutpoint contained in a small fraction of the initial gases.

The smallest particle size (cutpoint) that can be collected by a virtual impactor is linked to the design and operating parameters of the impactor through the Stokes number (Hinds, 1982),

$$\mathrm{St} = \frac{\rho_p C_c d_p^2 U}{9\mu D_j} \tag{24.1}$$

where D_i is the acceleration nozzle size (e.g., the diameter or width of the acceleration nozzle, for round or rectangular geometry nozzles, respectively), U is the average velocity of the jet ρ_p is the particle density, μ is the dynamic viscosity of the air, and C_c is the Cunningham slip correction factor.

The slip correction factor is given by the following equation (Hinds, 1982):

$$C_c = 1 + \frac{2}{Pd_p}\left[6.32 + 2.01\ \exp\left(-0.1095 Pd_p\right)\right] \tag{24.2}$$

where P is the absolute pressure in cmHg and d_p is the particle diameter in micrometers. The Stokes number value corresponding to the 50% cutpoint in conventional impactors is in the range 0.2 to 0.3 (Marple and Liu, 1974). However, in the case of virtual impactors, the value of the 50% Stokes number is greatly affected by the minor-to-total flow ratio, as will be discussed in the next few paragraphs. Thus, the cutpoint prediction from the Stokes number equation becomes more qualitative than quantitative.

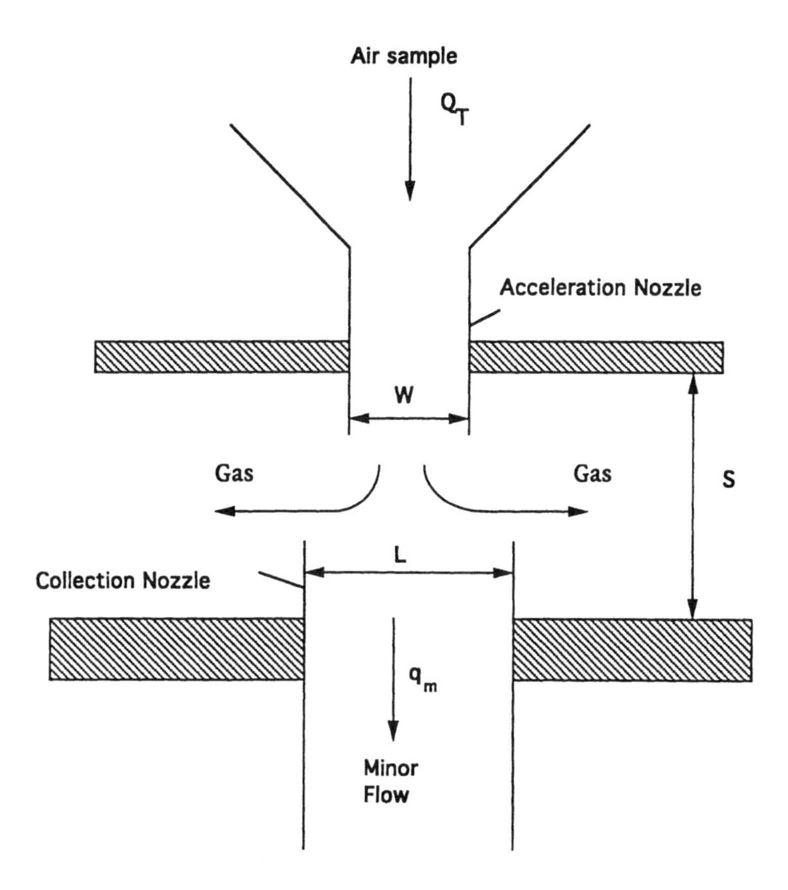

Figure 24.1 Schematic of a virtual impactor.

The technical challenges and limitations related to the development of small-cutpoint conventional impactors apply to small-cutpoint virtual impactors. To achieve a cutpoint of about 0.1 μm, and assuming that a Stokes number of about 0.3 corresponds to the 50% cutpoint, velocities higher than 150 m/s and nozzle sizes smaller than 0.04 cm are required, as predicted by the Stokes number equation. Nozzles in sizes smaller than 0.03 cm are difficult to manufacture by conventional means, and may result in excessive particle losses due to particle impaction onto the lateral nozzle walls (Fuchs, 1978). This sets an upper limit to the sampling flow of the aerosol entering the virtual impactor; even when the velocity of the aerosol equals the speed of sound (approximately 300 m/s at standard temperature and pressure), the total flow rate will be on the order of 1 l/min. In addition, operation of an impactor at sonic or compressible flow conditions may result in the aforementioned distortion of the physiochemical properties of the sampled aerosol (Biswas et al., 1987). Thus, developing virtual impactors operating at sampling flow rates in the range 100 to 1000 l/min would require the construction of a large number of acceleration and collection nozzles. Potential misalignments between the acceleration and collection nozzles, along with the aforementioned jet interactions, may increase substantially the internal particle losses, which are a typical shortcoming of even single-nozzle virtual impactors (Loo and Conk, 1988).

By designing virtual impactors with slit-shaped acceleration and collection nozzles, shown in Figure 24.2, sampling flow rates as high as 1100 l/min were achieved (Sioutas et al. 1994c; 1995a and b). The width of the acceleration nozzle was chosen to satisfy the Stokes equation, whereas the length was chosen to match the desired flow rate. Some of the most crucial design and operating parameters of those impactors are summarized in Table 24.1.

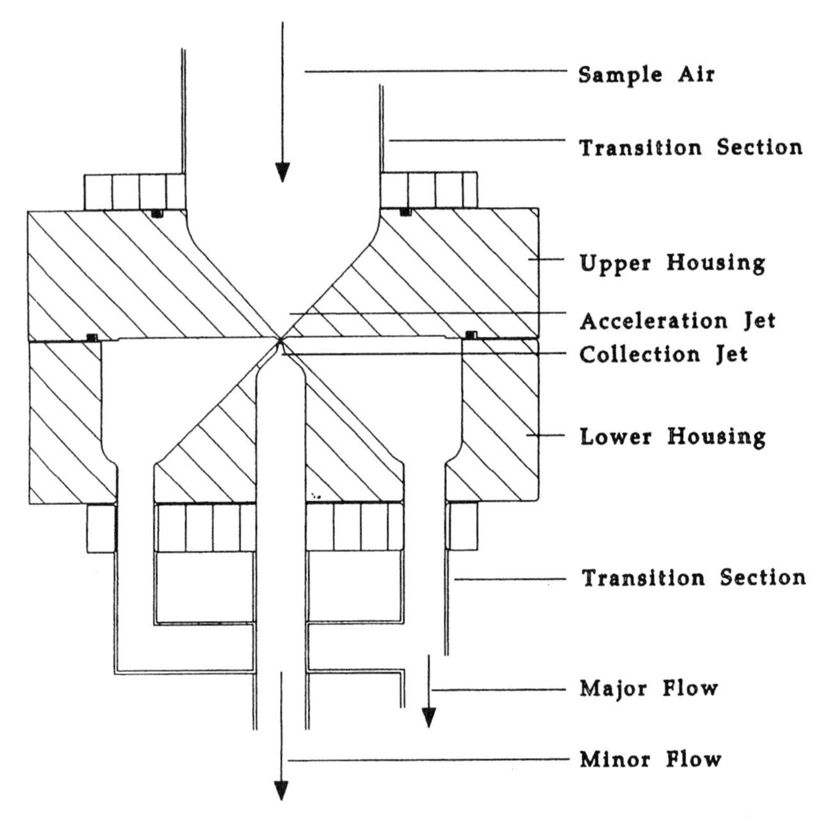

Figure 24.2 Schematic diagram of a slit-nozzle virtual impactor. (From Sioutas, C. et al., *Environ. Health Perspect.,* 103, 172–177; 1995. Reproduced with permission, Copyright © Environmental Health Perspectives.)

Figure 24.3 shows a typical particle collection efficiency and loss curve of a high-volume virtual impactor. Collection efficiency is defined as the fraction of particles of a certain size that were drawn into the minor flow of the virtual impactor from the total of particles collected in both minor and major flows. Thus, this definition does not account for particles lost in the impactor, but only considers particles that have penetrated either one of the major or minor flows. Particle losses are defined as the fraction of particles deposited in internal areas of the virtual impactor (e.g., areas other than the minor and major flows). They can be calculated by dividing the number of particles that have deposited in internal areas to the total number of particles sampled by the virtual impactor throughout the sampling period.

The specific curve corresponds to the virtual impactor operating at 225 l/min with a minor-to-total flow ratio of 0.2 (Sioutas et al., 1994c). The 50% cutpoint of the impactor determined in the experiments is 0.12 μm, and the particle losses range from 5 to 18%. Particle losses tend to increase slightly with particle size, probably due to inertial deposition of large particles on the top of the narrow collection slit. The collection efficiency curves are not as steep as those typically observed in conventional impactors (Wang and John, 1988; Marple et al., 1991). This is mainly due to the inherent differences in the collection mechanisms between virtual and conventional impactors. Particles impacting onto the tip of the collection nozzle of a virtual impactor are considered part of the losses and are not counted as part of the collection efficiency, since they have not penetrated the collection nozzle. These particles, however, would have impacted onto the plate in a conventional impactor, and thereby would have been effectively collected. The fact that the slip correction factor of particles smaller than 1.0 μm is not a constant, and increases significantly with decreasing particle

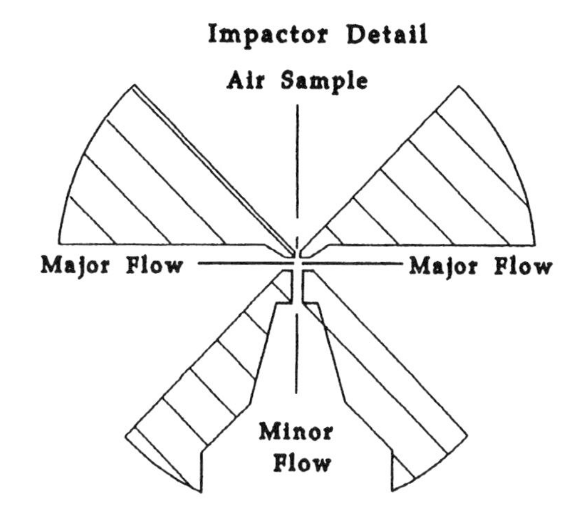

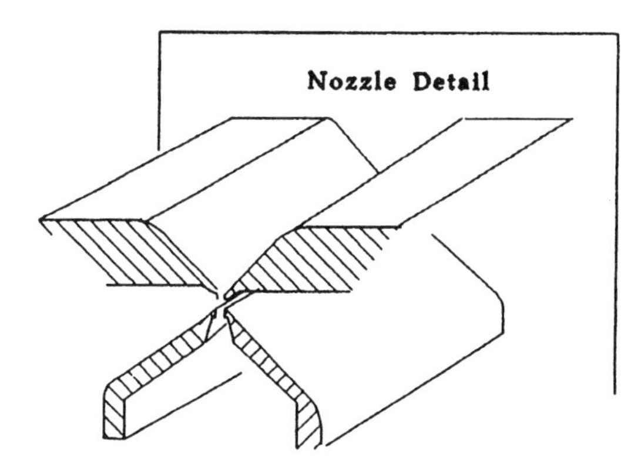

Figure 24.2 (continued)

Table 24.1 Design and Operating Parameters of High-Volume, Small-Cutpoint Virtual Impactors

Sampling Flow Rate (l/min)	Acceleration Nozzle Width (cm)	Collection Nozzle Width (cm)	Length of Nozzles[a] (cm)	Pressure Drop[b] (kPa)
15	0.023	0.036	1.5	8.5
24	0.023	0.036	1.5	18.1
32	0.033	0.036	1.5	12.2
225	0.033	0.050	5.6	28.2
1130	0.033	0.050	28.0	28.2

[a] Both acceleration and collection nozzles have equal lengths.
[b] Pressure drop across the acceleration jet of the impactor.

size (Equation 24.2), further decreases the steepness of the collection curve. Figure 24.4 shows a comparison between the experimentally determined collection efficiency and losses of the high-volume virtual impactor used in Figure 24.3, and those predicted by a theoretical study conducted for two-dimensional virtual impactors by Forney et al. (1982). In the latter study, theoretical solutions to Navier–Stokes equations were used to determine the particle trajectories and thereby

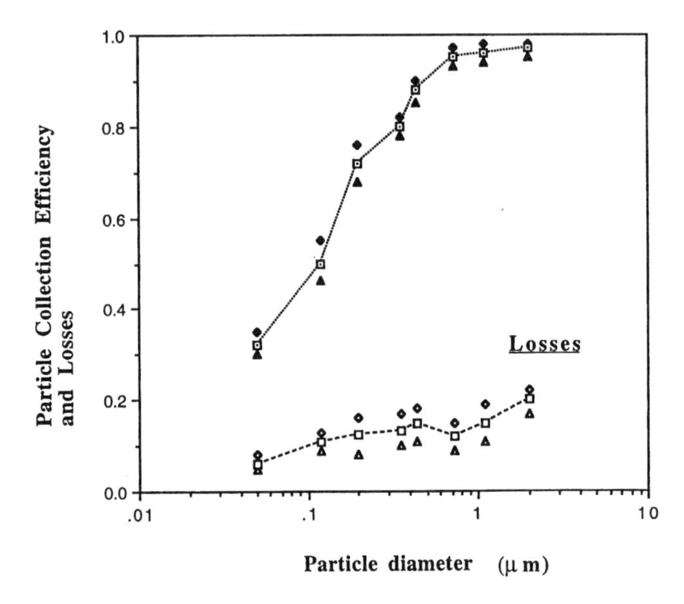

Figure 24.3 Collection efficiency and particle losses of a high-volume small-cutpoint virtual impactor. The total flow is 225 l/min and the minor flow ratio is 0.2. (From Sioutas, C. et al., *Particulate Sci. Technol.*, 12(3), 207–221, 1994. Reproduced with permission, Copyright © Taylor & Francis.)

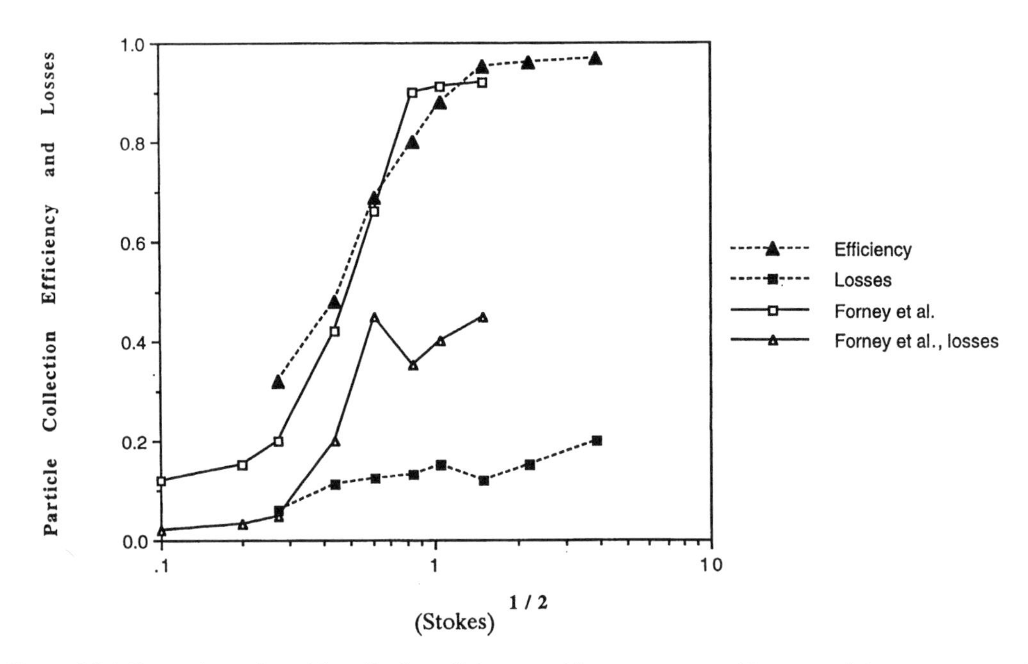

Figure 24.4 Comparison of particle collection efficiency and losses, expressed in terms of the square root of Stokes number, between the high-volume small-cutpoint virtual impactor and a slit-nozzle virtual impactor developed by Forney et al. (1982). (From Sioutas, C. et al., *Particulate Sci. Technol.*, 12(3), 207–221, 1994. Reproduced with permission, Copyright © Taylor & Francis.)

predict particle losses and collection efficiency of the slit-nozzle impactor. Instead of particle diameter, collection efficiency and losses have been expressed in terms of the square root of the Stokes number, $St^{1/2}$. The collection efficiency trends for both impactors appear to be quite similar. The increases value of q_m/Q_T (0.2) in the study by Sioutas et al. (1994c) compared with a value of

0.1 used by Forney et al. (1982) accounts for the relatively higher efficiencies observed at $St^{1/2}$ smaller than approximately 0.5.

Several design and operating parameters have been found to affect the impactor performance, e.g., the collection efficiency and particle losses. The most important parameters and their effect on the performance of small-cutpoint virtual impactors are summarized in the following paragraphs.

24.3.2 Effect of Minor-to-Flow Ratio

The minor-to-total flow ratio, whichs typically ranges between 0.05 and 0.4, affects dramatically the performance of the virtual impactor in terms of particle collection efficiency and losses. Figure 24.5 (Sioutas et al., 1994a) shows how the collection efficiency and particle losses are typically affected by the ratio q_m/Q_T. The 50% cutpoint decreases as the minor flow ratio increases, since the minor flow is contaminated by a larger fraction of fine particles. The increase in the collection efficiency with an increase in the minor flow ration (q_m/Q_T) is due to the fact that a higher flow ratios, more streamlines pass through the collection probe, as demonstrated in the theoretical analysis of particle flow in virtual impactors by Marple and Chien (1980). The square root of the Stokes numbers corresponding to the exprimentally determined 50% cutpoints ($St_{50}^{1/2}$) range from 0.38 to 0.55, depending on the ratio q_m/Q_T, as shown in Table 24.2.

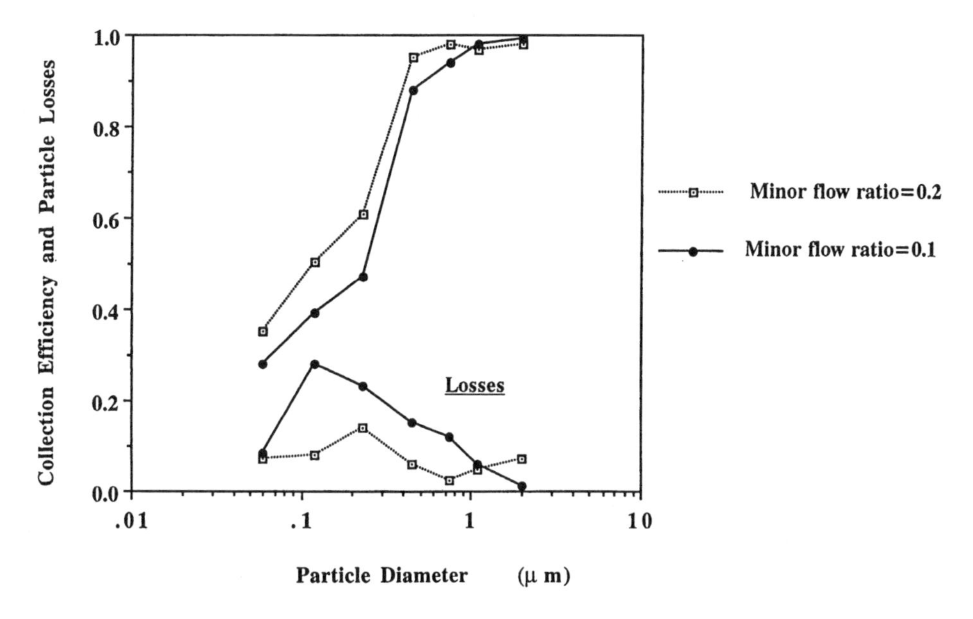

Figure 24.5 Effect of minor-to-total flow ratio on the particle collection efficiency and losses of small-cutpoint virtual impactors. (From Sioutas, C. et al., *Aerosol Sci. Technol.*, 21(3), 223–236, 1994. Reproduced with permission, Copyright © Elsevier Sciences, Inc.)

Table 24.2 Effect of the Minor-to-Total Flow Ratio on the Stokes Number Corresponding to the 50% Cutpoint

Minor-to-total Flow Ratio	$St_{50}^{1/2}$ [a]
0.1	0.55
0.15	0.48
0.2	0.35

[a] $St_{50}^{1/2}$ is the Stokes number corresponding to the 50% cutpoint.

Particle losses are also significantly affected by the minor-to-total flow ratio. The effect of q_m/Q_T on particle losses is shown in Figures 24.4 and 24.5. Depending on the size distribution of the sampled aerosol, the minor-to-total flow ratio can affect to a large extent the fraction of particles drawn into the minor flow. Figures 24.4 and 24.5 show that particles with size comparable to the 59% cutpoint are particularly affected by this ratio. In experiments conducted with indoor air as the test aerosol, Sioutas et al. (1995a) studied the effect of the minor-to-total flow ratio on the performance of high-volume small-cutpoint virtual impactors. The size distribution of the indoor aerosol was determined with the TSI Scanning Mobility Particle Sizer (SMPS Model 3934, TSI Inc., St. Paul, MN). Typically, the mass median diameter ranged from 0.15 to 0.3 μm with a geometric standard deviation of about 2. Thus, the majority of the particles were comparable in size to the cutpoint of the impactor (0.15 μm). The collection efficiency was evaluated by dividing the amount collected in the minor flow to the sum of the amounts in the minor and major flows. Particle losses were determined by comparing the total concentration on minor and major flows on the virtual impactor with the ambient fine particle concentration determined with a Harvard–Marple impactor (Marple et al., 1987). Operating at 4 l/min, the Harvard–Marple impactor has a 50% cutpoint at 2.5 μm and negligible (e.g., <2%) particle losses. By weighing the amount collected on the filter placed downstream of its impaction plate, the concentration of fine indoor particles was determined. Results from this study are summarized in Table 24.3. Decreasing the minor-to-total flow ratio results in a dramatic decrease in the collection efficiency and an increase in the particle losses.

Table 24.3 Characterization of a High-Volume (225 l/min) Small Cutpoint (0.15 μm) Virtual Impactor Using Indoor Aerosols

Minor Flow Ratio (%)	Efficiency (%)	Particle Losses (%)
20	80.6 (±0.8)	12.3 (±2.8)
10	61.7 (±2.5)	32.3 (±6.5)
7	55.6 (±3.1)	38.8 (±5.1)
5	51.1 (±1.2)	52.8 (±3.7)

24.3.3 Effect of the Ratio of the Collection-to-Acceleration Nozzle Size

The ratio of the width (or diameter for round nozzles) of the collection nozzle to the acceleration nozzle was also found to affect the performance of small-particle-cutpoint virtual impactors. Particle losses as a function of particle diameter are shown in Figure 24.6 for three different collection-to-acceleration nozzle ratios (Sioutas et al., 1994a). When this ratio is close to 1.0, particle losses occur primarily as a result of impaction on the tip of the collection nozzle. As expected, particle losses increase rapidly with increasing particle diameter because of the effect of inertia and become as high as 50% for particles in the micron-size range.

As this collection-to-acceleration nozzle size ratio is increased, losses are minimized because the streamlines of particles penetrating into the collection nozzle become less crowded. However, when this ratio becomes about 2, there is sufficient space in the collection nozzle for a complete flow reversal to occur, and for particle losses to occur on both the tip of the collection nozzle and the back of the acceleration nozzle. These experimental results agree well with a study by Masuda and Nakasita (1988), in which flow visualization techniques were used to examine the problem of particle losses in a slit-nozzle virtual impactor. Those experiments indicated the formation of a turbulent periodic eddy in the collection probe. The eddy residence time increased with increasing collection-to-acceleration slit widths, resulting in a more unstable flow. This eddy is formed by flow instabilities arising from the reduction of the flow velocity and the abrupt pressure recovery in the enlarged collection probe. The separation efficiency in those experiments became smaller

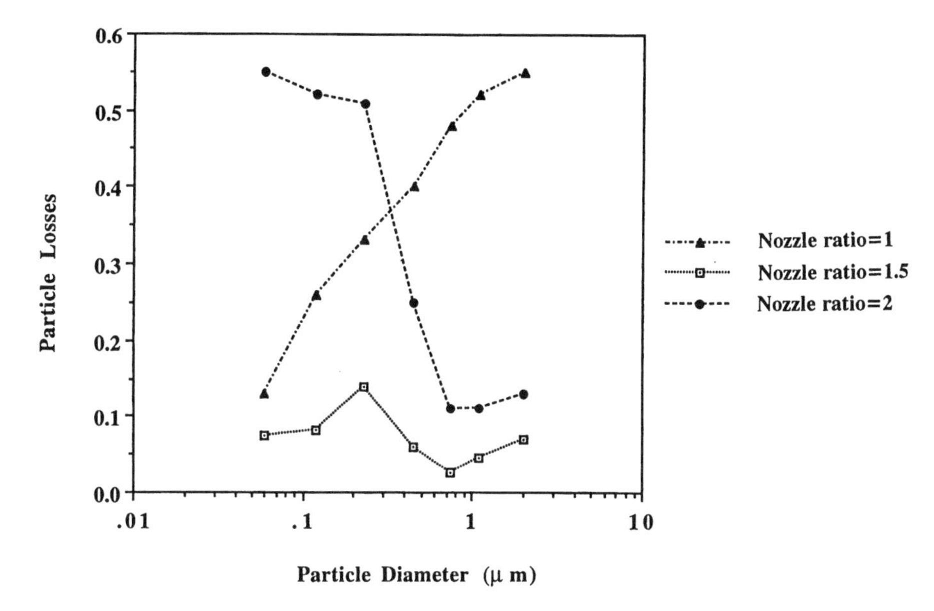

Figure 24.6 Effect of the ratio of the collection-to-acceleration nozzle size on particle losses in small-cutpoint virtual impactors. (From Sioutas, C. et al., *Aerosol Sci. Technol.*, 21(3), 223–236, 1994. Reproduced with permission, Copyright © Elsevier Sciences, Inc.)

than the theoretical prediction resulting in a larger cutpoint. Thus, a collection-nozzle-to-acceleration-nozzle ratio of 1.4 to 1.5 appears to be minimizing particle losses, as seen in Figure 24.6.

24.4 DEVELOPMENT OF AMBIENT FINE PARTICLE CONCENTRATORS FOR INHALATION STUDIES

The ability of small-cutpoint virtual impactors to separate fine ambient particles from gases can be used in several areas of air pollution and environmental health. This separation makes it possible to characterize these particles chemically and therefore improve our understanding on their toxicity. Of particular interest is the ability of small-cutpoint virtual impactors to also concentrate these fine particles while keeping them in the airborne state. Subsequent to being concentrated, fine particles can be supplied to human or animal exposure chambers to conduct inhalation studies. The concentration of particles larger than the cutpoint of a virtual impactor is ideally increased by a factor of Q_T/q_m, where Q_T is the total flow entering the virtual impactor and q_m is the minor flow. However, particle losses and deviations from 100% collection efficiency tend to decrease the concentration factor. The minor-to-total flow ratio, which typically ranges between 0.05 and 0.4, affects dramatically the performance of the virtual impactor and thereby the concentration factor, as shown in Table 24.3. The overall increase in the concentration enrichment becomes higher at lower flow ratios; it increases from a value of about 3.5 to 4.8 as the minor-to-total flow ratio decreases from 0.2 to 0.05, respectively. Nevertheless, the performance of the virtual impactor as a concentrator is deteriorated due to the increased losses and decreased collection efficiency, both of which tend to distort the size distribution of the sampled ambient aerosol. A minor-to-total flow value of 0.2 seems to optimize the operation of small-particle-cutpoint virtual impactors because it increases the collection efficiency, reduces particle losses, and maximizes the concentration of the aerosol in the minor flow with a minimal distortion of the original size distribution. Higher minor flow ratios (>0.2) presumably increase the collection efficiency of the impactor but do not result in concentration factors higher than 3.5.

To achieve higher concentration factors, which may be required for human or animal inhalation studies, Sioutas et al. (1995a and b) employed a series of virtual impactors operating at a minor flow ratio of 0.2. Although the pressure drop across the impactor nozzle is typically on the order of 0.3 atm (30 kPa) to achieve a cutpoint as small as about 0.1 μm, an almost complete pressure recovery occurs into the collection nozzle (Sioutas et al., 1995a). This is due to the conversion of kinetic energy to pressure as a small fraction of the flow enters the collection nozzle, resulting in an absolute pressure on the order of 0.99 atm (about 5 in. H_2O). This pressure recovery immediately downstream the tip of the collection nozzle is very important because exposure studies typically are not conducted under a negative pressure below 0.9 atm and techniques that could be used to transmit particles from a substantial vacuum to an atmosphere (for example, a compressor) could lead to substantial particle losses. Thus, several stages in series can be used, each of them concentrating ambient particles by a factor of approximately 3 to 3.5, to achieve the desired value of concentration for conducting inhalation studies. Since the concentrated aerosol is drawn through the collection nozzle, connecting more than one virtual impactor in series does not subject the concentrated particles to a substantial vacuum at the end of the series. In addition, no distortion of the physiochemical characteristics (e.g., particle growth or shrinkage) of the concentrated aerosol occurs, a frequent shortcoming of low-pressure sampling systems.

During controlled human exposure studies, it is desirable to have a system that supplies aerosols at higher than ambient concentrations during short-term experiments. For such experiments it is also preferable to have flow rates that exceed the average human inhalation flow rate (about 50 l/min) to avoid dilution. The ambient particle concentrator for human exposures developed by Sioutas et al. (1995a) consists of the following components (Figure 24.7):

1. A high-volume conventional impactor with a 2.5 μm cutoff size (separator).
2. Two virtual impactors with a 0.15 μm cutoff size (stages I and II).

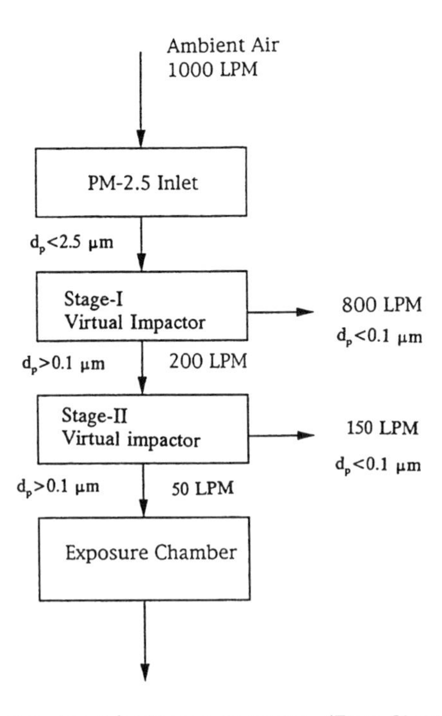

Figure 24.7 Ambient particle concentrator for human exposures. (From Sioutas, C. et al., *Inhalation Toxicol.*, 7(5), 633–644, 1995. Reproduced with permission, Copyright © Taylor & Francis.)

The first impactor is a high-volume conventional impactor (fractionating sampler, Andersen samplers) and removes particles larger than 2.5 μm operating at 1000 l/min, while particles smaller than 2.5 μm escape collection. The impactor has been characterized in detail by Burton et al. (1973). The deflected flow of the first impactor, containing particles smaller than 2.5 μm in aerodynamic diameter (1000 l/min), is drawn through a series of two virtual impactors. The first virtual impactor (stage I), shown in Figure 24.7, has a slit-shaped acceleration nozzle which is 28 cm long and 0.03 cm wide, and a collection slit-shaped nozzle, 0.045 cm wide and 28 cm long. The ratio of the collection to acceleration slit widths was chosen equal to 1.5 to minimize particle losses typically occurring at the tip of the collection slit. The impactor operates at 1130 l/min, with a minor-to-total flow equal to 0.2. The minor flow of stage I (225 l/min) is drawn through a second virtual impactor (stage II), with the same particle cutpoint size. The virtual impactor consists of a slit-shaped acceleration nozzle, 0.03 cm wide and 5.6 cm long, and a slit-shaped collection nozzle, 0.045 cm wide and 5.6 cm long. The minor flow, equal to 50 l/min, contains the concentrated ambient aerosol consisting of particles in the size range 0.1 to 2.5 μm.

The two-stage virtual impactor concentrator was evaluated by comparing the concentrations of fine mass and particulate sulfate to those determined with a Harvard–Marple impactor (Marple et al., 1987). The cutpoint of the Harvard–Marple impactor at 4 l/min is 2.5 μm. Furthermore, internal particle losses in the impactor have been shown to be negligible (e.g., less than 2 to 3%. Measurements were conducted by placing a 47-mm Teflon filter downstream of the minor flow of stage II and an identical Teflon filter downstream of the Harvard–Marple impactor and comparing the fine particulate mass and sulfate concentrations ($d_p < 2.5$ μm) determined by both sampling methods. The ratio of the concentration in the minor flow of stage II to that of the Harvard–Marple impactor gives an estimate of the particle concentration enrichment, based on either fine mass or particulate sulfate, are shown in Table 24.4.

Table 24.4 Characterization of the Concentrator Using Ambient Indoor Particles as the Test Aerosol

Exp. No.	Ambient Fine Mass (μg/m³)	Concentr. Fine Mass (μg/m³)	Concentr. Factor (fine mass)	Ambient Sulfates (μg/m³)	Concentr. Sulfates (μg/m³)	Concentr. Factor (sulfates)
1	17.3	122.3	7.2	4.0	24.6	6.2
2	13.0	122.9	9.5	3.3	28.1	8.6
3	11.3	114.1	10.1	3.4	27.8	8.2
4	11.6	102.6	8.9	2.6	21.1	8.2
5	18.1	151.3	8.4	3.0	28.2	9.4
6	20.8	186.4	9.0	4.3	30.2	7.1
7	22.1	200.6	9.1	4.8	40.7	8.5
8	8.9	81.9	9.2	2.2	20.4	9.8
9	10.9	115.5	10.6	2.5	22.8	9.1
10	15.1	135.4	9.0	2.9	25.1	8.7
11	28.2	268.1	9.5	3.1	24.6	8.0
12	28.1	309.5	11.0	2.8	31.1	11.1

The experimental results indicate that the enrichments in the concentrations of ambient fine mass and ambient sulfates achieved in the concentrator agree well with each other in every experimental run. The ambient fine mass concentration is increased by an average factor of 9.3 (±0.8), whereas the ambient sulfate concentration is increased by a factor of 8.6 (±1.3). The overall enrichment in the concentration depends on the average size distribution of the sampled aerosol, which is typically affected by factors such as the ambient temperature, relative humidity, and chemical composition (Koutrakis et al., 1989). Table 24.5 shows the concentration enrichment as a function of particle size for the two-stage system. Measurement of the ambient and concentrated aerosol were conducted using the TSI Scanning Mobility Particle Sizer (SMPS Model 3934, TSI Inc., St. Paul, MN). The results show a relatively uniform increase in the concentration for particles

Table 24.5 Concentration Enrichment as a Function of Particle Size in the Two-Stage Concentrator for Human Exposures

Particle Size Range (μm)	Concentration Enrichment
0.1–0.2	5.8
0.2–0.3	8.5
0.3–0.4	9.1
0.4–0.5	10.4
0.5–0.7	11.7
0.7–1.0	10.5

larger than about 0.2 μm in diameter. This concentration enrichment drops rapidly as particle size approaches the cutpoint due to the decrease in the collection efficiency of the virtual impactors.

A system for animal inhalation exposure studies was also developed and evaluated (Sioutas et al., 1995b). Ambient aerosols were concentrated by a factor of approximately 30 using a series of three virtual impactors, and introduced to an animal exposure chamber at 10 l/min. This is an adequate flow, since animal exposures require lower flow rates than human exposures. Again, due to pressure recovery occurring inside the collection nozzle, the measured absolute pressure inside the exposure chamber was 0.94 atm (equivalently, the overall pressure drop in the three-stage system was 24 in. of water). The performance evaluation of the three-stage system was similar to that of the two stage system described above. Ambient fine mass and particulate sulfate levels were determined using the Harvard–Marple impactor and compared to those in the exposure chamber. The results of the three-stage virtual impactor system are shown in Figure 24.8 and 24.9. The

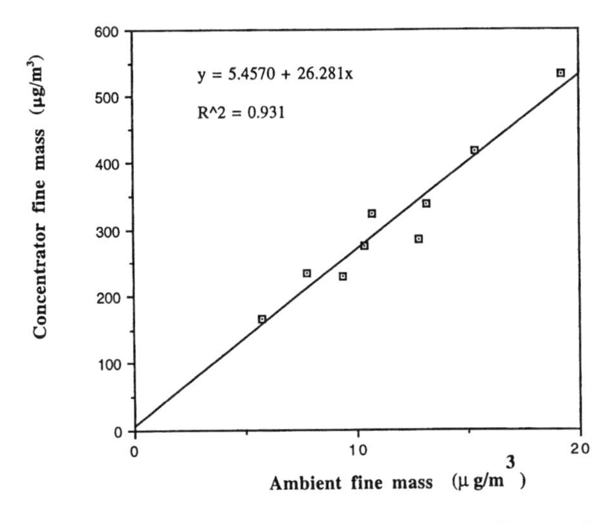

Figure 24.8 Ambient and concentrator fine mass concentrations. (From Sioutas, C. et al., *Environ. Health Perspect.,* 103, 172–177, 1995. Reproduced with permission, Copyright © Environmental Health Perspectives.)

concentrations of fine particulate mass and sulfates for a minor flow ratio (q_m/Q_t) of 0.2, shown in Figure 24.8 and 24.9, respectively, indicate that the fine mass concentration was increased by a factor of 26.3 (±2.7), whereas the fine sulfate concentration was increased by a factor of 28.7 (±3.4). The concentrator and ambient fine mass and sulfate concentrations were highly correlated (R^2 = 0.93 and 0.91, respectively). Thus, the addition of a third stage further increased the concentration of fine particle by a factor of approximately 3. The gradual increase in concentration through the

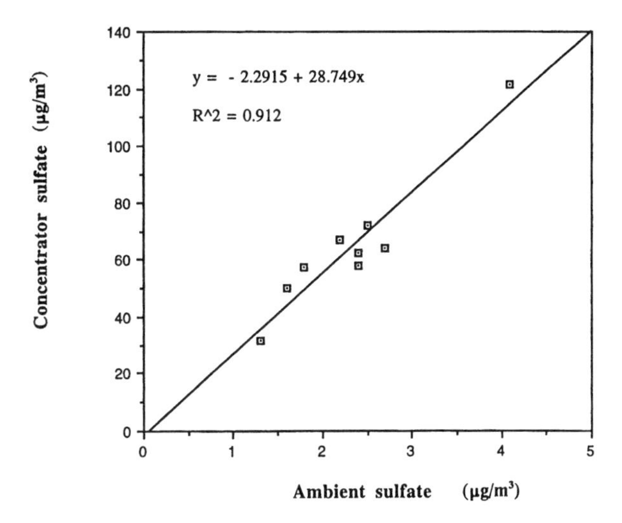

Figure 24.9 Ambient and concentrator sulfate concentrations. (From Sioutas, C. et al., *Environ. Health Perspect.*, 103, 172–177, 1995. Reproduced with permission, Copyright © Environmental Health Perspectives.)

different stages could result in some particle coagulation, which would tend to increase particle deposition on the internal surfaces of the virtual impactors and transition pieces. The coagulation effect may become more pronounced in the lower stages, especially in the third stage, where the lower flow rates increase the particle residence time in the various components of the system. The combined effect of the increase in the concentration and residence time may therefore decrease the overall concentration factor. The experimental results demonstrated that the overall concentration factor may be weakly affected by the increase in particle concentration. Since two virtual impaction stages increased the concentration by a factor of approximately 10, the placement of a third stage should have increased the concentration by a factor of approximately 32 (e.g., $10^{1.5}$). Nevertheless, the overall increase in the concentration, based on fine mass and particulate sulfate, was by a factor of 26.3 (±2.7) and 28.7 (±3.4), respectively, suggesting that the increase in particle concentration also increases particle losses in stage III. A decrease in the concentration factor due to particle coagulation could be more pronounced by adding a fourth virtual impaction stage, and could set an upper limit to the maximum concentration that can be achieved using multistage virtual impaction. The results from the evaluation of the concentrating systems described above demonstrated the feasibility of employing small-cutpoint high-volume virtual impactors to separate fine ambient particles from gases. Currently, these concentrating systems are being used by the U.S. Environmental Protection Agency for exposing animals and humans to fine ambient aerosols to determine ambient particulate toxicity.

24.5 APPLICATIONS OF SMALL-CUTPOINT VIRTUAL IMPACTORS IN OTHER AREAS OF ENVIRONMENTAL HEALTH

24.5.1 Artifact-Free Sampling of Semivolatile Organic Compounds

The ability of the newly developed virtual impactors to separate fine ambient particles from the majority of the surrounding gases can also be used in areas other than inhalation studies. A large number of ambient pollutants are distributed in both gas and particulate phases. Of particular interest in public health are semivolatile organic trace pollutants, such as polycyclic aromatic hydrocarbons and polychlorinated biophenyls formed during the incomplete combustion of coal, oil, wood, gasoline, and diesel fuel. The lifetime and fate of semivolatile compounds depends on their distribution between gas and particulate phases. Furthermore, control and removal of such compounds greatly depends on their partitioning between particulate and gas phases. Hence, accurate determination of the phase distribution is essential in assessing health effects related to exposures to semivolatile compounds as well as in designing control strategies. As discussed in the introductory part, currently, available technologies, including filter packs and diffusion denuders, collect the two phases sequentially, but chemical and physical interactions occurring during sampling result in sampling artifacts (overestimation or underestimations of the concentration of either the gaseous or the particulate phase). The advantage using virtual impaction is that the two phases can be separated within a few microseconds and collected concurrently instead of sequentially. As long as the cutpoint of the virtual impactors is small enough (e.g., on the order of 0.1 μm) to ensure that almost all of the particulate phase is contained in the minor flow, small-cutpoint virtual impactors offer a promising alternative to the existing particulate separation methods. Some polycyclic aromatic hydrocarbons have been found in significant amounts (e.g., 20 to 40%) in ambient particles smaller than about 0.1 μm (Venkataraman and Friedlander, 1994). The size distribution of these compounds was obtained with low-pressure conventional impactors which, as discussed in previous sections, may be subject to large artifact errors. However, the possibility of further decreasing the cutpoint of virtual impactors to the range 0.05 to 0.07 μm needs to be addressed for sampling polycyclic aromatic hydrocarbons more efficiently. Reducing the pressure in the impactor or the width of the acceleration slit-nozzle can further reduce the cutpoint, as predicted by the Stokes number equation. Particle residence time in the low-pressure area of the virtual impactor is on the order of few microseconds. Furthermore, once separated in the virtual impactor, particles can be collected on an appropriate medium (for example, a filter followed by a sorbent) instead of a flat surface exposed to a low pressure. Consequently, the cutpoint could be reduced without affecting the performance of virtual impactors and their ability to efficiently separate even polycyclic aromatic hydrocarbons associated with particle finer than 0.1 μm.

24.5.2 Replacement of High-Volume Filter Samplers

The use of high-volume small-cutpoint virtual impactors could also be extended to areas in the field of aerosol science which require higher concentrations than are normally present. Examples include radiochemical and X-ray fluorescent analysis (XRF) of collected particulate matter, as well as sampling of airborne allergens. Typically, a high-volume filter sampler has been used, which employs a 20.3 × 25.4 cm fiber filter connected to a vacuum pump sampling at 1130 l/min. Subsequent to sampling, the entire filter (or part of it) is extracted with an appropriate solvent to perform the required analysis of the collected matter. Although the particle collection efficiency of such filters is presumably 100%, there are several problems associated with the use of such filters; recovery of viable matter (microorganisms and allergens) is complicated because of the difficulties in extracting it from the filter surface (Chatigny et al., 1989). Furthermore, particles are collected over a large surface area, and the resulting low surface density of the collected matter decreases the sensitivity of the XRF analysis. Finally, background levels of ambient radioactivity can be very high (on the order of 750 fCi), thus making measurements of actual particulate-bound radioactivity difficult to perform (Cohen, 1989).

Replacing the traditional high-volume filters with high-volume small-cutpoint virtual impactors could solve several of the aforementioned problems. Although sampling can still be conducted at high flow rates (e.g., 1000 l/min or higher) to collect enough material within a short period, all particles can be contained within a small fraction of the original gas volume by being drawn through a virtual impactor. This can significantly reduce the surface area of the collection medium, thereby increasing the sensitivity of the analytical method. Furthermore, small-cutpoint virtual impactors also facilitate sampling and analysis of airborne biological particles because they combine high particle collection efficiency with the ability to maintain the micro-organisms airborne. This allows the collection of particles on the appropriate medium after they have been substantially decelerated, thus preserving their viability and minimizing the risk for damage. The feasibility of replacing high-volume filter samplers by high-volume small-cutpoint virtual impactors is currently being investigated by researchers at the Harvard School of Public Health.

REFERENCES

Appel, B. R., Tokiwa, Y., and Kothny, E. L. (1993). Sampling of carbonaceous particles in the atmosphere, *Atmos. Environ.,* 17, 1787–1796.

Appel, B. R., Tokiwa, Y., Haik, M., and Kothny, E. L. (1984). Artifact particulate sulfate, and nitrate formation on filter media, *Atmos. Environ.,* 18, 409.

Berner, A., Lurtzer, C. H., Pohl, L., Preining, O., and Wagner, P. (1979). *Sci. Total Environ.,* 13, 245–261.

Biswas, P., Jones, C. L., and Flagan, R. C. (1987). Distortion of size distributions by condensation and evaporation in aerosol instruments, *Aerosol Sci. Technol.,* 7, 231–246.

Burton, R. M., Howard, J. N., Penley, R. L. et al. (1973). Field evaluation of the high-volume particle fractioning cascade impactor, *J. Air Pollut. Control Assoc.,* 23, 277.

Chatigny, M. A., Macher, J. M., Burge, H. A., and Solomon, W. R. (1989). Sampling airborne microorganisms and aerollergens, in *Air Sampling Instruments,* S. V. Hering, Ed., American Conference of Governmental Industrial Hygienists, Inc., Cincinnati, OH.

Cohen, B. S. (1989). Sampling airborne radioactivity, in *Air Sampling Instruments,* S. V. Hering, Ed., American Conference of Governmental Industrial Hygienists, Inc., Cincinnati, OH.

Evans, J. S., Tosteson, T., and Kinney, P. L. (1984). Cross-sectional mortality studies and air pollution risk assessment, *Environ. Intern.,* 10, 55–83.

Fang, C. P., Marple, V. A., and Rubow, K. L. (1991). Influence of cross-flow on particle collection characteristics of multi-nozzle impactors, *J. Aerosol Sci.,* 22(4), 403–415.

Forney, J. L., Ravenhall, D. G., and Lee, S. S. (1982). Experimental and theoretical study of a two-dimensional virtual impactor, *Environ. Sci. Technol.,* 16, 492–497.

Forrest, J., Spandau, D. J., Tanner, R. L., and Newman, L. (1982). *Atmos. Environ.,* 16, 1473–1485.

Fuchs, N. A. (1978). Aerosol impactors: a review, in *Fundamentals of Aerosol Science,* D. T. Shaw, Ed., John Wiley & Sons, New York.

Hering, S. V., Flagan, R. C., and Friedlander, S. K. (1978). Design and evaluation of a new low pressure impactor-1, *Environ. Sci. Technol.,* 12, 667–673.

Hinds, W. C. (1982). *Aerosol Technology,* John Wiley & Sons, New York.

Kinney, P. L. and Ozkaynak, H. (1991). Associations of daily mortality and air pollution in Los Angeles County, *Environ. Res.,* 54, 99–120.

Koutrakis, P., Wolfson, J. M., Spengler, J. D., Stern, B., and Franklin, C. A. (1989). Equilibriumn size of atmospheric aerosol sulfates as a function of the relative humidity, *J. Geophys. Res.,* 94, 6442–6448.

Koutrakis, P., Thompson, K. M., Wolfson, J. M., Spengler, J. D., Keeler, G. J., and Slater, J. L. (1992). Determination of aerosol strong acidity losses due to interactions of collected particles: results from laboratory and field studies, *Atmos. Environ.,* 26, 987–995.

Koutrakis, P., Sioutas, C., Ferguson, S., Wolfson, J. M., Mulik, J. D., and Burton, R. M. (1993). Development and evaluation of a glass honeycomb denuder/filter pack system to collect atmospheric particles and gases, *Environ. Sci Technol.,* 27, 2497–2501.

Krieger, M. S. and Hites, R. A. (1992). Diffusion denuder for the collection of semivolatile organic compounds, *Environ. Sci Technol.,* 26, 1551–1555.

Lave, L. B. and Seskin, E. P. (1997). *Air Pollution and Human Health,* Johns Hopkins University Press, Baltimore.

Loo, B. W. and Cork, C. P. (1988). Development of high efficiency virtual impactors, *Aerosol Sci. Technol.,* 9, 167–176.

Marple, V. A. and Chien, C. M. (1980). Virtual impactors: a theoretical study, *Environ. Sci. Technol.,* 8, 976–985.

Marple, V. A. and Liu, B. Y. H. (1974). Characteristics of laminar jet impactors, *Environ. Sci. Technol.,* 8, 648–654.

Marple, V. A., Rubow, K. L., Turner, W., and Spengler, J. D. (1987). Low flow ratge sharp cut impactors for indoor air sampling: design and calibration, *J. Air Pollut. Control Assoc.,* 37, 1303–1307.

Marple, V. A., Rubow, K. L., and Behm, S. M. (1991). A micro-orifice uniform deposit impactor (MOUDI), *Environ. Sci. Technol.,* 14, 434–441.

Masuda, H. and Nakasita, S. (1988). *J. Aerosol Sci.,* 19, 243–252.

Ozkaynak, H. (1993). Reviedw of recent epidemiological data on health effects of particles, ozone, and nitrogen, paper presented at the *National Association for Clean Air Conference,* Nov. 11, 12, Dikholo Game Lodge, South Africa.

Ozkaynak, H. and Thurston, G. D. (1987). Associations between 1980 U.S. mortality rates and alternative measures of airborne particle concentrations, *Risk Anal.,* 7, 4.

Ozkaynak, H., Kinney, P. L., and Burbank, B. (1990). Recent epidemiological findings on morbidity and mortality effects of ozone, paper presented at the 83rd Annual Meeting and exhibition of Air & Waste Management Association, Pittsburgh, PA, June 24–29.

Pope, C. A., Dockery, D. W., Spengler, J. D., and Raizenne, M. A. (1991). Respiratory health and PM_{10} pollution. A daily time series, *Am. Rev. Resp. Dis.,* 144, 211–216.

Pope, C. A., Schwartz, J., and Ransom, M. R. (1992). Daily mortality and PM_{10} pollution in Utah Valley, *Arch. Environ. Health,* 47, 211–217.

Possanzini, M., Febo, A., and Liberti, A. (1983). New design of high performance denuder for the sampling of atmospheric pollutants, *Atmos. Environ.,* 17, 2605–2610.

Schwartz, J. and Dockery, D. W. (1992). Increases mortality in Philadelphia associated with daily air pollution concentrations, *Am. Rev. Respir. Dis.,* 145, 600–604.

Sioutas, C., Koutrakis, P., and Olson, B. A. (1994a). Development of a low cutpoint virtual impactor, *Aerosol Sci. Technol.,* 21(3), 223–236.

Sioutas, C., Koutrakis, P., and Burton, R. M. (1994b). Development of a low cutpoint slit virtual impactor for sampling ambient fine particles, *J. Aerosol Sci.,* 25(7), 1321–1330.

Sioutas, C., Koutrakis, P., and Burton, R. M. (1994c). A high-volume small cutpoint virtual impactor for separation of atmospheric particulate from gaseous pollutants, *Particulate Sci. Technol.,* 12(3), 207–221.

Sioutas, C., Koutrakis, P., Ferguson, S. T., and Burton, R. M. (1995a). Development and evaluation of an ambient particle concentrator for inhalation exposure studies, *Inhalation Toxicol.,* 7(5), 633–644.

Sioutas, C., Koutrakis, P., and Burton, R. M. (1995b). A technique to expose animals to concentrated fine ambient aerosols, *Environ. Health Perspect.,* 103, 172–177.

Stevens, R. K., Dzubay, T. G., Russwurm, G., and Rickel, D. (1978). Sampling and analysis of atmospheric sulfates and related species, *Atmos. Environ.,* 21, 589.

Thurston, G. D., Ito, K., Kinney, P. L., and Lippmann, M. (1992). A multi-year study of air pollution and respiratory hospital admissions in three New York State metropolitan areas: results from 1988 and 1989 summers, *J. Expo. Anal. Dis. Environ. Epidemiol.,* 2, 429–450.

Van Vaeck, L., Van Cauenberghe, G. C., and Janssens, J. (1984). The gas-particle distribution of organic aerosol constituents: measurement of the volatilization artifact in hi-vol cascade impactor sampling, *Atmos. Environ.,* 18, 417–430.

Venkataraman, C. and Friedlander, S. K. (1994). Size distribution of polycyclic aromatic hydrocarbons and elemental carbon-2. Ambient measurements and effects of atmospheric processes, *Environ. Sci. Technol.,* 28, 563–572.

Wang, H. C. and John W. (1988). Characteristics of the Berner impactor for sampling inorganic ions, *Aerosol Sci. Technol.,* 8, 157–172.

Whitby, K. T., Husar, R. B., and Liu, B. Y. H. (1972). *J. Colloid Interface Sci.,* 39, 177–204.

CHAPTER 25

Dry Aerosol Removal Capacity of a Newly Developed Plate Cyclone Filter

Rein André Roos, Alexander M. Möllinger, and Mark Stoelinga

CONTENTS

Abstract — Cyclonic particle separation and deposition is an interesting technique because of the almost infinite life of such a filter. However, efficiency of interception fine micronic particles was very low with the classic cyclone concept. A new approach is the use of a cascade of very small cyclones by using special profiles, giving the cyclone an appearance of a plate. The efficiency of these plate cyclone filters (PCF) for liquid aerosols is quite well known; however, its reaction with dry powder is missing. The tests carried out showed very good results under such conditions and although not yet optimalized it seems that one more step has been made in the direction of a filter with an indefinite service life. When it comes to the filtering of certain liquid aerosols, like oil mist, it has to be assumed that the PCF is a very good alternative to electrostatic precipitation methods whose efficiency can be affected by so-called electrospraying.

25.1 INTRODUCTION

The most common method used for air cleaning is filtration, which is in general simple, versatile, and economical. At low particle concentration fibrous filters provide for high-efficiency collection of near micrometer-size particles. What is the mechanism that lies behind the possibility of a filter

0-87371-830-5/98/$0.00+$.50

to remove particles from an airstream? In fact, it is not one, but quite a number of mechanisms that are at the origin of it. The five basic mechanisms by which an aerosol particle can be deposed on a fiber of a filter are, according to Hinds, 1982:

Interception — Here the particles that follow the air streamlines touch the surface of the filter fiber and become attached to it.

Inertial Impaction — This occurs with a quite large particle, which because of its inertia is unable to adjust quickly enough to the abruptly changing streamlines in the vicinity of a fiber, crosses those streamlines and hits the fiber to which it remains attached.

Diffusion — Fine aerosol particles do not follow the air streamlines very well although they are light enough to do so. On the contrary, they tend to move in various directions due to so-called Brownian motion, which is sufficient to greatly enhance the probability of hitting the fiber and remaining there.

Gravitational — This filtering mechanism is, as its name says, indicates the settling of particles under the influence of gravity. In general, it plays an important role only when the particles are heavy and the airflow is low.

Electrostatic — This is the case when the fibers of the filter are produced in such a way that strong electrostatic fields are obtained. Because almost all particles carry an electric charge or show an electric dipole, enhanced deposition will follow due to the coulomb forces exercised between particles and the fibers of the filter.

Quite an arsenal of mechanisms, so why look for other methods? The reason is quite simple: filters have to be changed regularly because the intercepted particles cannot be removed easily even if the aerosols involved contain more or less liquid particles. A good example is tobacco smoke; if you smoke, for example, a number of cigarettes through a cigarette holder, you then notice that its filter is already clogged after a few cigarettes. So in spite of their liquidity, meaning that they could leak away, cigarette smoke particles remain trapped inside the filter. One of the causes is the capillarity found between the crossing of two fibers (see also Figure 25.1) that will be filled by the aerosol.

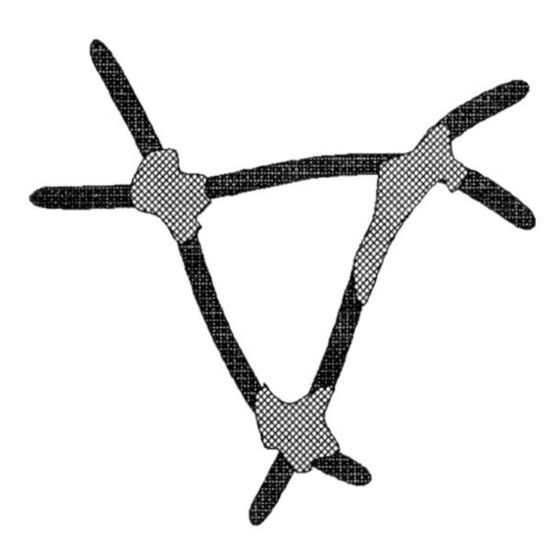

Figure 25.1 Capillarity filling with fibers.

25.2 BASIC PRINCIPLE

However, if we were able to construct a filter capable of leaking away the intercepted particles, then it could be used indefinitely and become even more economic than fiber filters. Such a type of filter exists and is based on an intercepting effect not mentioned before, namely,

Cyclonic — Coming from the Greek work *kuklos*, which means circle, here the particles are forced to rotate, thus increasing the coagulation between smaller and larger particles making them heavier. This causes the larger ones to be removed due to wall deposition caused by centrifugal forces; see also Figure 25.2.

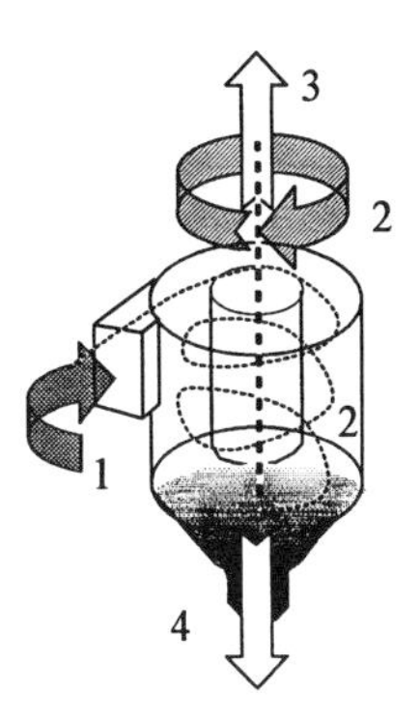

Figure 25.2 Typical cyclone; see also text.

The method used to separate or filter particles of different sizes from an airflow by means of curvilinear motion has been known for a long time in the industry under the name: cyclones. These are centrifugal dust separators in which the apparatus itself is fixed and circulation of gas around the axis of the cyclone is produced. See also Figure 25.2. The airflow to be treated centers through inlet 1 and passes through the cylindrical part of the cyclone. Here it requires a spiral motion 2, descending along the outer spiral and ascending by the inner spiral, to leave through exhaust tube 3. Particles precipitated on the walls of the cyclone by centrifugal force move downward along the wall and leave the cyclone through opening 4.

The simplicity and low cost of construction and maintenance, together with the high-volume treatment potential, have led to cyclones being used everywhere where particles have to be removed from airflows. However, mainly coarse particles are retained in this type of cyclone. An approximate analysis of the particle motion and cyclone performance has been carried out by Friedlander (1977), but he concludes that the aerodynamic pattern is too complex for an exact analysis and he refers to the semiempirical expressions used by Fuchs (1964). Another attempt to describe theoretically the aerosol collection in cyclones comes from Bürkholz (1984), while the flow along the walls is highlighted by Ranz (1984).

Concerning the smallest particle diameter that can be removed by the cyclone, Friedlander (1977) proposes the following equation:

$$d_p \min \sim [\mu(b - a)/\rho NU]^{0.5}$$

where

U = the average velocity of the airstream in the inlet tube
N = the number of turns the particles make
b = the radius of the outer wall
a = the radius of the exit tube of the cyclone
μ = the viscosity of the air
ρ = the density of the particle

From the relationship can be seen that, for a fixed velocity, performance improves as the diameter of the cyclone is reduced, because the distance the particle must move for collection decreases. It can also be seen that performance increases with the number of turns the airflow makes.

The effect of the increase in the number of turns becomes clear when we take a look at the cascade cyclone as described by Liu and Rubow (1984). This consists of a five-stage cyclone of the "axial flow" type, meant for size classification of airborne particulate matter. The principal advantage of this cyclone concept is its ability to collect a relatively large amount of material without particle bounce and reentrainment, making its performance superior to that of a cascade impactor in the case of a dense particle cloud with high mass loading.

The collection efficiency of this cyclone was determined as a function of particle size using solid, monodisperse aerosols of ammonium fluorescein. The experiments were performed with particles ranging in aerodynamic diameter from 0.8 to 23 μm. The obtained results are intercepted in Figure 25.3.

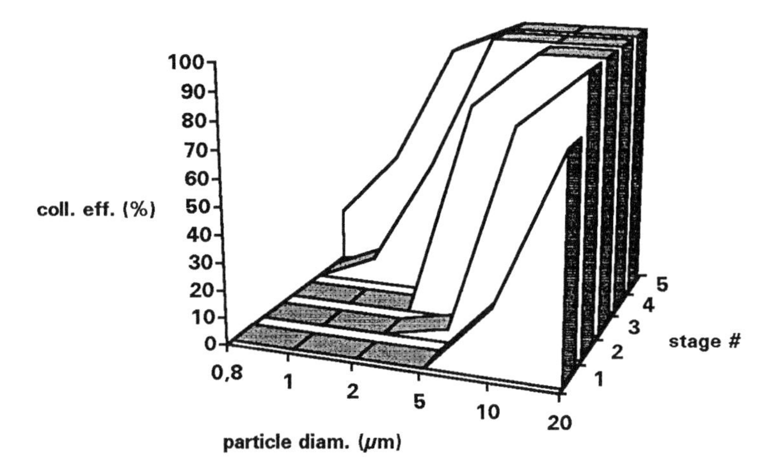

Figure 25.3 Collection efficiency after Liu and Rubow (1984).

Let us return to the relationship as proposed by Friedlander concerning the cutoff particle diameter, according to which efficiency should go up with decreasing dimensions and an increasing airflow speed. This can be verified with the small cyclone having a diameter of only 10 mm, which is used for studies on particle deposition in the respiratory system.

Tests carried out with such a device, using a liquid aerosol of silicon oil, show indeed that it is only smaller particles that are able to penetrate this device and that its performance increases with an increase in flow rate (l/min) and thus the speed of the airstream (Liden and Kenny, 1986). See Figure 25.4 for interception of the results. So if we want to create a filter with good cutoff characteristics for fine particles based on this cyclonic effect, it should have the following characteristics.

- Small-sized cyclones
- Multiple stages
- High airspeed

By using particularly shaped aluminum profiles, a number of companies are currently producing platelike filters based on cyclonic deposition, thus plate cyclone filters (PCF). Some of the profiles are shown in Figure 25.5, from which can be seen that they consist indeed of a succession of minicyclones with high airflow speeds, which should give a good efficiency in particle cutoff diameter.

However, there is a fundamental difference with this approach and that of the cyclone of Figure 25.2; namely, the airflow does not move downward in order to promote particle deposition at the bottom of the cyclone. So the application of this type of filter seems to be restricted to those aerosols difficult to intercept in the classic fiber filters, namely, the liquid ones like oil and acid mists. However, the virtually unlimited life of the PCF makes it very interesting if it also could be used for certain types of solid aerosols. A series of tests was carried out at the Laboratoire d'Aéroactivité in order to measure the behavior of such a PCF of brand "B" in respect to a solid aerosol, see also Figure 25.6.

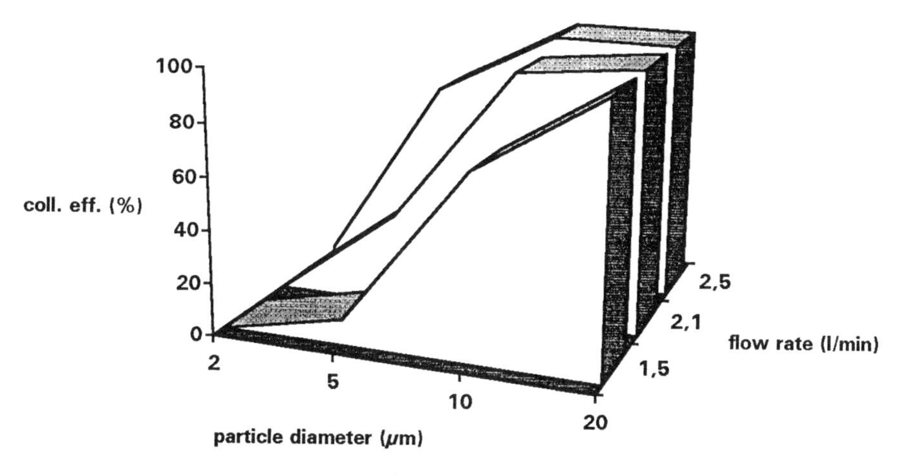

Figure 25.4 Collection efficiency after Liden and Kenny (1986).

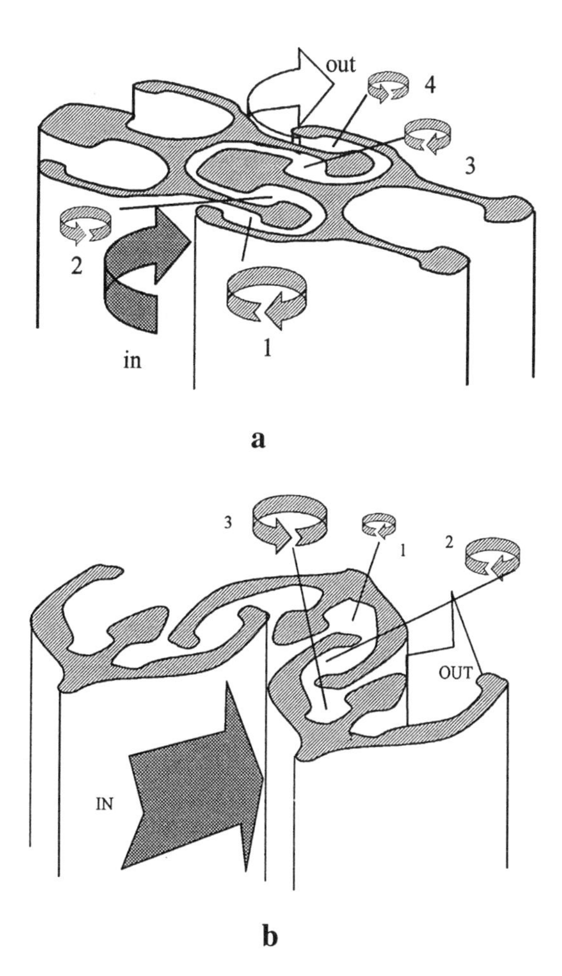

Figure 25.5 (a) Two sections of the profile of brand "B," 1 to 4 minicyclones. (b) Three sections of the profile of brand "F," 1, 2, 3 minicyclones.

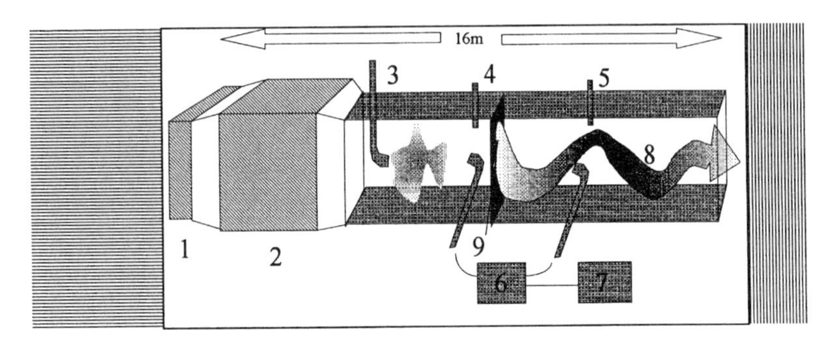

Figure 25.6 (1) Ventilator 11520 m³/h, (2) straighteners, (3) van der Wel aerosol generator, (4 and 5) pressure sensors, (6) particle counter (PMS), (7) computer, (8) wake, (9) filter under test.

25.3 EXPERIMANTAL

For the test we used a large wind tunnel. This wind tunnel consists of the following parts: a large radial blower which blows the air in a high-volume chamber (2×2 m and 2.5 m long) with filters and flow straighteners in it. This chamber is connected to a measuring section of 0.6×0.6 m and 7 m long. It is in this measuring section that the filter under test was placed 3.5 m from the entrance.

At the entrance of the measuring section of the wind tunnel the air velocity was measured. Directly before and after the filter pressure holes were made in order to measure the pressure drop over the filter. Pressure measurements were carried out with standard water manometers having an accuracy of 0.2 mm water column. For the measurement of the air velocity a fan-type anemometer was used with an accuracy of 0.1 m/s and an inferior airspeed limit of 0.4 m/s.

The aerosol particles used in the experiment were spherical alumina oxide (Al_2O_3). The particles were brought into the measuring section by means of a van der Wel generator. This generator consists of a powder supply feeding by means of a rotating disk bringing a constant amount of particles to a fluidized bed inside the measuring section. The particles emerging from this bed were intercepted by the airstream and blown toward the filter. The particles were measured just before and after the cyclone filter by means of probes sampling the aerosol stream. The particles with a diameter less then 3 μm were measured with an active scattering aerosol spectrometer from Particle Measuring Systems, Inc. (PMS).

25.4 RESULTS

The wind tunnel was able to generate a pressure difference of more than 1000 Pa over the filter and a speed in excess of 3 m/s. The leakage of air was reduced to almost zero by taping all seams. The pressure drop over the filter as a function of the velocity was measured and is shown in Figure 25.7. The results were corrected for the difference in cross section between the measuring section and the cyclone filter.

Cross section of the wind tunnel	605×604 mm
Cross section of filter "B"	560×560 mm

This gives a correction factor of 1.165 which has been taken into account in the graph shown. The function between pressure drop (P) and velocity (V) is

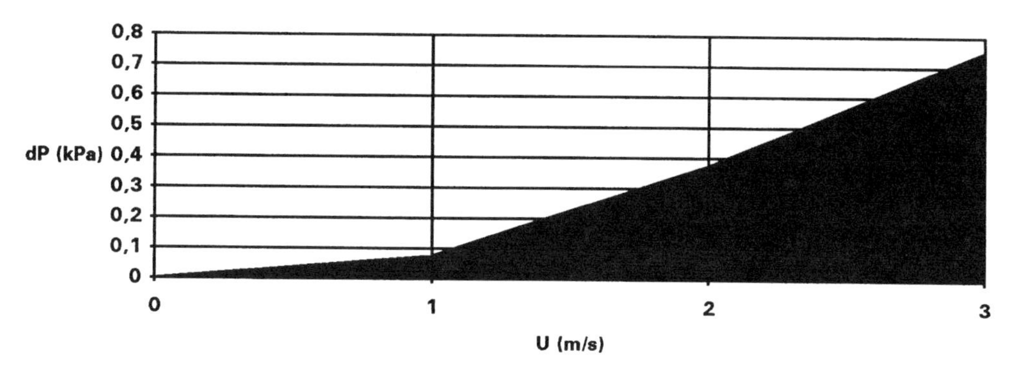

Figure 25.7 Pressure drop as function of velocity.

$$P = 3.28 + 31.25U + 71.88U^2$$

where pressure drop is in Pa and velocity is in m/s.

The particle size distribution (PSD) of the alumina oxide powder was determined by means of a sedigraph using water as liquid and stirring the particles for 15 min in an ultrasonic bath. Figure 25.8 shows the obtained PSD, with the PMS the number of particles before and after the filter were measured. Combination of a number of measurements resulted in the efficiency of the cyclone filter at three different wind speeds: 0.5, 2.4, and 3.0 m/s. The results are shown in the Tables 25.1 to 25.3.

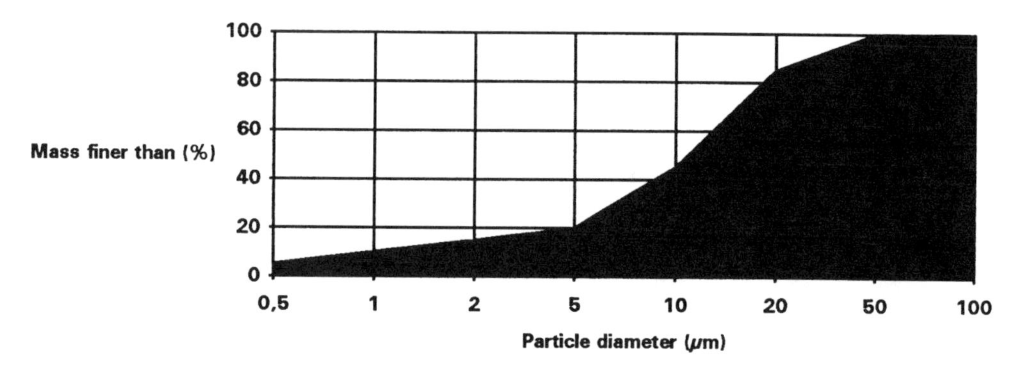

Figure 25.8 Particle size distribution; see text.

Table 25.1 Airspeed 0.5 m/s

Particle Size (μm)	Concentration before Filter (mg/m³)	Concentration after Filter (mg/m³)	Efficiency
0.24–0.60	0.0128	0.0097	0.24
0.60–1.08	0.0293	0.0163	0.44
1.08–1.59	0.0597	0.0266	0.55
1.65–2.04	0.0993	0.0384	0.61
2.04–2.52	0.1285	0.0645	0.49
2.52–3.00	0.1369	0.0474	0.65
0.24–3.00	0.4665	0.2029	0.57

Table 25.2 Airspeed 2.4 m/s

Particle Size (μm)	Concentration before Filter (mg/m^3)	Concentration after Filter (mg/m^3)	Efficiency
0.24–0.60	0.0087	0.0053	0.39
0.60–1.08	0.0184	0.0030	0.83
1.08–1.59	0.0360	0.0043	0.88
1.65–2.04	0.0595	0.0041	0.93
2.04–2.52	0.0893	0.0057	0.94
2.52–3.00	0.0773	0.0101	0.87
0.24–3.00	0.2892	0.0325	0.89

Table 25.3 Airspeed 3.0 m/s

Particle Size (μm)	Concentration before Filter (mg/m^3)	Concentration after Filter (mg/m^3)	Efficiency
0.24–0.60	0.0104	0.0077	0.26
0.60–1.08	0.0092	0.0044	0.52
1.08–1.59	0.0145	0.0031	0.79
1.65–2.04	0.0211	0.0052	0.76
2.04–2.52	0.0388	0.0040	0.90
2.52–3.00	0.0617	0.0071	0.88
0.24–3.00	0.1557	0.0315	0.80

25.5 DISCUSSION

From the tables it follows that the initial concentration decreased as the airspeed increased. This is due to the fact that the particle generator produced a constant amount of particles which is diluted into a larger amount of air. Similar to the cyclone, the PCE increases its efficiency with the windspeed, but seems to pass through a maximum. This is also the case for the particle cutoff size which when operated at the highest efficiency is on the order of 0.6 μm. The use of the PMS meant that the filter efficiency could be determined only for relatively low dust concentrations.

A large wake was present behind the tested filter; in fact, the whole airstream was blown to the wall of the wind tunnel. This flow pattern was made visible by means of a number of small flags. The presence of the wake made it difficult to measure the number of particles in the air. Concentrations were measured before and after the filter at the same position. The probe after the filter was situated in such a way that it intercepted very well the streamlines coming out of the filter. This is different from the generally used method using a grid of measuring points before and after the filter. It is our opinion that by using this classical approach, we should have obtained a number of unrealistic values due to the presence of the wake.

We obtained a lower pressure drop than was obtained in another filter test. It is thought that that can be caused by the fact that in our case the air was blown through the filter instead of being sucked through it. Suction means that the wake cannot escape as it did on our case, but that it remains blocked between the filter and the sucking ventilator, leading probably to turbulent phenomena and unusual flow patterns with possible consequences for the pressure drop.

25.6 CONCLUSIONS

Although probably not fully optimalized, the tested PCF showed remarkably good efficiencies even for fine submicron dry particles. Particle deposition on the walls will modify the shape of the cyclones in the PCF and influence the efficiency of the filter, so they have to be cleaned. The extent to which gravitation can play the role of "auto cleaner" as it does with liquid aerosols has not been investigated, but it seems that the PCF could have this characteristic when confronted with certain dry aerosols. The almost unlimited life of these filters makes them very attractive from an ecological point of view.

When it comes to the filtering of liquid aerosols, like, for example, an oil mist, the future role of these filters should not be underestimated. Recent investigations (Marijnissen, et al., 1993) have shown that certain electrostatic precipitators instead of removing these aerosols are capable by generating them in large inhalable quantities through the mechanism of electrospraying. So in this case PCF are superior from filtering and energetic points of view.

ACKNOWLEDGMENTS

We wish to thank the Aerosol Laboratory of the Technical University of Delft in The Netherlands and especially Dr. J. C. M. Marijnissen for the help we got during this research.

REFERENCES

Brand "B"; Brumex, Apogee, Z. I. du Tronchon, 45 Chem. du Moulin Carron, 69570 Dardilly, F.

Brand "F"; Fluidex, S. A. Charles Desbonnets, B. P. 24, 59051 Roubaix Cedex, F.

Bürkholz, A., Approximation formulas describing aerosol collection in cyclones, in *Aerosols,* Liu, B. Y. H. et al., Eds., Elsevier, New York, 1984, 647–649.

Friedlander, S. K., *Smoke Dust and Haze,* Wiley, New York, 1977, 109–112.

Fuch, N. A., *The Mechanics of Aerosols,* Pergamon Press, Oxford, 1964.

Hinds, W. C., *Aerosol Technology,* Wiley, New York, 1982, 164–178.

Liden, G. and Kenny, L. C., The performance of personal cyclones for the sampling of respirable dust, in *Aerosols, Formation and Reactivity,* Pergamon, Oxford, 1986, 579–582.

Liu, B. Y. H. and Rubow, K. L., A new axial flow cascade cyclone for size classification of airborne particulate matter, in *Aerosols,* Liu, B. Y. H. et al., Ed., Elsevier, New York, 1984, 115–118.

Marijnissen, J. C. M., Mollinger, A. M., and Vercoulen, P. M., Generation of droplets in electrostatic precipitators by electrospraying, *J. Aerosol Sci.,* Suppl. 39–40, 1993.

Ranz, W. E., Wall flows in a cyclone separator; a description of internal phenomena, in *Aerosols,* Liu, B. Y. H. et al., Eds., Elsevier, New York, 1984, 631–634.

CHAPTER 26

A Length-Selective Technique for Fibrous Aerosols

Toshihiko Myojo

CONTENTS

26.1 INTRODUCTION

Asbestos is widely used in various industrial products, such as cement slate, heat insulation, and brake linings, although its carcinogenicity has been pointed out in many studies. Researchers are also concerned about the carcinogenicity of other natural and anthropogenic fibrous materials. The toxicity of fibrous materials is an essential problem not only for biological researchers, but also for engineers who are searching asbestos substitutes.

When these fibrous materials are crushed during any handling process, small portions of fine fibers are suspended and fibrous aerosols are generated. It is important to examine the chemical and mineralogical characteristics of fibrous materials. Meanwhile, biological and medical studies

0-87371-830-5/98/$0.00+$.50

on the interaction of fibrous materials with the human body have been widely conducted in the past few decades.

In order to combine the results from material studies and the biological effects of fibrous materials, the techniques of generation and classification of the fibrous aerosols are needed. In this chapter, I will briefly review the characteristics of fibers and generation methods for aerosols, then describe a length-selective technique developed for measuring the length distribution of fibrous aerosols and the preparation of asbestos samples for animal studies.

26.1.1 Motion of Fibers

The dimension of fibers, their length and diameter (width), is important to quantify the transportation process of fibrous materials. Research work on fibrous aerosols began as a part of a study on the shape of suspended particles, for example, which we can find in *The Mechanics of Aerosols* by Fuchs (1964). However, there are still many unsolved problems concerning fibrous aerosols.

Figure 26.1 shows the dimensions of the three types of airborne particles: (a) spherical, (b) ellipsoidal, and (c) cylindrical. The size of the spherical particles (a) is just diameter Dp, but length Lf and diameter Df are needed to show the size of cylinder (c). The orientation of cylinder ϕ, which is the angle between the long axis of the cylinder and the flow direction, is very important to depict the motion of a cylinder in fluid. It is possible to solve the trajectory of ellipsoids (b) theoretically (Jeffrey, 1922). The trajectory of cylinders was solved using numerical methods (Gallily and Cohen, 1979; Gallily and Eisner, 1979; Gallily et al., 1986; Foss et al., 1989).

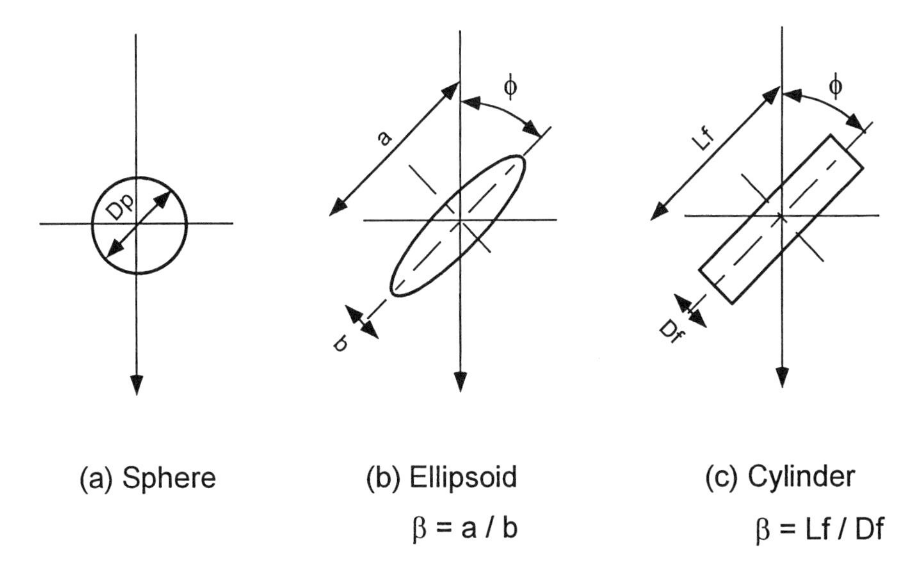

Figure 26.1 Dimensions and orientation of (a) sphere, (b) ellipsoid, and (c) cylinder.

Theoretical and experimental results for the aerodynamic diameter D_{av} of fibers are summarized in Table 26.1. Figure 26.2 shows the relationship between the aspect ratio and D_{av}. The density of fibers ρ_p is 2.5 g/cm^3. The theoretical equations for an ellipsoid were shown for three cases: parallel to the flow direction (1), perpendicular (2), and random orientation (3). The random orientation obtained from parallel and perpendicular orientation added to the weight of position.

Asbestos and/or glass fibers were introduced into a Timbrell aerosol spectrometer (settling, Timbrell, 1965), a centrifuge (centrifugal force, Stober et al., 1970) and a cascade impactor (inertia, Burke and Esmen, 1978). The dimension of the fibers collected on these instruments was compared with the aerodynamic diameter measured by these instruments. The observed data in original papers scattered, but the lines in Figure 26.2 show the average of the data.

Table 26.1 Geometric Dimesions and Aerodynamic Dimension of Fibrous Aerosol

No.	Equation	Conditions	Ref.
(1)	$D_{av} = 1,5D_f\sqrt{\rho(\ln(2\beta) - 0.5)}$	Parallel with flow direction	Harris, 1972
(2)	$D_{av} = 1.5D_f\sqrt{\rho\dfrac{(\ln(2\beta) + 0.5)}{2}}$	Prpendicular to flow direction	Harris, 1972
(3)	$D_{av} = 1.5D_f\sqrt{\dfrac{\rho}{\dfrac{0.385}{(\ln(2\beta) - 0.5)} + \dfrac{1.230}{(\ln(2\beta) + 0.5)}}}$	Randomly oriented prolate spheriod	Harris, 1972
(4)	$D_{av} = 66D_f\left(\dfrac{\beta}{2 + 4\beta}\right)^{2.2}$	Glass fiber with aerosol spectrometer	Timbrell, 1965
(5)	$D_{av} = 2.18D_f\beta^{0.116}$	Amosite with the spiral centrifuge	Stober et al., 1970
(6)	$D_{av} = D_f\{1 + 0.013(\ln \beta)^3\}\{0.71 + 0.91 \ln \beta\}^{0.5} \cdot \sqrt{\rho}$	Glass fiber with cascade impactor	Burke and Esmen, 1978

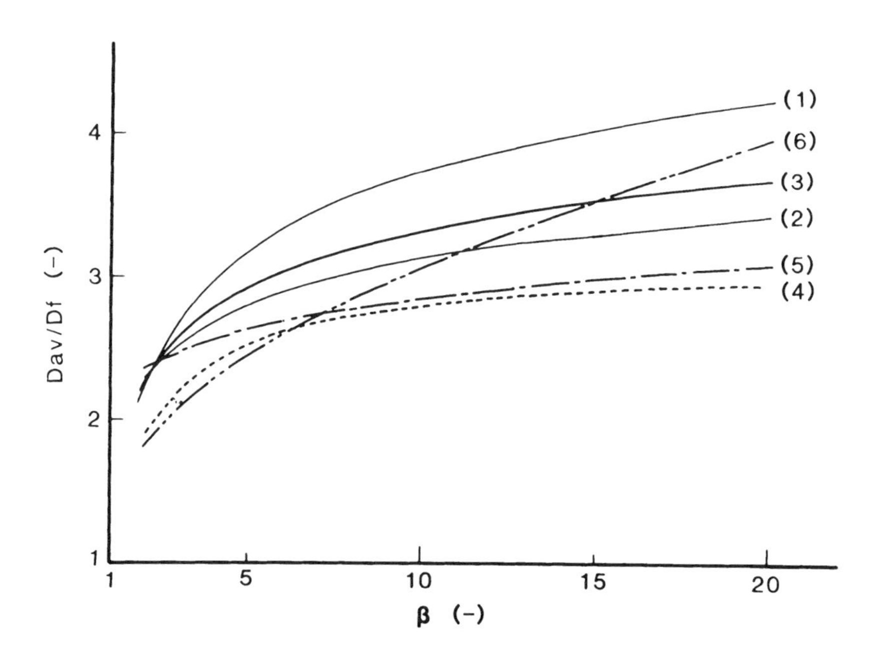

Figure 26.2 Relation between aspect ration β and ratio of fiber diameter and aerodynamic diameter. Lines from (1) to (6) correspond with the numbers in Table 26.1.

Table 26.1 and Figure 26.2 suggest that the diameter or the width of the fibers is more influential an the aerodynamic diameter than the fiber length or aspect ration β. The orientation of fibers against the flow direction also caused the experimental data to scatter. Consequently, it is not easy to classify fiber length by means of the method using settling, centrifugal force, and inertia.

The orientation of fibers is not stable in a nonuniform flow, such as flow in a pipe. Jeffery (1922) presented a theoretical description of the behavior of small ellipsoids in viscous fluid. The simplest case calculated from the equations is shown in Figure 26.3. Consider an ellipsoid having a long axis on the *X-Z* plane. The long axis forms an angle ϕ with the *X* axis. When the initial position is $\phi = 0$, an equation to describe the motion of the cylinder is derived from Jeffery's equations as shown below.

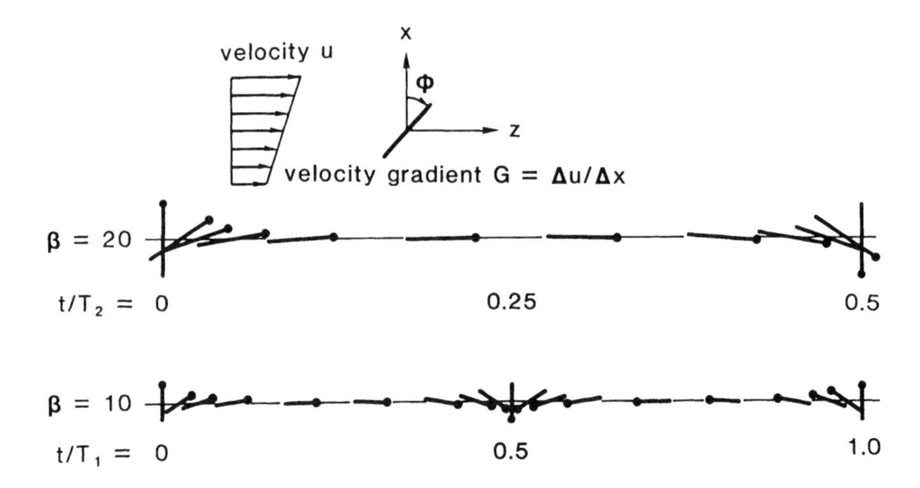

Figure 26.3 Rotation of ellipsoid in shear flow.

$$\tan\phi = \beta \tan\left(\frac{2\pi t}{T}\right) \tag{26.1}$$

where T is the orbit period, that is, the time required for one complete revolution of the ellipsoid around the Y axis,

$$T = \frac{2\pi t}{G} \tag{26.2}$$

for aspect ratio $\beta >> 1$. The velocity gradient G is set to a uniform value like Couette flow.

The angle ϕ in Figure 26.3 shows the position of a cylinder during periodic motion. The cylinder rotates rapidly when its angle ϕ is near 0 and rotates slowly when it is almost aligned with the streamlines of the fluid.

Harris (1972) and Gallily and Eisner (1979) performed experiments to determine the motion of rods in Couette and Poisouille flows. They observed that the motion of rods in liquids was very similar to the motion estimated by Jeffery.

26.1.2 Generation of Fibrous Aerosols

Monodisperse fibers on both length and diameter are preferable in research on aerodynamic motion and the light-scattering property of fibers. Studies on the techniques generating monodisperse fibrous aerosols are still ongoing (Esmen et al., 1980; Loo et al., 1982; Hoover et al., 1990; Baron et al., 1994).

Powder-dispersion-type aerosol generators have been widely used to study the inhalation toxicology of fibrous materials because of their easy operation and ability to generate high concentrations. The geometric standard deviation of the length distribution of generated fibrous aerosols used to be more than 2.0. Figure 26.4 summarizes the method for generating fibrous aerosols. A fluidized bed and/or a pulverizer are used to disperse the fibrous materials. The fluidized beds are classified as either one-component type or two-component type.

Spurny et al. (1975; 1976) developed a fibrous aerosol generator using a one-component fluidized bed with a vibration source. This generator enabled the size distribution to be charged by adjusting the frequency or amplitude of the vibration. However, the aerosol concentration generated by the device was relatively low.

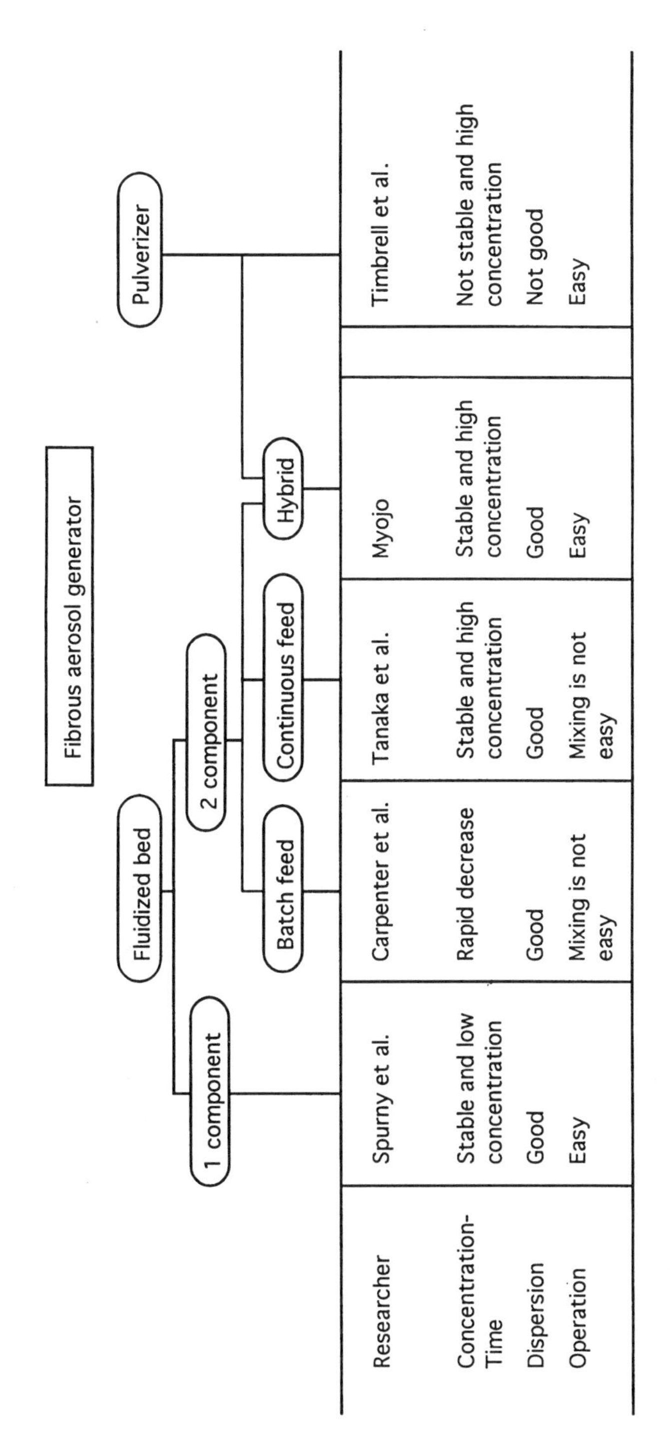

Figure 26.4 Fibrous aerosol generator.

A two-component fluidized bed consists of fibrous material to be aerosolized and a bed material that is larger in size than the fibers. After the fluidized action disperses the mixed powder, the aerosolized fibers are carried away from the bed. The bed material is separated in the bed column by the difference in settling velocities of the two powders.

Carpenter and Yerkes (1980) and Carpenter et al. (1981) designed a fibrous aerosol generator using a two-component fluidized bed. Since the system was operated as a batch-feed mode, it was necessary to use two systems alternately for continuous operations. The advantages of this method are the simple structure of the generator, and that it is possible to obtain good dispersion and high concentration. On the other hand, the disadvantages are a decrease in the concentration exponentially with the batch-feed mode and attrition of the bed material, as well as entrainment of fine particles. In order to generate fibrous aerosol continuously, Tanaka and Akiyama (1983; 1984) added a powder feeder to provide two components during aerosol generation.

On the other hand, the fibrous aerosol generator that was developed by Timbrell et al. (1968) and modified by Oritiz et al. (1977) was commonly used for inhalation toxicology of asbestos programs. The generators have two parts: a fiber-feed mechanism and a dispersing chamber using a pulverizer. These generators produced high mass concentration but less-stable generation than the fluidized bed aerosol generators.

Myojo (1983) modified the disadvantages of a fluidized bed aerosol generator. The generator enabled continuous generation of fibrous aerosols rather than a batch-feed-type generator.

26.2 INTERCEPTION EFFECT OF FIBROUS AEROSOLS

The interception effect is predominant at the penetration of fibrous aerosols through holes or other collectors. Spurny et al. (1979) suggested that it was most effective to use the interception effect for classifying the length of fibers. They tried to use a microsieve with mesh openings of between 5 and 50 μm.

Gentry et al. (1980) measured the collection efficiency of Nuclepore filters for fibrous aerosols. Irrespective of the flow velocity, the collection efficiency was constant. They concluded that the fibers were collected by the effect of the interception.

Gallily et al. (1986) and Foss et al. (1989) precisely calculated the flight trajectory of the cylindrical particle around the collectors, that is, sphere, cylinder, or hole. Gallily et al. calculated the collection efficiency of fibers ($Lf = 10$ μm and $Df = 1$ μm) through Nuclepore filters (pore size: 8 μm). The collection efficiency is the function of the initial orientation angle between the fibrous particles and the flow direction and Stokes number.

The collection efficiency of randomly oriented fibers is presented as a parameter of Stokes number. Foss et al. (1989) changed the parameters of the spherical collector and performed calculations similar to Gallily and Eisner (1979). The collection efficiency was changed by not only the initial orientation of the fiber but also the initial position. Their calculation results showed that collection efficiency depends on initial orientation. However, the correlation between initial orientation and the collection efficiency is not large.

26.2.1 Collection Model of Fibrous Particles through Wire Mesh Screens

Wire mesh screens are suitable for collecting long fibers by the interception effect. The wire screens of sieves can be used to separate the fibers. Compared with the microsieves or Nuclepore filters used by Spurny et al. (1979), the screens are stronger and enable the fibers to be shaken off more easily. Another advantage of wire mesh screens is a higher porosity than that of Nuclepore filters. Mesh openings of 19 μm (635 mesh) is the finest screen available. As yet, however, the relationships among fiber length, mesh opening, and penetration of fibrous aerosols through the mesh screens have not been measured quantitatively.

Myojo (1991) simulated the collection process of the fibers through a wire mesh screen using a Monte Carlo method. The simulation results are compared with experimental data using glass fiber aerosols. To simplify the collection process, we assumed that the orientation of the fibers at the initial position does not largely influence the collection efficiency of fibrous aerosols through the wire mesh screen. The collection process of fibrous aerosols can be depicted as follows. The conceptional diagram shown in Figure 26.5 will aid understanding of the concept.

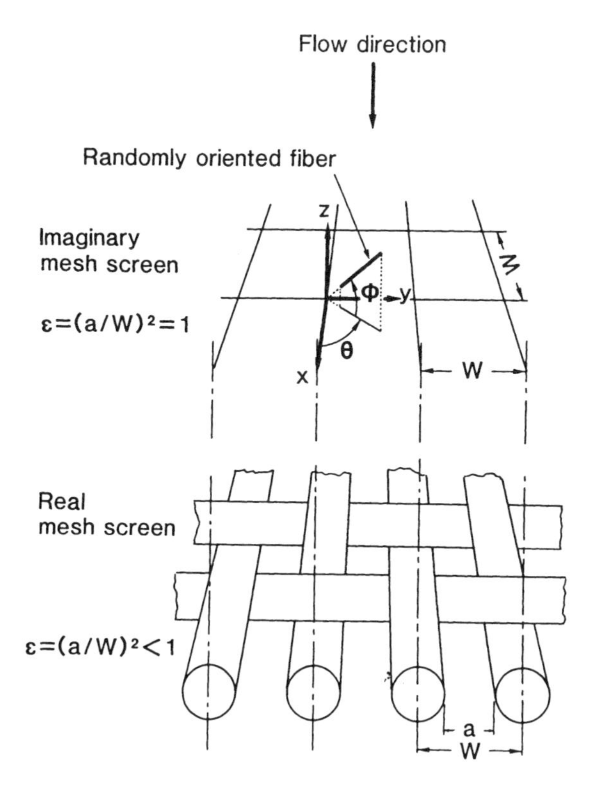

Figure 26.5 Conceptual diagram of a fiber introduced into a wire mesh screen.

1. The fibrous particles are randomly distributed at the infinite front of a mesh screen.
2. An imaginary mesh screen without wire thickness is introduced at the infinite front. As the porosity of the mesh screen is 100%, the flow near the screen is not disturbed by the wires.
3. The fibers interrupted by this imaginary wire mesh screen are collected in a real wire mesh screen.
4. The diameter of the fibers *Df* is small enough for the inertia effect and collection by interception to be negligible.
5. Collection is independent at each wire mesh screen. Hence, when the penetration of a wire mesh screen is *P*, the penetration of *N* pieces is *pN*.

The penetration of fibrous particles passing through the imaginary wire mesh screen ($\varepsilon = 1$) in Figure 26.5 was calculated using the Monte Carlo method. The following assumptions were used.

1. Fibers of length *Lf*, the center of gravity point (x, y), and angles θ and ϕ shown in Figure 26.5 are distributed uniformly and randomly. The range of distribution is $0 < x < W/2$, $0 < y < W/2$, $0 < \theta < \pi$ and $0 < \phi < \pi/2$. The uniform random numbers are generated by the RND function of Quick Basic (Microsoft, Redmond, WA). We used ten kinds of initial values as argument of the RND function.

2. If the projection image to the X-Y plane of a fiber touches or crosses the mesh screen, that fiber is collected. The above-mentioned assumption is correct for the imaginary mesh screen in Figure 26.5. The distance W between the center of the wires was chosen as the representative length of this system. We assigned nine different values to Lf/W between 0.15 and 5.0. We performed 5000 trials and calculated the penetration $P(Lf/W)$. The values of W, and a, and ε of the mesh screen used in this study are shown in Table 26.2.

Table 26.2 Dimensions of Wire Mesh Screen

No.	Mesh No.	W (μm)	a (μm)	ε
1	100	257 ± 12	162.4 ± 10.8	0.40
2	200	129 ± 6.5	78.3 ± 6.3	0.37
3	325	78.9 ± 6.7	44.2 ± 5.9	0.31
4	500	52.0 ± 2.7	26.7 ± 2.8	0.26
5	635	49.1 ± 3.6	19.5 ± 2.4	0.24

26.2.2 Measurement of the Penetration Rate through Wire Mesh Screens

In this investigation, we confirmed the separated fibrous samples passing through the wire screen by measuring the fiber sizes with a scanning electron microscope (SEM).

We used a fluidized bed aerosol generator to generate glass fiber aerosol as mentioned above. Glass fibers have simple shapes and are less harmful than asbestos. The aerosols were passed through ^{85}Kr neutralizer to neutralize the electrostatic charge of the particles.

Figure 26.6 shows the experimental setup used to facilitate penetration through the wire mesh screen. The wire mesh made of stainless steel was used as the test filter (4). Four layers of mesh screen with a diameter of 40 mm were set up at intervals of 5 mm. The approaching velocity was set at 0.27 m/s to reduce the inertia effect. Aerosols at the inlet and the outlet of the test filter were sampled with a Nuclepore filter (6) (pore size: 0.2 μm) at a flow rate of 16.7 cm^3/s.

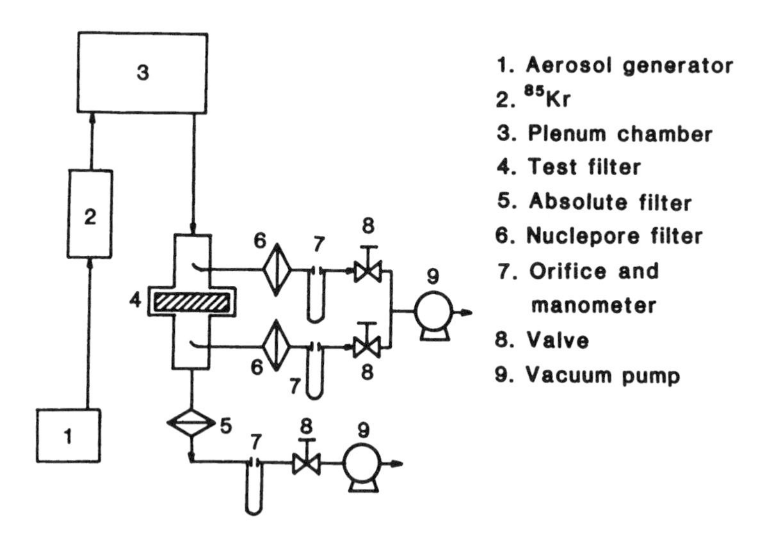

Figure 26.6 Experimental apparatus.

In order to determine the concentration and the size distribution, we observed the fibers collected on the Nuclepore filter using an SEM. The total number of fibers at the outlet of the mesh screen was divided by the number at the inlet for each size range to obtain the penetration.

26.2.3 Comparison between Simulation and the Experimental Results

Figure 26.7 shows the simulation results of the relationship between dimensionless fiber length *Lf*/*W* and the penetration. The dotted line in the figure indicates the penetration *P* passing through one screen. The solid line is the penetration of fibrous aerosols passing through four pieces, i.e., the fourth power of *P*.

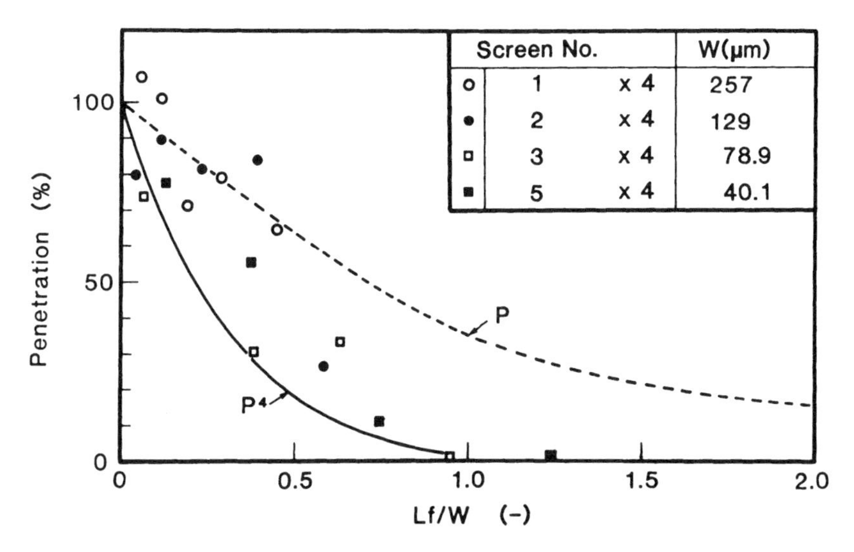

Figure 26.7 Penetration of fibrous aerosols through wire mesh screens. Dashed line is simulation result for one mesh screen. Solid line is for four mesh screens.

Figure 26.7 also shows the data of the penetration measured for each fiber length. We used four wire mesh screens with four distances between wires *W* in this experiment. The number of fibers in each range of length was counted. The arithmetic average of these ranges was used as the measured fiber length.

Generally, these data points are in agreement with the simulation results, P^4. However, the results of penetration are unstable because the number of fibers for each size range was too small. The porosity of these wire mesh screens was fairly small, ranging from 0.4 to 0.23 as shown in Table 26.2. However, penetration does not depend on porosity and is the function of dimensionless fiber length *Lf*/*W*. For a small *Lf*/*W* (less than 0.3), the experimental results show higher penetration than the simulation results. This may be because the flow field around the wire, which was ignored in this simulation, affected penetration.

According to Cheng et al. (1985), the collection efficiency of a similar wire mesh screen to the one used in this work was about 1% for spherical particles with a diameter of 1 μm. Therefore, the interception effect is predominant during the collection process of fibrous aerosols through wire mesh screens.

26.2.4 Length Distribution through Wire Mesh Screens

In order to calculate the length distribution of fibrous aerosols passing through wire mesh screens, we determined the penetration rate of fibers *P*(*Lf*/*W*) using the third spline interpolation method with nine values obtained in the above simulation. When a fibrous aerosol with a probability density function *f*(*Lf*) on the length distribution passes through *N* wire mesh screens, a cumulative penetration of less than *Lf* is expressed as follows:

$$y(Lf, W) = \int_0^{Lf} P(Lf/W)^N f(Lf) d(\log Lf) \tag{26.3}$$

If inlet fibers are units, the total amount of outlet fibers is shown in Equation 26.4.

$$Y(W) = \int_0^{\infty} P(Lf/W)^N f(Lf) d(\log Lf) \tag{26.4}$$

where $Y(W)$ is the penetration of fibrous aerosol through wire mesh screens. From Equations 26.3 and 26.4, a cumulative length distribution of less than Lf is as follows:

$$F(Lf, W) = \frac{y(Lf, W)}{Y(W)} \tag{26.5}$$

Equations 26.3 and 26.4 were solved by numerical integration. The probability density function $f(Lf)$ was determined from measured data using the third spline interpolation.

In this experiment, we measured more than 300 fibers to determine separately the cumulative length distributions of aerosol sampled at the inlet and the outlet of screens. The fibrous aerosols at the inlet and the outlet were not sampled simultaneously to collect enough fibers on the Nuclepore filters.

The solid lines in Figures 26.8 and 26.9 indicate the cumulative length distributions of the glass fiber aerosol introduced into the wire mesh screens. The dotted lines show calculated distributions using Equations 26.3 and 26.4 and solid lines.

Figure 26.8 shows the difference in mesh size for screen No. 2 ($W = 129$ μm) and No. 4 ($W = 52$ μm). The keys, ● and ■ for screens No. 2 and No. 4 show the measured length distribution. There were four mesh screens in total. Figure 26.9 shows the length distribution of aerosols passing through a different number of screens: two and eight. The keys, ● and ■ for two and eight mesh screens also show the measured length distribution. As mesh screen No. 2 was used in this figure, the case of four mesh screens is shown in Figure 26.8. The calculated lines in both figures agree with the experimental results for different cases.

These results suggest that the length-selection technique mentioned above is suitable for fibrous aerosols. The next section discusses one application of this technique.

26.3 APPLICATION OF THE SIZE-SELECTIVE TECHNIQUE USING WIRE MESH SCREENS

The hypothesis that "fibers which are long and thin, and remain in the lungs unaltered, are carcinogenic whether they are asbestos or not" was presented by Stanton et al. (1977; 1981), Pott (1978), and Pott et al. (1989). Stanton et al. emphasized that fibers with a diameter of 0.25 μm or less and a length of 8 μm or more exhibit high carcinogenicity. Pott also asserted that fibers with a diameter of 0.5 to 2.5 μm and a length of 3 to 20 μm are particularly carcinogenic.

A biological experiment was carried out by the National Institute of Industrial Health in Japan to evaluate the toxicity of short fibers that are supposedly less harmful. Chrysotile samples of various fiber lengths classified from the U.I.C.C. standard reference chrysotile B and the bulk sample were listed as target samples of the experiment.

Our biological experiment required about 1 g of separated fiber sample for the animal experiment (*in vivo*) and about 0.1 g for the cytotoxicity test (*in vitro*). To satisfy these requirements, we used a fluidized bed to aerosolize the samples, stainless steel wire screens as sieves to separate the fibers by length, and a virtual impactor to separate fibers of small diameter as an application of the previous section. This system can continuously classify asbestos aerosols to obtain size-selected asbestos samples without any change in chemical or physical properties.

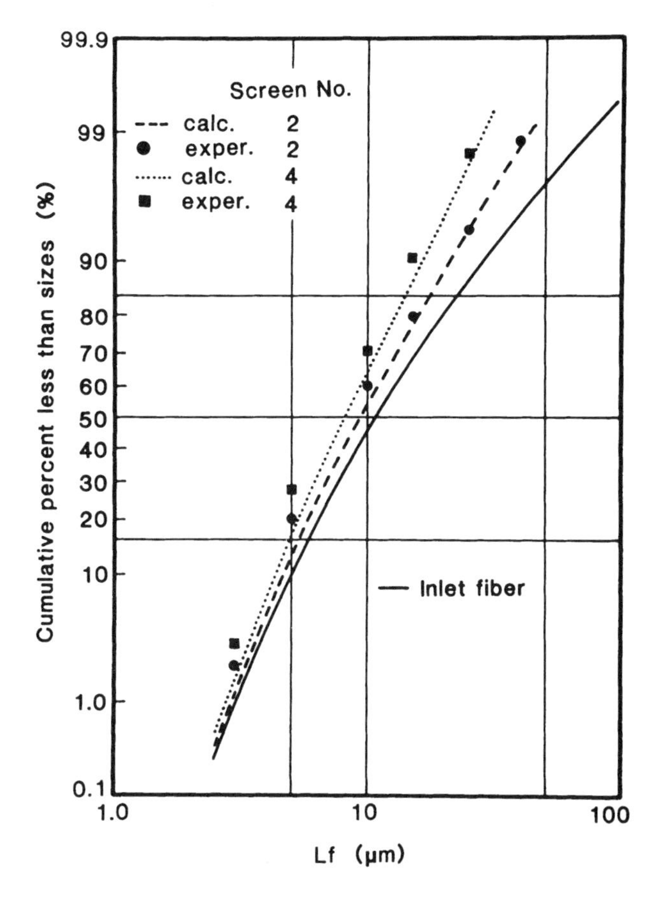

Figure 26.8 Comparison between calculated and experimental cumulative length distributions of fibrous aerosol. Four layers of mesh screens are used.

26.3.1 Evaluation of Length Classification System

(1) Classification Apparatus and Method

As shown in Figure 26.10, the continuous classification apparatus consisted of a fiber feeder (1), fluidized bed (2), wire screen for separation (4), and a virtual impactor (3). The feed rate was controlled by turning the motor on and off, which pushed out the fibers toward the grinder. We used a 50 mm fluidized bed, made of glass with a diameter of 50 mm and a height of 500 mm. For the fluidizing particles, we chose 100 g of 80 to 120 mesh glass beads as the fluidized material. The glass beads were washed sufficiently with water to remove small debris. A wire screen (4) with openings of 78 μm (200 mesh) or 19 μm (635 mesh) was inserted into the upper part of the fluidized bed for separation.

The aerosol was then introduced into a virtual impactor (3) (Variable Impactor manufactured by Sankyo Dengyo, Tokyo; hereinafter called "V.I."), which was developed by Masuda et al. (1979). The flow rate of the V.I. was adjusted so as to obtain particles of 2 μm in diameter for the 100% cutoff point with compact particles having a density of 2.5 g/cm^3. These separated fibrous aerosols were collected by membrane filters (8) and (9). Filter (8) corresponds to the side of the finest fibers. The experimental conditions of the V.I. are also indicated in Table 26.3.

(2) Cleaning of the Wire Mesh Screen

Since the fibers accumulating on the wire screens eventually cause clogging, the removal of such accumulated fibers is necessary to keep the wire screens in working condition. To clean the

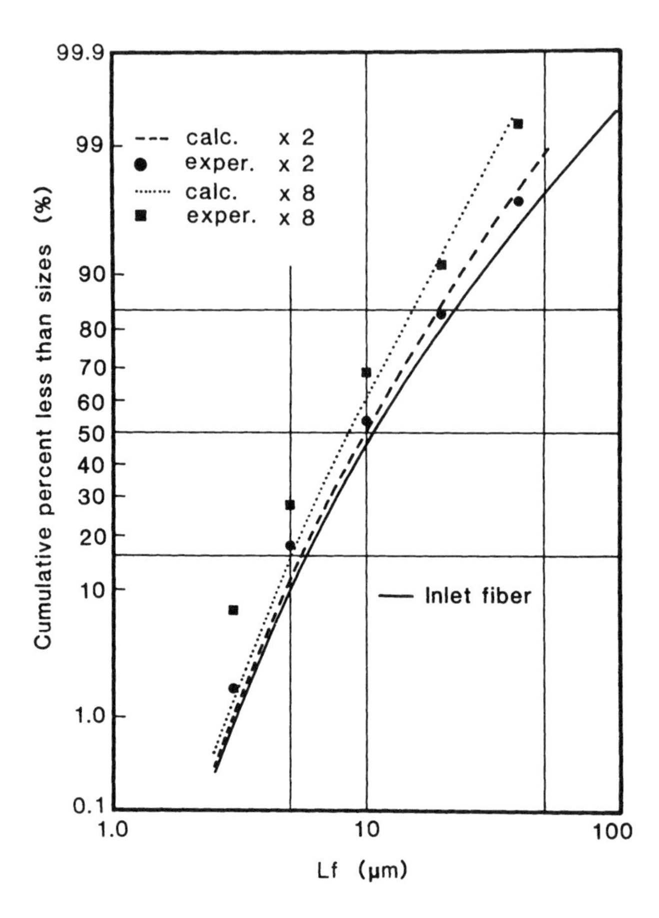

Figure 26.9 Comparison between calculated and experimental cumulative length distributions of fibrous aerosol. Mesh screen No. 2 was used to separate the fibrous aerosol.

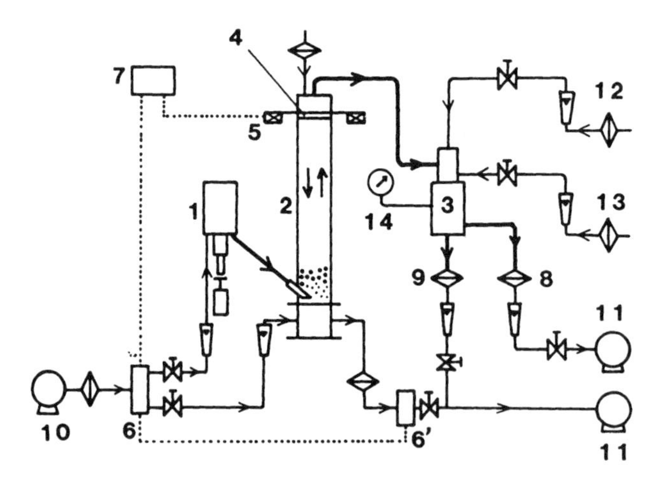

Figure 26.10 Schematic diagram of classification apparatus for asbestos fibers. (1) Fiber feeder, (2) 50-mm fluidized bed, (3) virtual impactor, (4) wire screen, (5) vibrator, (6) and (6′) solenoid valve, (7) controller, (8) filter (fine), (9) filter (coarse), (10) compressor, (11) vacuum pump, (12) core air, (13) sheath air, (14) pressure gauge.

Table 26.3 Experimental Conditions

Fluidized Bed (F.B.)	I	II
Bed diameter	50	70
Airflow into F. B. (dm^3/min)	10	17.0
Airflow into feeder (dm^3/min)	6	7.5
Superficial velocity in F. B. (m/s)	0.14	0.11
Glass bead (g)	100	200
Asbestos feed rate (mg/min)	1.8	1.4
Switching cycle of feeder	5 s on/15 s off	5 s on/20 s off
Fluidizing and cleaning cycle	110 s/10 s	110 s/10 s
Virtual Impactor		
Nozzle diameter (mm)	4	
Nozzle gap (mm)	1	
Aerosol flow rate (1/min)	26	
Core flow rate (1/min)	29	
Sheath flow rate (1/min)	9	
Fine flow rate (1/min)	58.5	
Coarse flow rate (1/min)	5.5	
100% cut diameter (μm)	2.0	

screens, the solenoid valve (6) on the clean-air supply side in Figure 26.10 is closed and the valve (6′) is opened to change the direction of airflow in the fluidized bed, and to suck air into the bed. At the same time, the frame of the wire screen is vibrated using a vibrator (5) to shake off accumulated fibers onto the fluidized particles. The apparatus repeats a cycle of generation for 110 s and shake-off for 10 s. The cyclic operation and monitoring of the whole apparatus was done using a microcomputer.

(3) Results and Discussion

The SEM micrographs of asbestos fibers in Figure 26.11 were obtained using this apparatus. Figure 26.11a shows asbestos fibers flowing out from the fluidized bed without a wire mesh screen. Considerably long fibers or bundle-form fibers were seen; the long fibers showed several similarities to the fibers contained in the bulk U.I.C.C. chrysotile B. It was difficult to measure the length of these fibers because the difference of the fiber length was very wide.

Figure 26.11b shows fibers that passed through wire screens with openings of 19 μm and are trapped on the coarse particle wide of the V.I. Some fine fibers can be seen, but fibers with large diameters and spherical particles are predominant.

Figure 26.11c is a photomicrograph of the fine side of the V.I. Most of the fibers here are shorter than 30 μm in length. A comparison between Figure 26.11b and c indicates that V.I. separation greatly depends on the fiber diameter. Figure 26.11d shows fibers that passed through the wire mesh screen with openings of 78 μm and collected on the fine side of V.I. The micrographs indicate the existence of fibers longer than 100 μm but there are no fibers or fiber aggregates having both a large length and a large diameter.

Figure 26.12 shows the fiber length distributions of the fibers in Figures 26.11c and d. The fibers in Figure 26.11c passing through the 19-μm-opening wire screen are close to the desired size, of which a fiber length of less than 5 μm accounts for 81% and that of less than 10 μm for 96%.

The fiber length distribution of the U.I.C.C. chrysotile B measured by Timbrell (1970) is also shown in the figure. These results were obtained using an optical microscope for bulk asbestos samples; the fibrous aerosol samples were generated by an aerosol generator (Timbrell et al., 1968). The length distribution of the bulk samples was wide and decreases slightly after aerosolizing. This distribution is similar to that of the fibers passing through the 78-μm wire screen in Figure 26.11.

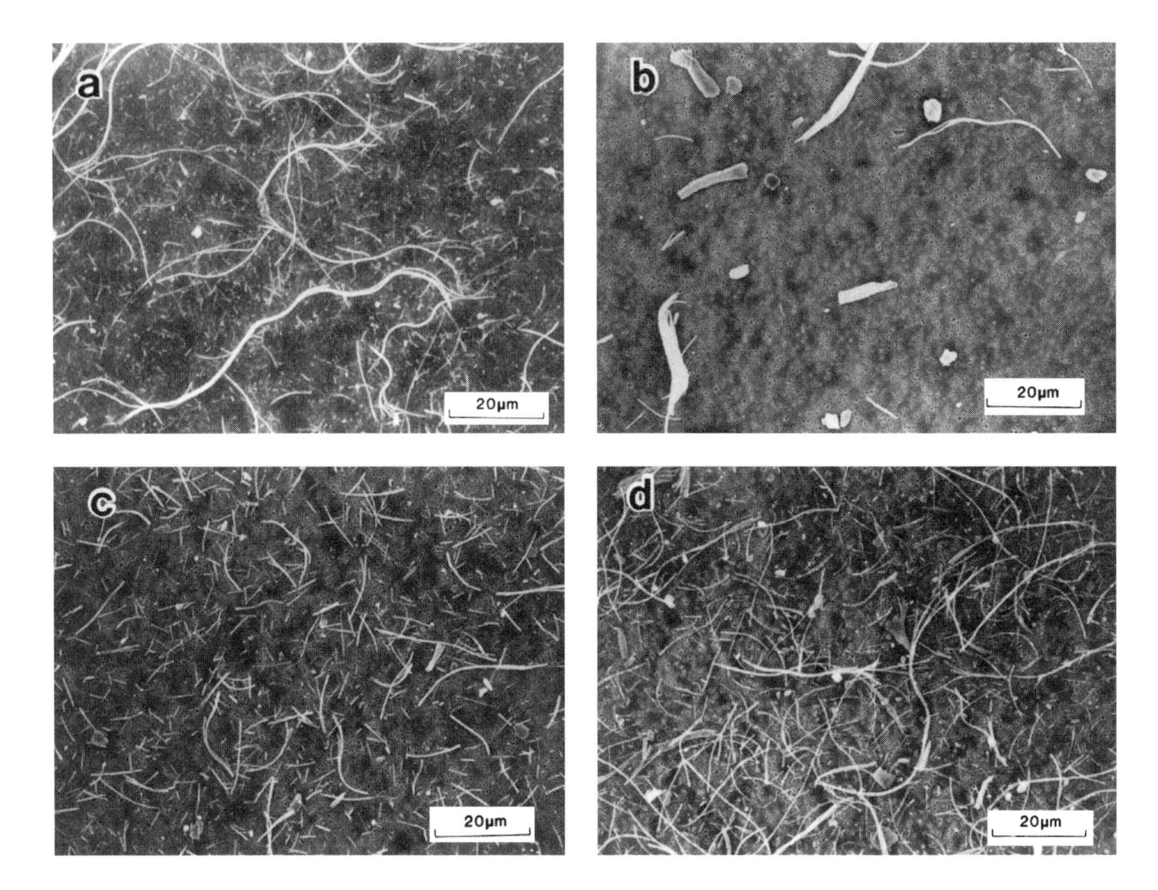

Figure 26.11 SEM micrographs of asbestos (U.I.C.C. chryosotile B). (a) Fibers flow out of fluidized bed. (b) Fibers penetrating 19-μm wire mesh screen and V.I. coarse side. (c) Fibers penetrating 19-μm wire mesh screen and V.I. fine side. (d) Fibers penetrating 78-μm wire mesh screen and V.I. fine side.

The distribution of the fibers passing through the 78-μm wire screen apparently differed from the fibers passing through the 19-μm wire screen.

In order to estimate the separation performance using this apparatus, we performed aerosol generation for 3 h and measured the amount of asbestos passing through the wire screen. The separation performance is determined from the ratio of the amount of collected asbestos to the amount fed (0.331 g) for 3 h as estimated from the feed rate. The results are shown in Table 26.4. Performance without a wire screen has not yet been measured, but the separation performance may be assumed to be about 10% in considerable of the results for glass fibers (Myojo, 1983). The amount of fibers passing through the wire screens was extremely small.

26.3.2 Preparation of Size-Selected Asbestos Samples for Animal Studies

(1) Apparatus and Method

We prepared another fluidized bed for continuous separation of chrysotile fibers. The inside diameter of the fluidized bed was increased to 70 mm to provide a larger filtering area for the wire mesh screen. In order to save time, the fluidized bed was designed to produce two types of separated samples simultaneously. We used the virtual impactor to separate fibers passing through 19-μm mesh screen.

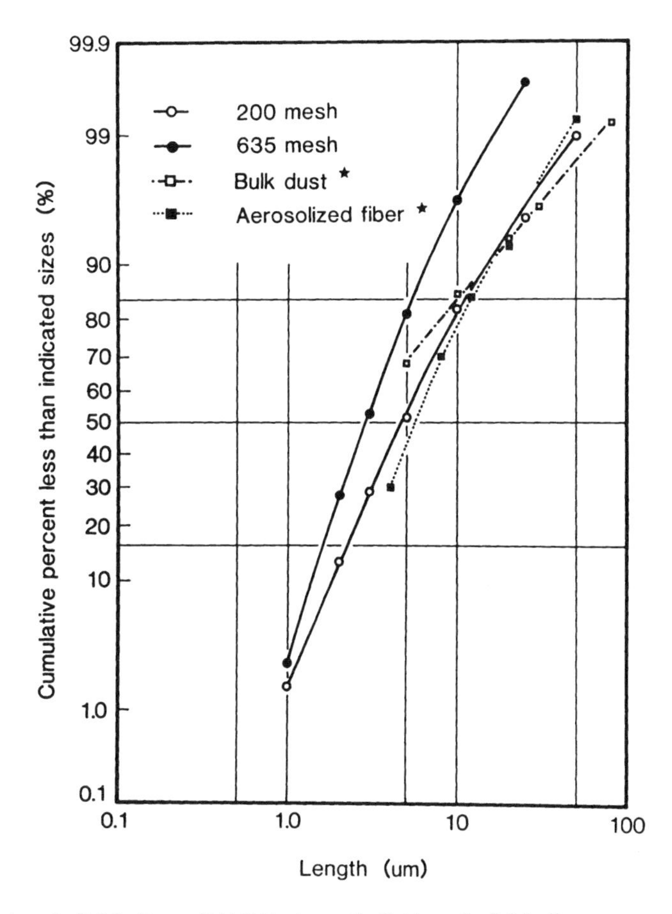

Figure 26.12 Fiber length distributions of U.I.C.C. chrysotile B. * Length distribution measured by Timbell (1970).

Table 26.4 Generation Efficiency of U.I.C.C. Chrysotile B

No.	Screen Size, Mesh	Generated Fiber (mg/3 h)	E %
1	200	6.25	1.9
2	200	6.50	2.0
3	200	5.68	1.7
4	635	1.31	0.40
5	635	1.70	0.51

Feed rate: 0.042 mm/min. Fiber mass fed for 3 h: 0.331 g.

Figure 26.13 is a schematic diagram of the 70-mm fluidized bed. This fluidized bed is made of glass with a height of 500 mm and contains 200 g of 80 to 120 mesh glass beads. The stainless steel wire screen (12) has openings of 78 μm, and the downstream wire screen (2) has openings of 19 μm.

In the 70-mm fluidized bed, fibers that accumulated on the 200-mesh wire screen (1) were shaken off by vibration onto the fluidized material by the airflow in the reverse direction. Fibers on the 635-mesh wire screen (2), which was cylindrical and rotated by a motor, were removed by the flow provided through the suction port (10) — fiber outlet (10) in Figure 26.13. Next, the fibers on the wire screen (2) were collected in the filter holder. The fibers caught at this point were used

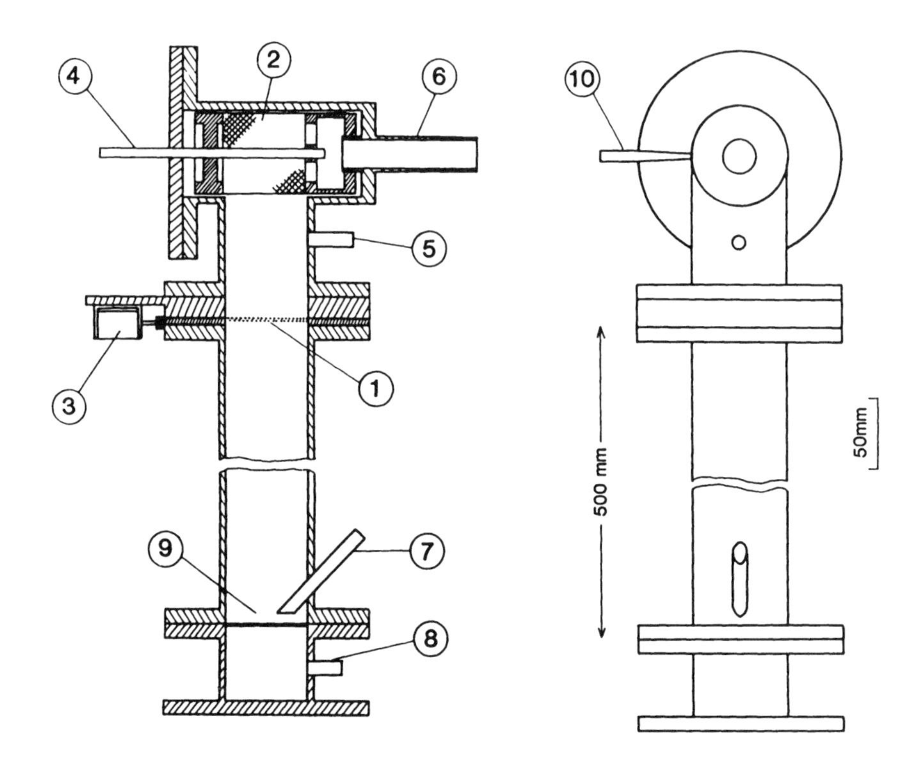

Figure 26.13 Detail of 70-mm fluidized bed. (1) Wire screen (200 mesh), (2) wire screen (635 mesh), (3) vibrator, (4) motor shaft, (5) clean air inlet, (6) aerosol outlet, (7) fiber inlet from feeder, (8) clean air inlet for fluidizing, (9) glass beads, (10) fiber outlet.

as a sample pass-through wire screen (1) but do not pass through the wire screen (2). The other overall configuration, such as the connection between the solenoid valves and the V.I., were the same as those shown in Figure 26.10.

Figure 26.14 shows the airstream direction at the time of generation (fluidizing) and shake-off (cleaning) of the 70-mm fluidized bed. In the fluidizing process, chrysotile fibers disperse on the bed and rise with the airflow. The fibrous aerosol pass through the 78-μm wire screen (1) from the bottom, then pass through the cylindrical 19-μm-opening wire screen (2) from the outside to the inside. Finally, the aerosol is sucked into the V.I. at a rate of 26 l/min.

In the cleaning process, the flow direction was reversed by the solenoid valve, and clean air was introduced into the bed through the clean air inlet (5) (see Figure 26.13). The clean air flowed in two directions: toward the bottom of the fluidized bed and toward the fiber outlet (10) on the side of the cylindrical wire screen. During this process, the frame holding the wire screen (1) was shaken by the vibrator. The fibers that accumulated on the wire screen (1) dropped onto the fluidizing particles. The fibers that accumulated on the cylindrical wire screen (2) were collected on the filter. The wire screens were cleaned as described above and the fluidizing process was started again.

(2) Results and Discussion

During continuous operation of the 70-mm fluidized bed, we obtained about 0.1 g of asbestos fibers which passed through the 19-μm wire screen and were caught on the fine side of the virtual impactor with a cutoff point of 1 μm and 1.5 g of asbestos which passed through the 78-μm wire screen but were unable to pass through the 635-mesh screen from approximately 20 g of U.I.C.C. standard reference asbestos (chrysotile B). The former, which consisted of short fibers, was an insufficient amount for use in injection or inhalation experiments on animals but was offered for cytotoxicity tests.

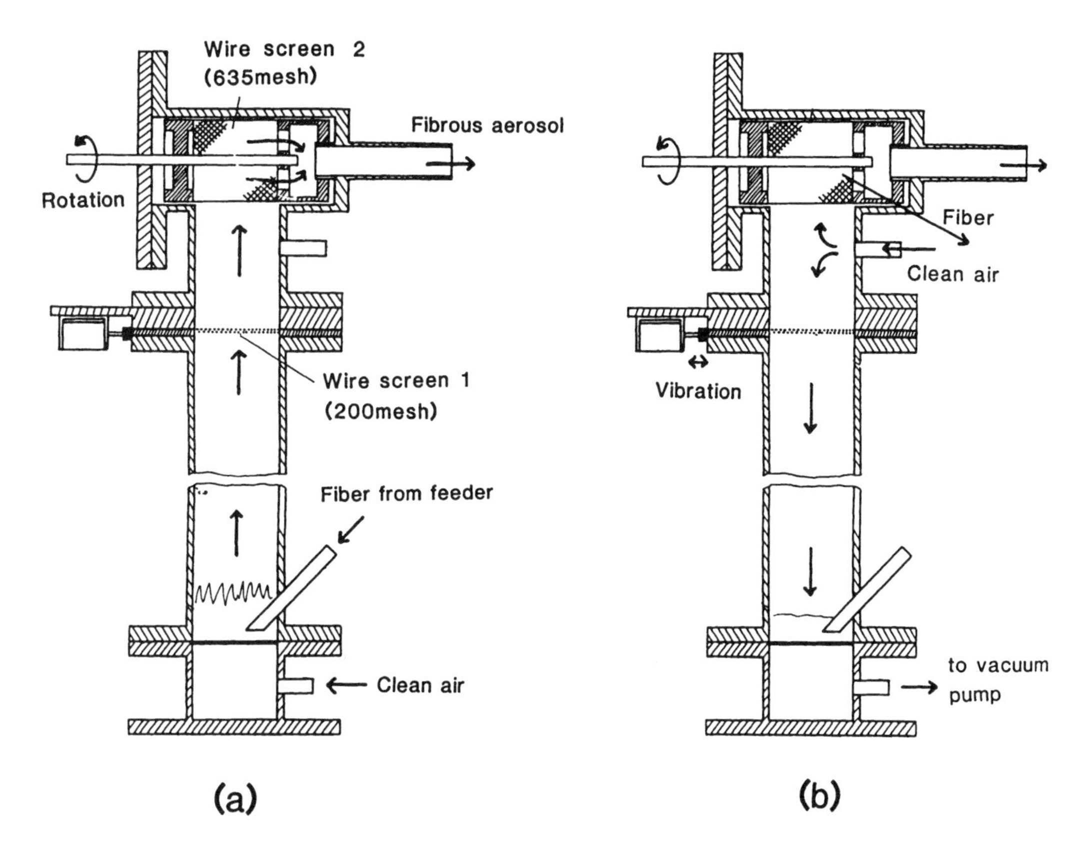

Figure 26.14 Schematic diagram of fluidizing process (a) and cleaning process (b).

26.4 CONCLUSION

We simulated the penetration of fibrous aerosols through wire mesh screens by means of the Monte Carlo method. The interception parameter was defined as the ratio of fiber length to wire distance of the mesh screen. The relationship between the interception parameter and the penetration of fibrous aerosols was blunt, but longer fibers were eliminated by the wire mesh screens. The experimental results for penetration were in agreement with the simulation results.

As an application of the results, a classification apparatus was built and operated automatically for a total of about 350 h. The fibers obtained by this method were not subjected to alteration of their chemical composition since separation was carried out in a dry state, and X-ray diffraction results did not indicate any obvious degradation in crystallinity. The process is a superior separation method for providing fibrous samples for biological experiments. The results of the cytotoxicity test using these samples revealed that short fibers were less harmful than medium-sized or bulk samples (Koshi et al., 1991).

ACKNOWLEDGMENT

This paper was rewritten on the basis of four papers published in Japan (Myojo, 1988, 1991; Myojo and Kohyama, 1990, 1992). The author greatly appreciates the courtesy of the publishers.

REFERENCES

Baron, P. A., Deye, G. J., and Fernback, J. (1994), *Aerosol Sci. Technol.,* 21, 179–192.
Burke, W. A. and Esmen, N. (1978), *Am. Ind. Hyg. Assoc. J.,* 39, 400–405.
Carpenter, R. L. and Yerkes, K. (1980), *Am. Ind. Hyg. Assoc. J.,* 41, 888–894.
Carpenter, R. L., Pickrell, J. A., Mokler, B. V., Yeh, H. C., and DeNee, P. B. (1981), *Am. Ind. Hyg. Assoc. J.,* 42, 777–784.
Cheng, Y. S., Yeh, H. C., and Brinsko, K. J. (1985), *Aerosol Sci. Technol.,* 4, 165–174.
Esmen, N. A., Kahn, R. A., LaPietra, D., and McGovern, E. P. (1980), *Am. Ind. Hyg. Assoc. J.,* 41, 175–179.
Foss, J. M., Frey, M. F., Schamberger, M. R., Peters, J. E., Leong, K. H. (1989), *J. Aerosol Sci.,* 20, 515–532.
Fuchs, N. A. (1964), *The Mechanics of Aerosols,* Pergamon Press, Oxford, 34–49.
Gallily, I. and Cohen, A. (1979), *J. Colloid Interface Sci.,* 68, 338–356.
Gallily, I. and Eisner, A. D. (1979), *J. Colloid Interface Sci.,* 68, 320–337.
Gallily, I., Schiby, D., Cohen, A. H., Hollander, W., and Schless,s D. (1986), *Aerosol Sci. Technol.,* 5, 267–286.
Gentry, J. W., Spurny, K. R., Opiela, H., and Weiss, G. (1980), *Ind. Eng. Chem. Prod. Res. Dev.,* 19, 47–52.
Harris, R. L. (1972), Ph.D. thesis, University of North Carolina, Chapel Hill.
Hoover, M. D., Casalnuovo, S. A., Lipowicz, P. J., Yeh, H. C., Hanson, R. W., and Hurd, A. J. (1990), *J. Aerosol Sci.,* 21, 569–575.
Jeffrey, G. B. (1922), *Proc. R. Soc. London,* A102, 161–179.
Koshi, K., Kohyama, N., Myojo, T., and Fukuda, K. (1991). *Ind. Health,* 29, 37–56.
Loo, B. W., Cork, C. P., and Madden, N. W. (1982), *J. Aerosol Sci.,* 13, 241–248.
Masuda, H., Hochrainer, D., and Stober, W. (1979), *J. Aerosol Sci.,* 10, 275–287.
Myojo, T. (1983), *Ind. Health,* 21, 79–89.
Myojo, T. (1988), *J. Aerosol Res. Jpn.,* 3, 60–63.
Myojo, T. (1991), *J. Soc. Powder Technol. Jpn.,* 28, 495–500.
Myojo, T. and Kohyama, N. (1990), *J. Soc. Powder Technol. Jpn.,* 27, 804–810 (in Japanese).
Myojo, T. and Kohyama, N. (1992), *KONA Powder Particle,* 10, 184–191 (in English).
Oritiz, L. W., Black, H. E., and Coulter, J. R. (1977), *Ann. Occup. Hyg.,* 20, 25–37.
Pott, F. (1978), *Staub Reinhalt. Luft,* 38, 486–490.
Pott, F., Roller, M., Ziem, U., Reiffer, F. J., Bellman, B., Rosenbruch, M., and Huth, F. (1989), Non-occupational exposure to mineral fibres, Bignon, J., Peto, J., and Saracci, R., IARC Scientific Publications, Lyons, France, 173–179.
Spurny, K. R., Boose, C., and Hochrainer, D. (1975), *Staub Reinhalt. Luft,* 35, 440–445.
Spurny, K. R., Boose, C., Hochrainer, D., and Monig, F. J. (1976), *Ann. Occup. Hyg.,* 19, 85–87.
Spurny, K. R., Stober, W., Opiela, H., and Weiss, G. (1979), *Am. Ind. Hyg. Assoc. J.,* 40, 20–38.
Stanton, M. F., Layard, M., Tegeris, A., Miller, E., May, M., and Kent, E. (1977), *J. Natl. Cancer Inst.,* 58, 587–603.
Stanton, M. F., Layard, M., Tegeris, A., Miller, E., May, M., Morgan, E., and Smith, A. (1981), *J. Natl. Cancer Inst.,* 67, 965–975.
Stober, W., Flachsbart, H., and Hochrainer, D. (1970), *Staub Reinhalt. Luft,* 30, 227–285.
Sussman, R. G., Gearhart, J. M., and Lippmann, M. (1985), *Am. Ind. Hyg. Assoc. J.,* 46, 24–27.
Tanaka, I. and Akiyama, T. (1983), *Am. Ind. Hyg. Assoc. J.,* 44, 875–877.
Tanaka, I. and Akiyama, T. (1984), *Am. Ind. Hyg. Assoc. J.,* 28, 157–162.
Timbrell, V. (1965), *Ann. N.Y. Acad. Sci.,* 132, 255–273.
Timbrell, V. (1970), *Pneumoconiosis,* H. Shapiro, Ed., Oxford University Press, New York, 28.
Timbrell, V., Hyett, A. W., and Skidmore, J. W. (1968), *Ann. Occup. Hyg.,* 11, 273–281.

Part III

APPENDIX

Nikolai Albertowich Fuchs: (1985–1995) Aerosol Scientist and Humanist
Rememberance to the 100th Anniversary of His Birthday

Kvetoslav R. Spurny

CONTENTS

1 INTRODUCTION

Professor N. A. Fuchs was the founder of aerosol science in Russia and an important cofounder of this scientific discipline worldwide. He was born on July 32, 1895 at Lantvarovo, a small town in Lithuania. He lived his life in Moscow — except the years he spent as an accused, but innocent, political dissident and persecuted democrat in Stalin's prisons in banishment. He passed away on October 10, 1982 in Moscow.

I met Prof. Fuchs several times between 1956 and 1970 — in Moscow, Tiflis, Odessa — and he visited me in Prague in June 1966. I had the opportunity to speak with him about science, life, hobbies, and sometimes about books that were suppressed or forbidden in Russia, about philosophies, as well as about politics. The latter discussions could be accomplished in "special environments" only — like during walks in woods, in theaters, etc. — where it could be assured that official and secret natural or electronic ears were not present.

I was therefore partly informed about the difficult life of Fuchs in Russia. Perhaps he had survived all these inequities and inhuman treatment because of his very good health and physical

0-87371-830-5/98/$0.00+$.50

condition, his modesty, admirable optimism and, last but not least, because of the great and permanent support of his family, mainly his wife, Marina.

Nevertheless, for a long time, we have had no full and realistic picture about his life. During the period of the Gorbatchov "Perestroika," Marina Guseva, Fuchs's wife, wrote a true and almost complete biography of her husband. My wife Drahuse and I were very good friends and letter friends of Nilolai Albertowich and his family. This friendship continues also with Prof. Michael Gusev, Fuch's son, and his family, who live in the same flat in Moscow. I very much appreciate that Marina has acknowledged our friendship and help during the saddest times in her memories.

2 AEROSOL SCIENCE

Fuchs was one of a select group of scientists who combined a brilliant experimental technique with the ability to make theoretical interpretation from experimental results. He was also an accomplished classical theoretical physicist (Editorial Board, 1970; 1975; 1976; 1983). He was one of a group of scientific giants which included Victor La Mer and Irving Langmuir, a Nobel Laureate, who contributed to the fundamental understanding of aerosol phenomena over roughly the same time period. Fuchs was unique among that group in his vision of aerosol science as a scientific discipline (Friedlander, 1990).

Fuchs attached great importance to the activities of scientific organizations. He was a member of the editorial board in Russia of the *Journals of Colloid Science* (*Kolloidnyi Zhurnal*). He joined the Board of Overseas Editors in the *Annal of Occupation Hygiene* (England) in 1963 and was a founding member of the *Journal of Aerosol Science,* since 1970. He gave this journal considerable support. He had been an honorary Member of the Gesellschaft für Aerosolforschung since 1982. He was happy about that and wrote us that in order to accept he had to wait, unfortunately, for the agreement of his government. He died unexpectedly the same year — without obtaining this permission. So he is the GAeF honorary member without such permission — and this is his best recommendation.

Fuchs began his scientific activity in the laboratory organized by the famous Russian physicist P. N. Lebedev. He started his independent scientific activity in the chair of the second Institute of Chemical Engineering in Moscow. His work during that period was devoted to the investigation of capillary equilibrium and the interface between two liquid phases and a single vapor phase, to the development of a method for investigating surfaces, and to obtaining two-dimensional crystals. He was a pioneer in the introduction of separational chromatography in Russia. He developed a method for determining the active gamma isomer of hexachlorocyclohexane, and a number of other chromatographic methods of chemical analysis (Petryanov et al., 1966).

He began his aerosol studies in 1932 in the L. Ya. Karpov Physicochemical Institute in Moscow, where he organized an aerosol laboratory. The organization of such a laboratory was proposed and supported by the academician A. N. Bach; A. V. Frunkin, I. Petryanov, B. Rotzeig, H. Rosenblum, and P. Lisovskii were his first co-workers.

At that time the study of aerodisperse systems worldwide was still in an embryonic state, and a necessary condition for the development of this field of science was the creation of precise methods for measuring the size and electric charge of particles. This problem was solved by the construction of equipment in which electric fields were photographed in an ultramicroscope. It was the so-called particle-oscillating method (Figure 1) that was published in 1933 (Bibliography 11).

At the same time he was involved with questions of the mechanisms of the formation of aerosols. He developed the theory of transfer phenomena in aerosols on the basis of the method of the boundary sphere. A great deal of work, carried out together with his students in the 1930s, has become classic. In 1934 he published his first theory of aerosol coagulation. He later considered his coagulation theory formula his chief scientific achievement (Figure 2). This field of aerosol coagulation was the basis of a theoretical disagreement between Norman Davies and Fuchs (Davies,

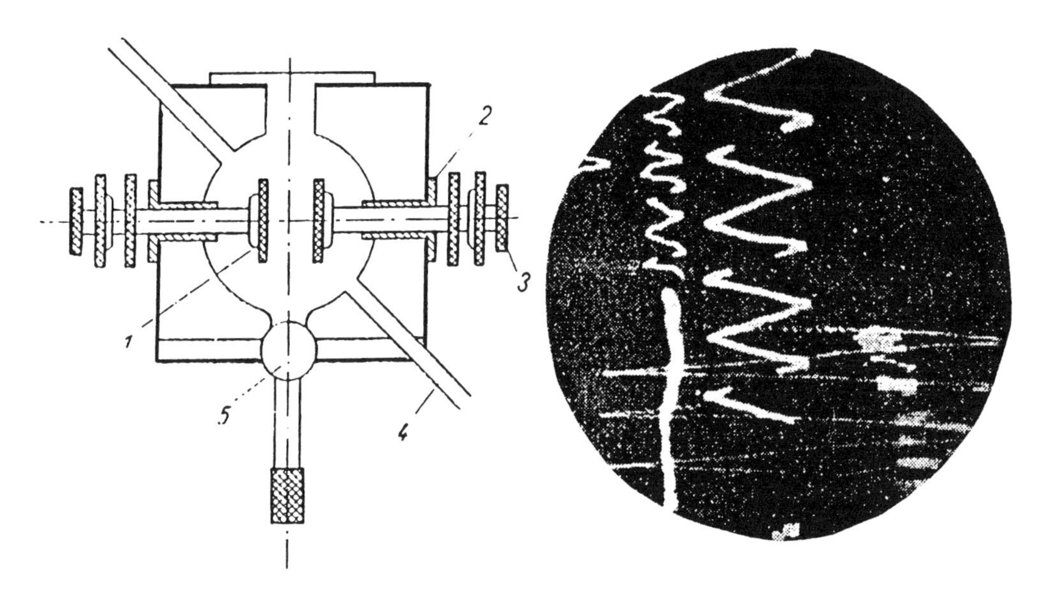

Figure 1 Schematic picture of the head of the ultramicroscope for measuring size and electric charges of single aerosol particles (left). A photograph of the particle movement (right).

1979; Fuchs, 1980). Davies was 68 years old and Fuchs 84 at that time. As it is well known, Fuchs had extrapolated the Smoluchowski coagulation theory in liquids for the gas phase as the basis for his aerosol coagulation theories (Fuchs, 1934; 1964). He developed and described a mode of the "boundary sphere" — for the case of intermediate values of Knudsen numbers Kn: $Kn = l/r$, where l is the mean free path of gas molecules and r is the particulate radius. This sphere is concentric with the particle, and the gap between is on the order of magnitude of the free path of gas molecules. He introduced also the "aerosol Knudsen number" $Kn_p = l_p/r$, where l_p is the "apparent free path of the Brownian particles" with the radius r. For this reason Fuchs introduced also a correction factor β_F for calculation of the actual value of the coagulation constant: $K = K_o \cdot \beta_F$.

Davies criticized this correction factor β_F as an "erroneous" one. Based on experimental results of several publications, Davies proposed that the correction for a concentration jump near the particle surface not be used. In his opinion the Smoluchowski formula with the Cunningham–Millikan slip correction is valid at $Kn < 15$. For cases of $Kn > 15$ he proposed an empirical formula.

Fuchs showed in his reply (Fuchs, 1980) — Davies had not allowed this reply to be published in the *Journal of Aerosol Science* — that Davies had ignored his considerations, by the use of different experimental data, the fact that K_o depends strongly on the polydispersity of the investigated model aerosol. And in the majority of cases of these published data, this factor was greater or much greater than 1. By considering this important fact, Fuchs could get very good agreement between his theory and the newly published data, which had also been partly used by Davies. Nevertheless, it seems to me that this "competition," which is not unusual among scientists, did not seriously touch the very good friendship between Davies's and Fuchs's families.

In the same year — 1934 — Fuchs proposed and developed, in cooperation with his co-workers Petryanov and Rozenblum, the first fibrous aerosol filters (see also the Kirsch paper). He was the author, or part author, of 109 publications (see also the Bibliography, collected by C. N. Davies and A. A. Kirsch). However, there were no publications between 1937 and 1947, because of Fuchs's "most difficult time" (Spurny, 1983; 1984). The year 1955 saw the publication of his monograph *Mechanika Aerosolei* (*Mechanics of Aerosols*), in Russia. It was the first book on this field of physics and is still one of the most important (Figure 3). This book brought together, for the first time, an enormous range of phenomena related to aerosol systems. Fuchs succeeded in taking set of diverse, apparently unrelated experimental observations by a multitude of investigations and

Moscow, April 26, 1979

Dear Drahuse,

Some days ago I received 3 discs with records of "Saoul" and yesterday - 9 disks with Bach's cantatas. All records are quite good. I am immensely grateful to you for all your troublesome and time-consuming work of bying and sending me the disks. How much joy they bring to me ! Now I have got a good headphone and can listen without disturbing Marina who does not like classical music. So I am quite happy.

We have had an exceptionally severe winter and fear that all our fruit-trees have perished from the frost. The April was also uncommonly cold and the real spring began only two days ago. Marina went with our pet - Airdale-Terrier dog Johnny to stay at our cottage , Mike and I will spend our rest-days there. In a week we shall see what plants have survived in our garden.

As I could not persuade C.N.Davies to abstain from publishing his paper in which he maintains that my coagulation formula - my chief scientific achievment - is quite wrong, I am forced to publish critical comments on this paper and this will probably end our friendship lasting more than 20 years. Nothing doing !

Did Kvetoslav get a better post? I am anxious to know it. My best regards to him.

With best wishes
Yours very grateful N.Fuchs

Figure 2 Fuchs's letter from 1979 confirming the receipt of new LP records and his remark on his coagulation formula.

organizing it in a coherent and comprehensive fashion. He showed that many of these phenomena could be treated within the same framework of theoretical concepts and mathematical equations. Fuchs's monumental work on the *Mechanics of Aerosols* has given us a solid foundation on which to build aerosol science and technology.

The first half of *Mechanics of Aerosols* book deals primarily with deterministic analysis of what we could call "coarse" particle transport, based on the equations of particle motion. The second half deals with various stochastic processes including Brownian and turbulent diffusion and coagulation (Friedlander, 1990). This book remains the most cited one in the world of aerosol literature. Indeed, this is the first fundamental monograph in which the basic information concerning almost all properties of aerosols is carefully summarized and presented in such a way that this book can be used both as a manual and as a reference book. It is still the actual handbook for all aerosol scientists.

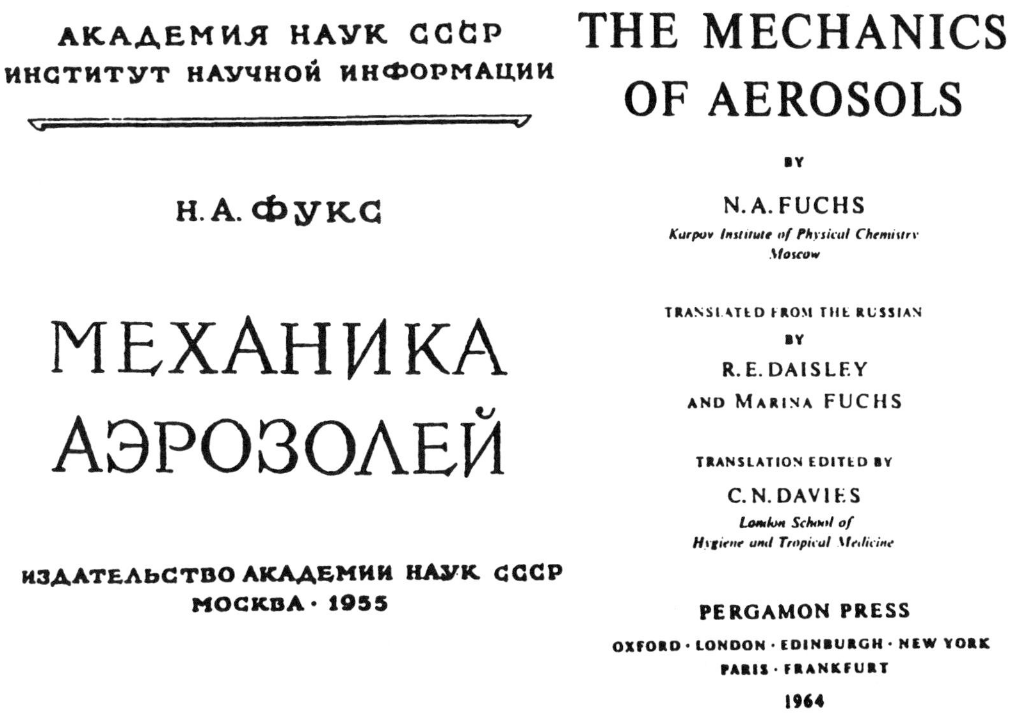

АКАДЕМИЯ НАУК СССР
ИНСТИТУТ НАУЧНОЙ ИНФОРМАЦИИ

Н. А. ФУКС

МЕХАНИКА
АЭРОЗОЛЕЙ

ИЗДАТЕЛЬСТВО АКАДЕМИИ НАУК СССР
МОСКВА · 1955

THE MECHANICS
OF AEROSOLS

BY

N. A. FUCHS
Karpov Institute of Physical Chemistry
Moscow

TRANSLATED FROM THE RUSSIAN
BY
R. E. DAISLEY
AND MARINA FUCHS

TRANSLATION EDITED BY
C. N. DAVIES
London School of
Hygiene and Tropical Medicine

PERGAMON PRESS
OXFORD · LONDON · EDINBURGH · NEW YORK
PARIS · FRANKFURT
1964

Figure 3 The title pages of the Russian and English versions of the book *The Mechanics of Aerosols.*

Fuchs had “invented” the designation “mechanics of aerosols,” and he defined it as the “study of the motion and precipitation of aerosol particles.” The first English translation was made in 1958 in the U.S. under the auspices of the U.S. Department of the Army. An “official” English edition was prepared by C. N. Davies (Vincent et al., 1994). I had obtained the Russian version of *Mechanics of Aerosols* as a gift from N. A. Fuchs in 1961 during my visit in Moscow (Figure 4). Returning to Prague, I sent this book immediately to the publisher of Pergamon Press in England, Mr. Maxwell, recommending to him that he prepare and edit an English version. Mr. Maxwell was from Czechoslovakia, hence my “countryman.” In the same letter I also recommended to Mr. Maxwell that he plan the edition of a new journal — the *Journal of Aerosol Science.* Both my proposals were accepted and realized by the great support and help of Davies. During the “translation” time, a friendship between the Davies and the Fuchs families began. Davies had also initiated and supported other Fuchs’ publications. This was at the time a great help for Fuchs to overcome his isolation in Russia. The second English edition of *Mechanics of Aerosols* appeared in 1989 published by Dover Publications, Inc., in New York, and was initiated by the American Association for Aerosol Research (AAAR). In 1958 there followed the book *Evaporation and Growth of Particles in a Gas Medium* and *Progress in the Mechanics of Aerosols.* The latter book was then included in the Pergamon Press edition of *Mechanics of Aerosols* in 1964.

The next important monograph, written by Fuchs and his co-worker A. G. Sutugin, was published in Russian in 1969 and in English in 1970 (Figure 5). In this book their research results as well as the state of art in the field of high-dispersed aerosols (HDA) were described. The HDA with particle sizes less than and much less than 100 nm found broad research interest and applications later in the 1980s and 1990s. They cover three orders of magnitude in particle diameter, equivalent to nine in particle mass. In 1969 the monograph of Green and Lane, *Particulate Clouds,* originally in English in 1957, was published in Russian (Figure 6). Fuchs was the editor. He had

Figure 4 My visit in Fuchs's laboratory in Karpov Institute in Moscow in 1961 (Prof. Fuchs on the left).

the book substantially expanded and modernized. Fuchs was also engaged in organizing the aerosol literature. In 1971 he succeeded in publishing the first part of his *Collected Abstracts* (Figure 7). This is a unique collection of the whole aerosols literature from 1850 to 1953. When you go through this book you will often discover that what many of the contemporary publications in aerosol science describe as "news" had been previously published many years before.

He also prepared the second part of this collection with abstracts published after 1954. George Hidy, myself, and other American friends, looked for a publisher in the U.S., unfortunately, without success, which was a great pity. In the 1980s, Fuchs was engaged in major work on the "Methods of Measuring Aerosol Concentrations," which was planned to be printed in Russian, in 1984. In his last letter to me (Figure 8) he had mentioned this book.

Besides his several monographs, Fuchs and his co-workers (Figure 9) made many new investigations during the period between 1960 and 1980. They were very busy. They continued in the research of HDA, developed and described an excellent method for measuring the size distributions of HDA, investigated the formation process of HDA, and the coagulation of aerosols in sonic and ultrasonic fields (e.g., F. I. Murashkewich), etc. Fuchs and his co-workers, mainly A. A. Kirsch and I. B. Stechkina, devoted many years to the study of aerosol filtration processes. It was shown that their "fan filter model" could well represent commercial filters well. Furthermore, in several papers, published mainly in the *Journal of Aerosol Science,* they described the filtration theory for a combination of several partial mechanisms. They also found good agreement of their theory with experimental results (see section by Kirsch and Stechkina in this book).

3 FRIENDS AND FRIENDSHIPS

Prof. Fuchs exchanged letters with many colleagues in Europe, the U.S., Japan, etc. As mentioned in his wife's memoirs (to be published by University Press, Delft, The Netherlands, 1998),

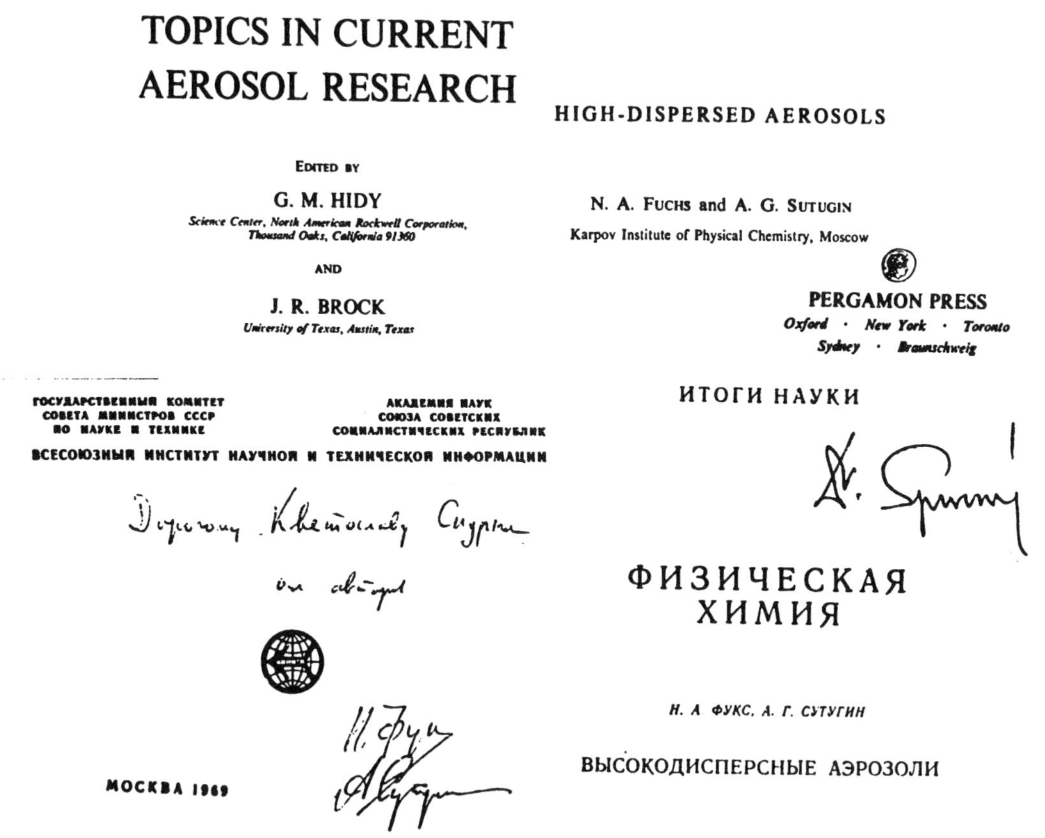

TOPICS IN CURRENT
AEROSOL RESEARCH

EDITED BY

G. M. HIDY
Science Center, North American Rockwell Corporation,
Thousand Oaks, California 91360

AND

J. R. BROCK
University of Texas, Austin, Texas

HIGH-DISPERSED AEROSOLS

N. A. FUCHS and A. G. SUTUGIN
Karpov Institute of Physical Chemistry, Moscow

PERGAMON PRESS
Oxford · New York · Toronto
Sydney · Braunschweig

ГОСУДАРСТВЕННЫЙ КОМИТЕТ
СОВЕТА МИНИСТРОВ СССР
ПО НАУКЕ И ТЕХНИКЕ

АКАДЕМИЯ НАУК
СОЮЗА СОВЕТСКИХ
СОЦИАЛИСТИЧЕСКИХ РЕСПУБЛИК

ВСЕСОЮЗНЫЙ ИНСТИТУТ НАУЧНОЙ И ТЕХНИЧЕСКОЙ ИНФОРМАЦИИ

МОСКВА 1969

ИТОГИ НАУКИ

ФИЗИЧЕСКАЯ
ХИМИЯ

Н. А ФУКС, А. Г. СУТУГИН

ВЫСОКОДИСПЕРСНЫЕ АЭРОЗОЛИ

Figure 5 The title pages of the Russian and English versions of the book *High-Dispersed Aerosols.*

he had correspondence with almost all the well-known aerosol scientists. He did meet some of them in Moscow — Viktor La Mer in 1964, Shal Friedlander in 1970, David Shaw in 1977, and a few others, as mentioned in Marina's memoirs. He met Irving Langmuir in the 1930s; he admired him greatly. The friendship between Davies and Fuchs lasted over 30 years, until Fuchs's death (Figure 10). It seems that the criticism of Fuchs's coagulation formula in 1979 (Figure 2) had no greatly negative impact on these relations. On the other hand, in my opinion, there was no reason for such criticism; the formula was correct.

My contacts and friendship with Fuchs and his family began in 1956 during my visit in Moscow and while attending the Russian Colloid Conference in Tiflis (Caucasus). This friendship lasted until Fuchs's death and, of course, continued with Marina and then with his son, Michael, and his family.

As a "people enemy" Fuchs was never allowed to visit "capitalistic" countries. Therefore, he was very happy to be able to spend a short time with me in Prague in 1966. He had presented several lectures in Prague (Figure 11). One of them was recorded. He was very excited to see Prague and its environment. He was, for example, surprised that private houses even in small towns and villages were built of bricks and not of wood as in Russia. He wondered at the well dressed people, etc. He suspected that I was showing him "Potemkin Villages" — that is, some small towns and villages that were well built to serve as communistic propaganda for capitalistic visitors. The Czech communists used a lot of propaganda and had "Potemkin Villages" themselves at that time; what we showed Fuchs was not the case.

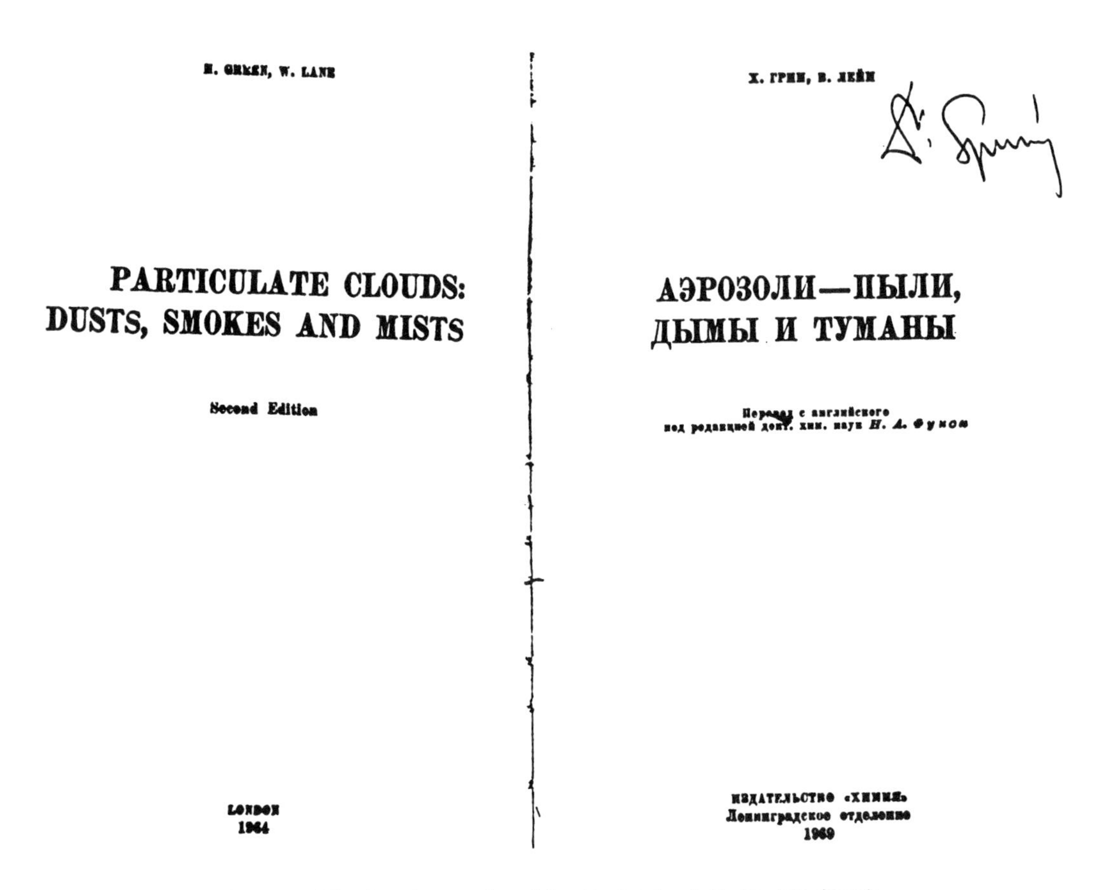

H. GREEN, W. LANE

PARTICULATE CLOUDS: DUSTS, SMOKES AND MISTS

Second Edition

LONDON
1964

Х. ГРИН, В. ЛЕЙН

АЭРОЗОЛИ—ПЫЛИ, ДЫМЫ И ТУМАНЫ

Перевод с английского
под редакцией докт. хим. наук Н. А. Фукса

ИЗДАТЕЛЬСТВО «ХИМИЯ»
Ленинградское отделение
1969

Figure 6 The title page of the Russian version of the English book *Particulate Clouds.*

As Marina mentions in her memoirs, Fuchs also had many good friends among scientists in Russia. His pupil and colleague Kirsch (Figure 9) was his devoted friend. It was he who prepared for publication the first three chapters of Fuchs's unfinished book. Unfortunately, he did not achieve official and broad recognition as a scientist — not during his life and not even after his death in Russia. Marina mentions in her memoirs that in a Russian publication remembering the 50th anniversary of aerosol science in Russia, the name Fuchs was "omitted."

4 FAMILY AND HOBBIES

The three members of the Fuchs's family were also happy together, remembers Marina in her memoirs. The family was, for all three, the base of freedom and hope and support for overcoming a very difficult life in difficult times. In addition to the family, his hobbies provided important support in his life.

He had several interests besides science. He very much loved nature — birds, flowers, trees. But most of all he loved mountains. As to botany, which was his chief hobby, there was not a shadow of doubt in the minds of botanists who met him that he was not one of them. For example, during a visit in the botanic garden, Pruhonice, near Prague in 1966, he informed the director of this garden — a botany professor — about several mistakes in the designated names of some wild flowers growing there.

Fuchs's other hobby besides botany was music. He often went to concerts of classical music. He was a good musician himself and played the violin very well. He liked classical music, and his favorite composers were Mozart, Bach, Beethoven, and Händel. He collected records, which he

ФИЗИКО-ХИМИЧЕСКИЙ ИНСТИТУТ ИМ. КАРПОВА

KARPOV-INSTITUTE OF PHYSICAL CHEMISTRY

ВСЕСОЮЗНЫЙ ИНСТИТУТ НАУЧНОЙ И ТЕХНИЧЕСКОЙ ИНФОРМАЦИИ

ALL-UNION INSTITUTE OF TECHNICAL AND SCIENTIFIC INFORMATION

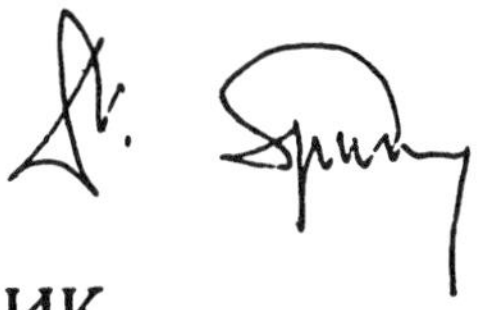

СБОРНИК

РЕФЕРАТОВ ПО АЭРОЗОЛЯМ

Том I, охватывающий литературу, опубликованную до 1953 г.

полутом I

под редакцией Н. А. ФУКСА

COLLECTION

OF AEROSOL ABSTRACTS

Volume I covering literature published before 1953.

Firs half

Edited by N. A. FUCHS

Figure 7 The title page of the monograph *Collection of Aerosol Abstracts.*

played over and over again. He loved best the cantatas and oratorios by J. S. Bach. He was therefore very happy when his professional friends in the U.S. and Western Europe continued to send him a large collection of LP records, including the sacred music of Bach. David Shaw and I organized a "Fuchs Music Foundation" in the 1970s. Several aerosol colleagues contributed to this foundation. My wife Drahuse and myself sent records from Germany to Moscow. Fuchs was very happy about that (Figure 12).

5 THE MAN NIKOLAI

His best friend, Prof. Efim Nesis, who got acquainted with Fuchs during his stay at Stavropol (according to Marina's memoirs) summarized his opinion: "Nikolai Albertowich was a man of great intellect and integrity. He remained such to the end of his days. He was devoted to science and his erudition was immense. But he was very modest and simple in his tastes and habits." He was well read in historical literature. He could read, write, and speak three foreign languages: English, French, and German. He was a well-educated man. He had a phenomenal memory and remembered all the scientific data of interest to him. He could communicate well at any time in his life. He could provide information on many subjects at short notice. He was always ready to help anyone who consulted him. He always maintained that his weak point was mathematics. And he was always studying it.

Moscow August 20, 19
1982

Dear Kristoslav

I send you herewith a reprint of my paper on thermophoresis published in „Fizika Atmosfery i Okeana". This paper in somewhat better English version will be also published in the J. of Aerosol Science but I have not received as yet any reprints from Dr. Davies.

Prof. Deryagin will never pardon me for my criticism of the numerous papers on this subject published by him and his pupils (Bakanov, Yalamov) but I hope that my paper will clear the situation in this very complex problem. If you think my paper is correct please write a review of it for Staub-Reinhaltung der Luft.

We are living with Marina the whole summer season at our cottage and growing berry-bushes and vegetables. I like this occupation very much. When I was a boy there were many kinds of vegetables in the market which have completely disappeared since and it is impossible to get even the seeds of them – e.g. artichokes, asparagus, topinambur, black root (schorzoner), Brussel sprouts etc. I would be very grateful to you or Drahuse if you could send me a catalogue of these seeds from a seed-growing German firm.

I reached on the 1/VIII the age 87 and decided to resign from Karpov Institute. My pension money will be 3 times less than my salary but I shall have more time and energy to finish my monograph on the methods of measuring aerosol concentration. I hope I shall do it next year and (have) publish it in 1984.

My health has not improved much after I left the hospital and I am still walking with a stick. The numerous drugs have not done much good to me; the best remedy is walking and I am doing it regularly every day for one hour (about 2 kilometers)

With best wishes to you and to Drahuse

N. Fuchs

Figure 8 Fuchs's letter written by his hand less than 2 months before his death.

Figure 9 The visit of Prof. La Mer from the U.S. in the Karpov Institute in Moscow. The picture is taken in the garden of the Karpov Institute. From left to right: Fuchs's co-worker F. Murashkewich, three unnamed men, Fuchs, La Mer and Fuchs's co-workers Kirsch and Sutugin.

He visited libraries regularly and worked there 4 to 5 hours on end. He had several notebooks where he stored his information. He had an immense capacity for work. He could work anywhere and all the time. He did not hear anything when he was working. He was not an ambitious man. He loved science for its own sake and worked for its advancement. Therefore, he did not resent the many injustices and unfairnesses coming from society and even from his colleagues and "friends."

Fuchs was also a good, strict, and critical teacher. But always a good father to his young co-workers. Several of them, such as Kirsch, Stechkina, Sutugin, Lushnikov, and others, became world-recognized aerosol scientists.

6 HIS LAST YEARS

In 1975, when Fuchs was 80 years old, he decided to leave the post of the chief of the LPAs (Laboratory of Physics of Aerodisperse Systems) of the Karpov Institute. He remained in the laboratory as a consultant until the middle of 1982. In his last letter to me (Figure 8) he mentioned that he decided to retire (he was 87) and his pension monthly would be reduced to three times less than his salary. But he would like to have more time and energy to finish his monograph on the methods of measuring aerosol concentration. Unfortunately, he did not finish this book — his time was over.

Nevertheless, he suggested and recommended his successor. His proposal was a very good one. The new head of LPAS became Prof. Lushnikov, who had joined the Fuchs laboratory in 1971. Lushnikov is an excellent aerosol theoretician, and was already known and recognized worldwide.

c/ Herrn Dr. Spurny

Journal of Aerosol Science

Chairman of the Editorial Board
Dr. C. N. Davies
Department of Chemistry
University of Essex
Wivenhoe Park, Colchester, Essex.

Telephone : OCO6 862286
Ext : 2307

CND/AJS

3 November 1982

Prof. Dr. W Stöber
Fraunhofer-Gesellschaft
Institut fur Aerobiologie
Nottulner Landweg 102
4400 Munster-Roxel
F R Germany

Dear Professor Stöber

I have just had a letter from Marina and Michael telling me that Sandy Fuchs passed away on October 10 of heart failure. "It was quite unexpected, he didn't suffer from any pains and didn't think he was going to die. He died a happy man."

Marina asked me to spread the news around. Undoubtedly this marks the end of the first era of aerosol science, with the passing of its leading figure.

We are on the point of sending the text of J. Aerosol Sci. 14(2) to the printers and I hope very much to include the obituary notice in this issue. Such notices are unsigned. I should be most grateful if you can lay your hands on any instances of personal contact and let me have any details which might be included in the obituary notice and any other material which you think is relevant.

There is a notice of his 75th birthday in J. Aerosol Sci. 1, 171-173 (1970) and of his 80th birthday in 6, 383 (1975).

Yours sincerely

Norman Davies

C N Davies

Figure 10 Letter of C. Norman Davies informing colleagues of the death of Prof. Fuchs.

7 FUCHS MEMORIAL AWARD

The GAeF in Germany, like AAAR, as well as the Japan Association of Aerosol Science and Technology (JAAST) decided to found the Fuchs Memorial Award (FMAW) in aerosol science in the middle of the 1980s. This award is conferred every 4 years during the International Aerosol Conferences to very respected and internationally recognized aerosol scientists. The first FMAW winner was Friedlander, from UCLA in Los Angeles, a meritorious aerosol scientist in the U.S. and worldwide in 1990. The second winners were Othmar Preining from Vienna University and Benjamin Y. H. Liu from the University of Minnesota in Minneapolis in 1994.

Figure 11 N. A. Fuchs during his seminar presentation in Prague in in 1966 (he was 71 years old at that time).

Moscow, 28 November, 1978

Dear Kvetoslav,

What a wonderful present you have sent me - all of Bach's organ compositions, and in such a fine interpretation ! I am so grateful to you. But it distresses me very much that I do not know what a present I can make to you or to Drahuse. Please write and tell me this.

I received also a disk from Mrs. Hochreiner. As I do not know her address, please give her my thanks.

My wife and I wish you and your whole family a merry Chris tmas and a very happy New Year !

With best wishes sincerely yours

N.Fuchs

Figure 12 Fuchs's letter of 1978, in which he acknowledges the receipt records of classical music.

8 FUCHS STREET IN HANNOVER

Fuchs's scientific merits were also acknowledged by some general expressions. In Germany, in the town of Hannover, the street on which the Fraunhofer Institute for Toxicology and Aerosol Research is situated, was named Nikolai Fuchs Strasse (Figure 13).

Figure 13 Photographs showing the designation of Nikolai Fuchs Street in Hannover in Germany. (This street nomination had been done with the proposal and support of Werner Stöbber, at that time director of the Frauhofer Institute for Toxicology and Aerosol Research).

9 SIMILARITIES

Fuchs was an important pioneer and founder of aerosol science in Russia and in the former U.S.S.R. But it is clear that he was one of several important aerosol scientists in this country (Grishpun, 1993). Therefore, when I was sent to Moscow in 1956 and in following years, I contacted other aerosol scientists, too, mainly those in the Institute of Physical Chemistry of the Academy of Sciences. So I have consulted, e.g., B. V. Deryaguin, L. V. Raduskiewich, Ya, I. Kogan, G. Ya. Vlasenko, P. Prochorov, Y. I. Yalamov, S. P. Bakhanov, D. V. Fedoseev, etc. (Spurny et al., 1961; Spurny, 1965). Later, during the Colloid Chemistry Conferences in Tiflis and in Odessa, etc., I had the opportunity to meet Russian and other aerosol physicists, chemists, and meteorologists. With several of them I had conducted extensive written correspondence.

Nevertheless, since the first contact with Nikolai Albertowich I had the feeling that our souls, our way of thinking, life, and political philosophies, etc. were very similar. And so he became my example and my first aerosol teacher. He was also the first aerosol teacher of my closest co-worker in Prague who later became a well known theoretician in aerosol physics. Of course, I was 28 years younger than Fuchs and was no equal partner in aerosol science. Nevertheless, as Marina mentions in her memoirs, Fuchs liked and appreciated my research ideas, my experimental research approaches, newly developed equipment, aerosol generation and radio-labeling methods, my pioneer work in pore filters, etc. He congratulated me when I succeeded in organizing a technical production of membrane filters and aerosol filters made of thin polymer and glass fibers in Czechoslovakia. Our Aerosol Laboratory in Prague, as well as our publications, were very positively accepted among the international aerosol science community, and several outstanding aerosol scientists visited our laboratory between 1960 and 1970 (Figure 14).

There were a lot of common interests in languages, history, literature, politics, and human relations. I had also been a persecuted person, in communistic Czechoslovakia. I was a "non-progressive" intellectual, who was not a "party" member and who had immediately recognized the tragic impact of communistic ideas and practices as well as that of the dictatorship over nations and individuals. This was the reason for my intensive engagement in the democratic movement of

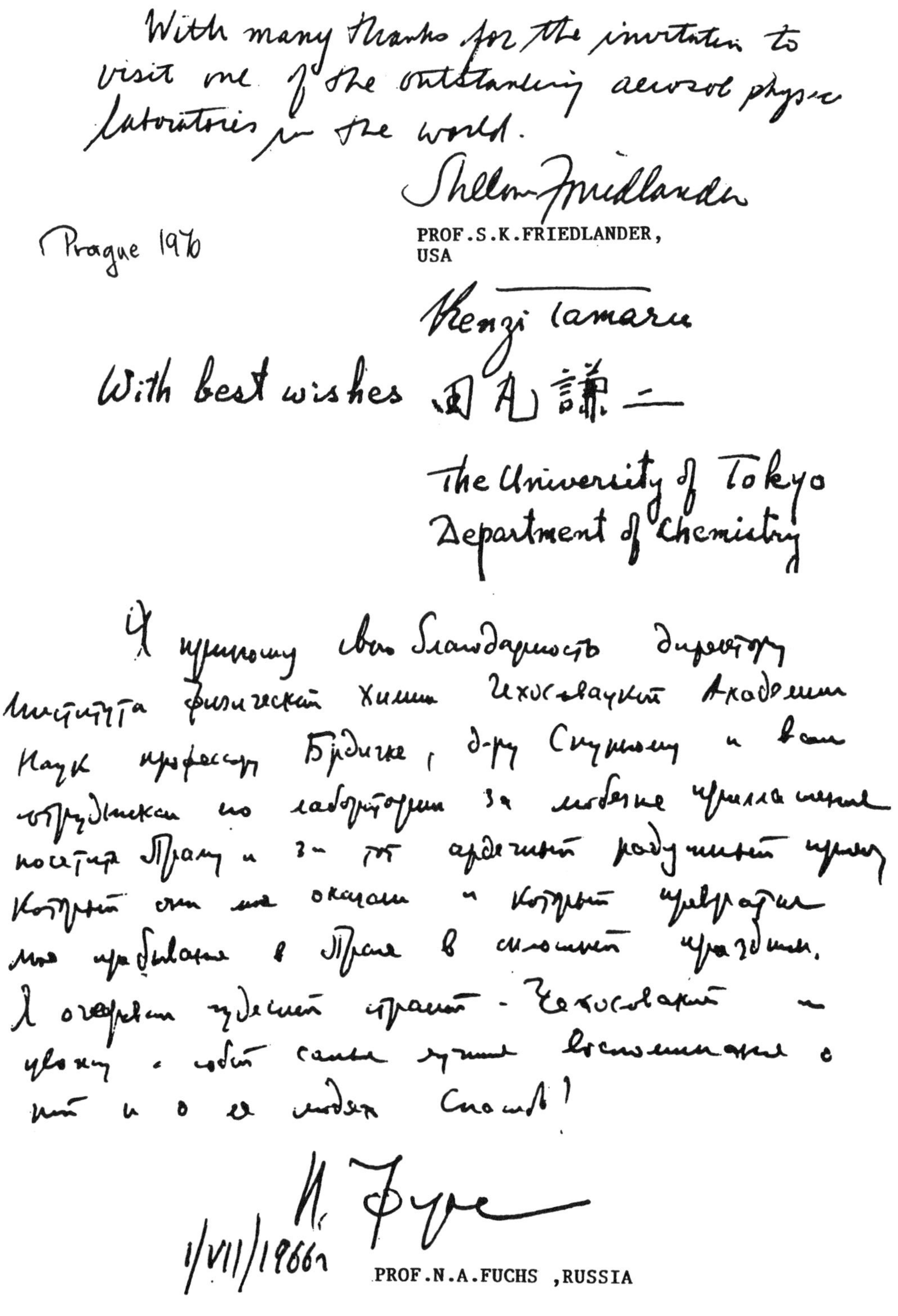

With many thanks for the invitation to visit one of the outstanding aerosol physics laboratories in the world.

Sheldon Friedlander

Prague 1976

PROF.S.K.FRIEDLANDER,
USA

Kenzi Tamaru

With best wishes 田丸謙二

The University of Tokyo
Department of Chemistry

Я приношу свою благодарность директору института физической химии Чехословацкой Академии Наук профессору Брдичке, д-ру Спурному и всем сотрудникам по лаборатории за любезное приглашение посетить Прагу и за тот сердечный радушный прием, который они мне оказали и который превратил мое пребывание в Праге в сплошной праздник. Я очарован чудесной страной - Чехословакией - и увожу с собой самые лучшие воспоминания о ней и о ее людях. Спасибо!

Н. Фукс

1/VII/1966г

PROF.N.A.FUCHS ,RUSSIA

Figure 14 An example of the remarks in the visitor's book of our laboratory in Prague.

the "Prague Spring" revolution in 1968. After this democratic period was stopped and the movement destroyed, I was — like Fuchs — labeled as a "people's enemy" and "counterrevolutionary person." In 1972 I was not allowed to continue my research work. My former research in the Department of Physical Chemistry of Aerodisperse Systems, which I founded and headed since 1956, was disturbed and my co-workers were dispersed. The worldwide recognized aerosol physicist J. Pich was isolated and could continue his aerosol research only under covert operation. "Aerosols" was a nondesirable word in communistic Czechoslovakia in the 1970s and 1980s because of the connection with my name. I had to emigrate abroad and I started again in Germany, being almost 50 years old.

Many similarities did exist between Fuchs and me without thinking about them. Therefore, I was very glad I was able to help in the recognition and rehabilitation of such a persecuted person and scientist. He, unfortunately, did not live to the end of the communistic dictatorship in Eastern Europe and in the U.S.S.R. But the Lord has made it possible that his wife Marina and I and some aerosol scientists were able to present to our aerosol friends worldwide the truth about the life and research conditions of Fuchs. Maybe, when a young aerosol scientist will read *The Mechanics of Aerosols,* in the future, he will remember the inhuman conditions under which this book was born.

REFERENCES

Davies, C. N. (1979) Coagulation of aerosols by Brownian motion, *J. Aerosol Sci.,* 10, 151–161.

Editorial Board (1970) Dr. Nicolas Fuchs. Seventy-fifth birthday, *J. Aerosol Sci.,* 1, 171–173.

Editorial Board (1970) A Tribute to Professor Nikolai Albertowich Fuchs on his 75th birthday. *J. Colloid Interface Sci.,* 32, iii.

Editorial Board (1975) Dr. Nicolas Fuchs. Eightieth birthday, *J. Aerosol Sci.,* 6, 383.

Editorial Board (1975) Nikolai Albertowich Fuchs on his 80th birthday, *Kolloidnyi Z.* (*J. Colloid Sci.,* Moscow), 38, 411–412.

Editorial Board (1975) Nikolai Albertowich Fuchs. In memoriam, *J. Aerosol Sci.,* 14, 1–3.

Friedlander, S. K. (1990) Fuchs Memorial Award. Acceptance talk, *Int. Aerosol Conference,* Kyoto, Japan.

Fuchs, N. A. (1934) *Physikalische Zschr. U.S.S.R.,* 6, 224–231.

Fuchs, N. A. (1934) *Z. Phys. Chem.,* 171, A, 199–209.

Fuchs, N. A. (1964) *The Mechanics of Aerosols,* Pergamon Press, Oxford.

Fuchs, N. A. (1980) On the Brownian coagulation of aerosols, *J. Colloid Interface Sci.,* 73, 248–249.

Grishpun, S. A. (1993) The state of aerosol research in the former Soviet Union, *J. Aerosol Sci.,* 24, 563–579.

Petryanov, I. V., Dubinin, M. M., Kogan, Ya. I., Raduskiewich, L. B., and Tchmurov, K. V. (1966) Nikolai Albertowich Fuchs. On his 70th birthday, *Zh. Fiz. Khim.* (*J. Phys. Chem. Moscow*), 40, 1428–1429.

Spurny, K. R., Jech, C., Sedlacek, B., and Storch, O. (1961) *Aerosoly* (*Aerosols*), SNTL, Publ. House of Technical Literature, Prague, 1–342.

Spurny, K. R., Ed. (1965) *Aerosols, Physical Chemistry and Applications,* Publ. House Academia, Prague, 1–943.

Spurny, K. R. (1983) Nikolai Albetowich Fuchs. In memoriam, *Aerosol Sci. Technol.,* 2, 301–302.

Spurny, K. R. (1983) Nikolai Albetowich Fuchs. In memoriam, *J. Colloid Interface Sci.,* 93, 597–598.

Spurny, K. R. (1984) Aerosol Conference — opening address, *J. Aerosol Sci.,* 15, 193–194.

Vincent, J. H., Kasper, G. et al. (1994) Charles Norman Davies 1911–1994. In memoriam, *J. Aerosol Sci.,* 25, 1253–1254.

BIBLIOGRAPHY OF THE WORK OF N. A. FUCHS, 1918–1970

Collected by C. N. Davies and A. A. Kirsch

1. Étude de movement des corps dans un milieu resistant par la methode photographique, *Arch. Sci. Phys. Inst. Sci. Moscou,* 1, 40, 1918.
2. Kinetics of hydration of meta- and pyrophosphoric acids, *J. Ruiss. Phys. Chem. Soc., Part Chem.,* 61, 1035, 1929.
3. Die Regel von Antonoff und die Molekulorientierung, *Kolloid Z.,* 52, 262, 1930.
4. Über die Realität des Neumannschen Dreiecks, *Z. Phys.,* 65, 262, 1930.
5. Über zweidimensionale Kristalle, *Z. Phys. Chem.,* 14B, 285, 19231.
6. Technisch-chemische Anwsendungen der "Taumethode." Paraffin und Erdwachs, *Z. Angew. Chem.,* 44, 969, 1931.
7. Zur Kinetik des Auystrocknens der pfänzlichen Öle, *Kolloid Z.,* 61, 365, 1932.
8. Mikroskopische Prüfung des Leinöls auf Minedalölgehalt, *Farben-Zig.,* 38, 217, 1933.
9. Zur Theorie der monomolekularen Adsorptionsschicht, *Z. Phys. Chem.,* 21B, 235, 1933.
10. Über die Oberflächenkondensation, *Phys. Z. Sovjetunion,* 4, 481, 1933.
11. (with I. Petryanov) Über die Bestimmung der Grösse und der Landung der Nebeltröpfchen, *Kolloid Z.,* 65, 171, 1933.
12. Zur Theorie der Koagulation, *Z. Phys. Chem.,* 171A, 1199, 1934.
13. Über die Verdampfungsgeschwindigkeit kleiner Tröpfchen in einer Gasatmorphäre, *Phys. Z. Sovietunion,* 6, 224, 1934.
14. On the activation energy of evaporation and condensation, *Dokl. Akad. Nauk. SSSR,* 3, 335, 1934.
15. Über die Stabilität und Aufladung der Aerosole, *Z. Phys.,* 89, 736, 1934.
16. Über die Bilding von Aerosolen. I. Methodik. Schwefelsäurenebel, *Acta Physiochim URSS,* 3, 61, 1936.
17. (with A. Frunkin). Über den Dampfdruck kleiner Tröpfchen und Kristalle, *Acta Physiochim, URSS,* 3, 783, 1935.
18. Über die Stabilität und Aufladung der Aerosole, II. Experimenteller Teil (with I. Petryanlv), *Acta Physicochim, URSS,* 3, 819, 1955.
19. Über die Effektivität der Zusammenstösse von Aerosolteilchen mit festen Wänden, *Acta Physiochim. URSS,* 3, 819, 1935.
20. Über die Taubildung, *Kolloidn Z.,* 71, 145, 1935.
21. On the nucleation of crystals, *Usp. Fiz. Nauk.,* 15, 496, 1935.
22. (with I. Petryanov and B. Rotzeig). Rate of charging of droplets by an ionic current, *Trans. Faraday Soc.,* 32, 1131, 1936.
23. Über die Fallgeschwindigkeit von überstokesschen Teilchen, *Techn. Phys. USSR,* 3, 255, 1936.
24. Die Bestimmung der Tröpfchengrösse in Wassernebein, *Phys. Z. Sovjetunion,* 10, 421, 1936.
25. (with B. Rotzeig). Zur ultramikroskopischen Grössenbestimmung von Aerosolteilchen, *Acta Physicochim, USSR,* 5, 893, 1936.
26. Dispersitätsmessung von Aerosolen, *Acta Physicochim URSS,* 6, 143, 1937.
27. (with I. Petryanov). Microscopical examination of fog-, cloud- and rain-droplets, *Nature* (London), 139, 111, 1937.
28. Determination of droplet size in sulfuric acid mists, *Zavad. Lab.,* 6, 210, 1937.
29. Stabilization of aerosols of NH_4Cl and HgI_4, *Zh. Fiz. Khim.,* 9, 294, 1937.
   [Gap of 10 years in publications (prisons and banishment)]
30. The value of charges on the particles of atmospheric aerocolloids, *Izv. Akad. Nauk SSSR. Ser. Geogr. Geosfiz,* 11, 341, 1947.
31. Use of distribution chromatography in the chemical analysis. Analysis of hexachlorocyclohexane, *Zh. Anali. Khim.,* 3, 320, 1948.

32. Separation of ammonia, methyl-, dimethyl- and trimethylamine by means of distribution cheomatography, *Dokl. Akad. Nauk SSSR.*, 60, 1219, 1948.
33. Distribution chromatography and its use in analytical chemistry, *Usp. Khim.*, 17, 45, 1948.
34. Progress of chromatographic method in organic chemistry, *Usp. Khim.*, 18, 206, 1949.
35. Determination of droplet size in oil fogs, *Kolloidn. Zh.*, 11, 286, 1949.
36. Preparation and standardisation of alumina for the chromatographic analysis of organic substances, *Zavod. Lab.*, 878, 1950.
37. Progress in aerosol studies and practical achievements in this field, *Usp. Khim.*, 19, 175, 1950.
38. Method of Tsvet in organic chemistry, in *Reactions and Methods of Investigation of Organic Compounds,* Vol. 1, Akademizdat, Moscow, 1951, 179.
39. Effect of dust on the turbulence of a gas stream, *Zh. Tekhn. Fiz.*, 21, 704, 1951.
40. A contribution to the theory of precipitation of "warm clouds," *Dokl. Akad. Nauk SSSR,* 81, 1043, 1951.
41. Observations on the wind-breaking action of shelterbelts, *Les Step,* 51(2), 1953.
42. *Mechanics of Aerosols,* Akademizdat. Moscow, 1955 (Am. ed., 1958).
43. Gas-liquid chromatography, *Usp. Khim.*, 25, 845, 1956.
44. (with P. Lesovskii). The charging of aerosols by ionic diffusion, *J. Colloid Sci.*, 11, 107, 1956.
45. Contribution to the theory of evaporation of small droplets, *Zh. Tekhn. Fiz.*, 28, 159, 1956.
46. *Evaporation and Growth of Particles in a Gas Medium,* Akademizdat, Moscow, 1958 (English ed., Pergamon Press, 1959).
47. (with S. Yankovskii). Thermophoresis in an aerosol stream, *Dokl. Akad. Nauk SSSR,* 119, 1177, 1958.
48. (with S. Yankovskii). A technique for depositing aerosols in a thermal precipitator for electron microscopic examination, *Kolloidn. Zh.*, 21, 133, 1959.
49. *Progress in the Mechanics of Aerosols*, Akademizdat, Moscow, 1961.
50. Deposition of aerosols on the waslls of chembers, *Izv.. Akad. Nauk SSSR Ser. Geofiz.*, 142, 1962.
51. On the vertical particle distribution in a turbulent stream; *Zh. Tekhn. Fiz.*, 32, 255, 11962.
52. (with I. Stechkina and V. Staroselskii). On the determination of particle size distribution in polydisperse aerosols by the diffusion method, *Br. J. Appl. Phys.*, 13, 280, 1962.
53. (with I. Stechkina). Resistance of a gaseous medium to the motion of a spherical particle of a size comparable to the mean free path of the gas molecules, *Trans. Faraday Soc.*, 58, 1949, 1962.
54. (with A. Sutugin). Droplet size distribution in dibuthylphtalate mists prepared with the use of fireign nuclei, *Kolloidn. Zh.*, 14, 519, 1963.
55. (with A. Sutugin). Generation and investigation of high-dispersed sodium chloride aerosols, *Br. J. Appl. Phys.*, 14, 519, 1963.
56. (with I. Stechkina). A note on the theory of fibrous aerosol filters, *Ann. Occup. Hyg.*, 6, 27, 1963.
57. On the stationary charge distribution on aerosol particles in bipolar ionic atmosphre, *Geofis.,Pura. Appl.*, 56, 185, 1963.
58. *Mechanics of Aerosols,* Revised and enlarged edition, Pergamon, Oxford, 1964.
59. (with A. Selin). Pneumatic dispersion of powders, *Inzh. Fiz. Zh.*, 7, 122, 1964.
60. "Aerosols" and "Coagulation of aerosols," articles for the *Physical Dictionary,* 2nd ed., Sovetskaya Entsiklopedia, Moscow, 1964.
61. (with A. Sutugin). High-dispersed aerosols, *Kolloidn. Zh.*, 26, 110, 1964.
62. (with A. Kirsh). Effect of condensation of a vapour on the grains and of evaporation from their surface on the deposition of aerosols in granular beds, *Chem. Eng. Sci.*, 20, 181, 1965.
63. (with A. Sutugin). Coagulation rate of highly dispersed aerosols, *J. Colloid. Sci.*, 20, 492, 1965.
64. (with V. Gubenskii). Determination of sizes and charges of individual particles formed by electrostatic atomization of liquid systems, in *Electrostatic Coating of Manufactured Articles,* Moscow, 1966, 35.
65. (with I. Stechkina). Studies on fibrous aerosol filters. I. Calculation of diffusional deposition of aerosols in fibrous filters, *Ann. Occup. Hyg.*, 9, 59, 1966.
66. On the Brownian coagulation of aerosols, *J. Colloid. Interf. Sci.*, 21, 110, 11966.
67. (with S. Yankovskii). Granulometric analysis of industrial dusts according to Stokes' diameters, *Zavod. Lab.*, 32, 811, 1966.
68. (with A. Sutugin). The generation and use of monodispersed aerosols, in *Aerosol Science,* C. N. Davies, Ed., Academic Press, London, 1966.
69. (with A. Kirsh). Studies on fibrous aerosol filters, II. Pressure drop in systems of parallel cylinders, *Ann. Occup. Hyg.*, 10, 23, 1967.

70. (with A. Kirsh). The fluid flow in a system of parallel cylinders perpendicular to the flow direction at small Reynolds numbers, *J. Phys. Soc. Jpn.*, 22, 1251, 1967.
71. (with A. Sutugin). On the production of high-dispersed powders via aerosol state, *Zh. Prikl Khim.*, 42, 567, 1967.
72. (with A. Sutugin). High-dispersed aerosols, *Usp. Khim.*, 37, 1965, 1968.
73. (with A. Kirsh). Studies on fibrous aerosol filters. III. Diffusional deposition of aerosols in fibrous filters, *Ann. Occup. Hyg.*, 11, 299, 1968.
74. (with A. Sutugin). Formation of condensation aerosols at high vapour supersaturations, *J. Colloid Interface Sci.*, 20, 216, 1968.
75. Versuch einer Klassifikation der Aerosolliterature, *Staub,* 28, 349, 1968.
76. (with A. Stechkina and A. Kirsh). Studies on fibrous aerosol filters. IV. Calculation of aerosol deposition in model filters in the range of maximum penetration, *Ann. Occup. Hyg.*, 12, 1, 1969.
77. (with A. Sutugin). *High-Dispersed Aerosols,* Akademizdat, Moscow, 1969.
78. Recent progress in the theory of transfer processes in aerosols at intermediate values of the Knudsen number, *Proc. 7th Int. Conf. Condensat. Nuclei. Prague and Vienna,* 1969, 10.
79. (with A. Kirsh and I. Strechkina). Studies on fibrous aerosol filters. Experimental determination of the filter efficiency in the range of maximum penetration, *Kolloidn. Zh.*, 31(2), 1969.
80. Tentative classification of aerosol literature [in Russian], *Kolloidn. Zh.*, 32, 471, 1969.
81. (with Sutugin). On the formation of aerosols at quickly changing conditions. Theory and method of calculation, *Koll. Zh,* 32, 212, 1970.
82. (with Murashkevich). Laboratoriums — Pulverzerstäuber — Staubgenerator, *Staub Reinhalt. Luft,* 30, 447, 1970.
83. (with Stechkina). Influence of inertia on the capture efficiency of aerosol particles on cylinders at small Stokes numbers, *Kolloid. Zh.*, 32, 467, 1970.
84. (with Sutugin). Formation of condensation aerosols under rapidly changing environmental conditions: theory and method of calculation, *J. Aerosol Sci.*, 1, 287, 1970.
85. (with Sutugin and Kotsev). Formation of condensation aerosols under rapidly changing environmental conditiokns: non-coagulated high-dispersed aerosols, *J. Aerosol Sci.*, 2, 361, 197.
86. (with Sutugin). High-dispersed aerosols, in *Topics in Current Aerosol Research,* Hidy, G. M. and Brock, J. R., Eds., Pergamon Press, Oxford, 1971.
87. (with Kirsch and Stechkina). Effect of gas slip on the pressure drop in a system of parallel cylinders at small Reynolds numbers, *J. Colloid Interface, Sci.*, 37, 458, 1971.
88. Some new methods and deviceds for aerosol studies, in *Assessment of Airborne Particles*, Mercer, T. T., Morrow, P. E., and Stöber, W., Eds., Thomas, Springfield, IL, 1972, Chap. 11, 200–211.
89. (with Kirsch and Stechkina). The capture of submicron particles by single fine fibres, *Atmos. Environ.*, 6, 73, 1972.
90. Latex aerosols — caution! *J. Aerosol Sci.*, 4, 405, 1973.
91. (with Kirsch and Stechkina). Effect of gas slip on the pressure drop in fibrous filters, *J. Aerosol Sci.*, 4, 287, 1973.
92. (with Kirsch and Stechkina). A contribution to the theory of fibrous aerosol filters, *Faraday Symposia of the Chemical Society,* 7, 143, 1973.
93. (with Kirsch and Stechkina). Pressure drop and aerosol deposition in a polydisperse fan model filters, *Kolloidn, Zh.*, 35, 971, 1973.
94. (with Kozhenkov, Kirsch and Simonov). On the mechanism of formation of monodisperse fosgs in electrostatic dispersion of liquids, *Dokl. Akad. Nauk. SSSR.*, 212, 879, 1973.
95. On aerosol chambers, *Kolloidn. Zh.*, 36, 1183, 1974.
96. (with Kirsch and Stechkina). Gas flow in aerosol filters made of polydisperse ultrafine fibres, *J. Aerosol Sci.*, 5, 39, 1974.
97. (with Kozhenkov and Kirsch). Investigation of the process of formation of monodisperse aerosols in electrostatic dispersion of liquids, *Kolloidn. Sh.*, 36, 1168, 1974.
98. Sampling of aerosols, *Atmos. Environ.*, 9, 967, 1975.
99. (with Kirsch). Air flow in a porous membrane, *Teor Osn. Khim. Tekhnol.*, 9, 311, 1975.
100. (with Kirsch and Stechkina). Efficiency of aerosol filters made of ultrafine polydisperse fislters, *J. Aerosol Sci.*, 6, 119, 1975.

101. (with Kozhenkov). Determination of the droplet size distributions in coarse fogs by means of the microdiffraction method, *J. Colloid Interface Sci.,* 52, 120, 1975.
102. (with Kozhenkov). Electrodynamical dispersions of liquids, *Usp. Khim,* 45, 2274, 1976.
103. Aerosol impactors (a review), in *Fundamentals of Aerosol Science*, Shaw, D. T., Ed., John Wiley, New York, 1978, 1–83.
104. On the Brownian coagulation of aerosols, *J. Colloid Interface, Sci.,* 73, 248, 1980.
105. (with Stulov and Murashkevich). The efficiency of collision of solid aerosol particles with water surfaces, *J. Aerosol Sci.,* 9, 1, 1978.
106. (with Belov, Datskevich, and Karpova). Formation of breakdown plasma in aerosol under the action of CO_2 laser radiation, *Zh. Tekh. Fiz.,* 49, 333, 1979.
107. High efficiency filtration of gases and liquids by fibrous materials, *Khim. Prom.,* 11, 668, 1979.
108. Thermophoresis of aerosol particles at small Knudsen numbrs: theory and experiment, *J. Aerosol Sci.,* 13, 327, 1982.
109. Methods for determining aerosol concentration, *Aerosol Sci. Technol.,* 5, 123, 1986.

INDEX

Index

F

G

H

I

J

K

L

P

Q

R

S

T